GEOTECHNICAL SPECIAL PUBLICATION NO. 120

GROUTING AND GROUND TREATMENT

PROCEEDINGS OF THE THIRD INTERNATIONAL CONFERENCE

Volume Two

February 10-12, 2003
New Orleans, Louisiana

SPONSORED BY
The Geo-Institute of the American Society of Civil Engineers
Deep Foundations Institute

EDITED BY
Lawrence F. Johnsen
Donald A. Bruce
Michael J. Byle

American Society
of Civil Engineers
1801 ALEXANDER BELL DRIVE
RESTON, VIRGINIA 20191–4400

Abstract: *Grouting and Ground Treatment: The Third International Conference on Geotechnical Grouting* was sponsored jointly by the Geo-Institute of the American Society of Civil Engineers and the Deep Foundations Institute. The proceedings includes 127 papers by authors from 21 countries. A segment of the conference, devoted to deep mixing and jet grouting, attracted 44 papers. Keynote lectures were presented by Goran Holm, Stuart Littlejohn, Giovanni Lombardi, Mitsuhiro Shibazaki, Massaaki Terashi, and James Warner.

ISBN 0-7844-0663-4

Preface

The 2003 Specialty Conference on Grouting is the third in a series of international grouting conferences organized by the ASCE/Geo-Institute Committee on Grouting. The first conference was held in New Orleans in February of 1982 and was chaired by Dr. Wallace Hayward Baker and Dr. Reuben Karol. It attracted 64 papers of which 29 were by authors from 11 countries outside the U.S. The proceedings, "Grouting in Geotechnical Engineering," were edited by Dr. Baker and published by ASCE. The conference discussions were published and distributed to the conference attendees.

The second conference, held in New Orleans in February of 1992, was chaired by Dr. Loren Anderson. This conference included papers on soil improvement, geosynthetics and grouting. Of the 106 papers presented, 49 covered grouting. The 49 grouting papers included 23 by authors from 10 countries outside of the U.S. The proceedings, "Grouting, Soil Improvement and Geosynthetics," were edited by Dr. Roy Borden, Dr. Robert Holtz, and Dr. Ilan Juran and published by ASCE. The conference proceedings were also supplemented by a volume containing discussions of conference papers.

This conference, held in New Orleans in February of 2003, was chaired by Mr. Michael J. Byle. The conference attracted 127 papers of which 61 were by authors from 20 countries outside of the U.S. The proceedings, "Third International Conference on Grouting and Ground Treatment," were edited by Lawrence F. Johnsen, Dr. Donald A. Bruce, and Michael J. Byle, and published by ASCE. The conference discussions will be published and distributed to the conference attendees. Reflecting the growing use of Deep Mixing methods in certain parts of the world, this conference included 20 papers on the technology from Japan, the Nordic countries and the United States

The organizing committee for the conference included:

Michael Byle	Gannett Fleming, Inc.
Richard Berry	REMBCO Engineering Corp.
Gordon Boutwell	Soil Testing Engineering, Inc.
Donald Bruce	Geosystems, L.P.
Francis Gularte	Hayward Baker, Inc.
Lawrence Johnsen	Heller and Johnsen
Ilan Juran	Polytechnic University
Kenneth L. McManis	University of New Orleans
Steven Scherer	Hayward Baker, Inc.
Cumaraswamy Vipulanandan	University of Houston
James Warner	Consultant
Joseph Welsh	Consultant

All papers were peer reviewed and are available for discussion in the ASCE Journals. The due date for any discussion is the same as that for a February 2003 journal paper. All of these papers are eligible for appropriate ASCE and Geo-Institute awards. The reviewers included:

Peter Aberle
Abir Al-Tabbaa
A Sheikh Bahai Ameli
Andy Anderson
Fioravante A. Bares
Jim Belgeri
Kenneth R. Bell
Stanley Bemben
Richard M. Berry
Tom Billups
Roy Borden
Andrew F. Brengola
Donald A. Bruce
George K. Burke
Michael J. Byle
Allen Cadden
Dave Campo
Joe Cavey
Marcelo Chuaqui
Patrick Crowell
Umakant Dash
John R. Davie
James Davies
Jim DeStefano
Ata G. Doven
Trent Dreese
Adel M. El-Kelesh
Melvin I. Esrig
Tao-Wei Feng
Joseph A. Fischer
Mathew Francis
Patricia M. Gallagher
Jesus E. Gomez

Deborah Goodings
Jim Graham
Donald Gray
Clay Griffin
Francis Gularte
Jacco Haasnoot
Tarek Haider
Douglas M. Heenan
Wolfgang F. Heinz
Chu Eu Ho
Jozef Hulla
Jon Jagello
Michael Jefferies
Stephan Jefferis
Lawrence Johnsen
Reuben Karol
Edward Kashdan
Masaki Kitazume
Raymond Krizek
Clemens Kummerer
James R. Lambrechts
Wei F.Lee
Daniel R. Lees
San-Shyan Lin
Justice J. Maswoswe
Gordon Matheson
Ross McGillivray
John Meyers
Paul Michaels
Michael J. Miluski
Jens Mittag
Russ Morgan
Brennon Morioka

Saiyouri Nadia
Lou Narduzzo
Alex Naudts
James C. Ni
J. Pestana-Nascimento
Ali Porbaha
Ellen Rathje
Chris Ryan
Lois Schwarz
Shui-Long Shen
J.N. Shirlaw
Dawn Shuttle
Orjan A. Sjostrom
G. Smoak
Staffan Swedenborg
Masaaki Terashi
S. Thevanayagam
A. E.C. van der Stoel
Janne Vataja
C. Vipulanandan
John Volk
Zhao Wang
David A. Ward
James Warner
Stuart T. Warren
Kenneth D. Weaver
Mitch Weber
David Wilson
Michael Xu
David S. Yang
Hiroshi Yoshida

Steve Scherer organized the grouting committee's first Grouting GREATS presentations. During the Monday luncheon Edward Graf, Reuben Karol, James Warner, Kenneth Weaver, and Joseph Welsh were honored.

We wish to thank the staffs of DFI, the Geo-Institute, and ASCE. This conference marked the first that the Geo-Institute jointly sponsored with another organization. Geordie Compton, Executive Director of the DFI and Theresa Rappaport proved invaluable in organizing the conference and covering the daily details. Carol Bowers of the Geo-Institute and Donna Dickert guided us through ASCE and G-I, and the publishing of the proceedings. DFI set up a pre-conference web site in which all attendees could read the papers prior to the conference. This led to well-considered questions for the discussions.

The ASCE/G-I Grouting Committee intends to continue the tradition begun by Wally Baker and Reuben Karol of hosting international geotechnical grouting conferences every ten years. Visit our web site at www.newhaven.edu/grouting to

iv

see our other activities. Give our webmaster the web address for your non-profit grouting site and we will post a link.

Lawrence F. Johnsen, Principal   Heller and Johnsen

Dr. Donald A. Bruce, President   Geosystems, L.P

Michael J. Byle, Geotechnical Manager   Gannett Fleming, Inc.

Contents

Volume One

KEYNOTE LECTURES

Fifty Years of Low Mobility Grouting..1
James Warner

The State of Practice in Deep Mixing Methods...25
Masaaki Terashi

The Development of Practice in Permeation and Compensation Grouting:
A Historical Review (1802-2002) Part 1: Permeation Grouting............................50
Stuart Littlejohn

The Development of Practice in Permeation and Compensation Grouting:
A Historical Review (1802-2002) Part 2: Compensation Grouting.......................100
Stuart Littlejohn

State of Practice in Dry Deep Mixing Methods...145
Göran Holm

Grouting of Rock Masses..164
Giovanni Lombardi

State of Practice of Jet Grouting...198
Mitsuhiro Shibazaki

JET GROUTING

A Ten-Year Perspective of Jet Grouting: Advancements in Applications
and Technology...218
Gary T. Brill, George K. Burke, and Alan R. Ringen

QA/QC for Jet Grouting in Deep Boston Blue Clay Central Artery/Tunnel Project 236
Justice J. G. Maswoswe

Specialist Foundation Construction Techniques used in the Reconstruction of the
University Library "Bibliotheca Albertina" in Leipzig, Germany.......................248
Wolfgang G. Brunner

Jet Grouting Soft Clays for Tunnelling and Deep Excavations—Design
and Construction Issues...257
J. Nick Shirlaw

Stabilization of Deep Open Excavations in Soft Soil by Jet Grouting.................269
Chu Eu Ho and Chin Gee Tan

Construction and Quality Control of Jet Grouting Applications in Turkey..................281
Rasin Düzceer and Alp Gökalp

Development Oversized Jet Grouting...294
    Mitsuhiro Shibazaki, Mitsuru Yokoo, and Hiroshi Yoshida

Microtunneling and Horizontal Directional Drilling (HDD) Performance
in Jet Grouted Soil...303
    James Kwong and Mathew Francis

Jet Grout Stabilization of Steeply Excavated Soil Slope.................................318
    John Meyers, Tim Myers, and Kerry Petrasic

SuperJet Grouting Repairs and Extends the Life of Ailing Coastal
Front Structure...330
    Dennis W. Boehm and Thomas A. Posey

Ground Treatment Associated with the Construction of Cross-Passages
for the Kowloon-Canton Railway Corporation (KCRC) West Rail Phase I,
Kwai Tsing Tunnels in Hong Kong..342
    R. B. Storry, A. Richely, E. Nelson, and D. O. Licuanan

SuperJet Grouting Reduces Foundation Settlement for La Rosita Power Plant
in Mexicali, Mexico..354
    Kenneth R. Bell, José L. M. Clemente, Francis B. Gularte,
    and Roberto A. Lopez

Jet Grout Columns Partially Support Natural Draft Cooling Tower...............365
    John Davie, Mehmet Piyal, Armagan Sanver, and Bahattin Tekinturhan

Case History for Soil Improvement of SETAT 2002 High Rise Residential
by Jet Grouting in Istanbul..377
    H. T. Durgunoglu, H. F. Kulac, S. Yilmaz, and D. Kocak

Jet Grout Columns in Mixed Profile to Control Foundation Settlement:
Gerald Ratner Athletics Center..389
    Raymond J. Franz and Kyle E. Camper

Case Histories of Ground Treatment with Vertical Jet Grouting Solutions....401
    Alexandre Pinto, João Falcão, Carlos Barata, Sandra Ferreira, Duílio Cebola,
    and Joana Pacheco

Jet Grouting Experience at Posey Webster Street Tubes Seismic Retrofit Project........413
    Umakant Dash, Thomas S. Lee, and Randy Anderson

Jet Grout Foundations to Resist Compressive, Uplift and Lateral Loads
at an Operational Power Plant...428
    Andrew F. Brengola and Bradford W. Roberts

Improving Deep-Seated Soft Clays Using Super-Jet Grouting........................440
    H. Senapathy, J. R. Davie, and D. Boehm

A Case History of Ground Treatment with Jet Grouting Against Liquefaction
for a Cigarette Factory in Turkey..452
    H. T. Durgunoglu, H. F. Kulac, K. Oruc, R. Yildiz, J. Sickling, I. E. Boys,
    T. Altugu, and C. Emrem

North Airfield Drainage Improvement at Chicago-O'Hare International Airport:
Soil Stabilization using Jet Grouting..464
    Dwayne A. Lewis and Martin G. Taube

DEEP MIXING

The Practitioner's Guide to Deep Mixing...474
Donald A. Bruce and Mary Ellen C. Bruce

Deep Mixing: An Owner's Perspective 489
David P. Shiells, Thomas W. Pelnik III, and George M. Filz

Ground Stabilization in the United States by the Scandinavian Lime Cement
Dry Mix Process..501
Melvin I. Esrig, Peter E. Mac Kenna, and Edward P. Forte

The Application of Various Deep Mixing Methods for Excavation
Support Systems 515
Kenneth B. Andromalos and Eric W. Bahner

Guidelines for Design and Installation of Soil-Cement Stabilization............527
David L. Druss

Design and Construction Aspects of Soil Cement Columns
as Foundation Elements 540
Gyimah Kasali and Osamu Taki

Mass Stabilization of Organic Soils and Soft Clay............................552
Nenad Jelisic and Mikko Leppänen

Effects of Lime-Cement Soil Stabilization Against Train Induced
Ground Vibrations...562
Mehdi Bahrekazemi and Anders Bodare

Soil Mixing to Stabilize Organic Clay for I-95 Widening, Alexandria, VA....575
James R. Lambrechts, Margaret A. Ganse, and Carrie A. Layhee

Laboratory Tests on Long-Term Strength of Cement Treated Soil..............586
Masaki Kitazume, Takeshi Nakamura, Masaaki Terashi, and Kanta Ohishi

Field Observation of Long-Term Strength of Cement Treated Soil 598
Hirochika Hayashi, Jun'ichi Nishikawa, Kanta Ohishi, and Masaaki Terashi

Measured and Predicted Five-Year Behaviour of Soil-Mixed Stabilised/Solidified
Contaminated Ground 610
A. Al-Tabbaa, B. Chitambira, R. Perera, and N. Boes

Measured Permeabilities in Stabilised Swedish Soils 622
Helen Åhnberg

Evaluation of Property Changes in Surrounding Clays due to Installation
of Deep Mixing Columns...634
Shui-Long Shen, Norihiko Miura, Jie Han, and Hirofumi Koga

Strength Properties of Soil Cement Produced by Deep Mixing.....................646
Osamu Taki

Prediction Method for Ca Leaching and Related Property Change
of Cement Treated Soil 658
Takahiro Nishida, Masaaki Terashi, Nobuaki Otsuki, and Kanta Ohishi

Coring Soil-Cement Installed by Deep Mixing at Boston's Central Artery/Tunnel
(CA/T) Project...670
James R. Lambrechts and Scott Nagel

Column Penetration Tests for Lime-Cement Columns in Deep
Mixing—Experiences in Sweden...681
Morgan Axelsson and Stefan Larsson

In Situ Techniques for Quality Assurance of Deep Mixed Columns.............695
Ali Porbaha and Anand J. Puppala

GROUTING FOR ENHANCEMENT FOR PILES AND PIERS

Study on Bearing Capacity of Bored Cast-in-Situ Piles by Post Pressure Grouting.....707
Xudong Fu and Zhengbing Zhou

Base Grouted Bored Pile on Weak Granite..716
Chu Eu Ho

Pile Foundation Improvement by Permeation Grouting..............................728
Almer E. C. van der Stoel

Underpinning of a Pier by Microfine Cement Grouting
and Compensation Grouting..740
Georg Breitsprecher and Paul Stefan Tóth

GROUTING FOR MICROPILES AND ANCHORS

The Basics of Drilling for Specialty Geotechnical Construction Processes 752
Donald A. Bruce

Reliability of Estimated Anchor Pullout Resistance.....................................772
Yasser A. Hegazy

Grouting of Micropiles in Scandinavia...780
Jouko Lehtonen and Stefan Aronsson

Rehabilitation of Union Pacific Railroad Tunnel, Ryndon, Nevada.............791
Francis B. Gularte and Gerry Millar

Volume Two

COMPENSATION GROUTING

Railroad Embankment Stabilization Demonstration for High-Speed
Rail Corridors...803
Andrew Sluz, Theodore R. Sussmann, and Gopal Samavedam

Active Settlement Control with Compensation Grouting—Results from
a Case Study..813
Clemens Kummerer, Helmut F. Schweiger, and Reiner Otterbein

Frac Grouting—A Case History...824
Douglas M. Heenan, Janne W. Vataja, and Trent L. Dreese

Compensation Grouting to Reduce Settlement of Buildings During an Adjacent
Deep Excavation..837
Jinyuan Liu

Effect of Injection Rate on Clay-Grout Behavior for Compensation Grouting............845
K. Soga, S. K. A. Au, and M. D. Bolton

A Retrospective on the History of Dam Foundation Grouting in the U.S. 857
K. D. Weaver

Compaction Grouting for Sinkhole Repair at WAC Bennett Dam 869
James Warner, Michael Jefferies, and Steve Garner

GROUTING FOR WATER RETAINING EMBANKMENTS

Quantitatively Engineered Grout Curtains...881
David B. Wilson and Trent L. Dreese

California Aqueduct Foundation Repair Using Multiple Grouting Techniques...........893
Timothy M. Wehling and David C. Rennie

Flexibility in Grouting: Solutions for Old Dams...905
Allen Cadden, Jesús Gómez, Graham C. G. Smith, and Robert Traylor

Curtain Grouting for the Antamina Dam, Peru: Part 1—Design and Performance.....917
T. G. Carter, F. Amaya, M. G. Jefferies, and T. L. Eldridge

Curtain Grouting for the Antamina Dam, Peru: Part 2—Implementation
and Field Modifications..929
D. G. Ritchie, J. P. Garcia, F. Amaya, and M. G. Jefferies

GROUTING IN KARST

Shallow Foundations in Karst: Limited Mobility Grout
or Not Limited Mobility Grout..941
Jesús E. Gómez and Allen W. Cadden

Grouting in Karst Terrane—Concepts and Case Histories 953
Joseph A. Fischer, Joseph J. Fischer, and Richard S. Ottoson

Grouted Seepage Cutoffs in Karstic Limestone...967
Arthur H. Walz, Jr., David B. Wilson, Donald A. Bruce, and James A. Hamby

Nittany Lions' New Convocation Center: Rock Solid with Limited Mobility
Grouting (LMG) 979
Allen W. Cadden and Richard A. Wargo

COMPACTION GROUTING AND LIMITED MOBILITY GROUTING

Using the Grouting Intensity Number (GIN) to Assess Compaction Grouting
Performance...991
Steven W. Perkins and Joe Harris

Case History: Broadcast Tower Anchor Stabilization, Portland, Oregon 1010
Rajiv Ali and Jeffrey Geraci

Grouting and Ground Treatment Case Studies in Applications of Grouting
and Deep Mixing Use of Compaction Grout Columns to Stabilize Uncontrolled
Loose Fill and to Lift a Settled Tunnel: A Significant Case History.............................1020
   Ray (Alireza) Boghart, Paul S. Hundley, Jeffrey R. Hill, and
   Steven D. Scherer

Compaction Grouting Used for a Water Treatment Plant Expansion..........................1032
   Michael W. Oakland and Michael L. Bachand

Low Strain Testing of Compaction Grout Columns...1044
   Lawrence F. Johnsen, Andy Anderson, and John J. Jagello

Effect of Soil and Grouting Parameters on the Effectiveness
of Compaction Grouting 1056
   Adel M. El-Kelesh and Tamotsu Matsui

Design Considerations for Inclusions by Limited Mobility
Displacement Grouting..1071
   Michael J. Byle

GROUTING IN ROCK AND MINING

Characterization of Fractured Rock for Grouting Design
Using Hydrogeological Methods...1082
   Åsa Fransson and Gunnar Gustafson

Rock Mechanics Effects of Cement Grouting in Hard Rock Masses...........................1089
   S. Swedenborg and L-O. Dahlström

Subsidence Mitigation Using Void Fill Grouting 1103
   Darrel V. Holmquist, Damon B. Thomas, and Kent Simon

Mining Grouting: A Rational Approach...1115
   W. F. Heinz

Innovative Grouting Solves Geotechnical Issues: Five Case Histories 1130
   H. Clay Griffin and Richard M. Berry

GROUT MIXTURES AND MATERIALS

Long Term Performance of Grouts and the Effects of Grout By-Products.................1141
   Stephen A. Jefferis

Mix Design and Quality Control Procedures for High Mobility
Cement Based Grouts..1153
   M. Chuaqui and D. A. Bruce

Fly Ash Utilization in Grouting Applications...1169
   Ayse Pekrioglu, Ata G. Doven, and Mehmet T. Tumay

Additives and Admixtures in Cement-Based Grouts..1180
   Alex Naudts, Eric Landry, Stephen Hooey, and Ward Naudts

Evaluation of Fly Ash and Clay in Soil Grouting..1192
   S. Akbulut and A. Saglamer

New On-Site Wet Milling Technology for the Preparation of Ultrafine
Cement-Based Grouts 1200
    Alex Naudts and Eric Landry

Characterization of a Non-Shrinkage Cement Grout Used for Water Pipe Joints......1208
    C. Vipulanandan and Y. Mattey

Experimental Investigation of Factors Affecting the Injectability
of Microcement Grouts 1221
    M.C. Santagata and E. Santagata

Treatment of Medium to Coarse Sands by Microcem H900 as an Alternative
Grouting to Silicate-Ester Grouts...1235
    Murat Mollamahmutoglu

Formulation of High-Performance Cement Grouts for the Rehabilitation
of Heritage Masonry Structures 1243
    S. Perret, G. Ballivy, D. Palardy, and R. Laporte

Sealing of Dilatation Joints with Polyurethane Resins 1254
    Jozef Hulla, Peter Slastan, and Drahomir Janicek

Irreversible Changes in the Grouting Industry Caused by Polyurethane
Grouting: An Overview of 30 Years of Polyurethane Grouting....................1266
    Alex Naudts

Wet-Dry Cyclic Behavior of a Hydrophilic Polyurethane Grout 1281
    C. Vipulanandan, Y. Mattey, David Magill, and Steve Hennings

Hot Bitumen Grouting: The Antidote for Catastrophic Inflows 1293
    Alex Naudts and Stephen Hooey

Liquefaction Resistance of a Colloid Silica Grouted Sand............................1305
    H. J. Liao, C. C. Huang, and B. S. Chao

A Study on the Optimal Mixture Ratio for Stabilization of Surface Layer
on Ultra-Soft Marine Clay 1314
    Byung-Sik Chun and Jin-Chun Kim

GROUTABILITY AND GROUT EVALUATION

A Method for Measuring and Evaluating the Penetrability of Grouts 1326
    Magnus Eriksson and Håkan Stille

Selection Criteria of Polyurethane Resins to Seal Concrete Joints
in Underwater Road Tunnels in the Montreal Area 1338
    J. P. Vrignaud, G. Ballivy, S. Perret, and E. Fernagu

Soil Grouting: Means, Methods and Design..1347
    Daniel Lees and Marcelo Chuaqui

Soil Solidification with Ultrafine Cement Grout...1360
    James Warner

The Groutability of Sands—Results from One-Dimensional and Spherical Tests 1372
    Jens Mittag and Stavros A. Savidis

GROUTED CUT-OFFS

Grouted Cofferdam for an Intake Structure in Mixed Rock
and Gravel Environment..1383
K. Ramachandra, A. Wern, R. J. Kapadia, and S. K. Shim

Humboldt Bay Nuclear Power Plant Repair Project.....................................1392
Monica M. Rourke

Grout Curtain Effectiveness in Fractured Rock by the Discrete Feature
Network Approach..1405
D.A. Shuttle and E. Glynn

Performance Monitoring of Grout Curtains in Slovakian Flysh
and Volcanic Rocks..1417
Jozef Hulla, Dusan Chlapik, and Robert Hok

Seepage Control by Grouting Under an Existing Earthen Dike....................1429
Hasan Abedi, Gary Simard, and David P. Lohman

MODELING AND VERIFICATION

State of the Art in Computer Monitoring and Analysis of Grouting............1440
Trent L. Dreese, David B. Wilson, Douglas M. Heenan, and James Cockburn

Numerical Simulation of Chemical Grouting in Heterogeneous Porous Media 1454
Tirupati Bolisetti and Stanley Reitsma

Some Aspects on Grout Time Modeling..1466
Thomas Dalmalm and Håkan Stille

Model Testing of Passive Site Stabilization: A New Grouting Technique 1478
Patricia M. Gallagher and Alyssa J. Koch

Geophysical Investigations to Assess the Outcome of Soil Modification Work............1490
Bruno Gemmi, Gianfranco Morelli, and F. A. Bares

Electro-Osmotic Grouting for Liquefaction Mitigation in Silty Soils.........1507
S. Thevanayagam and W. Jia

GROUTING FOR TUNNELS

Ground Treatment for Tunnel Construction on the Madrid Metro...............1518
Pedro R. Sola, A. Sarah Monroe, Lucas Martin, Miguel Angel Blanco,
and Raúl San Juan

Grouting Techniques as Part of Modern Urban Tunneling in Europe 1534
Eduard Falk and George Burke

Grouting to Minimize Settlements Prior to Tunnel Excavation—A Case Study..........1546
Douglas M. Heenan and Michael Xu

Design of Grouting Procedures to Prevent Ground Subsidence
over Shallow Tunnels...1557
Ross T. McGillivray

Use of Grouting to Reduce Deformations of an Existing Tunnel Underpassed
by Another Tunnel 1570
 S. A. Mazek, K. T. Law, and D. T. Lau

Soil Stabilization Grouting Under a Railway for Micro-Tunneling
for a Sewer Crossing 1582
 M. Chuaqui and R.P. Traylor

Principles of Ground Water Control Through Pregrouting in Rock Tunnels 1594
 Orjan A. Sjostrom

Injection of a Ventilation Tower of an Underwater Road Tunnel Using Cement
and Chemical Grouts 1605
 Danielle Palardy, Gérard Ballivy, Jean-Philippe Vrignaud,
 and Caroline Ballivy

The Toronto Transit Commission's Subway Tunnel and Station Leak
Remediation Grouting Program 1617
 L. Narduzzo

Large-Scale Field Investigation of Grouting in Hard Jointed Rock 1628
 Thomas Dalmalm and Thomas Janson

Long-Distance Grouting, Materials and Methods 1640
 Christopher R. Ryan, Steven R. Day, and Donald W. McLeod

The Effect of TAS Method by a Supplementary Method to Tunnel 1652
 Byung-Sik Chun and Yoo-Hyeon Yeoh

Subject Index I-1

Author Index I-5

Railroad Embankment Stabilization Demonstration for High-Speed Rail Corridors

Andrew Sluz[1], Theodore R. Sussmann[2], and Gopal Samavedam[3]

Abstract

The development of high-speed railroad corridors in the United States is being considered by Congress as a fuel efficient and economical alternative to air or highway passenger travel. The existing infrastructure is, in many cases, suitable for freight traffic but not for the more exacting geometry standards of high-speed rail passenger trains. In many cases the proposed passenger service would use existing trackage heretofore carrying only slower moving freight trains (e.g., the newly opened service on the Northern New England Corridor (The Downeaster) between Boston, Massachusetts, and Portland, Maine). Instability in the roadbed can cause changes in track geometry at a rate unacceptable for safe or economical high-speed operation over existing lines. This project was conducted to demonstrate that existing ground stabilization techniques could be utilized to economically improve track performance for high-speed service.

Rail traffic and the resulting limited track time available for maintenance in high-speed corridors dictate that embankment stabilization methods must be employed with minimum traffic disruption. The Federal Railroad Administration (FRA) Office of Railroad Development initiated a demonstration project to identify an unstable railroad embankment and effect a remedy. The purpose of the project was to develop experience with and demonstrate the capabilities of ground improvement techniques for reducing track maintenance requirements.

The line segment selected for demonstration had a history of track settlement that continued after the line was rehabilitated for passenger service. After only a few years of renewed service, it became evident that the embankment was still subject to chronic settlement that required frequent resurfacing. A sub-surface investigation determined that a variable-thickness peat layer underlying the embankment caused the settlement.

---

[1] Civil Engineer, John A. Volpe National Transportation Systems Center, 55 Broadway, Cambridge, MA 02142, phone 617-494-2276; sluz@volpe.dot.gov

[2] Civil Engineer, John A. Volpe National Transportation Systems Center, 55 Broadway, Cambridge, MA 02142, phone 617-494-3663; sussmannt@volpe.dot.gov

[3] Engineering Manager, Foster-Miller, Inc., 350 Second Avenue, Waltham, MA 02451, phone 781-684-7275; gsamavedam@foster-miller.com

Based on the information from the site investigation, a remedial program was devised to minimize the track settlement by improving the stability of the peat layer. Grout pipes were installed from the side of the embankment with little disruption of rail service. A cement grout was pressure-injected into the embankment, targeting the peat layer. After grouting, the track elevation was monitored periodically to determine whether the program had stabilized the track geometry.

This paper describes the site, investigation procedures, rehabilitation, and post-stabilization monitoring of the embankment as an example of a method to economically address problematic track conditions with minimal disruption of rail operations. The program was successful and it is believed that similar strategies can be employed to fix a variety of embankment problems and reduce the cost of maintaining the high quality track geometry necessary for high-speed service.

Introduction and Background

There are many corridors in the United States connecting major population centers that are being considered for high-speed rail passenger service. The benefits of the service to ease highway congestion and to benefit conservation and the environment are quite clear (U.S. DOT, 1996). Most of the new high-speed tracks will take advantage of existing railroad track upgraded for higher speeds. The great majority of these lines were constructed over 50 years ago, many over unsuitable foundations.

Typical railroad embankment construction in the early twentieth century minimized haul by borrowing material from within the right-of-way. The existing ground was only minimally prepared for placement of the embankment. Often, the embankments were allowed to "winter" (i.e., sit untouched over the winter months, to "settle" in). After commencement of service, if the track continued to settle, ballast (the angular crushed rock surrounding the ties on top of the embankment) was added and the track resurfaced and raised (literally pulled up to proper elevation and cross-level). For most embankments, settlement would gradually slow and the track would become relatively stable. With a stable embankment to provide track support, track geometry degradation would be caused primarily by the service live loads.

Where soft or weak soil underlay the embankments, track geometry would continue to deteriorate because of bearing capacity or slope failures. The railroads would try various measures, even reconstruction, to solve this problem. Tracks that were built over swamps with embankments underlain by layers of fine-grained or organic material remained problems. Consolidation of the fine-grained layers could continue over many decades. Constant addition of ballast to compensate for the settlement would add to the total stress on the peat layer increasing the rate of consolidation and exacerbating the problem.

A railroad could usually abide long-term settlement with periodic resurfacing. For ordinary freight traffic, speed restrictions can be employed until geometry is brought back within tolerable limits. High-speed passenger service requires tighter tolerances on geometry and better on-time performance than freight service, making this strategy impractical. Frequent interruptions of service negate the advantages of the higher speeds, while lowering track geometry standards jeopardizes safety with potentially disastrous results. To make high-speed rail corridors practicable, it is necessary to develop a means of stabilizing the embankments to reduce settlement rates. To this end, the FRA Office of Railroad Development sponsored a demonstration stabilization project.

Site Description

The site chosen for the demonstration project was the newly reopened Plymouth/Kingston commuter line of the Massachusetts Bay Transportation Authority (MBTA) near South Weymouth, Massachusetts. The 3 m (10 ft) high embankment had been constructed over soft marshland in the early 1900s as part of the Old Colony Line. The line was abandoned for approximately 35 years and then reconstructed and reopened for service in 1996.

The embankment had settled over the 35 years of abandonment and was reconstructed by adding approximately 1 m (3 ft) of new fill and ballast to bring the rails to proper profile. After service resumed, the track had to be resurfaced once or twice a year to maintain Class 5 service. Amtrak, responsible for maintaining the track, sought a solution that would inhibit settlement and reduce maintenance requirements for the track.

Site Reconnaissance/Local Geology. The area traversed by the railroad at the problem site was a marshy lowland. Upon initial investigation, the presence of surface water to the west of the railroad embankment (see Figure 1) was noted. The Wisconsin era glacial retreat (8,000 to 15,000 years ago) eroded the Dedham Granite bedrock and the draining meltwater deposited the eroded material in stratified glacial deposits of sand and gravel over the bedrock in the lowlands (Kauffman and Trepanowski, 2000 and Peragallo, 1989). Repeated vegetation of the areas has since filled in many of the remaining smaller surface depressions (e.g., glacial lakes and ponds) making swamps, bogs, and marshes as observed during the initial site reconnaissance. The surface soil at the site was identified as Freetown Muck by Peragallo (1989) consisting mainly of very poorly drained soils formed of highly decomposed organic material and silty alluvium that developed near local rivers. The Freetown Muck is poorly suited for most uses because of the seasonal high water table, flooding, and low strength (Peragallo, 1989).

In the early 1900s, the original single line railroad embankment was constructed over the Freetown Muck. Typical construction of early railroad embankments consisted of building a roadbed of locally available granular soil at the approximate grade and alignment required for the line. Local railroad authorities report that this particular section of the railroad has a long history of settlement and repeated maintenance required to keep the line operational. The line remained in service until the 1960s, after which the line was abandoned until 1996. In 1996, the track was reconstructed for passenger service operations by the MBTA. Reportedly, the track at the site had settled up to 0.9 m (3 ft) during the 35 year abandonment.

During reconstruction, it is likely that the old track was removed, the surface of the existing embankment graded, and new fill placed to the approximate height of the present line. After service was reestablished, additional settlement of up to 0.3 m (1 ft) was observed between 1996 and 1999. Maintenance personnel estimated that the maximum settlement rate in the test zone was 8 to 15 cm (3 to 6 in.) per year.

Figure 1. Standing water at the base of the embankment.

The overall stratigraphy at the site consists of a track surface of ballast, followed by granular fill (placed during the 1996 reconstruction), over granular fill from the original embankment construction, a variable thickness of organic Freetown Muck, a glacial sand and gravel deposit, and Dedham Granite bedrock at a depth of approximately 13.7 m (45 ft) (Kauffman and Trepanowski, 2000). The mechanism causing the reported 0.9 m (3 ft) of settlement during the 35 year abandonment was, most likely, compression or consolidation of the Freetown Muck.

Site Investigation. The site investigation was conducted to confirm the source of the embankment settlement: either in the embankment fill or in the foundation soil. Three test borings, numbered B1-B3, were conducted using a hi-rail truck-mounted drill rig to provide on-track access to the site. The test borings consisted of advancing the drill hole with hollow stem augers and performing nearly continuous standard penetration tests (SPTs), conducted in general accordance with American Society for Testing and Materials (ASTM) D1586 (ASTM, 2000). The split spoon sampler was driven with a 63.5 kg (140 lb) hammer raised using a rope and cathead. A test boring made in 1993 during the design of the line reconstruction for the present passenger service was used to guide the investigation. The embankment was underlain by peat (Freetown Muck) over sand and gravel. The blow count (N) from the SPT ranged from 27 to over 75 for the original fill and from 37 to over 100 for the newly placed embankment layer indicating dense and well-compacted soil in each layer.

The four test borings, three from the testing conducted for the stabilization, and one from the design of the reconstruction, were made along the eastern track and indicate a relatively uniform embankment. A longitudinal section along the eastern track is presented in Figure 2, which shows the relatively uniform embankment construction, but a variable thickness peat layer. The thickest portion of the peat layer approximately corresponded to the location of the maximum track settlement, which provided an indication that the peat layer in the foundation soil was the source of the settlement. A thin clay layer was identified in test boring B1 underlying the peat.

One undisturbed 0.076 m (3 in.) diameter, thin walled Shelby tube sample was obtained from the Freetown Muck/Peat subgrade material in Test Boring B1 at a depth of 4.3-4.9 m (14-16 ft). The sample was obtained for laboratory testing and appeared to be fine-grained, cohesive soil, but contained a large percentage of fibrous organic material.

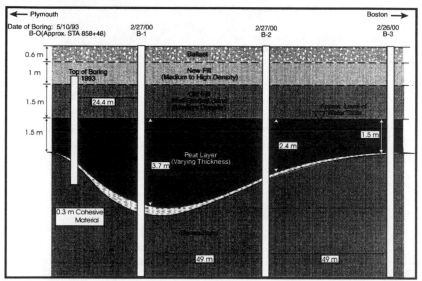

Figure 2. Variation in the subsurface conditions in the demonstration area.

Laboratory Testing. Laboratory testing consisted of grain-size distributions on the existing embankment fill and on the sand and gravel underlying the Freetown Muck/Peat, as well as an organic content test and a consolidation test on the Freetown Muck/Peat.

The Freetown Muck/Peat consisted mainly of loose black fibrous organics, with fine sand in some borings. The consistency of the peat varied with depth from mainly fibrous organics near the top of the deposit to what was classified as fine-grained peat at the bottom. In test boring B1, a layer of clay was found to underlie the peat. The organic deposit consists of a mixture of fine sand, silt, and clay with a large amount of fibrous organic material. The organics test on the peat sample indicated approximately 86 percent organic material.

A consolidation test was conducted on a portion of the undisturbed Shelby tube sample due to the appearance of some of the peat samples as a fine-grained soil and the presence of clay under the peat,. The results from the consolidation test are shown in Figure 3. The soil was found to be normally consolidated clay soil with a preconsolidation pressure of approximately 57.5 kPa (8.3 psi). The preconsolidation pressure was approximately equal to the current stress state of the soil at the surface of the peat deposit, indicating normally consolidated conditions (Craig, 1998).

Since the peat was found to behave similar to a normally consolidated fine-grained soil and since the location of the maximum settlement coincided with the location of the thickest layer of peat (Freetown Muck), it was concluded that consolidation of the peat layer was responsible for the track settlement.

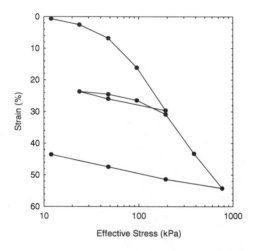

Figure 3. Consolidation test results.

Grout Application

A grout program was developed to mitigate the settlement of the embankment due to compression and consolidation of the organic peat layer. The grouting concept was to compress, compact, or consolidate the peat layer by adding grout under pressure (the actual mechanism depending on the local consistency of the peat which varied from the top to the bottom of the layer). The specific grout program could be termed compensation grouting, which is typically applied to compensate for loss of material volume. Since the embankment settlement was previously compensated for during track maintenance such as ballast dumping and track surfacing, the goal was to install the grout and maintain the grade of the track. The goals of this grout project fall within the goals of the general category of limited mobility displacement (LMD) grout, as described by Byle (1997 and 2000).

The procedure was to install sleeve port pipes through the embankment and the peat layer. Cement grout mixed in a 1:1 water-cement ratio by volume was pumped at a pressure between 1 MPa (150 psi) and 2 MPa (300 psi) through specific ports using nitrogen filled double packers (shown in Figure 4) to confine the grout to the desired port. During grouting, the track elevation was monitored to ensure the track surface did not move appreciably during grouting. The largest track movement recorded was 6.1 mm (0.02 ft).

Sleeve port pipes with grout injection ports spaced approximately every 0.9 m (3 ft), Figure 4, were installed from the side of the track at angles of 50°, 30°, and 20° from horizontal, as shown in Figure 5. To install the sleeve port pipes, a 0.13 m (5 in.) diameter casing was advanced past the bottom of the peat layer with a Davey Kent DK620 drill rig. The drilling was accomplished from the side of the track, as shown in Figure 6, to minimize operational delays. The casing was then washed out to remove any soil and the sleeve port pipe was inserted and the casing was removed. Thirty six sleeve port pipes were installed to treat the 76 m (250 ft) of the embankment where the peat thickness exceeded approximately 1.3 m (5 ft). The average spacing was approximately 2.1 m (7 ft), with sleeve port pipes spaced more closely where the peat layer was thickest.

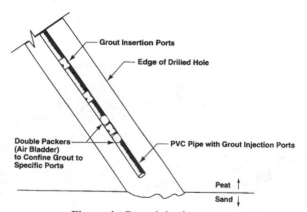

Figure 4. Grout injection concept.

The 50° sleeve port pipe was grouted first. After the grout cured in this location, it would provide confinement for the grouting in the 30° and 20° pipes. In all pipes, the ports near the ground surface were grouted first to seal the lower ports so that the grout would penetrate the peat layer instead of propagating to the surface. The initial grouting of the upper 50° pipe and the upper grout ports was done to seal the grout in the peat layer and prevent flow of the grout to the ground surface or into the adjacent layer. During grouting, volume was measured using a flow meter and the pressure was recorded on the grout pump. On average, 284 L (75 gallons) of grout was used for each port in the peat layer. The approximate total volume of grout pumped was 61,000 L (16,000 gallons). The sleeve port pipes were left in the embankment and the grout was washed from the pipes to accommodate future grouting, if any further settlement was observed.

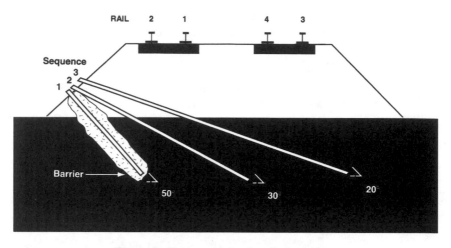

Figure 5. Sleeve port pipe installation plan.

A) Rig placement							B) Drilling grout holes

Figure 6. Drilling for grout injection pipe installation.

Post-Grout Monitoring and Maintenance

Track settlement was monitored by surveying the track elevation prior to and after grouting, directly before and after tamping, and over the 6 months following surfacing. The elevations were measured with a transit on the top of each rail every 13 m (50 ft) originating at the 1993 borehole (as shown in Figure 2). Figure 7 shows the measured changes in elevation. The track settled at an approximately uniform rate (5 mm/yr, 0.2 in./yr) for about 1 year after grouting, until the track was lifted and aligned. Track surfacing maintenance was conducted to lift and align the track in July 2001 and resulted in the 10-15 mm (0.4-0.6 in.) increase in elevation. Following surfacing (track lift and tamp), the behavior of the 2 tracks (rail 1 and 2 compared to rail 3 and 4) was different, likely due to rearrangement of the ballast, which was loosened by the surfacing operation, differently under each track.

After monitoring the track for 1 year after grouting, the maximum movement, 8.9 mm (0.35 in.), was seen on the west rail of the west track. The other three rails moved 6.4 mm (0.25 in.) or less. The total movement was probably not the result of consolidation settlement of the peat alone, but probably included movement in the ballast due to train action. During resurfacing, the rail was raised an average of 13 mm (0.5 in.). The observed movements over 6 months after surfacing was small and included heave on the west track, indicating that train loading may have had a larger impact on elevation changes than settlement. The railroad's conservative estimate of settlement prior to grouting was between 76 to 153 mm (3 to 6 in.) a year. After grouting, the settlement rate was reduced by more than an order of magnitude.

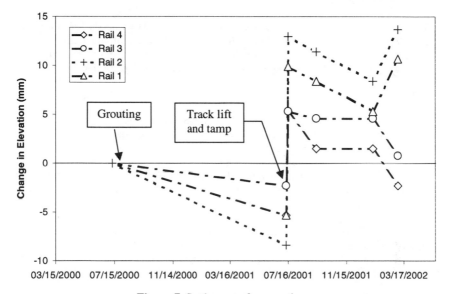

Figure 7. Settlement after grouting.

Conclusions

An investigation of a railroad embankment with a chronic settlement problem indicated that the source of the settlement was an organic peat layer underlying the embankment. The investigation consisted of several test borings using standard penetration tests to obtain samples and characterize the soil and one Shelby tube sample obtained from the fine-grained peat layer for laboratory consolidation testing. The embankment consisted mainly of dense sand, underlain by an organic peat deposit. A grout program was developed to compress, compact, or consolidate the peat layer.

Results from continued monitoring indicate that the settlement in the stabilized segment was reduced greatly after grouting and could have been eliminated altogether. For this scenario where the embankment was underlain by a relatively

shallow layer of consolidating peat, early findings indicate that the compensation grouting has been effective. The MBTA has chosen to employ the method at other sites.
Future development of high-speed rail corridors will be needed to relieve congestion on highways and air routes between major urban centers. These developments require cost-effective and innovative techniques to improve existing infrastructure. In the example presented in this paper, a technique to identify the cause of an embankment settlement problem and to improve the condition of the embankment was demonstrated.

Acknowledgements

This project was funded under the FRA Next Generation High Speed Rail Program. John Kidd, formerly of Foster-Miller Inc. conducted much of the fieldwork and managed the program. GZA Geoenvironmental conducted the site investigation, laboratory testing, and grouting program under contract to Foster-Miller, Inc.

References

ASTM (2000). "Standard Method for Penetration Test and Split-Barrel Sampling of Soils," Standard D1586, American Society for Testing and Materials, *Annual Book of ASTM Standards*, Vol. 04.08.

Byle, M. J. (1997). "Limited Mobility Displacement Grouting: When 'Compaction Grout' is Not Compaction Grout," Geotechnical Special Publication 66, *Grouting: Compaction, Remediation, and Testing*, ASCE.

Byle, M. J. (2000). "An Approach to the Design of LMD Grouting," Geotechnical Special Publication 104, *Advances in Grouting and Ground Modification*, ASCE.

Craig, R. F. (1998), *Soil Mechanics*, Sixth edition, Spon Press, London.

Kauffman, M. D. and J. Trepanowski (2000). "Basewide Groundwater Flow Assessment Phase II Remedial Investigation South Weymouth Naval Air Station, Weymouth, Massachusetts," Submitted to Northern Division Environmental Branch, Code 18 Naval Facilities Engineering Command, Lester, PA, by Tetra Tech NUS, Inc., King of Prussia, PA. Contract Number N62472-90-D-1298, Contract Task Order 0310, December, 2000.

Peragallo, T. A. (1989). *Soil Survey of Norfolk and Suffolk Counties, Massachusetts*. United States Department of Agriculture, Soil Conservation Service, In Cooperation with the Massachusetts Agricultural Experiment Station.

U.S. DOT (1996). *High-Speed Ground Transportation for America, Overview Report*, U.S. DOT, FRA, August 1996.

ACTIVE SETTLEMENT CONTROL WITH COMPENSATION GROUTING - RESULTS FROM A CASE STUDY

Clemens Kummerer[1], Helmut F. Schweiger[2], Reiner Otterbein[3]

ABSTRACT: Near surface tunnelling in built-up urban areas has lead to the development of special geotechnical measures to protect buildings from damage resulting from undue (total and differential) settlements. In contrast to passive ground improvement techniques, compensation grouting is an active method applied to counteract subsidence induced by tunnel excavation. Compensation grouting is done in two stages: In the first stage grouting between the ground surface and the tunnel is performed for "conditioning" the soil. After the immediate response of the system is ensured, settlements monitored with accurate measurement devices are compensated in the actual grouting phase. In this paper compensation grouting operations for a tunnel excavation underneath a station building are described in a case study. The efficiency of compensation grouting is discussed for this practical example. To show the basic effects of compensation grouting, finite element calculations are provided for different stages of the grouting process and compared with in-situ measurements.

INTRODUCTION

The construction of shallow tunnels in urban areas requires special protective measure to prevent the structures within the zone influenced by the excavation from damage. To overcome problems associated with (total and differential) building movements, a variety of protective systems can be applied. After Harris (2001) protective measures can by divided into ground treatment measures, in-tunnel measures and structural measures.

[1] Research Assistant, Institute for Soil Mechanics and Foundation Engineering, Graz University of Technology, Rechbauerstrasse 12, A-8010 Graz, Austria;
[2] Professor, Institute for Soil Mechanics and Foundation Engineering, Graz University of Technology, Rechbauerstrasse 12, A-8010 Graz, Austria; email: helmut.schweiger@tugraz.at
[3] Branch Manager, Keller Grundbau GmbH, Mausegatt 45-47, D-44866 Bochum, Germany; email: R.Otterbein@KellerGrundbau.com

Most of these methods are passive, which means that after their implementation no modification can be made. If an active control of settlements is required, "observational" measures such as compensation grouting or structural jacking have to be implemented. In the course of the last decade, compensation grouting has increasingly gained in significance and proved to cope with difficult geotechnical problems.

It is common with this type of grouting to distinguish two different grouting stages, firstly the conditioning phase and secondly the actual heaving phase.

Prior to the tunnel excavation, the ground has to be treated during the conditioning phase to ensure immediate response of grouting when settlements occur. During this pre-treatment phase, the ground is locally displaced and thereby compacted. In normally consolidated soils, the lateral stress increases caused by the displacement of the ground. When the horizontal stress approximately equals the vertical one, horizontal fractures will occur resulting in heave (Raabe and Esters 1993). Loose zones are compacted and the original stress state in the soil that was disturbed during drilling and implementation of the TAMs (Tubes á Manchette; grouting pipes) is restored. The conditioning is finished when heave is observed at the surface. Due to the tightening and strengthening of the subsoil during this stage, a considerably stiffer response of the treated soil compared to the natural soil is achieved. Chambosse and Otterbein (2001a) report on a settlement reduction of 25 to 50% depending on the soil condition. As a consequence, less grouting is necessary in the actual compensation phase.

The grout volume can hardly be estimated for the conditioning phase, as many influences, e.g. loose zones and inhomogeneities, are not explored at that stage. Chambosse and Otterbein (2001a) investigated a number of major projects in Germany. They concluded that the grout quantities range from approx. 40 to 115 litres per m² for medium to dense soils, mainly depending on the stress level.

Simultaneously with the excavation, the real compensation (or heaving) phase is performed on basis of building movements observed. If long-term movements are expected (especially when excess pore pressures have developed due to injections), grouting is necessary even after the excavation is finished. The distribution and quantities of grouting are selected according to the measured displacements. Recently very precise water lever systems have been developed for this purpose.

In the compensation phase, the efficiency of grouting, defined as (average) heave divided by injected volume, is investigated. According to Chambosse and Otterbein (2001a) the efficiency of grouting at that stage is typically between 5 and 20%. These efficiency values vary considerably with stress level, i.e. the efficiency for heaving high loaded foundations is lower than for lightly loaded ones. Moreover, the grouting efficiency is not constant during the grouting process, but increases considerably from the initial to the final stages (see Watt 2002).

NUMERICAL MODELLING OF COMPENSATION GROUTING

As compensation grouting is increasingly specified as a complementary measure in urban infrastructure projects, there is a need to develop a tool to predict the effects of grouting at a preliminary stage. In the proposed paper the application of the finite element method for the design of grouting problems is examined.

Compensation grouting is performed by injecting small quantities at different valves in several passes. From the practical point or view, it is therefore impossible to simulate this process in great detail. A global approach seems to be more appropriate to capture the characteristics of grouting. At the global level a grouting zone around the injection points has to be specified containing all the fracs that develop during grouting. In the proposed approach the grouting zone consists of two regions. In the outer zone the modifications of the soil properties are accounted for. The dimensions of the grouting zone have to be chosen according to the circumstances of the project considered. The actual injections are simulated in elements within the treatment zone, representing the zone adjacent to the valves. The volume change is realized by means of applied volume strains.

The way of realizing the volume increase can by validated by means of a closed form solution developed by Sagaseta (1987). In order to compare the finite element model with this analytical solution (valid for circular injections under undrained conditions in an elastic half space), one single injection at 7.5 m depth below the ground surface was simulated.

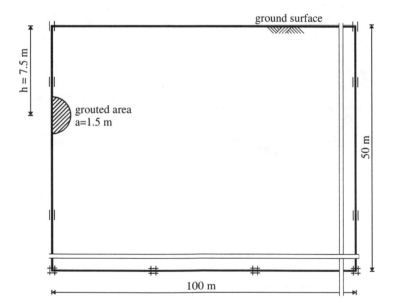

FIG. 1. Geometry for single injection comparison

A volume increase of 5% in the treated cluster was applied representing the injection of 0.35 m³ of grout (solid content) per metre in the treated area. The grouted zone was assumed to be initially 3.0 m in diameter that is a typical value observed in practice (see Fig. 1). For reason of simplicity only one half is modelled.

The soil was assumed to behave as an elastic material with a Young's modulus E=10.000 kPa and a Poisson ratio ν=0.495. During the injection phase the Poisson ratio in the grouted area is reduced to an artificial value.

Fig. 2 shows the results of the comparison with respect to vertical (+ve) and horizontal displacements at ground surface. The calculated maximum heave above the injection axis for both the numerical and the analytical solution was about 15 mm. The numerical result matches the analytical solution very well. With increasing distance from the injection point the results differ slightly because of different boundary conditions. This also applies to horizontal displacements. With increasing distance from the injection point differences between both approaches occur because in the finite element analysis horizontal displacements are zero at the remote boundary, whereas in the analytical solution zero displacements are assumed at infinity.

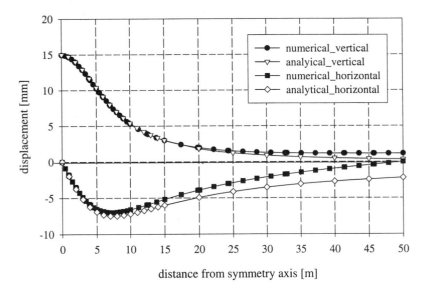

FIG. 2. Vertical and horizontal displacements at surface for single injection

DESCRIPTION OF COMPENSATION GROUTING WORK AT ANTWERP CENTRAL STATION

As part of the 3.8 km long "Antwerp North-South Link" the classified monumental Central Station of Antwerp in Belgium is reconstructed at present by order of Belgian Railways NMBS.

Built as a dead-end station between 1899 and 1905, the Antwerp Central Station (see Fig. 3) could not meet the requirements of the important high-speed railway link between Brussels and Amsterdam, which is in fact one of the major European priority transport projects. The construction of a new 80 m long tunnel underneath the historic station will allow high-speed trains to pass through the station at the second underground level. In 2006, the aforementioned construction activities on the whole railway section will be finished.

FIG. 3: Antwerp Central Station

Although there was a lot of experience with the Belgian Tunnelling Method incorporating a pipe roof, there were concerns at the preliminary stage of the project as actual settlement calculations for the designed construction steps predicted subsidence reaching from 60 to 120 mm. Cracks with more than 15 mm were expected. Moreover, the construction of an old Metro tunnel below the edge of the Central Station yielded to excessive settlements, which had to be counteracted with hydraulic jacks. Especially, in the vicinity of the main piers which transfer the high loads of the dome to large footings (12 x 27 m) settlement problems were expected. A cross section through the main piers of the building is depicted in Fig. 4. To limit the building strains to 0.05%, compensation grouting was specified by Eurostation (the engineering division of NMBS) as an additional measure to protect the Central Station from damage. Contractor Keller Grundbau was charged to perform the grouting operations.

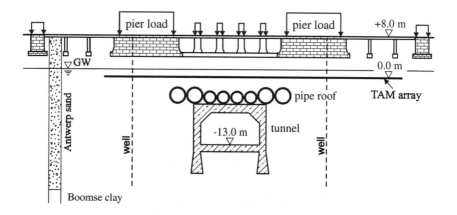

FIG. 4. Cross section through the main piers

The first construction phase was the lowering of the ground water table by about 15 m. The excavation of two shafts on both sides of the building followed this dewatering. After that, the drilling and installation of the TAMs from both shafts was made (Chambosse and Otterbein 2001b). About 3500 m grouting pipes with a diameter of 50 mm and a valve spacing of 0.5 to 1.0 m were put in place 3.5 m below the bottom of the foundations. The position of approx. one third of the TAMs was controlled with inclinometers. The drilling deviation in the sand layer was about 1%. Once the TAMs were installed, the conditioning phase was started. This phase was considered to be completed when the columns within the zone of influence of the works had been heaved by 2 to 5 mm. Underneath the grouted "slab" the jacking of two pipes $\varnothing$2.96 m at each side and four inner pipes $\varnothing$2.47 m was carried out. The outer pipes provided the access for the hand excavation of the trenches, which were stabilized with prefabricated concrete panels. After the tunnel walls and the pipe roof were completed by reinforcing and filling with concrete, the tunnel was excavated in sections with a maximum length of 6 m. Compensation grouting was made after each pipe jacking operation and during the construction of the tunnel walls and excavation of the tunnel.

As the settlement prevention is done in an "observational" manner, adequate measurement systems form the basis of all the grouting work. In the actual project eight different systems were applied (Otterbein 2000) of which the water level system was the most important one. The GeTec-system with pressure sensors applied in Antwerp records changes of pressure instead of changes the water level itself. In contrast to conventional water level systems this principle is more robust and an overall accuracy of 0.3 mm can be guaranteed. Settlement data are available every 30 seconds. In addition, 93 water levels (including 9 temperature sensors) were installed on columns at different levels. Beside these relative settlement measurements precise levelling was used as a control. In addition, 67 crack devices, 5 vertical extensometers, 3 horizontal and 4 vertical inclinometers were installed.

Watt (2002) investigated the efficiency of compensation grouting for various stages of the construction process (Tab. 1). In total, about 750,000 litres of grout were injected during the whole activities. More than 40% of the total volume was needed for the pre-treatment yielding a relatively low efficiency of 10%. During the grouting progress the efficiency increased considerable to 35%, whereas the average injected volume decreased. The total grout efficiency was about 20 % which lies within the range usually observed for similar conditions.

TABLE 1. Grout efficiency after Watt (2002)

phase	grout volume (litres)	average per injection (litres)	efficiency (%)
pre-treatment	330,000	50	10
pipe jacking and part of tunnel walls	200,000	25	20
rest of tunnel walls and the preparatory work for the tunnel excavation	50,000	15	30
excavation of the tunnel and deflection of the roof slab	170,000	15	35
total	**750,000**	**33**	**20**

FINITE ELEMENT CALCULATIONS

Finite element calculations were made in order to investigate the effects of grouting for different construction phases. The finite element package PLAXIS 3D Tunnel (Brinkgreve and Vermeer 2001) was utilized for this purpose. As the cross section through the main piers was significant for the construction works (see Fig. 4), this part was considered in the calculation. Some simplifications were introduced in the model: plane strain conditions were assumed which were archived by a slice of unit thickness in the three-dimensional domain. Moreover, only one half of the cross section was modelled and the effects of the construction of the Metro tunnel were neglected. The finite element mesh consisting of about 2000 elements is given in Fig. 5.

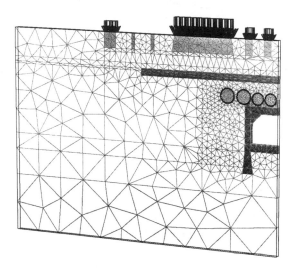

FIG. 5: Finite element mesh for construction at the Antwerp Central Station

For the description of the Antwerp sand an advanced elastic-plastic constitutive law, the so-called Hardening Soil model, was employed. The basic characteristics of this model are:

- stress dependent stiffness according to a power law
- hyperbolic stress-strain relationship.
- distinction between primary deviatoric loading and unloading/reloading.
- failure according to Mohr-Coulomb criterion

In addition to shear hardening this model also accounts for compression hardening and can therefore be applied to both soft soils and stiff soils. The relevant mechanical properties are given in Tabs. 2 and 3, respectively. E_{50}^{ref} and E_{oed}^{ref} represent the stiffness for primary loading, E_{ur}^{ref} for unloading/reloading at a reference stress p^{ref}. In areas which were influenced by the dewatering a cohesion of 15 kPa due to capillary suction was assumed.

TABLE 2. Stiffness parameters for Antwerp sand

E_{50}^{ref} (kPa)	E_{oed}^{ref} (kPa)	E_{ur}^{ref} (kPa)	p^{ref} (kPa)	m (-)	v_{ur} (-)
50,000	50,000	150,000	100	0.5	0.2

TABLE 3. Strength parameters for Antwerp sand

c (kPa)	φ (°)	ψ (°)
1	37	7

The load cases were considered as follows:

- Initial state of stress (the coefficient of lateral earth pressure K_0 was chosen according to $K_0 = 1-\sin \varphi$)
- Backfill of foundations and application of loads
- Groundwater lowering (displacement reset to zero)
- Contact grouting for soil conditioning
- Pipe jacking B-D-A-C with grouting after each jacking operation
- Excavation of the side walls and the tunnel

The vertical movements observed at two water level points (P455 and P470) of the main footing were compared with calculated displacements. In Fig. 6 the result of the comparison for point P455 is shown. After the model was calibrated for the first load case (contact grouting), the pipe jacking with subsequent grouting was modelled. A good agreement of the measured and calculated displacements was found. This also holds for the last load case, where the tunnel excavation with concurrent grouting was modelled.

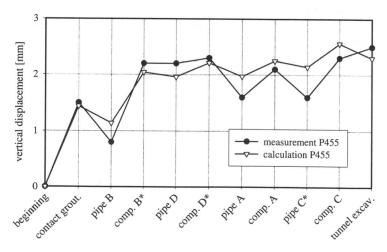

FIG. 6. Comparison measured - calculated settlements for water level P455

Similar results were obtained for point P470 (Fig. 7). Again, the main construction phases were captured with the finite element model.

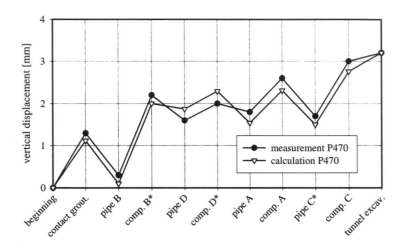

FIG. 7. Comparison measured - calculated settlements for water level P470

In order to quantify the effects of compensation grouting, the load cases described above were calculated without considering grouting. A maximum settlement of about 60 mm was obtained which is within the range of expected settlements. It has to be emphasized though that this value is at the lower end of the range because the numerical model does not account for construction specific uncertainties that are included when settlements are estimated from past experience.

Differences occurred between the grout volumes injected in practice in comparison with the volume expansion applied in the finite element model. For the conditioning phase it is obvious that the grout volume in reality is strongly influenced by loose zones in the soil, inhomogeneities and the local stress concentrations. These influences can hardly be taken into account, as they are usually unknown in the design phase. For subsequent phases it turns out that the grout intake in practice is still higher than in the calculation. However, as observed in practice, the grout efficiency is lower for the settlement compensation of highly loaded than for lightly loaded foundations. Moreover, the grout efficiency is not constant during the grouting operations, but increases with the grouting progress (Watt 2002). This phenomenon is captured in the finite element model, too. Further investigation on the correlation between injected and calculated volumes have to be made.

CONCLUSIONS

For the transition of the Central Station of Antwerp into a modern station the construction of a tunnel underneath the monumental building was necessary. To protect the station from damage due to settlements, compensation grouting was applied in all relevant construction phases as a protective measure. It was proved that settlements due to excavation were controlled by means of grouting, so that only minor damage occurred. The efficiency of the grouting activities varied considerably during the progress.

A finite element model was proposed in this paper that utilizes volume strains to simulate the soil displacement associated with grouting. It also accounts for the modification of the mechanical properties in the zone influenced by grouting.

In finite element calculations the main construction phase were back analysed. The model could reproduce the construction sequence in a very reasonable way. Similar tendencies between the real grouting operations and the calculations were observed, still differences between the grout volume used in practice and the calculation differed.

REFERENCES

Brinkgreve, R.B.J., and Vermeer, P.A. (2001). *PLAXIS 3D Tunnel-Version 1. User-Manual*. A.A. Balkema Publishers, Rotterdam.

Chambosse, G., and Otterbein, R. (2001a). "Central Station Antwerp. Compensation grouting under high loaded foundations". *Proc. Conf. on building response on tunnelling. Case studies from Construction of Jubilee Line Extension*, Burland et al. (eds.). Th. Telford, London.

Chambosse, G., and Otterbein, R. (2001b). "State of the art of compensation grouting in Germany". *Proc. 15th Int. Conf. on Soil Mechanics and Geotechnical Engineering*, A.A. Balkema Publishers, Rotterdam, 1511-1514.

Harris, D.I. (2001). "Protective measures". *Proc. Conf. on building response on tunnelling. Case studies from construction of Jubilee Line Extension*, Burland et al. (eds.), Th. Telford, London, 135-176.

Otterbein, R. (2000). "Einsatz neuer Vermessungssysteme bei der Soilfrac-Sicherung Centraal-Station, Antwerpen", *Beitraege zum Studiedag KIvI*, Breda 2000 (Keller Report 61-59 D, in German).

Raabe, A.W., and Esters, K. (1993). "Soilfracturing techniques for terminating settlements and restoring levels of buildings". *Ground improvement*, Moseley (ed.), Chapman & Hall, London, 175-192.

Sagaseta, C. (1987). "Analysis of undrained soil deformation due to ground loss". *Géotechnique*, London, 37(3), 301-320.

Watt, A. (2002). "Structural works under Antwerp Central Station completed successfully". *J. Cement*, Vol. 1.

FRAC GROUTING – A CASE HISTORY

Douglas M. Heenan[1], P.Eng., Member, ASCE, Janne W. Vataja[2], Trent L. Dreese[3], P.E., Member, ASCE

ABSTRACT: The United States Postal Service's (USPS) Eastpointe Facility in Clarksburg, West Virginia is located at the site of a former strip mine. Reclamation activities prior to construction were accomplished by backfilling the stripped area with the excavated materials after the coal seam had been removed. Subsurface investigations performed for the design of the facility identified settlement concerns due to loose fill and the existence of nested cobbles and boulders. To address this concern the site was reportedly over excavated and replaced with a controlled engineered fill. Placement and compaction methods utilized during placement of this controlled fill are unknown. Constructed in 1989, the 9,300 m^2 single story steel frame structure is supported by columns founded on spread footings. The floor of the building is a grade supported concrete slab isolated from the primary foundation system. Differential settlement of up to 150 mm has occurred causing structural damage and serviceability problems. This paper details, from a construction perspective, the applied grouting technology utilized to stabilize the structure, along with an analysis of the performance achieved. The work was substantially completed in the fall of 2001.

---

[1] Vice President, Advanced Construction Techniques Ltd., 10495 Keele Street, Maple, Ontario L6A 1R7
[2] Project Manager, Advanced Construction Techniques Ltd., 10495 Keele Street, Maple, Ontario L6A 1R7
[3] Senior Geotechnical Engineer, Gannett Fleming Inc., P.O. Box 67100, Harrisburg, Pennsylvania 17106-7100

FRAC GROUTING BACKROUND AND THEORY

Frac (hydrofracture) grouting is a relatively new grouting technique that originated in Europe. The process is applicable to stabilization and compensation grouting of fine grained soils with low hydraulic conductivity where permeation grouting is not effective or is limited. The original development of Frac grouting was to compensate for soil losses and subsequent settlement during tunneling or excavation activities. The process evolved into a soil stabilization technique for the purpose of preventing future settlement due to an unstable soil mass. This latter application was the purpose intended for the Clarksburg project.

The premise of Frac grouting is to inject a grout under pressure to hydrofracture or open pre-existing fractures within the insitu soil. The grout filled fractures create grout lenses which densify and reinforce the soil mass. Grout is initially injected at pressures required to create soil failure, which in turn fills the fracture. The magnitude of the Frac, with regard to length, width and volume is a function of hydraulic gradient established by the grout injection pressure and insitu soil overburden pressures.

In theory, if a soil is normally consolidated, the first grout lens produced will propagate in the vertical direction (direction of major principal stress), which will cause horizontal pressures to increase and hence cause compression in the soil (Raabe and Esters, 1990). If further injections are undertaken until the principal stresses are reoriented, horizontal fractures will be propagated and eventually (sometimes rapidly) lead to heave. In practice, the direction of the initial fracture is often controlled by pre-existing defects, or weak zones within the soil or weathered rock mass, and not the existing state of stress (Rawlings et.al., 2000).

Initially the grout forms a bulb that continues to grow, until the principal stresses of the soils are overcome. At this point a pressure drop occurs, and flow remains either constant or increases, depending on the fracture that has developed. The dynamics of this process are complicated and often difficult to predict. The process, therefore, requires continuous real-time monitoring of structural movement and the grouting parameters to adequately and safely control the Frac grouting operation.

The grout delivery system typically involves the installation of sleeved pipes called tube-a-manchette's (TAM's) that are drilled, driven or jacked into the soil in some predetermined array designed to provide adequate grout zone coverage of the soil mass. TAM's consist of steel or plastic pipes with one-way sleeved valves spaced at 0.3 to 1.5 meters along the pipe alignment through which grout is injected and if necessary re-injected. A double pneumatic packer assembly is used to seal off the individual sleeved valves, through which grout is injected for the purpose of treating the surrounding soil.

CONTINUOUS ELECTRONIC MONITORING

Soil fracture grouting by nature is a technique that has the potential to cause significant structural damage to adjacent structures and tunnel linings. Once Frac grouting is initiated the insitu soil is subjected to internal pressure, and soil displacement will eventually propagate into surface heave. Controlled fracturing of soil requires real-time electronic monitoring of the pressure and flow rate during grouting operations as well as the real-time electronic monitoring of the ground surface or buildings to detect movement. Based on the monitoring data, informed decisions can be made and the fracturing of the foundation soils can be performed with confidence. The data collected during the project is electronically stored within a relational data base that may be used for quantity reconciliation and execution verification along with post analysis of grouting work.

DESIGN AND CONSTRUCTION EXECUTION

The USPS Eastpointe Facility located in Clarksburg, West Virginia, had experienced differential movement since construction of up to 150 mm. As a result, structural damage to interior floors and walls in the form of cracking was evident; as well as some serviceability issues of doors not closing properly. Several elevation surveys were conducted all indicating continual differential movement.

Soils investigations at the site, prior to the building construction in 1989 found that settlement concerns existed due to the presence of nested cobbles and boulders. To address this concern the site was reportedly over excavated and replaced with a controlled engineered fill. Excavation was apparently performed to a depth of 3 to 4.5 meters, but the placing techniques as well as the origin and quality of the controlled fill are unknown.

Subsurface conditions at the site prior to grouting indicated that the fill material was variable (soils investigations were performed by a soils testing company). Boring logs and laboratory test results indicated that subsurface conditions consisted of silty clay with varying amounts of rock fragments, and sand. Site testing consisted of Standard Penetration Testing, with the resulting N values ranging between 1 and 30, with an average N value of 8.

It was found during shaft excavation that the insitu soil comprised of a gray to brown clayey, gravelly silt. Numerous large sandstone and limestone boulders (±

1m x 1.25m x 1.25 m) with clusters of smaller, densely packed sandstone and limestone boulders were also uncovered. In addition, ground water seepage in the order of 20 liters/minute was encountered during the shaft excavation.

DESIGN

The grouting approach (original design by an engineering company overseeing the project) included the construction of four vertical shafts, with radiating fans of sleeved pipes originating from these four shafts and dividing the building into four separate grouting quadrants (refer to Figure 2). From these four shafts, 4 layers of pipes were installed at locations of 4.56 meters below the building slab (Bottom Layer, Mid Layer, and Intermediate Layer), and 3.05 meters below the building slab (Top Layer)(refer to Figure 3). These were subsequently classified as lower level and upper level holes. This exterior approach, utilizing access shafts, was adopted due to the facility being in use 24 hours a day, 7 days a week. Any work that was to be performed from the inside was considered to be too obtrusive.

The contract specifications indicated the following:
- Installation of pipes to be either drilled or driven.
- Grouting flow and pressure to be continuously monitored in real-time.
- Ground monitoring to be capable of reading and processing movement in real-time during grouting operations with an accuracy of 1.5 mm.

INSTALLATION OF SLEEVE PIPES

Inducing further settlement of the existing fragile soil during sleeve pipe installation was a major concern when determining the sleeve pipe installation method. Conventional drilling methods utilize pressurized air or fluid and have the potential to create further damage to the existing fragile soil beneath the USPS Facility. Jacking of the sleeve pipes would provide the least disruption to the existing soils and eliminate any soil losses during sleeve pipe installation while providing some degree of soil compaction in the process. As a result, custom designed and fabricated steel sleeve pipes were jacked into place utilizing a custom designed hydraulic jacking unit.

This process involved precise initial sleeve pipe alignment with continual monitoring of jacking pressures as well as sleeve pipe alignment profiling during installation. The Jacking unit (refer to figure 1) was custom manufactured to meet contractor requirements and was used to hydraulically jack sleeved pipes under the facility.

Figure 1. Custom manufactured Jacking Unit.

The unanticipated presence of boulders, deflection and early termination of some sleeve pipes presented a constructability issue. Jacking pressures were continually monitored and a refusal criteria of 45.36 metric tons jacking force was established as a maximum termination pressure. In two separate instances, boulders encountered during the jacking operation resulted in abrupt deviations and caused sleeve pipes to penetrate the floor slab of the facility. As a result of these two events, more stringent refusal criteria was established for deviation control based on both jacking pressure and sleeve pipe alignment profile. The sleeve pipes were periodically surveyed using a liquid settlement profiler, and the jacking pressures were continuously monitored. A 1.5 meter buffer zone was established from the tip of the pipe to the top of the building floor slab, and in the event that either this 1.5 meter buffer was breached or jacking pressures exceeding 45.36 metric tons, installation of the pipe was terminated. This revised refusal criteria was successful in completing all remaining sleeve pipe installations without further incident.

Upon the successful installation of a sleeve pipe an optical borehole survey instrument was used to establish the final sleeve pipe location, for verification of grout zone coverage. Additional sleeve pipes were installed when early termination occurred or the sleeve pipe was determined to have excessive deviation that would create a situation of inadequate grout zone coverage.

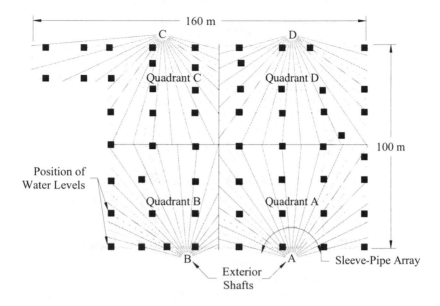

Figure 2. Plan View of Sleeve Pipe Layout.

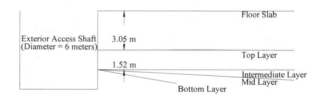

Figure 3. Elevation View of Sleeve Pipe array from one shaft.

GROUTING

The mix design chosen was a slow setting, stable cementitious grout mix with the following formulation and physical properties.

- Mix Design Formulation
  - o Water / cement ratio: 2.25:1 (by weight of cement)
  - o Pre-hydrated Bentonite: 8% (by weight of cement)
- Physical Properties
  - o Compressive Strength: 3.5 MPa
  - o Specific Gravity: 1.29 kg/L
  - o Marsh Funnel: 80 seconds
  - o Bleed: 5%
  - o Initial Set: ± 8 hours
  - o Final Set: ± 12 hours

This above mix was chosen to provide a reasonably long initial set time to allow time for grout delivery and complete injection and re-injection as required. The strength parameter is based on final shear strength requirements and insitu soil parameters.

Grout was mixed at a central batching plant using high-speed colloidal mixers and was delivered to the point of injection via helical screw type pumps connected to a header and return lines. Grouting within the sleeve pipes was performed via a double packer arrangement used to isolate the individual sleeve ports. All grouting operations were electronically monitored for real-time pressure and flow rate.

The following general guideline specification was adopted for this project:

- Maximum production grouting pressure: 550 KPa
- Maximum target volume per sleeve: 500 Liters
- Maximum movement: 1.5 mm
- Maximum flow rate: 20 Liters/min

ELECTRONIC MONITORING

The grouting operation was continuously monitored for pressure, flow rate and volume parameters. A flow meter and pressure transducer unit was located within the shaft at the point of injection and real-time information was electronically transmitted to a central monitoring control center. The information was visually displayed for the operator and stored electronically for future reference. The grouting technician was in constant radio contact with the header man and batch plant operator, and completely directed and controlled the operation based on the project requirements and real-time monitoring data. The monitoring provided information on the physical injection process and was correlated with building movement to provide meaningful control. Based on the requirements outlined in the specifications, options were given to the selection of electronic ground monitoring devices, with the selection being the electronic liquid level system.

A total of 24 electronic liquid levels were used per grouting quadrant, the locations of which are indicated in Figure 2. These levels were attached to the building columns. However, with the building columns being spaced approximately 23 meters apart additional monitoring points were required at closer intervals. To bridge this gap and to provide monitoring of the lighter floor slab, 2 levels were mounted on moveable carts. The liquid levels mounted on the building columns were referenced to a location outside the potential movement zone and the differential movement between the liquid levels and reference point for all locations was electronically transmitted to the central monitoring control station. The grouting technician could then correlate very discrete building movements with the grout injection process and have full control of the operation. Figure 4 shows a schematic of the grouting operations and the electronic monitoring utilized on this project.

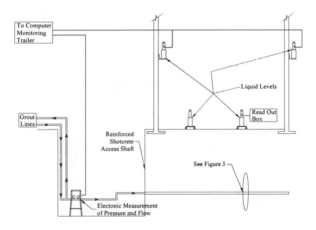

Figure 4. Cross sectional view of grouting operations.

SLEEVE PIPE INSTALLATION ANAYLSIS

The jacking system that was utilized preformed remarkably well from both a production and deviation standpoint. Without the introduction of traditional drilling

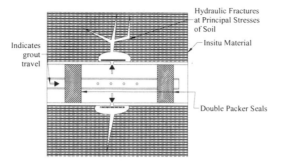

Figure 5. Close-up of sleeve during grouting operations.

fluids for the flushing of the drill cuttings, the in-place soil remained intact and actually received a small degree of compaction from the driving of the sleeve pipes. The presence of numerous cobbles and boulders did provide for a challenging situation that required close monitoring and operator judgment. When a boulder was encountered, early termination and excessive deflection did occur, and additional sleeve pipes were installed in order to attain adequate grout zone coverage. The jacking operation was continually monitored for deviation and jacking pressure as shown in Figures 6 & 7, which represents a typical jacking operation.

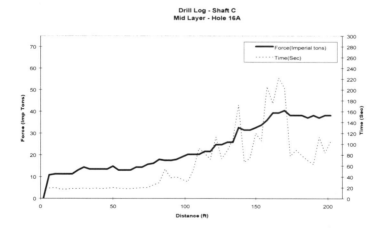

Figure 6. Typical results detailing the jacking pressures and times

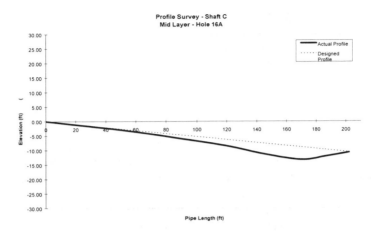

Figure 7. Typical results detailing the profile alignment survey of sleeve pipes.

Deviation of the sleeve pipes exceeded original contractual specifications due to the unexpected frequency of boulders; however grout zone coverage was not compromised due to the installation of additional sleeve pipes to compensate for the early termination, or the excessive deviation of installed sleeve pipes. The following table (Table 1) indicates the average deviation of the sleeve pipes jacked from Shaft A, which is typical of the other three shafts.

PIPES JACKED	AVERAGE DEVIATION (METERS)	DIRECTION
40	1.47	Up
28	1.00	Down
31	1.30	Right
37	1.25	Left

Table 1. Shaft A summary of deviations.

A total of 2,813 meters of pipe was installed at 136 locations from Shaft A with an average deviation of 4%. The magnitude of the deviation is attributed to the presence of the cobbles and boulders within the insitu soil combined with jacking lengths exceeding 60 meters.

GROUTING ANALYSIS

The grouting operation for this project was carefully monitored and was considered to be successful based on post survey data. Analysis of the survey results indicated that subsidence had ceased, thereby indicating that the Frac grouting was a success.

Due to the extensive amount of instrumentation utilized on this project, the primary refusal criteria was based on the electronic liquid level heave monitoring system. This was the only true measure of effectiveness of the grouting operations in terms of real numbers. Detection of very small movements was capable in seconds, which is required in this type of work. The grouting proceeded from the lower level sleeve pipes (Bottom, Mid, and Intermediate Layers, collar elevation at 4.57 meters below the facility floor slab) and proceeded to the upper levels (Top Layer and any subsequent additional pipes, collar elevation at 3.05 meters below the facility floor slab). As expected, the lower zones were able to accept considerably more grout material than the upper due to proximity of the facility. The table (Table 2) below indicates the percentage of the total grout placed per layer on the entire project.

Upper Levels (Top Layer, and any subsequent additional pipes)	28.8 %
Lower Levels (Bottom, Mid, and Intermediate Layers)	71.2 %

Table 2. Project total grout volumes.

Table 2 is an indication that the grouting was being successfully applied. Since the approach was to begin with the lower level holes and to proceed to the upper level, the results indicate that the soil matrix was becoming increasingly compacted, and consolidated.

Figure 8 indicates the responsiveness of the liquid level system to surface heave. As indicated, Sleeve 19 took a considerably greater amount of grout and caused one of the building columns to heave slightly. When grouting resumed on Sleeve 17, the same building column was immediately affected by the influx of a minimal amount of material. Since this heave monitoring system was so responsive, this system was used almost entirely as refusal criteria to ensure that hydrofracturing of the ground was being successful, leading to improvements in the soil characteristics.

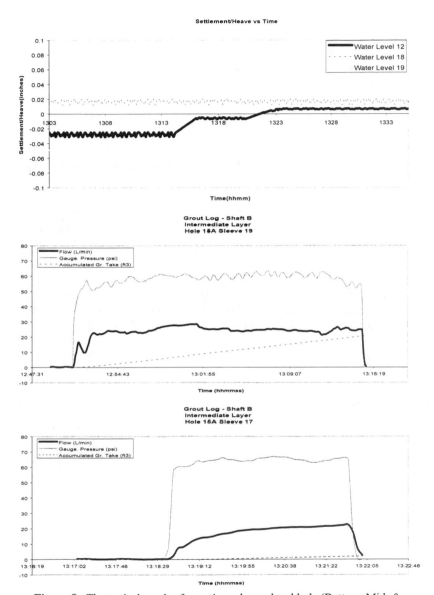

Figure 8. The typical result of grouting a lower level hole (Bottom, Mid, & Intermediate Layers were 1.52 meters lower than the Top).

SUMMARY AND CONCLUSION

Frac grouting on this project was performed to mitigate settlement, and was performed under very stringent operational and performance criteria. A total of 12,800 meters of sleeve pipe was installed and 680 m^3 of grout was injected during the Frac grouting operation over an 18-month period. Early survey results indicate good performance with ongoing surveys being conducted semi-annually.

Frac grouting together with extensive electronic monitoring equipment can alleviate either the anticipated settlement due to tunneling or stop continuing settlement due to poor foundation conditions as was demonstrated on this project.

REFERENCES

Raabe, A.W., Esters, K., (1990). "Soil fracturing techniques for terminating settlements and restoring levels of buildings and structures." *Ground Improvement.* pp. 175-192.

Rawlings, C.G., Hellawell, E.E., and Kilkenny, W.M., (2000). "Grouting for Ground Engineering." *CIRIA Publication C515, Construction Industry Research and Information Association*, Basingstoke Press, London. Chapter 7.

COMPENSATION GROUTING TO REDUCE SETTLEMENT OF BUILDINGS DURING AN ADJACENT DEEP EXCAVATION

Jinyuan Liu[1], Ph.D., Student Member, ASCE

ABSTRACT: Compensation grouting used to protect seven masonry buildings near deep excavation in Shanghai, China is presented in this paper. Grout was injected between the diaphragm wall and the buildings to compensate for ground loss and stress relief caused by the excavation. The design and construction of compensation grouting during adjacent deep excavation are detailed. This project demonstrates that compensation grouting is a cost-effective and promising method to protect properties near deep excavations in soft clay.

INTRODUCTION

Deep excavations will induce a substantial ground settlement behind the retaining wall in soft ground (Peck 1969). Ground surface settlement can cause damages, even structural collapse, to the buildings within the influenced zone. There is an acceptable settlement limit for the building according to a building's structural type (Bjerrum 1963). A successful project using compensation grouting to protect seven six-story masonry buildings during an adjacent deep excavation in soft Shanghai Clay is introduced in this paper. Grout was injected between the diaphragm wall and the buildings to compensate for ground loss and stress relief caused by the excavation. This project demonstrates that compensation grouting is a cost-effective and efficient method to protect properties near deep excavations in soft clay.

COMPENSATION GROUTING

Grouting has a history of nearly 200 years. With the development of material and equipment, many grouting techniques have evolved. Grouting mechanisms are difficult to verify and geotechnical information of sites varies project-by-project.

---

[1] Research Assistant, Department of Civil Engineering, Polytechnic University, 6 MetroTech Center, Brooklyn, NY 11201, liu@poly.edu

Therefore grouting often introduces a grout volume, which is more than the ground loss. The consolidation associated with the dissipation of the pore pressure caused by higher grout volume will reduce the protection result substantially, especially in clay. Compensation grouting is a relatively new technique. It is being used increasingly to control ground settlement during tunneling, where grout is injected between the tunnel and the building foundations to compensate for ground loss and stress relief caused by tunnel excavation (Mair et al. 1994). This technique was used successfully in Baltimore, USA to protect about 40 masonry buildings (Baker et al. 1983). This paper is to present a successful project in Shanghai, China, using compensation grouting to protect buildings during an adjacent deep excavation.

BACKGROUND

The excavation had plan dimensions of 20 m by 277 m (Fig. 1). The excavation depth was 15 m in the middle and 17 m at two ends. A 0.6 m thick diaphragm wall with a depth of 26 m was used as the retaining structure, which is supported by four levels of steel braces with preloaded forces varying from 500 to 2180 kN. The braces were located at depths of 1.2, 5.2, 9, and 12.5 m below the ground surface (Fig. 2) The horizontal spacing between the braces is 3 m.

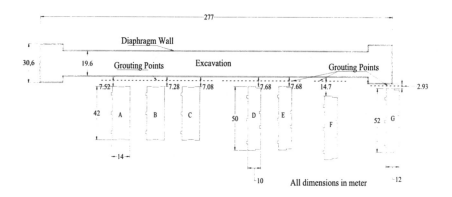

FIG. 1. Layout of Excavation and Buildings

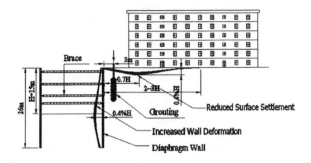

FIG. 2. Principle of Compensation Grouting Near A Deep Excavation

Shanghai Clay is soft clay. The annual average water level is about 0.5–1 m below the ground surface. Typical properties of soils within 30m below the ground surface are shown in Table 1. Properties of Shanghai Clay are detailed in Liu (1999). Most underground structures, for example, basements and tunnels, are located within 20 m below the ground surface. Within this depth, gray clay is the main stratum because of its depth and its characteristics. Gray clay has a water content near its liquid limit. The shear strength of gray clay, however, is very low. It also exhibits an apparent creep property.

TABLE 1. The Physical and Geotechnical Properties of Shanghai Clay

No	Description	Bottom depth	Unit weight	Water content	Void ratio	Modulus	Shear Strength	
			γ	ω	e	E	C	ϕ
		m	kN/m^3	%	--	MPa	kPa	°
(1)	(2)	(3)	(4)	(5)	(6)	(7)	(8)	(9)
2	Yellow clay	2.4~4	18.15	30.2	1.03	7.81	9.6	19
3	Gray silty clay	8.5~10	17.36	42.1	1.30	3.34	18.8	12
4	Gray clay	17~19	17.07	48.8	1.32	1.96	19.5	12
5	Gray silty clay	19~21	17.85	37.6	0.99	3.93	21.0	14
6	Green silty clay	26~29	19.82	22.3	0.68	3.57	26.2	16

There were seven six-story masonry buildings 2.9 to 14.7 m away from the excavation (Fig. 1). Based on large amount of well-documented field monitored data, the maximum settlement is located approximately 0.7 times the excavation depth behind the wall with a magnitude of around 0.4 percentage of the excavation depth. The influenced zone is about two to three times the excavation depth (Fig. 2). More descriptions of underground construction in Shanghai Clay can be found in Liu and Hou (1998).

The proposed excavation was expected to induce a substantial settlement behind the wall. The maximum ground settlement in this project would be about 10 m away from the wall, which would be under the buildings. The maximum ground settlement would be much more than 6 cm considering the surcharge from the buildings. The differential settlements induced by the excavation would be expected to exceed the limits for the masonry structure. It was necessary to take steps to ensure the buildings' safety during excavation. Compensation grouting was selected among many possible methods based on its proposed budget.

DESIGN AND CONSTRUCTION OF COMPENSATION GROUTING

Principle

Ground loss from the deformation of the diaphragm wall will transfer through the soil to the ground surface behind the diaphragm wall. This transmission takes place gradually over a period of time. Compensation grouting compensates for ground loss before the deformation can transfer to the building. Thus it limits the buildings' settlement. Theoretically speaking, grouting simultaneously with the excavation will produce the best result. The shaded areas in Fig. 2 illustrate that compensation grouting not only reduces the ground surface settlement but also increases the deformation of the diaphragm wall. In practice, however, grouting starts after the placement of braces for the safety of the excavation. The grout gel time is limited so as to be as short as possible, and the grout volume is tightly controlled to limit the pore pressure build-up and the potential damage to the diaphragm wall.

Design

The design and construction of compensation grouting followed the construction of the excavation. The total excavation depth was divided into five layers. The first layer of excavation was followed by the placement of first level of braces. The same procedures were followed until the total excavation depth was reached.

In order to reduce the pressure on the diaphragm wall, the grouting points were located 3 m away from the diaphragm wall and spread parallel to the diaphragm wall as shown in Fig. 1. The horizontal spacing between the grouting points was 3 m.

One grouting point consisted of three plastic pipes used for the three stages of grouting. The spacing between plastic pipes was 25 cm. The first stage grouting started from 4 m deep up to 1 m following the second layer excavation (Fig. 3a). The second stage was from 8 m deep up to 4 m for the third layer excavation (Fig. 3b). The third stage grouted from 11 m deep up to 8 m for the fourth layer excavation (Fig. 3c).

Time control in compensation grouting includes both grout gel time and elapsed

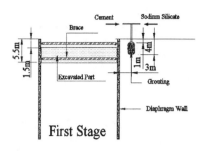

First Stage

(a)

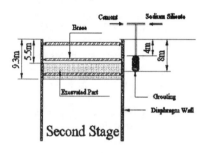

Second Stage

(b)

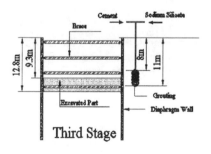

Third Stage

(c)

FIG. 3 Construction Procedure of Compensation Grouting

time between the placement of the brace and grouting. Chemical grouting using cement and sodium silicate was selected for its gel time of one minute. The grouting started within half an hour after the placement of the brace to ensure a better result.

The grouting volume for each stage was designed based on the equation:

$$V_{grout} = \alpha . V_{groundloss} \tag{1}$$

Where V_{grout} was designed grout volume, $V_{groundloss}$ was predicted ground loss from the deformation of the diaphragm wall, and α was the parameter, which was specified as 2 at the beginning and was adjusted based on in situ test results.

Equipment

The equipments used consisted of a driller, two screw mixers, two hydraulic piston pumps, plastic pipes, high-pressure hoses, and steel injection pipes. Measuring equipments including pressure gauge, timer, earth pressure cell, inclinometer, and optical surveying equipment were used to ensure the safety of excavation and buildings.

Preparation

Plastic pipes needed to be installed before the excavation. Holes were drilled using 5 cm header to the designed depth. Plastic pipes with a diameter of 4 cm were then placed in position. Plastic pipes were cemented at the top 1.5 m deep to prevent them from being pumped out during later construction.

Each plastic pipe piece was 33 cm long and connected to the others to the desired length. Predrilled holes were located at both ends of plastic pipe piece to facilitate grouting. Plastic covering was used to cover the predrilled holes to prevent them from being plugged-up by the slurry and also from being broken by the nearby grouting.

Construction

After the brace was placed in position, the corresponding grouting stage needed to start within half an hour. The injection pipes with a diameter of 2.5 cm were put inside the plastic pipe. Each injection pipes was 1 m long and connected to the others and lowered to the position. A 50 cm long special head with holes along its length was used. There were two plastic O-rings mounted at both ends of the head to ensure that the grout flowed horizontally.

A three-way T-shape connector was used to connect the pressure hoses for the two slurries and the injection pipes. The ratio between the two slurries was 1:1 by volume. The grout volume is critical in compensation grouting. Under a constant pumping rate, the grout volume is controlled by the grouting time. A pumping rate of 20 strokes per minute with a flow rate of 20 liters per minute was used in this project.

The gel time was checked before grouting. Regular checks were needed during grouting. If the gel time got too long, the work was stopped until the material had met the requirement.

A staging-up method was used in each grouting stage. The stage length was 15cm. Time for lift-up was determined based on the reading of the pressure gauge placed at the collar of the injection pipe.

Safety protections must be implemented during this type of construction. Grouting personnel must be trained and telecommunication equipments were required for a smooth construction.

RESULT ANALYSIS

In situ testing, including the ground surface settlement, the brace force reaction, and earth pressure response, were performed. The force in the brace was found to be proportional to the grout volume. The distance between the grouting point and the diaphragm wall minimized the grouting pressure influence. Load increments from 250 to 400 kPa were measured in the brace.

The change in the ground surface settlement during grouting is presented in Table 2, where the grouting was the third stage. The ground surface change due to grouting was like a cone. The influence radius was about 1.6 times the depth of grouting.

The volume of surface change calculated based on the monitored data was about half of the grout volume.

Because both the gel time and the grouting pressure were tightly controlled, no sudden settlements were observed in either the ground surface or the buildings throughout the project.

In addition to the tests shown above, the earth pressure change due to grouting was also monitored and analyzed, shown in Fig. 4. The earth pressure cells were located 2 m away from the grouting point. The earth pressures were increased instead of reduced due to grouting, especially around the depths of grouting. It showed that the grout compensated for ground loss in the region around the grouting point.

The construction of compensation grouting started from Building A, and then moved from B and G toward the middle of the excavation to Building D and Building E. At the final stage of compensation grouting, the grout volume was increased from two to three times ground loss and the limit for grouting pressure was also raised from 0.25 to 0.5

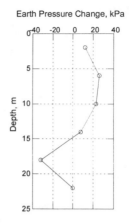

FIG 4. Earth Pressure Change Due To Grouting

MPa. The total settlement of seven buildings throughout the excavation is listed in Table 3. The duration of the corresponding excavation is also listed along with the settlement.

Table 2   The Ground Settlement Change Due To Grouting

Monitoring points	G2	G3	G4	G5	G6	G7	G8	G9	G10
(1)	(2)	(3)	(4)	(5)	(6)	(7)	(8)	(9)	(10)
Distance from grouting (m)	0.2	3	6	9	12	15	18	21	24
Before grouting (mm/day)	1.89	2.04	2.11	1.73	1.87	1.57	0.87	0.68	0.59
During grouting (mm/day)	0.46	1.02	0.92	0.96	0.93	0.51	0.65	0.80	0.54
After grouting (mm/day)	1.37	1.29	1.25	1.01	0.54	1.04	0.94	0.86	0.72

Table 3.   Settlements of Seven Buildings During Excavation

Building	A	B	C	D	E	F	G
(1)	(2)	(3)	(4)	(5)	(6)	(7)	(8)
Distance from wall (m)	7.52	7.28	7.08	7.68	7.68	14.68	2.93
Duration of excavation (day)	106	61	43	40	61	81	37
Total settlement (mm)	51.24	34.90	23.95	34.50	41.88	24.30	20.48

CONCLUSION

Compensation grouting is a feasible technique to reduce the buildings' settlement during an adjacent deep excavation. The projects showed that compensation grouting would have reduced the settlement of those buildings approximate 3 cm. The protection results were very encouraging. Not even tiny cracks were observed in the buildings, except in Building A, where only the third stage grouting was applied and the time control was not very tight near Building A. It shows that the time control is also critical in compensation grouting.

Compensation grouting was proved to be a cost-effective technique. The cost of compensation grouting was only one tenth of the budget for the proposed underpinning method.

Compensation grouting, however, should not be regarded as a panacea for all potential settlement problems. The practice of underground construction in Shanghai Clay demonstrated that well-controlled construction can efficiently reduce the ground loss and the risk of damage to surrounding properties.

ACKNOWLEDGEMENT

The writer wants to thank Mr. Hou Xueyuan and Liu Guobin of Tongji University and Mr. Liu Jianhang of Shanghai Metro Corporation. The cooperation from Shanghai Tunnel Engineering Co., Ltd. and China Coal Construction Group Corp. is also greatly appreciated. The writer is also grateful to Ms. Pat George and Ms. Chen Zhan for their editorial helps.

REFERENCES

Baker, W., Cording, E., and MacPherson, H.(1983). "Compaction grouting to control ground movements during tunneling." *Underground Space*, 7(3), 205-212.
Bjerrum, L.(1963)."Allowable settlement of structures." *Proc. European Conf. on Soil Mech. & Found. Eng.*, Wiesbaden, Vol.2, 135-137.
Liu, J and Hou, X. (1998). *Excavation engineering handbook.* Chinese Building Construction Press (In Chinese).
Liu, J. (1999). "Centrifugal modeling of excavation in soft clay." Ph.D. dissertation, Tongji University, Shanghai, China (In Chinese).
Mair, R.J., Harris, D., Love, J., Blakey, D., and Kettle,C.(1994). "Compensation grouting to limit settlements during tunnelling at waterloo station, London." *Proc. 7th Int. Sym. on Tunnelling'94*, London, 279-300.
Peck, R.B. (1969). "Deep excavations and tunneling in soft ground." *Proc. 7th Int. Conf. on Soil Mech. & Found. Eng.*, State-of-the-Art Volume, 225-290.

Effect of Injection Rate on Clay-Grout Behavior for Compensation Grouting

Soga, K.[1], Au, S.K.A[2] and Bolton, M.D.[3]

Abstract

Laboratory grout injection tests were performed in clay to investigate the effect of grout injection rate on soil-grout behaviour during and after injection for both 'compaction' type and 'fracture' type grouting, with particular emphasis on compensation grouting in clays. Injection of low mobility grout was simulated by expanding a latex balloon, whereas injection of high mobility grout was simulated by injecting epoxy resin directly into clay specimens. For balloon expansion case, there was a minor effect of injection rate on pressure-injection volume and grout efficiency-time relationships. For epoxy injection case, on the other hand, fracture initiation and propagation mechanisms were affected by the magnitude of injection rate and hence the pressure-volume and grout efficiency-time relationships were injection rate dependent.

Introduction

Compensation grouting is becoming a popular method to offset subsidence caused during underground excavation (Mair and Hight, 1994). The basic principle is that grout is injected between the excavation and building foundations to 'compensate' for the ground loss and stress relief caused by underground excavation. Grout injection is often undertaken simultaneously with excavation in response to detailed observations so that settlements and distortions are limited to specified amounts (La Fonta, 1998; Buchet et al., 1999).

Compensation grouting includes grouting processes such as compaction grouting, intrusion and fracture grouting. Although the common practice of injecting low mobility grout (commonly referred to as compacting grouting) is to increase the

---

[1] Senior Lecturer, University of Cambridge, Engineering Department, Trumpington Street, Cambridge CB2 1PZ, UK; phone +44-1223-332713; ks@eng.cam.ac.uk
[2] Lecturer, City University of Hong Kong, Department of Building and Construction, Kowloon, Hong Kong; bcskaa@cityu.edu.hk
[3] Professor, University of Cambridge, Engineering Department, Trumpington Street, Cambridge CB2 1PZ, UK; mdb@eng.cam.ac.uk

strength of the soil, it has been successfully used as a method of compensation grouting in tunnel construction (Baker et al., 1983). Low mobility grouts are always used in compaction grouting so that the grout permeation into the matrix is limited and the grout does not fracture the soil matrix. The grout is injected through open-ended steel tubes, which are inserted down into boreholes. However, when it is required to react to additional settlements, it becomes difficult to re-inject from the same position. The high injection pressure is also a concern when applied in close proximity to an excavation face.

In contrast to compaction grouting, injection of high mobility grout (commonly referred to fracture grouting) results in penetration of grout fingers, thin sheets or lenses through the ground. It is common to use a sleeved tube known as a Tube a Manchette (TAM). Its advantages include possible reinjection from the same position and the relatively low injection pressure compared to compaction grouting. The reported case studies show that it is a more popular method for compensation grouting (e.g. Pototschnik, 1992; Droof et al., 1995; Harris et al., 1996; Harris, et al., 1999; Sugiyama et al., 1999). However, the uncertainty in fracture direction and distribution can lead to some difficulty in achieving precise settlement control.

Due to lack of defined criteria for implementation of compensation grouting under various conditions, as well as of understanding of grout behavior during injection in various types of soils, compensation grouting has been developed mainly by trial and error in the field. The reported injection volumes are often several times greater than the heaved volume, particularly in soft clays (Au et al., 2002). Moreover, the compensation effect deteriorates critically as the excess pore pressures generated during injection in the surrounding soil dissipate and the soil consolidates (Shirlaw et al., 1999; Komiya et al., 2001). In order to control the settlements precisely and to achieve high compensation efficiency, there is a need to find controlling factors that allow optimization of the method. Although modification of grout properties is certainly one of the possibilities, the grout injection process also affects the soil-grout behavior. The main objective of this study is to understand the manner in which the surrounding soils react to the introduction of a given grout in a particular way (Jafari et al., 2000; Au, 2001, Au et al., 2002 and Soga et al., 2002). This paper reports the effect of injection rate on soil-grout behavior for a very low mobility 'compaction' type grout and a high mobility 'fracture' type grout, with particular emphasis on compensation grouting in clays.

Laboratory Grout Injection Tests

Grout injection tests were performed on laboratory reconstituted E-grade kaolin clay specimens placed in a modified consolidometer. A schematic diagram of the experimental set-up is shown in Figure 1. Two modified consolidometers with different diameters (50 mm and 100 mm) were used and an injection tube, which is a 4 mm OD and 3 mm ID copper needle, was placed at the centre of the consolidometers. The total length of the needle was 130 mm and the height above the bottom porous plate was 50 mm.

E-grade Kaolin (LL = 72, PL = 38) was used as an injected soil medium. Clay slurry was prepared by mechanically mixing dry E-grade kaolin powder with de-aired water under vacuum, giving a water content of 120%. The slurry was placed in the modified consolidometers and the volume of slurry was determined based on the condition that the specimen height after consolidation will be approximately 100 mm. The samples were prepared in different overconsolidation ratios ranging from 1 to 5. The vertical effective stress before the injection tests was fixed at 140 kPa.

Two types of injection were performed; (a) injection of water or epoxy resin into a latex balloon attached at the tip of the injection point to simulate 'compaction' type grouting and (b) injection of epoxy resin directly into soil specimens to simulate 'fracture' type grouting. The use of latex balloon idealises 'compaction' type grouting by removing the effect of possible permeation. The purpose of using epoxy resin rather than a more realistic grout (such as cement bentonite mixtures) was to reveal the fracture patterns after the tests by exposing the hardened epoxy. A ratio of 10:6 of resin and hardner by weight was used to obtain the highest strength of epoxy resin.

Injections were made using a pressure/volume controller, which can control the injection rate and volume. When epoxy resin was injected, a pressure interface chamber (PIC) was used to transmit the pressure from the hydraulic fluid in the pressure/volume controller to the grout during injection as shown in Figure 1. Injection pressures were measured using a pressure transducer positioned between the PIC and injection needle. Pressure corrections were made for both cavity expansion and fracture grouting tests. The surface displacement of the specimen was measured using a LVTD.

The injection volume was 5000 mm³. This small volume is to ensure that the grout will be contained within the specimen and will not migrate through fractures reaching the boundaries. It was found from the preliminary tests using the smaller 50 mm diameter consolidometer that, if the grout volume was larger than this volume,

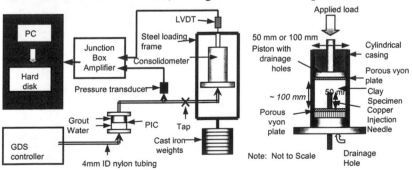

Fig. 1 Experimental set-up

the grout escaped from the testing boundaries through fractures, resulting in an artificial reduction in grouting efficiency. Further experimental details can be found in Au (2001).

Experimental Results and Discussion

'Compaction' type grouting by balloon expansion
Injection of a very low mobility 'compaction' type grout was simulated by expanding a latex balloon placed at the tip of an injection tube. The tests were performed at different injection rates (41.7, 83.3 and 500 mm^3/s) on normally consolidated clay specimens placed in the 100 mm diameter consolidometer. Figure 2 shows the measured relationships between normalized pressure (injection pressure/vertical effective stress) and normalized volume (injection volume/initial cavity volume). The initial cavity volume is calculated based on the outer diameter of the balloon. The peak pressure in the slow injection test (Q = 41.7 mm^3/s) is slightly larger than the peak pressures in the faster injection tests (Q = 83.3 and 500 mm^3/s). As the slowest injection test took 2 minutes to complete the injection, it was considered that the soil around the injection point started to consolidate during the injection process, giving a slightly higher peak injection pressure than the faster injection tests. Pore pressures were not measured inside the specimen and hence it was not possible to confirm this. However, the measured settlement curves after grout injection indicate that soil consolidation is occurring during injection for the slow injection tests, as shown next.

The change in specimen volume during and after injection was measured by the LVDT placed at the loading plate (see Figure 1). The changes are plotted in Figure 3 as grout efficiency-time curves. The grout efficiency is defined as the ratio of heaved volume to injection volume (5000 mm^3 in this case). If injection is performed rapidly, the soil deformation is undrained and therefore the grout efficiency should become one. As shown in Figure 3, this is not the case even for the very high injection rate of 500 mm^3/sec. This is because tiny air bubbles trapped in

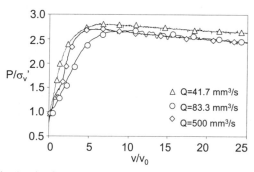

Fig. 2 Injection pressure vs. volume for compaction grouting

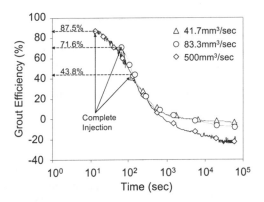

Fig. 3 Grout efficiency – time curves for compaction grouting

the injection system were compressed so that less than the exact 5000 mm^3 injection was achieved. This was confirmed by the initial zero heave at the beginning of the injection. In general, the volume loss due to air bubble compression was measured to be approximately 500 mm^3 (10% of the total injection volume).

The initial grout efficiency, which is measured immediately after the injection, decreased with the decrease in injection rate (87.5% for 500 mm^3/sec, 71.6% for 83.3 mm^3/sec and 43.8% for 41.6 mm^3/sec). As grout permeation is not possible, this decrease is due to soil consolidation during the injection stage.

The grout efficiency-time curves after injection were similar among the three tests; they overlap each other at least during the first 700 seconds. At the end of the test, the grout efficiency dramatically reduced with time to less than zero efficiency due to consolidation. Extensive shearing during the expansion phase will have generated excess pore pressures in the normally consolidated clay, leading to a subsequent reduction in volume during consolidation. The final grout efficiencies for the slower injection tests (Q = 41.67 and 83.3 mm^3/s) were approximately 10% higher than that of the fast injection test (Q = 500mm^3/s). However, this difference is rather small considering the variation of initial grout efficiency loss due to compression of air bubbles.

It was therefore concluded that the effect of injection rate was relatively minor when a perfect low mobility grout is used for settlement control purpose in the long-term. However, for a given volume of injection, the heaved volume achieved immediately after injection will depend on the injection rate due to soil consolidation occurring during the injection stage.

'Fracture' type grouting by epoxy injection
Epoxy resin was injected into clay specimens with different overconsolidation ratios (OCR = 1.5, 2 and 5) to examine the soil-grout behaviour in 'fracture' type grouting.

Two different injection rates (500 mm³/sec and 16.7 mm³/sec) were used. The tests were performed using the 50 mm diameter consolidometer.

The measured injection pressure-volume curves are shown in Figures 4 and 5 for 500 mm³/sec and 16.7 mm³/sec, respectively. The fracturing pressures in the fast injection tests were larger than those in the slow injection tests; the increase in the fracturing pressure of 500 mm³/sec was about 50% from that of 16.7 mm³/sec for a given overconsolidation ratio. The volumes at peak fracturing pressure were larger for the high injection rate. It should be noted that this increase in peak pressure with injection rate was opposite to the observation made in the compaction grouting cases described in the previous section.

It is hypothesized that the difference in fracturing pressure is attributed to different fracture initiation and propagation mechanisms for the two rates. When grout is injected under a high rate, it is more likely for the surrounding soil to fail in

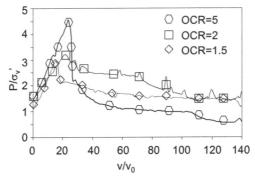

Fig. 4 Injection pressure vs. volume for injection rate of 500mm³/sec

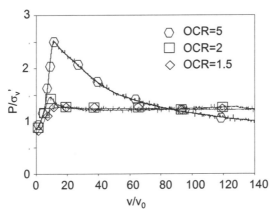

Fig. 5 Injection pressure vs. volume for injection rate of 16.7mm³/sec

shear. This is because the injection duration is too short for the grout to penetrate into micro fissures creating tensile cracks and/or to permeate into the soil pore to weaken the strength of the soil. This concept was proposed by Mori and Tamura (1987), who concluded that shear failure occurs around the borehole rather than tensile failure as long as the injection rate is high. When grouting is performed under a low injection rate, the grout can penetrate into micro-cracks or imperfections to initiate tensile failure, leading to a smaller peak injection pressure.

The measured grout efficiency-time curves are plotted in Figure 6. In the fast injection tests, the initial efficiency loss immediately after the injection stage was about 10% to 20%. In the slow injection tests, the initial grout efficiencies were smaller than the fast injection tests due to soil consolidation during the injection process. The grout efficiencies immediately after injection varied between 20% and 65% depending on the overconsolidation ratio of the clay specimens. This variation is due to the difference in the degree of consolidation occurring during the injection stage. In Figure 7, the measured heave during the injection stage is plotted against time for the slow injection tests. For the OCR=1.5 specimen, the rate of heaving decreased with increase in injection volume and was almost zero at the end of injection. As more grout was injected into the soil, the fractures extended quickly to increase the size of the excess pore pressures zone but the excess pore pressures quickly dissipated with time during the injection stage. Moreover, the highly pressurised grout penetrated into the clay at the grout-soil interface. For the OCR=5 specimen, there was no surface heave during the first 30 seconds due to compression of the trapped air. As more grout was injected into the soil, the surface heave increased linearly at a constant rate, and finally levelled off at 2mm when grout injection was completed. This linear increment of heave took place because the net effect of consolidation was small compared to the OCR=1.5 specimen. As a result, the effect of consolidation during the injection was very small for the heavily overconsolidated case.

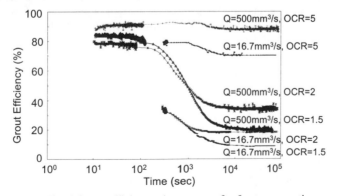

Fig. 6 Grout efficiency-time curves for fracture grouting

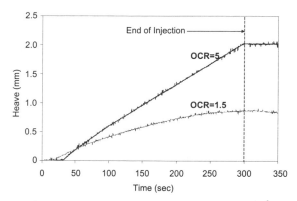

Fig. 7 Specimen heave with time during injection (16.7 mm^3/sec injection)

The effect of injection rate on final grout efficiency was found to be larger for 'fracture' type grouting than for 'compaction' type grouting. As shown in Figure 6, the final grout efficiencies in the faster injection tests are approximately 20% larger than those in the slow injection tests. The difference is primarily due to relatively large consolidation occurring during the injection stage for the slow injection tests; the grout efficiencies immediately after injection for the slow injection tests is smaller than that of the equivalent time for the fast injection tests as shown in Figure 6. The efficiency curves at different injection rates do not overlap each other for a given OCR in contrast to what was found in the compaction type grouting tests (see Figure 3). Hence, the differences in fracture pattern and propagation during the injection stage influenced the extent and magnitude of the excess pore pressure generated in the soil specimens, leading to a larger consolidation effect in the slow injection tests.

Although it was not possible to investigate the differences in fracture propagation during injection for different injection rates, the fracture pattern was examined by sectioning the soil specimens after the tests. It was found that vertical fractures were formed in the soil sample with OCR =1.5 and 2 in the low injection rate tests and a horizontal fracture was formed in the other soil specimens. It is thought that this difference in fracture direction is due to the effect of initial stress condition and the boundary confinement in the consolidometer (further details can be found in Au (2002) and Soga et al. (2002)). Figure 8 shows a typical horizontal fracture formed in the fast injection tests, whereas Figure 9 shows vertical fractures formed in the slow injection tests. When vertical fractures were initiated, the pressurised grout penetrated into the fracture and pushed the soil from the crack in the lateral direction. Hence, the measured surface heave was not directly caused by the 'jacking-up action' (i.e. the injected grout displacing the soil vertically towards the ground surface.). It is possible that excess pore pressures were generated in a larger volume and hence the grout efficiencies for OCR 1.5 and 2 samples in the slow injection tests were lower than those in the fast injection tests.

In summary, for 'fracture' type grouting, the fracture initiation and propagation were different depending on the injection rate. This difference led to further loss in grout efficiency during the injection stage for the slower injection case.

Fig. 8 Horizontal fracture   Fig. 9 Vertical fracture

Effect of specimen diameter
Assuming symmetry of the radially fixed outer boundary condition, the single injection test can be considered to be equivalent to injecting grout simultaneously with spacing of 113 mm and 56.5 mm for 100 mm and 50 mm diameter specimens as shown in Figure 10. When grouting is performed in a regular geometrical array with injections of the same volume at each port, the behaviour of any grouting unit can be considered to be similar to that of a single injection confined in an equivalent radial fixed boundary (Jafari *et al.*, 2000). This symmetry hypothesis for multiple simultaneous injections was confirmed in the multiple injection tests (both balloon

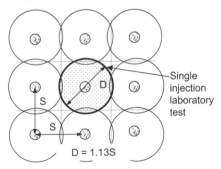

Fig. 10 Symmetry for multiple simultaneous injections

expansion and epoxy injection) performed by Au (2001).

In order to investigate the effect of radial boundary condition on soil-grout behaviour for different injection rates, both balloon expansion and epoxy injection tests were performed on normally consolidated clay specimens with different injection volumes and different diameter consolidometers. Since the radial boundary effect is relative in terms of injected volume against radial boundary size, it is assumed that the effect can be normalised by the radial boundary ratio, which is defined as the radial boundary distance (r_b) to the equivalent radius (r_g) of the injected grout $(n = r_b/r_g)$. The equivalent radius was calculated assuming that the grouted volume is spherical. For the investigation, two different injection rates were used; Q = 500 and 83.3 mm^3/sec for epoxy injection tests and Q = 500 and 16.67 mm^3/sec for balloon expansion tests.

The final grout efficiency against the radial boundary ratio n is plotted in Figure 11. Similar to the findings described earlier, the injection rate effect was found in the 'fracture' type grouting cases; the final grout efficiencies of the slow injection tests were about 5 to 10% smaller than those of the fast injection tests. No injection rate effect was observed in the 'compaction' type grouting cases.

For a given injection rate, the final grout efficiency decreased with the n value; more consolidation settlement was induced with a larger spacing of larger injection than with closely spaced injections of small grout volume. This is primarily due to overlapping of excess pore pressure zones generated by each injection. For the small spacing condition, an injection is influenced by the neighboring injections and the total amount of excess pore water pressure generated is affected by this overlapping condition. It is hypothesized that this overlapping suppressed the amount and extent of excess pore pressure generation, leading to less consolidation

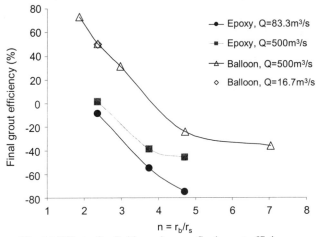

Fig. 11 Effect of radial boundary on final grout efficiency

settlement and a better final grout efficiency for the smaller spacing condition. This hypothesis has been verified by the coupled consolidation finite element analysis of multiple grout injections tests (Au, 2001).

Conclusions

Laboratory grout injection tests were performed to investigate the effect of grout injection rate on clay-grout behaviour during and after injection. Injection of a very low mobility grout ('compaction' type grouting) was simulated by expanding a latex balloon, whereas injection of a high mobility grout ('fracture' type grouting) was simulated by injecting epoxy resin directly into the soil specimens. For balloon expansion case, the test results showed a minor rate effect on the pressure-injection volume relationship and the grout efficiency-time relationship. On the other hand, for epoxy injection case, the effect of injection rate on the grout efficiency-time relationships was significant. Larger grout efficiency loss was observed when epoxy was injected at a slower rate. Based on the difference in the measured injection pressure-volume curves as well as in the observed fracturing patterns, this rate effect was attributed to the difference in fracture initiation and propagation mechanisms during the injection stage, which resulted in complex patterns of excess pore pressure generation and consolidation in the soil. The radial boundary condition also affected the grout efficiency. More consolidation settlement was induced with larger spacing of larger injections than with closely spaced injections of small grout volume.

In summary, a better grout efficiency was obtained when grout was injected rapidly and the injections are closely spaced. It should be noted that this finding is based on small-scale laboratory tests and the impact of injection rate on grout efficiency should depend on injection volume and the drainage condition. Larger scale tests are necessary to confirm this finding for possible field application.

References

Au, S.K. (2001) : "Fundamental Study of Compensation Grouting in Clay," PhD thesis, University of Cambridge

Au, S.K.A., Soga, K., Jafari, M.R., Bolton, M.D. and Komiya, K. (2002) : "Factors affecting long-term efficiency of compensation grouting in clays," accepted for publication in the *Journal of Geotechnical and Geoenvironmental Engineering*, American Society of Civil Engineers

Baker, W.H., Cording, E.J. and MacPherson, H.H. (1983) : "Compensation grouting to control ground movements during tunnelling," *Underground Space*, Vol. 7, pp. 205-212

Buchet, G., Soga, K., Gui, M.W., Bolton, M.D. and Hamelin, J.P. (1999) : "COSMUS; New methods for compensation grouting," Association Francaise des Travaux en Souterrain (AFTES) International Conference "UNDERGROUND

WORKS - Ambitions and Realities," 25-28 October 1999, pp. 131-137

Droff, E.R., Travares, P.D. and Forbes, J. (1995) : "Soil fracture grouting to remediate settlement due to soft ground tunnelling," *Proceedings of Rapid Excavation and Tunnelling Conference*, pp. 21-40.

Harris, D.I., Pooley, A.J., Menkiti C.O. and Stephenson, J.A. (1996) : "Construction of low level tunnels below waterloo station with compensation grouting for Jubilee Line Extension," *Geotechnical Aspects of Underground Construction in Soft Ground*, Mair, R.J. and Taylor, R.N. (eds), Balkema, pp. 361-366

Harris, D.I.. Mair, R.J., Burland, J.B. and Standing J.R. (1999) : "Compensation grouting to control tilt of Big Ben Clock Tower", *Geotechnical Aspects of Underground Construction in Soft Ground*, Kusakabe, O., Fujita, K. and Miyazaki , Y. (eds), Balkema, pp. 225-232

Jafari, M.R., Au, S.K.A., Soga, K., Bolton, M.D. and Komiya, K. (2001) : "Fundamental laboratory investigation of compensation grouting in clays," *Geotechnical Special Publications No. 113*, American Society of Civil Engineers, pp. 445-459

Komiya, K., Soga, K., Akagi, H., Jafari, M.R. and Bolton, M.D. (2001) : "Soil consolidation associated with grouting during shield tunnelling in soft clayey ground," *Geotechnique*, Vol. 51, No. 10, pp. 835-847

La Fonta, J. (1998) : "Puerto Rico real-time control of compensation grouting with the cyclops system", *Geotechnical News*, Vol. 17, part 2, pp.27-32.

Mair, R. J. and Hight, D.W. (1994) : "Compensation grouting", *World Tunnelling*, pp. 361-367

Mori, A. and Tamura, M. (1987) : "Hydrofracturing pressure of cohesive soil", *Journal of the Soil and Foundations, Japanese Society of Soil Mechanics and Foundation Engineering*, Vol. 27(1), pp. 14-22.

Pototschnik, M.J. (1992) : "Settlement reduction of soil fracture grouting", *Proc Conference on soil grouting, Soil improvement and Geosynthetics*, ASCE, Vol. 1, pp.398-409

Shirlaw, J.N., Dazhi, W., Ganeshan, V. and Hoe, C.S. (1999) : "A compensation grouting trial in Singapore marine clay" *Geotechnical Aspects of Underground Construction in Soft Ground*, Kusakabe, O., Fujita, K. and Miyazaki, Y. (eds.), Balkema, pp. 149-154

Soga, K., Au, S.K.A., Jafari, M.R. and Bolton, M.D. (2002) : "Laboratory investigation of multiple injection into clay," submitted to *Geotechnique*

Sugiyama, T., Nomoto, T., Nomoto, M., Ano, Y., Hagiwara, T., Mair, R.J., Bolton, M.D. and Soga, K. (1999) : "Compensation grouting at the Docklands Light Lewisham Extension project", *Geotechnical Aspects of Underground Construction in Soft Ground*, Kusakabe, O., Fujita, K. and Miyazaki , Y. (eds), Balkema, pp. 319-324.

A Retrospective on The History of
Dam Foundation Grouting in the U.S

K.D. Weaver[1], Aff. M. ASCE

Abstract

The history of dam foundation grouting in the U.S., which began with a project in New York in the late nineteenth century, is - to some extent - one of objectives not fully achieved. It also is one of innovative procedures and insightful ideas only some of which were applied, and of questionable procedures that look all too familiar to today's grouting practitioners. An early suggestion that a closely adjacent two-row grout curtain consisting of closely-spaced grout holes might be preferable to a three-row curtain clearly was not incorporated in the design of Teton Dam, but has been incorporated in the design of a few dams constructed in recent years. The early twentieth century concept of injecting essentially endless volumes of high w:c ratio grouts survived unto the late twentieth century, despite a realization in some quarters that such grout would travel far beyond the area requiring treatment and would either not set up at all or would merely form "films." By the time Boulder Dam was constructed, the design of grouting programs was considered to have become "systematic". However, in this case, remedial grouting entailing deepening the curtain and injecting very substantial volumes of grout subsequently was found to be necessary. There have since been many other cases in which the initial grouting was done "systematically" (using now outmoded concepts and procedures) and in which remedial grouting ultimately proved to be required.

Introduction

It is potentially instructive as well as interesting to review the history of dam foundation grouting in the U.S. That history is - to some extent - one of objectives not fully achieved. It also is one of innovative procedures and insightful ideas only some of which were applied, and of questionable procedures that look all too familiar to today's grouting practitioners. Thus,

---

[1] Consultant. 40442 Valencia Court, Fremont, CA 94539-3625: phone 510-657-5127; kengrout@earthlink.net

we find an early suggestion that a closely adjacent two-row grout curtain consisting of closely-spaced grout holes might be preferable to a three-row curtain. It is noteworthy in that respect that construction of two-row grout curtains has become relatively common practice in recent years. On the more negative side, a review of U.S. grouting history reveals the advocacy of injecting grouts with a water:cement ratio as high as 30:1 in essentially endless volumes. Although such extremely high water:cement ratios may no longer be used, and at least one dam-building agency took steps in the early years to avoid excessive grout travel, the use of super-plasticized 5:1 grout mixes in large volumes on one relatively recent project arguably are the modern-day equivalent. By the time of construction of Hoover Dam on the Arizona-Colorado border in the 1930's, the design of grouting programs was considered to have become "systematic". However, "systematic" is not necessarily a good thing if all of the geologic conditions that might contribute to reservoir leakage, and the depths to which those conditions might be present, are not fully taken into account and if inappropriate materials and procedures are used. Thus, at Hoover Dam and at many other dams built prior to the 1970's, a remedial grouting program subsequently has been found to be necessary.

The First Half Century

The history of grouting in the United States dates back at least to 1893, when cement grout was injected into the foundation of a dam in the New Croton Project, in New York (Franklin and Dusseault 1989). Glossop (1961) reported that the objective of this grouting was to reduce uplift pressures, and that no attempt was made to construct an impermeable cutoff. However, Verfel (1989), citing an earlier report, stated that joints in the rock at this 91m (297 ft)-high concrete gravity dam were systematically sealed by drilling and flushing holes up to 33m (100 ft) deep and grouting with a cement suspension. This would appear to be a reasonable depth for grout holes for a dam of this height, suggesting that U.S. grouting procedures were off to a good start. However, published records indicate that the depth to which grouting was done on some other dams during the early years may have been overly shallow compared to the heights of the dams.

Arrowrock Dam, in Idaho, which was the worlds tallest dam when it was constructed during the period 1911-1915, is an example of an early project in which the grout holes were perhaps overly shallow. The grout holes at this 107m (350 ft) - high concrete arch structure penetrated only 8m (26 ft) into the reportedly "sound but seamy" granite porphyry bedrock. On the positive side, however, the grout curtain consisted of two rows of holes - perhaps as a result of one of the lessons learned at Estacada Dam. The rows were 2.4m (8 ft) apart, and the grout hole spacing was 3m (10 ft) - comparable to row and hole spacings used in late 20th century practice.

Warm Springs Dam in Oregon, which was constructed on a series of olivine basalt flows during the period 1918-1919, is another case in point regarding dams at which the depth to which grouting appears to be quite shallow compared to the dam height. Following stripping to remove seamy rock, holes were drilled only 4.6m (15 ft) into the bedrock below just a part of this 32.3m (106 ft) - high thin arch structure and were "filled with cement grout to intercept any fine seams that might have been overlooked." (It is not clear whether or not this grout was injected under pressure.)

The first major application of cement grouting for construction of a grout curtain in the United States was done in 1910 at Estacada Dam, a 34m (110 ft)-high concrete buttress structure on the Clackamas River in Oregon Although a three-row curtain consisting of 10,400m (34,000 ft) of grout holes was constructed in the highly permeable volcanic rock foundation, the seepage pressures and underflows were not reduced to the extent expected (Rands 1915). Interestingly enough, a consultant to the project (Rippey 1915) indicated that it was remarkable that the results were as good as they were, considering management changes, the resistance and lack of cooperation by the local authorities, and the resultant difficulties experienced in causing the work to be done as recommended. (These sorts of problems are not absent in current practice; to them can be added the potential adverse effect of budget constraints on the ability to achieve the project objectives.)

Rands' conclusions regarding the lessons learned included the following observations, which remain wholly correct:

1. "Air following the grout into the hole is apt to make trouble."
2. "To force charge after charge of thin grout into a hole probably means in great measure a wastage of cement."
   "The writer's opinion, now, is that either a single row of holes with close spacing or two rows of holes very close together in an up-and-down-steam direction, with casings staggered, is preferable to the triple line used at Estacada."
3. "The proper diffusion of the grout can be secured only when the concrete of the cut-off closes surface seams and confines the pressure to a depth at which it may be effective in tightening the underlying material."

A few comments on Rands' observations appear to be in order:

- The first observation indicates that compressed air "pressure pots" (rather than pumping) was used to inject the grout at Estacada Dam. Although that type of equipment is no longer in use, air ingestion is possible (and not desirable) when open-throat pumps are used to inject grout under pressure.
- Rands' observation regarding "forcing charge after charge of thin grout into a hole ...." should not be open to serious question. Nonetheless, there has been one recent case in which grout containing more than 17,000 bags of cement was injected into a single stage of a single hole (Sherwood, 1992).
- It can be inferred from this comment that none of the three rows was grouted to closure. Additionally, one may speculate that all of the grout holes may have been vertical, and thus parallel to a major joint set. The failure of Teton Dam, in Idaho, illustrates the point that a closely-spaced double-row curtain (although not a single-row curtain) might be preferable to a three row curtain in which the outer rows are

not tightly closed An investigation of the causes of failure at that site indicated that the three-row curtain had actually functioned only as a single-row curtain (Boffo 1977).

• Rands' fourth observation would appear to lend support to the use of grout caps constructed in trenches rather than grout caps that are no more than sidewalks pinned to the ground surface. By extension, this observation appears to lend support to the relatively recent practice of applying shotcrete or "regularizing concrete" over extensive exposures of fractured rock in embankment dam foundations.

Publication of the details of the grouting program at Estacada Dam gave the engineering profession as a whole the opportunity to discuss the procedures, problems and results of the project. Comments made by one reviewer (Hulse 1915) of Rands' paper provide an interesting insight on early United States grouting practice, but demonstrate some questionable reasoning of a type which, unfortunately, has persisted for decades:

1. "The first grout introduced should be quite thin - say, 1 part cement to 30 parts of water..."

2. "The writer once grouted for eight continuous days and nights on a hole that took somewhat more than 20 tons of cement. For several days, thin grout was literally poured into that hole - much of it by gravity - and the grout was kept thin until the hole showed signs of closing up."

3. "There is good evidence that Portland cement sets very slowly - if at all - after it has been introduced underground in this fashion, but, should the process of grouting necessarily entail an assumption of the necessity for the setting of the cement used?"

Each of Hulse's comments warrants further comment in the light of current knowledge:

• The reasoning behind using grout with high water:cement ratios was that this would keep the grains of cement separated, hence unable to prematurely clog fine fractures. By the time construction of dams for the California State Water Project got underway in the 1950's grout with a w:c ratio of 7:1 (still scarcely more than "dirty water") was being specified as a starting mix. Eventually, it was found that superplasticizers place electrostatic charges on grout particles, causing them to repel one another. This effect, of course, negates the previously perceived need to use large volumes of water to separate the grout particles. Nonetheless, old practices tend to cling to life. For example, although the results of laboratory tests performed by the Bureau of Reclamation (Smoak and Mitchell, 1993) demonstrated the potential effectiveness of superplasticizers in substantially reducing (or even eliminating) any need for high w:c ratios, extensive use was made of superplasticized grout with a 5:1 w:c ratio at the Bureau's New Waddell Dam. However, where institutional constraints have been absent or have been overcome, the European practice of using stable grouts (essentially zero bleed) has finally been adopted in the most recent U.S. practice. Unfortunately, Hulse's experience of injecting "20 tons" (18 metric tons) of cement into a single grout hole does not represent a record. His experience is dwarfed by that of the grouting crew at New Waddell Dam, where (as has been reported orally at an annual short course on "Fundamentals of Grouting") grout containing in excess of 10,000 bags (426 metric tons) of cement was injected into a single stage of a single hole! Clearly, in each case, much of the grout injected must have traveled far beyond the zone in which a relatively impermeable barrier to seepage was needed. Arguably, any grout hole or stage that will accept such large volumes of grout should very readily - and more effectively - accept an appropriately formulated stable grout.

• Hulse's suggestion that it is not necessary for cement injected in grout to set at all is truly astounding. Suffice to say that hydrated cement forms soluble compounds that dissolve over time as seepage passes through the fissures in which these compounds are deposited. This effect has been observed in the form of carbonate mineral deposits in drains in concrete dams. Grout, to be effective, should completely fill the openings into which it is injected, rather than simply leaving a soluble coating on the fissure walls. Optimally, realizing that some seepage flow may reach the solidified grout, a chemically stable (i.e., relatively insoluble) grout should be used - hence the recent practice of including a pozzolan such as fly ash or silica fume in grout mixes.

The grouting program performed at Hoover Dam (also known as Boulder Dam) has been said to mark the beginning of systematic design of grouting programs in the United States (Glossop 1961). This 221m (726 ft)-high concrete arch dam was constructed between 1932 and 1936 on the Colorado River, where that river forms the border between Nevada and Arizona. The initial curtain and consolidation grouting operations began following completion of foundation and abutment excavations in 1933. The usual depth of the curtain holes, which were drilled using air- driven diamond drills, was 30 to 46m (100 to 150 ft) - relatively shallow considering the height of the dam. Holes less than 15m (50 ft) deep were drilled using pneumatic percussion drills. The grouting operations in the foundation and abutment areas entailed drilling 143 grout holes to a cumulative depth over 1,650m (5,300 ft), and injecting 181m^3 (6,395 ft^3) of grout. Most of the grouting operations were accomplished using compact grout plants mounted on flatbed trucks that were picked up by a cableway and set down were they were needed (W.J. Simonds undated).

The grouting program for Hoover Dam, like the one performed at Estacada Dam, has been described in some detail (Simonds 1952). The design of the initial grouting program reportedly was based in large part on the results of a review of grouting programs on fifty high concrete dams. However, numerous problems were experienced during the course of construction of the grout curtain, and excessive seepage occurred during the first filling of the reservoir. Therefore, extensive remedial grouting quickly became necessary. The remedial grouting operations performed during the period from 1938 to 1947 included deepening the grout curtain from the original 21% to 41% of the height of the dam, and regrouting construction voids around one of the penstock tunnels. It involved injection of more than 10,660 metric tons (250,000 bags) of cement grout into about 80,500 lineal meters (264,000 lineal ft) of boreholes. The presence of hot alkaline groundwater led to experimentation with a number of special cements and additives - principally oilwell cements and retarders. The w:c ratios of the grout mixtures used were in the range of 20:1 to 7:1.

Some of the points made in the discussion of Simonds' (1952) paper on Hoover Dam are particularly instructive and continue to be broadly applicable to the design, execution and evaluation of dam foundation grouting programs. For example:

"... the science of geology has made great strides in the field of foundation exploration. This has taken much of the guesswork out of foundation grouting and drainage. Modern practice consists of a program whereby the original grouting design, based upon the general information obtained during preliminary investigations, is modified in accordance with information which becomes available as construction progresses. Thus, after the site has been dewatered and the foundation rock exposed by stripping, the surface is cleaned meticulously so that surface manifestation of subsurface defects can be seen, measured, mapped, and studied."

" It has been found that, if these data are collected and analyzed as they become available, any engineering geologist who is 'worth his salt' will have a detailed knowledge of subsurface conditions and is in a position to give valuable advice to the construction forces. This goes far in eliminating the necessity of making important decisions on a trial and error basis."

"There is another phase of the grouting problem that merits careful consideration. It is the necessity for the services of an experienced geologist in the early stages of the investigation and also during the grouting operations. A geologist who has had considerable experience in dam foundation treatment can contribute much to the success of the job by virtue of his knowledge of geological formations. Such knowledge is not usually possessed by civil engineers."

Contemporaneously with the construction of Boulder Dam, the Tennessee Valley Authority (TVA) was building the 81m (265 ft)-high Norris Dam on the Clinch River in Tennessee. Realizing the desirability of limiting grout travel beyond the desired treatment zone as well as producing a relatively permanent filling, TVA was performing laboratory and field tests on grout mixtures with low w:c ratios. Kennedy (1961), citing an earlier paper, reported a finding that grout consisting of equal parts of cement and rock flour and a w:c ratio of 1:1 was preferable to neat cement grout because it penetrated the cracks and did not set up in the pipeline. It was noted that the rock flour was finer than the cement, and that it had a retarding effect. Concern that the retarding effect would allow the grout to travel excessive distances beyond the desired treatment area led to addition of a 3% solution of $CaCl_2$ by weight of cement to the grout. The results of laboratory tests made on cores of grouted seams while grouting was underway demonstrated a satisfactory compressive strength. Attempts to economize by adding sand to the grout were unsuccessful: the pump and lines became solidly plugged, cores of the sanded grout taken from holes near the injection location showed a tendency toward segregation, and the material in the cores was lean and crumbly. Therefore, it was concluded that it would be unwise to use a sanded grout mix in seams that would be subjected to a high hydraulic head.

Lewis (1940) reported that grout communication to holes previously grouted to refusal was observed, and speculated that this may have been due to grout "shrinkage" and to washing of seams at higher pressures in the later holes. In that latter respect, extensive washing back and forth from hole to hole with air and water under pressure was done at this site. Observations in tunnels driven into "seams" revealed that the seams were not continuously open: Lewis (1940) reported that "openings frequently vanished entirely, only to reappear further along." (This observation clearly is a powerful argument in favor of a general practice of intersecting known or suspected "seams" with grout holes at several locations, as some intersections may likely occur at locations where the "seams" are closed..) Other observations revealed that grout built up in layers in the seams and concentrically in grout holes, leading to an assumption that a chemical affinity within the grout could lead to clogging of a hole or seam before full grouting could occur. Although successive depositions of grout in incompletely (or inadequately) grouted seams during sequential grouting from two or more holes is perhaps a more likely explanation for the noted layering, it is worth noting that lower w:c ratios were employed at this site than were in common use by dam construction agencies other than TVA. In a procedure that has been revived in recent years, pre-grouting water tests were done to gain an idea of the amount of grout that might be taken, and to assist in the selection of an appropriate w:c ratio for the starting grout mix. The w:c ratio used for "tight" holes was 3; for holes offering "a fair degree of resistance" it was 1.5; for average conditions the normal mix was 1.0; and the grout mix for very open conditions had a w:c ratio of 0.66. Interestingly, grout holes were explored with a mechanical "feeler" with spring-actuated legs and with a periscope. (Comparably, in more recent years, caliper logging of grout holes was done at the Eastside Project in Southern California, and Houlsby (1990) mentioned the use of a periscope during grouting operations for dams in Australia.)

The ultimate in close spacing of grout holes in a single-row curtain was approached during construction of Chicamauga Dam, a 39m (129 ft)-high concrete gravity structure in Tennessee, starting in 1936. As described by Hayes (1940), 6.35 cm (2.5 in) diameter holes were first drilled on 1.2m(4 ft) centers, and the spacing was sequentially reduced to a final spacing of 0.3m (12 in) Shallow cavities encountered by these holes were opened to the surface and backfilled with concrete; deeper cavities were intersected with 0.9m (36 in) calyx holes, then cleaned out and backfilled with concrete.

Consolidation grouting prior to the curtain grouting reportedly having filled all of the larger and more extensive seams, only 35 metric tons (830 cu ft) of cement was accepted in 3,290 lin m (10,800 lin ft) of grout holes.

The Second Half Century

Kennedy (1961) reported that "Thin mixes injected at pressures of 500 psi or slightly greater formed excellent grout films." The concepts that grout should fill joints and fractures and that it should bond to their walls rather than be deposited on them as "films" evidently had not taken hold.   In another context, Thompson (1954) pointed out a number of facts relating to embankment dam foundation treatment, including the following:

1. "The design of grout curtains should be based in part on results of pressure tests made in exploratory boreholes;
2. Grouting specifications should allow increasing or decreasing the grouting program based on conditions actually encountered during construction of the curtain; and
3. The factors that grouting specifications should take into account include the proper angle of grout holes needed in order to achieve the greatest number of intersections with groutable "seams."

Attention has, for many years, been given to the first two of the facts listed above. Thus, despite an unfortunate tendency to use a formula based on the height of a dam when selecting a grout curtain depth, pressure test results started being used to ascertain grout curtain depths that would be appropriate in consideration of the foundation conditions, and grout curtains have in some cases been deepened locally based upon grout takes. For example, the grout curtain at the 91m (300 ft)-high Ruedi Dam in Colorado - which was completed in 1968 - is deeper in the abutments than in the channel section, and was extended below the planned maximum depth of 79m (260 ft) to a final depth of 110m (360 ft) locally (Walker and Bock, 1972). However, it can be inferred from the fact that the grout holes were drilled normal to the foundation surface that attention was not given to the third fact on the list presented by Thompson (1954), i.e., achieving the maximum number of intersections with groutable "seams."

The unfortunate practice of constructing a grout curtain with vertical holes without regard to the orientations of prominent "seams" continued at least until 1956, when construction started on City of Bethlehem's original Penn Forest Dam. Schweiger and others (1999) reported that "turbid seepage" developed downstream from this 145 foot high embankment dam during first filling of the reservoir, and that a sinkhole developed in the embankment itself. Repair attempts - and emergency measures other than drawing down the reservoir - were unsuccessful, leading to the dam being replaced with a 180 foot high RCC structure. The design features of the new dam, which was completed in 1998, included a three-row curtain constructed with stable grouts.

Walker and Bock (1972) indicated that most of the grout used in constructing the Bureau's 537.5 ft high Trinity Dam in California, which was completed in 1962, had a w:c ratio of 5:1. Although this represents some improvement over the 7:1 ratio that previously had been used for a number of projects, it was still quite thin by contemporary European standards and by modern U.S. standards. More commendably, however, the grout curtain consisted of two rows of holes - the minimum number that ordinarily should be considered for use in conservative grout curtain design practice. Walker and Bock's description of grouting operations at the 390 ft-high Blue Mesa Dam, competed in Colorado in 1967, indicated that the w:c ratio of some of the grout used in construction of the grout curtain at that location had retrogressed to 7:1, but that w:c ratios as low as 1:1 were used.

Grout with a w:c ratio as low as 8:1 was used in the center row of the 3-row grout curtain at the USBR's Teton Dam, in Idaho (Aberle, 1976). This 93m (305 ft)-high embankment dam, constructed during the period 1972 - 1976, failed on first filling (Arthur, 1976). Photos taken as the failure developed indicate that the failure initiated with major seepage at the base of the right abutment (Olson 1976). Large near-vertical "fissures," partially filled with rubble, had been observed during excavation of the key trench, and low-angle joints - some also partially filled with silt and rubble - were discovered in the post-failure exploration. In retrospect, the presence of these infillings probably was responsible for the inability (as reported by Aberle, 1976) to inject grout into intervals that accepted less than 19 liters (2 cubic feet) of water in a 5 minute period during pre-grouting water tests. Also in retrospect, it seems likely that piping developed in at least some of the infillings of silt and rubble during filling of the reservoir. Neither Aberle (1976) nor Arthur (1976) mention what, if any, surface treatment was done in the core trench to protect the embankment. One might speculate that an application of shotcrete, which is now common practice prior to grouting, might have reduced the potential for piping to migrate from the fissures into the embankment.

Appropriately, blanket grout holes were drilled at angles designed to intersect defects observed in the foundation excavation (Aberle 1976). The grout holes in the downstream curtain row were vertical and the grout holes in the upstream and central rows of the curtain were inclined, thus intersecting reported flat-lying and vertical joints at favorable angles. In view of post-failure findings that the 3-row curtain had functioned only as a single-row curtain, it is noteworthy that - as reported by Aberle (1976) - the spacing of the grout holes in outer rows was not reduced to less than 6m (20 ft) regardless of the grout takes in adjacent holes in those rows. Although a deficiency in the grout curtain was not identified as the sole possible factor in the failure of the dam, one lesson that - in retrospect - should be learned is that all rows of holes in a multi-row grout curtain should be closed to appropriately tight standards. However, it is unlikely that any amount of grouting could have rendered the reported joint infillings impermeable. thus it is appropriate that one of the lessons learned by the USBR was that design changes should include "redundant measures to control seepage and prevent piping" (Pedde 2001).

Late 20th Century Advances

Dam foundation grouting practice in the U.S. had to wait for visionaries such as Dr. Wallace Baker, and for the influx of foreign ideas and concepts which began in the early 1980's through the efforts of Dr. Donald Bruce, Dr. Don U. Deere, A.C. Houlsby and others, before standards, practices and attitudes changed. Construction of two-row grout curtains and grouting to standards became common practice, and the Swiss concept of multi-row grout curtains with holes at oppositely inclined orientations was adopted for major projects - perhaps most notably the Metropolitan Water District's Eastside Reservoir Project, in Southern California. Ultrafine cements, first introduced from Japan, came into common use for treating fractured rock foundations. Artificial pozzolans came to be standard ingredients in cement-based grouts, as did superplasticizers and - perhaps to a lesser extent - anti-bleed additives. The European concept of employing stable grouts gradually (if grudgingly) began to be accepted, and low mobility grouts were adapted for use in dam foundation grouting. Fear of applying injection pressures greater than "rule-of-thumb" began to subside. Due in large part to the efforts of innovative specialty contractors, equipment manufacturers, material suppliers, and assorted consultants with international experience, U.S. practice began to evolve rapidly and is internationally acknowledged as a source of innovation, accomplishment and expertise - especially in the remediation of grout curtains originally constructed between 1920 and 1970.

REFERENCES

Aberle, P.P. (1976). "Pressure grouting foundation on Teton Dam." *Rock Engineering for Foundations and Slopes*, ASCE. 245-263.

Arthur, H.G., (1976). "Teton Dam failure." *The Evaluation of Dam Safety: Engineering Foundation Conference Proceedings*, Nov. 28-Dec. 3, 1976. ASCE. 61-68.

Arrowrock Dam. http//www.usbr.gov.dams.arrowrock.html

Boffey, P.M. (1977). "Teton Dam verdict: a foul-up by the engineers." Science, V. 195, n. 4275, 270-272. January 21, 1977.

Franklin, J.A. and Dusseault, M.B. (1989). *Rock Engineering*. McGraw Hill, New York.

Glossop, R. (1961). "The invention and development of injection processes; part II: 1850-1960." Geotechnique, V. 11, 255-279.

Hayes, J.B. (1941). "Foundation treatment of Chicamauga Dam." Symposium on Foundation Experiences, TVA. Transactions, ASCE v. 106, 780-801.

Houlsby, A.C. (1990). *Construction and design of cement grouting.* Wiley, New York..

Hulse, S.C. (1915). Discussion in Rands, H. A., (1915).

Kennedy, T.B. (1961). "Symposium on grouting: Research in Foundation Grouting with cement." Journal of Soil Mech. and Found. Div., ASCE, v. 87, No. SM 2, Paper No. 2794. 55-81.

Lewis, J.E., Jr. (1940). "Foundation treatment and reservoir rim tightening of Norris Dam." Transactions, ASCE, Paper No. 2113. p. 685-721.

Minear, V.L. (1936) "Field methods for pressure grouting foundations of Boulder Dam and appurtenant structures." Bureau of Reclamation Technical Memorandum 535.

Olson, E. (1976). Teton Dam failure slide gallery. http://ucsb.edu/~arthur/Teton%20Dam/welcome_dam.html.

Pedde, K.. (2001). "The failure of Teton Dam." News release, USBR. http://www.pn.usbr.gov/news/01new/dcoped.html.

Rands, H. A. (1915). "Grouted cut-off for the Estacada Dam." Transactions, ASCE, v. 78, 447-482.

Rippey (1915). Discussion, in Rands, H. A., (1915).

Schweiger, P.G., and others (1999). "Replacement of a deteriorated earthfill dam with an RCC gravity dam: the Penn Forest Dam replacement project." Nineteenth Annual USCOLD Lecture Series, Dealing with Aging Dams. 359-374.

Sherwood, D.E. (1992). "Report on Session 3; Equipment, and control of grouting processes." *Bell, A.L., ed., Grouting in the Ground*; Conference Proceedings, Institution of Civil Engineers, Thomas Telford, London, p. 247-259.

Simonds, A.W. (1951). "Final foundation treatment of Hoover Dam." Transactions, ASCE, Paper 2537, Proc. Separate no. 109.

Simonds, W.J., (undated). "The Boulder Canyon Project, Hoover Dam." http//www.usbr.gov/history/hoover.htm

Smoak, W.G., and Mitchell, K.D. (1993). "Effect of high-range water reducers on cement grout." Concrete International, Jan. 1993, p. 56-61.

Souder (1915). Discussion, in Rands, H. A., (1915).

Thompson, T. F. (1954). "Foundation treatment for earth dams on rock." Proc. ASCE, Soil Mech. and Found. Div., Vol. 80, Separate No. 548.

Verfel, J. (1988), *Rock Grouting and diaphragm wall construction.* Elsevier, New York. 532 pages

Walker, F.C., and Bock, R.W. (1972). "Treatment of high embankment dam foundations." Journal of Soil Mech. and Found. Div., ASCE, SM 10, n. 9272. 1099-1113.

Warm Springs Dam. http//www.datawebusbr.gov/dams/oroo82.htm

Weber, A.H. (1950), "Correction of reservoir leakage at Great Falls Dam." Transactions, ASCE, Paper No. 2424, p.31-48.

COMPACTION GROUTING FOR SINKHOLE REPAIR AT WAC BENNETT DAM

By James Warner[1] F. ASCE, Michael Jefferies[2] M. ASCE, and Steve Garner[3]

Abstract

WAC Bennett dam is a zoned earth embankment slightly greater than 600 ft (183 m) in height. Two sinkholes, of which the deepest descended about 400 ft (120 m) from the crest, were discovered in the core of the dam in 1996. Compaction Grouting was chosen as the best remedial method, and was used. Because even the slightest chance of hydraulic fracturing of the core could not be tolerated, grouting was carefully designed and controlled including: precision drilling of the deep holes to within a verticality tolerance of one degree; continuous analytical evaluation of the grouting based on real time computer monitoring; careful control of the grout mix and especially its aggregate component; and, rigid control of the grout injection rate. The dam core was also monitored during grouting using a broad range of instrumentation, both on and within the embankment. Procedures were verified by a full-scale test before working on the dam.

INTRODUCTION

A sinkhole was discovered in the WAC Bennett Dam embankment in June 1996. During investigation of this sinkhole, a second one was discovered. Both sinkholes were about 10 ft in diameter, underlain by columns of loose ground within the impervious core of the dam that extended laterally to about 20 ft (6 m) in diameter and vertically to depths as great as 400 ft (120 m) below the dam crest. The goal of the remediation was to densify the loosened soils so that no further collapses would occur. A broad study of possible remedial methods, Garner et al (2000), established Compaction Grouting to be the most advantageous. However, because compaction grouting had neither been previously used to such great depths or to repair the core of a dam, a very cautious approach was mandatory. This included precision drilling to assure the holes were correctly placed, design of a highly stable grout mix and control of the injection parameters so as to preclude the risk of hydraulic fracturing or other damage to the embankment, and extensive instrumentation to provide constant knowledge of the in-place conditions.

---

[1]Consulting Engineer, Mariposa, CA, [2]Golder Associates, Nottingham, England, [3]B. C. Hydro, Vancouver, B.C., Canada

In order to qualify the proposed methods and criteria for the work, a field trial consisting of full scale test columns were drilled and grouted prior to any work on the dam. These holes were in an area south of Vancouver, B.C. that had soils approximating the various types and conditions of the loosened zones within the dam core. An important objective of the trial was to fully acquaint the contractor and the owner's inspection staff with the full range of injection behaviors that might be experienced. Paramount of these was the pressure increase that always accompanies an increase in the pumping rate and the sharp drop of injection pressure indicative of a hydraulic fracture. It was intended to intentionally hydraulically fracture the soil with a sudden increase in pumping rate and/or addition of clay to the grout, so the working personnel could visually see and become acquainted with the sudden pressure drop signature of such an event. Artificial creation of such an event was not needed however, as a perfect signature was observed within minutes of the start of injection of the first test hole, due to an excessive pumping rate caused by malfunction of the grout pump. Although the cost of the field trial was considerable, its value was solidly confirmed by the lessons learned and information gained.

REMEDIATION PLAN

An extensive investigation had resulting in mapping of the approximate locations of low density sinkhole materials. This work is described by Watts et al (2000) and Gaffran et al (2000) and comprised a mixture of sampled boreholes (including frozen samples to preserve water contents), penetration tests and geophysical methods including cross-hole shearwave tomography. Based on this work, it was found that the sinkholes were underlain by columns of loose soil about 20 ft (6 m) in diameter. These soils were much looser than the surrounding dense sandy silt core, as shown on Figure 1. The boundaries of the sinkhole were neither uniform nor followed a straight course, however, and the soils within it were grouped around their critical state void ratio, which was determined by testing reconstituted samples. This is consistent with their having been displaced large distances.

As can be seen on Figure 2, the deep but confined geometry of the loose sinkhole zone was quite different than the more commonly found large area and shallower depths typical in compaction grouting. The remediation was originally planned to use seven grout holes for each sinkhole: six equally spaced holes in a six-foot diameter circle, around a central grout hole. The design layout was adjusted on site to reflect the locations of existing instrumentation, drill rods and casing, as well as to account for new

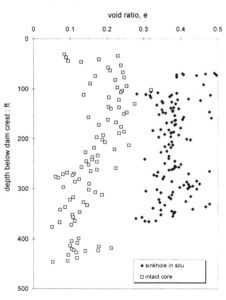

Figure 1 – Comparison of void ratio below sinkhole with undisturbed dam core.

information that was obtained as the grouting program progressed. Eventually, eight holes were grouted at Sinkhole 1 and seven at Sinkhole 2. These holes extended to depths as great as 380 ft (114 m) from the surface of the dam crest. The goal of the compaction grouting was to densify the loose soils to an initial average void ratio of less than 0.3, equivalent to an average induced in situ volumetric strain of $\varepsilon_v > 7$ %.

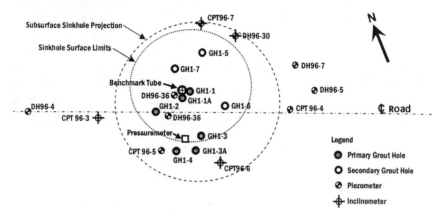

Figure 2 – Instrumentation and Grout Hole Layout at Sinkhole 1.

GROUT HOLES

Drilling of the holes to within one percent of verticality was required to ensure that the extent of the remediation zone was fully treated. To meet this provision, the drilling contractor Foundex Explorations, used a heavy and very powerful Foremost DR24 dual rotary drill rig, Figure 3, to install a relatively stiff six-inch diameter casing. The 20-ft long casing sections were carefully aligned and welded as each hole advanced. During grouting, the sections were cut with a pipe cutter as required during retrieval. In order to prevent drill fluid pressures from escaping at the bit and causing a hydraulic fracture, a 6 ft (1.8 m) length of tightly fitting flight auger, Figure 4, was used in advance of the flush ports in the drill string, so as to effect a tight soil plug between the drill flush and hole bottom.

Upon completion of the holes, a gyroscopic survey measured the accuracy. The initial sinkhole had been hastily filled with a mixture of sand, gravel, and cobbles, which extended to a depth of about 50 ft (15 m). To work through this fill an 8 in. (200 mm) casing was first place to that depth. The grout hole was then drilled to final depth using a 6 in. (152 mm) diameter casing. Initially, a 0.219-inch (5.56 mm) wall welded seam casing was used, which was combined with a variety of heavier lead sections. After problems with this casing bending and breaking in the first grout hole on the dam, a heavier 0.432-inch (11 mm) wall, seamless steel casing was used. This proved adequate and was employed on the remainder of the project without further difficulty. With but one exception where the drill is believed to have encountered abandoned drill string, the required verticality tolerance was readily met.

Figure 3 – Powerful dual rotary drill rig.   **Figure 4 – Lead auger section in casing.**

The initial proposal was to grout directly through the 6 in (152 mm) casing. There was concern however, the weight of the grout in the casing might provide sufficient head pressure to hydraulically fracture the core. Historically, the vast majority of compaction grouting has been performed through 2 in (51 mm) casing and a pressure loss of about 1 psi/ft (6.9 kPa/0.3 m) has been established for this size. No such criteria were available for the much larger 6 in (152 mm) casing however, so an evaluation was made during injection of the first test hole utilizing a down-the-hole pressure cell. A head *gain* of 0.46 psi /ft (3.2 kPa /0.3 m) of depth was found for the larger casing. As this could not be tolerated in the production work, a smaller three in. (76 mm) grout casing was telescoped inside the 6 in. (152 mm) drill casing. It resulted in a head *loss* of about 0.45 psi/ ft (3.1 kPa /0.3 m). This smaller casing was sealed at the bottom so that grout could not run up the annulus. In production grouting, 20 ft (6 m) long sections of both casings were removed as the work progressed.

GROUT MIX

From a geotechnical standpoint, it was desirable to match the properties of the grout as closely as possible to those of the core material. The inclusion of cement, which is normally used in Compaction Grout, was thus undesirable. However, as has been well recognized from research and experience, Warner (1992) and Warner et al (1992), control of the grout during injection, is greatly influenced by the grout rheology. Maximizing the coarse grain proportion, while providing just enough silt size particles to allow pumping, will result in the greatest control and minimize the likelihood of hydraulic fracturing, which were both absolute requirements for injection into the dam core. Because the work was to be done in the winter under extreme freezing conditions, any moisture in the finer aggregate fractions would be subject to freezing, and obviously, it would be extremely difficult to excavate and handle frozen material. Further, there were no natural deposits of satisfactory aggregate that could be located within a reasonable haul distance from the dam.

A custom site-blended mix was thus designed. It contained a slightly gap-graded concrete sand and a round-grained pea gravel, both of which remained in piles from the original dam construction. This was combined with natural occurring silt that existed on site adjacent to the dam. To preclude freezing of the silt, it was excavated in good weather, and stored in a tunnel leading to the powerhouse, for later use. These three components were precisely proportioned so as to result in a stable grout that would provide good controllability and yet be pumpable, even though no cement was included. And, it provided a grout with grain size distribution that closely matched the gradation of the dam core material, as can be seen from the curves in Figure 5. This aggregate blend was mixed simply with water, so as to result in a stiff, essentially no slump grout mixture.

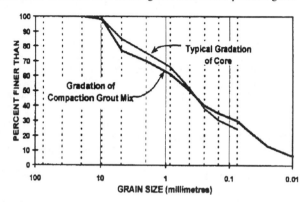

INJECTION PARAMETERS

Since the reservoir was to be maintained at a high level during the work, an unconditional project requirement was to prevent damage to the core during the grout injection. Parameters were thus established to assure the forces of the grout into the soil would be reasonably gentle. In this regard, an early decision was made to inject generally small grout quantities into many closely spaced holes, rather than the more typical larger quantities into a smaller number of more widely placed holes. On this basis, an initial volume limit of 4 ft³/ft (113 L/ 0.3 m) of hole was established. As the work progressed, based on analysis of retrieved data which will be detailed shortly, the volume limit was increased to 7 ft³/ft (198 L/0.3 m) of hole. The hole spacing varied, but was typically on the order of about three ft (0.9 m).

A further objective was avoidance of the development of excessive pore pressures, and continuous drained behavior of the core soils during the injection. Simply stated, the grout could not be injected faster than the water could drain from the area being densified. Based on the permeability of the core soil, an initial injection rate of 0.25 cfm (7 L/min) was selected. As planned, this resulted in essentially drained behavior. During the early stages of the work the contractor used a grout pump with a 6 in (152 mm) cylinder. Such a big pump was found to be virtually impossible to control at the very low pumping rates, and considerable variation of pump output occurred, within a range of 0.2 to 0.6 cfm (5.6 to 17 L/ min) or greater. During this early work, the piezometers indicated several sporadic episodes of reaching excessive pore pressure which resulted in pauses of the injection. Examination of the injection records found these excursions to have occurred only during the uncontrollably high pumping rates. Replacement of the original pumps with better controlled machines and smaller four in. (102 mm) cylinders enabled more uniform control of the injection rate.

Once the new pumps were operational, the optimal injection rate was established experimentally through careful manipulation within a range of 0.25 cfm and 0.5 cfm (7 and 14 L/min). The bottom portion of one grout hole was injected at 0.25 cfm (7 L/min) without incident. As the injection level came to within 15 ft (4.5 m) of adjacent piezometers, the rate was increased to 0.5 cfm (14 L/min). As much as 43 ft (13 m) in piezometric head increases resulted at this higher pumping rate. A production rate of 0.38 cfm (10.7 L/min) was then adopted for the work, which continued without incident.

The pressure levels that would be required for the work were not positively known as there had been no prior Compaction Grouting experience to such depths. The grout pressure is directly related to the injection rate, however, so use of a safe injection rate based upon the effect on pore pressures as above discussed, should result in a safe injection pressure. An initial pressure limitation of 1000 psi (6.9 MPa) at the hole collar was initially planned, with the requirement that the grout pump be able to operate at up to 2000 psi (13.8 MPa) should such become necessary. After allowing for the head losses within the grout casing, the actual pressure measured at the grout header was up to about 1400 psi (9.7 MPa) for the deeper portions of the holes. This reduced to about 500 psi (3.4 MPa) in the shallower portions as required to prevent any surface heave.

GROUTING EXECUTION

As previously mentioned, the work had to be accomplished during the winter in order to allow rising of the reservoir with the spring freshet. This required virtually all work to be performed within heated shelters as illustrated in Figure 6. The mast of the drill rig was enclosed in hoarding that extended out about 25 ft (7.5 m) to provide for the 20 ft (6 m) long sections of drill rod and casing. An additional tent was provided for the mixing and pumping equipment. Therein, a stockpile of aggregate that was supplied to the grout mixer by a skid steer loader was also enclosed. A separate trailer was provided for the computers and control operations, which can be seen to the right of the shelter in Figure

Figure 6 - Cold weather protection and grouting operation layout at Sinkhole 1.

FIELD MONITORING AND INSTRUMENTATION

Prior to the start of remediation, vibrating wire piezometers were installed in the 7 operational standpipes in and around Sinkhole 1 as shown on Figure 2. These provided readings at two minute intervals, which was considered adequate for the work. They were connected to and part of an existing automatic data acquisition system (ADAS) that included 147 other instruments and had alarming capabilities, as described by Scott and Hill (2000). Water levels in the nearby observation wells were also monitored by ADAS at 10 minute intervals. Readings in the inclinometers installed in the Sinkhole 1 area during the investigation were read periodically during the work, to observe deformations caused by the Compaction Grouting. A hose was also inserted into each inclinometer casing to facilitate sealing up of the hole, should the Compaction Grouting significantly deform it. After completion of the first grout hole (GH1-1), one inclinometer required sealing due to significant deformations below the 220 ft (66 m) depth.

Survey prisms were placed on the upstream and downstream slopes adjacent to Sinkhole 1. They were read from two survey booths at the edges of the crest. During the remediation, two sets of survey readings were taken per 12-hour shift to monitor any movements on the dam faces. Results of the surveys indicated that only negligible movement occurred during the remediation program.

All drilling parameters were constantly monitored. They included depth, rate of rotation, torque, and down pressure, all measured on the outer casing. Calculations using the parameters were made to determine Specific Energy, as was presented by DePaoli et al (1987) and is here illustrated in Figure 7. It is expressed as force/area and provided information as to the existing soil conditions prior to injection. The grout pumps had proximity switches installed on each cylinder to tally pump strokes. Pressure was input at the grout line near the pump and at the grout hole collar.

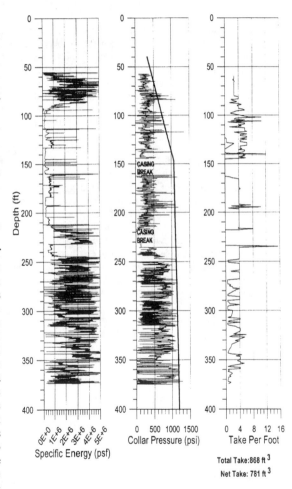

Total Take:868 ft³
Net Take: 781 ft³

Figure 7 – Drilling and grouting data for grout hole 1.

INJECTION ANALYSIS AND CONTROL

Compaction grouting, like most ground modification, developed from experience which has been supplemented by limited full-scale experiments. However, analysis of grouting allows an understanding of what can be achieved, gives guidance on how grouting can best be done, and allows evaluation of injections to see how well things have been accomplished. Analysis of compaction grouting is quite rare however, because it is very difficult. Sophisticated models are needed as the aim is to change the density of the target soil, but soil behavior is dependent upon its density which needs to be represented. Such models don't have simple solutions that can be put in a spreadsheet, and the analysis requires large strain as radial strains exceed several hundred percent.

The analysis used was simplified by adopting a cavity expansion approach which assumes either cylindrical or spherical movement. With such simplification, one must ask, which shape does the grout develop? This was resolved by early inclinometer data, however. There is a huge difference in the predicted displacement patterns between cylindrical and spherical development. The inclinometer data showed that the soil movements were very close to those expected from cylindrical symmetry, so it was adopted as the appropriate idealization. Details of the modeling approach and its implementation have been previously presented in Shuttle & Jefferies (2000) and Jefferies & Shuttle (2002).

Design Predictions: Analysis of compaction grouting at the outset of the project was aimed at confirmation that it could achieve the desired remediation and also to provide control values for pressure and injection to be used. An initial calculation was made to predict the distribution of void ratio in the ground after compaction to various treatment ratios. In response to concern that the grouting could loosen the dense intact core, numerical studies showed that to shear the dense soil required a pressure significantly greater than that required to densify the loose ground. In the sinkholes, a treatment rate of 7 ft³/ft (198 L/0.3 m) would accomplish the desired densification with few holes. However, there was a secondary concern that the grout might not displace the soil as desired with this quantity of injected volume. It was therefore decided to inject less per hole and use more holes. 4 ft³/ft (113 L/0.3 m) was adopted as the initial treatment protocol.

A further issue was the grout injection flow rate. Too high a flow rate would lead to an undrained response, with consequent loss of compaction efficiency. Conversely, a low flow rate had cost and operational implications. The numerical analysis was therefor extended to consider the effects of partial drainage (a 'Biot' type analysis). The simulations indicated that an injection flow rate of 0.25 cfm (7 L/min) was required for optimum compaction but 0.5 cfm (14 L/min) could be tolerated, as the moderately induced pore pressures would still allow effective compaction. The adopted treatment protocol and the design hole layout, then led to the predicted effect of the compaction grouting as shown on Figure 8 (compare this with the initial conditions of Figure 1). It was expected to comfortably attain the desired $e < 0.3$ target.

Control During Grouting: The availability of computer acquired data on the ground response to grouting, obtained for each 1 ft (0.3 m) grout stage, provided information on which to assess the accuracy of the design predictions and to evaluate the progress of compaction. Only three properties of the ground varied significantly from place to

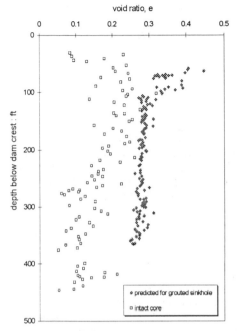

Figure 8: Predicted effect of adopted protocol.

place; void ratio, initial horizontal stress, and average distance from dense soil. Analysis proceeded iteratively by estimating the properties, computing the response, comparing this response to that measured, and then changing the estimated properties to improve the fit of the model to the data. When a reasonable fit was found, the parameters were used directly, to give information about the state of the ground; and the model itself contains the information on what was achieved.

Each individual grout stage was analyzed as starting from the in situ stress state. Grouting was carried out without stopping the injection pumps, however, and this resulted in many instances where the grout pressure did not fall to the expected in situ stress at the start of a grout stage. Here, the controlling factor was the speed of casing withdrawal achieved by the driller, compared to the pumping rate. The depth of the casing was

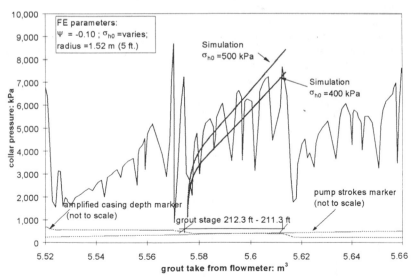

Figure 9 - Analyzed grout behavior for one stage of Grout Hole 1-4

recorded by the ADAS system and this signal has been amplified and shown on the pressure-take plots in Figure 9, so that the start and end of a grout stage can be seen, as well as the relative speed of casing withdrawal. As can be observed, the grout pressures often fluctuated substantially during any given grout stage.

Most of the fluctuation was caused by the inability of the grout pump to maintain a constant pumping rate at the very slow rates demanded. This was especially difficult when the pump approached the end of its stroke. The strokes were recorded and an amplified form of this record has been plotted on the pressure-take plots so that the effect of the pump stokes can be readily discerned. Putting the above factors together, it is clear that the pressure-take response during grouting can be used as a guide to the ground conditions, but it is not as accurate or precise as would be expected with, say, a pressure meter test. The approach adopted, therefore, was to evaluate the pressure-take response for consistency with that which was expected. The computed behavior is shown as a thick line on the pressure-take plots provided in Figure 9.

Grout takes were limited by the injection pressure limit in hole Grout Hole 1-4, such that the takes only exceeded 2 ft³/ft (57 L/0.3 m). on only a very few grout stages. The character of the low take responses is illustrated in Figure 9. This behavior corresponded to quite compact ground, more typical of normal (i.e. non-sinkhole) stress levels and with even denser ground within about 5 ft (1.5 m) of the grout bulb, a quite consistent result for the end of primary grouting. The secondary holes typically showed high pressures at low takes, with small distances to dense ground and further increased initial stresses. Grout takes were so low that the effects of compaction could not be distinguished from the effects of distance to dense ground and initial stress level.

The allowable grout injection rate for effective compaction was based on the finite element simulations. The accuracy of these simulations was confirmed with piezometers installed near the grout injection holes in order to measure the induced excess pore pressures. Figure 10 shows an example of the measured data as the grout injection zone approached and then moved past the piezometer. An excess head of approximately 5 ft (1.5 m) was induced in this instance, well within what was computed to be tolerable.

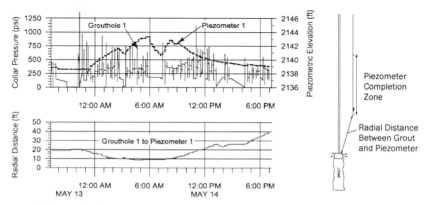

Figure 10 – Piezometric Response to Compaction Grout Injection.

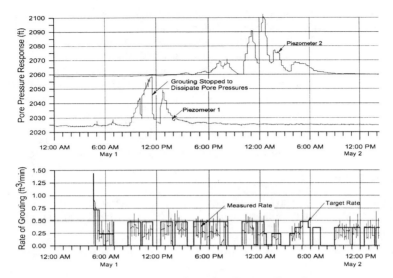

Figure 11 – Influence of Grouting Rate on Pore Pressure.

Going further, Figure 11 shows the response induced as the injection rate was slowed to 0.25 cfm (7 L/min) in the case of grouting the loose soils near piezometer DH96-36. As can be seen, reducing the injection rate lowered the excess head from about 35 ft to 15 ft (10.6 m to 4.5 m), again consistent with the expectations from the finite element analysis.

Achieved Compaction - The nature of the ground response changed as successive holes were grouted. Examples shown here are taken from the first grout hole 1-1, which is detailed on Figure 10, and the last primary hole injected, Grout Hole 1-4. These fits are fairly typical, however, they are more difficult to make when there is no clear differentiation between the grout stages.

In the first grout hole, grout takes reached the 4 ft^3/ft.(113 L/0.3 m) volume limit above about 270 ft (81 m) depth, but below this depth there were many instances where the pressure limit was reached at far less take. Analysis of ground response concentrated on the full take zones, as these clearly comprise the less dense sinkhole soils. As can be seen from Figure 9, which shows the pressure-take data from 211 ft (64 m), the measured data reasonably fits the expected grout pressure curve.

CONCLUSIONS

Compaction Grouting has been used to effectively repair extremely deep sinkholes in the core of a very large earthfill dam. Rigorous monitoring and near real-time analysis of the work, was effectively used to assure safety of the dam structure as well as to control the quality and cost of the very sensitive work. Further, the exceptional degree of control of the grout injection parameters coupled with the extensive instrumentation and continual numeric analysis set to rest some formerly contentious issues of Compaction Grouting technology. The grout mass was found to assume a basically cylindrical rather than spearical shape regardless of depth. Although this is in agreement with considerable research in which the specimens have been extricated there has been considerable question as to whether such was the case at depth. Compaction Grout can fall down larger diameter casings inparting positive gravity pressure at the base. The largest size casing that will result in a head loss at depth is about three inch (76 mm) internal diameter. Effective Compaction Grout mixes can be designed and blended with a combination of silt, sand and pea gravel. These can be pumped at pressures of at least 1500 psi (10.3 MPa) and to great depths.

REFERENCES

DePaoli, B. and Tomiolo, A., (1987) "The Use of Drilling Energy for Soil Classification", 2nd International Symposium, FMGM87, April 6-9, Kobe, Japan.

Gaffran, P.C. and Watts, B.D. (2000) "Geophysical Investigations at WAC Bennett Dam", Proceedings, 53rd Canadian Geotechnical Conference, Montreal, Quebec, Oct. 15-18.

Garner, S., Warner, J., Jefferies, M., and Morrison, N. (2000) "A Controlled Approach to Deep Compaction Grouting at WAC Bennett Dam". Proceedings, 53rd Canadian Geotechnical Conference, Montreal, Quebec, Oct. 15-18.

Jefferies, M.G & Shuttle, D.A. (2000), Discussion on the paper by Kovacevic et al "The Effects of the Development of Undrained Pore Pressure on the Efficiency of Compaction Grouting" Geotechnique.

Scott, D. and Hill D., (2000) "Surveillance Improvements at a Large Canadian Dam", Proc. Canadian Dam Association – 3rd Annual Conference, 16-21 Sept., Regina, Canada

Shuttle, D.A. & Jefferies, M.G. (2000) "Prediction and Validation of Compaction Grout Effectiveness". Proceedings, Wallace Baker Memorial Symposium, Geotechnical Special Publication 104, ASCE, Reston, Virginia.

Warner, J., "Compaction Grouting – Rheology vs. Effectiveness" (1992), Proceedings, Grouting, Soil Improvement and Geosynthetics, ASCE, New Orleans, Geotechnical Special Publication No. 30, Feb.

Warner, J., Schmidt, N., Reed, J., Shepardson, D., Lamb, R., and Wong, S. (1992) "Recent Advances in Compaction Grouting Technology", Proceedings, Grouting, Soil Improvement and Geosynthetics, ASCE, New Orleans, Geotechnical Special Publication No. 30, Feb.

Watts, B.D., Gaffran, P.C., Stewart, R.A. and Sobkowicz, J.C. (2000) WAC Bennett Dam – Characterization of Sinkhole No. 1", Proceedings, 53rd Canadian Geotechnical Conference, Montreal, Quebec, Oct. 15-18.

Quantitatively Engineered Grout Curtains

David B. Wilson, P.E.[1] and Trent L. Dreese, P.E.[2]

Abstract

It is now possible, under reasonably favorable geologic conditions, to design and construct grout curtains as fully engineered elements having specific design properties and achieving specific, end performance results. This paper summarizes the procedures, tools, and techniques to utilize the Quantitatively Engineered Grout Curtain (QEGC) Design Approach. Recommendations are provided for curtain design parameters based on evaluation of the set of integrated project characteristics that affect the outcome of grouting. The potential favorable impacts on technical and cost effectiveness afforded by a quantitative design approach are illustrated by simple examples.

Quantitatively Engineered Grout Curtain (QEGC) Design Approach

Prior to the 1980's, grout curtains were not assigned specific engineering properties that could be used in design and subsequently constructed to achieve specific end performance results. At the lowest end of design sophistication, the curtain configuration was selected only on the basis of empirical rules of thumb related to applied head and type of dam. Frequently, the curtain configuration had little relevance to the geologic conditions, and consequently they were often neither technically effective or cost effective. At the highest end of design sophistication, the curtains were designed on the basis of sound experience coupled with a good understanding of the geologic conditions.

Depending on where design practice fell within the continuum described above, some of the potential pitfalls and/or shortcomings for curtains designed in the era prior to the 1980's are listed below:

---

[1] Vice President and Practice Leader, Earth Science & Hydraulics Practice, Gannett Fleming Inc. P.O. Box 67100, Harrisburg, PA 17106-7100
[2] Senior Project Manager, Geotechnical Section, Gannett Fleming Inc., P.O. Box 67100, Harrisburg, PA 17106-7100

1. Subsurface investigations were performed to understand the stratigraphy and structure, but boring depths were often set at predetermined depths rather than being varied to define the relevant geologic conditions. Curtain configurations were often symmetrical across the site, which often bears little resemblance to the geologic conditions relevant to grouting. This resulted in curtains being deeper than necessary in some areas and of insufficient depth in others, thereby being neither technically nor cost effective.

2. Water pressure testing might have been incorporated in the program, but the results were rarely incorporated into design in a meaningful or rational manner.

3. Geologic investigations sometimes delineated fracture orientations, but less attention was normally paid to fracture spacing, fracture size, or fracture characteristics. Holes were sometimes not oriented properly to intercept primary fractures or the interval of penetration of critical fracture systems was inadequate.

4. Modeling of grouting was generally not performed, and specific design parameters and project performance requirements were not generally established (i.e. specific residual seepage rates and pressure distributions). Therefore, the grouting goals were often unclear and the programs subject to curtailment solely on the basis of economics when construction cost expectations were exceeded.

5. Grout curtain performance was assumed to not be affected by the quality of work, the contracting procedures, or the quality of inspection procedures.

6. Grouting was assumed to create an effective hydraulic barrier at the location at which it is constructed, often with relatively little regard to its interaction with adjacent materials.

Houlsby (1982) (1990) and others through the 1980's and into the 1990's published a wealth of information that promoted a much more rational approach to grouting based on careful site investigation and site characterization, matching high quality field techniques to the site conditions, and performing at least semi-analytical approaches to grouting design and analysis of field results. Wilson and Dreese (1998) first coined the term Quantitatively Engineered Grout Curtains (QEGC's) to describe a methodology that takes the design approach to an advanced level, in which all elements of the design are performed based on quantitative analyses and considerations. This paper summarizes the procedures, tools, and techniques to utilize the QEGC Design Approach. Figures and examples used to illustrate the procedures are from the Penn Forest Project, which was designed and constructed to meet specific performance parameters.

Site Characterization

In order to effectively utilize the QEGC Design Approach, it is necessary to have sufficient site information to confidently and quantitatively characterize the geologic units at the site. The basic tools for site characterization are geologic mapping, quality borings, and close interval water pressure testing to whatever depths are found to be necessary. The minimum end products needed from the site characterization program are illustrated in Figure 1 and include the following items:

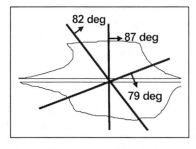

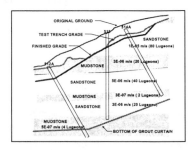

Fracture Orientations   *Quantitative Geologic Profile*

Figure 1. Site Characterization Products

1. Geologic profiles and cross-sections showing stratigraphy, structure, and weathering characteristics
2. Fracture orientations, widths, spacing, and condition (i.e. weathering and infilling)
3. Hydraulic conductivity values for the formations based on the most adverse testing orientations
4. Fracture behavior diagrams from stepped water pressure testing

The design element outcomes from the site characterization will be:
1. Identification of probable tie-in zones for the bottom of the curtain
2. Definition of the probable top of curtain elevation (usually considering the depth of weathering, in-filled material, and hydraulic conductivity)
3. Determination of the most advantageous orientation of grout holes
4. Identification of special or critical zones
5. Assignment of hydraulic conductivities for design modeling
6. Preliminary assessment of maximum safe grouting pressures

Project Performance Requirements

A critical element in the QEGC Design Approach is to establish specific, project performance requirements for residual seepage rates, flow paths of residual seepage, and/or seepage pressure distributions. A detailed discussion of factors to consider in establishing project performance requirements was presented by Wilson and Dreese (1997). In general, the following items should be considered in establishing those design requirements:
1. Dam safety implications
2. Impacts and value of pressure control
3. Value of lost water (i.e. via cost/benefit analysis)

4. "Political/public acceptability" of residual seepage

Commonly, an initial target value for acceptable residual seepage rates might be set based on an overall judgment of "tolerable seepage". That value is usually low, possibly on the order of 7.6×10^{-4} m^3/s to 7.6×10^{-3} m^3/s (10-100 gpm) for the entire dam. The grout curtain designer then begins a simplified assessment of various grouting alternatives to determine the intensity of grouting that might be required to achieve those results. If achieving those rates is found to be extremely costly or difficult, then the design requirements are revisited and reevaluated in more detail before proceeding further.

For the Penn Forest project, which was a new roller compacted concrete (RCC) dam approximately 55 m high and 610 m long, the project performance requirements were established through a process of preliminary analyses and discussions between the Designer, a Board of Consultants, and the Owner. A residual seepage performance requirement of 0.015 m^3/s (200 gpm) was established based primarily on consideration of the following factors:

1. Water loss rates were a critical issue due to the fundamental purpose of the dam, which is water supply, and the small drainage area
2. The prior embankment dam, which the new RCC dam was replacing, had failed by seepage
3. The prior embankment dam, located just downstream, contained a grout curtain, and residual seepage for the new dam was required to be substantially lower in order to prevent the old curtain from acting as an "underground dam" that would have a bathtub effect and create adverse uplift conditions for the new dam.

Grout Curtain Parameters

In the QEGC Design Approach, the grout curtain is not treated as a vague "impervious barrier". Rather, it is treated as an engineered structure with a specific geometry and specific hydraulic conductivity that interacts with the natural geologic materials having their own set of engineering characteristics.

The issue of assigning design parameters to a grout curtain is highly complex because the parameters are controlled by design, construction, and inspection factors. The achievable results are dependent on many elements of the grouting process including the grouting materials, grout mixes, construction equipment, field technique, number of lines utilized, hole spacing, hole orientation, experience and diligence of the contractor, experience and diligence of the inspection staff, the field monitoring and analysis techniques utilized in evaluating the completed work, contractual incentives and disincentives, climatic conditions, and other factors. While extensive information and discussion of the importance of each of these parameters is readily available, the factors have not, in general, been combined into specific design recommendations.

Table 1 contains the authors' recommendations for grout curtain permeability for use in grout curtain design. Table 1 is applicable for materials that are readily groutable (i.e. initial permeability greater than 1×10^{-4} m/s.) and which are inherently favorable for grouting. Features that make rock masses inherently favorable for grouting and for application of the QEGC Design Approach include highly fractured rock masses, well-connected fracture systems, and fracture systems of relatively uniform nature (i.e. spacing, size, and condition). The recommended values become progressively less applicable when these conditions are not present. These recommendations are based on consideration of the many relevant discussions of the factors and the authors' own experience on projects. Italics in Table 1 are used to highlight the differences in project characteristics between the various levels.

Table 1: Recommended Design Permeability for Grout Curtains

	Project Characteristics	Single Line Curtain	Triple Line Curtain
Level 1	Low bid contracting; inexperienced contractors Procedure specification – not engineer directed Payment provisions based on solids injected Thin neat cement grouts (thinner than 3:1) Minimal water pressure testing Final hole spacing on about 3 m centers Conservative, rule of thumb grouting pressures Dipstick & gage monitoring technology Holes not grouted to absolute refusal Inexperienced inspection staff Number of inspection staff inadequate to cover all operations full-time Non-existent understanding/analysis of results	Results Unpredictable Not Recommended	1×10^{-5} m/s (80 Lugeons)
Level 2	Low bid contracting *with experience requirement for grouting contractors* Primarily procedure specification – *limited engineer direction* Payment provision based on solids injected *Neat Type I or Type II cement grouts with mixes not thinner than 3:1* *Limited* water pressure testing *Final hole spacing on about 1.5 m centers* Conservative, rule of thumb grouting pressures Dipstick & gage monitoring technology *Holes grouted to absolute refusal with holding period after refusal* *Inspection staff with limited experience* *Number of inspection staff adequate to cover most operations full-time* *Limited understanding/analysis of results*	1×10^{-5} m/s (80 Lugeons)	1×10^{-6} m/s (8 Lugeons)

	Project Characteristics	Single Line Curtain	Triple Line Curtain
Level 3	*Pre-qualification of specialty grouting contractors and equipment & advance commitment by General Contractor* *Payment provisions based on fluid injection time* *Balanced stable grout mixes* *Extensive* water pressure testing *Engineer directed program* Final hole spacing 1.5 m *maximum* *Higher grouting pressures determined in the field as safe for each geologic unit* *Monitoring based on pressure transducers, magnetic flowmeters, and real-time display of results* Holes grouted to absolute refusal *Experienced inspection staff* *Number of inspection staff adequate to cover all operations full-time* *Good understanding/analysis of results*	1×10^{-6} m/s (8 Lugeons)	4×10^{-7} m/s (3 Lugeons)
Level 4	*Best Value Selection* or pre-qualification of specialty grouting contractors and equipment & advance commitment by General Contractor Payment provisions based on fluid injection time *Balanced stable grout mixes and special cements* Extensive water pressure testing Engineer directed program Final hole spacing 1.5 m maximum *and based on detailed closure analyses* *Highest grouting pressures determined in the field as safe for each geologic unit* Monitoring based on pressure transducers, magnetic flowmeters, real-time display of results, *and real-time analytical systems* Holes grouted to absolute refusal *Highly experienced inspection staff* Number of inspection staff adequate to cover all operations full-time *Comprehensive understanding/analysis of results*	4×10^{-7} m/s (3 Lugeons)	1×10^{-8} m/s (0.1 Lugeon)

The recommendations in Table 1 place considerable weight on Defect Theory considerations, since many of the factors are items that will contribute to development of small distributed defects throughout the grout curtain. Cedergren (1989) illustrates the importance of distributed small defects from mathematical analyses showing that well-distributed openings with an open space ratio of only 0.1 percent of the total cutoff area reduces the cutoff efficiency to only 29%, thereby allowing 71% of the flow to pass through. Based on examination of unsatisfactory grouting projects, the authors believe that one of the greatest problems with grouting has been distributed small defects, and the greatest errors in design are to assume that "grouting is grouting" and to overestimate the final effectiveness that will be achieved without regard to the entire design/construction/inspection process.   In picking a reliable design permeability, it is necessary to honestly evaluate the collection of factors that most closely matches how the project will, in reality, be executed.   Although specifying and achieving higher quality requirements might appear to slightly increase costs, experience has shown that major program cost savings actually occur with increased quality and use of advanced technology (Wilson and Dreese, 1998).

One source that at least lends some credibility to the recommended values in Table 1 is the report prepared by the Independent Panel to Review Cause of Teton Dam Failure (1976).   The outer rows were intended to be semi-pervious grout barriers intended to provide confinement for grouting an impervious barrier along the final, center row.   The final row of holes was spaced on 10-foot centers with split spacing wherever the primary holes did not indicate a tight curtain.   The grouting process utilized most closely resembles what is represented as Level 2 in Table 1.   Re-analysis of post-failure water testing data in the failure area suggests that Lugeon values within the completed grout curtain in excess of 100 were present in several locations and that Lugeon values between 10-100 were as common as Lugeon values less than 10.   The Independent Panel's conclusions were that (1) "a triple or 3-row grout curtain was not constructed. Instead it should be termed a single-row curtain", and (2) "the tests indicate that in the critical key-trench area the grout curtain was not fully closed."   A second thoroughly analyzed source supporting the Table 1 recommendations is from the Penn Forest Project.   That curtain was designed and constructed at Level 3 quality factors, and the back calculated performance indicates the performance of the 3-line curtain is at the 1 Lugeon level.

The remaining parameter for the grout curtain is selecting the effective thickness of the completed curtain.   While there are theoretical approaches to estimating the radius of grout travel, based on observations and analyses of single line and multiple line curtain closure, the authors recommend an effective design thickness of 1.5 m for single line curtains and 4.5 m for three-line grout curtains (assuming a 1.5 m spacing between each line).

Analytical Methods

General.   Grout curtain design in the QEGC Design Approach is modeled with the same techniques used for flow through porous media.   The validity of this approach obviously varies depending on the particular site conditions, but in general the authors believe that it is generally applicable except under extreme conditions such as in massive limestone heavily dissected by large open joints.   Porous media models for flow through a rock mass are clearly more valid when the rock is highly fractured.   It is also generally valid from a "bulk scale" consideration, where the fracture size and spacing relative to the dimensions of the areas being modeled are in reasonable proportion.

Preliminary Analyses. Having assigned quantitative parameters to the geologic units, the initial performance requirements, and the grout curtain, it is now possible to proceed with rational preliminary analyses of grouting alternatives. These preliminary analyses will provide a sound basis for determining the need for grouting, the value of alternative depths of grouting, the cost and schedule impacts to meet the desired performance requirements, and the relative benefits of performing the work at higher vs. lower levels of quality. The authors believe that the best tool for performing the preliminary analyses are flow nets prepared with a substantial amount of simplification. Flow nets are an incredibly useful tool, and any particular analysis can be performed with considerable accuracy in a matter of minutes if properly constructed. Cedergren (1989) provides an excellent basic reference on flow net construction.

A surprising degree of simplification without grossly distorting the results is possible in the Preliminary Analyses. In most cases, materials can be grouped by order of magnitude permeability for the preliminary analyses. For example, whether one unit has an assigned permeability of 1×10^{-5} m/s and another unit has an estimated permeability of 3×10^{-5} m/s has relatively little consequence. For practical purposes, they could be assigned the average of the two or the higher of the two values. Another major simplification based on the impact of relative permeabilities and the proportional geometries that exist for most dams is that it is only necessary to model the units as follows:

1. Units that are 2 orders of magnitude more permeable than the grout curtain or adjacent geologic units can be assumed to be infinitely permeable.
2. Units that are at 2 orders of magnitude less permeable than the grout curtain or adjacent geologic units can be assumed to be infinitely impermeable.
3. If the grout curtain permeability, relative to other units, is 2 orders of magnitude less permeable, it can be assumed to be infinitely impermeable or can be modeled as a separate component using Darcy's Law of q=kia.
4. Units that are either 1 order of magnitude more or less permeable than adjacent units will have an effect on head loss and flow. However, flow nets can easily be constructed for these intermediate materials by using rectangular elements having a 1:10 dimensional ratio, or a 10:1 dimensional ratio, as applicable. To construct these rectangular elements, it is only necessary to construct squares in all materials but to remember that the squares in the more permeable materials contain 10 flow tubes and the squares in the less permeable materials contain 10 equipotential drops. If less accuracy in the preliminary analyses can be accepted, the intermediate materials can simply be treated as infinitely permeable or impermeable.

An example of the maximum type of simplification is shown in the Figures 2 and 3. Also shown in Figure 4, for comparision purposes, is an "exact" solution based on finite element modeling that was performed later. With respect to the figures, the following should be noted:

1. The first step in the simplification was to reduce the permeabilities to simple order of magnitude values.

2. In each figure, units with a relative permeability one order of magnitude greater than the grout curtain or the lower permeability unit into which the grout curtain was connected were assumed to be infinitely permeable. Similarly, units with a relative permeability one order of magnitude less were assumed to be infinititely impermeable.

3. The grout curtain permeability was very low, but because it's geometry became substantial with depth, it was modeled separately from flow around the bottom of the curtain using Darcy's Law.

4. The extremely simplified analysis for a 24 m deep grout curtain yielded an estimated flow of 5.8×10^{-3} m³/s, while the "exact" analysis that accounted for the actual, more complex interaction of all the materials yielded a calculated flow of 7.6×10^{-3} m³/s.

While one way of looking at the comparative results is that the calculated flow differences are 30%, the reality is that because permeability and flow calculations are an order of magnitude analysis, the results are, in fact, very similar and are entirely adequate for preliminary evaluations of multiple alternatives The flow nets also provide a fundamental understanding of the flow regime to the designer that computers do not seem to impart.

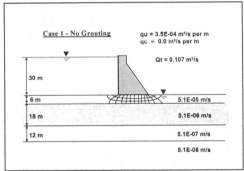

Figure 2. Simplified Analysis – Ungrouted Site

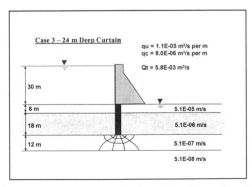

Figure 3. Simplified Analysis – 24 m Grout Curtain

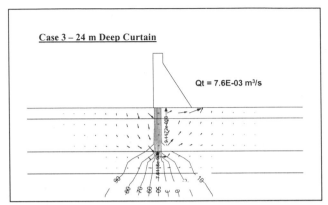

Figure 4. "Exact" Analysis – 24 m Grout Curtain

In this particular example, the results of the Preliminary Analysis for various curtain depths are shown in Table 2. Assuming that the performance requirement is 0.015 m^3/s (200 gpm) or less residual seepage, the analyses clearly show that it is necessary to grout to a depth of 24 m, and it is equally clear that grouting beyond that depth provides only a marginal improvement in performance at a 50% increase in cost and construction duration.

Table 2 – Results of Preliminary Analyses

Grout Curtain Depth (m)	Total Residual Seepage (m^3/s)	Grout Curtain Cost	Curtain Construction Duration
0	0.107	$ 0	0 Months
6	0.039	$ 600,000	2 Months
24	0.006	$ 2,400,000	8 Months
36	0.005	$3,600,000	12 Months

Final Design and Optimization. Many of the important basic design questions are either settled or at least partially answered in the Preliminary Design. However, the final design provides the opportunity to move forward with modeling all the complexities of the geology and topography. Grouting is an expensive and time consuming operation, and it is important that the design is optimized. The end result will be a grout curtain design that is tailored to the site, efficient, and which should reliably produce a residual seepage within the the performance requirements and the design estimates. Finite element modeling is normally utilized as the optimization tool since it can easily accommodate far more complex and subtle differences than can be effectively modeled with the simplified analytical procedures.

Matching Contract Requirements and Skills with the Design. One of the most critical elements of final design is to match the contracting mechanisms, the contract documents, the assignment of contract responsibilities, and the abilities of all parties with the results that must be achieved in construction of the grout curtain. Careful consideration is necessary to ensure that disincentives to quality are not built into the process and documents. Curtain efficiency requires that a substantial difference must exist between the permeability of the grout curtain and the materials in which it is installed. Typically, it will be found that a curtain that is only 1 order of magnitude less permeable than the material in which it is installed may have an efficiency on the order of only about 50%, whereas a curtain that is 2 orders of magnitude less permeable than the material in which it is installed will have an efficiency on the order of 99%. Therefore, the quality level that must be used in grouting can be dictated by the permeability of the natural materials. In simple terms, even relatively low quality grouting can reduce large flows by a dramatic amount. However, when control of medium flows is required, the grouting must be carried out in an exacting fashion to have a positive impact and to produce the desired result.

Penn Forest Case History Summary

All of the principles described herein were applied in the design and construction of the 3-line grout curtain for Penn Forest Dam:

1. Intensive investigation and testing programs were conducted to characterize the site conditions and assign quantitative design parameters.
2. A project performance requirement was set that the residual seepage not exceed 0.015 m^3/s (200 gpm) under full pool.
3. Grouting programs of various quality were considered along with their associated, applicable design parameters.
4. The grout curtain was designed using the QEGC Design Approach.
5. The program was optimized for depth along the length of the valley, which resulted in an asymmetric configuration. Geologic factors actually resulted in a substantially shallower curtain depth in the areas of highest applied head.
6. The contract documents, contract mechanisms, contractor requirements, field techniques, mix and monitoring technology, and the inspection staffing were at Level 3, which was consistent with achieving a 3 Lugeon triple line curtain.
7. Two years into service with a full pool, the total measured seepage at Penn Forest Dam is less than 0.008 m^3/s (100 gpm). The actual performance shows a curtain performing at the 1 Lugeon level.
8. A program cost savings of approximately 25% was attributed to use of Level 3 quality and technology factors.

References

Cedergren, H.R.. (1989). *Seepage, Drainage, and Flow Nets, Third Edition.* John Wiley & Sons, Inc., New York.

Houlsby, A.C. (1982). "Cement Grouting for Dams." *Proceedings of the Conference on Grouting in Geotechnical Engineering,* ASCE, New Orleans, Louisiana.

Houlsby, A.C. (1990). *Construction and Design of Cement Grouting,* John Wiley & Sons Inc., New York.

Independent Panel to Review Cause of Teton Dam Failure (1976). *Report to U.S. Department of the Interior and State of Idaho on Failure of Teton Dam,* U.S. Government Printing Office, Washington, D.C.

Wilson, D.B., and Dreese, T.L. (1997). "Grout Curtain Design for Dams." *Proceedings of the 1997 Annual Conference,* ASDSO, Pittsburgh, Pennsylvania.

Wilson, D.B., and Dreese, T.L. (1998). "Grouting Technologies for Dam Foundations." *Proceedings of the 1998 Annual Conference,* ASDSO, Las Vegas, Nevada.

CALIFORNIA AQUEDUCT FOUNDATION REPAIR USING MULTIPLE GROUTING TECHNIQUES

Timothy M. Wehling[1], Assoc. Member, and David C. Rennie[2], PE, Assoc. Member

ABSTRACT: On June 5, 2001, a major leak was discovered beneath the California Aqueduct embankment at Mile Post 4.25, discharging more than 4,200 L/min (1,100 gpm) at its peak. This situation prompted emergency action to control the leak to avert breaching the embankment. The leak was plugged after pumping 42 m^3 (55 yd^3) of concrete to fill the large piping void. Then a geologic exploration was conducted to delineate problematic zones beneath the embankment, followed by several grouting techniques used to investigate and improve the distressed foundation. The comprehensive grouting program included 27 compaction grout holes to explore and densify the soft alluvium foundation, 13 permeation grout holes to investigate and fill voids in the underlying fractured bedrock, and abandonment of an obsolete seepage collection system beneath the Aqueduct. This paper shows chronologically how data collected from the grouting program, geologic exploration, and construction records were used to better understand the mechanisms of the piping failure.

INTRODUCTION

A leak was discovered along the California Aqueduct, Mile Post (MP) 4.25, near Tracy, California on June 5, 2001 (FIG. 1). Seepage from a boil located 46 m (150 ft) downslope from the Aqueduct increased to approximately 4,200 L/min (1,100 gpm) over the next 36 hours as operations personnel, engineers, and geologists from the California Department of Water Resources (DWR) worked to locate and control the leak. DWR hired a grouting consultant, who was onsite to provide expert advice and assistance regarding all grouting methods used. Several emergency measures were undertaken to stop the piping failure before it could breach the embankment and release nearly 5.5×10^6 m^3 (4500 acre-ft) of water from 4.3 km (2.7 mi) of Aqueduct and Bethany Reservoir, located 0.4 km (0.25 mi) downstream. Once divers located the inlet of the leak near the Aqueduct invert, a quick-setting concrete was pumped through the cracked concrete canal liner to plug the leak.

By the late afternoon of June 6, the leak was plugged and DWR engineers began developing plans to conduct a permanent repair of the embankment, liner, and

---

[1] Water Resources Engineer, California Department of Water Resources, P.O. Box 942836, Sacramento, CA 94236, wehling@water.ca.gov
[2] Water Resources Engineer, California Department of Water Resources, P.O. Box 942836, Sacramento, CA 94236

893

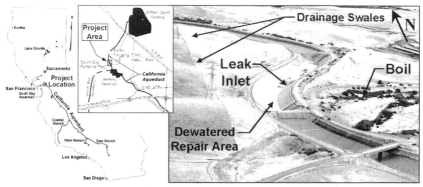

FIG. 1. Project Location Map and Aerial Photo of the Repair Site

distressed foundation near the leak. A section of the Aqueduct was dewatered to allow crews to replace damaged concrete liner panels and embankment fill near the leak inlet and to install an impermeable geomembrane liner along the channel section. Only 5½ weeks were allowed for the Aqueduct service outage since any additional downtime would have resulted in severe impacts to water users dependent on State Water Project (SWP) deliveries.

A key element of the final repair was the implementation of a comprehensive grouting program to investigate and improve the weakened foundation materials beneath the eastern Aqueduct embankment. The grouting program consisted of:

a) Compaction grouting to investigate and densify the weak alluvial foundation beneath the embankment; and

b) Permeation grouting to investigate and reduce the global permeability of the fractured bedrock where the piping may have developed.

In addition, pressure grouting was conducted beneath the canal invert to fill and abandon an obsolete seepage collection system.

SITE BACKGROUND

The 715-km (444-mi) long California Aqueduct is the main conveyance facility of the State Water Project, the largest state-built, multipurpose water project in the United States. The SWP delivers water to more than 23 million people and 400,000 hectares (1 million acres) of farmland in Central and Southern California.

The leak was discovered along the Aqueduct at MP 4.25, 2 km (1.25 mi) downstream from the Banks Pumping Plant and 26 km (16 mi) northwest of Tracy, California (FIG. 1). This reach of the Aqueduct crosses a series of broad, east-trending alluvium-filled swales and valleys (FIG. 1) along the eastern foothills of the Diablo Range. Three such valleys are located in the vicinity of MP 4.25. Beneath this reach, the Aqueduct foundation consists of up to 5.5 m (18 ft) of Quaternary Alluvial (Qal) deposits of soft, lean to fat clays containing varying amounts of sand and gravel. The underlying bedrock is part of the Cretaceous Panoche Formation (Kp) consisting of moderately to intensely weathered and fractured, often weak, sandstone with interbedded shale. The bedding strikes to the northwest and dips about

25 degrees to the northeast. The embankment fill (Qf) consists of compacted lean to fat clays with varying amounts of sand (DWR, 1970).

This reach of the Aqueduct is a cut and fill section approximately 1.9 km (1.2 mi) long between the Banks Pumping Plant and Bethany Reservoir. The channel invert is level throughout this reach at Elevation 65.0 m (213.4 ft) with an invert width of 12.2 m (40 ft) and 1.5:1 canal slopes approximately 10.7 m (35 ft) high. The canal prism is lined with 100-mm (4-in) thick unreinforced concrete panels with longitudinal and transverse contraction joints 3.6 m (12 ft) on-center. The design capacity of this channel is 292 m^3/s (10,300 cfs) at a normal depth of 9.1 m (30 ft). With no inlet control structure at Bethany Reservoir, the canal and reservoir are operated as a single pool controlled by upstream and downstream operations.

EMERGENCY RESPONSE

June 5, 2001

The leak was discovered the morning of June 5. Leakage from a boil was observed flowing into a drainage swale approximately 46 m (150 ft) east of the eastern Aqueduct embankment. DWR operations personnel quickly constructed a V-notch weir to monitor the flow rate from the boil. The initial head differential between the Aqueduct water surface and the boil was 18 m (59 ft). Dewatering of the Aqueduct pool and adjoining reservoir was initiated immediately at a rate that would prevent slope failures along the Aqueduct due to rapid drawdown. In addition, a sandbag chimney was constructed around the boil (FIG. 2) to reduce the head differential by roughly 2.4 m (8 ft).

Seepage at the boil was initially estimated at 3,600 L/min (950 gpm) and increased despite drawing down the Aqueduct pool (FIG. 3). The increasing flowrate and turbidity of water exiting the boil (FIG. 2) indicated that the piping condition was progressing, and the stability of the embankment was in jeopardy.

Divers conducted several underwater inspections before discovering the location of leak inlet, which was a 0.6-m (2-ft) long crack along a construction joint approximately 0.3 m (1 ft) above the concrete canal invert. The divers reported considerable suction as water flowed into the crack. Then based on the site geology, the pipe was believed to have developed either at the alluvium-bedrock interface or through the fractured bedrock formation.

Meanwhile, several emergency measures were undertaken during the first 24 hours of the response. Broadly-graded rock was stockpiled near the leak to be used as last-

FIG. 2. Sandbag Chimney at Boil

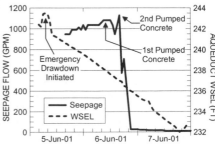

FIG. 3. Seepage Rate from Boil vs. Time

resort plugging material in the event flow began increasing beyond control. By late evening on June 5, a large, steel bulkhead had been fabricated and was lowered into the Aqueduct. This steel plate was designed to be placed directly over the leak to distribute stresses and avert a collapse of the liner panels. However, DWR divers were not able to guide the plate into place due difficulties. Then an unsuccessful attempt was made around 3AM to grout behind the misaligned plate with a cement/bentonite mix. Seepage at the boil was closely monitored throughout the early morning as plans were developed to plug the leak with injected concrete.

June 6, 2001

By mid-morning of June 6, hardhat divers and concrete pumping equipment were mobilized to the site, followed by the grouting consultant and a concrete boom pumper truck. Based on previous experience along the Aqueduct, the leak was initially believed to be comprised of numerous small seepage channels leading to a major pipe in the embankment material, alluvial foundation, and/or underlying bedrock. Thus, a minus 9.5-mm (3/8-in) pea-gravel concrete was ordered from a nearby ready-mix plant.

The objective of this small, rounded aggregate was to enable the concrete to flow through the small seepage channels to treat the primary void. A viscosity modifying admixture (VMA) was ordered to prevent cement washout while pumping underwater. However, no VMAs were available from the local plant, so silica fume was requested to increase the mix cohesion and reduce cement loss. However, this admixture was also unavailable locally. Finally, a 10-bag mix with 30% minus 9.5-mm (3/8-in) rounded pea gravel was ordered under the premise that the anticipated cement loss would be compensated for by the rich mix.

By noon, divers had cut a 150-cm (6-in) hole through the concrete canal liner near the leak to provide access for grouting. The concrete boom truck was positioned on the embankment crest to access the hole. Divers guided the pump nozzle into the hole and reported grout action at the inlet via two-way radio.

Two truckloads, 14.5 m^3 (19 yd^3) total, of the pea gravel mix were rapidly pumped into the grouting hole. Cement and aggregate were observed exiting the boil within 2-3 seconds after pumping began, demonstrating the direct connectivity between the leak inlet and the boil. Boil ejecta was compared to the concrete mix after pumping had ceased. Both included angular aggregate rather than the rounded pea gravel that was ordered, which was deemed more suitable to treat small channels leading to a major void. The transport of aggregate suggested the size of the pipe was significant and continuous.

This grouting attempt achieved a temporary 10% reduction in seepage flowrate (FIG. 3). However, the flowrate increased approximately 30 minutes after the grouting had ceased. In addition, the nature of the water exiting the boil began to change rapidly, becoming more turbid. It was theorized that grout may have partially filled the bottom of the pipe, forcing the route of piping to change. This change was likely due to the development of high velocities in the pipe and scour from the angular pea gravel, increasing the rate of erosion along the top of the pipe.

Given these conditions, the grouting consultant concluded that a major pipe had formed between the canal invert and the boil, and a grout with maximum plugging ability was needed. Thus, a second batch of concrete was ordered immediately. This was also a 10-bag mix with the maximum amount of minus 38-mm (1½-in) aggregate that could be pumped. Calcium chloride ($CaCl_2$; 3% by weight) was added as an accelerator to rapidly set the mix during injection. In addition, polypropylene fibers were added to reduce the slump and increase the plugging ability of the mix. The grouting consultant believed that the ability to pump concrete at a high enough rate to overwhelm the flow of water in the pipe was the most important factor in stopping this massive leak. Thus, the second grouting attempt was not initiated until a sufficient volume (*i.e.,* number of trucks) of concrete was onsite to inject continuously until the leak stopped.

A second hole was drilled through the concrete canal liner approximately 1.8 m (6 ft) downstream from the first access hole, which failed to take anymore concrete. By 3:00 P.M., the first load was injected without $CaCl_2$ to increase the mobility of the grout front. Three subsequent loads included $CaCl_2$ to reduce the potential for the cement to wash out. Four truckloads, approximately 29 m^3 (36 yd^3), were ultimately injected during this attempt. The maximum pumping rate was approximately 1.9 m^3/min (2.5 yd^3/min), or 4-5 minutes per truckload.

Divers reported that the hole stopped taking grout after most of the fourth load was injected. The flowrate at the boil almost immediately decreased from in excess of 4,200 L/min (1,100 gpm) to less than 8 L/min (2 gpm) (FIG. 3) and catastrophic failure of the embankment was averted. It was concluded that, while the accelerator played a significant role in plugging the leak, a continuous, high rate of pumping was more important than the precise nature of the mix. Most importantly, the ability to recognize changing field conditions and adjust the grouting program on the fly, as well as contributions from the ready-mix plant technicians, were critical to the success of this emergency grouting operation.

PERMANENT AQUEDUCT REPAIRS

Once the leak was stopped and the emergency averted, DWR personnel immediately developed a plan to permanently repair the Aqueduct while dewatering continued. The repair program included the following major work elements:

a) Conduct a geologic/geotechnical investigation to evaluate the condition of the embankment foundation;

b) Improve the foundation deficiencies discovered by the investigation;

c) Construct two rockfill coffer dams to fully dewater a 365 m (1,200 ft) reach of the Aqueduct (FIG. 1);

d) Repair the concrete canal foundation and lining near the leak inlet;

e) Construct a 243-m (800-ft) long impermeable barrier between the Aqueduct pool and distressed embankment and foundation; and

f) Construct a filtered toe drain near the location of the boil.

As an aside, following the completion of the grouting work, the impermeable membrane was installed along more than 12,000 m^2 (130,000 ft^2) of the channel. The membrane was ballasted with shotcrete and secured in anchor trenches that intercepted an existing invert drainline 85 m (280 ft) upstream from the leak. To reduce the conveyance of future seepage, 150 m (500 ft) of this 100-200 mm (4-8 in) diameter perforated drainpipe was abandoned using a standard flowable fill mix with 110 g/m^3 (3 oz/yd^3) of Darafill admixture. This admixture created a high fly ash, cement-sand mix with a high entrained air content that is almost self-leveling and yet free of bleed. Approximately 15 m³ (5 yd³) of grout was estimated for the drainpipe abandonment.

Construction crews were afforded only 38 days to complete the repairs so as not to impact valuable SWP deliveries scheduled for the following month. To further complicate matters, almost 2 weeks would be required to safely dewater this portion of the Aqueduct without inducing slope failures caused by rapid drawdown. Crews completed the entire repair in only 25 days. In fact, the geologic investigation and additional foundation grouting activities (described below) were completed before the work area was completely dewatered.

INITIAL GEOLOGIC INVESTIGATION

A geologic investigation was initiated immediately following the emergency grouting and included hollow-stem auger and mud rotary drilling, cone penetration testing (CPT), and trenching. The primary objective of the investigation was to evaluate the deterioration of the embankment foundation. These drill holes and CPTs were performed along the embankment crest, the lower access road, and at the boil location. In all, 31 CPTs and 10 boreholes were drilled in 3 and 12 days, respectively, to provide a clear depiction of the embankment, alluvial, and bedrock formations described earlier (FIG. 4). Seven boreholes were later converted to observation wells, and a trench was also excavated just west of the boil.

The key findings from this initial investigation are summarized by a profile along the embankment crest (FIG. 4) and a cross section perpendicular to the embankment (FIG. 5). The CPT was an especially insightful tool that provided immediate interpretation of the foundation materials prior to the foundation grouting program. The CPT soundings along the crest in FIG. 4 indicate a well compacted embankment, very soft alluvium (CPT tip resistances as low as 0 tsf), and hard bedrock. The alluvium is thickest in the middle of the swale (hole DH-2), which is also where the piping occurred. The cross section in FIG. 5 demonstrates that the alluvium and the bedrock are continuous and slope gently from the leak inlet to the boil.

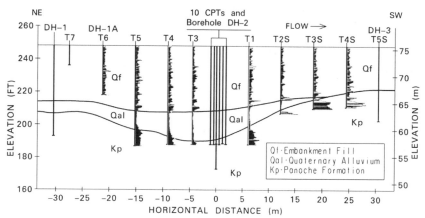

FIG. 4. Geologic Profile of the Eastern Embankment Crest at the Leak

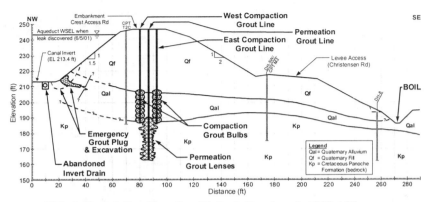

FIG. 5. Foundation Grouting Methods Employed at MP 4.25

Some additional findings supported the concept that the piping most likely occurred through the fractured bedrock or at the alluvium/bedrock interface. For example, despite the extremely soft nature of the alluvium, none of the CPT soundings intercepted any pockets of hardened grout from the emergency grouting. Furthermore, at a depth of 4.8 m (15.8 ft), borehole DH-4 contained a 5-mm (0.2-in) seam of grout, which was identified as grout from a nearby CPT hole (M5) that took five times more grout than was expected to backfill the CPT hole. This grout seam occurred at the alluvium-bedrock contact, where voids or fractures in the bedrock are expected to provide a pathway for fluid to travel horizontally. Similarly, during mud rotary drilling of hole DH-3, more than 380 L (100 gal) of bentonite drilling fluid was lost in the bedrock. This indicates the nature of the fractured bedrock is highly variable.

FOUNDATION GROUTING PROGRAM

The initial geologic investigation, outlined above, identified the weak alluvium and fractured bedrock foundation materials, both of which contributed to the piping condition. Therefore, a phased grouting program was implemented to investigate and improve the embankment foundation. Compaction grouting was used to densify the soft alluvium beneath the embankment, and permeation grouting was used to reduce the hydraulic conductivity of the fractured bedrock (FIG. 5). The extent of these two methods are shown in FIG. 6 and discussed below. In addition, a ready-mix flowable-fill grout was used to abandon an existing underdrain system beneath the Aqueduct invert.

Grouting Program Objectives

The primary goals of the grouting program were to:

- Collect grouting data as a tool to investigate the subsurface conditions;
- Densify the soft alluvium foundation and close voids;
- Fill joints to reduce the global permeability of the fractured bedrock; and,
- Fill the perforated drain pipes beneath the canal invert to block conduits for future seepage.

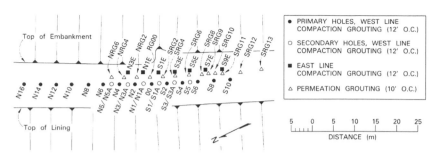

FIG. 6. Compaction and Permeation Grouting Along Embankment

Compaction Grouting

Compaction grouting was performed by Geo Grout, Inc. of San Francisco from June 15-20, 2001. About 20 m³ (710 ft³) of 25-mm (1-in) slump, low-mobility grout was pumped into 27 holes along two parallel 53 m (175 ft) long rows centered on the leak (FIG. 6). The depths of these holes varied from 12.8-18.9 m (42-62 ft) below the embankment crest. Drilling was accomplished using two track mounted, pneumatic rotary (Air Trac) rigs driven by one 21 m³/min (750 cfm) compressor. Upstage compaction grouting was performed in the following sequence:

- Predrilled hole by air-percussion 7.6 m (25 ft) into the compacted fill;
- 38-mm (2.0-in) inner diameter casing (with plugged end) was inserted down the hole and driven through the alluvium and an additional 0.3 m (1 ft) of bedrock;
- Grout was pumped into the alluvium in 1-m (3-ft) stages starting at the bottom of the hole;
- Grout was injected at each stage under constant flow of 0.02 m³/min (1 ft³/min) while monitoring grout take, pressure, and time for each stage;
- Each grouting stage ceased when wellhead pressure reached 3500 kPa (500 psi) or 1 mm (1/16 in) of ground heave was observed; and
- Grout pipe was raised 0.9 m (3 ft) to begin the next grouting stage and the procedure repeated until the entire thickness of the alluvium was treated.

Compaction grouting began with 14 primary grout holes along the embankment crest spaced 3.7 m (12 ft) on-center. These holes are designated on FIG. 6 as the West Line Primary holes (even numbered holes). Primary holes near the center and south of the leak were then split-spaced with secondary compaction grouting holes (odd numbered holes) based on high takes in these areas.

The effectiveness of the primary holes was evident by two observations. First, the secondary hole casings were much more difficult to install, as evident by the broken casings and difficult predrilling. (Note that the hole numbers followed by the letter "A" in FIG. 6 show the locations of the broken casings.) This suggested that the compaction grouting from the primary holes had densified the alluvium, making the secondary holes more difficult to install. Second, grout takes in the secondary holes were considerably lower than those in the primary holes, suggesting that the range of compactive influence of the primary and secondary holes had overlapped. Therefore, to facilitate installation of the secondary holes, guide holes were predrilled to bedrock. The grout pipes were then driven into the predrilled hole down to the top of bedrock, and grouting continued as previously described.

Compaction grouting results along the West Line are shown in FIG. 7. Based on these results, an additional row (N3E through S9E) of compaction grouting was completed 3 m (10 ft) east of the West Line. Seven holes were grouted along this East Line of compaction grouting, as shown in FIG. 6, and the grouting results are shown in FIG. 8. These seven holes were installed in the same manner as the secondary holes along the West Line.

These figures show that grout takes were larger in the primary holes than in the secondary holes. Also, holes further north from the leak inlet generally took less grout, which confirmed the CPT interpretation of stiffer alluvium north of the leak. Given that, the highest grout takes in the alluvium were relatively small, about 0.57 m³ (20 ft³) per stage. For example, assuming a 1.5-m (5-ft) radius of influence and a 0.9-m (3-ft) stage, a grout take of 0.57 m³ (20 ft³) would result in a replacement ratio of 8.5%. Moseley (1993) suggests that replacement ratios during compaction grouting usually range between 3% and 12% for soils free of large voids. Therefore, with one exception, the small compaction grouting takes show that the alluvium was free of large voids and was compacted by particle displacement. The exception occurred in hole S3A, which took 0.76 m³ (27 ft³) near the alluvium-bedrock contact. The alluvium at the bottom of this hole showed very little resistance during drilling and some wet material near the alluvium/bedrock interface. Therefore, the high grout take most likely filled a piping void at the contact between the alluvium and bedrock, where the pipe was suspected to have occurred.

Permeation Grouting

Upon completion of the compaction grouting, permeation grouting was conducted by Geo Grout, Inc. to investigate and fill joints in the upper 7.6 m (25 ft) of fractured bedrock. Approximately 4.6 m³ (164 ft³ or 1230 gal) of cement slurry was injected into 13 holes spaced 1.5-3 m (5-10 ft) apart along a 33.5-m (110-ft) long line situated between the compaction grout lines (FIG. 6). After completion of the initial nine holes, four additional holes (SRG-9, 11, 12, & 13) were drilled and grouted between July 1 and July 3, 2001. This additional work was conducted due to a large grout take,

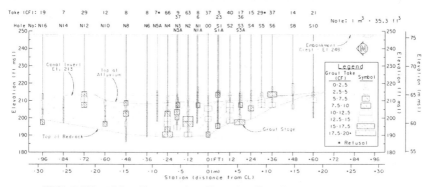

FIG. 7. West Line Compaction Grouting Results

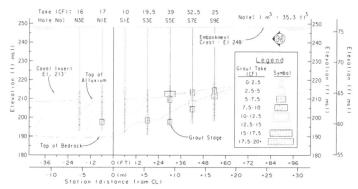

FIG. 8. East Line Compaction Grouting Results

227 L (60 gal), in the southern most hole (SRG10) completed in June. The following procedures were used in completing the permeation grouting:

- Hole was predrilled by air percussion 0.3-0.6 m (1-2 ft) into the bedrock;
- 51-mm (2.0-in) inner diameter casing driven into the predrilled hole;
- Approximately 7 L (2 gal) of quick-setting cement grout was poured into the casing to seal the connection between the casing and the bedrock;
- After the grout set, a 47.6-mm (1 7/8-in) diameter drill bit (AX) was lowered inside the casing and drilled 7.6 m (25 ft) into the bedrock with water circulation to flush the cuttings;
- Grout was injected with a recirculating system using a manual valve to control the flow of grout into the casing;
- Grout volume and gage pressure versus time were monitored and recorded during pumping.

Grout mix and pumping parameters were adjusted throughout the permeation grouting process. An ultrafine grout consisting of Nittetsu Super-Fine cement at a water-cement (w/c) ratio of 1.25 (by weight) was initially injected into each hole. In holes that readily took the initial 230-380 L (60-100 gal) batch, the mix was changed to type I-II portland cement mix with a w/c ratio of 0.83, reduced to as low as 0.67 for very large takes. The objective of the mix design was to develop a very fine, mobile grout front capable of filling fine bedrock fractures ahead of the thicker grout mass filling larger voids. The recirculating valve was continually throttled to maintain a maximum gage pressure of 172 kPa (25 psi) at the top of casing. Grouting was ceased when the injection rate decreased to approximately 7.6 L (2 gal) per 5 minutes.

Four primary holes (NRG6, NRG2, SRG2, and SRG6) spaced 6 m (20 ft) apart were drilled and casings set before grouting began. This minimized grout communication between holes. The secondary holes (NRG4, RG00, SRG4, and SRG8) were split-spaced between the primary holes (FIG. 6) after the primary grout had set. This resulted in a final spacing of 3 m (10 ft) between permeation grout holes. The final boring drilled and grouted during the initial phase was SRG10. This hole was added to the final work due to the relatively large grout take at SRG8 (550 L, or 145 gal), the southern most hole in the first series, to see if additional voids continued south of

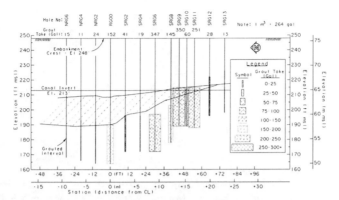

FIG. 9. Permeation Grouting Results at the Embankment Crest

SRG8. Holes SRG-9, 11, 12 and 13 were drilled and grouted based on the high takes in holes SRG8 and SRG10.

Grout takes for the permeation grouting holes are shown in the cross section on FIG. 9 where five holes took more than 380 L (100 gal) of grout each. Holes placed just south of where the leak occurred took the majority of the grout. For example, the five holes between SRG6 and SRG11 took 75% of the permeation grout injected at the site. This indicates that the bedrock fractures can transmit significant flows. Grout takes for the southern most holes, SRG-12 and 13, were much smaller and, thus, illustrate the variability in the bedrock, including rock fracture frequency, orientation, and thickness.

MONITORING AND FUTURE ACTIONS

Continuing seepage from the boil was monitored daily during the embankment repair. The flowrate remained nearly constant at approximately 20 L/min (5 gpm) and slowly increasing to roughly 40 L/min (10 gpm) as the Aqueduct was refilled in early July 2001. As a defensive measure, a filter drain and berm were constructed along the downstream toe to intercept and measure seepage. In the 14 months following completion of the canal repairs, the flowrate and seepage pressures remained relatively constant and approximately equal to conditions prior to the leak. Seepage is likely sourced from Bethany Reservoir and portions of the Aqueduct outside this repaired area. Future monitoring and evaluation will be necessary to verify the long-term viability of designs and decisions made under emergency conditions.

SUMMARY AND CONCLUSIONS

The MP 4.25 leak was plugged by pumping fast-setting concrete into the piping void. Recognition of changing field conditions and timely modification to the grouting program using available resources were critical to the success of the emergency action. After recognizing that the piping mechanism was a single large conduit, the strategy shifted to overwhelming the flow of leakage with a continuous, high-volume grout injection. In addition, the grouting consultant compensated for the lack of locally available admixtures by increasing the cement content of the concrete to offset the expected cement washout.

Due to the inability to obtain critical admixtures for the emergency grouting at MP 4.25, DWR has implemented a simple solution to prevent a repeat of this scenario in the future. The grouting consultant, in cooperation with Surecrete, Inc., has developed a dry powder admixture that has a long shelf life and is easy to administer in the field. This VMA with high-range water reducer (for dispersal) is to be used in conjunction with polypropylene fibers (to decrease excessive slump) to create an effective anti-washout mix. The product is sealed in five-gallon buckets for ease in handling and dosing by DWR personnel. Stockpiles of the product have been distributed to DWR field divisions throughout the state that may encounter a similar emergency in the future. The availability of this product should improve the Department's ability to respond to future emergencies, independent of material availability from local vendors.

Monitoring and real-time interpretation of key parameters including grout take, injection rate, and pressure vs. time during compaction and permeation grouting were successful in investigating and improving the distressed foundation materials beneath the Aqueduct embankment. The low grout takes and high injection pressures during the compaction grouting confirmed the absence of significant voids in the alluvium. A single, large take at the bottom of the alluvium suggested that the piping occurred near the bedrock-alluvium interface. Then, high grout takes and injection rates during permeation grouting demonstrated the capability of the fractured bedrock to transmit the large flow rates necessary to trigger internal erosion of the overlying alluvium. Repair of this reach of Aqueduct has been achieved through densification of the soft alluvium by compaction grouting, reduction of the mass permeability of the bedrock by permeation grouting, and installation of the impermeable liner along the canal to reduce seepage losses from the Aqueduct.

After completion of grouting, the emergency concrete bulb was excavated at the leak inlet. The excavation revealed a large, funnel-shaped concrete mass immediately behind the Aqueduct panels. The size and symmetry of this concrete plug confirm that the leakage path was a large, continuous void rather than a system of small channels leading to a major pipe. Also, the concrete plug leading to the boil entered the soft alluvium in a southeast direction approximating the strike of the underlying bedrock, which helps confirm the findings from compaction and permeation grouting that the bedrock had a major role in triggering the piping conditions at MP 4.25.

ACKNOWLEDGEMENTS

The authors thank the following people who contributed personal accounts of field activities, drawings, photos, and geologic data used in this paper: DWR's grouting consultant, Jim Warner of Mariposa, California, who led the grouting program and provided details about the grouting activities; and DWR staff including Les Harder, Mike Inamine, Mike Driller, Don Walker, Steve Belluomini, and Jeff Van Gilder.

REFERENCES

DWR (1970), Project Geology Report No. C-53, *Final Geologic Report, Delta Pumping Plant to Chrisman Road*, California Department of Water Resources.

DWR (2002), Project Geology Report No. C-128, *Final Construction Geology Report, Emergency Repair: Mile 4.25*, California Department of Water Resources.

Moseley, M. P., ed. (1993), *Ground Improvement*, CRC Press, Boca Raton, FL.

Flexibility in Grouting: Solutions for Old Dams

Allen Cadden[1], P.E.; Jesús Gómez[2], Ph.D., P.E.; Graham C.G. Smith[3];
and Robert Traylor[4]

Abstract

The Ivex Packaging Corporation purchased a paper-recycling mill in Chagrin Falls,
Ohio. The 100-year old dam was constructed of masonry and timber cribbing, and
had shown a history of wear in recent years. Seepage flows through the embankment
and right abutment, and into the process building were a steadily increasing problem
requiring constant attention. To combat this problem, a combination of grouting
techniques was used, including sleeve port pipe grouting, and limited mobility
grouting. Difficulties encountered during construction required close coordination
with the plant personnel, grouting specialist contractor, and the engineers to adjust the
hole pattern, grout materials, and injection methods as work proceeded. Following
four weeks of grouting, seepage flows were stopped through the right abutment. This
paper describes the process for development of a grouting plan adapted to the
particular site conditions, and the field adjustments made during construction. These
adjustments included modification of type of grout and grout placement methods,
especially where significant flows were found to be untreatable using low viscosity
grout.

Introduction

Chagrin Falls Dam is located on the Chagrin River, less than 800 m (½ mile)
upstream of the town of Chagrin Falls, in Cuyahoga County, Ohio (see Figure 1).

The dam was built over one hundred years ago, and is presently used as water
supply for a paper mill owned by IVEX Packaging Corporation. The dam consists of
an earth embankment, and a masonry arch spillway running approximately north to
south, as illustrated in Figures 1 and 2. The paper processing building is a three-story
structure with perimeter masonry walls. The east wall of the building acts as a
retaining structure for the adjacent earth embankment, and has a height of
approximately 6 m (20 ft) as measured from the crest of the embankment. The
embankment is composed of a heterogeneous mix of soils, wood cribbing, and other
materials, which were originally placed directly against the building wall. The

---

[1] Member ASCE, Principal, Schnabel Engineering Associates, Inc., 510 East Gay Street, West
Chester, PA 19380; acadden@schnabel-eng.com
[2] Member ASCE, Associate, Schnabel Engineering Associates, Inc., 510 East Gay Street, West
Chester, PA 19380; jgomez@schnabel-eng.com
[3] Project Manager, Structural Preservation Systems, Inc., 39 Utter Avenue, Hawthorne, NJ 07506
[4] President, Traylor, LLC, 17204 Hunter Green Road, Upperco, MD 21155

spillway and abutments consist of rock blocks of various sizes and composition. A concrete facing covers the exposed portions of the masonry spillway and abutments.

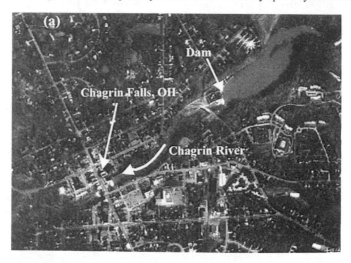

Figure 1. (a) Aerial view of the vicinity of the dam (Source: USGS); (b) The dam as seen from the north embankment (Source: IVEX).

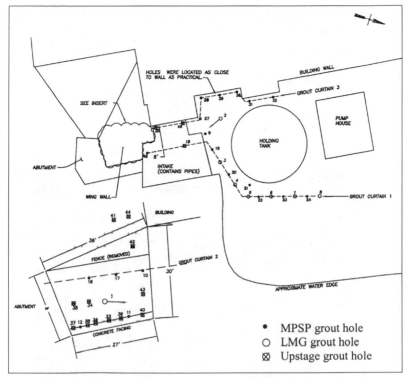

Figure 2. Site plan showing general layout of the site and grouting plan.

A set of intake pipes is located north of the right abutment and carry the water used for paper processing into the building. The intake pipes are encased within a 3 m (10 ft) thick concrete block (see Figures 1 and 2). A masonry wing wall spans the distance between the right abutment and the intake structure.

Originally, Chagrin Falls Dam consisted of two ponds. The upper pond dam (upstream of the site) collapsed in 1994. According to information from personnel in the paper plant, the apparent cause for the collapse was failure of one of the abutments of the spillway. The collapse caused debris to accumulate throughout the lower pond, and directly behind the existing spillway. The depth of the existing lower pond now ranges from about three to seven feet behind the spillway.

As illustrated in Figure 3, water was infiltrating from the pond into the process building and through the abutment. The infiltration rate had been increasing with time, and was a cause of concern for the owner, especially considering the preceding failure of the upper dam, and the location of the town of Chagrin Falls directly downstream of the dam. Seepage into the building was so severe that drainage channels were excavated in the basement to catch this flow and direct it

the flow of water from the pond into the intake structure made it difficult to interpret the results of the Self-Potential tests.

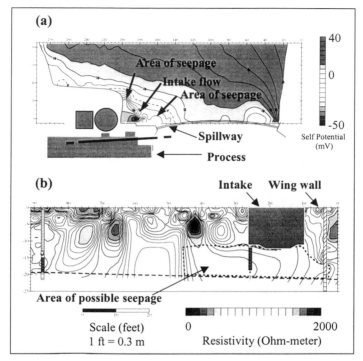

Figure 4. (a) Self Potential test results - plan view;
(b) Cross-hole resistivity test results - section.

Figure 4 shows the results of the geophysical surveys. These results, together with the information from exploratory borings and observations on site, indicated that significant water flow was occurring through the right abutment. In addition, a potential flow path was detected directly north of the intake structure, and there was evidence of possible flow taking place under the intake structure and through the wing wall.

The fact that significant water was flowing through the abutment and the embankment, and that some of the flow was fast, was of great concern. It was inferred from the data that relatively large voids existed within the embankment, with the consequent potential for increased washout of the embankment material. These voids may have been produced by gradual erosion around and inside the cribbing. Most importantly, the existence of large voids within the abutment could develop into potential instability or collapse of the abutment over time.

back into the river channel downstream. Several attempts for remediation had been undertaken in the past ten years using various synthetic and cement grouts, each with little success.

Figure 3. Observed leak paths.

A geophysical study was carried out to determine the flow path of the water seeping though the earthen embankment and wall. Based on this information, a grout treatment plan was designed to reduce the rate of water inflow into the process building, while limiting the environmental impact of potential washouts of grout into the river.

Geophysics for Detection of Leak Paths

A geophysical survey was conducted in the pond, dam embankment, and abutment areas. Two methods were used for this survey: Self-Potential and Cross-Hole Resistivity. The Self-Potential method was employed in the western extent of the pond along the upstream face of the dam in five linear traverses extending north-south at 6 m (20 ft) intervals. The data collected was utilized to determine what areas of the pond were losing water through its banks or bottom.

The Cross-Hole Resistivity survey was conducted along the embankment and right abutment wall to detect wet or saturated conditions within the embankment and wing wall. The survey was conducted using cross-hole techniques to gain additional depth than would have been obtained from a surface survey, given the site constraints. The boreholes also provided quality assurance for the geophysical results by correlating visual descriptions of material types and conditions.

Interpreting the results of the geophysical exploration was challenging due to the configuration of the plant. The presence of the intake structure, utilities, and other buried obstacles was a challenge for the interpretation of resistivity data. Similarly,

Grouting Program

Grouting was carried out in two phases. Phase 1 was performed to control seepage through the embankment, wing wall, and intake structure areas. Phase 2 focused on eliminating seepage though the deteriorated west abutment. Each grouting phase is discussed below.

Phase 1. The designed grouting program for the embankment, wing wall, and intake structure involved the installation of three vertical grout curtains constructed from ground surface to top of bedrock, as illustrated in Figure 2. Grout Curtain 1 started in the embankment area and continued across the intake structure and the wing wall. It consisted of primary Low Mobility Grout (LMG) holes spaced approximately 2.44 m (8 ft) on center. Between the LMG grout holes, secondary Multiple Port Sleeved Pipes (MPSP) were installed. The purpose of the LMG grout holes was to displace softer materials and fill any major voids and seepage paths by the injection of low-slump cement/sand grout mix. Potential damage to the surrounding structures or ground heave was not a major concern in this area. The purpose of the MPSP grout holes was to provide final closure to the LMG grout curtain, and to fill low-velocity seepage paths upstream and downstream of the curtain through the injection of high-viscosity neat cement grout.

Originally, for the wing wall area of Grout Curtain 1, only MPSP holes were proposed at approximately 1.83 m (6 ft) on center. Significant seepage in this area was not anticipated and potential damage to the abutment was a concern if LMG grouting was implemented.

Grout Curtain 2 extended from the abutment to the north end of the intake structure and consisted of MPSP grout holes spaced at 1.83 m (6 ft) on center along the wing wall, and LMG holes in the intake area. To treat the zone under the intake structure, LMG holes were drilled at an angle from both sides of the intake. In addition, holes were drilled through the intake structure away from the encased pipes. As discussed later in this paper, the grouting methods considered during design for Grout Curtains 1 and 2 were significantly modified to adapt to the soil conditions encountered during construction.

Grout Curtain 3 ran along the upstream wall of the process building and consisted of MPSP grout holes located 2.13 m (7 ft) on center. LMG grouting was not considered due to the potential for damage to the adjacent building wall.

The grout hole layout for the curtains was intended as a guideline for construction. Field adjustments to the locations and quantities of the grout holes were anticipated due to the varied fill material and access constraints. Therefore, it was important to bring the tools and equipment to perform both LMG grouting and MPSP grouting interchangeably.

Drilling. The drill rig used was a Klemm 704 electric hydraulic rotary, track-mounted rig designed for limited access work (see Figure 5). A 102 mm (4 inch) diameter, heavy wall casing fitted with a carbide-tooth shoe was advanced to bedrock with external water flush. Once the casing was installed, either LMG was injected through

Figure 5. View of the drilling rig.

this casing following an upstage grouting method, or the MPSP was installed and the casing was extracted.

In the area of the concrete intake structure, a down hole hammer (DHH) with drill rods was used to advance through the concrete. After the DHH penetrated through the concrete, the drilling method was switched to rotary flush technique to complete the hole.

LMG Grouting. The drill rig was equipped with a grout-through head. This head allowed the grout casing to be extracted in a single motion without any plumbing disconnection. This setup proves to more productive than a standard top of casing grout header, and allows instant relief of excessive grout pressures in the ground by extracting the casing a short distance.

The mix proportions of the LMG grout were 435.5 kg (960 lb) flyash, 65.3 kg (144 lb) cement, 997.9 kg (2200 lb) sand, and 0.15 m^3 (40 gallons) of water per 0.76 m^3 (cubic yard). The slump of the grout was between 25 mm and 76 mm (1 and 3 inches), and its unconfined compressive strength was at least 6900 kPa (1000 psi). The grout was delivered to the site and pumped using a high-pressure swing tube grout pump. Two grout pressure gauges were installed: one located at the pump, and the other located just above the drill head.

LMG grout holes were grouted to refusal pressures between 345 kPa and 2410 kPa (50 and 350 psi) depending on the slump of the mix, and stage depth. Holes were upstage grouted in 0.61 m (2 ft) lifts.

MPSP Grouting. MPSP grouting was performed using 38 mm (1.5 inch) diameter PVC grout pipes with sleeved ports located every 381 mm (15 inches), as shown in Figure 6a, and capped at the bottom. The pipe was supplied in 3 m (10 ft) sections, and was joined with standard PVC and glue. Single- and double-fold, double-port

Figure 6. (a) View of multiple port sleeved pipes; (b) View of packer.

packers were utilized to isolate and inject grout through the sleeved ports (see Figure 6b).

For most of the MPSP grouting, multiple line grouting was used to simultaneously grout two holes at one time. The grout plant consisted of a high-speed colloidal mixer, an agitated holding tank, and a Moyno-type pump. Magnetic flow meters and pressure gauges were used to monitor grout pressures, flows and volumes. Pressures were manually controlled with a return line valve. An inline screen was used to capture any unwanted solids before the grout was circulated and injected in the hole.

Once the MPSP pipe was installed and the casing removed, grout was injected to fill the approximately 32 mm (1.25 inch) annulus between the MPSP and the drill hole wall. A weak (less than 690 kPa [100 psi]) bentonite/cement grout was used. A geotextile barrier bag located on the first sleeve below ground surface was inflated with the same grout to provide additional confinement (see Figure 7a). After the casing grout set for 24 hours, each port was fractured open with water to establish access to the formation.

Grout mixes with varying viscosity and set times were formulated to both stop water seepage and fill fine apertures. Two base cement suspension grout mixes were established using Type 3 cement. The first mix was a 1.5:1 water-cement ratio by weight. Other constituents included prehydrated bentonite (3.75% by weight of cement) and superplasticizer (1.25% by weight of cement). This base mix had a

Figure 7. (a) View of grout bag after deployment test; (b) MPSP grouting throughout the embankment after completion of primary LMG grouting (note flowmeters in the background).

viscosity, as measured using the Marsh Funnel, of 34 seconds (28 seconds for water), a specific gravity of 1.31, and less than 3% bleed. The Marsh Funnel time could be increased to 45, 55 and 70 seconds by adding 1.25%, 2.5% and 3.75% by weight of cement, respectively, of anti-washout additive.

The second base mix had a water-cement ratio of 1:1. Doses of bentonite and superplasticizer were similar to those in the first base mix. Again, the viscosity was increased using higher proportions of anti-washout additive, and it was possible to achieve a mix with very large Marsh time. Also, for cases of extreme washout, an accelerated mix was tested with small doses of diluted sodium silicate. In the field, the accelerator would be injected inline at the header using a small dosage pump equipped with a check valve. All the grout mix formulations were injectable through the small ports in the MPSP.

Field Adjustments to the Grouting Plan. MPSP grouting started in Grout Curtains 1 and 3 in the area of the embankment, after completion of LMG grouting (see Figure 7b). Each pipe was grouted from the bottom sleeve working upwards. All sleeves were fractured open with water and injected at least twice to refusal.

During injection of many of the MPSP holes of Grout Curtain 3, grout connection to the inside of the building wall was observed in previous seepage areas and also through dry, open joints. In these instances, hydraulic cement and oakum were used to point the masonry from the inside as grouting continued.

Seepage through the building wall, north of the intake structure, decreased significantly as grouting progressed towards the wing wall. However, seepage through the wing wall, south of the intake, appeared to be increasing. This was likely caused by a rise of the groundwater "pool" behind the building wall, as grouting sealed the existing flow paths through the embankment and wall.

During MPSP grouting of the wing wall area, south of the intake, severe grout washout was observed through the wing wall and abutment. Interestingly, grout washout was also observed at one location on the downstream face of the spillway, which was a clear indication that voids within some areas of the masonry were large and interconnected. Unsuccessful attempts were made to prevent grout washout by accelerating the "thickest" mix using sodium silicate.

Because there was concern about the rising water table within the wingwall, a hydrophobic, single-component polyurethane resin was injected through the MPSP holes in this area. The resin was injected through sleeves that were known connections to seepage paths. However, washout of the urethane was observed downstream, even after the amount of urethane accelerator was increased. Eventually, after 0.095 m³ (25 gallons) of urethane were consumed with limited results, this operation was abandoned.

It was decided that a grout containing solids, such as sand, and injected at high flow with minimal pressure was required to grout the wing wall. Control of grouting pressure was important to reduce the potential for damage to the structure. In addition, low pressures were needed to keep grout takes under control given the permeability of the wall.

The chosen method was to inject a high-slump, sand/cement mixture under gravity through the end of the 102 mm (4 inch) diameter casing following an upstage procedure. Grout takes were significant; however, grout washout was minimal and significant improvement was observed as grouting progressed. After 12 holes were completed, seepage through the wing wall and adjacent building wall was eliminated. Figure 8 illustrates the reduction of leaks through the basement wall during an intermediate grouting stage. It can be noted that the large seepage flows had been eliminated, and that only minor seepage through the wall remained. This minor flow was eventually eliminated as well.

Phase 2. As mentioned earlier, the abutment consisted of a masonry wall with concrete facing. The concrete facing extended from the upstream to the downstream side of the abutment. The downstream side of the facing showed spalling and evidence of long-standing seepage. It was known that large open voids existed within the abutment material, especially considering the interconnection between the wing wall and the spillway observed during grouting. However, it was expected that flow velocities were relatively slow due to the confinement provided by the concrete facing, and the effect of the grouting previously performed in the wing wall.

In total, six holes were drilled and grouted. Drilling was performed by directly spinning the casing through the masonry and flushing with water until bedrock was reached at a depth of approximately 6.7 m (22 ft) from the top of the abutment. Each

Figure 8. (a) Leaks through basement wall before grouting;
(b) Leaks significantly reduced during grouting of wing wall area.

hole was gravity-grouted through a tremie tube installed to the bottom of the borehole using a neat cement-water mix with a water-cement ratio of approximately 0.5 by weight.

The grout take per hole decreased steadily as grouting progressed. Interconnection between holes also decreased as confirmed by dye tests performed at different depths during drilling. After completion of the last hole, leaks on the downstream face of the abutment had essentially stopped, and the grouting operation was considered complete.

Conclusions

The grouting program was successful and exceeded the initial expectations. There were several reasons for this success. First, it was recognized from the start that the site presented significant challenges for exploration and grouting. The exploration plan was conceived to provide the most adequate information with as little cost as possible, given the site constraints. The geophysical exploration techniques used were an important part of the investigation, and aided in the definition or confirmation of potential leak paths. They provided the necessary resolution in difficult areas, and were complemented by direct exploration techniques.

In addition, the initial grouting plan was developed considering the variety of materials, space constraint, and unknown characteristics of the dam construction. The plan was the result of a combined effort between the contractor, the owner, and the engineer, in which both engineering and construction aspects were thoroughly considered.

Finally, the presence of the designer's representative in the field during construction allowed a rapid response to deviations from the design, which were proposed by the contractor in view of the encountered field conditions. This reduced the time required for the work, and cost for the owner.

Although it is possible that leaks will recur over time as new flow paths are created within the embankment and wing wall, the experience gained during this work will facilitate the correction of future leaks.

Acknowledgments

We would like to thank the owner of this project, IVEX Packaging Corporation, and the Ohio Department of Natural Resources for working closely with us as the remediation plan was developed and implemented successfully. Additional thanks must also go to the significant resources provided by both Structural Preservation Systems, Inc., and Schnabel Engineering, who provided the information and support for this work. In particular, we would like to thank Mr. Mark Dunscomb, P.G., of Schnabel Engineering, who was responsible for the successful application of innovative geophysical tools to this unusual challenge.

CURTAIN GROUTING FOR THE ANTAMINA DAM, PERU
Part 1 – Design and Performance

T.G. Carter[1], F. Amaya[2], M.G. Jefferies[3], and T.L. Eldridge[4]

ABSTRACT
A deep grouted cut-off was constructed as part of the 140 m high, concrete-faced rockfill dam at Antamina in the high Andes of Peru. Foundation conditions comprised upstream-dipping, contorted argillaceous rocks in the left abutment, varying to limestone with karstic solutioning in the right abutment. This potential for karst was of concern for ensuring the integrity of the dam cut-off and governed much of the rationale for defining curtain depths and for ensuring flexibility of grouting procedures. This, the first of two papers on the dam, concentrates on the design aspects of the curtain and foundation grouting philosophy. Curtain performance on impoundment to full height is then reviewed. The accompanying paper, (Ritchie et al., 2003 - this conference) describes the curtain construction.

Figure 1: Dam during first filling.
For scale note identical240 t haul trucks on access road and crest.

[1]Principal, Golder Associates Ltd., Mississauga, ON, Canada; (001-905-567-4444; tcarter@golder.com)
[2]Partner, Ingetec S.A., Bogota, Columbia; (57-1-287-7759; famaya@ingetec.com.co)
[3]Associate, Golder Associates Ltd., Vancouver, B.C., Canada; (001-604-296-4200; mjefferies@golder.com)
[4]Prinicpal, Golder Associates Ltd., Vancouver, B.C., Canada; (001-604-296-4200; teldridge@golder.com)

INTRODUCTION

The 140m high, 650m long, concrete-faced rockfill dam constructed at the Antamina copper mine in Peru, is located in a steep sided rocky valley in the high Andes at an altitude of 4000 m (Figure 1). It is located on a regional watercourse, with towns downstream. Mine development was funded by international loans, which required a *fast-track* start-up sequence that imposed two operational modes:

- *Phase 1 - Initial Operation*: For two years, the dam stores water for mill start-up, concentrator operation, and export of product in a slurry. The cut-off, designed to water supply retention requirements will allow operation through dry periods.

- *Phase 2 – Progressive Tailings Deposition and dam raising over 25 years*: Environmental requirements dictated a tight cut-off to control seepage release.

A flexible design process was required to suit the project schedule which had impoundment commencing before dam completion with no provision for controlling the filling rate or lowering the impoundment for remedial grouting. These factors led to decisions to: (a) ensure that appropriate technical staff were present on site to make necessary design modifications; and, (b) provide significant flexibility in the contract to allow changes to grouting procedures without creating contractual complications. It further justified adopting a stringent target curtain permeability criterion of achieving less than 2 Lugeons.

SITE GEOLOGY

River erosion at the site has created an incised valley through the complex geology characterized by a series of steep, spiny ridges of competent limestone between intermediate gullies within more shaly and argillaceous beds (Figure 1).

Lithological Conditions

The rock units forming the dam foundation comprise siltstones, mudstones and minor limestones in the left abutment and limestones with minor shales in the right abutment. As indicated on Figure 2, the rock units are from two formations, from north to south and in decreasing geological age: the Jumasha and the Celendin. The Jumasha (zone 1 on Figure 2) consists almost entirely of limestone. The Celedin is divided into two segments: the Lower Celendin (zone 2 on Figure 2), where the rocks are 75% limestone and 25% siltstone and mudstone; and, the Upper Celendin with 25% limestone and 75% siltstone. The Lower Celendin is further subdivided into a lower unit predominantly composed of calcareous mudstone, a generally more fractured and disturbed middle unit, and a distinctly interbedded limestone and mudstone upper unit (zones 3, 4 and 5 respectively on Figure 2).

Structural Conditions

Major Structures

Near the upstream toe, a downstream-dipping backthrust occurs comprising two to three sub-shear bands. As shown on Figure 2, the beds are intensely folded adjacent to this backthrust and wrap around such that part of the most disrupted zone intersects the plinth. At this crossing, the rock is extensively disturbed and altered, with a 2 m wide clay-shear present. In contrast, at the top of the left abutment, the

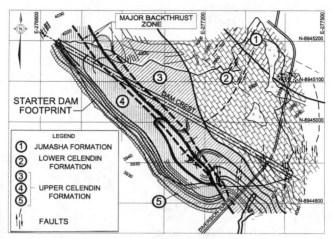

Figure 2: General as-mapped geology of the dam footprint.

backthrust, although steeper, comprises a 5 - 10 m wide deformation zone within which two or three mylonitized sections exist, but with only minor clay-shear zones.

The downstream part of the foundation is dissected by several 30° to 60° downstream-dipping backthrusts that parallel the major thrust at the upstream toe. Eight intersections with the plinth alignment were identified on the right abutment, and four on the left. Each comprised a 0.3 – 0.5m wide shear zone, with associated more fractured ground each side of the shear plane. In addition, dissecting the foundation near the downstream toe, close to the contact with the Jumasha, a NW trending sub-vertical fault was encountered .

Jointing

As a result of the complex folding and faulting, several well-developed, subordinate structural fabrics exist. In general, three joint sets exist, the most prominent of which (set J1) parallels the backthrusts, with the J2 and J3 sets aligned sub-parallel to the valley. Occasionally, sub-vertical joints of the J2 and J3 sets were open due to relaxation and localized solutioning.

HYDROGEOLOGY

Precipitation is seasonal as shown on Figure 3. In the wet period, extending

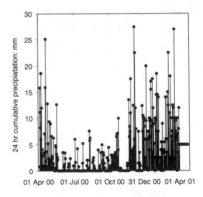

Figure 3: Rainfall - Apr.'00 – Mar.'01

from December to April, daily precipitation in excess of 25mm can occur. By contrast, dry season flows are small. Inflows to the creek are dominated by local catchments, which respond rapidly to rainfall events.

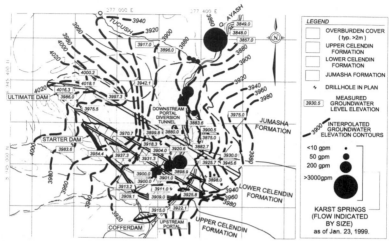

Figure 4 : Interpolated pre-construction groundwater surface contours (dry-season)

Groundwater Levels

Figure 4 shows the dry season groundwater table prior to construction. The levels in the left abutment, within the Upper Celendin, were found to be relatively high and sympathetic to topography, i.e. a normal condition. This contrasts markedly with the right abutment and the downstream part of the left abutment within the karstic Lower Celendin and Jumasha units. In these rock units, groundwater levels were found to be significantly lower (by as much as 90m relative to elevations in the Upper Celendin).

Wet season levels fluctuate. Figure 5 shows the readings in well MW99-3, located in the karstic Jumasha (Figure 6) over the same time period as the rainfall data of Figure 3. In wet periods, all the piezometers within MW99-3 showed water tables near surface. A

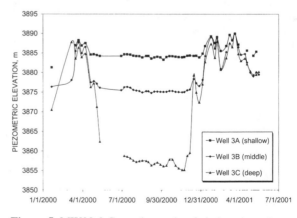

Figure 5: MW99-3 Groundwater levels in karstic rocks

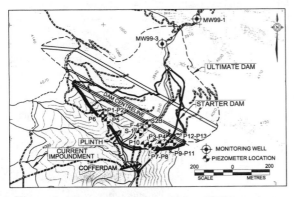

Figure 6: Piezometer and Monitoring Well Locations

strong, >0.4 downward hydraulic gradient was established within two weeks of seasonal rainfall diminishing, consistent with karstic under-drainage.

Direct observation of the depressed water table were made during the Diversion Tunnel excavation (Figure 2). Minimal groundwater inflows were seen during the dry season in the Jumasha and Lower Celendin, despite the fact that solutioned joints were present. Drillhole data showed however, that groundwater tables were compartmentalized within the Lower Celendin, as sharp changes in water level depth were measured between observation wells and artesian pressures caused by interbedding of the limestones with the less-pervious siltstones and shales were noted.

Hydraulic Conductivity

Comparative conductivity plots, such as Figure 7, based on data from some 50 investigation drillholes, indicated substantive differences by formation. The Upper Celendin was found to be markedly less pervious than either the Lower Celendin or the Jumasha. Distinct trends with depth and structural complexity are also evident from Figure 7. While the mean conductivity for all tests is in the 10^{-3} to 10^{-4} cm/sec range, some much higher permeabilities were found with two distributions evident: the very fractured zones; and, the more average rock mass. A break in slope of the upper limit line for the average rock mass suggests a difference in conductivity with depth. Near surface, the Upper Celendin rocks are of the order of 1×10^{-2} cm/s, somewhat higher than measured for either of the other

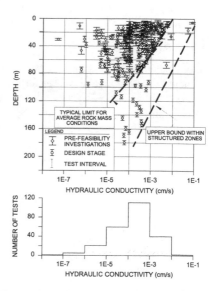

Figure 7: Hydraulic conductivity vs. depth.

formations. By contrast, at depth, the highest permeability values (about 2×10^{-3} cm/s) occur in the Lower Celendin. These values are approximately two orders higher than seen in the other two formations, which typically showed mean permeabilities of about 6×10^{-5} cm/s for the same depth range. Interestingly, the lowest conductivities were found in zones of massive to medium fractured siltstone and/or limestone, and not in the argillaceous units

The upper 50 m of the Upper Celendin rocks zone was found consistently more permeable than at depth. This contrasts with the more structured or karstic zones of the right abutment, which showed distinctly higher conductivity at depth where solutioned joints were found in the drill core. Test sections close to the backthrusts exhibited values in the order of 10^{-3} cm/s, either reflecting zones of more brittle fracturing in the margins or washout of clay infill from within the shear zones.

Karst Considerations

The depressed phreatic surface in the Lower Celendin and Jumasha rocks, and the high conductivities measured close to the depth of the inferred deep karstic groundwater table, suggested karstic flow control, but no major voids or cavities were encountered during exploration. Multi-point piezometers were installed in the right abutment to allow definition of potential flow paths based on measured vertical gradients and phreatic surface levels. Analyses of these suggested that karsticity at the base of the Upper Celendin was limited to localized solutioning. In the Lower Celendin, karsticity was found more developed as reflected by the presence of several springs at the toe of the slopes (indicating underground flows exiting from the higher mountain areas). Only within the Jumasha Formation, immediately downstream of the dam, were karstic caverns evident, some several metres across and with permanent flow. Most of the spring locations (see Figure 4) were where prominent limestone beds intersected a major discontinuity or a backthrust.

CURTAIN DESIGN

Focused grouthole layout and injection control was required to achieve results without hydro-jacking the weaker formations. Seepage analyses dictated the curtain criterion of less than 5 Lugeons, with a target of 2 Lugeons.

Key requirements were therefore:
  (i)   to cut-off steep, valley-parallel jointing noted to exhibit karstic solutioning;
  (ii) to ensure adequate intersection of the curtain with major structures;
  (iii) to balance injection methodology with rockmass conditions, recognizing lithological differences and different zones within the abutments;
  (iv) to incorporate adequate flexibility for sealing potential karst ; and
  (v)   to ensure interconnection of the plinth, rock mass and tunnel plug.

Three zones were identified as being of most concern for layout of the curtain:
  (a)   the plinth intersection with the major backthrust shear and fold zone,
  (b)   the contact between the Lower and Upper Celendin; and,
  (c)   around Elevation 3900 m in the right abutment (river level) where the potential for solutioning existed.

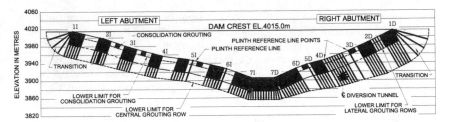

Figure 8: Design Curtain Layout (at Tender stage)

As cave development and solutioning in karstic rocks preferentially occurs close to and just above historic prevailing groundwater tables, the Ayash Valley was mapped downstream of the dam site to verify and record the elevation of all the spring and sink sources. All features were found at higher elevations than the major spring identified at the structural boundary between the Jumasha and the next lower (Ayash) Formation, some 4 km downstream from the dam site. This established the most likely hydraulic grade line of the karst system. Projecting this line inferred that the karst system in the right abutment historically would not likely have been lower than about elevation 3890 m at the plinth, reasonably defining the required depth of the curtain. The curtain was therefore configured for a minimum 20 m penetration below this elevation, based on the following design precepts:

- The curtain depth to be one-third of the maximum final head, except where groundwater tables were proven to be near the rock surface. The resulting depth thus varied between 50 m and 85 m, the greater depth corresponding to the upper right abutment where the karstic groundwater table was very deep (Figure 8).
- A multi-row curtain to provide adequate cut-off through zones with high degrees of fracturing and alteration: a deep central row and two 30 m deep lateral rows.
- Additional 10 m deep consolidation rows in areas where fracturing and alteration of the rock was especially pronounced.
- Grout rows at 2 m spacing within the plinth, (up to 10 m width at the creek bed).
- Drilling and grouting sequence from the outer rows towards the inner rows.
- All primary holes, spaced at 6 m centres, were to be grouted, with split-spaced higher order holes optional, based on previous and/or surrounding grout takes.
- Extensive water pressure testing verification to check that final overall mass permeability were always less than 5 Lugeons.

As most of the rock mass beneath the plinth appeared non-karstic, the specification used GIN principles (Lombardi and Deere, 1993) and flow controlled injection with a low viscosity grout mix matched to the median permeability rock mass. Maximum pressures, GIN envelope and injection volume limits were selected based on classic European grouting guidelines of 1 to 2 bars/metre depth and total cement volumes equivalent to $0.5 - 1.0\%$ fracture porosity of the injected rock mass (assumed applicable within a 3 m radius around each grouthole).

Because the specified grouting pressures and flowrates could need modification to avoid hydrofracture, and because the GIN control method is not appropriate in karst, the specifications were written with considerable flexibility. Provisions were built in for changing not only mixes and pressure/flow envelope control criteria, but also to change the grouting method from GIN to the Australian method of pressure control and formally sequenced mix thickening, (Houlsby, 1990). Changes from routine super-plasticized portland cement mixes to clay-cement, sand-cement, or even, if necessary, bitumen or pump concrete mixes were allowed for in the documents.

In order to optimize not only the grouting itself but also when it would be most appropriate to make changes to techniques, state-of-the-art electronic monitoring of injection behaviour was specified. This required that experienced geotechnical staff be present throughout the grouting operation to verify actual conditions relative to design assumptions, and introduce appropriate changes if and as needed. In turn, this required incorporating special contractual arrangements to avoid unreasonable situations for either owner or contractor. These requirements were met by the developed specification and contractual basis, which employed 'method-related charges' to capture costs with an incentive for efficient working while allowing the engineer to guide the works.

Curtain Configuration Modifications

Because of the valley width upstream and topographic limitations relative to the geology, it was impracticable to found the plinth totally on the Upper Celendin Formation. In the upper right abutment it had to be founded on the Lower Celendin.

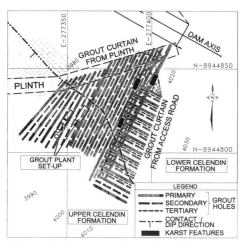

Figure 9: Grouting at Upper Celendin contact

Field mapping and plotting procedures were used for hole layout in the abutments. Holes were oriented parallel to the dam axis, typically dipping 60° and 70° from horizontal. In general, such hole layout arrangements were relatively straightforward except in the upper right abutment where grout holes were initiated in potentially karstic Lower Celendin rocks.

The geological complexity and local karst features within the Lower Celendin revealed during foundation excavation required modification of the tendered curtain arrangement (Figure 8), so as to create a fan of grout holes that crossed the karstified contact between the Upper and Lower Celendin. Field mapping defined the contact, and stereonets of the main structures resulted in the optimized layout shown on Figure 9. This layout was developed based on three-dimensional plotting of the geometry of the contact zone. Two fans of holes were laid out to intersect within the Upper Celendin upstream (south-west) of the

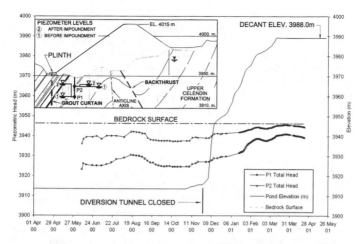

Figure 10: Left Abutment Piezometric Response (P1 and P2)

formation contact boundary. In this way, a positive stratigraphic cut-off could be created, isolating the karst-prone zones.

GROUTING EFFECTIVENESS

The reservoir was impounded between November 2000 and April 2001, immediately following substantial completion of the curtain. [Note – procedures for curtain installation are described in a companion paper at this conference (Ritchie et al., 2003)]. In parallel with the curtain construction, as part of the dam monitoring instrumentation, piezometers were installed into the dam foundation and rock mass downstream of the site. Based on likely geological controls, three nests were installed at: the mid left abutment; the lower right abutment; and, high right abutment. For brevity, only the Left Abutment response at piezometers P1 and P2 will be discussed in any detail (ref. inset diagram on Figure 10, for location in section). Comparable results were found in the right abutment.

In addition, some comparison will be made between rainfall records and monitored responses in well MW99-3, located just downstream of the dam, (refer to plots in Figures 3 and 5 respectively).

Expected Piezometric Behaviour

Figures 3 and 5 suggest that both rainfall and reservoir level may exert some influence on the piezometers downstream of the grout curtain. Rainfall could influence the water table in the abutment, thereby establishing a "tailwater" elevation for the phreatic surface passing through the dam. In addition, reservoir level could also influence the piezometers, by seepage. Based on these assumptions, as a first approximation, the expected piezometric head response, h, would be:

$$h = \alpha \Delta h_r + \beta \sum \Delta q + h_0 \qquad [1]$$

where h_o would be the initial reading, Δh_r defines the change in reservoir elevation, $\Sigma \Delta q$ is the rainfall accumulated over some period prior to the head measurement, and α and β are influence factors. The influence factor α is related to grout curtain performance, while the factor β more reflects rock mass response to rainfall and includes storativity and geometric factors. As α is of particular relevance to curtain performance, this parameter was examined closely during impoundment.

For assessing the response factor α, the position of a piezometer along a streamline from the reservoir to tailwater and the head drop across the grout curtain must be defined. For a grout curtain centrally located on the seepage path, α can be estimated from the hydraulic impedances along the streamline passing through the

$$\alpha \approx \left(\frac{x \cdot I}{I_g + I} \right)$$ [2]

piezometer as:

where, I is the hydraulic impedance (=length/permeability) of the streamline outside the grout curtain, I_g is the impedance in the grout curtain, and x is the piezometer location expressed as a fraction of the distance between downstream tailwater control and the grout curtain. Strictly, [2] needs to be modified for head loss upstream of the grout curtain, but this complexity is not warranted here as there are only clusters of two piezometers in a borehole, not instrumented planes through the dam.

In the case examined here, the lower piezometer, P1, is at a depth of 23 m and some 20 m downstream of the grout curtain. For P1, the streamline over which reservoir head drops is at least 75 m (measured from the upstream face of the grout curtain) with 10 m in the grout curtain. Thus $x=1-[20/(75-10)]=0.7$. For the ungrouted average hydraulic conductivities of about 35 Lugeons (because the Left Abutment was in better ground and hence to the lower end of the pre-grouting range from the site investigation), the impedance is $I = (75-10)/35 = 1.3$ m/Lugeon . Regarding the grout curtain, water pressure tests in the final stage holes indicated that an average permeability of less than 1.6 Lugeons had been achieved across at least a 10 m width. (Ritchie et al, 2003, this conference), giving $I_g=10/1.6=6.3$ m/Lugeon. Substituting these values into [2] gives the expectation $\alpha \approx 0.16$ for P1. Similar calculations could be done for P2, however P1 and P2 responses were expected to be similar and not be further distinguishable in the absence of 3-D seepage modelling.

Measured Left Abutment Response

Actual responses measured in piezometers P1 and P2, plotted as a function of reservoir elevation are shown on Figure 11, based on the raw response data and reservoir elevation curve plotted on Figure 10. As reservoir filling only started in late November, 2000, the piezometric fluctuations shown in the early part of the graphs, can rightly only be attributed to rainfall or to changes in the groundwater regime caused by construction (e.g. diversion of streams into lined ditches). The expected behaviour of increased rainfall leading to increased piezometric elevation is not seen, and this is thought caused by grading and redirection of streams during the final steps of dam construction influencing the flow regime more than the rainfall. In addition, the initial rise in piezometric elevation shown on Figure 10 for the first few readings after the start of impoundment is considered to be due to initial saturation of the piezometer tips and the surrounding sandpacks.

For both P1 (the lower piezometer) and for P2, it is clear from inspection of the early data on Figure 10 that the reservoir water level had no significant effect on piezometric elevations until reservoir levels reached elevation 3935 m (approx.). As is evident from the inset cross-section on Figure 10, this elevation is slightly below ground surface, but above the elevation of the groundwater table in P1 prior to impoundment. The best-fit linear response to these data plots, which are shown as dotted lines on Figure 11, correspond to $\alpha_1 = 0.25$ for P1 and $\alpha_2 = 0.18$ for P2. Further examination of the figure highlights a significant change in piezometric response is observable at an approximate reservoir elevation of 3980m. This change corresponds to the period 14 – 24 February, 2001, when the intensity of rainfall markedly diminished (see Figure 3). This change in piezometric response is considered to be a localized effect on the water table within the abutment.

The measured response coefficients, α_1 and α_2 for P1 and P2 respectively are comparable to those expected for the estimated rock mass and curtain permeabilities. However, these calculations are only approximate and assume a specific tailwater control. More important is the fact that even on reservoir impoundment to 3990m (some 75m of retained water) there is a very clear trend across the site that a strong downward hydraulic gradient still exists in the karstic Lower Celendin and Jumasha rock units downstream of the dam face and grout curtain. This downward gradient has not been affected by construction of the dam, suggesting that leakage is well controlled. The fact that the observation wells downstream of the dam show that this strong downward gradient still develops in the foundation when there is only limited infiltration (less than about 2mm/day equivalent precipitation) is consistent with earlier site investigation data and reflective of predominating karstic groundwater

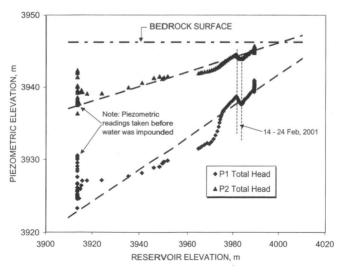

Figure 11: Response of Left Abutment Piezometers to Reservoir Filling

Substantial completion of the work took place between March and December, 2000. The total hole length drilled and grouted was 38,400 m, and 2,867 m^3 of grout was injected. Grouting was performed from a reinforced concrete plinth at the toe of the dam. The plinth is anchored to the bedrock by grouted dowels providing the ability to resist 2.5 MPa of grout pressure at the bedrock contact. For drill traction, timber grids were bolted to the plinth. On the steeper slopes, anchored steel ropes and "come-alongs" were used to support the drills (Figure 1).

The curtain comprises 3 to 5 rows of holes, based on the retained head and rock conditions: two 10 m deep consolidation rows; two 30 m deep lateral rows; and a central row as deep as 95 m. Primary holes are spaced at 6 m in each row, with split-spacing to 3 m for secondary holes and so forth. The rows are spaced at 2 m with the outer rows dropping out as the plinth narrows towards the dam crest. Holes were collared in the plinth, except for a transverse fan required in the karst zone.

The outer consolidation rows were grouted first, followed by the lateral curtain rows, split-spaced as necessary to achieve the target 2 Lugeon conductivity

Figure 1: *Steel cable drill to anchoring*

in water pressure testing. Only then was the deep central row grouted. The drills were worked in panels, choreographed so that no drilling took place within 12 m of a hole injected in the previous 12 hours to avoid disturbing the setting grout.

Grouting started in the valley section as the flat plinth was convenient for sorting out procedures. Work then moved to the lower abutments and progressed thereafter on two fronts with between two and four drilling and grouting plants in operation. Mixing and injection dictated progress, with few occasions of idle grout plants due to a lack of holes. Worked was performed around the clock 7 days per week.

Grouting was primarily by the upstage method. Exceptions to this were areas of the lower right abutment and upper left abutment where rock conditions dictated downstage working for hole stability, and in the karst zone where drill fluid loss was so great that holes could only be completed downstage.

Drilling was by three top-mounted, 2.75" (70 mm nominal) diameter hammer rotary percussive rigs and one downhole, 3" (76 mm nominal) diameter hammer rotary percussive rig. All were track-mounted and used water flush, supplemented with compressed air. The curtain holes were drilled vertically in the valley and inclined at 60° and 70° from horizontal in the left and right abutments, respectively. Drill hole deviation was measured by Sperry-Sun survey. Replacement of several holes was required because of potential gaps in the curtain at depth due to hole drift.

Progressing cavity pumps (i.e. Moyno) were specified, however, the contractors were committed to piston pumps. Some low displacement piston pumps were operated at constant speed on a single injection line under computer control for automatic GIN termination. Higher displacement piston pumps were operated on a re-circulation line system with injection controlled by a diversion valve. Pressure fluctuations were controlled by the use of in-line dampers downstream of the pumps, re-circulation lines, and diversion valves adjacent to pressure transducers under the supervision of a grouting technician (who was also responsible for the data acquisition).

Each water test or grout stage was monitored for flow and pressure at the hole collar, largely following Jefferies et al. (1982). Laptop computers were used for display and archiving of data.

Once the deficiencies of the GIN method became clear (as discussed below), grout injection reverted from the flow-control protocol of the GIN method to the pressure-control protocol of the Australian Method (Houlsby, 1990). Injection criteria were changed and/or the grout viscosified to suit the observed ground response.

INITIAL GROUTING METHOD
Initially one grout mix was used for all stages, in accordance with the GIN method. The mix had a water cement ratio of 0.7:1 by weight with the addition of 0.9% of superplasticizer (Euco 357). This mix had an average Marsh Cone time of 34 seconds (for reference water is 32 seconds), with about 5% bleed in 3 hours.

The initially specified GIN parameters varied by section of the dam as shown in Table 1. The specified injection rate was 10 L/min for a 5 m stage length.

Table 1: *Specified GIN parameters*

Zone	Limiting Values		
	Pmax (bars)	V max (L/m)	GIN (bar L/m)
Abutments: 0 - 15 m depth	15.0	150	500
Abutments: depths >15 m	22.5	200	1,000
Valley Floor: 0 - 15 m depth	30.0	250	1,500
Valley Floor: depths > 15 m	40.0	350	2,000

In view of the complex geologic conditions and uncertainties regarding karst the specifications provided flexibility for adjusting the grouting procedures.

GROUND RESPONSE TO SPECIFIED GIN GROUTING

Grout Acceptance Measure
The extent to which stable grout fills apertures in the rock does not depend on the water:cement ratio directly; each litre of grout injected sets to fill a 1 litre volume of rock aperture. The water:cement ratio only determines the strength of the grout, with grout viscosity being controlled by admixtures. Correspondingly, the most relevant measure of grout acceptance is quoted in terms of litres per metre of grout hole.

Hydrojacking

The maximum injection pressures of Table 1 are far higher than would be used for the rocks being grouted using the *Australian Method* (Houlsby, 1990). This is a known feature of the GIN method, however, it does not prevent formation damage. This was evident in early grouting of an area of sound rock. The pressure versus injected volume relationships for four stages on hole 437BP are shown on Figure 2.

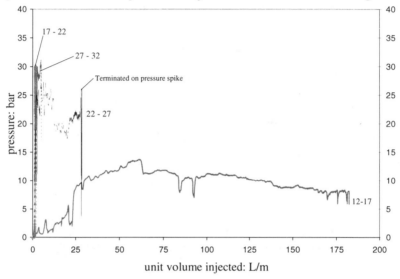

Figure 2: Pressure versus Injected Volume for grouthole 437BP.

Two of these stages, 17–22 m and 27–32 m, showed tight ground conditions. The specified refusal pressure of 30 bar was obtained and held for ten minutes with grout acceptance declining rapidly as illustrated on the penetrability plot, Figure 3.

More interesting behaviour was seen in the two higher take stages. The 12–17 m stage had the greatest take (183 L/m). Progressively declining penetrability was seen until about 140 L/m when the onset of refusal becomes apparent; grouting was terminated on the GIN criterion (Figure 3). This is consistent with the behaviour of a simple fracture grouted with a Bingham fluid (a conventional idealization of cement grout).

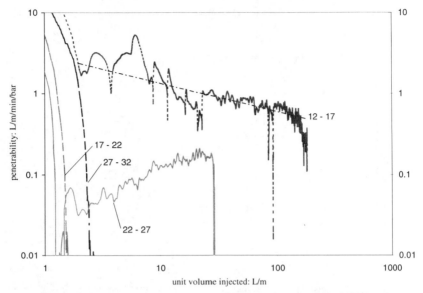

Figure 3: *Penetrability versus Injected Volume for grouthole 437BP*

The other higher-take stage, 22–27 m, initially showed the same response as the low-take stages: a rapid rise to the desired target pressure with only 2 L/m injected. However, during the specified 10 minute pressure hold, the flow began to increase and the stage was eventually terminated on the GIN criterion. The penetrability plot (Figure 3) shows the rapid decrease to a low penetrability followed by a subsequent increase. This indicates hydrojacking, with continuing formation damage evidenced by the increasing grout flow rate.

Injection Rate Effects

The GIN system does not consider the effects of flow rate, merely specifying a "slow" rate. This allows the operator to arbitrarily determine how the GIN curve is reached – the flow rate directly changes the pressure (i.e. doubling the flow rate doubles the injection pressure) so that the injected grout volume at the specified GIN essentially depends on the operator or engineer. What has the appearance of a precise specification is in fact no such thing without a prescription of how the flow rate is to be set (and varied) during the injection. To illustrate this point, Figure 4 shows the pressure against injected volume for a specific hole [234bp, 27.6-32.6 m] where it the GIN curve was met several times prior to terminating injection, points A, B, C and D. Similar responses were obtained on many occasions.

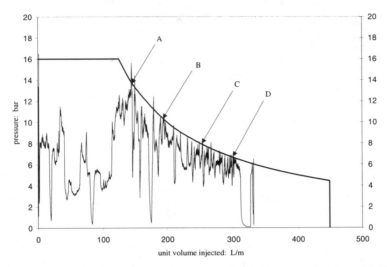

Figure 4: *Example P vs. V curve showing several GIN closures [234bp 27.6-32.6m]*

The flow characteristics were investigated for each point as the GIN curve was approached. Grout acceptance and injection rate are essentially inversely related, consistent with the grout behaving as a viscous fluid.

REVISIONS TO INJECTON PARAMETERS

Limitations of the GIN Method

In addition to the pressure-related concerns discussed above, a deficiency of the GIN method is its reliance on the apparent cohesive strength of the grout. GIN closure theory is only technically correct if the grouting is terminated on the GIN curve simultaneously with ground refusal to grout injection (zero flow rate), and only then for grout behaving as a Bingham fluid. However, the same radius of grout penetration can arise with a finite grout flow with greater injection pressure. The situation is not unique as implied by GIN proponents. A proper understanding of ground response to grouting requires tracking (i.e. integrating) of the entire grout penetrability history, not reliance on a GIN.

With time and experience on the project it became clear that GIN procedures were not achieving the desired control of the grouting in the sedimentary rock at Antamina. The detailed electronic monitoring identified four detrimental factors:

- Hydrojacking of the formation due to excessively high injection pressures;
- Dilation following hydrojacking, shown by non-reducing curtain permeability and the lack of curtain closure;
- High take grout stages terminated at the specified volume limit while still accepting grout at high flow rates, raising concerns that important features were not being treated; and

- No relationship between the GIN criterion and the total volume injected, with GIN providing no control for adequate treatment of the fracture system.

These deficiencies were identified in the relatively good rock of the valley section. In the more fractured rock and faulting in the right abutment, these deficiencies magnified. Ongoing analysis of the data and subsequent changes to the injection criteria was essential to complete the works to the specified standard. Each section of the site was treated in light of its own response to grout

Basis of Changed Grouting Criteria

A useful contribution of the GIN method is that it provides a basis for high injection pressures. If properly controlled, higher pressures lead to more thorough treatment and less cost. However, the essential step is to use electronic systems and analysis of injection transients to determine just how much pressure the ground can take. Jefferies et al. (1982) provide a methodology to do this, and it amounts to no more than treating each and every grout stage as a pressure test.

The GIN doctrine also introduced stable, strong grouts using admixtures to control viscosity. The advantages include simplicity in batching operations and rational understanding of ground response, as injection involves a fluid rather than 'washed in' cement particles.

Following on from these views, the revised grouting procedures were based on:

- Retaining the GIN concept of allowable injection pressures decreasing with injected volume;
- Field defining the maximum injection pressures to measured hydrojacking values;
- Estimating volume limits from inferred fracture aperture and target hole spacing;
- Calculating the GIN to give traditional *Australian Method* injection pressures at the target injection volume limit; and
- Viscosifying grout if penetrability was greater than 0.2 litres/m/min/bar on reaching the GIN envelope.

Required Grout Volumes

Understanding required injection volumes depends on the attributes of the discrete fracture network to be treated. In particular it is necessary to evaluate fracture apertures. Ideally, fracture apertures should be directly measured using calibrated borehole camera methods. These techniques are commonly used for contaminant transport studies involving nuclear waste but are uncommon for most grouting works. Antamina was no exception in this regard. Fracture apertures were therefore estimated from zone transmissivities. A common approach is the 'cubic law', which is based on the analogy of flow in a fracture being equivalent to viscous fluid flow between smooth parallel plates. However, this often turns out to be a poor idealization for real fractures. Detailed modelling in discrete fracture network studies, including the correlation of fracture aperture measurement with borehole cameras to distributions of hydraulic transmissivity, (Uchida et al., 1996) has shown that a relationship holds of the form:

$$h = \varepsilon \, T^n \qquad (1)$$

where: h is fracture aperture (mm); ε is a coefficient; T the hydraulic transmissivity (m^2/s) of the grout stage; and, n a dimensionless exponent in the range 0.5 to 0.75. Based on prior experience a cautious estimate of $\varepsilon=5000$ and $n=0.6$ was used so as to not over-estimate grout penetration. For the average regional pre-grouting conductivity of about 50 Lugeons this gives an upper size for the expected apertures of about 1 mm/m of hole.

The required injection volumes were then calculated assuming a uniform radial flow along fractures of constant aperture. This introduces a further complication in that fractures actually exist in inter-connected networks with fractal-like characteristics. A simplification is the generalized radial flow model (Barker, 1988) which envisages the area contacted by the flowing grout A as progressively increasing with distance r of the grout front away from the borehole (of radius r_0):

$$A = h\ 2\pi r_0\ r^D \qquad (2)$$

The exponent D is sometimes referred to as the fractal dimension. Many aspects of fracturing within rock massed appear to have a fractal dimension of $D \approx 1.7$ (e.g. Turcotte, 1989). A more conservative value of $D = 2$ was adopted for control of the Antamina grouting, again to avoid overestimating grout penetration.

For the specified primary spacing of 6 m, a grout penetration radius of 5 m provides reasonable overlap between holes, giving closure for uniform grout travel. Figure 5 shows the calculated radius of grout travel as a function of injected volume, with the results for four Lugeon values being shown.

Criteria for Higher-Order Holes

Grout does not travel uniformly and therefore using the volumes in Figure 5 does not ensure closure. Secondary holes were added if the pressure on completion of a stage was less than 60% of the required target pressure. Holes were put down either side and taken one stage below the stage in question. Grouting proceeded upstage provided that the achieved injection pressure was at least 60% of target; otherwise, grouting was stopped for 12 hours to allow the grout in the stage to set.

Comparing actual with expected takes in the initial secondary grouting showed that the 60% of target pressure criteria produced many secondary holes with little or no take. But, most importantly, the criteria failed to indicate secondary holes in several stages with much greater than 2 Lugeon conductivity. The completion penetrability in primary holes was therefore used to decide on secondary holes. Detailed analysis of the data showed that primary grouting terminating at less than 0.1 L/m/min/bar

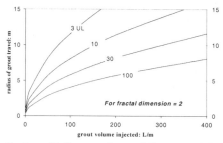

Figure 5: *Radius of influence of grouthole versus injected volume*

penetrability for standard grout produced less than 2 Lugeons in secondary hole water tests, regardless of the primary hole take. The same approach was therefore followed for determining tertiary holes from secondary injection data.

Viscosifying Sequence

To expedite grouting and seal high take areas, a sequence of viscosifying was used to achieve pressure and therefore allow upstage grouting to continue. Table 2 provides a mix summary. A different sequence was used in the karst zone. In the other areas, viscosifying was implemented based on stage penetrability. Two decision points were used: 200 L/m; and 400 L/m. If the measured pressure-volume trend indicated that the GIN envelope would be intersected with greater than the target penetrability then the grout was thickened.

Table 2: Grout Mix Viscosifying Sequence

Mix Identification	Water Cement Ratio (by mass)	Superplasticiser (%)	Bentonite Mass (%)	Marsh Cone Time (s)
A	0.7 : 1	0.9 %	0 %	30 – 35 secs
B	0.7 : 1	0 %	0 %	35 – 40 secs
C	0.7 : 1	0 %	0.6 %	40 – 50 secs

It was found that grout takes varied with local geology, except within the karst region where primary takes were nearly double other areas. The pattern encountered in the "normal" areas is illustrated on Figure 6 using data from the left abutment; the consolidation holes are neglected in this discussion.

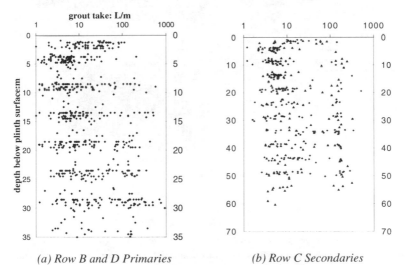

(a) Row B and D Primaries   (b) Row C Secondaries

Figure 6: *Grout takes in Left Abutment*

The primary grouting of the outer rows showed almost a bi-modal distribution with no trend by depth or position within the abutment. Most stages took less than 20 L/m but the remaining 20% or so took much more, typically 200–300 L/m. This is consistent with the distribution of fracture apertures inferred from the distribution of hydraulic conductivity, and was encouragingly similar to the expectations shown on Figure 5. As noted earlier, the C row was grouted only after all grouting had been completed in the B and D rows, so that the secondary grouting in the central row is the last stage of work for the curtain over much of its length. The takes in the secondary C holes are also shown on Figure 6. The grout takes continue with the bi-modal pattern, but now shifted to lower values with the majority of stages having less than   10   L/m   with   about   the   remaining   10%   averaging   120   L/m.

Figure 7 summarizes the grout takes for the deep central curtain row with an inset showing an enlargement of the take records within the lower right abutment zone.

Zones of high relative grout take can be identified by the increased hole density and the heavier relative shading (reflecting the higher injection takes). All of these zones correspond with areas of known complex geology, where fracturing or karsticity is more pronounced, or zones of shearing and folding cross the curtain alignment. Figure 7 shows only the deep central row (Row C) which was grouted after completion of the upstream and downstream curtain rows (Rows B and D). Therefore, the plot reflects "primary acceptance" below the lateral row depth (30 m).

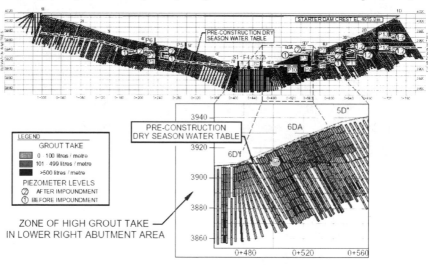

Figure 7: *Grout Acceptance, Central Row C*

Table 3 summarizes the grout takes in the upper 30 m of the site where three rows exist. The values and primary/secondary ratios are different below 30 m depth where the curtain is a single row and where grout hole deviation was also an issue (some secondary holes were in effect primary once their deviation was accounted for). As can be seen from Table 3 the average grout takes are very comparable to the expectations from Figure 5 for the chosen 6 m primary hole configuration. Overall, the grout take for the project was 75 L/m. The project had been bid based on the design estimate of 80 L/m.

Table 3: *Summary of Grout Acceptance, Triple Row Zone 0 – 30m depth*

Area	Grout Take	
	Primary Holes (L/m)	Secondary Holes (L/m)
Left Abutment	63	35
Valley	60	7
Lower Right (Fracture Zone)	140	27
Middle Right	107	37
Upper Right (Karst)	175	27

POST-GROUTING WATER PRESSURE TEST DATA

Adequacy of the grout curtain was established by water pressure testing. This avoided subjective judgments at the early stage of specific site experience and checked that the grout travel and curtain closure estimates were reasonable.

Water pressure testing by dedicated crews independent of the grout injection used similar electronic data acquisition equipment so that tests could be analysed in detail. Procedures for this approach are also given in Jefferies et al. (1982).

Figure 8 shows the water pressure test data in a 90 m long part of the curtain in the Left Abutment (the whole curtain was tested, with each geologic zone individually assessed). Primary grouting reduced the permeability of most zones to less than 5 Lugeons, but about 10% of the rock mass was more pervious. The identified secondary grouting further reduced the hydraulic conductivities of these more pervious zones. Treating each test as independent, the effective conductivity of this 90 m panel of the curtain was 0.9 Lugeons after limited secondary grouting, comfortably below the 2 Lugeon target

Water testing is readily carried out at small incremental cost in the early stages of the work when secondary holes are put down anyway to assess how closure is developing. However, in the later stages of the project when grouting procedures have become sufficiently reliable to predict closure from the ground response to grout it is desirable to limit holes drilled just for water pressure testing. The methodology adopted for the later stages of the works was to water pressure test every other secondary hole in the central C row as a statistical sampling.

At completion, the overall 780 m long curtain has average central-row closure conductivities of less than 1.5 Lugeons (including the karst region). Many zones indicating around 0.1 Lugeons. The excellent results are attributed to the high injection pressures relative to conventional grouting, the use of stable mixes, and detailed electronic monitoring.

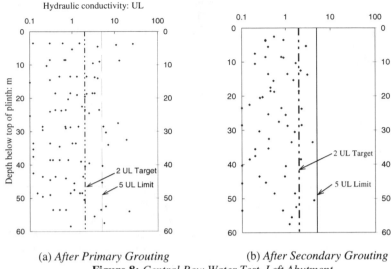

(a) *After Primary Grouting*   (b) *After Secondary Grouting*
Figure 8: *Central Row Water Test, Left Abutment*

ACKNOWLEDGEMENTS

Many individuals and companies contributed. In particular, colleagues within Golder Associates and Ingetec. Also individuals from Bechtel, Parina-Andina, Geotechnica and Soletanche Bachy are noted. Special thanks to Compania Mineria Antamina for permission to publish this account of the works.

REFERENCES

Barker (1988)."Generalized Radial Flow". *Journal of Water Resources And Hydrology*.ASCE.

Carter, T.G., Amaya F., Jefferies, M.G. and T.L. Eldridge. (2003). "Curtain Grouting For The Antamina Dam, Peru - Part 1 – Design And Performance. 12 pp. (Proceedings, this conference).

Houlsby, A.C. (1990). *Construction and Design of Cement Grouting*. Wiley, ISBN 0-471-51629-5.

Jefferies, M.G., Rogers, B.T. and D.W. Reades. (1982). "Electronic Monitoring of Grouting". *Proc ASCE Specialty Conference On Grouting In Geotechnical Engineering*, New Orleans.

Lombardi, G. and D. Deere. (1993). "Grouting and Control Using the GIN Principle". *Water Power and Dam Construction*. June 1993. pp.15-22.

Turcotte, Donald L (1997). *Fractals and Chaos in Geology and Geophysics*. Cambridge UP. ISBN: 0521567335.

Uchida, M., Doe, T., Dershowitz, W., Thomas, A., Wallman, P., and A. Sawada (1994). "Discrete-fracture modelling of the Aspo LPT-2, Large Scale Pumping and Tracer Test". *Report ICR 94-09*, published by SKB (Svensk Karnbranslehantering AB, the Swedish Nuclear Fuel & Waste Management Company).

Shallow Foundations in Karst: Limited Mobility Grout or Not Limited Mobility Grout

Jesús E. Gómez, Ph.D., P.E.,[1] and Allen W. Cadden, P.E.[2]

Abstract

Construction of shallow foundations in karstic geology always carries risks. In many cases, the perceived risks are high enough for the engineer to recommend alternative deep foundations or ground modification. The difficulties associated with construction of deep foundations in karst have led to the development of Limited Mobility Grouting (LMG) methods to improve the subsurface conditions and, in many cases, to permit the use of shallow foundations. This paper reviews the aspects to be considered for selection of foundation alternatives in karst, and presents three case histories where LMG has allowed the use of shallow foundations for major structures.

Introduction

Karstic geology is the term used to describe the remnant subsurface formations left behind by weathering of carbonate rocks. These include an irregular rock surface marked with pinnacles, troughs, cutters, and caverns (Figure 1). The overlying soils can be almost as irregular. Weathering of carbonate rocks is dissimilar to most rock formations. Often, the overburden materials closest to the rock surface are the softest, and consist of voids and remnants of the solutioned rock, which are softened by water flowing along the rock surface. These remaining soils (epikarst) typically have very low shear strength.

The irregularities of the rock surface and heterogeneity of the overlying soils create significant difficulties for the foundation engineer. Where very soft soils or irregular depths to rock occur near foundation grades, the design of spread footings is problematic. Bearing capacity, total settlement, and differential settlement may become significant issues in this environment.

The use of deep foundations comes with its own difficulties in karstic areas. A driven pile system often encounters irregular depths to rock, sloping rock surface, and "floating boulders;" all creating difficulties and unreliability. The authors have utilized driven steel H-piles in karstic geology, and experienced highly irregular driving depths varying from 15 to 30 m (50 to 100 ft) in a single pile cap. Where the piles encounter such crevices in the rock or the sloping rock in general, piles often kick out of plumb or twist before reaching driving refusal. The irregularities in length

---

[1] Member, ASCE, Associate, Schnabel Engineering Associates, Inc., 510 East Gay Street, West Chester, PA 19380; 610-696-6066; jgomez@schnabel-eng.com

[2] Member ASCE, Principal, Schnabel Engineering Associates, Inc., 510 East Gay Street, West Chester, PA 19380; 610-696-6066; acadden@schnabel-eng.com

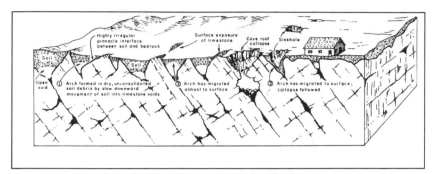

Figure 1. Karstic conditions (Geyer and Wilshusen, 1982).

and final configuration can leave the engineer with concerns about reliability of the piles, and will likely result in very conservative capacity estimates and additional redundancies in the foundation system.

Drilled shafts can overcome many of these issues. The size of the excavation can allow inspection of the bearing rock or, where highly irregular rock is present, the shaft can be designed to carry the load in side friction through relatively sound rock. However, the cost of penetrating irregular, sloping, cavernous, and inconsistent rock is high. The presence of voids or loose soil pockets above the bearing grade of the shaft can also be highly problematic since casing will become necessary for drilling and concreting. Where voids are present below the groundwater level, permanent casing may be necessary to ensure continuity of the shafts. Each of these factors adds to cost and very often construction delays.

The LMG technique, often referred to as compaction grout, can help meet the needs of the project and the engineer in a cost-effective manner. The key factors for selection of LMG are discussed in this paper. In addition, case histories where LMG was successfully used are presented.

The Decision Process

When are the conditions associated with karst too severe to utilize spread footings? This is not an easy question to answer. Every engineer has preferences and "gut feelings." The decision will involve local knowledge, and a thorough review of the subsurface data. Often the field data includes test borings, test pits, air percussion drill holes, geophysical surveys and occasionally, in situ test data. The analysis of this information will consider overall variability of the karstic features, percentage of exploration points with problems, type of structure, foundation loads, depths of cuts and fills, and groundwater levels. The alternative of finding a safer location for the structure needs to be evaluated as well (DeStephen and Wargo, 1992).

These issues are weighted differently by each engineer. Furthermore, each factor may carry different weighting on different projects. It is beyond the scope of this paper to detail a method of evaluating a site for the use of spread footings.

Conditions dictating that a site is unsuitable for spread footings include the presence of very soft soils or voids within or near the influence zone of the foundations, or drastic irregularities in the rock surface that would cause differential movements. In both cases, it is possible to improve the conditions of the site to support the foundation elements as uniformly as possible. This improvement would come in the form of densification or displacement of the very soft soils above and near the rock surface.

Often, treatment of the upper rock materials may provide additional benefits. Filling of voids and cavities improves the consistency of the rock mass and limits the available avenues for water flow and ongoing solutioning of the rock, or erosion of soils into rock cavities or joints.

LMG Background

LMG has a broader application than traditionally associated with what is referred to as "compaction grout;" thus, the more behavior-specific name. Since compaction grout is a type of LMG, densification of soils would be the first logical application of this material. Where the site conditions for new construction so dictate, ground improvement can be performed to improve soil conditions and, therefore, to allow smaller foundation elements or eliminate the need for deep foundations and structurally supported floor systems.

The broader sense of LMG would be its application to projects where the main goal is void filling and displacement of soft soils. In these cases, the resulting soil or rock properties are not necessarily the controlling factor. Where the goal is to fill voids, control of the grout material properties to limit the unrestricted flow of grout in the ground is the primary consideration.

Case Histories

Three examples are provided to give an understanding of the variety of cases where LMG has been applied successfully in karstic areas and the benefits realized. The first case is a project in Conshohocken, Pennsylvania, where LMG was used beneath the garage foundations to provide suitable conditions for spread footings supported on a combination of soil and rock.

The second case, the Bryce Jordan Center at Penn State University, State College, Pennsylvania, is a general example of the use of LMG for rock-supported spread footings. Although a little unusual, the application provided assurance that the foundation rock was consistent throughout the structure.

The third case is a more traditional example where the overburden soils were of significant depth and variable consistency. LMG was used to improve the ground conditions beneath the strip footings of a six-story assisted living care facility in Lebanon, Pennsylvania.

In each of these case histories, the basic geologic features, building foundation requirements, and special constraints are presented along with some of the lessons learned during construction. A discussion is also presented of alternative foundation systems considered for each project. It is the hope of the authors that the readers will find anecdotal information in these examples that can be useful in making future decisions on where LMG is appropriate in karstic geology.

Case 1. Mercy-Keystone Headquarters, Conshohocken, Pennsylvania. The new corporate headquarters were planned at a site along the Schuylkill River near the Fayette Street Bridge (Reith et al., 1999). This complex consisted of one five-story, and one six-story office building along the north side of the site, and a five-level parking deck along the south side (Figure 2). The first phase of the exploration was developed to provide basic information for foundation design, and to meet local building code requirements. Fifty-four traditional soil test borings with split spoon sampling and limited rock coring were completed. The results of this program revealed that the site was at the contact of the Wissahickon Schist Formation and the Conestoga Limestone Formation. A diabase dike and sills also crossed this site (Figure 3).

The northern portion of the site was underlain by firm consistency residual schist throughout the depth of the exploration. Other than a pinnacle of limestone encountered between the buildings, no rock was encountered. However, the southern side of the site was highly variable with pinnacled limestone, where rock levels varied from above the foundation bearing grade to depths of over 30 m (100 ft). The eastern central portion of the parking garage was further complicated with the intrusion of a diabase dike and thin layers 0.3 to 4.5 m (1 to 15 ft thick) of diabase sills near the surface. Beneath these sills, soft soils were present above the limestone rock. To the far east of the site, very soft residual diabase and schist soils were present to depths of over 30 m (100 ft).

Shallow spread footings were the easy choice for the office buildings where residual soils were well suited for the maximum 1,800 to 2,700 kN (400 to 600 kip) column loads. However, given the complex site conditions, further study was necessary to define the most efficient foundation system for the garage where, due to the long spans between columns, the loads approached 7,100 kN (1600 kip). Numerous alternatives were considered that included a mat foundation, various

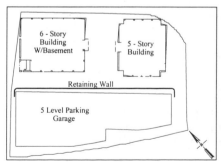

Figure 2. Mercy Keystone schematic parking garage.

Figure 3. Diabase dike exposed in site plan.

driven piles, caissons, auger cast piles, pressure-injected footings, micropiles, and various ground improvement methods. The options were narrowed to the use of LMG for ground improvement based on technical feasibility and overall costs. The construction manager also wanted to have a drilled shaft alternate bid, since it appeared to have a favorable initial cost, and was a more familiar system to the contractors in the area.

A second subsurface exploration program was therefore conducted in the parking garage area. This program generally consisted of drilling one test boring to bedrock, and 3 m (10 ft) of rock coring at every column location, and one air rotary percussion hole drilled 1.5 m (5 ft) from each column location. This program was developed to evaluate the quality of the underlying rock, to provide estimates of bearing grades at each drilled shaft location, and to evaluate the sloping rock conditions. In this way, a lump sum price bid could be obtained for the drilled shafts.

Ninety-seven test borings and 67 air rotary percussion holes were drilled for the second subsurface exploration. The air rotary percussion drilling was performed using a 101-mm (4-inch) diameter down hole hammer with a button bit in an uncased borehole. The drilling encountered problems with obstructions and sloping rock in almost every hole, which resulted in numerous off-sets and re-drills, and breakage or loss of drilling equipment in the hole. Running sand was also encountered in several instances. During drilling of the air rotary holes, air/mud/water was often expelled from several of the surrounding holes, revealing intricate interconnected paths within the formation. The rock surface was observed to be extremely variable, changing as much as 3 to 9 m (10 to 30 ft) between the test boring and the percussion hole.

The rock cores recovered indicated that the limestone and diabase were highly fractured across most of the garage footprint. The orientation of bedding planes and fractures in the limestone was nearly vertical. This resulted in a highly pinnacled rock surface with troughs, ledges, and several deep soil-filled crevices At least four deep zones of soft soil and highly weathered rock were identified by the test borings.

Given these highly variable site conditions, drilling and construction of a drilled shaft foundation were considered impractical. LMG was thus selected to provide ground improvement and allow construction of shallow spread footings for the garage. This method of ground improvement was also considered the least susceptible to cost overruns and time delays for several reasons. The grout probe holes are made using rotary percussion methods that can easily penetrate overburden, man-made obstructions, and variable depths of weathered or fresh rock. In addition, the drilling costs per unit length do not vary with the material encountered as is typical with drilled shaft construction. Since the system relies on composite ground improvement rather than discrete foundation elements, the locations are more flexible should difficulties be encountered during construction. If the subsurface conditions are worse than anticipated in some areas, elements can easily be added until the required level of ground improvement is achieved. Finally, the work is generally not weather dependent.

The LMG system was designed to create a mass of soil and grout with increased stiffness and strength within the zone of stress influence of the foundations. The goal was not to create grout columns that would act as individual foundation elements, but rather to create uniformly distributed inclusions within the soil mass. In such a system, the interaction between the soil and the grout inclusions permits transmission of the foundation loads to the underlying rock.

An added benefit was lateral improvement beyond the foundations consisting of increased stiffness and reduced potential for subsidence feature development in the surrounding area. Such supplemental improvement cannot be achieved with traditional deep foundation systems.

Continuous strip foundations were supported on a crushed stone mat to provide stress dissipation and redistribution over the grout-improved soil and rock. The thickness of the mat ranged from 0.3 to 1.4 m (1.0 to 4.5 ft) depending on the width of the foundation and the depth to rock. The strip footings were designed using a maximum net design bearing pressure of 290 kPa (6,000 psf), and typically ranged in width from 1.5 to 1.8 m (5 to 6 ft). A minimum of two rows of grout elements was specified for the exterior footings, and three rows for the interior footings. The grout holes were required to extend at least 0.6 m (2 ft) into the underlying rock, but should be at least as deep as three times the width of the strip footings.

The LMG specification called for a minimum compressive strength of 6.9 MPa (1,000 psi) after 28 days. LMG was mixed on-site using two 11.5 cubic meter (15 cubic yard) capacity mobile batch plants. The contractor used a mixture of Type I Portland Cement, screened topsoil, No. 10 limestone screenings, and 6 mm (¼ inch) open graded stone to produce the grout (Figure 4). The approximate proportions were 20 percent cement to 20 percent topsoil, and 60 percent of a mixture of two parts limestone screenings to one part stone by volume. The actual mix was adjusted throughout the project as delivered materials varied slightly. Laboratory testing of the basic grout mix used (without the open graded stone) indicated that the blend of materials classified as silty sand having 27 percent by weight passing the No. 200 sieve, 100 percent passing the No. 4 sieve, and was non-plastic. Compressive strength testing of grout cylinders indicated an average 28-day compressive strength of 19 MPa (2,800 psi), which was significantly greater than the strength specified.

The grouting was performed in 300 mm (1 ft) stages using duplex piston positive displacement pumps. Measured pressures during injection ranged from essentially zero, indicating no resistance from the surrounding soils; to over 4.1 MPa

Figure 4. LMG extruding from an injection pipe.

(600 psi), which was considered refusal or maximum resistance from the surrounding soils. The criteria utilized to evaluate the injection process and determine when a grout stage was complete included: grout flow ceases at a maximum pressure of 4.1 Mpa (600 psi); ground surface movement was observed; a minimum injected volume of 0.23 m^3/m (2.5 cf/lf) at a pressure of at least 0.70 MPa (100 psi); a maximum of 0.38 m^3/m (13 ft^3) was injected in a stage. During production, the injection criteria were modified to account for several site-specific effects observed. In order to account for the high water table and frictional losses within the grout pipes, the minimum pressure to allow a cutoff of 0.07 m^3 (2.5 ft^3) for any stage was increased from 0.70 to 1.25 MPa (100 to 180 psi) for grouting performed at a depth greater than 12 m (40 ft).

A total of 1,201 m^3 (42,418 ft^3) of grout was injected over 3,956 m (12,982 ft) in the 550 original design grout locations. A total of 14.7 m^3 (591 ft^3) of grout was injected over 60.7 m (199 ft) in the nine additional grout locations. Given these data, an average grout take of 0.3 m^3/m (3.3 ft^3/lf) was achieved.

Case 2: Bryce Jordan Convocation Center, Penn State University, State College, Pennsylvania *(Major Reference Cadden & Wargo, 2003).* Penn State University required the construction of a 16,000 seat convocation center on campus. At the time of its construction, this was one of the largest on-campus facilities of its kind. Given its size, column loads ranged from about 1,100 kN (250 kips) to over 11,200 kN (2,500 kips). Based on mapping by the Pennsylvania Geological Survey, the site is located near the contact of the Stonehenge and Nittany Formations within the Ridge and Valley Physiographic Province.

Subsurface conditions were explored using electromagnetic geophysics and 78 test borings. The exploration data revealed highly variable rock conditions ranging from near-surface, massive material to highly fractured rock, and deep soil filled troughs.

One of the key features determined from this exploration was that the rock formation, although highly variable in quality, was present at or near the bowl level (arena floor grade) of the center. The main foundations of the structure were located at this level. The traditional foundation selected by state agencies in this area has been drilled shafts end bearing on relatively massive rock. This would have required coring the caissons into the variable rock, which is difficult and costly. Therefore, placing the foundations directly on rock at nominal depths seemed like a logical solution.

In order to apply a shallow foundation solution to this structure, the consistency and quality of the bearing material had to be improved. To accomplish this, two solutions were developed. First, the major foundation elements (primary oval line of foundations around the stadium) were designed as foundation ring elements; thus, making them continuous, providing some ability to span inconsistencies. Second, LMG was selected as a means of improving the consistency of the rock mass, densifying weathered rock, as well as filling some joints and voids. Following LMG improvement of the rock, a bearing pressure of 575 kPa (12 ksf) was used for the foundation elements.

Figure 5. LMG injection at Bryce Jordan Center.

A basic pattern of drilling and grouting was planned and completed beneath all new foundation elements (Figure 5). Beneath the ring foundations, primary and secondary holes were completed resulting in a 2.1 m (7 ft) center-to-center spacing. Hole layout patterns for individual foundation elements were based on the size of the element and ranged from 4 to 5 holes per footing, and a net spacing of about 1.5 m (5 ft).

The drilling and grouting of each of these locations were used as a means of further exploration. Monitoring for drilling resistance, grout pressure, and volume takes were direct indications of the consistency of the material beneath the foundation element. Based on this data, tertiary holes and subsequent split spacing were completed in many areas to develop "closure."

Rock material was not present in some areas. Where rock was not present at or near the bearing grade, a modification of the grouting intent was developed. The revised goal was to provide closely spaced holes that would become inclusions in the soils, similar to Case 1. These inclusions provided some form of densification or displacement of the soils, resulting in a composite mass that was reinforced with column-like grout elements to create a stiffer bearing mass. Thus, the load was dissipated and/or transferred to the underlying rock.

A total of 9,609 m (31,527 lf) of drilling was completed for grouting. A total of 2,288 m³ (2,992 yd³) of LMG was injected in these holes, resulting in an average grout takes of about 0.23 m³/m (2.6 ft³/ft) of injected hole. As would be expected, the actual injection ranged from "no take" to the maximum allowed in a stage. No large open voids were encountered in this program.

Case 3: Six-Story Housing Structure, Palmyra, Pennsylvania. The site for a new six-story, assisted living housing structure located in Palmyra, Pennsylvania, was underlain by dolomite and limestone of the Epler Formation within the Ridge and Valley Physiographic Province. Test borings revealed that the subsurface conditions varied from soft to stiff silt and clay, and had highly variable underlying rock profiles. Several exploration programs were performed at this site including test borings, test pits, and air track probes which were originally intended to be pilot holes for pipe pile installation. Upon drilling the pile holes and finding further evidence that voids and poor quality rock were present to depths in excess of 58 m (190 ft), a redesign of the foundation was commissioned. Additional test borings supplemented the original studies, and led to the recommendation for the use of LMG to improve the near surface soil and rock so that spread footings could be used for support of the building. Compilation of the exploration information revealed that numerous probes encountered voids or very soft zones as thick as 9.1 m (30 ft). However, given the high variability of the geology, and considering the tight spacing of the probes, these zones were believed to be vertical seams and troughs, rather than large open caverns.

The revised foundation/ground improvement scheme was selected as a means to limit the exposure to extra and unforeseen costs during construction. The potential costs associated with additional drilling and injection of grout in relatively shallow conditions were considered significantly less than those of drilling piles into the highly variable rock.

Given the existence of fine-grained soils at this site, densification was not believed to be a realistic approach to ground modification. However, the process of injecting the stiff LMG at relatively high pressures did allow voids and the very soft zones to be filled, and material displaced to create a stiffened composite mass beneath the foundations.

Approximately 200 grout injection holes were drilled, resulting in a total of about 2,377 m (7,800 lf), of which about 2,133 m (7,000 lf) were grouted. A total of 777 m^3 (27,452 ft^3) of grout was injected. Thus, the resulting grout take at this site was about 0.35 m^3/m (3.9 ft^3/ft). The load bearing wall foundations with loads of up to 175 kN/m (12 kips/ft) were supported on spread footings with a net bearing pressure of 191 kPa (4 ksf).

Summary

The selection of the proper foundation system in karstic environments is a matter of engineering principles, geologic understanding, and professional judgement. However, several overriding factors should be considered, particularly in the application of LMG.

Exploration. The consideration of the proper solution will require complete understanding of the site conditions. Many tools are available including traditional test borings and test pits. Drilling techniques such as air percussion drilling may provide cost-effective means of expanding the site understanding. It must be recognized that percussion drilling is well suited for defining the top of rock and relative changes in drilling consistency; however, no samples or accurate measurements are obtained with the percussion drilling method.

Other available tools include geophysical methods. Non-intrusive techniques, such as resistivity testing among others, have been used very successfully to assess subsurface conditions in karstic areas. Experienced operators who can apply the correct technique or combination of techniques will determine the success of a geophysical survey.

Structure. The first stage of the foundation decision process may be to address the potential impacts of variable subsurface conditions on the structural system. The use of combined or continuous foundation elements can often limit the risk or impact of possible differential settlements. As described in both Case 1 and Case 2, continuous foundation elements were used to uniformly spread the loads. In these cases, the karstic conditions were conducive to differential settlements beneath a portion of the structure. However, the loads would tend to redistribute to other portions of the foundation strip or at least limit deflections until repairs can be made.

It is up to the designer to establish the desired ability of the foundation to span subsurface problems. Often, selection of span lengths on the order of 3 to 4.5 m (10 to 15 ft) may be considered. Given the loads of the foundations in Cases 1 and 2, continuous foundations with thicknesses on the order of 1 to 2 m (3 to 6 ft), and reinforced top and bottom, were required to provide spanning capabilities. These dimensions are significantly greater than those of typical shallow foundations.

Ground Improvement. If ground improvement is deemed necessary to facilitate the use of spread footings, then the decisions will turn to the extent of the improvement. Extent involves the limits of the treatment within the plan area of the structure (i.e., whether treating the areas beneath the foundations is sufficient, or if treatment of the floors will be required), and the maximum depth to treat.

In general, goals of the karst LMG program include void filling and displacement of very soft soils. Densification of the soils is also likely to be achieved, although not to the extent possible when compaction grouting granular soils. The end result is an improved, less compressible mass consisting of soils reinforced with cementicious bulbs and column-like elements. Where grouting is completed in a relatively continuous process, each individual grout column could be regarded as vertical reinforcement, or even as a deep foundation element. However, such grout columns should not be considered as load carrying elements for design unless reinforced and constructed in such a manner to ensure that they are continuous and would meet all applicable codes.

Consideration of the extent of the grouting program must be based on the site conditions and the stress distribution beneath the foundation elements. Grouting depths should be based on the location and extent of the very soft soils and voids as a primary consideration. Although no guarantees can be made to the value of the LMG improvement, efforts should be made to address areas likely to result in excessive differential movements either from foundation loads, new fill placement weight, or potential collapse of existing conditions.

Extending the improvement to floor, pavement, or general site areas can result in significant cost implications. Loading conditions, traffic flow, site conditions, risk to safety, impact on facility use; all must weigh in the decision to treat areas beyond foundations.

One additional zone of consideration would be water storage and conveyance areas. It is a well-recognized fact that changes to groundwater and surface water flows have an impact on the risk of karst feature development (sinkhole formation). Although LMG can be used to improve subsurface conditions for foundation applications, it should not be expected to create conditions suitable for water storage.

LMG Design Considerations. Once the area and depth to be treated are established, the layout and grouting criteria must be established. Often the plans are developed to provide successive levels of improvement. This would entail working to create a perimeter or base grouting (primary injection), then filling between this initial work with secondary and subsequent holes.

Each additional hole serves two roles. Additional improvement can be achieved until the drilling and resulting grout volume and pressures meet the desired results. Closure is assessed by reaching a desired pressure and resistance to grout volume injection. Every effort should be made to avoid the acceptance criteria involving costs. The second role is to provide additional subsurface understanding which is achieved from the drilling and grout records.

The projects described above, as well as other experiences of the authors, have shown that primary grout hole spacing on the order of 3 to 4.5 m (10 to 15 ft) is generally appropriate. Split spacing with secondary and subsequent holes will reduce the final spacing to dimensions of about 1.2 to 1.5 m (4 to 5 ft) in most cases. Although geologic conditions vary considerably in karstic environments, volumes of grout injected are often in the range of 0.3 to 0.45 m^3/m (3 to 5 ft^3/ft) of injected depth. However, volumes as great as 1 to 1.3 m^3/m (10 to 15 ft^3/ft) or more have been reported. The size of the project often has a significant impact on this final injection volume. Larger projects with significant split spacing will generally fall within the range experienced by the authors. Where small projects focusing on remediating known soft zones or voids are completed, or where systematic split spacing is not completed, the volumes can be much larger. It should also be recognized that additives in the grout mix can result in significantly larger grout takes, since the mix is easier to pump, and anticipated cutoff criteria pressures may not be realized.

Quality Control. Much has been written about verification of grouting. A good resource is the ASCE Special Publication 57, "Verification of Geotechnical Grouting." A key consideration of quality control in a karstic grouting program is that the grouting is a tool to develop further understanding of the site. Establishing reasonable criteria to monitor and control the grouting program is quality control in itself. Drilling resistance, grout take, pressure, ground response, and relative improvement between primary and subsequent injections all serve as measures of the effect of the grouting.

Often, efforts to use standard exploration measures for determining the level of ground improvement are applied to karstic sites. Although this can provide some data, use of tools such as Standard Penetration Test (SPT) is not recommended as a suitable sole measure of successful ground improvement. The initial use of SPT tests at the sites described previously proved that the site was highly variable in the first place. Often, borings less than 0.9 m (3 ft) apart show significantly different blow count results. Therefore, determining a baseline measure to compare with is nearly impossible. Another difficulty with SPT measurements is that the LMG generally displaces the fine-grained soils associated with older karstic geologies. The nature of these materials, low permeability and the rate of grouting, does not allow densification that can be measured by SPT tests.

There are numerous reports of projects where the SPT measurements were the sole acceptance criterion. In one case on a project in York, Pennsylvania, the contractor completed a triangular test pattern of grout injection, with a good mix design and slow injection rate that resulted in an actual decrease in SPT values when measured between the injection points (Miluski, 2002).

Grout strength requirements are another area of concern. Unless LMG piles are being created, the required grout strengths are not as large as in other grouting applications.

The projects discussed above were completed utilizing the grouting process and experienced field inspection as the primary quality control. No distress attributable to settlement or karstic feature development has been reported in these structures to date. For structures of a critical nature, or where specific uncertainties exist, monitoring points can be established on foundation elements, and a program of long-term inspection implemented throughout construction and the life of the structure.

Acknowledgement

We would like to thank Schnabel Engineering Associates, Inc., for making their files available for review and for providing administrative resources that were essential in completing this paper.

References

Byle, M., Border, R. "Verification of Geotechnical Grouting," ASCE Geotechnical Special Publication No. 57, October 1995.

Cadden, A., Wargo, R. "Nittany Lions' New Convocation Center: Rock Solid with LMG," 3rd International Conference on Grouting and Ground Treatment, New Orleans, LA, 2003.

DeStephen, R., Wargo, R. "Foundation Design in Karst Terrain," Bulletin of the Association of Engineering Geologists, Vol. XXIX, No. 2, 1992.

Geyer, Alan R., and Wilshusen, Peter J. (1982). Engineering Characteristics of Rocks of Pennsylvania, Pennsylvania Geological Survey, Environmental Geology Report No. 1, Second Edition.

Miluski, M. Personal Communications, 1999.

Reith, C., Cadden, A., Naples, C. "Engineers Challenged by Mother Nature's Twist of Geology," Seventh Multidisciplinary Conference on Sinkholes and the Engineering Environmental Impacts of Karst, Hershey, Pennsylvania, April 1999.

GROUTING IN KARST TERRANE – CONCEPTS AND CASE HISTORIES

Joseph A. Fischer[1], P.E., Member ASCE, Joseph J. Fischer[2] & Richard S. Ottoson[3]

ABSTRACT

Throughout the United States and especially on the east coast, development pressures have increased construction atop carbonate rocks. A myriad of concepts and techniques for identifying and remediating karst hazards has been proffered, many reasonable, some not so. Only remediation by grouting is discussed herein.

The difference between effective and ineffective techniques for both the identification and remediation of karst are related to the nature of the subsurface, the likely failure mechanism, and the type of construction. Bedrock character, bedding orientation, tectonic alteration, glacial activity, nature of the overburden, as well as the location and size of soil or rock cavities, will influence the failure mechanism in some way. In addition, the nature of any planned or existing construction can modify the manner, extent and procedures of the grouting operations.

Any cost-effective grouting program in solutioned carbonates must consider all of the above concerns. Grouting can be as simplistic as filling the "throats" of isolated sinkholes with an *appropriate* "flowable fill" to a full-fledged exploratory/grouting program using compaction and/or slurry grouting techniques below a major structure. Developing an *accurate* cost estimate varies from quite difficult to impossible.

Case histories of successful remediation operations that consider these concepts are presented herein.

INTRODUCTION

Karst is an international problem, although only one location truly typifies it - the Karst province on the Adriatic coast of what was formerly Yugoslavia. Areas of the United States and China, for example, certainly have similar conditions; a pinnacled rock surface that often protrudes above the general ground surface. However, in the United States, the term "karst" has been used for almost any area underlain by solutioned carbonate rocks whether it is a "sinkhole plain" or pinnacled rock protruding from the surface. Unfortunately, this use of the term masks the fact that differences in character and failure mechanisms will pose different investigatory and remediation concerns for areas underlain by:

---

[1] President, Geoscience Services, 3 Morristown Rd., Bernardsville, NJ 07924, geoserv@hotmail.com.
[2] Vice-President, Geoscience Services.
[3] Associate, Geoscience Services.

- The geologically recent (Cenozoic) sediments of, for example, the Caribbean, Bahamas Platform and Florida;
- The older (Mesozoic and Paleozoic) and harder limestones and dolomites of the folded and faulted valleys and ridges of, for example, both the eastern United States and the intermontane valleys of the western United States.
- The equally old, but flat-lying carbonates more typical of the mid-continent;
- The metamorphosed (but solutioned) Proterozoic carbonates (marble and skarn) found, for example, in the Appalachian Mountains.

As a result, developing appropriate remedial measures must include an understanding of the potential failure modes of the solutioned carbonates below a particular site as well as the more conventional engineering parameters of building/facility type, loadings and potential remedial alternatives. Of course, a powerful player in any investigatory and/or remedial effort is economics in relation to safety. Not only should consideration be given to cost versus remedial effectiveness, but the owner must also balance the overall expected costs related to construction in "limestone" against abandoning the facility.

ENGINEERING GEOLOGY

First, one must recognize that some carbonates are hard and some are soft. Some are old, faulted, folded and weathered. Some are old, hard and flat-bedded. Some are recent, weak and flat. For purposes of this paper, we have considered failure in the structural sense (i.e. loss of foundation support), not in the environmental sense (i.e. contaminant flow through highly porous and/or solutioned carbonates). Such foundation failures can occur in a variety of ways, as noted below:

- Soft, young karst is often characterized by "cover collapse" (solution) or "subsidence" sinkholes (e.g., Figures 1A and 1B) where loose sandy materials fall or erode into generally near-vertical solution features within the underlying, relatively flat-lying "limerock". As can be seen on Figures 1A and 1B, "solution" and "subsidence" sinkholes are essentially the same. "Cave collapse" sinkholes (Figure 1C) can also occur in these materials. Foundation failures can occur through the loss of support under all or part of a structure. This type of sinkhole formation is often seen in the carbonates of areas such as Florida and the Caribbean.
- An actual structural failure within the flat-lying, but hard Paleozoic rocks of the central United States generally occurs through "cave collapse" (e.g., essentially the same as shown on Figure 1C for "young" carbonates). The roof of a linear solution feature (cave) that has formed, generally along bedding, may collapse as a result of continued solutioning (rare) or increased loading (more common). The best examples of this type of failure are the sinkhole plains of the Mammoth Cave area in Kentucky. Cave collapse is not as common in folded (e.g., Valley and Ridge) terrane, but can occur (e.g., Figure 1 D).
- Conversely, a raveling-type of failure generally occurs as a result of the erosion of overburden materials into solution features along inclined bedding or shear zones in, for example, the folded and faulted rocks of the

Valley and Ridge Physiographic Province of the eastern United States (e.g., Fischer, et al, 1996). These sinkholes generally form from the rock surface upward and may not be evident until the soil arch (Figure 2) collapses. Foundation concerns are obvious as the soil "arch" can fail from either overloading or roof material removal. Small cavities or water-softened materials often lie atop the rock because of surface water infiltration or open-channel water movement from, for example, precipitation.

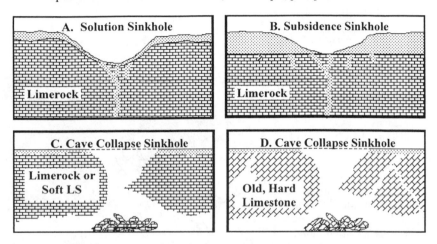

Figure 1. Common failure modes in karst (Figures 1A, 1B and 1C from Chen & Beck, 1989; Figure 1D from Fischer, et al, 1996).

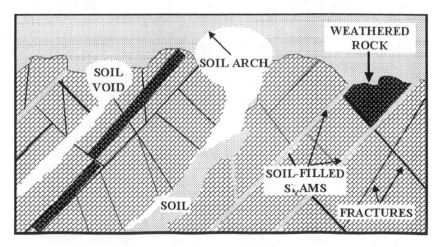

Figure 2. Common failure modes in "bent rock" karst (from: Fischer, et al, 1996).

- Solutioned cave features are often found in hard metamorphics of, for example, the eastern United States. Failure ordinarily requires the overloading of a cavity roof (cave collapse sinkhole Figure 1D) or cutting into a cave rather than persistent dissolution or by erosion of the surficial soils into ancient cavities. These solution features may follow ancient zones of weakness resulting in odd shapes (Figure 3).

These two-dimensional representations of possible failure mechanisms yield a clue to what the geotechnical engineer planning the remediation project must consider in three dimensions. Variations in the surface of flat-lying sedimentary rocks often result from the existence of near-vertical solution channels ("cutters") found within these rocks. Cavities (or caves) can extend hundreds of meters laterally. Erosion occurs along bedding and often at different elevations because of ground water level changes occurring over long periods. Different rates and/or quantities of precipitation can change the elevations of such "conduit" flow phenomenon.

In contrast, folded rock

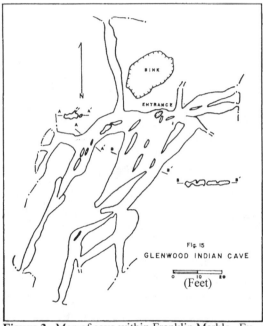

Figure 3. Map of cave within Franklin Marble. Extent is 493 feet (from Dalton, 1976).

terranes exhibit significant strength and solubility variations across bedding strike with, perhaps, down-dip continuity in physical properties. As fracturing can enhance permeability, variations in previous and current stress magnitude and direction will further exacerbate the problem of trying to understand the variations in physical properties, dissolution, and bedrock elevations for purposes of designing a remediation program.

Even glaciation has an effect. Are the landforms across the site a result of glacial action or carbonate geomorphology? They can appear quite similar. Are the near-surface soils till with numerous carbonate fragments or residual soils? Did a glacier override the site removing weathered rock, flattening pinnacles, and crushing thin cave roofs? Have low permeability glacial deposits protected the upper carbonates from post-glacial water inflow? Geologic maps can be useful in this regard, but are usually not site specific.

Thus, it becomes obvious that to develop an appropriate site remediation scheme (whether it be grouting, deep foundations or dynamic compaction), an understanding of the nature of the subsurface and its likely failure mode (or modes), as well as the more conventional engineering concerns of loading magnitude and grading (cuts and fills) must be considered. One would hope to have the data from a well-designed and properly executed field investigation on hand to develop an appropriate geologic model of the subsurface prior to planning a remedial program. However, from a practical standpoint, there is never enough funding for a full investigation of a karst site even in its most simplistic manifestation. Alternately, post-construction problems often occur at sites where no or limited geotechnical information is available. When such subsurface information is not available, aerial photography, site reconnaissance, and knowledge of the regional geology will at least allow a general understanding of the problem (e.g., Fischer, et al, 1989).

Whatever the nature of the grouting program selected for eventual use (slurry, modified or hybrid slurry techniques, or compaction grouting), the use of exploratory holes (air-track probes or test borings) to continually modify the geologic model is necessary. The probes or test borings should be used to:

1) Develop subsurface information at each planned grouting location during the progress of the (exploratory/grouting) program;

2) Evaluate the preliminary geologic model in light of the data from each exploratory hole;

3) Select the grouting method(s) including the location of the next exploratory/grout hole, as well as the type of grout and procedure for placement to be used subsequently; and

4) Continually develop a more refined geologic model of the area of concern as the work progresses.

ECONOMICS AND SAFETY

All clients are interested in the economics of a project. However, the cost of a project is dependant upon a number of factors. What method of grouting is most appropriate; how long will it take; how can a contract be let that is fair to both sides as well as realistic; and how "safe" should the end product be? Any cost estimate must consider the grouting procedure, the subsurface conditions in the area of interest, contracting concerns, as well as the degree of conservatism and/or level of safety acceptable to both the engineer and owner. Ready answers to this conundrum are not available. Obviously, experience and judgment are important ingredients. Some of the possible considerations are listed in Table 1. The list is not complete, nor will all of the items be of concern to all projects.

Obviously, there is no firm answer to the safety/cost aspect of a karst grouting project. The problem can be explained to an owner, but they in turn are faced with financial pressures whether it is from a bank looking over a developer's shoulder or the County Commissioner reviewing the Public Works budget.

Even with a reasonable understanding of the site subsurface, cost estimates are often scientific guesses at best. With a site that has little to no subsurface information at hand, but with the existence of (or potential for) sinkhole formation, the scope and cost of a remedial program cannot be accurately estimated.

TABLE 1 – COST AND SAFETY CONSIDERATIONS	
ITEM	**COMMENT**
Geology	Depth to bedrock, strike and dip of bedding, size and frequency of solutioned zones, potential cavity filling materials, ground water depth and variations, potential for additional solutioning and/or erosion of surficial soils.
Loading	Intensity. Dimensions and importance of facility. Significance of failure of facility.
Solution Feature	Sinkholes. Excavation of feature. Size and presence or absence of throat.
Existing or Planned Land Use	Will the repair of the feature of concern reroute subsurface water movement to nearby areas, thus essentially moving the sinkhole?
Drainage Area	Will the sinkhole closure cause areal flooding?
Grading	Increase in loading upon, or a decrease in soil cover above a subsurface soil void? Is grading necessary to redirect surface water flow from area of concerns.
Nature of "removed" materials	Was material used to backfill old sinkhole a potential contaminant?
Remediation Materials	Availability of cement, fly ash, sand, water, etc. Suitability of materials for remedial program (e.g., flowability, strength, shrinkage, water quality). The cost of performing chemical grouting usually eliminates it from the list of alternatives.
Access	For equipment, materials and operating personnel. Lateral and overhead constraints. Residual soils from older crystalline carbonates often turn to sticky, slick mud when wet.
Safety/Conservatism	Level required, number of "check" holes, type of grouting, temporary or permanent repair, level and experience of technical supervision.
Type of Contract	Time and expense, fixed cost with materials, or quantity pricing.

REMEDIATION ALTERNATIVES

Remediation in solutioned carbonate rocks can be accomplished by a variety of schemes, but the scope of this paper is grouting. Grouting is often the procedure of choice because of its flexibility and relatively non-intrusive nature. Grout movement can be controlled by using thicker mixes or quick-setting agents, even with a high mobility grout. When used appropriately, compaction grouting can densify loose granular soils, fill voids or compress soft cohesive materials. Larger voids can be filled with lower cost aggregate or transit mix grouts. On larger projects, on-site batch plants can be used to place large quantities of variable-mix grouts. Air- or hydro-track drilling of grout holes is relatively inexpensive and can provide useful subsurface data if drilled by knowledgeable technicians and monitored by an engineering geologist/geological engineer with experience at karst sites.

Sinkhole backfilling is unfortunately considered a simple procedure. Often, "remediation" is performed by the mason or construction supervisor calling for a truckload of "5,000-psi" concrete and "chuting" it into the sinkhole that opened during the previous night's rain, or the soil void revealed during excavation. The concrete is certainly strong enough, obviously much stronger than the surrounding soils that support most structures. If the throat to the subsurface is permanently closed and water no longer erodes material into the subsurface, the "patch" will likely work (e.g., Figure 4A). However, if the subsurface looks more like that of Figure 4B, the patch will likely not succeed. The writers have seen both "fixes" in the field.

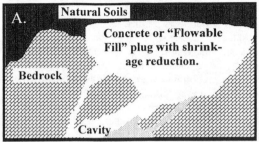

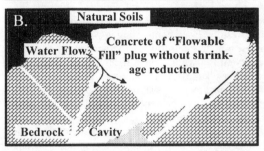

Figure 4. Examples of possible "sinkhole plug" remediation results.

Placing transit mix grout, "flowable fill", "sinkhole mix", or ready-mix structural concrete in a sinkhole is often the first thought when an owner or constructor finds a sinkhole after a rainy night. Commonly, the wrong grout mix is used with "transit mix fills". The mix is too thick or coarse to flow through the existing subsurface channels; there may be too little cement to bond the aggregate and reduce permeability and erosion; or it does not have a shrinkage-limiting agent added. Flowable fill has a variety of meanings, but conventionally, none of it is readily flowable through small subsurface openings. Without a shrinkage-limiting agent, surface water can move around the edges of the now-contracted concrete mass, find a passage to the causal cavity and erode surficial soils into unfilled subsurface cavities as illustrated on Figure 4B. Often the sinkhole is merely relocated a few feet away and discovered after the next heavy rain.

It is possible to improve the effectiveness of the "quick fix mix" by using the appropriate proportions of cement, water, fine aggregate (such as mason sand), combined with an appropriate anti-shrinkage agent. However, unless a definitive "throat" leading to the rock cavity can be revealed (generally by washing after excavation), even an appropriate grout will merely fill the soil void, not the causative cavities or fractures. In the authors' experience, even with the best of transit mix grouts; the chance of a sinkhole reopening nearby in the future is problematic.

In any case, one can decrease the chances of surface water infiltration with the placement of properly compacted, low permeability materials atop and adjacent to any repaired sinkholes. Often, on-site clayey, residual carbonate materials are adequate for such purposes.

Similar geotechnical concerns exist when one performs downhole grouting. As the concrete mass increases, so does the amount of potential shrinkage. Obviously, shrinkage must be controlled, but a number of agents exist to reduce shrinkage and increase flowability (e.g., bentonite, Aero 2000™, Darafill™). Again, one must consider the size and nature of the voids to be filled or sealed, the grouting procedure to be used, the grout mixes to be used, the aggregate size to be used, and the nature of the structure or facility to be located above or near the remediated area. Also of concern can be what may to nearby facilities if all of the subsurface voids in an area are not successfully closed and/or sealed?

For purposes of this paper, compaction grouting is defined as the injection, under pressure, of a bulb of low-mobility grout into a zone of loose or soft materials and/or voids with the intention of forming a suitable bearing zone or perhaps columns of strengthened materials. Slurry grouting or a modified slurry grouting program is crudely defined herein as the injection of a high-mobility grout with the primary intent of filling voids. In karst terrane, essentially using modified slurry grouting techniques, the authors often use additives (e.g., bentonite, calcium chloride, polymers, and sodium silicate) to limit mobility to the area of interest (e.g., Fischer, et al, 1992). A combination of slurry and compaction grouting can often be effective.

When selecting the grouting procedure, both geotechnical and cost concerns are important. Compaction grouting can be most useful if heavy loads are to be supported. However, is the added strength that results from compacting low slump concrete necessary to support roadways, homes, transmission towers and similar? Not much compaction is needed to turn soft cavity filling within sound rock into something that can withstand the loading of even a very large home supported 12 meters above solutioned bedrock. Generally, it is only necessary to seal the entrance(s) to the rock cavity and fill any surficial soil voids (Figure 2) to prevent further soil migration into pre-existing cavities if the structure or facility imposes only light loads

Hence, from a geotechnical standpoint, one should consider the intensity of the loads being imposed upon the remediated feature(s) when selecting a grouting program. From a cost standpoint, the need to install steel casing and have an air-track with crew standing by to remove it usually negates the lower material costs that result from compaction grouting's somewhat more limited grout movement and higher aggregate content. However, the lower mobility of the grout can also be a concern in filling or sealing small, interconnected voids in the area of concern. A sand, cement and water grout mix (1:1:1 by volume) usually travels the 1½ to 2 meters distance between holes drilled on a three-meter grid pattern. The authors have seen compaction grout hole spacing of 0.6 meters needed in an effort to fill a void or compact soft soils below a structure. Borden and Ivanetich (1997) noted that compaction grout mixed "with significant proportions of coarse aggregate have [internal friction] values of 40 to 47 degrees". As a result, high pressures and/or close grout hole spacing would likely be necessary to fill a void. There is also the concern of creating high pore pressures and/or hydrofracturing while grouting within the residual, generally cohesive soils of the Mesozoic and Paleozoic rocks of the eastern United States. Both slurry and compaction grout procedures can be successfully used in karst, but each provides its own advantages depending upon the mix of numerous considerations and concerns previously covered. However, without awareness of the subsurface variability and the mobility of the grout being used, a karst grouting program is likely to be either a failure or a costly exercise in the placement of grout with, perhaps, limited effectiveness. Hiring a grouting contractor without a karst-experienced grouting supervisor usually leads to cost overruns. An understanding of both the subsurface and the inherent difficulties are imperative in achieving a cost-effective solution. Even well prepared and karst-seasoned professionals are frequently surprised by how far off their original cost-estimate was.

CASE HISTORIES

Case 1 - This project was to remediate existing and possible future sinkhole formation at a high-tension power transmission tower that lies on a fault-bounded block of Allentown Formation dolomites. The Allentown rocks in the region are known to be solution-prone and have a history of sinkhole occurrences. Two depressions were located at tower legs and a large depression was centrally located below the tower.

Initially, the client had planned to use their civil engineer/designer to direct the remediation, but decided that their karst grouting experience was limited. A grouting contractor only experienced in compaction grouting was the low bidder. The contract for the grouting equipment and personnel was let on a daily price plus material costs (by the yard) basis. Subsequently, the authors' firm was chosen to direct the remediation upon the basis of their experience in the region. Several pre-grouting test pits revealed loose soils, boulders (some with voids below them), a pinnacled rock surface, and a healthy groundhog lair, but no obvious sinkhole throats or bedrock openings.

The grouting program originally proposed by the authors was an exploration/grouting procedure using air-track (pneumatic percussion drilling equipment) probe holes and the previously mentioned modified slurry grouting techniques. However, the drilling/grouting firm contracted by the client did not have the appropriate equipment for, and was not familiar with, slurry grouting procedures. As a result, a program of grouting at low to moderate pressures with a moderately high mobility grout mix delivered to the site was performed for the remediation program (a hybrid slurry grouting program?). The air-track probe holes were filled with a somewhat viscous mix of cement, flyash, sand, retardant/anti-shrink agent, and water introduced under variable pressures. The sand used in the mix was as fine as could readily be provided by the supplier. Grouting pressures were intentionally limited to avoid, as much as possible, potential lifting of the tower legs.

Originally, exploration/grouting procedures were planned for the two tower legs (A and B) where sinkholes were located. In addition, one or two exploratory probe holes were planned at each of the other Legs (C and D). An effort would also be made to isolate the central sinkhole from the legs. However, the probe holes drilled for the legs of apparent lesser concern (C and D) indicated suspect conditions as well. Therefore, additional probe holes were drilled and grouted at these locations. Each leg received treatment intended to isolate their foundation from future sinkhole occurrences. Therefore, only cursory remedial action was taken on the central sinkhole (i.e., costs were mounting).

A roughly three-meter pattern of primary grout holes was drilled some five feet into the rock for each leg of the tower. Secondary "check" holes were then drilled between the holes that had the larger grout-takes to evaluate the efficacy of the program during the progress of the work. The center sinkhole was remediated last.

A summary of the probes drilled and their grout-takes by area/leg is as follows:

Leg A – 8 probes, 97.0 cubic meters (m³) of grout.

Leg B – 9 probes, 27.3 m³ of grout.

Leg C – 7 probes, 19.2 m³ of grout.

Leg D – 6 probes, 21.4 m³ of grout.

Center Sinkhole – 3 probes, 1.5 m³ of grout.

As can be seen by the grout takes, the surface depressions were sinkholes that resulted from the solutioned nature of the bedrock.

Overall, we feel that the project would have been less expensive to complete if handled somewhat differently. Large quantities (although less than expected) of grout were wasted when a full eight cubic yard (6.1 m³) load of transit mix grout could not be used. The unit price for grout was per delivered (not in place) yard. In addition, no drilling was being performed while the grout pump was operating and much time was lost in the transition between equipment and the changes in operator duties. Also, grout deliveries could not keep pace with the drilling. If the losing contractor (using site-mixed grout) had performed the work on their time and material bid, job operations would likely have been much more efficient (thus less expensive), even though the perceived cost of two crews, extra equipment and a higher unit price for grout made it appear otherwise.

Case 2 - This site is a toll road service plaza in eastern Pennsylvania. The site area lies in a carbonate-floored valley where the valley itself is an inlier of folded and faulted Paleozoic-age sedimentary rocks. The valley is underlain by Cambrian and Ordovician limestone and dolomite, as well as by Cambrian quartzite and quartz schist. The susceptibility to chemical solutioning and erosion of the carbonate rocks is responsible for the subdued topography of the valley. The service plaza is on a fault-bounded block of Ledger Formation dolomite.

A geotechnical study was performed after a sinkhole (or sinkholes) occurred at a fuel storage area adjacent to the service plaza. The study included a review of aerial photography, data from previous investigations within the service area, and three rotary wash test borings using techniques designed for investigation in carbonate rock areas (e.g., Fischer and Canace, 1989).

This study revealed that the bedrock had been subject to fracturing and thus, increased solutioning in the past. Numerous open channels and soil-filled cavities were noted during the drilling and coring operations. Unexpectedly, the total of four test borings encountered rock at the same depth (about 7.6 meters). Rock depths found during the exploratory/grouting program ranged from 7.6 to 19.8 meters below grade. Upon the basis of the recovered core, a small fault was believed to trend through the westerly portion of the proposed tank site.

Prior to the initiation of remedial work at the site, sheet piles were installed in a 12.5-by 11-meter rectangular area just outside the proposed tank locations. The existing tanks were removed and the area backfilled with 1-centimeter pea gravel. Both the placement of the stone backfill and site work being performed on nearby storm sewers complicated drilling and grouting operations at the tank site. The pea gravel backfill resulted in changes in the drilling and grouting procedures and increased costs as a result. It was originally planned to use one rotary-wash drilling rig and one air percussion (air-track) rig for the drilling operations. However, as result of the placement of the gravel backfill prior to and during the beginning of the drilling and grouting operations, it was necessary to install drill casing to limit grout flow into the stone (and stone into the borehole), a function that is difficult and slow for an air-track. As a result, two conventional truck-mounted drilling rigs were used. In addition, owing to the weight of the drilling equipment and the sloped nature of the uncompleted stone fill, it was necessary to occasionally employ heavy site equipment to facilitate the movements of the drilling rigs.

A primary grout hole grid of 3 meters was established for the slurry grouting operations. A grout perimeter was constructed below the sheet piling through the use of a setting agent (calcium chloride in this case) to limit grout movement outside the area of interest. Secondary grout holes exhibited little grout take indicating the effectiveness of the lean cement, mason sand, water, and bentonite grout mix used. A total of 43.6 m³ of grout were placed within the 12.5- by 11-meter fuel storage area. The remediation costs totaled about one-half to two-thirds of the estimated cost of the test boring/seismic tomography investigation proposed by the initial site evaluation team.

As predicted, the completion of a grout wall with internal cavity filling moved the formation of sinkholes to the parking area outside the remediated tank area. Eventually, remediation of these areas through slurry grouting procedures and the repair of a leaking water line completed the project.

Case 3 - The third site is a 98-hectare, 240-lot residential subdivision located in one of northwestern New Jersey's carbonate valleys. This large site is underlain by several carbonate rock formations in a repeating sequence. At least two faults and a synclinal fold traverse portions of the site. A cost-effective site planning and geotechnical study program was performed for this site prior to any development work. The program consisted of a review of the available geologic data and aerial photography, a geotechnical reconnaissance and widely spaced test pits and test borings performed using special procedures and equipment as outlined in Fischer and Canace, 1989. These data were utilized to develop a "geologic model" of the site subsurface. This model resulted in dividing the site into six segments that attempted to anticipate the conditions in areas of similar subsurface hazard. Prior to development in each of these segments, a more complete geotechnical investigation was to be performed in order to increase the understanding of each segment's specific subsurface conditions and design/construction concerns. Central to the planned program was to be ongoing construction inspection of critical areas by personnel with broad experience with development in karst. The developer agreed with the Township to comply with this program and to utilize the recommended construction procedures for karst terrane including minimal grade changes, rock excavation by hydraulic hammer or controlled blasting for minimal vibration/disturbance, site grading with appropriate drainage to prevent surface water ponding, and the use of compacted, low permeability trench backfill. The construction process started in full accordance with these concepts.

For the first phase of the build out, development proceeded as expected which included a study of the subsurface, the use of karst-friendly construction procedures and

964 GROUTING AND GROUND TREATMENT

the foundation inspection/remediation program as described above. The work was planned and accomplished with only one house foundation requiring sinkhole remediation. However, the market for housing in western New Jersey was skyrocketing and the developer was able to sellout the homes faster than he could construct them. Given these sales conditions and the perceived need to complete and close on as many homes as possible before the market cooled, the developer dropped all proactive karst-related procedures.

The developer started work in future sections using uncontrolled rock blasting for utility and foundation installations. Trenches were not drained to prevent ponding of rainwater during utility installations. Backfill was poorly compacted resulting in utility "swales" rather than firm, level trench areas. Drainage for housing foundation excavations was not provided, permitting water to pond in open foundation areas. The developer was advised of the likely effects of his chosen construction procedures on the formation of sinkholes in the future. The decision made was always to continue development using the procedures that appeared to speed construction and reduce construction costs with little apparent concern for sinkhole remediation costs and delays.

As expected, large numbers of sinkholes developed throughout the site during the construction. Sinkhole occurrence was directly related to the extent of blasting for utility and basement construction, lack of backfill compaction, the amount of rainfall during the construction period, and the proximity of a thrust fault.

Most of the sinkholes that developed during construction were not overly large nor had open "throats" to the cavities within the rock. However, a sinkhole took out the only road leading to the next section to be developed. As a result, the most common method of remediation (i.e., cleaning out loose soil, running water to open a "throat" and filling with transit-mix grout to seal off the rock opening) was not particularly effective. Where an open "throat" was unearthed this method was used, but for the most part transit-mix grout was only used to fill a sinkhole void to permit construction, such as paving, to continue prior to performing a final remedial grouting program. Exploration/grouting procedures were used continuously throughout the project.

At this site, sinkholes generally resulted from many small open fractures in the bedrock near blasted areas, not large open voids in the rock. In many cases the fractures extended great distances laterally and vertically from the intended boundary of the rock excavation. To effectively seal off the multiple, generally small, openings into the rock, numerous probe holes were drilled surrounding the sinkhole to intercept as many of the fractures as possible. Plastic "tremie" pipes were placed in the probe holes and a site-mixed, very high mobility, cement/sand/water/bentonite grout (slurry grouting method) was pumped to the bottom of the probe holes through the "tremie" pipes. Where very small sinkholes were found, plastic "tremie" pipes were often jetted into the sinkhole to reach underlying fractures/void systems and then grouted with site-mixed grout. These methods allowed the maximum penetration of the grout into the small fractures and open joints in the rock and appeared to completely seal off the rock from surface water inflow. The site-mixed grout permitted rapid variation in the grout mixture to match the "flowability" required to effectively seal the rock surface.

The drilled probes consisted of both rotary wash-drilled borings using only water as a drilling fluid and pneumatic percussion- (air track) drilled probes. The rotary-drilled borings were utilized at the start of remediation efforts in each segment to obtain soil samples and rock cores to better identify the subsurface conditions in the area. Air track probes were utilized for the majority of probe holes once the subsurface conditions of each segment and the causes of sinkholes were better understood.

Initial remediation efforts began in August 1997. After the acceleration of the site earthwork operations, remediation was performed semi-continuously from mid 1998 to early 1999, then continuously through to June 2000. A total of 561 drilled probes and 72 jetted probes were installed and grouted. A total of 2,202 m³ of grout was placed.

CONCLUSIONS

Some form of grouting is often a suitable remediation procedure at a karst site. Different remedial concepts are available, but they all depend upon having or developing an understanding of the subsurface, the nature of the structure or facility of concern, the degree of conservatism desired, scheduling requirements, and economics. One must be aware of the potential consequences of remediating one area and forcing sinkholes to open nearby.

Often, a flexible or hybrid slurry grouting operation is the most economical solution for remediating karst. However, establishing a realistic and firm contract is not easy, generally requiring trust between the owner, contractor, and grouting engineer.

REFERENCES

Borden, R.H., & K.B. Ivanetich. (1997). "Influence of fines content on the behavior of compaction grout". *Grouting: Remediation and Testing*, Proc. of Geo-Logan '97 Conf., (Logan, UT).

Byle, M.J. (2000). "An approach to the design of Low Mobility Displacement (LMD) grouting". *Advances in Grouting and Ground Modification*, ASCE (Special Publ. No. 104).

Chen, J. & B.F. Beck.. (1989). "Qualitative modeling of the cover-collapse process". *Engineering & Environmental Impacts of Sinkholes & Karst*, Proc. of 3rd Multidisciplinary Conf. on Sinkholes and the Engineering and Environmental Impacts of Karst, (A.A. Balkema).

Dalton, R.F. (1976). "Caves of New Jersey". NJ Geological Survey (Bul. 70).

Fischer, J.A., T.C. Graham, R.W. Greene, R.J. Canace, & J.J. Fischer. (1989). "Practical concerns in Cambro-Ordovician karst sites". *Proc. of 3rd Multidisciplinary Conf. On Sinkholes and the Engineering & Environmental Impacts of Sinkholes & Karst*. (A.A. Balkema).

Fischer, J.A., R.W. Greene, J.J. Fischer, & F.W. Gregory. (1992). "Exploration Grouting in Cambro-Ordovician Karst". *Grouting, Soil Improvement and Geosynthetics, V. 1*. ASCE (Geotech. Publ. No. 30).

Fischer, J.A., J.J. Fischer, R.F. Dalton. (1996). "Karst site investigations". *New Jersey and Pennsylvania sinkhole formation and its influence on site investigation: Karst Geology of New Jersey and Vicinity*, Proc. of 13th Annual Mtg. of Geological Assoc. of NJ, (Whippany, NJ).

Fischer, J.A., J.J. Fischer and R.J. Canace. (1997). "Geotechnical constraints and remediation in karst terrane". *Proc. of the 32nd Symposium on Engineering Geology and Geotechnical Engineering*, (Boise, ID).

Fischer, J.A., J.J. Fischer & R.S. Ottoson. (1998). "Geotechnical investigation in karst: bent rock versus flat rock". *Proc. of 49th Highway Geology Symp.* (Prescott, AZ).

Nichols, S.C., & D.J. Gooding. (2000). "Effects of grout composition, depth and injection rate on compaction grouting". *Advances in Grouting and Ground Modification*, ASCE (Special Publ. No. 104).

Shuttle, D. & M. Jeffries. (2000). "Prediction and validation of compaction grout effectiveness". *Advances in Grouting and Ground Modification*. ASCE (Special Publ. No. 104).

Grouted Seepage Cutoffs in Karstic Limestone

Arthur H. Walz, Jr. P.E.[1], David B. Wilson, P.E.[2], Donald A. Bruce, Ph.D, C.Eng.[3,] and James A Hamby, P.E.[4]

Abstract

Four types of seepage cutoffs have been successfully utilized in limestone formations: open-cut excavated cutoffs, diaphragm wall cutoffs, secant pile cutoffs, and grouted cutoffs. After falling somewhat out of favor due to lack of success on some projects, recent advances in materials, procedures, and techniques have resulted in practitioners regaining confidence in grouting.

This paper examines cutoff methods with respect to geologic compatibility and the issues, problems, and limitations for each type of cutoff. Four recent case histories of successfully grouted seepage cutoffs are discussed: the reservoir rim cutoff at Tims Ford Dam, Tennessee, the Patoka Lake spillway in Indiana, Wujiangdu Hydroelectric Project in China and a multi-material cutoff in an operating quarry in West Virginia.

Introduction

Any construction in or on karstic geology is very complex and the desired results are frequently difficult to achieve. When voids and cavities are encountered in the excavation operations for foundation preparation in the initial construction, the conventional method for providing the necessary cutoff is by open cut excavation with conventional earth and rock excavating equipment. This has proved to be a successful method. In the past, many reservoir projects in the United States have been constructed on karstic limestone foundations with only limited grouting and foundation treatment. Over time, the head created by the reservoir can cause increased seepage by continuing to wash out the clay or soil filled joints and cavities. Traditionally, when this happened seepage cutoffs were installed using the diaphragm wall and secant pile methods of construction. Grouting methods had

---

[1] Senior Water Resources Engineer, Gannett Fleming, Inc., Harrisburg, PA and former Chief, Geotechnical and Materials Branch, Headquarters, U.S. Army Corps of Engineers, Washington, D.C.
[2] Vice President and Manager, Geotechnical Section, Gannett Fleming, Inc., Harrisburg, PA.
[3] President, Geosystems, L.P., Venetia (Pittsburgh), PA.
[4] Civil Engineer, TVA, Chattanooga, TN

fallen out of favor due to lack of success on several projects. However, recent advances in grouting technology, materials and grouting procedures have resulted in the profession regaining confidence in grouting as a cost effective method of constructing cutoffs in karstic limestone.

Karstic Topography and Geology

Karst topography is the landform that develops in areas of exceptionally soluble carbonate rock (limestone and dolomite). It is characterized by the presence of sinkholes, closed depressions, enlarged fracture systems, caves and underground rivers. The sinkholes and depressions allow soil to fill or partially fill some of the voids. Unique surface and subsurface drainage patterns develop and large quantities of water flow beneath the surface.

The major problem for seepage control is the presence of red clay from the chemical degradation of the rock. This clay usually occupies large cavities and is highly erodible. As the erosion of the clay or soil material continues, these conditions become more extensive and present a real challenge for the construction of infrastructure projects.

Cutoff Types

Excavated Open Trench: This method consists of excavating and removing the rock and joint materials to the required depth and length. The excavation is then backfilled with the appropriate concrete mix to form a thick and continuous wall. It results in a positive cutoff since the termination points can be visually inspected before backfilling operations begin. This method is applicable when the voids are discovered during the initial construction, when the overburden or fill above the rock surface is shallow and the required depth is easily obtained by conventional drilling, blasting and excavation equipment. This is a traditional method used for many dams worldwide.

Diaphragm Wall: This method consists of excavating a narrow trench that is temporarily stabilized by a slurry fluid in a series of continuous panels. After the excavation for a primary panel is completed, the excavation is then backfilled with concrete. Reinforcing is added if required. After completion of several primary panels, secondary panels are then excavated between the existing primary panels and backfilled with concrete to form a continuous wall. It is important that the completed diaphragm wall have adequate structural strength and integrity as well as to provide a positive cutoff for differential hydraulic head.

Secant Pile Wall: This method consists of drilling large primary and secondary piles, in sequence, and then backfilling them with tremied concrete. After two primary piles have been drilled and the concrete has cured, a secondary pile is drilled between them and backfilled with concrete to form a continuous concrete cutoff wall. This is an effective method where the depth of overburden is shallow and the depth of penetration into the rock is significant.

Grouting: Historically, grouting operations in limestone have not been very successful because of our lack of understanding of the characteristics of limestone foundations and with the conventional equipment and procedures, it was difficult to control the grout mixes and injection pressures needed to ensure that the grout filled all of the voids. Monitoring of grout takes and the results were accomplished manually after a zone had been grouted. After a hole was completed an assessment was made. This process made it difficult to measure the degree of success of grouting an area or the project as a whole. While grouting was successful in stopping muddy water from flowing from the toe of the Wolf Creek Dam between 1968 and 1970, additional studies concluded that because of the high head on the foundation, grouting could not be considered as permanent long term treatment and a diaphragm cutoff wall was then installed.

Considerations for Selecting the Type of Cutoffs in Karstic Limestone

In the evaluation of alternative methods for constructing cutoffs, there are many factors to consider. A basic and sound characterization of the site and foundation is very important and many times difficult to obtain due to the nature of karstic limestone. It is often difficult to determine the locations and extent of the voids and cavities. The physical properties of the limestone or dolomite will have a significant impact on the type of cutting heads used to cut through the rock. Selection of the wrong type of head and equipment can result in major change orders or termination of the contract.

An example of the difficulty in assessing the insitu properties of the rock foundation occurred at Beaver Dam, AK. The contractor started constructing a diaphragm cutoff using a hydrofraise built especially for the job. A large sample of the rock from a quarry was tested to refine the cutting head design. The hydrofraise had two cutter heads with 32 cutting picks per head and cut a 9-foot by 33-inch panel. After 500-600 picks were used and only a small fraction of the wall was excavated, the contractor stopped work after less than a month and the contract was terminated.

In the evaluation and design phase, it can be difficult to establish the depth and length of a cutoff wall to achieve the desired reduction in seepage or flow of groundwater. This can have a significant impact on the investigation costs. Mobilization and construction costs can be high for a wall constructed by the diaphragm or secant methods. For wall construction the alignment is critical to ensure continuity of the wall and at times can be costly to accomplish.

Today, the use of computer monitoring systems clearly provides an improvement and is characterized by totally integrated systems of data analysis and management and real time monitoring and adjusting the grouting operations. Recent experience has shown that properly designed and constructed high pressure grouting can successfully provide a seepage cutoff in karstic limestone.

Recent Developments and Current Grouting Methods

Grouting is an especially unique type of construction. It is the process of injecting mixes of water, cement, fillers and additives into open or soil filled cracks and voids in rock with the intent of stopping the movement of water or to improve the physical characteristics of the rock. It involves managing and performing many simultaneous operations, each of which requires a high degree of care. Grouting is further complicated by the fact that we cannot see the formation to be grouted or see grout permeating the voids or fractures. In the past, this has led people in the profession to commonly describe grouting as an "art", based on rules of thumb.

Recognition of the potential benefits and experimentation with "automated" monitoring or data recording systems for grouting started in the 1960's. Use of electronic measurement devices mated with computers was recognized as having significant potential almost as soon as desktop computers came into being in the late 1970's. Since the mid-1990's, there have been dramatic improvements in both the number and type of flow and pressure measuring devices, computer hardware, data acquisition software, and data management and display software. The dipstick and pressure gage method is becoming a practice of the past. Using proper investigation, evaluation, design and construction techniques, we now have the ability to design and build Quantitatively Engineered Grout Curtains (QEGC) for seepage control with a high degree of confidence and reliability. This system is now sufficiently reliable and user friendly that grouting can now be considered as a very efficient and cost effective method to provide a cutoff for future projects.

Karst Grouting Issues, Theory and Methods

There are a number of issues in confidently constructing a durable and reliable grout curtain in karst formations. However, successful procedures have evolved that address these problems. These items are discussed in subsequent paragraphs.

Irregular Top of Rock Surface. The first problem that must be dealt with is the enlarged joints, seams and soil filled voids that create an irregular top of rock surface. For new dams, this problem is usually solved by variation in the depths of foundation excavation to expose a groutable rock surface. The excavation depth is established based on the experience and judgment of the engineer or geologist, and the seam is backfilled with concrete before grouting.

Water Loss During Drilling. The recommended practice for drilling grout holes in rock is to use water for removal of cuttings (Houlsby, 1990 and Weaver, 1991). When open joints or voids are encountered during drilling, water loss will occur. This is a frequent occurrence in grouting karst foundations, particularly in the early stages of the program. The frequency of occurrence will decrease dramatically as the grouting progresses. When water loss occurs, the drilling is suspended a short distance past the point of loss and the fracture is washed and grouted before continuing with further drilling of the hole (Weaver 1991). There are multiple reasons for this procedure:

1. The hole has encountered precisely the type of feature being sought, and a significant opening has been encountered that deserves special treatment.

2. Continuing to drill without having cuttings brought to the surface means that the cuttings are entering the fracture and potentially reducing the ability to later inject grout into the fracture.

3. Drilling difficulties are likely to occur without return of water including possible damage to the bit, losing the hole, or losing drilling equipment in the hole.

4. Continuing past that point and grouting later is highly likely to result in grouting in of packers resulting in loss of equipment and likely the drill hole itself.

Large Voids. When a void is encountered during drilling, the size of the void is unknown. Rod drop may have little meaning, since it is not known whether the center or edge of the void has been penetrated or perhaps the drill has encountered an enlarged vertical seam of limited horizontal dimensions. A downhole camera system is of great value in assessing the conditions, and it is recommended that one be kept onsite full time for that purpose. In the absence of a camera, and sometimes even with image information, the recommended procedure is to still begin grouting at a relatively fluid mix, but to proceed through a sequence of thicker grouts relatively quickly. The grouting should be brought to refusal with whatever grout mix is found to be necessary. Low mobility grouts (LMG) will usually perform adequately when large voids are encountered. The use of sanded mixes and concrete mixes should be limited to those voids where the condition is known by camera inspection.

Large Water Flows in Fractures. When fractures with large water flows are encountered, the flows must be reduced by whatever means necessary to allow later high quality grouting to be performed. A staggering variety of materials have been used for this purpose including any and all materials readily available (Bruce et al

1998). After the flow has been initially blocked, extensive grouting is performed with proper materials.

Soil Filled Fractures. Clay, silt or sand filled fractures and solution features are extremely common and have always been the major source of concern for the quality of grout curtain construction in karst conditions. The concern is heightened by the fact that these infilled materials are frequently soft and loose and are clearly erodible under even moderate flow conditions. Houlsby (1990) and others, including the authors, have concluded that removal of all of these infilled materials is neither possible nor necessary. Simply from an intuitive standpoint, it is clear that the relatively small diameter of the water washing pipes makes it impossible to inject enough water through the drilling system to clean out a major infilled seam. At best, it might erode an opening through the material of sufficient diameter to accommodate the inflow rate. Accordingly, it is recommended that washing continue until the return flow becomes clean, but it should not be assumed that the infilling has been removed to a significant degree. In the absence of any return flow, the fracture should be washed for a pre-determined and limited amount of time, typically 5 to 10 minutes.

A critical question remains, however, regarding how an effective grout curtain can possibly be constructed in the presence of such materials. In fact, it is precisely this question and the combination of unsatisfactory grout curtain performance on some projects that has given rise to the loss of confidence in grouting as a solution. However, despite the "horror stories" about grouting limestone sites, there are a larger number of success stories, including effective grouting of foundations with major cave systems for dams with applied heads as high as 165 m (540 feet) (Weaver, 1991; Zuomei and Pinshou, 1982; Houlsby 1990). Why grouting works under these circumstances is one of the most intriguing and least understood technical issues because it is such a deviation from the normal concepts of what is required for successful grouting. Clearly, there must be alternate mechanisms involved that allow successful grouting of karst despite these conditions. These alternative mechanisms are discussed in the following case histories.

Patoka Lake Dam, Indiana

The recent seepage remediation project at the Corps of Engineers Patoka Lake Dam required grouting of a vertically fissured limestone unit with considerable soil infillings. A 3-line grout curtain was used, and real-time computer monitoring of grouting was employed. As the grouting progressed, it was clear that in these enlarged, infilled fractures a common pattern was a gradual build-up of pressures followed by one or more episodes of hydrofracturing of the clays. The grouting pressure being used was well in excess of the head that would eventually be applied to the completed curtain, and each stage was brought to refusal at that pressure.

Water pressure testing was performed at a lower, uniform pressure adequate to locate fractures and to calculate the Lugeon (Lu) value for the stage.

Early in the project, the theory of successful grouting at this site was based only on a containment concept. A 3-line curtain would be constructed, but the primary mechanism for success would be that a wide zone of rock would have all the open fractures filled, thereby keeping gradients relatively low and simultaneously protecting and containing the infilled materials left in place. However, as the program evolved, extensive observations and analyses of the developing behavior were performed. In particular, the effects related to the hydrofracturing were considered, and a revised closure criterion was developed. Previously, closure of a line was considered to occur when analysis of series of grouted holes suggested line closure had occurred and when the lower pressure water tests in supplemental holes indicated that open fractures had been completely filled. When that condition was found to exist, the supplemental holes were simply gravity grouted since they were absolutely tight as per the water testing. As the program progressed, the criterion for closure of a line was modified such that drilling and grouting of supplemental holes continued until closure had been obtained. When the lower water pressure tests of supplemental check holes indicated absolutely tight conditions, and when grouting of supplemental holes showed no hydrofracturing. This modified closure criterion resulted in total confidence that a fully effective grout curtain had been created, because it now showed that not only had containment been achieved, but also that it was not possible to hydrofracture through the infilled materials even at elevated pressures.

Wujiangdu Hydroelectric Project, China

The work by Zuomei and Pinshou (1982) has added more information on the mechanisms related to hydrofracturing and other factors. The Wujiangdu Hydroelectric Project in China required effective grouting of extensive karstic caves filled with soft clay to be able to withstand heads of up to 165 m (540 feet). Grouting was accomplished under high pressures (60 kg/cm^2), and hydrofracturing was observed. After grouting, specimens were removed from the grouted caves in the curtain line. Based on examination and testing of soil samples from the voids, they concluded that multiple mechanisms were at work that allowed the curtain to be successful under very adverse conditions. Specifically, they concluded the following:

1. Hydrofracturing occurred in a radial pattern that created an extensive network of cement veins within the clay. There was also evidence that the grouting created circumferential fracturing. This resulting network acts as an effective barrier to confine the clay from erosive flows. In order to verify the effectiveness of the grouting, specimens from the voids were obtained and tested. A secondary benefit of the hydrofracture cement network was a chemical alteration and hardening of the clay from calcium carbonate formed

by the grout. This produced a measurable increase in strength and slaking resistance that improved the fundamental erosion resistance of the infilled materials.

2. While the hydrofracturing effects are the most important items, two other beneficial effects were also observed. The first is that grouting did displace some of the softer materials by extrusion under the high grouting pressures, thereby increasing the efficiency of the grouting. Secondly, by using the natural water content and density of the clay and the volumetric ratio of the cement grout in the specimens, consolidation of the clay in the voids was calculated to be 6.5% of the grout volume.

Combining the information from these two projects provides both a reasonable explanation of the mechanisms (hydrofracturing, displacement and consolidation), as well as the importance of adopting a dual closure criteria for the curtain lines based both on demonstration of tight stages by water pressure testing and by cessation of hydrofracturing at elevated grouting pressures.

Tims Ford Dam, Tennessee

The Tennessee Valley Authority's (TVA) Tims Ford Dam is a 175-foot high rock-fill dam, located on the Elk River, Tennessee. The rock at the Tims Ford site consists of thinly bedded, nearly horizontal layers of limestone and shale. The reservoir rim near the right abutment of the dam is comprised of red residual clay from the Fort Payne formation underlain by three limestone formations - the Brassfield limestone containing numerous solution cavities along bedding planes and near-vertical joints; the Fernvale limestone, also containing solution cavities along bedding planes and joints; and the Catheys-Liepers limestone, containing shale lenses, fossilized layers and displaying a prominence of clay seams.

During the initial planning of the dam project, it was decided to treat only those portions of the rims that showed leakage after impoundment. In March 1971, when the pool reached approximately 865 feet in elevation, leakage developed in both left and right reservoir rims, as well as in the left abutment adjacent to the dam. Over time leakage from the Fernvale limestone on the right reservoir rim increased steadily to about 4,000 gallons per minute (gpm) at maximum pool by 1995. Later that year the rate of flow increased dramatically to just less than 8,000 gpm.

This leakage prompted TVA to evaluate the need for remedial action at the right rim of the reservoir. Several alternatives were evaluated. Grouting from the top of the rim was selected. The advantages of this scheme were: the area to be treated could be fairly well defined; the likelihood of success was good; the work could be done with the reservoir a few feet below normal minimum pool elevation; standard drilling and grouting equipment could be used; and the effectiveness of the grouting could be monitored by observing the leakage.

Results from a site investigation indicated that a multi-row, remedial grout curtain having a length of 800 feet would be adequate. The holes penetrate all three limestone formations, were inclined at 30 degrees to the vertical to ensure intersection of sub-vertical features and were oriented in opposite directions in the two outside rows. Primary holes in each row were located at 40-foot centers, with conventional split spacing methods to be employed (to reduce interhole spacing to 10-foot centers). The central tightening row was vertical.

TVA's goals were to reduce the peak seepage to about 1,000 gpm and to focus only on the major features. Holes that did not encounter voids or active flow were to be grouted with fluid, cementitious grouts. The grouting was designed to be performed using upstage methods, although it was anticipated that poor foundation conditions could require localized utilization of downstage methods in conjunction with polyurethane resin. However, because larger than anticipated were encountered, low mobility grout (LMG) was used to fill the major voids in lieu of the polyurethane resin.

Actual field conditions varied from what was anticipated. The work progressed as scheduled, but several major modifications were made.

1. This project had to comply with a restricted schedule due to reservoir drawdown constraints. The LMG (slump 2 to 6 inches) was batched onsite, using a two-conveyor, three-component, trailer-mounted batch plant, with hydraulic-driven mixer/conveyor auger.

2. When the reservoir was drawn down to elevation 859 feet, the flow from the major seepage exit point stopped. This "no flow condition eliminated the need for the polyurethane grouts and extending the applicability of cement-based formulation (including LMG).

3. Larger-than-anticipated open or clay-filled features were encountered, especially in the upper 20 feet of the curtain. For technical, commercial, environmental, and scheduling reasons, such features were treated with LMG.

4. A suite of four cement-based grout mixes were developed to permit the appropriate match of mix design and "thickening sequence" to the particular stage conditions as revealed by drilling and permeability testing (both multi- and single-pressure tests). In summary, about 2,000 cy of LMG, 400 gallons of polyurethane, and 790 cy cement-based grouts were injected into a total of 250 holes (comprising 11,000 linear feet of rock drilling)

Real-time performance monitoring during the grouting operations included the results of drilling, water pressure tests, calculating reduction ratios and dye testing. This monitoring allowed onsite engineers to track the development of the integrity of the grout curtain and focus grouting efforts on specific zones along the grout rows. Monitoring also included data from 1) discharge from the rim leak; 2) groundwater elevations down gradient from the grout curtain; and 3) headwater elevations. Engineers used results from water tests to evaluate permeability of the rock in Lugeon (Lu) values. The water tests confirmed more open void stages in two certain areas, and another order holes were added to these zones. Grout takes closely followed trends observed in the water test data. To evaluate grouting progress, reduction ratios were calculated by dividing the average take of one order of holes by the average take of the previous order of holes. By completion of the program, total seepage at full head had been reduced to less than 300 gpm and has remained at that level since completion of the grouting.

Limestone Quarry, West Virginia

A large operational dolomitic limestone quarry is situated in West Virginia less than 1,500 feet (460m) from the Shenandoah River. In April 1997, a major sudden inflow developed into the southwest corner of the quarry pit following production blasting activities and several abnormally severe precipitation events that caused flooding of the river and nearby sinkhole formation. An observed vortex in the river appeared to be the point source of the flow. The initial magnitude of the flow, estimated at over 35,000 gpm (132,500 L/min) was far greater than the capacity of the existing pit pumping facilities.

The new inflow posed a severe threat to both the current and future viability of the quarry. Several unsuccessful attempts were made to construct a cofferdam with sandbags on and around the location of the vortex. In May 1997, pumping operations were discontinued, and the quarry water level was allowed to rise. Extensive investigations were conducted to determine the source and extent of the inflow. Prior to the design and construction of the remediation, it was agreed to "baseline" the hydrogeologic situation as closely as possible. Wells with deep piezometers were located between the river and the quarry to evaluate the water level, pH, conductivity, and temperature. This monitoring continued during and after the remediation.

The owner's goal of the remedial program was to reduce the total inflow into the quarry to a flow of 8000 gpm, (30,300 L/min) with the quarry completely dewatered. Later data would indicate this would require reducing the flow from the river to below 3000 gpm (11,400 L/min). Three specific options were considered:

1. Identify the specific solution cavities in the river and seal them.
2. Construct an intercepting cut off at some appropriate location between the river and quarry.
3. Treat the problem close to the quarry.

Option 2 was clearly favored, on logistical, technical, and environmental grounds, and it was decided to locate the cut off on a convenient road side location about 50 feet from the river bank

The main challenges were: the very high velocity and rate of the flow through potentially multiple conduits; mud filled karstic features, creating the possibility for erosion, piping, and "blow out" after curtain placement when the hydraulic gradient increased; and the possibility of grout migration "upstream", into the river. Several grouting technologies were studied to provide the curtain, in part or in whole jet grouting; polyurethane injection; LMG; hot bitumen injection; accelerated cement based slurries; use of the multi packer sleeve pipe (MPSP) system; and geotextile grout-filled bags. For the very severe geological and hydrogeological regimes to be accommodated, each technique was assessed based on technical feasibility, likelihood of successful treatment of the inflow in both short and long terms, and cost. Grouting was accomplished in nine phases. Throughout the grouting operation, several modifications were made to enhance control and responsiveness and allow simultaneous injection of both bitumen and slurry into the same hole. For example, stringers were used to allow the simultaneous injection of both slurry and bitumen into the hole. It was decided to first treat the "Cold Karst' zones (open voids without flowing water) with LMG and slurry grout via the MPSP system and then treat the "Hot Karst", i.e., the zones were water flowed, with hot bitumen from the downstream row of holes, backed up by slurry grouts simultaneously injected from the upstream row via further MPSP locations.

Monitoring of groundwater wells, water levels in the quarry, flow and visual observations of the river eddy indicated that the program was successful. By the end of the grouting, the flow from the river into the quarry had essentially stopped. This success of this case history clearly illustrates many important features, but three are particularly noteworthy. Firstly, this is an illustration of how contemporary grouting technology can be used, if correctly designed, implemented, analyzed, and closely monitored, to provide a successful result in even the most adverse conditions. Secondly, is that all sources of information must be studied before and during the operation in order to gain the best possible "picture" of what is really happening in the ground and the incremental changes actually brought about the grouting itself and changes in. the hydrogeological regime. Thirdly, and perhaps most importantly, this project illustrated the need for all stakeholders (owner, designer, consultants and the contractor) to partner fully and openly, and to provide mutual support at all times and in all aspects. In such circumstances, patience and trust are vital ingredients to successful teamwork in arduous and stressful conditions.

Summary:

Any type of construction in karstic geology is very complex and it is sometimes difficult to achieve the desired result. Establishing permanent seepage control can be difficult, expensive and requires monitoring and future evaluation. The recent advances in the approach to grouting now enable geologists and engineers to make significant improvements in the grouting practices and procedures to control seepage or the flow of groundwater. In summary, these advances in technology, materials and grouting methods have made grouted cutoffs in karstic limestone a reliable, efficient and cost effective method for permanent seepage control

References:

Bruce, D.A., G.S. Littlejohn, and A. Naudts. (1997). "Grouting Materials for Ground Treatment: A Practitioner's Guide," Grouting – Compaction, Remediation, and Testing, Proc. of Sessions Sponsored by the Grouting Committee of the Geo-Institute of the American Society of Civil Engineers, Logan UT, Ed. by C. Vipulanandan, Geotechnical Special Publication No. 66.

Bruce, D.A. and Gallavresi, F. (1988). "The MPSP System: A New Method of Grouting Difficult Rock Formations." ASCE Geotechnical Special Publication No. 14, "Geotechnical Aspects of Karst Terrains.".

D.A. Bruce, J.A. Hamby, and J.F. Henry. (1998). "Tims Ford Dam, Tennessee: Remedial Grouting of Right Rim." Proceedings of the Annual Conference, Dam Safety '98, Association of State Dam Safety Officials, Las Vegas, NV.

Bruce, D.A., Taylor, R.P. and Lolcama, J. (2001) "The Sealing of a Massive Flow through Karstic Limestone." ASCE, Blacksburg, VA, June 9 – 13, Geotechnical Special Publication No 113, pp. 160 – 174.

Hamby, J.A. and Bruce, D.A., Monitoring and Remediation of Reservoir Rim Leakage at TVA's Tims Ford Dam, Proceedings, United States Society on Dams, Meeting and Lecture, July 2000

Houlsby, A.C. (1990). Construction and Design of Cement Grouting, John Wiley & Sons, New York.

Weaver, K. (1991). Dam Foundation Grouting, ASCE, New York, New York.

Zuomei, Z. and Pinshou, H. (1982). Grouting of the Karstic Caves with Clay Fillings, Proceedings of the Conference on Grouting in Geotechnical Engineering, ASCE, New Orleans, Louisiana.

Nittany Lions' New Convocation Center: Rock Solid with LMG

Allen W. Cadden, P.E.[1], and Richard H. Wargo, P.E.[2]

Abstract

Limited Mobility Grouting (LMG) has been a relatively recent technology that began in the United States. Applications of this method of ground improvement are still evolving. The Bryce Jordan Center, Penn State University, State College, PA, is an example of the use of LMG in karst for rock supported spread footings. Although a deviation from the traditional compaction grout roots of LMG, the application provided assurance that the foundation rock was consistent/continuous. This paper provides a brief discussion of the exploration methods used to assess the site variability, and the development of the foundation design. This approach was developed as a cost saving alternative to drilled shafts, and represents the first recent structure where the State owner of this facility had allowed foundations other than drilled piers for buildings in karst. The paper includes discussion of the design methodology and estimating procedures, as well as construction experience. Conclusions related to the construction activities, difficulties, and results are included. The design methodology builds on the work of Schmertmann and Henry to develop a strength-based approach to unifying the rock mass to limit settlements under foundation loadings. The success of the method has been demonstrated by six years of satisfactory performance. As a unique application of LMG to rock foundations, particularly on a structure of this magnitude with column loads up to 2500 kips (11,000 kN), this case study serves as an example of successful innovation with an evolving technology.

Introduction

Penn State University required the construction of a 16,000 seat convocation center on campus. At the time of its construction, this was one of the largest on-campus facilities of its kind. The new convocation center was to take the place of an existing parking lot for Beaver Stadium that was located adjacent to this site. Recent construction of major structures at the University generally involved the use of drilled shafts. Given the difficulties associated with drilled shaft construction in karstic geology, the Pennsylvania Department of General Services (DGS) required test borings at each foundation location in an effort to manage the potential costs and construction delays. For a structure of this magnitude, such an approach carried with it a significant up front cost.

---

[1] Member ASCE, Principal, Schnabel Engineering Associates, Inc., 510 East Gay Street, West Chester, PA 19382, 610-696-6066, acadden@schnabel-eng.com
[2] Member, ASCE, Principal, Schnabel Engineering Associates, Inc., 104 Corporate Boulevard, Suite 420, West Columbia, SC 29169, 803-796-6240, rwargo@schnabel-eng.com

Modifications to the brute force approach for this project involved the exploration of the site geologic structure, trends and anomalies through the use of standard test borings, and geophysical surveys. This approach was further deemed reasonable because suitable bearing material (that which would meet the strict quality requirements of the DGS) would likely require significant excavation into the underlying limestone. Costs associated with constructing numerous drilled shafts into this rock were likely to be prohibitive. Therefore, the goal was to develop a feasible shallow foundation system for this structure.

Structure Features

The convocation structure consists of a Concourse Level at the road grade of EL 358.5 m (1176 ft), where the main entrance and ticket areas are located. The site slopes significantly from north to south, which accommodates a depressed bowl for the event floor. Thus the bottom of the bowl, the Event Level, is located about 10.4 m (34 ft) below the Concourse Level, EL 348 m (1142 ft). To the south, significant fill was placed to allow room for practice courts and locker rooms, as well as ramps and an access road. The practice levels are located at EL 348 m (1142 ft). Entrance ramps along the south side rise to the Concourse Level. The total plan event area is 97.5 x 122 m (320 x 400 ft). With the addition of the ancillary practice areas and locker rooms, the total plan area of the entire facility is about 27,870 m² (300,000 ft²).

Four oval rings of foundations surround the event floor with the most significant loads being carried on the outermost of the rings, Ring D. Foundation loads vary considerably around the structure. The main superstructure over the event area was supported by a ring of 32 columns located at the back of the seating area. These loads were typically about 6670 kN (1500 kips) with the maximum loads of about 11,000 kN (2500 kips) occurring near the four corners of the oval. Between the event floor and the main ring of foundations, two rings of columns were used to support the seating area loads. These foundations ranged from about 2200 to 3300 kN (500 to 750 kips). Most of the foundation grades were consistent with the event floor level, except to the north; where the lower levels were not present, foundations stepped up about 6 to 9 m (20 to 30 ft). The remainder of the structure was more typical masonry wall and concrete column construction with loads typically less than 146 kN per meter (10 kips per foot) for walls and 2200 kN (500 kips) for columns.

The majority of the structure and grandstands was cast-in-place concrete with precast concrete seating. However, the structure supporting the saddle-shape roof was long span steel trusses. Span lengths of 109.7 m (360 ft) were required to maintain a clear event floor. Given the type of structural elements, the designers required settlement of less than 25 mm (one inch), and angular distortions of less than 0.001 m/m.

Site Geology

This site is located within the Nittany Valley of the Ridge and Valley Physiographic Province, and is believed to lie near the contact of the Nittany Formation and the interfingered Stonehenge and Larke Formations. The contact between the Nittany and the Stonehenge/Larke Formations has been mapped by the USGS along the north side of the project. The Nittany Formation is described by the Pennsylvania Geological Survey as a "fine to coarsely crystalline dolomite" which is sandy or cherty in areas. The Stonehenge Formation consists of "fine crystalline limestone and dark gray laminated limestone" containing breccia and shaley interbeds (Geyer, 1982).

Investigation

Although the typical requirements of the DGS is to complete a test boring at each column location, given the size of this facility and the layout of the structure, this was not considered practical. Alternatively, subsurface conditions were explored using a combination of electromagnetic geophysics and 78 test borings. This approach utilized the electromagnetic survey to gain an overall image of the subsurface variability, and to identify potential anomalies. The borings then focused on major foundation loading locations, and on identified anomalies and locations where gaps remained in the exploration plan.

Terrain conductivity (electromagnetic) surveys were performed at the site. Two different conductivity meters were used. The Geonics EM31 is a one-man unit with a fixed separation 3.65 m (12 ft) between the transmitter and receiver coils. The measured apparent conductivity values represent contributions of conductivity primarily from the 4.6 m (5 to 15 ft) depth range. The Geonics EM34XL is a two-man unit. The separation between the receiver and the transmitter coils was set at 20 m (63 ft). With this configuration, the measured apparent conductivity values represent contributions of conductivity primarily from the zero to 15.2 m (50 ft) depth range in the horizontal dipole mode, and 9 to 21 m (30 to 70 ft) in the vertical dipole mode. Several anomalies were observed in these surveys which were indicative of the variable quality of the underlying rock. Fracture zones and poor quality rock areas were detected; however, a large portion of the center of the site was obscured by the influence of a steam line running north-south through the site.

The borings generally ranged in depth from 7.6 to 30.5 m (25 to 100 ft), with the maximum depth of a test boring reaching 39.9 m (131 ft). Within the limits of the main structure, all borings extended to depths where suitable rock was found, should deep foundations have been selected for the project. This suitable rock was defined as Rock Quality Designation, RQD, of 60 percent for at least 3 m (10 ft). Borings also extended to at least 6 m (20 ft) below the anticipated nominal foundation bearing grade. Although most borings were completed in accordance with these requirements, one boring was extended to over 39.6 m (130 ft) to meet this requirement. This included coring over 36.6 m (120 ft) of rock.

The exploration data revealed highly variable rock conditions ranging from near-surface, massive material to highly fractured rock, and deep soil-filled troughs. A review of the rock qualities near the foundation grade indicated that the average rock core recovery (REC) value across the building within about 20 ft (6 m) of the bearing grade was 72%. The range of REC was 0 to 100% in this zone. The Rock Quality Designation (RQD) values ranged from 0 to 93% with an average of 38% near the bearing grade. RQD is a ratio of the summation of the length of rock core segments greater than 100 mm (4 inches) in length to the total length cored. Figure 1

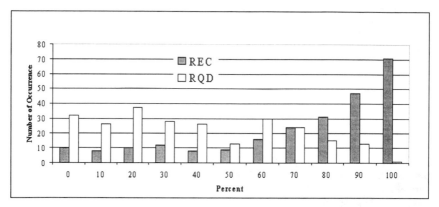

Figure 1. Rock core summary near foundation grade.

shows the frequency of REC and RQD measured within about 6 m (20 ft) of the bearing grades.

The soil samples obtained from the exploration were subjected to physical testing including gradation and Atterberg Limits. The fine grain portions of the samples tested classified as low to moderately high plasticity clay. Within the samples tested, the gradation analysis indicated that the samples had about 70 percent fine grain material (smaller than the No. 200 Standard Sieve).

Point load uniaxial compressive tests and unconfined compression tests were performed on selected samples of limestone and dolomite rock at varying degrees of weathering. These tests indicate that the average compressive strength of the intact rock is on the order of 138 MPa (20,000 psi). Lower values were obtained on several samples generally where vugs were present in the rock core, or the break occurred on a pre-existing fracture plane.

Foundation Alternatives

Foundation design began prior to the field exploration, with a preliminary comparison of drilled shafts and spread footings. As the structure design developed and the field data was obtained, it was confirmed that the drilled shaft system would be prohibitively expensive. The design grades of the building indicated that many of the main foundation elements would be located below the surface of the rock encountered in the borings. Therefore, the use of high bearing capacities that capitalized on the strength of the rock, and limited the size of the foundation elements was desired.

Design of spread footing systems must consider the available bearing capacity of the foundation material as well as total and differential settlement under the design building loads. In the case of karstic geology, the design must also consider the likely subsurface variability that will be encountered during construction and system performance, should features such as sinkholes develop post construction.

The analysis of bearing capacity and settlement of foundations on rock is well documented by many sources. In the case of this structure, reference was made to the guidelines developed by the Pennsylvania Department of Transportation for foundations on rock (PADOT, 1993). A design bearing pressure of 0.6 MPa (12,000 psf) was selected. However, given the variability of the formation detected by the borings, and the critical use of this structure, the rock present at the foundation level was not believed to provide the level of confidence desired. Several steps were taken then to address the potential variability both during construction and to limit post construction risk.

A common procedure to limit the effects of future sinkhole development on a foundation element is to combine multiple individual foundations into a single larger element. Thus, the likely size of the sinkhole will be small relative to the foundation element, and the foundation will span the problem with limited movement. In the case of this structure, this was accomplished by combining the main oval lines of foundations into continuous rings that were designed to take the individual column loads, as well as to span voids that may occur beneath the foundation. A void diameter of about 3 to 4.5 m (10 to 15 ft) was considered when designing the ring. The final design of the ring further provided for stability of the dead load of the structure, if support for one column was lost completely, 18.3 m (60 ft) span.

The second step taken was to require limited mobility grouting (LMG) of the foundation bearing material to improve the consistency and reliability of the formation near the bearing grades. Where rock was not readily present (isolated areas of the building), the LMG was recommended to improve the overlying soil and rock matrix and help transfer the loads to the deeper rock. Thus, eliminating the requirements for undercutting and replacement, or lowering foundation elements to the rock. There were some areas, however, that did require undercutting during final construction, either due to the very poor quality of the subgrade, or to expedite construction when the grouting crews were focused in other areas of the site.

Grouting Design

LMG was selected over a low viscosity material to limit the flow of the grout material within the rock. It has been the experience of the authors that using a low viscosity grout provides little control of the improvement area and final results. LMG involves the injection of a very stiff (low slump) mortar-like grout in stages directly beneath the area of concern. The grout is injected under pressures typically between 0.3 to 3.5 MPa (50 and 500 psi) to densify and displace material immediately adjacent to the point of injection. Due to the stiff nature of the grout, the relative travel of the material is limited. Where soft soils are present, the material often remains in a bulb-like shape around the injection point. Where open voids are encountered, significant volumes of the grout can be injected; however, due to its stiffness, when pressure is stopped, the grout does not continue to flow. This is advantageous since it will leave a mound of grout that future injections can use as a reaction. Thus, closure can be achieved directly below the area of interest at significantly smaller volumes of grout.

This material is not well suited for filling fine fractures in the rock mass. However, these small fractures are not the primary concern for a spread footing support. The grout does, however, create a continuous column through the rock mass, expanding in the larger open or soil-filled fractures, and assisting the rock mass to transfer load past smaller fractures.

Similar improvement occurs in the soil formations where the LMG creates bulbs and columns through the soil mass. The bulbs provide densification and displacement of the soils while creating a composite mass with a net increase in stiffness. This mass then carries increased loads, and transfers that load to the underlying rock, or dissipates it laterally through the improved mass.

The design of the grout considers the spacing of the elements, as well as the anticipated volume of grout required to reach the desired goals. Where the underlying materials are uniform, this would consider the relative starting density, void ratio, and strength, as well as the target strength and stiffness requirements. Several papers have discussed design methods and theories of LMG including Graf (1992), Schmertmann (1992), Warner (1997), and Byle (2000).

However, due to the variability of karst, it is more difficult to determine the spacing and volume requirements. The LMG process in this case is looked at as an investigation as well as improvement tool. As each hole is drilled, valuable information is collected on the underlying conditions based on drill cuttings, rate and equipment performance. Then, as the grout is injected in each stage, the starting pressure, changes in pressure during injection, and final pressures provide information about the stiffness and consistency of the formation. Both drilling and grouting must be monitored. Often where drilling seems to be relatively uniform, significant grout takes occur at isolated locations. Likewise, where drilling indicates soft zones, the grouting process can be modified to address these zones, as appropriate. Where grout takes do not match indicated soft or void zones identified by the initial subsurface exploration and geophysics, additional holes may be required to ensure that suspect problem areas are addressed.

As a starting point in karst, the hole spacing is estimated to provide a reasonable coverage beneath the area to be improved. Hole layout should allow logical split spacing to secondary and tertiary levels. In the case of the main foundation line (D-Ring), the primary holes were located on the inside and outside of the ring at 4.2 m (14 ft) center to center. The inside and outside lines were located 0.6 m (2 ft) inside the edge of the foundation, and 1.8 m (6 ft) between the lines. The two primary grout hole rings were staggered. Secondary holes were required between each primary hole, with tertiary holes located down the center of the ring at about 3 m 10 ft) spacing. The tertiary holes were used for verification. Where the volumes and pressures were not considered suitable, additional holes were added beneath the foundation until suitable results were obtained. Figure 2 depicts the general foundation layout and grout hole locations.

Other areas of the foundation system were handled in similar ways. Smaller rings and wall foundations used this philosophy with modified spacing. Where individual column elements were used, the grout layout was generally 2 to 5 holes completed in a primary and secondary pattern, with a tertiary center hole used for verification.

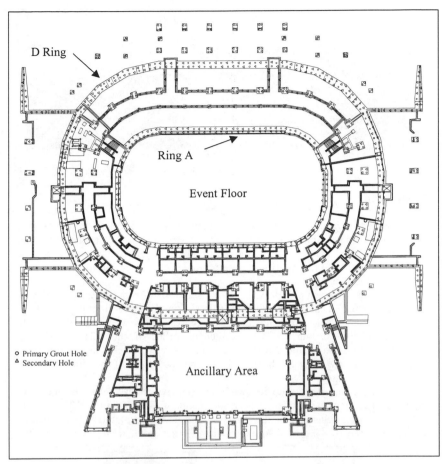

Figure 2. Foundation plan and grout hole layout.

The depths of the grout holes were based on the stress zone of the spread footings, and the conditions encountered during drilling. Once the base depth (1.5 times the maximum width of the foundation element) was reached, the drilling was terminated when at least 0.6 m (2 ft) of competent rock were present at the bottom of the hole.

Where lightly loaded foundations were present or significant depths of soil were expected, a reduced bearing pressure of 0.2 MPa (4000 psf) was utilized on LMG improved subgrade materials.

Figure 3. LMG operation.

Construction

A basic pattern of drilling and grouting was planned and completed beneath all new foundation elements. Hole layout patterns for individual foundation elements were based on the size of the element and ranged from 2 to 5 holes per footing, and a net spacing of about 1.5 m (5 ft). Drilling and grouting each of these locations were used as a means of further exploration, Figure 3.

Drilling was completed with air powered drill rigs utilizing top drive percussive drilling techniques. Grout holes extend from the top of foundation grade to at least the nominal design depths. In many cases, holes were extended due to unsuitable conditions, with the maximum depth drilled of about 21.3 m (70 ft). Following completion of the drilling, a 50 mm (2 inch) steel casing from grout injection was inserted into the hole, and the annulus space filled with stone dust where necessary.

Grouting was completed with swing tube positive displacement piston pumps. The rated capacity of each stroke of the piston ranged from 0.006 to 0.018 m^3 (0.2 to 0.65 ft^3) per stroke depending on the pump being used. Pumping rates were generally held to less than 0.14 m^3 (5 ft^3)/minute where larger grout takes were occurring.

GROUTING AND GROUND TREATMENT 987

Where higher pressures and lower volumes of grout were being injected, the rate of pumping was less than 0.06 m³ (2 ft³)/minute, and often less than 0.03 m³ (1 ft³)/minute. Figure 4 is an example of the typical grout injection setup on this site.

Monitoring of drilling resistance, and grout pressure and volume were direct indications of the consistency of the material beneath the foundation element. This monitoring was accomplished by experienced full time field personnel under the direction of the Design Geotechnical Engineer. Based on this data, tertiary holes and subsequent split spacing were completed in many areas to develop "closure." Completion of each grout stage was defined using the criteria outlined in Table 1.

Table 1. Grout stage refusal criteria.

Closure Criterion	Limit
Grout Flow Ceased	Maximum Pressure 3.5 MPa (500 psi)
Ground Surface Movement	2.5 mm (0.1 inch)
Injected Volume/Pressure	
Grout Slump = 50-75 mm (2-3 inches)	0.4 m³ and >0.7 MPa (15 cf and >100 psi)
Grout Slump = 25-50 (mm 1-2 inches)	0.6 m³ and >0.55 MPa (20 cf and >80 psi)
Grout Slump = <25 mm (<1 inch)	1.1 m³ or >0.55 MPa (40 cf or >80 psi)

A total of 800 m (31,527 linear feet) of drilling was completed for grouting. A total volume of LMG of 2288 m³ (2,992 yd³) was injected in these holes, resulting in an average grout injection rate of 0.07 m³ (2.6 ft³) per linear foot of injected hole. As would be expected, the actual injection ranged from "no take" to the maximum allowed in a stage. No large open caverns were encountered in this program.

A closer look at the grouting records of the primary, secondary, and tertiary/subsequent holes for the D-Ring indicates that the grout take per foot of hole decreased steadily until closure was reached. For the primary holes, the average take was 0.12m³/300 mm (4.3 ft³/ft); and 0.07m³/300 mm (2.4 ft³/ft) for the secondary holes. The final holes were completed with an average of 0.5m³/300 mm (1.8 ft³/ft). This is as would be expected as the intended closure was being achieved. Table 2 provides a detailed breakdown of the grout results across the site.

A finite definition of "closure" was not formulated for the project, given the highly variable conditions at the site. The Geotechnical Engineer's discretion was utilized to evaluate when an area or foundation was complete. This was based on a review of the conditions encountered during drilling and grouting, the relative improvement observed between the primary, secondary and tertiary holes in the area, the overall grout volume injected, and resulting pressures. Conducting such a review occurred regularly on the site and could only be completed by relying on the detailed and accurate records of the engineer's field representatives.

Table 2. Site LMG grout summary.

Location	D-Primary	D-Secondary	D-Proof	C- Footing	C-Wall	B-Footing	B-Wall	A-Ring	Auxiliary Area	Sinkholes	Total
No Holes	180	178	179	186	55	155	38	295	868	7	2141
Hole Depth, m	5.2	5.3	5.8	5.2	4.1	5	2.8	4	3.9	7.6	4.6
Total Footage, m	933	947	1031	960	224	775	108	1190	3386	54	9609
Total Grout, m³	373	209	175	262	73	238	30	176	738	13	2288
Grout Take, m³/300 mm	0.12	0.07	0.05	0.08	0.10	0.09	0.08	0.05	0.07	0.07	0.07

Further evaluation of the data in relation to the rock cores obtained can provide some measure of comparison for the effective impact of the grout injections. Considering that the average void space of the rock was equivalent to the material not recovered by the coring process, then the grout would have to replace approximately 28% of the volume beneath the foundations (average recovery was about 72% within 4.6 m (15 ft) of the bottom of the foundations). Looking only at the D-Ring of foundations, these numbers would be slightly different, with an average recovery of 66%, and an assumed void space of 34%. Although the assumed void space inferred in this method is far from accurate, it provides a means of comparison when evaluating the grout takes.

The area of the D-Ring was about 1161 m² (12,500 ft²). With an average depth of grouting of about 5.3 m (17.5 ft), the total volume of the treated area is about 6200 m³ (219,000 ft³). Given the total grout volume injected beneath the D-Ring of 759 m³ (26,800 ft³), this would indicate that about 12% of the void space was filled. Therefore, only about one third of the available space assumed above from rock core information (34%) was filled.

An alternative means of evaluating these results may be to determine a minimum equivalent radius of the grout elements, assuming that the grout is fully effective at reaching the available void space. Therefore, the equivalent radius could be calculated as:

$$radius = \sqrt{\frac{T_{avg}}{(1 - REC)\pi}}$$

where T_{avg} = average grout take, REC = rock core recovery.

Interestingly, utilizing this method, when evaluating the aggregate of the D-Ring holes as well as total site numbers, both indicate that the equivalent minimum

Figure 5. D-Ring foundation excavation. (Inset - reinforcing steel mat.)

radius was about 0.5 m (1.7 ft), or an effective diameter of at least 1 m (3.4 ft). Given that the final hole spacing was less than about 1.5 to 2 m (5 to 7 ft), and it is highly unlikely that we were able to reach all of the available void spaces (1-REC) with this grout, it may be inferred that the majority of the area beneath the foundations was treated with this hole spacing and refusal criteria.

As foundation work was being completed in areas, construction of the footings and superstructure began. In general, this proceeded with little incidence. Scheduling of the general construction was modified to accommodate the grouting process, as was the grouting process adjusted and accelerated in some instances to meet the general construction needs (Figure 5).

Construction of the utilities in the area of the bowl presented a significant challenge. Many of the utilities for the structure, as well as a major steam line for the University, traversed the event floor area, and this had to be constructed within the rock foundation materials. Due to concerns with the damage that may occur to completed grout elements (particularly in soil), strict blasting controls were imposed on the contractor should he elect to blast in the building. This created difficulties and delays for the foundation and utility excavation.

The project was bid based on a unit rate contract with estimated quantities. The base bid for the project grouting was \$665,000 with add/deduct rates obtained at the start of the grouting of $13/m^3$ (\$10/yd^3) of grout, \$18/m (\$5.50/ft) additional drilling, and \$8.20/m (\$2.50/ft) reduced drilling. Given these numbers, it was evident that the contractor anticipated that the total grout volume would not reach the contract volume. More typical grout costs experienced by the authors would be on the order of \$265 to $325/m^3$ (\$200 to $250/yd^3$) on a large production project.

Conclusions

This project was completed successfully because the design team focused on site understanding, and gathering proper field data: performing a thorough evaluation of the needs of the structure, looking beyond the traditional foundations used in the area, and providing necessary field observation under the direct review of the Geotechnical Engineer during construction.

Since this project was completed in 1995, the Convocation Center has been successfully utilized for a variety of campus functions including concerts, sporting events, and graduations. Similar ground improvement was used for the library expansion on the campus, which is also performing well. The LMG applications in karst have been proven to effectively improve sites to the point where shallow spread footings can be used even for high sensitivity structures such as the Bryce Jordan Center.

Acknowledgements

The authors would like to recognize the design team of Haas/Rosser Fabrap/Brinjac Kambic Joint Venture. The grouting contractor on the project was Hydro Group.

References

Byle, Michael J. (2000). "An Approach to the Design of Limited Mobility Displacement (LMD) Grouting," Advances in Grout and Ground Modification, ASCE, Denver, CO, p. 94.

Geyer, Alan R., and Wilshusen, Peter J. (1982). Engineering Characteristics of Rocks of Pennsylvania, Pennsylvania Geological Survey, Environmental Geology Report No. 1, Second Edition.

Graf, E. (1992). "Compaction Grout," Grouting, Soil Improvement & Geosynthetics, ASCE, New Orleans, LA, p. 275.

Pennsylvania Department of Transportation (PADOT) Design Manual Part 5. (1993). PDT-Pub No. 15, Table 4.4.8.1.2, pg B.4-16.

Schmertmann, J.H., and Henry, J.F. (1992). "A Design Theory for Compaction Grouting," Grouting, Soil Improvement & Geosynthetics, ASCE, New Orleans, LA, p. 215.

Warner, J. (1997). Compaction Grouting Mechanism - What do We Know," Grouting: Compaction, Remediation & Testing, ASCE Geo-Logan '97, Logan, UT, p. 1.

Using the Grouting Intensity Number (GIN) to Assess Compaction Grouting Performance

By Steven W. Perkins[1] and Joe Harris[2]

1 Introduction

Compaction Grouting is a method of ground improvement using the injection of low slump (<25 mm) soil cement into weak soils to compact the soil by lateral displacement. This technique, as practiced by the Contractor, uses a 50-mm diameter steel casing driven to the bottom of the weak zone and withdrawn in 300-mm stages. Each stage receives a specified volume of grout, typically 145l/0.3 m (5 cf/lf), which creates a bulb of grout that locally compacts the soil to some radius around the hole. Compaction Grouting has been discussed in detailed by Warner (1982) and others. This paper describes the development of the GIN concept's application to compaction grouting of loose sand embankments.

The Grouting Intensity Number (GIN) concept has been previously used in consolidation grouting to standardize the level of grouting effort conducted at each grouted hole. On this project the GIN concept was applied to compaction grouting to provide real-time quality control (QC) to ensure that the construction work was being successfully completed. This approach reduced the contractor's risk of failure to meet the design requirements for no increase in construction cost or schedule. QC for acceptance testing was provided by periodic test borings using conventional Standard Penetration Test (SPT) at four progress milestones. The work was executed in US customary and has been converted to SI units for this presentation.

The Consumers Energy (Consumers) owned and operated Croton Dam on the Muskegon River was constructed in 1906. Hydraulic fill placement methods used during construction created zones of loose saturated sand adjacent to the powerhouse and other structures. Analysis of the left earth embankment found potential for liquefaction during the design earthquake. The site is generally uniform sand and gravel placed at a loose density over a dense original ground. In situ soil density varied from very loose (WOR) to medium density (Dr = 40%) in the embankment while the density of the original ground was very dense. The post construction results were uniformly medium dense sand (40%<Dr<60%) reflecting the rapidly increasing incremental difficulty of compacting sand as the density increases. Therefore, the same level of effort will increase the density of loose sand more than medium dense sand leading to more uniform results at completion.

Based on the analysis of test panel grouting results, the GIN concept was applied to provide a real-time method to monitor the degree of compaction achieved for each 300-mm (1-ft) stage of each grouted hole. The concept was later expanded to cover additional conditions as more data was gathered. This method provided QC analogous to the slump test used for concrete placement (i.e. indicative and immediate rather than precise and after the fact). This method was used in combination with conventional in situ test methods such as SPT or cone penetrometer.

[1]Geotechnical Engineer, Acres International Corp, 140 JJ Audubon Parkway, Amherst, NY 14228-1180
[2]Project Manager, Denver Grouting – A Division of Hayward Baker, 11575 Wadsworth Blvd. Broomfield, CO 80020-2752

The conceptual justification for this empirical method is based on doing a consistent minimum level of work effort to each volume of soil being compacted and then correlating it to the required density. This approach is similar to Proctor density test methods where a specified amount of compaction effort would compact a specific soil to a predictable density assuming the soil starts from a loose condition.

The conventional GIN approach was modified for compaction grouting as follows. The product of the peak injection pressure (psi) and the injected volume (cf/lf) divided by square of the hole separation distance was used to predict the SPT "N" value taken at the center of the four surrounding holes. This approach treats each grouted hole as a "test" hole. Testing for acceptance was based on the standard SPT "N" value measurements (uncorrected for overburden) conducted by an independent testing firm. A soil constant, C, correction factor and US customary units matching the Contractor's equipment were used so that the GIN and the specified N value were of similar numeric value for ease of field use.

The development of the GIN concept follows.

2   Test Panel

The test panel selected by the contractor was 9 meters by 12 meters and had 19 grouted holes at 2.4-m centers on a square grid (Figure 1). This panel included slightly more than 5% of the work area. Two test borings, B-1 and B-2, were drilled inside the test area to document the initial conditions and to confirm that a representative area had been selected. The area was then grouted using the criteria selected by the Contractor. On completion of the grouting, three confirmation borings, B-3, B-4 and B-5 were drilled within the test area and as far from the grouted holes as possible to measure the areas likely to have the least compaction. Testing results indicated that adequate compaction was achieved in the test area and only minor changes to the grouting program were required to compact the loosest areas.

3   Construction Method

The contractor's initial methodology for construction was based on their experience at similar sites. The grout holes were spaced at 2.4-meter centers in a square grid (Figure 1). 50-mm diameter (ID) steel pipe casing was driven through the loose sand of the embankment approximately 900-mm into the original ground; the interface was obvious as the driving difficulty increased substantially in the original ground. The casing was kept free of soil by a 60-degree conical end cap that was knocked out of the end prior to grouting and abandoned at the base of the hole. The hole was then grouted from the bottom up in 300-mm stages.

The grout mix used for the test panel was primarily a blend of locally available soils proportioned to create sandy silt of very low plasticity; about 15% round fine gravel were then added. The mix was screened on site to remove any coarse material greater than 10-mm. Cement and water were added in the concrete batch truck to create a low slump (<25-mm), non-plastic grout. The grout was then pumped to the individual holes. The truck operator adjusted the water content based on a visual assessment of the material; adjustments were minor since the source material was uniform and maintained in dry condition.

The crew foreman remotely controlled the pumping rate, withdrew the casing and recorded the data controlled the grout injection at the header. Grout injection pressure was read directly from a 100-mm diameter gage mounted at the header. The quantity of grout injected was measured by counting the strokes of the positive displacement pump; each stroke delivered ~7 liters (0.25 cf) of material. The stroke on the jack that withdrew the casing was 300 mm and the casing sections were 900 mm long, which aided in record keeping.

The contractor initially selected the following procedures based on their previous experience: the grout take limit was 145 liters per 300 mm (5 cf/lf) of hole and a maximum injection pressure of 5.5 MPa (800 psig) below 6 meters and 4 MPa (600 psig) to 3 meters. Grouting from 3 meters to 2 meters was subject to a limit pressure of 2.5 MPa; grouting above 2 meters depth was conducted at a nominal pressure sufficient to fill the hole to minimize ground heave. The original ground typically accepted only a nominal volume of grout before the limit pressure was reached while the embankment soils reached the volume limit prior to developing the maximum allowable pressure. This procedure was modified during construction as the GIN concept was developed.

Hole 1 was grouted first. Grouting was then conducted sequentially through Hole 19. This approach was used to provide confinement for subsequent holes by grouting on a 4.8-meter spacing and then split space grouting the remaining holes. The success of this approach became evident when the contractor drove the casings for the intermediate holes; the drive rate for the primary hole casings was approximately 1-2 minutes per 900 mm segment while the rate for the secondary holes was approximately 3-5 minutes per segment. Drive rates in the original ground were typically 15 minutes per section and were not noticeably affected by grouting. Grouting and driving adjacent holes was not attempted during the test program as it was the contractor's experience that this caused the casing to bind and slowed the drive rate.

4   Test Panel Results

The compaction grouting was successful below 2 meters but had little effect near the ground surface (Figure 2). Figure 2 shows the comparison of the initial conditions (B-1, B-2) to the final conditions (B-3 to B-5) and the specified requirements. Only 2 of 40 "N" values within the compaction zone (below 2 meters) failed to meet the specified criteria; these two failures were close to the acceptance criteria (N=13 vs. 15) and the probable cause was addressed in the production grouting program.

The compaction grouting's effect on the soil density was two fold; the soil became denser as indicated by the increased "N" values and the density became more uniform throughout the soil column in the compaction zone based on the work presented by Bowles (Figure 3). The less dense sands compacted more easily than did the denser layers resulting in greater increases in relative density of the loose areas than in the dense layers. This reduced the overall range of the density, and made the embankment both denser and more uniform than its initial condition (Figure 4).

A quantitative measurement of the level of grouting effort was developed from the test panel results to provide a real-time assessment of the grouting program based on the volume of grout injected and the maximum injection pressure. For purposes of this discussion, this measurement is referred to as the Grouting Intensity Number (GIN).

5   The GIN Concept

The GIN concept was devised as a rational method to correlate grouting effort with soil density, as measured by the Standard Penetration Test (SPT) (ASTM D-1586) using US customary units. The GIN is defined as follows:

GIN = Take x Pressure /(C x Distance^2)   Where:
                                            Take   = grout take (cf.)
                                            Pressure = Peak injection pressure (psi)
                                            Distance = center to center distance
                                            between holes (ft)
                                            C = 0.75; Site specific constant.

The numeric values of the GIN are similar to the "N" values from the SPT for the embankment sands tested. This relationship does not apply to the original ground "N" values, as these soils are initially dense, yielding low GIN values but high "N" values (Figure 2). This relationship is based on data from square grids ranging from 2.4 meters down to 1.2 meters which covers the range likely to be used in practice. The constant of 0.75 in the denominator of the GIN value is used to empirically account for the effect of head losses, soil parameters, rationalize the measurement units and other unmeasured factors. The effect of the grout bulb diameter is expected to be significant at small radii (<1 meter) but the present data are insufficient to resolve this issue.

Development of the GIN Concept

The sequence of hole grouting in the test panel created a grouting pattern where the initial holes were essentially uninfluenced by earlier grouting due the hole separation (~4.8 meters). As the grouting progressed, each successive hole grouted became more influenced and the grouting effort confined by the work that had been done previously. To quantify this observed effect a rating scale was developed to group holes with similar degrees of confinement by neighboring holes. Holes in a square grid pattern have eight holes adjacent to them; this allows each hole to receive a *neighbor rating* percentage based on the percentage of the hole's neighbors that were grouted previous to its grouting. After some trial and error analysis, the holes were divided into three groups (Figure 5). Holes in Group 1 had the least confinement by neighbors while holes in Group 3 were typically confined on at least part of three sides. The effect of this confinement on the GIN can be seen in Figures 6, 7 and 8 which show increasing correlation between GIN and "N" as the degree of confinement approaches the conditions under which the SPT testing occurred. The Group 3 configuration is the condition that was actually tested as the testing was conducted after all the holes were completed.

The low level of confinement provided by adjacent holes to group one holes may contribute to the high degree of scatter in the GIN value range for these holes (Figure 6). Group 2 holes show a similar pattern of variation (Figure 7). The greater confinement of the Group 3 holes may result in less scatter of the GIN values (Figure 8) as the soil density increases and grouting pressures rise for the same volume of grout injected. This trend toward higher and more uniform GIN values for the Group 3 holes is of interest because it is possible to arrange the order of the production grouting holes such that nearly half of the production grouting holes meet the Group 3 confinement criteria and can be used as an indicator of the minimum level of compaction achieved. A strong correlation between the GIN value of the Group 3 holes and the SPT test results in the verification borings will allow extrapolation between borings and greater confidence that proper compaction is being achieved.

There were four possible combinations of grout take and peak pressure ranges and resulting GIN values that were likely to occur at site. These combinations were:

-Low grout take (< 30 liters/300 mm) at a high injection pressure (>5.5 MPa) which typically occurred in the dense original ground;

-Moderate grout take (~ 145 liters/300 mm) at moderate pressure (2.7 to 4 MPa) which typically occurred in the loose sand (20% - 40% relative density). The Test Panel and most of the site were in this category;

-Moderate grout take (~145 liters/300 mm) at low pressure (0.7 to 1.4 MPa) which typically occurred in loose to very loose sand (< 40% relative density) in areas with a high water table (low effective overburden stress). The Remedial Grouting area at the toe of the dam was in this category;

-High grout take (200 to 350 liter/300 mm) at low pressure (~1.4 MPa) which typically occurred in very loose sand (0 - 20% relative density) with significant overburden pressure. The area adjacent to B-3-89 at the top of the dike near the corewall and powerhouse was in this category.

Each of these conditions except the grouting of the original ground is discussed below.

GIN and "N" value Correlation in the Test Panel

The relationship between the "N" value of each of the confirmation borings and the GIN of the adjacent grouted holes was determined. The relationship between the GIN and the measured "N" value was strongest for the Group 3 holes.

Confirmation boring B-3 and its adjacent grouted holes are shown in Figure 9. All the SPT "N" values within the compaction zone exceed the specified requirements. This data set has two Group 3 holes diagonally opposite to each other and the overall level of grout effort as defined by the GIN is high. The major exception is in Hole 9 between El 692 and 696 where the maximum injection pressure was reached with only nominal grout take yielding a low GIN number.

Confirmation boring B-4 was located on one of the grouting lines so the nearest six grouted holes are presented in Figure 10. The location of B-4 was chosen along a grouting line instead of in the center of a square to examine if there was any visible effect. Any difference from the other two borings is not readily apparent. The overall level of grouting effort was high with the GIN generally greater than the required "N" value. All the SPT results were above the specified requirement.

The location of confirmation boring B-5 was selected to be between the set of four grouted holes with the lowest average group number and where the GIN appeared to be the lowest to test the GIN concept. Figure 11 presents the GIN values for each of the grouted holes as well as the average GIN for each elevation. Low GIN values for Hole 8 was primarily due to low pressure while the low GIN values for Hole 13 were due primarily to low grout take; these areas of low GIN values correspond to the regions of the compaction zone that did not achieve the specified "N" value albeit by only 2 blows each (13 vs. 15). This leads to the observation that additional grouting effort (either higher pressure or more volume) in Holes 8 and 13 would have compacted these zones to within the specifications.

GIN and "N" value Relationship in the Remedial Area

It was observed that the GIN values and peak injection pressures were unusually low in the portion of the dam toe below Pz #3 (Figures 12, 13 & 14) indicating that this area would not have acceptable "N" values when tested. The low peak injection pressure was due the ground heaving at relatively low pressure; this tendency to heave at low injection pressure was attributed to low effective confining pressure due to the high water table (~1 m). This suspicion was confirmed by test boring B-7-98 in which all the test samples in the compaction zone failed to meet the required specifications. A successful remedial grouting program was conducted by the Contractor after assessment of the production grouting program to determine the extent of the problem area using the GIN concept as discussed below.

Assessment of the Remedial Area

The production grouting phase of the remedial area (Figure 12) was characterized by distinctly poor compaction results (interpreted from low GIN values and peak pressure readings) for 13 of the holes. Three typical holes are shown in Figure 15 and there were six additional holes with marginally poor compaction (Figure 14 is typical). The remedial area was surrounded by well-compacted holes (Figures 15, 16 and 17) typical of the test panel and other grouted areas of the site. The extent of the remedial area was initially defined by the examination of the grouting logs and later confirmed by test borings 9 and 9A. A remedial grouting program was then developed and conducted based on the assumption that the standard 2.4 meter square grid was too coarse to provide sufficient overlapping radii of compaction at the center points between the holes. This interpretation was based on the observation that only limited compaction was measured at the test point (B-7-98) when compared with the pre-construction test borings (Figure 18).

Figure 18 shows that measured relative density prior to construction was very loose (<20%) to loose (20% to 40%) for most of the area; the production grouting program marginally improved the relative density to a consistently loose state. The relative density further improved after the remedial grouting (as measured by B-8-98) to a uniform medium dense compaction (40% to 60%) at or slightly above the specified requirements.

Remedial Grouting Phase

The remedial grouting program consisted of 20 additional grouted holes on a 2.4 meter grid split spaced around the poorly compacted holes (Figure 12). The grouting criterion for the remedial holes was the same as for the production holes.

The GIN values for the remedial holes were higher than the values for poorly compacted holes adjacent to them but were typically lower than the well-compacted holes outside of the remedial area. The remedial area was expected to be sufficiently compacted because the combined effects of the production and remedial grouting programs provided nominally twice the compaction effort as the production grouting alone and the center to center hole separation was approximately half the production distance. Test Boring B-8-98 was consequently closer to the sources of compaction than the other test borings. The effect of the remedial grouting program is shown in Figures 19 and 20.

Test Boring B-8-98 was drilled to confirm that adequate compaction had been achieved in the test area. All but one of the "N" values in the compaction zone were within the specified requirements; the one test that was lower than the specifications was taken at the interface with the original ground. Two additional holes, B-9-98 and B-9A-98, were drilled in the area adjacent to the remedial area to confirm that the extent of the remedial work was correct (Figure 17). These test borings had acceptable "N" values throughout the compaction zone and the GIN values of the adjacent holes were typical of well-compacted areas in other parts of the site.

7   Project Use of the GIN Concept to Reduce Testing Failures

The area adjacent to the core wall and the powerhouse near pre-construction borings B-3-89 and B-7-90 (Figure 1) was constructed of very loose sand (N=WOR) and compacting this area to achieve the specified compaction was expected to be more difficult than for other areas of the site. The GIN values for this area initially achieve low values but rose to acceptable values when the grouting was continued until the GIN value was in the expected range. The Contractor was able to pump up to 350 liters / 300 mm of grout into this zone as the effective overburden stress was much higher than at the toe due to the lower water table (~ 8 m). The decision to continue pumping at any stage was made during that stage by the grouting technician at the header. This area was subsequently tested (B-11-98) and nearly all the "N" values were above the specified requirements; the low "N" value at 13.5 meters does not fit the pattern (Figure 21).

8   Use of the GIN Method

Use the following steps to apply the GIN concept to compaction grouting:

1) Determine that the existing site conditions are loose cohesionless soils (sand and silts) more than 2 meters below the ground surface. Identify the required degree of compaction to be achieved by the compaction grouting program in terms of SPT "N" values (Dr < 60%).
2) Select a representative area of the project site for a test panel and conduct the planned grouting program. Analyze the data to determine the local soil constant "C" and assess the hole spacing.
3) Adjust the hole spacing and grouting limit pressures and volume if required.
4) Monitor the GIN values during production grouting and authorize the grouting foreman to increase grouting effort at specific stages where the GIN value predicts low future N values.
5) Prediction is more accurate for grout holes with higher *neighbor ratings,* which occurs as the grid is filled (Figure 8).

9   Conclusions

The Croton site had the advantage of being uniform sand and gravel and was sufficiently large (338 plus holes) that considerable data was available. While the in situ soil density varied from very loose (WOR) to medium density (Dr = 40%), the final results were uniformly medium dense (40%<Dr<60%) reflecting the rapidly increasing incremental difficulty compacting sand as the density increases. The same level of effort will increase the density of loose sand more than medium dense sand leading to more uniform results at completion.

The intent of GIN concept is that it can be used as a "real-time" indicator of compaction success for holes at various grid spacing (1.4 m to 2.4 m) when an initial calibration test panel is constructed to determine the value of "C". It is analogous to the concrete slump test which is used as a predictor of the concrete's future strength based roughly on a specific mix's slump; in both cases, the intent is to provide a minimum threshold value above which the material will behave acceptably. The GIN concept can also be used to identify and delineate insufficiently compacted areas within a site to minimize the extent of remedial grouting and re-testing should the primary compaction grid be too coarse for some areas. These two uses combined should allow for fewer test failures, more efficient remedial grouting and greater confidence that the compaction grouting program has achieved the intent of the work as it allows interpretation of the compaction between soil verification borings. Verification borings can located in more convenient locations if soil conditions hole to hole can be interpreted with confidence.

References:

Bowles, Joseph E., <u>Foundation Analysis and Design</u>, 4th Edition, 1988. Table 3-4.

Warner, James, "Compaction Grouting - The First Thirty Years", <u>Grouting in Geotechnical Engineering</u>, ASCE, Feb 10-12, 1982. p 694 - 707.

Acknowledgements:

1) We would like to thank Consumer's Energy Company for allowing the data to be published.
2) We would also like to thank Denver Grouting of Denver, CO for their assistance as the contractor with the execution of this work.

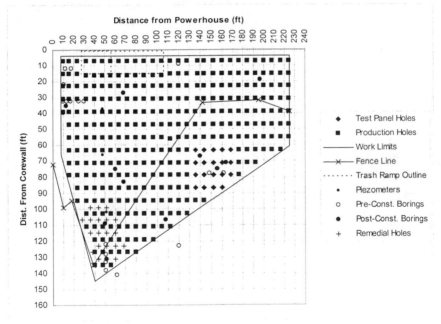

Figure 1: Locations of Grouted Holes at Completion

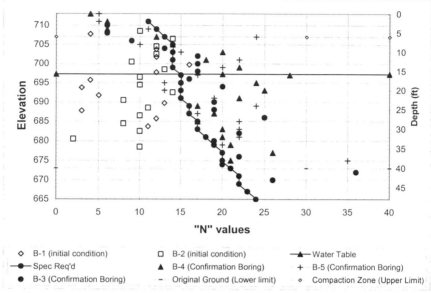

Figure 2: Test Panel SPT Results Comparing Initial Conditions to Final

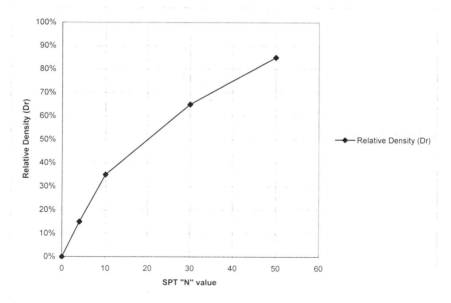

Figure 3: SPT Values vs Soil Density (Ref. Bowles)

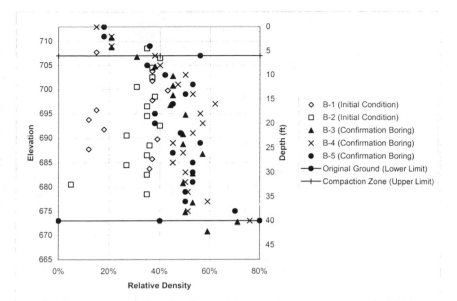

Figure 4: Test Panel Density comparing Initial Conditions with Final after Compaction Grouting showing both increased density and uniformity

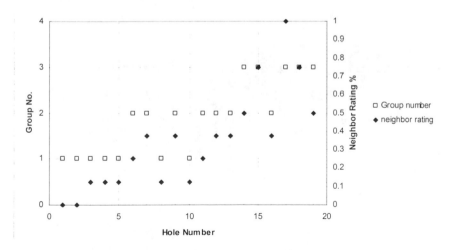

Figure 5: Neighbor Rating Number for Test Panel Grout Holes for grouping holes with similar confinement conditions

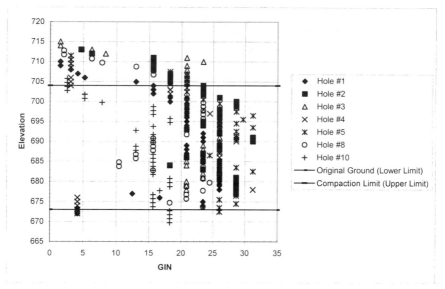

Figure 6: Test Panel; Group 1 Compaction Grouted Holes showing relatively higher scatter of GIN values with Depth in the Compaction Zone

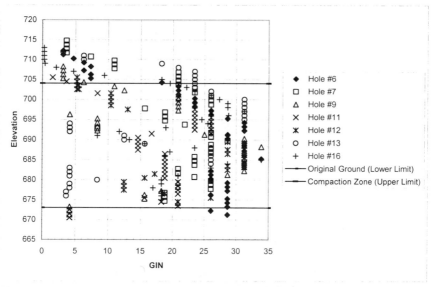

Figure 7: Test Panel; Group 2 Compaction Grouting Holes showing high scatter of GIN with Depth in the Compaction Zone

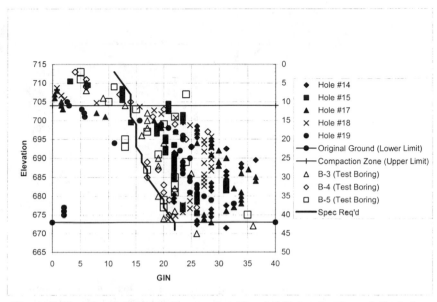

Figure 8: Test Panel; Group 3 Compaction Grouted Holes showing relatively low scatter of GIN values with Depth in the Compaction Zone

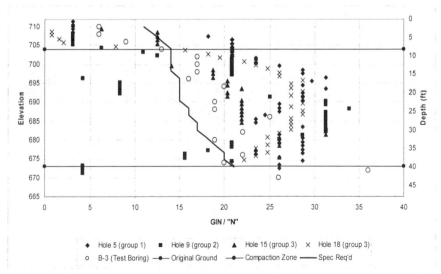

Figure 9: Compaction Grouted Holes Adjacent to Confirmation Boring B-3 showing relationship between GIN, "N" and specified requirements by Hole Group Number

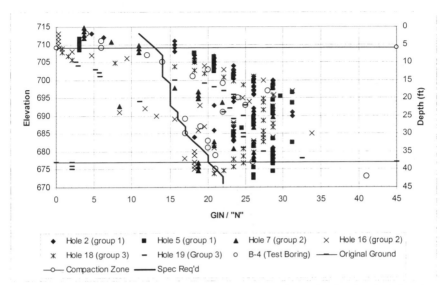

Figure 10: Compaction Grouted Holes Adjacent Confirmation Boring B-4 showing relationship between GIN, "N" and specified requirements by Hole Group Number

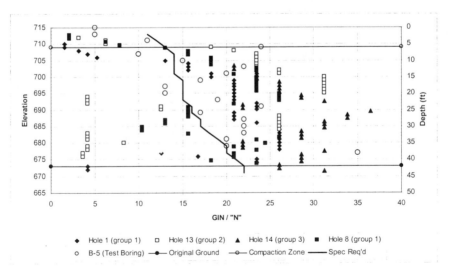

Figure 11: Compaction Grouted Holes Adjacent Confirmation Boring B-5 showing relationship between GIN, "N" and specified requirements by Hole Group Number

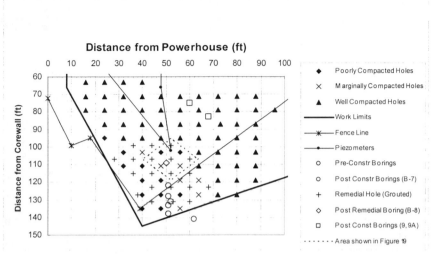

Figure 12: Remedial Compaction Grouting Area at toe of Dam. GIN concept used to define limits of Remedial Work Area based on GIN value relationship to "N" for comparison to specified requirements

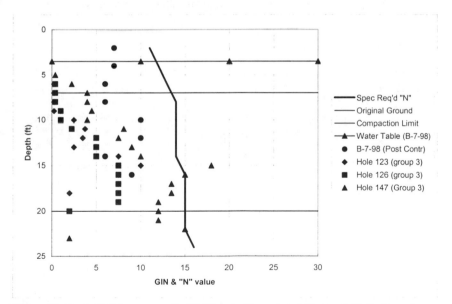

Figure 13: Remedial Area near Pz #3 (Poorly Compacted Group 3 Holes have low GIN values that predict low "N" values)

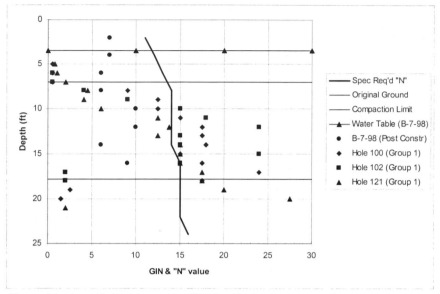

Figure 14: Remedial Area near Pz #3 (Marginally Compacted Group 1 Holes have GIN values that predict marginal but failing "N" values)

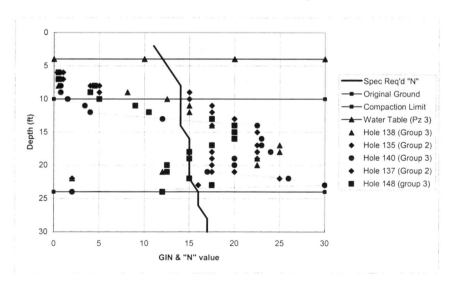

Figure 15: Remedial Area near Pz #3 (Well Compacted holes; Row 95 showing limit of poorly compacted area) Row 95 beyond limit of Remedial Grouting Area

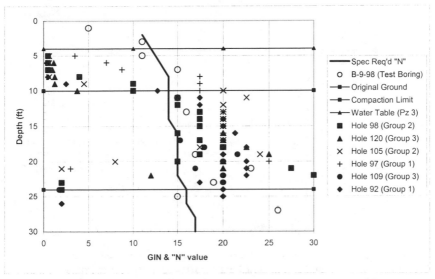

Figure 16: Remedial Area near Pz #3 (Well Compacted Holes in all Groups;
Columns 64, 72 beyond limit of Remedial Grouting Area)

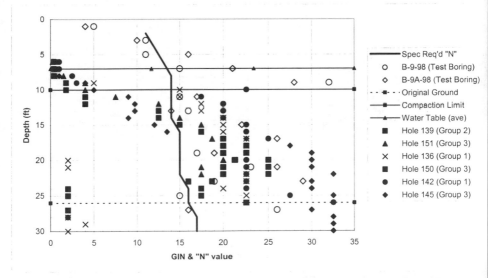

Figure 17: Remedial Area near Pz #3 (Well Compacted Holes in all Groups; Row 87
beyond limit of Remedial Grouting Area)

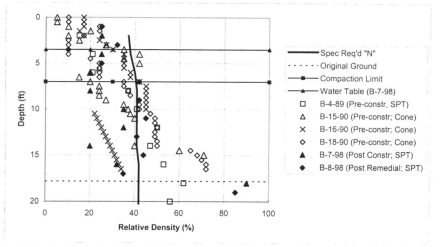

Figure 18: Density History in Remedial Area showing increase in density from Initial Pre-construction Borings and Post Construction Boring to Post Remedial Boring and increase in uniformity of density with depth

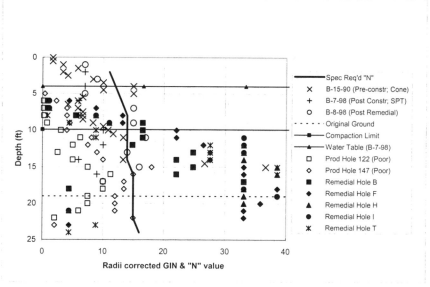

Figure 19: Effect of Remedial Grouting on "N" and GIN values showing low initial GIN and "N" values compared with higher Remedial Hole GIN values and "N" values

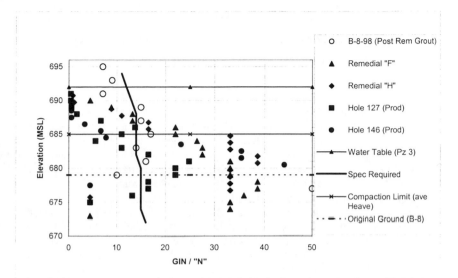

Figure 20: Post Remedial Grouting GIN values predict acceptable "N" values within compaction zone

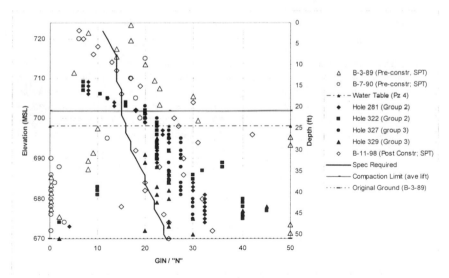

Figure 21: GIN Concept used to avoid Testing Failure in Initially Loose Area near Powerhouse by increasing grout take in Grouted Holes to achieve Acceptable Test Results

Case History:

Broadcast Tower Anchor Stabilization, Portland, Oregon

Rajiv Ali, PhD[1], Associate Member, ASCE, and Jeffrey Geraci[2], Member, ASCE

ABSTRACT

On Superbowl Sunday 2000 an extreme windstorm with sustained gusts of 80 km/hr induced random and potentially damaging cable motions in the supporting cables for a 350 m high, 680 tonne broadcast tower. Observed amplitudes in the 50 mm diameter support cables were reported to be on the order of up to 10 m. The tower is supported at the base on a spread footing and by three cable anchor systems placed 120 degrees apart. Relatively high moisture contents in the soft silt and clay were encountered within the foundation zone for the piles at the northeast anchor. Similar conditions were observed in the backfill soils covering the deadman anchor on the south. A simple pullout analysis indicated that the northeast and south anchor would fail under the design loading. Selection of the appropriate remedial option was primarily driven by the need for uninterrupted broadcast operations. The support cables and anchors are essential for tower support, and could not be dismantled without disassembling the tower itself. Consequently, consideration of remedial options required the ability to effect repairs without disturbance to the tower support system. A remedial system consisting of a precise compaction grouting pattern for *in-situ* ground improvement combined with load transfer by pin piles to bedrock was implemented at the northeast anchor. Compaction grouting was used at the south deadman anchor to add weight and increase soil strength properties within the active earth fill wedge.

INTRODUCTION

The KPDX Fox 49 (FOX) Transmission Tower is located in the Portland Hills on the northeasterly side of NW Miller Road in Portland, Oregon. This 350 m high transmission tower was constructed in 1983 and is supported on a 5 m square concrete mat foundation. Five 50 mm diameter support cables on each of three sides project to anchor stanchions at 120 degrees apart. Figure 1 shows the locations of the tower and support anchors superimposed on the local site topography. The northwest and northeast anchors are each supported by three 1.5 m diameter drilled piers, 10.5 m deep, including a 1.8 m thick by 6.0 m long pile cap at top. The south anchor is supported by a concrete deadman, 10.5 m long by 3.6 m wide by 1.8 m thick, installed at a depth of 5.5 m. The vertical load on the tower is on the order of 680 tonnes. Design loads imposed on each support anchor are 260 tonnes uplift and 150 tonnes lateral.

The support cables were originally tensioned in 1983 and were retensioned in 1992. Unusual movements were observed in these cables beginning in about the summer of 1998.

----

[1]Associate, GeoDesign, Inc., 14945 SW Sequoia Parkway, Suite 170, Portland, OR 97224; phone 503-968-8787; rali@geodesigninc.com
[2]Project Manager, Moore & Taber, 1290 North Hancock Street, Suite 202, Anaheim, CA 92807; phone 714-779-0681; j.geraci@mooreandtaber.com

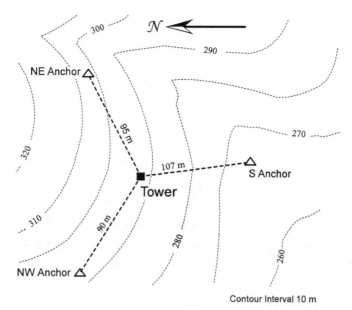

Figure 1. Configuration of Tower and Anchors in Plan View

An intense windstorm had developed on January 30, 2000, producing sustained gusts of 80 km/hr from the southwest. Studio engineers reported observing cable oscillations on the order of 10 m in amplitude during this event. Consequent evaluation of the tower's anchor foundation systems allowed for development of an appropriate remedial option. A design/build approach was then applied, with construction completed late autumn of 2000.

EVALUATION AND REMEDIAL DESIGN

Five borings were advanced to depths varying from 6.0 m to 15.7 m below ground surface (BGS) near the support anchors and the tower foundation using mud rotary drilling techniques. Subsurface conditions are summarized in Table 1 with respect to each structure location, and described below.

Table 1. Summary of Conditions at Each Anchor Location

Anchor	Condition	Remedial Design
S Anchor (deadman)	Wet, poorly compacted backfill $FS_{uplift} < 1.0$	Compaction grouting
NE Anchor (pier)	Wet, soft alluvium $FS_{uplift} < 1.0$	Micropile system with compaction grouting
NW Anchor (pier)	Moist, stiff alluvium $FS_{uplift} > 2.5$	None recommended

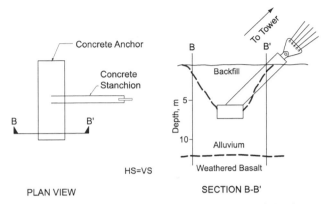

Figure 2. Pre-Construction Conditions at the South Anchor

Selection of appropriate remedial options required that the broadcast tower remained continuously operational. Given this constraint, complete reconstruction or relocation of the anchors was not an option. Consequently, remedial options were selected that provided *in-situ* ground treatment at the existing anchors.

South Anchor

The South Anchor is of a "deadman" type, relying on the weight of its concrete mass and surrounding soil reaction zone to resist lateral and uplift loads. Figure 2 is a depiction of the subsurface condition at the south anchor. Native alluvium consisted of soft to medium stiff clayey silt to a depth of 7.9 m, underlain by stiff to very stiff silt over weathered basalt.

Backfill above the anchor block consisted of medium stiff clayey silt with fine sand, and was determined to be approximately 5.5 m thick. Moisture contents were in the range of 25 to 28 percent, indicative of wet conditions and poor soil compaction. The backfill was reported to have been placed at a moisture content of 10 percent above the optimum and at a density of 80 to 82 percent as referenced to the ASTM D1557 Modified Proctor test.

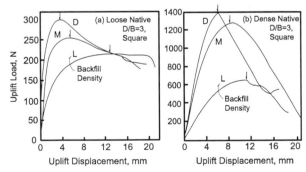

Figure 3. Influence of Backfill Density on Load Displacement Response
(from Kulhawy, et. Al., 1987)

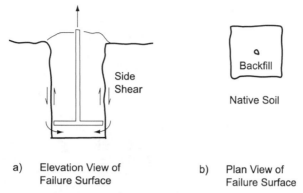

a)   Elevation View of   b)   Plan View of
     Failure Surface   Failure Surface

Figure 4. Side Shear Failure Mode Observations for Poor Soil Backfill
(from Kulhawy, et. al., 1987)

The capacity of the deadman concrete anchor had been reduced significantly by poor backfill compaction. Figure 3 shows the effect of backfill density and consequent strength reduction on the uplift capacity of a deadman-type anchor (Kulhawy, et.al, 1987). The contribution of soil friction, i.e. a conical zone of failure, had been significantly reduced, as the failure zone was limited to the soil zone immediately above the anchor (Figure 4). The reduced capacity likely resulted in horizontal and vertical movements in the anchor under repeated cyclic loadings, creating an even more weakened condition between the deadman anchor and the surrounding soil. An analysis of the south anchor revealed "incipient failure" conditions (Factor of Safety ≤ 1) under the design load of 260 tonnes uplift. Consequently, this anchor was determined to be in "incipient failure", and required remedial stabilization.

Compaction grouting was selected as a remedial treatment measure for the south anchor to increase the mass above the deadman and also to increase frictional resistance along the postulated nearly vertical failure surfaces. The recommended treatment zone included the backfill above the anchor block and a 3-meter zone surrounding the foundation. Close spacing of 1.2 meters center-to-center was recommended to maximize grout take in the relatively shallow target treatment interval. The nearest row of grout injection points along the anchor sides was recommended at a distance of 0.5 meters to fill potential voids between the buried anchor and surrounding soil.

Northeast Anchor

The top of the pile cap is buried 2.0 m into the ground. Subsurface conditions consisted of soft silt with clay underlain by medium dense silty sand with clay (alluvium). Weathered basalt was encountered at a depth of 12.5 m BGS. Pre-construction subsurface conditions at the northeast anchor are shown graphically in Figure 5. Laboratory testing indicated unusually high moisture contents of between 45 to 74 percent indicating a soft or loose condition. The intermittent stream on the northeasterly side of this anchor had undercut its bank and had migrated toward the structure, creating extremely wet conditions along that side of the anchor foundation system.

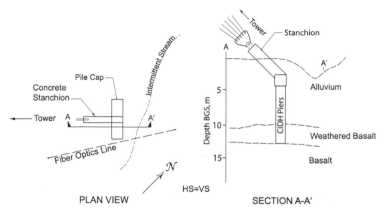

Figure 5. Pre-Construction Conditions at the Northeast Anchor

The softened soil conditions were not considered suitable for supporting the anchor under the design load condition. It appeared that poor initial construction practice combined with repeated cyclic loadings had weakened the anchor/soil interface. The contribution of skin friction was determined to be close to zero. Thus, the forces resisting uplift at this anchor were the buoyant weight of the anchor system due to saturated soil conditions, with minor contributions from the reduced soil friction and embedment into the local weathered rock. Analysis of the uplift capacity indicated that the factor of safety against uplift was significantly < 1.0 under the design loads. Lateral load analysis using the computer program "Lpile" produced a factor of safety for lateral movements of 12.5 mm on the order of 1.0 or less. This anchor also required remedial stabilization.

Initial recommendations for stabilization of the northeast anchor were for a combination of high capacity helical anchors and soil stabilization using compaction grouting. Eight 45 tonne capacity helical anchors with grouted collars were designed to transfer uplift loads from the pile cap to the deep zone of weathered rock. These anchors were to be attached to the pile cap with engineered structural connections. During construction, however, it was discovered that field conditions precluded the helical anchor concept. A micropile system was subsequently implemented to address field constraints. This is discussed in greater detail in the "Construction" section herein.

Compaction grouting was recommended to stabilize the soils around the existing piers. Based on preliminary analyses, recommendations provided for a target vertical treatment zone of 13.7 m, extending laterally to 3.0 m beyond the pile cap. The purpose of the grouting was to provide enhanced lateral support for the foundation system, and as a means of reducing local subsurface hydraulic conductivity in the pile/soil reaction zone. Grout injection points were designed for a maximum horizontal spacing of 1.8 m.

An average "displacement ratio", or soil density improvement as a result of injected grout, in the range of 5 percent was anticipated to meet project objectives. Low-mobility sand/cement grout with a minimum 28-day unconfined compressive strength of 2.75 MPa was recommended. Upon completion of the grouting operation, the anchors would be connected to the existing pile cap and tensioned to 80 percent of their design capacity.

Northwest Anchor and Tower Foundation

Analysis of uplift capacity for the northwest anchor confirmed a factor of safety against uplift of greater than 2.5, and a factor of safety on the order of 3 or greater for lateral movement of 12.5 mm. Bearing capacity and settlement analyses of the tower foundation showed adequate support under the design vertical load. Consequently, no further remedial recommendations were provided for these structures, with the exception of minor drainage improvements.

CONSTRUCTION

Compaction Grouting

Compaction grouting is a ground improvement method whereby very stiff, low-mobility grout is injected through pre-placed casings, targeting discrete soil zones (Warner, 1997; Brown and Warner, 1993, 1994). The preferred slump for this type of work is generally less than 25 to 50 mm. The grouting process translates pump pressure to the expanding grout bulb, displacing and densifying surrounding soils. The degree of compaction for a treated soil mass is typically expressed in terms of its bulk volumetric strain resulting from grout injection. Displacement ratios were calculated in terms of a hypothetical radius of influence as a function of injection point spacing.

Casing tubes for each injection point consisted of steel tube segments of 60 mm outer diameter (OD) and 48 mm inner diameter (ID). The tubes have flush joint threaded couplings in lengths that varied from 0.9 m to 1.5 m. Compaction grout injection points were drilled with rotary methods using water to remove cuttings, or by driving with a heavy pneumatic hammer. The drill rig consisted of a hydraulic drill with a direct drive motor producing 66 m-kg at 125 rpm.

The compaction grout mix consisted of natural silty sand, combined with approximately 10 percent Portland cement, and sufficient water to produce a 0 mm to 50 mm slump in accordance with the ASTM C-143 standard slump test. The silty sand grain size distribution conforms to a range established through years of empirical research (Warner, et. al., 1992; Warner, 1997; and Bandimere, 1997). Compaction grout was mixed and injected using an S-tube type pump, capable of producing 10.6 MPa at the faceplate, and equipped with an integral batch mixer.

Compaction grouting was performed by the split-spaced ascending stage method in vertical intervals of up to 1.5 m. A hydraulic straddle jack and injection header was then attached to the port to prepare an open hole interval for starting the grout injection. Typical configuration of the grout injection header is shown in Figure 6. The first injection was started with the tip of the injection port 0.3 m to 0.6 m above the maximum depth at which the casing was installed.

Figure 6. Typical Configuration at Grout Injection Point

Each stage was grouted to refusal. Stage termination criteria for this project consisted of any measurable movement at the adjacent structure, ground uplift of 2.5 millimeters, achievement of sustained high pumping pressure (location-specific), or grout return to the surface

During initiation of grout emplacement for any stage, grout pressure was allowed to rise as needed, up to the maximum pump output, in order to overcome static friction of the grout in the system. When the pressure stabilized, it was recorded as the starting pressure for that stage. The injection pressure recorded was taken during a pump stroke (not the start or end of the pump stroke).

As grouting in any stage continued, the operator monitored the surface level readings and injection pressures. Injection continued until refusal to take grout was reached. Then the casing was lifted to the next stage and the cycle repeated. Ground surface level in the area of injection was monitored with water manometers. Structural uplift of the anchor foundations was monitored with optical levels targeting the adjacent concrete stanchions.

South Anchor

The compaction grouting pattern for the south anchor is shown in Figure 7. A total of 117 grout injection points were advanced in a triangular pattern at a spacing of 1.2 m on-center, including 25 points on top of the concrete deadman anchor. A total of 9 m^3 of grout was injected above the concrete anchor, accounting for an average displacement ratio of 5 percent, and equating to approximately 18 tonnes of additional mass directly above the anchor. The other 92 injection points were advanced to a depth of 7.6 m.

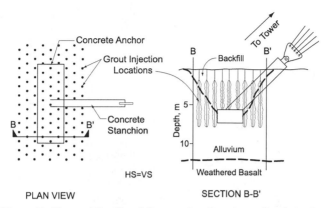

PLAN VIEW   SECTION B-B'

Figure 7. Plan and Profile of Compaction Grouting at South Anchor

A row of grout points was advanced at a distance less than 0.6 m from the edge of the concrete deadman anchor to fill potential voids between the soil and the concrete block along the sides and bottom. The grout injection was started at a pressure of up to 2.1 MPa in the lowermost stages, and was terminated at approximately 1.5 m below the ground surface at an average pressure of 0.35 MPa. Grout quantities injected into the zone surrounding the plan-view footprint of the concrete anchor amounted to 32.7 m³, corresponding to a displacement ratio of approximately 6 percent.

Northeast Anchor

Approximately 2.0 m of soil cover was removed to expose the pile cap and provide access for equipment. Two significant problems were noted during excavation, namely the presence of local scattered boulders in the alluvial/colluvial soils, and the existence of a proximal buried fiber optics line. The existence of the boulders required a change in design of the supplemental anchorage system from helical anchors to micropiles. In addition, the compaction grouting injection pattern was modified to avoid conflict with the buried fiber optics. Battered grout injections were incorporated into the design to avoid working within the adjacent intermittent stream channel. Additional grout injection locations were positioned to the southwest of the pile group to strengthen the distal portion of the inbound wedge of soil subject to cyclic lateral loading.

Compaction grouting was performed at 36 locations in the vicinity of the Northeast anchor to a typical depth of 10.7 m below the pile cap's upper surface. Figure 8 shows the as-built grout injection layout and generalized cross section. Refusal to casing installation was met at varying depths on the basalt surface and its overlying weathered zone. Relatively high grout pressures (on the order of 5.5 MPa to 6.9 MPa) were often recorded when pumping in the weathered basalt, and resultant grout takes in this horizon were typically lower than for overlying alluvial horizons.

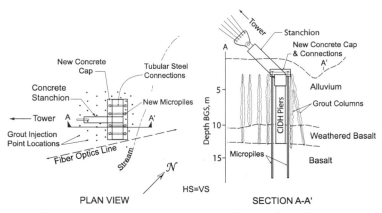

Figure 8. Plan and Profile of Remedial Construction at Northeast Anchor

Eight micropiles were installed to a depth of 15.9 m below the pile cap by drilling a 152 mm diameter boring with a VD–10 air-percussion rig with a 200 mm down-hole hammer. Due to extremely wet, soft soil conditions in the upper profile, a majority of cuttings were not recovered. Most of the cuttings were forced laterally into the formation. A 32 mm-diameter Williams R 71 all thread bar (Grade 150) was installed with centralizers. Neat cement grout with a 28-day unconfined compressive strength of 2.75 MPa was installed with a tremie tube starting from the lowest point of the micro pile upon installation. Each micropile was stressed after the grout had cured, and was locked-off at 80 percent of the 45 tonne design capacity in tandem with tubular steel connections. A new reinforced concrete cap was installed over the pile cap and micropile connections prior to grade reconstruction. A specially designed bearing pad was installed between the existing and the new pile cap to prevent concrete crushing.

CONCLUSIONS

The flexibility of the design/build approach was tested as varying conditions were encountered during construction. A quick and efficient design change was made from helical anchors to micro piles due to the presence of boulders within the zone penetrated. In addition, the grouting pattern was quickly modified to take into account the presence of a deeper than expected fiber optic line, steep slopes and the adjacent stream.

Compaction grouting at the south tower anchor has provided additional mass above the deadman anchor concrete block and failure zone and has also increased the shear resistance along the failure wedge. This has increased the calculated factor of safety against uplift and lateral load to the target values of > 2.5.

The majority of uplift loads for the northeast anchor have been transferred to underlying volcanic bedrock with new micropiles. Passive resistance for the northeast anchor's foundation system has been improved by the emplacement of compaction grout in the reaction zone. Preliminary data from tilt-meters placed at the stanchions at each anchor indicate that the project goals to mitigate adverse ground movements have been met.

ACKNOWLEDGEMENTS

We would like to acknowledge Mr. Edward Williams and Steve Benedict of Fox 49 KPDX to provide us with this opportunity to lead the Design/Build Tower Stabilization Project Team. Our sincere appreciation goes to Dr. Rick Thrall, PhD., P.E., Karen Geraci, P.E., and John Tims for providing valuable comments and review of the manuscript. We would also like to acknowledge Ms. Kelly Perkins who graciously assembled all figures and text for this paper.

REFERENCES

Bandimere, S. W., 1997, "Compaction Grouting State of Practice, 1997", Grouting: Compaction, Remediation and Testing, ASCE, Geotechnical Special Publication No. 66, July, 1997, pp. 18-31.

Brown, D. R., and Warner, J., 1973, "Compaction Grouting", Journal of the Soil Mechanics and Foundations Division, ASCE, Vol. 99, No. SM8, Proc. Paper 9908, August 1973, pp. 589-601.

Kulhawy, F.H. and Charles H. Trautman, 1987, "Spread Foundation in Uplift: Experimental Study", Foundations For Transmission Line Towers, Geotechnical Special Publication No. 8, April 87, pp. 96-109.

Warner, J., 1982, "Compaction Grouting – The First Thirty Years", Proceedings on the Conference on Grouting in Geotechnical Engineering, ASCE, February, 1982, pp. 694-707.

Warner, J., 1997, "Compaction Grouting Mechanism – What do we Know?", Grouting: Compaction, Remediation and Testing, ASCE, Geotechnical Special Publication No. 66, July, 1997, pp. 1-17.

Warner, J., and Brown, D. R., 1974, "Planning and Performing Compaction Grouting", Journal of the Geotechnical Engineering Division, ASCE, Vol. 100, No. GT6, Proc. Paper 10606, June 1974, pp. 653-666.

Warner, J., Schmidt, N., Reed, J., Shepardson, D., Lamb, R., and Wong, S., 1992, "Recent Advances in Compaction Grouting Technology", Grouting, Soil Improvement and Geosynthetics, ASCE, Geotechnical Special Publication No. 30, February, 1992, pp. 252-264.

Grouting and Ground Treatment
Case Studies in Applications of Grouting and Deep Mixing Use of Compaction Grout Columns to Stabilize Uncontrolled Loose Fill and to Lift a Settled Tunnel: A Significant Case History

Ray (Alireza) Boghart[1], Paul S. Hundley[2], Jeffrey R. Hill[3], Steven D. Scherer[4]

Abstract

Since the early 1960's, compaction-grouting techniques have been widely accepted as a means to densify cohesionless soils and to lift settled structures. Compaction grouting techniques have been used less frequently in silts and cohesive soils, primarily due to the slow dissipation of pore water pressure. When used in silts and clay, careful monitoring of the water level is required. This paper documents the use of compaction grouting to densify uncontrolled loose saturated fill and to lift a settled structure.

Introduction

This paper represents the conditions of the overburden materials under a reclaim tunnel. The tunnel was constructed with corrugated steel on an old filled quarry. The concrete mat foundation for this corrugated steel tunnel is approximately 7.6 meters (25 feet wide) and 0.6 meters (2 feet) thick. At the locations where the corrugated steel ends, the reinforced mat foundation is thickened with two 0.6-meter (2-foot) wide and about 0.5-meter (1.5-foot) high sections at each end of the tunnel. There is a notch at the middle of these sections, where the corrugated steel ends are secured. The height of the tunnel is about 4.3 meters (14 feet) and the backfill on the two sides of the tunnel in the initial construction was between about 1.8 meters (6 feet) to 3 meters (10 feet) above the mat bottom.

----

[1]Geotechnical Department Manager, PSI, Columbus, OH
[2]District Manager, PSI, Columbus, OH
[3]Project Manager, Hayward Baker, Inc., Buffalo Grove, IL
[4]Vice President, Hayward Baker, Inc., Buffalo Grove, IL

Approximately 90 meters (300 feet) of this 304-meter (1000-foot) long tunnel had settled. A geotechnical study was conducted to determine the condition of the underlying materials below the existing tunnel. After correctional procedures by compaction grouting, another geotechnical exploration was performed to find the effects and changes after grouting.

This case history presents the subsurface conditions of the site, structure and foundation design criteria and the compaction grouting operation. Specifically, the paper addresses the success of compaction grouting in densification of the soil profile beneath the tunnel and the construction techniques utilized to obtain this densification.

Compaction-grouting techniques have been widely used since the early 1960's. This technique will densify the surrounding soils to the desired depth and also can uplift the settled structures. In this technique, a very stiff grout material is injected into the soils.

The ASCE Grouting Committee defines compaction grouting as follows (1980): "Grout injected with less than 1 inch (25mm) slump. Normally, a soil-cement with sufficient silt sizes to provide plasticity together with sufficient sand sizes is utilized to develop internal friction. The grout generally does not enter the soil pores, but remains in a homogeneous mass that gives controlled displacement to compact loose soils, gives controlled displacement for lifting of structures or both." Brown and Warner (1973) stated that compaction grouting is most effective in the weakest portion of the soils and also is very effective in fine-grained soils that were considered ungroutable in the past. Also, Brown and Warner (1992) mentioned that the grout mass usually has an irregular shape, but in uniform soils it is uniform and can become approximately spherical/columnar in shape. The soils close to the mass of the injected grout undergo large amounts of plastic deformities, whereas at the farther distance, the deformations could be considered elastic. El-Kelesh, Mossaud and Basha (2001) state that the grout injection in the soil creates two mechanisms. First, the growing grout bulb is like a cavity expanding in the soil mass. And second, a conical shear failure occurs above the grout bulb that creates an upheaval in the ground surface.

Nichols and Goods (2000), using a small-scale model compaction grouting, state that the preliminary results show that the shape of the injected grout is a function of overburden pressure. The grout bulbs in their experiments were either cylindrical or teardrop in shape. It should be noted that their tests were conducted on uniform cohesionless soils.

If the soil mass is not uniform as in an uncontrolled fill, the shape of the grout bulb will not be a uniform shape either. The injected grout will take the pass of least resistance (if existing) and will create a non-uniform grout bulb shape.

Henry (1986), in his sinkhole correction paper, states that the compaction grouting is a reliable method that can reduce or even eliminate the risk of damage to the structures due to the slow or sudden collapse of voids in the soil mass.

Schmertmann and Henry (1992) presented a design procedure for using compaction grout "mat" to eliminate the subsidence of landfill liner system as a result of underlying local loss of support due to sinkhole actions.

Ivanetich, Gularte and Dees (2000) presented a case study for an economical solution to solve a liquefaction prone subsurface layer. The geotechnical study showed that a considerable portion of the underlying soils would liquefy. A part of the bridge structure was stabilized using drilled piers penetrating the liquefiable soils. However, at the pier location, this solution was not feasible. Therefore, at this location, compaction grouting was used to densify the soils around and under each pile group.

Project Description

A mix gradation horseshoe-shaped tunnel was built on 18 meters (60 feet) of loose silty fill material. Two conveyors run the length of the tunnel. Aggregates of various sizes are stockpiled on top of the tunnel. Gates open along the alignment of the tunnel to allow blending various aggregate stone sizes along the conveyor. During peak operation these stone piles can be as high as 20 meters (65 feet).

The conveying system collects aggregate from stockpiles situated above the tunnel. The tunnel is surrounded with fill material that was placed up to a height of about 2.1 meters (7 feet) to 3 meters (10 feet). The material was to consist of a well-graded granular backfill compacted to a specified degree of compaction described in the project specifications, ASTM designation A 807A 807M-97 "Standard Practice for Installing Corrugated Steel Structural Plate Pipe for Sewers and Other Applications."

While the stockpiles of aggregate were being placed over the tunnel, the steel arch section collapsed in an area near the north end of the tunnel. A section of the tunnel immediately south of this area has experienced significant settlement and water has accumulated on the slab in this area. The source of water is not known.

The slab and concrete walls in the tunnel showed signs of stress as a result of settlement. There were diagonal cosmetic cracks in the slab and on the walls on two sides. The cracks were not separated. There was water at the bottom of the slab in the damaged area.

Approximately 90 meters (300 feet) of the 304-meter (1000-foot) long tunnel had settled as much as 30 centimeters (12 inches). Settlement was due to changes in the uncontrolled loose fill materials lying beneath the tunnel. The 20 meters (65 feet) of stone covering the top of the tunnel surcharged the uncontrolled loose fill and caused settlement of the structure. This significant differential settlement was considered as a possible contributing factor to the collapse of the tunnel at one location and misalignment of the conveyor at other locations, halting operations. PSI through geotechnical exploration determined some of the causes of settlement. Also, after compaction grouting by Hayward Baker, Inc., another deep exploration operation was conducted to examine the results.

Compaction grouting from bedrock to the bottom of the tunnel was used to densify and displace the silt/clay and sand/silt mixtures. There were larger sized cobbles and boulders in the fill materials. The depth to the rock surface was about 18 meters (60 feet). During the compaction grouting process, increases in SPT counts were recorded along with increases in the pore water pressure. In addition to densification of the soil, the tunnel was lifted to near construction elevations at all points. A significant amount of data on soil density, water levels (pore pressures) and elevation of the tunnel were collected prior to grouting and immediately after grouting. Long-term settlement readings after grouting and under various loads applied by the stockpiles verify the success of the grouting program.

Scope

The scope of the job was divided into two phases. At phase one, the natural condition of the site was in question. For this purpose, twenty-two (22) test borings were drilled and standard penetration tests were performed using a CME automatic hammer. And, disturbed split barrel samples were obtained. After auger refusal occurred on the underlying bedrock, 1.5-meter (5-foot) rock cores were taken at twelve locations. The completion of this phase provided the information concerning the materials and their conditions under the existing tunnel. All the tests were performed in the northern 90 meters (300 feet) of the tunnel. The N-values, natural water content and soil types at the boring locations were determined. Also, sieve and hydrometer analyses were performed on selected samples. All the field and laboratory tests were conducted with respect to the latest ASTM procedures.

The second phase was conducted during the compaction grouting operation. At this phase, seventeen (17) test borings were drilled near the locations of the previous operation after the compaction grouting was over at a section. At each location, standard penetration tests were performed and disturbed samples were obtained. For consistency of the testing procedure, the same drilling equipment and crew were employed. The results of this phase of the operation were used to make a comparison between the soil conditions before and after grout injection.

Initial Field and Laboratory Test Results

There was a layer of crushed stone over the original fill at the boring locations to a depth that varies from approximately 0.6 meters (2 feet) to 3.4 meters (11 feet). At the southern section of the tunnel in the area of Borings B-2 through B-8, there are natural in-situ materials. These materials are mostly loose to medium-dense sand/silt mixtures. The limestone rock layer at this location is at a depth of about 6 meters (20 feet) to 6.7 meters (22 feet). North of Boring B-8, the rock elevation drops. Locations of Borings B-1 and B-3 were moved from the south to the north side. At the locations of Borings B-1, B-3 and B-9 through B-22, underneath the crushed stone, there are layers of non-uniform fill material. The test results show that the fill materials are mostly a lean clay or lean silt/clay mixture (ML) with regard to Unified Soil Classification. Also, the results of sieve analyses on the in-situ soil samples from Borings B-2, B-4, B-6 and B-8 indicate that they consist of approximately 4 percent gravel, 27 percent sand and 69 percent fines (silt and clay). The results of Atterberg Limit tests on selected samples indicate that the liquid limit of the fines ranges from 14 to 19 and plasticity index varies from 2 to 3.

The fill materials consist of firm to very stiff silt/clay mixtures and loose to very dense sand/silt mixtures. There are larger-sized cobbles and boulders in this fill material. The rock surface under the fill is at a depth of approximately 18 meters (60 feet). This is the area where the rock was excavated, and then the old quarry was filled with uncontrolled fill. The existing rock was cored to a depth of 1.5 meters (5 feet) at many borings after complete auger refusal. The percent recovery in the cores varies from 93 to 100 and the RQD varies from 0 to 100 percent. As an average value, 80% and 50% were assumed for recovery and RQD, respectively.

Groundwater was encountered in most of the borings at the end of the drilling operation at a depth that varies from about 0.8 meters to 17 meters (2.6 to 58.5 feet). Table-1 presents the results of field operations before compaction grouting. In this table, the final depth of each boring, the depth to rock surface, depth of water table during the drilling operation, the average N-value of the fill materials in blows per 300 millimeters, the average water content of fill materials, percent recovery of rock cores and RQD of cores are presented.

Based on the boring data, a portion of the south end is resting on the natural material and the remaining portion on a non-homogenous fill of approximately 18 meters (60 feet) in depth. At some locations, very unsuitable materials, such as tires and tree trunks, etc., were found during construction. They were removed and replaced with crushed stone. The fill material is a combination of very different materials, such as clay/silt, sand/silt, cobbles and boulder-sized pieces.

Boring No.	Boring Depth Meter, (ft)	Rock Depth Meter, (ft)	Water Table Meter, (ft)	Nave (blows/ 300 mm)	ω_n (%)	Core Recovery (%)	Core RQD (%)
B-1	17.7, (58.0)	17.6, (57.8)	12.4, (40.8)	9.67	16.50	-	-
B-2	8.4, (27.6)	6.9, (22.6)	1.2, (4.0)	17.00	13.20	93	13
B-3	18.9, (62.1)	17.8, (58.5)	12.9, (42.3)	8.63	16.20	37	100
B-4	8.5, (27.8)	7.0, (22.8)	-	16.80	16.20	100	0
B-5	6.5, (21.6)	6.5, (21.3)	-	17.20	13.20	-	-
B-6	7.7, (25.3)	6.2, (20.3)	0.8, (2.6)	18.00	15.33	100	0
B-7	8.0, (26.3)	6.5, (21.3)	3.4, (11.3)	23.50	16.33	100	0
B-8	7.0, (23.0)	7.0, (23.0)	-	20.20	9.20	-	-
B-9	20.0, (65.3)	18.4, (60.3)	12.4, (40.7)	11.00	16.10	100	78
B-10	17.8, (58.3)	17.8, (58.3)	-	12.00	13.80	-	-
B-11	20.1, (66.0)	18.6, (61.0)	17.8, (58.5)	11.25	12.55	100	87
B-12	15.2, (50.0)	-	-	8.83	16.25	-	-
B-13	19.9, (65.2)	15.3, (60.2)	8.6, (28.2)	9.10	15.08	97	85
B-14	15.2, (50.0)	-	-	10.70	13.60	-	-
B-15	17.7, (58.1)	17.6, (57.8)	-	7.50	14.70	-	-
B-16	18.0, (62.3)	17.5, (57.3)	2.2, (7.2)	7.00	15.50	100	70
B-17	16.2, (53.0)	14.6, (48.0)	-	6.29	16.38	90	13
B-18	17.6, (57.6)	16.0, (52.6)	2.4, (7.8)	8.10	20.73	100	97
B-19	15.9, (52.0)	15.9, (52.0)	7.0, (23.0)	9.10	16.61	-	-
B-20	17.7, (58.1)	16.2, (53.1)	1.3, (4.2)	8.86	17.73	100	100
B-21	13.4, (11.0)	2.4, (8.0)	1.7, (5.5)	14.50	19.00	-	-
B-22	17.7, (58.2)	16.2, (53.2)	6.1, (20.0)	7.50	17.44	100	87

Table 1. Field-testing results before compaction.

The N-values in the fill vary from about 2 to over 50 blows per 300 millimeters. The granular materials are in a very loose to very dense state of density. The consistency of the silt/clay mixtures varies from firm to hard. All of these findings indicate that the fill materials were not placed properly and did not have adequate compaction. The material would have settled under excess static loads, dynamic loading and vibration. The approximately 18-meter (60-foot) fill was not consolidated; therefore, the densification process was continued.

As can be seen in Table-1, the water table at the boring locations was different along the tunnel. This asymmetrical water level could have an adverse effect. As a result of different water elevations, a flow could occur. This could have resulted in the finer particles washing between the larger ones (piping effect) and causing further settlement. The piping action could occur in the unconsolidated fill materials if water is allowed to flow through them. The seepage force will remove the smaller particles. Therefore, the contact stress between larger particles will increase and cause more and accelerated settlement. This could have been one of the reasons for the sudden settlement and failure.

Compaction Grouting

Compaction grouting was used to raise the settled section and to fill the cavities and voids in the existing fill materials. Compaction grouting operation for this project had many advantages, which made it an attractive choice. Some of the advantages were:

1) The injection of three grout columns through the fill materials from the rock surface to the bottom of the slab created a vertical loading capacity. The grouting contractors usually do not rely on these vertical loading capacities through the created grout columns. But in reality, such a high-pressure grout injection from the bottom up and at stages will create a column-type vertical section.

2) The structure could be easily lifted up close to its original elevation at the end of the operation. Use of different grout columns will enable the contractor to monitor the rise and correct it from different sites by injecting additional grout.

3) The grouting operation was proceeding from within the tunnel without interfering with other operations.

4) To improve the fill materials under the tunnel, the density of the soils around the grout bulbs was increased and the cavities filled. The effective zone of the densification is about 2 to 3 times the diameter of the grout bulb. It was assumed that the effective area is about 1 meter.

The grout holes were drilled every 1.8 meters (6 feet) at the damaged and stressed location. Three rows of grout holes were drilled: one on the east side, one on the west side and one in the middle of the tunnel. All the grout holes were drilled through the 0.6-meter (2-foot) concrete slab. One-hundred-fifty-millimeter (6-inch) holes were cored through the slab and a 100-millimeter (4-inch) (minimum) ID steel pipe was placed through the cored hole to the existing ground surface. The treated section of the tunnel was about 76 meters (250 feet). The test drilling operation to monitor the result was performed within 0.3 meters on the outer edge of the slab.

Grout injection pipes were drilled, or in some cases pneumatically driven, to bedrock at a depth of about 18 meters (60 feet). Grout was then injected while the injection pipe was extracted in 0.3-meter (1-foot) lifts. Primary grout columns were spaced initially at 3.7 meters (12 feet). Secondary grout columns were then split spaced at 1.8 meters (6 feet). In a few cases, tertiary holes were then split spaced as well.

The grouting criteria defined by Hayward Baker is as follows: Inject grout at each stage until 4137 kN/m^2 (600 psi) is reached, 0.14 cubic meter (5 cubic feet) of grout is injected or unwanted ground movement is observed. If excessive grout takes are recorded with little or no increase of injection pressure, then additional compaction grout holes will be required in a given area. Grouting pressures reached as much as 2413 kN/m^2 (350 psi), in order to place the 0.14 cubic meters (5 cubic feet) in each stage. The attached graph (Figure 1.) illustrates a representative average grout injection pressure versus depth for primary and secondary holes.

Grout injection was completed with a 0.007 cubic meter (0.25 cubic feet) per stroke grout pump. During much of the project, two pumps were utilized to meet schedule demands. The compaction grout was injected at no more than 0.06 cubic meter (2 cubic feet) per minute. The slump at the grout injection point was 25 to 50 millimeters (1 inches to 2 inches). The following mix design was utilized:

Sand	9.3	kN (2100 pounds)
Fly Ash	3.2	kN (750 pounds)
Cement	1.33 kN (300 pounds)	

The primary holes were utilized to stabilize and densify a majority of the fill. Secondary holes were used to verify and complete the stabilization, and lift the structure. In some cases primary holes were then grouted a second time in order to more precisely lift the structure and insure any voids created beneath the tunnel were filled with grout.

Due to the irregular fill at the site the drilling and grouting conditions varied from location to location. One to 1.5-meter (three- to five-foot) deep areas of soft material were routinely encountered. In many of these areas the drill steel was able to advance without rotation.

Structure Monitoring

Early in the grouting work, monitoring was completed by utilizing piano wire stretched across the work area and site levels. Once the structure began to lift and monitoring was necessary, it was performed by the owner's registered surveyor at appropriate intervals.

Discussion and Analysis

Compaction grouting involved injecting low slump grout into soft and loose soil (uncontrolled fill) at a high pressure. In this type of grouting, the grout material does not permeate the voids in the soil, but expands and creates an assumed homogeneous mass, which causes densification of surrounding loose and soft materials. This type of grouting will also give controlled displacement to lift the structures.

The compaction grouting at the project site started with a 3.7-meter (12-foot) spacing to create primary grout columns. The secondary grout columns were installed at 1.8-meter (6-foot) spacings. The grouting operation was done through an approximately 18-meter (60-foot) injection pipe, from the bottom. The grout was injected while the pipe was extracted.

The grouting was done to a depth of −2.4 meters (-8 feet) below the ground surface. The final 2.4 meters (8 feet) was done at a later date. The grout pressure, volume and spacing were controlled so as not to create too much pore pressure or other adverse effects.

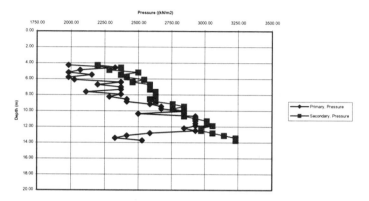

Figure 1. Average grout injection pressure vs. depth for primary and secondary holes.

To monitor the effects of grouting in soil compaction and densification, a set of new standard penetration test (SPT) operations were performed near the grout holes. The SPT procedure was started about a week after the first grout injection. The time lapse was needed to allow some of the excess soil pore pressure generated by the grouting operation to dissipate. The SPT operations were performed near the previous test location to create a base for comparison of the N-values before and after compaction grouting.

The tests were done on both sides of the tunnel. The tunnel did not have enough headroom to perform the SPT inside and to determine the grouting effects for the center holes. The center holes were done last. Therefore, the materials at this location were confined by the grout columns on either side. To model the confining effect of soil by the grout columns, a test section was prepared at one location. At this location, between E29 and E31 (E29 and E31 are grout hole numbers), three extra grout columns were created about 1.8 meters (6 feet) away from the first grout column. The effect of confinement could be checked between the two sets of grout columns.

Monitoring and Verification

As mentioned, a second drilling operation was performed to establish the effectiveness of compaction grouting under the tunnel slab. The holes were drilled close to the grouting holes, which were within the influence zone of the injected grout. These results are shown in Table-2. Compaction grouting will create a grout bulb, which densifies the surrounding soils and fills the existing cavities. These cavities were assumed to be present between large pieces that were bridged over softer materials. Therefore, it was expected that the geotechnical properties would improve, that is, larger N-values, denser soil, less permeable soil mass and so on. It should be noted that the drilling operation during this stage was much more difficult. The gas consumption of the rig was greater and the drilling was overheating the rig. These were signs that the materials were much stiffer with more friction that resulted in more difficult drilling.

The results of the initial and final test boring programs are presented in Table-2. When a new boring was between two of the initial borings, a weighted average of the N-values and water contents were used for comparison. In Table-2, the average N-values and water contents of the

Control Boring No.	Boring Depth Meter	Initial Borings Nearby	Nave (Initial)	Nave (Final)	Change (%)	ω_n (Initial)	ω_n (Final)	Change (%)
CB-1	19.4	B-19	9.10	9.86	+8.35	16.61	19.90	+19.81
CB-4	14.0	B-19	9.10	9.78	+7.5	16.61	19.00	+14.39
CB-5	19.6	B-17, B-19	7.65	9.11	+19.08	16.41	19.20	+17.00
CB-11	14.0	B-12, B-14	8.79	9.25	+5.23	13.98	17.10	+22.32
CB-13	16.1	B-14	10.70	10.91	+1.96	13.60	18.10	+33.09
CB-15	16.2	B-14	10.70	12.40	+15.89	13.60	16.30	+19.85
CB-17	16.1	B-14	10.70	10.27	-4.02	13.60	17.60	+29.41
CB-19	16.2	B-14	10.70	13.36	+24.86	13.60	16.20	+19.12
CB-21	16.2	B-14	10.70	11.36	+6.17	13.60	18.30	+34.56
CB-23	16.1	B-14, B-16	8.85	10.45	+18.08	14.55	17.50	+20.27
CB-25	15.8	B-16	7.00	7.60	+8.57	15.50	19.1	+23.23
CB-28	20.5	B-13	9.10	12.70	+39.55	15.08	16.50	+9.42
CB-29	15.9	B-16	7.00	6.30	-10.00	15.50	15.90	+2.58
CB-30	15.9	B-16	7.00	6.44	-8.00	15.50	17.30	+11.61
CB-30A	13.3	B-16	7.00	16.11	+130.14	15.50	17.60	+13.55
CB-32	15.8	B-16, B-18	8.00	8.00	0	17.75	18.70	+5.35
CB-39	16.4	B-20	8.86	9.57	+8.01	17.73	14.13	-20.30

Table 2. Field testing results and comparisons after compaction grouting.

uncontrolled fill material before and after the compaction grouting are presented. Also, the percent changes in N-values and water contents between the two boring operations are shown. The initial boring numbers are shown as "B" and the final boring numbers are presented as "CB," or "Control Boring." The control borings are numbered with respect to the location of the compaction grouting holes. If they are between the two compaction-grouting holes, the smaller number is used to number the boring. For example, CB-29, which is drilled between compaction grouting holes 29 and 30, is numbered CB-29. Analysis of the values in the presented table results in the following information:

1)   The N-values generally increased. If the extreme high and low values are omitted, the values of increase range from about 2 to 40 percent. The average value of increase is about 15 percent.

2)   The water content of the soils at the test locations generally increased from about 3 to 35 percent. The increase in water content is normal because compaction grouting displaced groundwater and increases the pore water pressure in the vicinity of the grouting, thereby increasing the moisture content of the soil.

3)   Generally, when the increase in water content of the material was larger than about 30 percent, the increase in the N-values showed the lowest numbers. This means that the soil has not consolidated to reach equilibrium. The compaction grouting will cause the moisture and pore pressure to increase initially. With time, the moisture and pore pressure will decrease (consolidation) and the strength (N-values) will increase.

4)   Confinement will have a great effect on the N-values and densification of the fill materials. CB-30 was drilled before the confining grout holes, and CB-30A was drilled after this effect. As shown, the average N-value increased tremendously because of confinement. Confinement will allow more densification of the loose and soft fill and, therefore, result in larger N-values.

5)   The test borings that were drilled at a much later date, after the completion of the compaction grouting (about 12 days), generally resulted in a much larger average N-value. This is the direct result of more consolidation time.

Conclusion

It is believed that the compaction grouting penetrated most of the cavities in the uncontrolled fill and resulted in a more uniform mass under the tunnel. A larger value for the average N-values after compaction grouting was expected. The present increase in N-values is acceptable and the soil will become stronger with time after all excess pore pressures are dissipated. At that time, the soil will have reached its final strength.

It is also believed that the center compaction grouting has created much more densification of the fill material than the two outside grouting columns. The reason for this is the confining effect, which was proven in one test section.

As the pore pressure diminishes with time, the soil consolidates and settles a little. Therefore, surveying the critical elevations after about three months to monitor conditions was recommended. Not much settlement is expected because most of the cavities are already filled with grout. It was discovered that water was flowing out from one of the grout holes on the west side for hours during the grouting operation. The site was visited the same day and no problem was found. The reason for the flow was that the compaction grouting had filled most of the cavities. The new grouts were pushing the in-situ water away and with no place for the water to go, it started flowing out of the grout hole. This process was normal and excess water flowing out actually resulted in more strength in the fill by displacing the water with grout.

References

Brown, D. R. and Warner, J. (1980). "Compaction Grouting". *Journal Soil Mechanics and Foundation Division,* ASCE, 99(8), 589-601.

El-Kelesh, A. M., Mossuad, M. E. and Basha, I. M. (2001). "Model of Compaction Grouting". *Journal of Geotechnical and Geo-environmental Engineering*, ASCE, Vol. 127, No. 11, 955-964.

Henry, J. F. (1986). "Low Slump Compaction Grouting for Correction of Central Florida Sinkholes". *Presented at a Conference entitled "Environmental Problems in Karst Terrain and Their Solutions".* Bowling Green, Kentucky. October 28-30.

Ivanetich, K., Gularte, F. and Dees, B. (2000). "Compaction Grout: A Case History of Seismic Retrofit." *Proceeding of "Advances in Grouting and Ground Modification."* Geo. Institute/ASCE Denver, Colorado. August 5-8, 83-93.

Nichols, S. C. and Goodings, D. J. (2000). "Physical Model Testing Compaction Grouting in Cohesionless Soil". *Journal of Geotechnical and Geo-environmental Engineering*, ASCE, Vol. 126, No. 9, 848-852.

"Preliminary Glossary of Terms Relating to Grouting". *Committee on Grouting, Journal of Geotechnical Engineering Division.* Proceedings of the ASCE, Vol. 106, No. GTI, July 1980. New York, New York, 803-815 (1973).

Schmertmann, J. H. and Henry, J. F. (1992). "A Design Theory for Compaction Grouting". *Proceedings of "Grouting, Soil Improvement and Geo-synthetics".* GT Div./ASCE. New Orleans, Louisiana. February 25-28, 215-228.

Compaction Grouting Used for a Water Treatment Plant Expansion

Michael W. Oakland, Ph.D., P.E.[1] and Michael L. Bachand[2]

Abstract

This paper discusses the selection, design and implementation of compaction grouting to densify a zone of loose sands and silts below a proposed expansion to the Adkins Water Treatment Plant in Six Mile, South Carolina. The water treatment plant is owned and operated by Greenville Water System. The expansion included construction of a duplicate set of sedimentation basins and filters adjacent to the existing basins. Each set of basins is about 183 m (600 ft.) in length and 34 m (110 ft.) in width spanning across a shallow natural valley.

Settlements of up to about 27 cm (10.5 inches) were observed following construction of the original basins. To avoid settlement of the new structures and provide a foundation compatible with the existing basins, compaction grouting was selected as a versatile remediation technique to densify areas of loose sands and silts that were identified in the center portion of the natural valley. Compaction grouting allowed remediation of only the problem areas while conventional foundation support could be used at the ends without special transition features.

Special design considerations included using compaction grouting in the silty environment. Slow rates of injection combined with high ultimate pressures were selected to avoid hydrofracturing or the need for wick drains. The high pressures were possible due to a thick layer of overburden from the previous site grading. However, sequencing considerations, allowing the contractor to begin foundation construction in remediated areas while the compaction grouting was still underway, limited the pressures that could be used in some areas of the program.

While the settlement of the existing basins appeared to have stabilized, as a precautionary measure, the grouting program was extended to stabilize the loose soils below the existing basins to limit potential future seismically induced settlement. Inclined injection holes were drilled at several angles below the existing basins. Careful monitoring and control of injection pressures and rates were essential to avoid heaving the basins.

The paper summarizes the history of the project from identifying the previous settlements, through selection of the remediation technology, to construction performance. The paper discusses evaluation of remediation alternatives, reasons for

---

[1] Senior Geotechnical Engineer, Camp Dresser & McKee, Inc., One Cambridge, Place, 50 Hampshire Place, Cambridge, Massachusetts 02139 Phone 617-452-6402, Fax 617-452-8402

[2] Geotechnical Engineer, Camp Dresser & McKee, Inc., 500 Laurel Street, Suite 400, Baton Rouge, Louisiana 70809 Phone 225-387- 3822, Fax 225-383-7735

selection of compaction grouting, construction monitoring and revisions of the program in the field to account for actual grout performance. The paper also presents a summary of grout takes, pay items and confirmatory testing program to assess the effectiveness of the compaction grouting.

Introduction

A new set of basins was planned as part of a proposed expansion to the Adkins Water Treatment Plant in Six Mile, South Carolina. The expansion included construction of a duplicate set of sedimentation basins adjacent to the existing basins. The new set of basins, as shown on Figure 1, was to be about 183 m (600 ft.) in length and 34 m (110 ft.) in width spanning across a shallow natural valley.

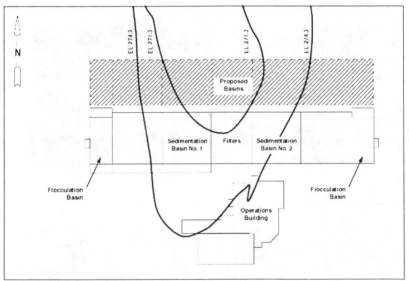

Figure 1: Location Plan of Existing and Proposed Construction

The original plant was constructed in 1982 with the operation building and original basins constructed at the head of the valley. The plant included an operations building, several storage and maintenance buildings and provisions for up to five rows of treatment basins. The plant was constructed with one row of basins. The existing row has flocculation basins at each end. Towards the center are a pair of sedimentation basins with a deeper common filter at the center of the row. Four additional rows of basins were planned, crossing the shallow valley parallel to the existing basins. The valley, however, at these future locations, is deeper and wider than at the existing plant. Some subsurface explorations had been conducted for the future basins during the original design.

The original basins were located about 3 to 4.5 m (10 to 15 ft.) below the existing grade. Excess soil excavate was used to construct a berm to the north of the original basins, covering about 75 percent of the footprint of the next row of basins. The berm was about 6 m (20 ft.) high in the center of the valley, decreasing in height towards the ends of the basin at the sides of the shallow valley where almost no fill was added. The outward slope of the berm was about 3 horizontal to 1 vertical. The shallow valley and existing berm are shown on Figure 2.

Figure 2: Photo of Shallow Valley and Existing Berm over Expansion Site

Subsurface Conditions

Test borings taken for the original structure indicated that the site is underlain by residual soils consisting of very loose to medium dense red-brown to black or white-tan silty sand or sandy silt with increasing amounts of gravel with depth. At a depth of about 20 m (65 ft.) below the ground surface, brown-black to gold partially weathered rock was encountered. The construction documents for the original plant identified the potential for loose subgrade soils and made provisions for removal of soft soils. However, the documents did not specify where the unsuitable soils may have existed and not limits were shown where the soft soils were to be removed. It was reported that some over-excavation was conducted, however, the location, depth and as-placed condition of the replacement soils were not recorded.

The basins displayed an excessive amount of cracking, especially near the deeper filter structure in the middle of the basin. Vertical cracks were observed in the basin walls, as well as, cracks and displacement of the beams supporting the discharge troughs within the basins. As a result of reviewing the structure to assess geotechnical performance, it was believed that the cracks and deformations were signs of significant settlement, up to about 27 cm (10.5 inches). A survey of the basin confirmed that it had settled towards the center by at least 15 cm (6 inches). An independent panel of experts was assembled who concluded that the cracking and depression was the result of settlement.

As shown in the subsurface soil profile on Figure 3, the loosest material found (shown in bold) was generally in the center of the valley between about El. 271.3 m (El. 890 ft.) and El. 265.2 m (El. 870 ft.). The low blow count material seemed to span a length of about 92 m (300 ft.). The soils appeared to become generally denser towards the sides of the valley.

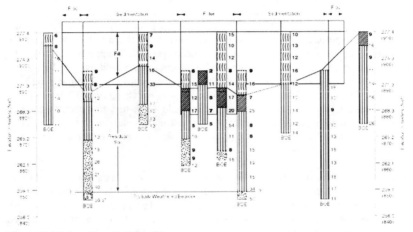

Figure 3: Subsurface Soil Profile

The location of the loose soil is consistent with the settlements observed at the existing basins.

Foundation Options

Prior to the geotechnical investigations, the designers anticipated that conventional mat foundations with no special subgrade preparation would be used to support the new basins. Several foundation alternatives were considered including:

Preloading/Surcharging. Either by relying on the existing berm or supplementing it with additional material, some benefit potentially could be gained by preloading the existing soils. However, the test borings did not indicate any noticeable gain in strength of the deep loose soils that had been overlain by the existing berm, therefore improvement under additional load seemed uncertain. In addition, the construction schedule would not allow for significant additional preloading or surcharging time.

Excavation and Replacement. Removal of the loose soils and replacement by compacted structural fill is often the most economical method of improving subgrade conditions. In addition, conformation of the subgrade is relatively conventional using compaction criteria to monitor the quality of the construction. However, in this case, excavation to the necessary depth would undermine the adjacent existing basins requiring underpinning or substantial lateral earth support to avoid further settlement. The difficulties imposed on the construction and operational risk to the existing basins were considered to be unacceptable.

Dynamic Densification. Dropping a weight or vibro-replacement techniques could be used to densify the soils in place. However, the impact of the vibration would likely cause more settlement to the existing structure.

Piles. Support of the new facilities on a variety of piles was technically feasible. However, this approach presented potential problems at the interface between the relatively rigidly pile supported basins and the original soil supported basins. To accommodate the differential settlement, the new structure would have to be moved away from the existing structure so that flexible connections could connect the two. In addition, the structural design of the foundations, which had been based on the original structure, was largely complete and would have had to be redesigned to accommodate the pile support. However, the cost and impact on the operations of relocating the facilities to allow for the flexible connections were considered unacceptable.

Soil Mixing/Jet Grouting. Soil mixing or jet grouting the soil is a means of forming a competent foundation by solidifying all the soils below the structures. The technique met all the requirements to support the foundations as originally designed, was considered to be compatible with the adjacent basins and could be used if needed below the existing basins by drilling through the adjacent basin floors. However, the cost and excessive amount of spoil generated was a negative consideration for the technique.

Compaction Grouting. Compaction grouting, a technique using pressure applied by a stiff grout to densify soils within a limited area of the borehole also met all of the requirements of the design. Compaction grouting does not result in as strong a foundation as the cement stabilized soil mixing, however, it is less expensive. In addition, the compaction grouting does not generate as much spoil as the jet grouting and can be conducted through inclined holes to stabilize soils below the existing basins without drilling through the basin floors.

Design Evaluation

To limit settlement of the new structures and provide a compatible foundation with the existing basins, compaction grouting was selected as a versatile remediation technique. Compaction grouting was proposed to densify areas of loose sands and silts identified in the center portion of the original valley. Compaction grouting allowed remediation of only the problem areas while conventional foundation support could be used at the ends without special transition features. Compaction grouting was also capable of densifying the soils while the berm soils remained in place. In fact, the berm soils would provide confinement, enhancing the effectiveness of the compaction grouting.

Compaction grouting was also proposed to densify soils below the existing distressed basins using angled drilling without having to drill through the basin floor. Densification of the soils below the existing basins by inclined drilling would be used to keep the existing basins in service during construction.

Compaction grouting was significantly less expensive than other alternatives being considered for the site. The primary concern was the silty nature of the subsurface soils. Silty soils restrict dissipation of pore pressures during densification, potentially resulting in hydrofracturing.

Hydrofracturing occurs when pore water pressures created during compaction grouting exceed the vertical overburden pressure causing a fracture which allows the grout to form a lens rather than compact the surrounding soil. Hydrofracturing is identified by a sudden drop in grout pressure and ground heave. To avoid hydrofracturing, grout pressures must be applied slow enough to allow pore pressures to dissipate. Application rates less than 0.055 m³/min (2 ft³/min) were considered impractical for this project. If this rate could not be met, wick drains would have to be considered to accelerate pore pressure dissipation. However, wick drains would require additional cost, construction time and equipment at this already confined site.

CDM consulted with Hayward Baker, Inc. of Odenton, MD, a specialty subcontractor experienced in the use of compaction grouting. While the typical gradation of the loose soils at the Adkins site was within the optimal zone to be compacted in accordance with literature provided by Hayward Baker, as shown on Figure 4, the soils were near the limit of silt which could be compacted without drainage. Hayward Baker confirmed that compaction grouting could be conducted and assisted with establishing the required grout hole spacing best suit the site.

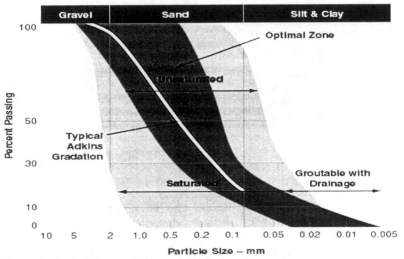

Figure 4: Typical Range of Compactable Soils

The zone of loose soil below the proposed basins, Phase I shown on Figure 5, was to be compaction grouted under the full weight of the existing berm to facilitate the high design pressures. Based on input from Hayward Baker, a 2 m (7 ft.) grid pattern was selected. The pattern was somewhat aggressive for the silty conditions, however, with the depth of the overburden placed over most of the site, relatively high grout pressures could be used to allow the wider spacing. Grouting was extended about 3 m (10 ft.) beyond the limits of the basin footprint. Beyond this limit, progressively shallower compaction grouting was included to create transitions zones to avoid the potential for abrupt changes in subgrade modulus below the mat foundations. To accommodate the construction schedule, it was agreed that the berm could be excavated in these areas prior to grouting.

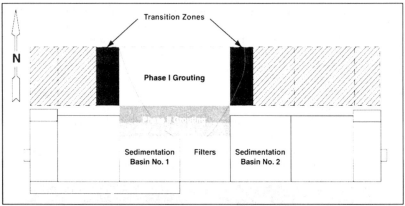

Figure 5: Area of Compaction Grouting

While the settlement of the existing basins appeared to have stabilized, as a precautionary measure, the grouting program was extended to stabilize the loose soils below the existing basins to densify a reasonable zone of influence below the foundations. Inclined grouting was used for the Phase II area as shown in Figure 5 to keep the existing basins in service during the work. Test borings indicated that the loose soils existed primarily along the northern side of the basin. The program included compaction grouting extending about 6 m (20 ft.) below the existing basins. The site would have to be excavated to the foundation level, removing the berm, to facilitate drilling for the Phase II compaction grouting. However, the excavation could not be made until completion of the Phase I that was to rely on this overburden to confine the high pressure grouting. The drill angles were left to the compaction grouting sub-contractor to coordinate with excavation levels of the general contractor, however, the grout pattern had to be capable of densifying the zone below the existing basins.

The objective of the compaction-grouting program was to increase the SPT N value of the loose soil to an average of 18 blows per 30.5 cm (1 ft.) with no individual N-values of less than 12. The compaction grouting specifications required as a minimum, the following criteria:

- a pressure refusal of 2758 kN/m^2 (400 psi) above the line pressure
- grout take of 0.17 m^3/vertical meter (6 ft^3/vertical ft.)
- heave of 0.32 cm (1/8-inch) at ground surface or 0.16 cm (1/16inch) at any structure.

Equipment

To achieve the compaction requirements, the compaction-grouting Contractor mobilized two compaction-grouting rigs, a cement silo, a mechanical screen, a pump, and a mixing truck to the site on January 20, 2001. Two different grout rigs were brought two the site; a wash drill rig (shown on Figure 6) and pneumatic drill rig that drove the grout pipe. The wash drill rig used a self-contained pump to conduct the wash bore, as well as, the compaction grouting and could conduct the grouting without physically switching lines. However, a separate grout line and pump had to be connected to the pneumatically driven grout pipe to conduct the compaction grouting.

Figure 6: Wash Drill Compaction Grout Rig

The mixing truck contained the grout, water, and sand that formed the low slump grout, which was mixed by a screw auger. The cement for the grout was stored in a silo on site. A mechanical screening belt was used to screen out large particles and debris from on-site material.

Compaction Grouting Procedure Overview

The Phase I compaction grouting included vertical holes drilled from 4.5 m (15 ft.) above the foundation level of the sedimentation basins through the existing confining berm. The zone to be compacted in Phase I was approximately 55 m (180 ft.) long by 40 m (130 ft.) wide requiring 667 grout locations. Each hole was drilled a total depth of about 12 m (40 ft.) with compaction conducted over the lower 6 m (20 ft.). The transition zones were also drilled with vertical holes over a combined area of approximately 12 m (40 ft.) wide by 40 m (130 ft.) long requiring 45 grout locations. This phase was largely drilled from the foundation level to depths of up to 6 m (20 ft.) with compaction occurring over most of the drilled length.

The Phase II compaction grouting used the inclined drill holes to densify soils beneath the existing sedimentation basins as shown on Figure 7. The work included grouting an area approximately 79 m (260 ft.) long by 40 m (130 ft.) wide. The zone of grouting was approximately 6 m (20 ft.) in depth below the bottom of the existing basins.

Figure 7: Drilling Angled Compaction Grout Holes Below the Existing Basins

Heave and settlements were monitored using a series of laser levels and weighted stands with rulers fixed to them. Rulers were also affixed to near by structures. Surface settlements were read at the beginning and end of each grout hole. Readings were taken every half hour on the existing structures and monitored constantly when grouting under or within 6 m (20 ft.) of an existing structure.

Compaction Grouting Construction

The compaction grouting generally proceeded from east to west so that the Contractor could excavate the 6 m (20 ft.) of fill material from the areas already treated and begin basin construction. The wash rig grouted the holes in 0.6 m (2 ft.) vertical lifts while the pneumatic rig grouted the holes in 0.3 m (1 ft.) vertical lifts. Only the wash rig was used to drill and grout locations within 12 m (40 ft.), in plan, of the exiting basins to avoid vibrations that may have caused settlement of the existing basins.

To prevent hydrofracturing while optimizing densification, the injection rate was initially specified to be no greater than 0.055 m³/min (2 ft³/min) based on advice from Hayward Baker. However, once in the field, it was agreed to try an increased injection rate of 0.11 m³/min (4 ft³/min) based on several initial grout locations that received maximum grout take at the higher injection rate. The program was further modified so that if the maximum grout pressures or heave criteria were the limiting criteria at an injection rate of 0.11 m³/min (4 ft³/min) for two consecutive grout increments in a single hole, the injection rate would be lowered back to 0.055 m³/min (2 ft³/min).

After approximately 12 percent of the Phase I grout holes were completed, six test borings were drilled to assess the performance of the modified program. The confirmation borings showed that the modified compaction grouting program only increased the loose soils to an average SPT N value of 16. The results were slightly less than the target compaction criteria. As a result, the maximum grout take was increased from 0.17 m³/vertical meter (6 ft³/vertical ft.) to 0.23 m³/vertical meter (8 ft³/vertical ft.) to achieve the required SPT N-value and the injection rate was reduced to the originally specified rate. Six additional confirmatory test borings were drilled after approximately 30 and 90 percent of Phase I grout holes were completed. The SPT data in these test borings confirmed that the modified procedures were achieving the target results under the adjustments made after the initial testing.

The Phase II grouting program was left largely to the contractor to specify based on his experience with grouting below existing structures. The Contractor proposed to install inclined grout holes from 45 locations along the edge of the existing basin. Each location four holes inclined holes were drilled at varied angles. The array of inclined drill holes were as follows: 0 degrees for 6 m (20 ft.), 26 degrees for 8 m (26 ft.), 38 degrees for 10 m (33 ft.) and 52 degrees for 14 m (47 ft.). Each grout hole was grouted in 0.7 m (2 ft.) lifts until the entire grout zone was grouted. The wash rig was used for the Phase II drilling to avoid vibrations below the structure. Due to the Contractor's concern for heaving the sedimentation basins, the following compaction grouting procedures were accepted:

- if the grout take was between 0.17 m³/vertical m and 0.23 m³/vertical m (6 and 8 ft³/vertical ft.), the maximum pressure would be 690 kN/m² (100psi) over line pressure.
- if the grout take was between 0.0.11m³/vertical m and 0.17 m³/vertical m (4 and 6 ft³/vertical ft.), the maximum pressure would be 1379 kN/m² (200 psi) over line pressure
- if the grout take was between 0.055 m³/vertical m and 0.11 m³/vertical m (2 and 4 ft³/vertical ft.), the maximum pressure would be 2069 kN/m² (300 psi) over line pressure
- if the grout take was between 0 and 0.055 m³/vertical m (0 and 2 ft³/vertical ft.), the maximum pressure would be 2758 kN/m² (400 psi) over line pressure

Compaction Grouting Summary

Phase I program of 667 grouting locations was completed in 40 days from the start of mobilization with two crews and rigs. To make up for lost time due to inclement weather and equipment malfunctions, one rig worked 24 hours a day for 8 days. The total length of holes drilled and volume of grout used in Phase I was 6895 lineal meters (22,621 lineal ft.) and 1266 m^3 (44,730 ft^3), respectively.

Phase II program of 133 grouting locations was completed in 15 days from the start of mobilization with two crews and rigs. The total length of holes drilled and volume of grout used in Phase II was 828 lineal meters (2715 lineal ft.) and 43 m^3 (1527 ft^3) respectively.

As discussed above, the compaction grouting was terminated either when a certain amount of grout had been pumped or a pressure refusal where the maximum pressure was achieved prior to pumping the specified maximum amount of grout. In general, grout quantity termination (referred to as grout refusal) indicated particularly loose soils that require compaction by multiple holes to densify the area while pressure refusal indicated initially denser soils that require limited further densification. Grout takes and pressure refusal are recorded for each lift in each hole.

While data from each lift would be too numerous to summarize simply, the predominate refusal, grout or pressure, was assessed for each hole. A plot of the mode of refusal within the Phase I area is shown in Figure 8. Grout take was primarily controlled by grout refusal within the center part of the valley where the lowest SPT N-values were observed. As anticipated, within the transition areas along the sides of the valley, pressure refusal generally controlled indicating that the limits of the grouting extended adequately into suitable foundation soils.

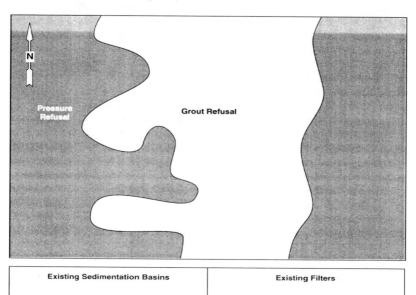

Figure 8: Plan of Pressure Refusal verses Grout Refusal

The results of the compaction grouting seem to agree very closely with the loose material found during the explorations and also seem to be consistent with the settlement observed in the existing basins. The slow grout rates resulted in no significant heave of the soils and with no erratic loss of pressure during grouting, it does not appear that hydrofracturing was a problem. In addition, with careful control of the grouting pressures used while compacting the soils below the existing basins, no movement of the existing structures was observed.

Conclusions

The compaction grouting program at the Adkins water treatment plant was performed on loose silty soils that required minimal improvement for the required bearing capacity. The silty nature of the soils was on the edge of being able to be compacted without the use of wicks drains or any other pore pressure dissipations methods. Thus, careful monitoring of the compaction grouting injection rates and pressure refusals are required to achieve good results. In conclusion, the following observations were made related to this project:

- Compaction grouting does not necessarily result in dramatic increases in soil densification; however, it is very useful where other methods may not be applicable.
- Compaction grouting seems especially applicable in controlling settlements or potential liquefaction rather than attempting to just increase bearing capacity.
- Compaction grouting offers versatility that may not be found in other remedial techniques.
- Adequate overburden soils must be in place to allow high grout pressures.
- Compaction grouting methods require constant adjustment to accommodate soil conditions.
- In silty conditions, the rate of injection is critical to achieving good results.
- Typically, the drilling costs are substantially more than the grouting costs so slower injection rates with higher maximum grout take is likely to be more economical than closer grout hole spacing.
- An experienced compaction grouting contractor is critical. If possible, the contractor should be retained during design to provide consultation on a design/build basis.

Acknowledgements

The authors wish to thank the following municipalities, companies, and individuals for their input and cooperation in writing this paper: Greenville Water System, owners of the Adkins Treatment plant who were actively involved in all aspects of the project and supportive of investigating application of this innovative technique for the treatment plant expansion, David Zimmer, Client Officer for CDM, Gabe Valdes, Project Manager for CDM, Hayward Baker for their valuable input during both the design and construction phases of the project, specifically Project Manager Mike Terry, and Bunnell-Lammons Engineering Inc. who conducted the test boring work at the site and assisted with field monitoring of the compaction grouting and subsequent subgrade preparation at the site.

Low Strain Testing of Compaction Grout Columns

By Lawrence F. Johnsen[1], Andy Anderson[2] and Jon J. Jagello[3]

Abstract

Low strain testing was performed on grout columns on three compaction grouting projects. The purpose of the testing was to investigate the capability of low strain testing to detect the depth, shape and continuity of compaction grout columns. The methods and objectives of the compaction grouting varied among the three projects.

Introduction

Low strain testing is a commonly used method to investigate the integrity of drilled piers, concrete piles and auger cast piles. It can detect poor quality concrete, broken piles, necking or bulging and provide an estimate of the pile length. Judgment is required both in the running of the test and the interpretation of the test data.

The test method was applied on three projects, discussed herein, to determine its applicability to compaction grouting.

Low Strain Test Method

This method of testing has many names including sonic pulse echo, pile integrity testing and low strain integrity testing, as ASTM refers to it in the standard for this method (ASTM, 1996). The testing described in this paper was performed using the Pile Integrity Tester manufactured by Pile Dynamics, Inc.

The method requires a clean and hard surface at the top of the grout column. This was achieved by removing loose material with a shovel, and grinding the top of the grout with a hand held grinder. An accelerometer was attached to the top of the column with a thin layer of either wax or lithium grease. The operator then hit the top of the grout column with a one or two pound instrumented hammer. The hammer strike generates accelerations in the 10 to 100 g range, strains less than 0.00001, velocities less than 0.1 fps and displacements less than 0.001 inches. The hammer contains a transducer to measure the impact force.

---

[1] Principal, Heller and Johnsen, Foot of Broad Street, Stratford, Connecticut 06615
[2] Area Manager, Hayward Baker, Inc., 17-17 Route 208, Fair Lawn, New Jersey 07410
[3] Project Manager, Heller and Johnsen, Foot of Broad Street, Stratford, Connecticut 06615

The hammer generates a low strain wave. The wave travels down the pile at a wave speed, c, which can be estimated by the following equation for concrete:

$$c = \sqrt{E/\rho}$$

where E is the elastic modulus of the pile and ρ is the mass density of the pile. The wave speed is often estimated by testing reference piles on a project, where the pile length is known. In this case, the wave speed is simply twice the pile length divided by the time for the toe reflection to be received at the pile top. With concrete piles, wave speeds typically vary up to 5% on projects where the speed is determined by reference piles, and up to 15% on projects where the speed is determined by equation.

As the wave travels down the pile, it is reflected by soil resistance and changes in impedance. The impedance changes are caused by changes in cross-sectional area, modulus of elasticity or grout strength. Upon reaching the toe of the pile, the wave will reflect back to the top of the pile. The toe reflection may be difficult or impossible to detect if the pile has a large amount of soil resistance. The typical limiting L/D ratio is 60:1. Also, large variations in impedance will tend to mask the toe reflection. The equipment allows the operator to exponentially amplify the reflection starting at the depth that the operator believes corresponds to the start of significant soil resistance.

Software is available to interpret the pile profile on the basis of the wave reflections. Since changes in impedance will either increase or decrease the reflected wave, a profile can be computed from the time integral of the velocity wave effects at the pile top. The following case histories will illustrate the method.

Westport, Connecticut

Compaction grouting was used to create a series of lightly loaded piles to underpin a two-story wood-framed duplex, and slab-on-grade. The subsurface profile consisted of granular fill to about 8 feet, a thin layer of peat grading to organic silt, which extended to 14 feet. The organic silt was underlain by medium dense to dense gravelly sand.

The piles were constructed by driving a 2 inch I.D. pipe a minimum of 5 feet into the gravelly sand. As the pipe was withdrawn in two foot stages, a stiff grout mix was injected under pressures of up to 500 psi. At the completion of grouting, a 1 inch diameter bar was driven into the grouted pile. The grout mix achieved 28 day breaks of 2150 to 5380 psi.

A test program was conducted in an open area adjacent to the duplex prior to the start of production piles. Four test piles were installed to depths of 15 feet. PIT testing was performed on two of the piles. Using an assumed wave velocity of 13,000 fps, Pile TP-2 showed a length of 13 feet with bulging beginning at 6 to 7 feet. A toe reflection was not detected at TP-1. Subsequent excavation of the piles showed that TP-1 had broken at a depth of 5 feet, and that TP-2 had a tear shaped bulge beginning at a depth of 9 feet. The full lengths could not be excavated. The relatively large area of the reinforcing bar in comparison to the grouted area may have affected the wave speed. The logged and tested depths are provided on Table 1. Velocity wave vs. depth plots are provided for each pile. Additionally, the computed profile of TP-2 is provided.

Allentown, Pennsylvania

Compaction grout columns were installed on a closely spaced grid to densify loose soil and fill potential cavities in karst terrain. The columns were formed by first predrilling a 7 inch diameter hole with a down-the-hole air hammer. The hole was taken about five feet into competent rock. Later, a 7 inch diameter steel pipe was driven into the hole with a small vibratory hammer. A grout hose was attached to the top of the steel pipe. The pipe was lifted in two foot stages and either 7 or 12 cf of grout was pumped unless heave was observed or a grout pressure of 400 psi was reached.

No test cylinders were broken for this project. On previous projects, compressive strengths ranged between 600 and 1000 psi for the grout mix used. A wave speed of 7500 fps was selected based on reference piles.

A total of 16 columns were tested. Each column had been grouted a minimum of one week earlier. Table 1 shows the results. Seven of the columns exhibited clear toe reflections. Five of the toe reflections were within 2% of the logged column depth. The other two were 18% and 27% less than the logged depth. Two columns were ambiguous in that they showed two wave forms that could be interpreted as toe reflections. Table 1 provides the logged and PIT estimated depths for the 16 columns. Velocity wave vs. depth plots are provided for each pile.

The possible reasons for the inconclusive results on the other seven columns include shallow horizontal cracks through the column or poor quality grout near the surface.

Reading, Pennsylvania

Sinkholes developed shortly after foundations had been constructed, and structural steel placed, for a building in Reading, Pennsylvania. Compaction grout columns were installed adjacent to footings to depths up to 86 feet.

The grout columns were constructed by driving a 2 inch diameter pipe to refusal, and then pumping grout in stages as the pipe was lifted one foot at a time. Grout takes were recorded for each stage. The grout had 28 day compressive strengths of 800 to 1200 psi.

PIT testing was performed on five columns that varied in length from 16 to 86 feet. Two of the five columns showed toe reflections at the logged depth. The wave speed of 9500 fps was obtained by the reference pile method. Since only two piles exhibited clear toe reflections, the variation between logged and tested depths is reduced by using the reference pile method. Table 2 shows the logged and PIT estimated depths. Velocity wave vs. depth plots are provided for each pile.

Acknowledgements

Hayward Baker, Inc. was the grouting contractor for the Westport, Connecticut and Allentown, Pennsylvania projects. Compaction Grouting Services was the grouting contractor for the Reading, Pennsylvania project. The PIT testing was performed by Jon Jagello of Heller and Johnsen.

Conclusions

1. The low strain method appears to have applications for compaction grouting. Potentially, the method may provide a means of verifying that compaction grout columns contain a continuous cross section of good quality grout to the bottom of the column, and identify the depths at which bulging, necking or poor quality grout occur.
2. Judgment is required in using the method. Since compaction grouting as commonly practiced will naturally form necking and bulging points, the need for this judgement is amplified. This would be minimized if grouting stages were performed smoothly and uniformly over the depth, instead of abruptly over short stages. Site conditions and project requirements will typically dictate this.
3. The method will not produce definitive results on all columns and has the following limitations:
   a. An average wave speed is required input to calculate the column length. The wave speed varies with the compressive strength of the grout. Variations in wave speed will result in inaccurate calculated depths.
   b. Under ideal conditions, the method is limited to a maximum length to diameter ratio of 60:1.
   c. Large variations in cross-section or grout quality will mask the underlying conditions.
   d. Horizontal cracks across the column will produce toe reflections, and may be interpreted as the bottom of the column.
4. Additional investigations are necessary to demonstrate the applications and limitations of the method as applied to compaction grouting. Ideally, these investigations will include subsequent excavation of the columns and compressive strength testing of the grout.

References

ASTM, (1996), Designation D 5882 –96, "Low Strain Integrity Testing of Piles".

Rausche, Frank, Likins, Garland and Hussein, Mohamad, " Formalized Procedure for Quality Assessment of Cast-in-Place Shafts Using Sonic Pulse Echo Methods", Transportation Research Record 1447.

TABLE 1
LOGGED VS. ESTIMATED DEPTHS

PROJECT	NUMBER	LOGGED DEPTH (ft)	ESTIMATED DEPTH (ft)	GROUT VOLUME (cy)
WESTPORT	TP-1	15		
C=13,000 fps	TP-2	15	13	
ALLENTOWN	31	87.5		20.8
C=7500 fps	50	30	22	2.5
	51	39.5	39	3.4
	60	58.5		8.4
	67	33.5	33	2.9
	68	42.5		5.1
	69	44.5		4.2
	80	45	45	3.3
	134	33		2.7
	139	35	35	6.4
	208	45		6.6
	209	45		5.6
	224	33	27	5.2
	302	26.5	20 or 40	1.0
	305	45	30 or 57	5.8
	317	25	25	1.0
READING	1A	32.5		
C=9,500 fps	6	20		6.6
	9	16	16	1.3
	13	86		
	14	23	23	3.4

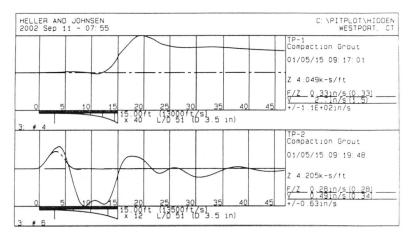

Figure 1. P.I.T. results, Westport, Connecticut

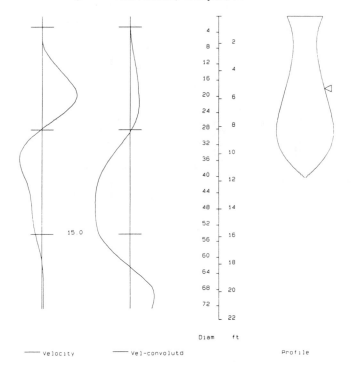

Figure 2. Profile of pile TP-2, Westport, Connecticut

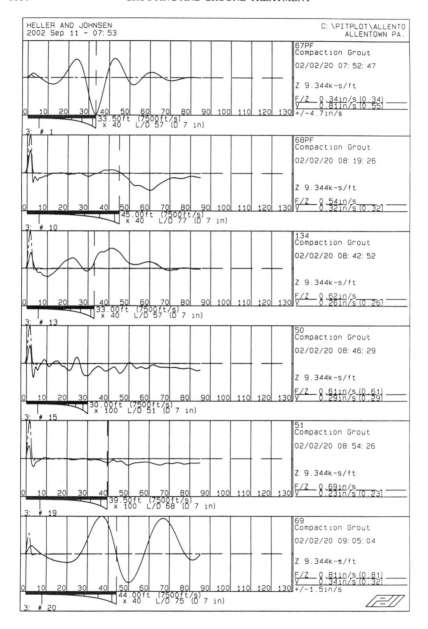

Figure 3. P.I.T. results, Allentown, Pennsylvania

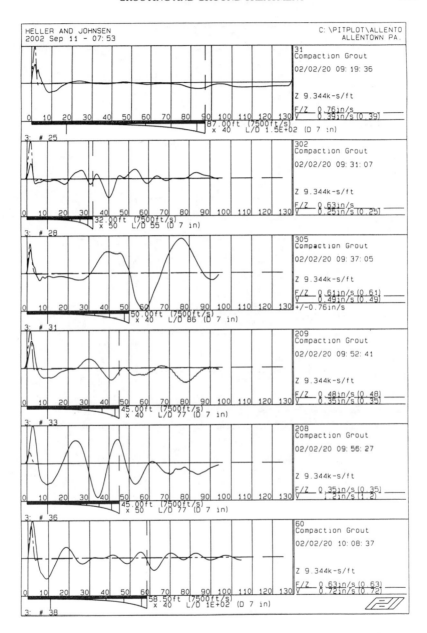

Figure 3 (cont.). P.I.T. results, Allentown, Pennsylvania

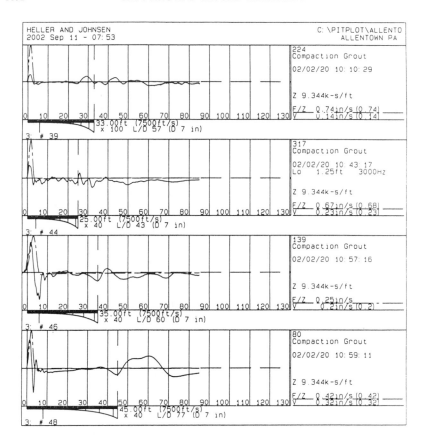

Figure 3 (cont.). P.I.T. results, Allentown, Pennsylvania

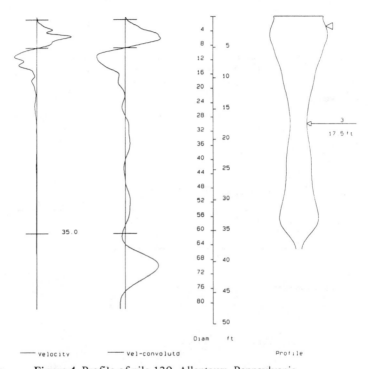

Figure 4. Profile of pile 139, Allentown, Pennsylvania

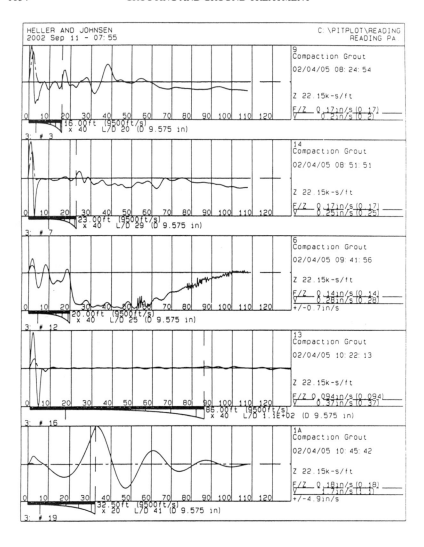

Figure 5. P.I.T. results, Reading, Pennsylvania

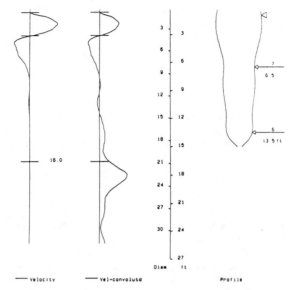

Figure 6. Profile of pile 9, Reading, Pennsylvania

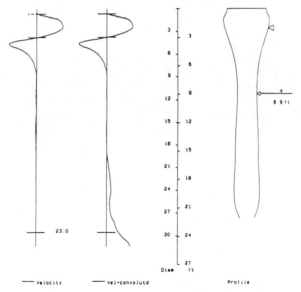

Figure 7. Profile of pile 14, Reading, Pennsylvania

Effect of Soil and Grouting Parameters on the Effectiveness of Compaction Grouting

Adel M. El-Kelesh, M.ASCE,[1] and Tamotsu Matsui, M.ASCE[2]

Abstract

Despite its many applications, compaction grouting has been suffering from the minimal empirical guidelines to predict the treatment effectiveness. This paper discusses the effects of soil parameters, soil compressibility, replacement ratio and injection sequence on the groutability and effectiveness of compaction grouting, through analyses of the results of a field test conducted under the old taxiway of Tokyo International Airport. The test consisted of eighty-seven compaction grout piles injected in three cases of different spacings, pile diameters and grouting procedures. Based on the analyses presented, unique relationships between the groutability of compaction grouting and the soil parameters are identified. Also, empirical correlations for estimating the treatment effectiveness, based on the initial soil properties, are presented. Another finding is that the injection sequence may be controlled to provide confining action for the subsequence injections. The confined injections can effectively improve the highly compressible soils much more than the unconfined ones. In addition, it is recommended to avoid the condition, beyond which particle crushing becomes the dominant mechanism of deformation.

Introduction

In compaction grouting, a very stiff grout material is injected into the soil. During injection the material increases in size and remains in a homogeneous state, with distinct grout-soil interface, resulting in displacement and densification of the surrounding soils. The current design practice depends on the assumption that the required change in void ratio, and consequently the required improvement, can be attained by volumetrically straining the treated soils. Therefore, the replacement ratio is used as an index of the volume of voids that is replaced by the volume of grout.

---

[1] Dept. of Civil Engrg., Osaka University, 2-1 Yamadaoka, Suita Osaka 565-0871, Japan; phone +81-6-6879-7626; adel@civil.eng.osaka-u.ac.jp
[2] Dept. of Civil Engrg., Osaka University, 2-1 Yamadaoka, Suita Osaka 565-0871, Japan; phone +81-6-6879-7623; t-matsui@civil.eng.osaka-u.ac.jp

In most applications compaction grouting has been effective (e.g., Brown and Warner 1973) and in some cases the improvements have been much larger than estimated (e.g., El-Kelesh et al. 2002b). However, in other cases the improvements have been less than estimated (e.g., Boulanger and Hayden 1995) or highly variable in the treatment zone (e.g., Salley et al. 1987). Another problem may be faced is the early refusal to grouting, and as a result, all the design grout volumes can not be injected. These problems are basically attributed to the little understanding of the grouting mechanisms and the effects of pertinent soil parameters and grouting variables on treatment effectiveness. Despite the many applications of the technique (e.g., Graf 1992; Boulanger and Hayden 1995) and its extensive literature, to the authors' knowledge, empirical correlations that account for the effects of soil parameters and grouting variables on the treatment effectiveness are not available.

This paper presents analyses of the results of a field test conducted under the old taxiway of Tokyo International Airport. In this test, eighty-seven compaction grout piles were injected in three cases of different spacings, pile diameters and grouting procedures. The analyses emphasize the effects of soil parameters, soil compressibility, replacement ratio and injection sequence on the groutability and treatment effectiveness. After brief descriptions of the test procedure and ground conditions, discussions and quantitative evaluations of the effects of soil properties and replacement ratio on the groutability and effectiveness of compaction grouting are presented. Then, the effect of injection sequence on treatment effectiveness for the cases of highly and less compressible soils is discussed.

Test Layout and Procedure

The field test consisted of eighty-seven compaction grout piles injected in three cases of different spacings and pile diameters, as shown in Fig. 1. The three cases are characterized in Table 1. The piles of each case were laid out on a triangular pattern of equidistant spacing. The spacings and diameters of the piles were selected to attain replacement ratios of 10, 15 and 20% for Cases-1, 2 and 3, respectively. The piles of Case-2 were injected first, then those of Cases-1 and 3. The arrows in Fig. 1 show the injection sequence for each case. The piles of Case-2 were injected in seven lines,

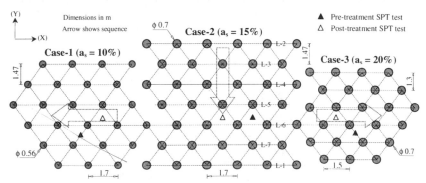

Figure 1. Layout of compaction grout piles and SPT borings.

Table 1. Characteristics of test cases.

Case No.	No. of Piles	Pile Diameter (m)	Spacing (m)	Depth (m)	Design Vol./Step (m³)	Design Replacement Ratio, a_s (%)	Schedule From	Schedule To
1	25	0.56	1.70	11.80-3.13	0.098	10	Aug. 17	Aug. 31
2	37	0.70	1.70	11.80-3.13	0.140	15	Jul. 16	Aug. 07
3	25	0.70	1.50	11.80-3.13	0.140	20	Aug. 21	Sep. 04

L-1 through L-7. It should be noted that L-1 and L-2 were injected first at the boundaries of treatment zone, then L-3 through L-7 were subsequently injected between L-1 and L-2. However, the piles of Cases-1 and 3 were injected while going outward form the zone of Case-2 in the negative and positive X-directions, respectively.

The piles were injected in steps using the bottom-up procedure form 11.80 to 3.13 m in depth. Each pile comprised 27 steps, with depth interval of 0.33 m between every two successive steps. Upon completion of the injection of a given step, the injection pipe was raised to the depth of the next step, by means of air-driven hydraulic jacking system. The grout material was mixed on site using auger mixer and injected using high pressure positive displacement pump. The pump strokes were counted so that volumes resulting in steps of assumed diameters of 0.56, 0.70 and 0.70 m were injected for Cases-1, 2 and 3, respectively. The grout material, a mixture of cement, silt, sand, gravel and water, had a slump ranging from 2.0 to 4.0 cm.

Grout is usually injected until a specified limiting criterion is reached. The limiting criteria may include pressure, volume and ground surface upheave or combinations. In this test, injections were limited by grout volumes of 0.098, 0.140 and 0.140 m³ for Cases-1, 2 and 3, respectively, or a maximum pressure of 8.0 MPa at the collar of grout pipe, whichever is attained first. Ground surface upheave was not included in the limiting criteria. To assess the effectiveness of treatment, pre- and post-treatment SPT borings were conducted for the three cases. The relative positions of these borings are shown in Fig. 1.

Ground Conditions

The results of exploratory SPT borings, whose positions are shown in Fig. 1, indicate the following general strata of soil profile: (1) Pavement: Green gray to black gray poorly graded gravel with sand overlain by asphalt of 0.15 m thick, extends from ground surface to a depth of approximately 1.80 m; (2) Fine Sand: Black gray to dark gray fine sand, fines content of 7-31%, extends from approximately 1.80 to 9.00 m in depth; (3) Silty Sand: Black gray silty sand, extends from approximately 9.00 to 10.25 m in depth; and (4) Silt: Black gray silt with partial amounts of sands and clays, appears at approximately 10.25 m in depth.

Groutability of Compaction Grouting

The current practice of designing compaction grouting operations is very simple and depends on attaining a given replacement ratio, a_s, to result in a given average improvement. Because this design approach does not account for the effects of soil properties, in some cases all the design volumes can not be injected due to the early

Attained a_s (%)

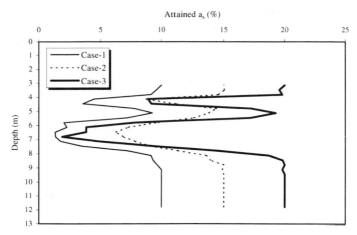

Figure 2. Variation of attained replacement ratio with depth.

refusal to grouting. During injections of the three cases, the design a_s was attained at some depths, while it was not at others. Figure 2 shows the variation with depth of the attained a_s for the three cases. It is seen that the three profiles are similar. However, the profiles of Cases-1 and 3 are better consistent with each other than with that of Case-2. This may be attributed to the sequence of cases injection, where the piles of Case-2 were injected first, then those of Cases-1 and 3. In other words, the primarily injected piles of Case-2 should have contributed to the improvements of the soils treated by the subsequently injected piles of Cases-1 and 3. By comparing the profiles of a_s in Fig. 2 for Cases-1 and 3 (cases injected under almost the same conditions), the following observations can be made (the results of Case-2 will not be discussed in this section, because of its dissimilar sequence of pile injection):

• The design a_s was attained for the soils below about 9.30 m, the highly compressible soils (soils of large fines content). However, for the above soils, there is a wide variation of the attained a_s with depth. This variation is most likely attributed to the variation with depth of the soil properties.

• Although the same limiting pressure of 8.0 MPa was considered for both cases, the attained a_s for Case-1 is always less than that for Case-3. For example, at the depths between 6.00 and 7.00 m, although a_s for Case-3 is less than 10% (design a_s for Case-1), a_s for Case-1 is less than that for Case-3. In addition, for the soils above about 9.30 m, it is observed that the variation of a_s for Case-3 is larger than that for Case-1. In other words, the difference between the design a_s and the attained a_s for Case-3 is larger than that for Case-1. These observations suggest that the design a_s also influences the attained a_s.

For better understanding of the effects of soil properties and design a_s on the attained a_s, the variation of the attained a_s normalized to the design one, $a_{s(Attained)}/a_{s(Design)}$, with pre-treatment fines content (Fc), mean particle size (D50) and N-value of SPT test, as shown in Fig. 3, may be considered. The Fc and D50 are considered as representatives of the soil intrinsic parameters, while N-value as an index reflecting the soil state.

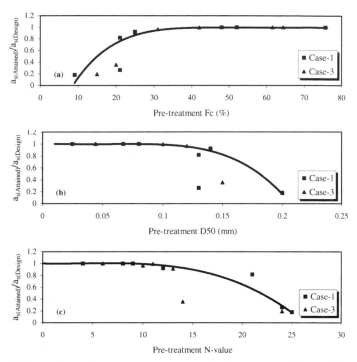

Figure 3. Correlation of $a_{s(Attained)}/a_{s(Design)}$, for soils deeper than 5.30 m, with: (a) Fc; (b) D50; (c) N-value.

El-Kelesh et al. (2001, 2002a) reported that the injections shallower than a given depth defined as the critical depth are usually characterized by upheave of ground surface, while the deeper ones are not. In this regard, it should be mentioned that for Case-2, significant upward displacements of ground surface were observed during the injections shallower than about 4.80 m. These displacements are reported by Yamaguchi et al. (2000). It was also reported (El-Kelesh et al. 2002a) that the critical depth increases with increasing the diameter of grout pile. Therefore, it is reasonable to assume that d_{cr1} ($d_{cr1} < 4.80$ m), 4.80 m and d_{cr3} ($d_{cr3} > 4.80$ m, because of the smaller spacing of Case-3 piles) are the critical depths for Cases-1, 2 and 3, respectively. Because of space limitations, the performance and effectiveness of the shallower injections (shallower than the critical depth) will be discussed elsewhere. However, those of the deeper injections will be considered in this paper. The results in Fig. 3 are those for the soils deeper than 5.30 m. The results in Fig. 3 show the following:

- The groutability of compaction grouting in terms of $a_{s(Attained)}/a_{s(Design)}$ has unique correlations with the pre-treatment Fc, D50 and N-value.
- For the soils of Fc > 40%, D50 < 0.10 mm and N < 9, all the design replacement ratios are attainable.

- For the other soils, for a given design a_s, the attainable a_s increases with increasing Fc, decreasing D50 and decreasing N-value.

The above discussions and correlations answer the questions "Why is the early refusal to grouting and what are the factors affecting the groutability of compaction grouting?". In addition, they provide reliable guidelines for predicting the attainable a_s, for a given design a_s, based on the initial soils properties, and as a consequence, predicting the attainable average improvement.

Effect of Initial Soil Properties on Treatment Effectiveness

Compaction grouting effectiveness is usually evaluated by comparing the pre- and post-treatment N-values of SPT test or tip resistances of CPT test. The post-treatment tests are usually conducted at the grid centers to evaluate the minimum improvement. Figure 4 shows comparisons between the pre- and post-treatment N-values of SPT test for the three cases. As for the improvements by the deeper injections (deeper than the critical depths), it is observed that although all the soils were improved due to grouting, there is a wide variation of the improvement throughout the treated soils. In addition, the improvement is not consistent with the attained replacement ratio. Because of the non-uniform diameters of the injected piles, the treatment effectiveness can not be assessed directly based on the comparisons in Fig. 4. However, there is a need for a criterion or an index that considers both of the improvement and the attained a_s. Therefore, the ratio $\Delta N/a_s$ is used henceforth as an index of the treatment effectiveness, where ΔN = the increase in N-value due to grouting; and a_s = the attained replacement ratio. Figure 5 shows the variation of $\Delta N/a_s$ with depth for the three cases. Because of the dissimilar sequence of pile injection for Case-2 and Cases-1 and 3, the results of Cases-1 and 3 will be discussed in this section, while those of Case-2 will be discussed in the next section.

The results in Fig. 5 reveal that the effectiveness of treatment by Case-1 is almost the same as that by Case-3 for most of the treated soils. Taking into

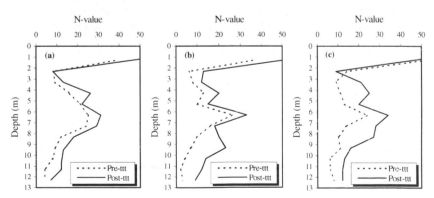

Figure 4. Comparison between pre- and post-treatment SPT N-values: (a) Case-1; (b) Case-2; (c) Case-3.

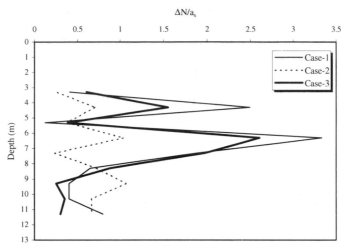

Figure 5. Variation of $\Delta N/a_s$ with depth.

consideration the variation of attained a_s with depth for Cases-1 and 3, as given in Fig. 2, it is concluded that, at a given depth, the improvement in terms of ΔN increases directly with increasing a_s. This implies that the improvement at a given depth can be reliably controlled by controlling the grout volume.

The results in Fig. 5 also show that the effectiveness of treatment in terms of $\Delta N/a_s$ varies significantly with depth. This variation is essentially attributed to the variation of soil properties with depth. For better understanding of the effects of soil properties on the treatment effectiveness, Figs. 6 through 8 show comparisons throughout the treated soils between $\Delta N/a_s$ and the pre-treatment soil properties in terms of 1/Fc, D50 and N-value, for the three cases. These comparisons reveal that the variation of $\Delta N/a_s$ is consistent with those of 1/Fc, D50 and N-value throughout the treated soils, and indicate that $\Delta N/a_s$ increases with decreasing Fc, increasing D50 and increasing N-value.

However, it is observed that the results for the soils shallower than about 5.30 m, for Cases-1 and 3 in Figs. 6 and 7, are less consistent ($\Delta N/a_s$ is lower) than those for the deeper soils. The more consistent results for the deeper soils indicate that the local soil properties are highly controlling the treatment effectiveness. On the other hand, the less consistent results for the shallower soils are explained by the mechanisms of the shallow injections that are characterized by associated ground surface upheave and are mainly controlled by overburden pressure and shear strength of the soils overlying the masses being injected (El-Kelesh et al. 2001, 2002a). As for the comparisons with N-value in Fig. 8, for the soils shallower than about 5.30 m, the results are better consistent than the corresponding ones in Figs. 6 and 7. This may be explained by the effect of overburden pressure that is mainly controlling the measured N-value (Meyerhof 1957) and the shallow mechanisms of compaction grouting (El-Kelesh et al. 2001).

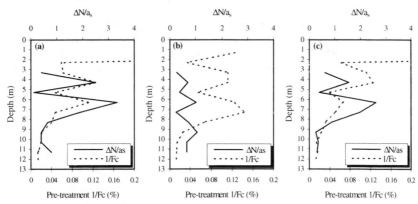

Figure 6. Variation of $\Delta N/a_s$ and 1/Fc with depth: (a) Case-1; (b) Case-2; (c) Case-3.

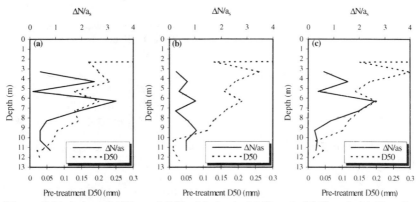

Figure 7. Variation of $\Delta N/a_s$ and D50 with depth: (a) Case-1; (b) Case-2; (c) Case-3.

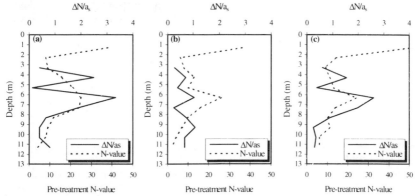

Figure 8. Variation of $\Delta N/a_s$ and N-value with depth: (a) Case-1; (b) Case-2; (c) Case-3.

For practical considerations, it is very important to quantify the effects of soil properties on the treatment effectiveness. Figures 9 through 11 show correlations of $\Delta N/a_s$ with the pre-treatment Fc, D50 and N-value, respectively, for the three cases. The data considered in these figures are those for the soils deeper than 5.3 m, to consider only the effectiveness of the deep injections. Correlation lines are also shown in the figures for the results of Cases-1 and 3. The correlations in Fig. 9 indicate that the treatment effectiveness in terms of $\Delta N/a_s$ is highly influenced by Fc and that $\Delta N/a_s$ decreases sharply with increasing Fc until reaching Fc of about 40-50%. Then, with more increases in Fc, $\Delta N/a_s$ decreases slightly. Despite the wide range of Fc considered, the correlation lines for Cases-1 and 3 are very close and similar. Therefore, an average correlation line (the solid line) for all the data of Cases-1 and 3 is also considered in Fig. 9. This line may be represented by the expression

$$\frac{\Delta N}{a_s} = \frac{47.09}{Fc^{1.17}} \tag{1}$$

The correlations in Fig. 10 indicate that $\Delta N/a_s$ increases linearly with increasing D50. The larger the particle size, the more the treatment effectiveness. The average correlation in Fig. 10 may be represented by the expression

$$\frac{\Delta N}{a_s} = 14.74\,D50 - 0.44 \tag{2}$$

Similarly, the correlations in Fig. 11 indicate that $\Delta N/a_s$ increases linearly with increasing N-value. The stronger the initial soil in terms of N-value, the more the treatment effectiveness. The average correlation in Fig. 11 may be represented by the expression

$$\frac{\Delta N}{a_s} = 0.12\,N - 0.40 \tag{3}$$

The correlations in Figs. 9 through 11 answer a question frequently asked in the practice of compaction grouting: "How do the initial soil properties affect the treatment effectiveness?". The expressions given by Eqs. 1 through 3 also give a satisfactory answer for the question: "How can the treatment effectiveness be predicted with accounting for the variation of initial soil properties?". Therefore, for design purposes, the ratio $\Delta N/a_s$ can be determined based on Fc, D50 or N-value. Then a_s is decided based on the required improvement in terms of ΔN.

Effect of Injection Sequence

Highly Compressible Soils. As for Case-2, Fig. 5 shows that the effectiveness of treatment by Case-2 is different from those by Cases-1 and 3. For the soils below about 8.5 m, the treatment by Case-2 is more effective, while it is less effective for the above soils. It is believed that the main factors contributing to this variation are as follows:

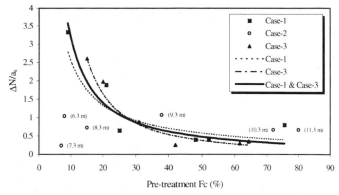

Figure 9. Correlation of $\Delta N/a_s$ with Fc for soils deeper than 5.30 m.

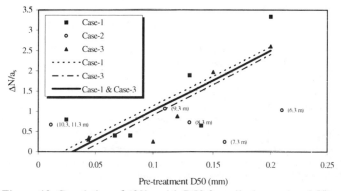

Figure 10. Correlation of $\Delta N/a_s$ with D50 for soils deeper than 5.30 m.

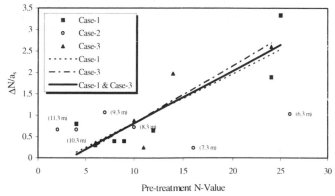

Figure 11. Correlation of $\Delta N/a_s$ with N-value for soils deeper than 5.30 m.

- Soil compressibility. The soils below 8.50 m are highly compressible in terms of Fc. Below 8.50 m, Fc ranges from 38 to 84%. However, above 8.50 m, Fc ranges from 1 to 31%.
- Sequence of injection. For Case-2, L-1 and L-2 were injected first at the boundaries of treatment zone. Then, the other lines (L-3 through L-7) were subsequently injected between L-1 and L-2 (see Fig. 1). For Cases-1 and 3, the lines were injected while going outward from the area of Case-2 in the negative and positive X-directions, respectively (see Fig. 1). In other words, the injection of Case-2 may be considered as confined, while those of Cases-1 and 3 as unconfined.

The effectiveness of treatment by Case-2 for the soils above 8.50 m will be discussed in the next sub-section. For the soils deeper than about 8.50 m (the highly compressible soils), the following observations however can be made:

- Because all the design replacement ratios were attained as shown in Fig. 2, the pre- and post-treatment N-values as given in Fig. 4 may be used to assess the treatment effectiveness. Figures 4-a and 4-c show that increasing a_s from 10% (Case-1) to 20% (Case-3) did not result in appreciable increase in the improvement. This observation might not be strange, because when treating highly compressible soils, significant improvement is usually unexpected.
- However, Figs. 4-b and 4-c show that although a_s of Case-3 (20%) is larger than that of Case-2 (15%), the improvement by Case-3, unexpectedly, is much less than that by Case-2.

Therefore, among the three cases, Case-2 resulted in the largest improvement, although its attained a_s was not the largest. The more effective treatment by Case-2 is shown in Figs. 6 through 8 (the horizontal scales of Case-2 figures are the same as those of Cases-1 and 3) by the larger values of $\Delta N/a_s$ (than those of the assumed $\Delta N/a_s$ profile that gives correlation factor similar to that of Cases-1 and 3). It is also shown by the larger correlation factors in Figs. 9 through 11, where the data points of the soils below 8.50 m are above the correlation lines (the depth of each data point of Case-2 is shown in the figures).

The main reason for the more effective treatment by Case-2 is undoubtedly the injection sequence, in which the injection of L-1 and L-2 provided effective lateral confinement and containment for the subsequent injections of L-3 through L-7. Very interestingly, a confined injection of 15% (Case-2) in replacement ratio is much more effective in treating highly compressible soils than an unconfined one of 20% (Case-3) in replacement ratio. Usually, the injection sequence is an important consideration when using compaction grouting to underpin or lift structures. However, for the conventional ground improvement works, the above conclusion on the effect of injection sequence provides new approach and guideline for improving the treatment effectiveness, especially for the highly compressible soils, whose effective treatment is usually unexpected. In other words, by controlling the injection sequence, larger improvements may be obtained without any cost increase or time delay; or alternatively, the target improvements may be obtained by smaller replacement ratios.

Less Compressible Soils. It is well known that the more compressible the soil, the less the improvement with respect to grout take. In addition, the discussion presented in the previous sub-section revealed that the confined injection is much more effective than the unconfined one. Nonetheless, it is seen in Fig. 5 that the treatment by Case-2 (confined injection) for the soils above 8.50 m (less compressible soils) is less effective than those by Cases-1 and 3 (unconfined injections). It should also be considered that, as shown in Fig. 2, the attained a_s for Case-2 is larger than that for Case-1. The less effective treatment by Case-2 is also shown in Figs. 6 through 8 by the smaller values of $\Delta N/a_s$ (than those of the assumed $\Delta N/a_s$ profile that gives correlation factor similar to that of Cases-1 and 3), and in Figs. 9 through 11 by the smaller correlation factors (the data points of Case-2 are below the correlation lines).

Therefore, the question that needs to be answered is: "Compared to those by Cases-1 and 3, why is the less effective treatment by Case-2 for the soils above 8.50 m?". It is believed that this question may be answered by considering the injection sequence and the deformation mechanisms. Fundamentally, there are two deformation mechanisms in granular soils: crushing of individual particles and relative displacement between particles as a result of sliding or rolling. The two mechanisms are seldom independent of one another. The deformation mechanisms may be understood by comparing the pre- and post-treatment soil intrinsic parameters.

Figures 12 and 13 show comparisons between the pre- and post-treatment Fc and D50, respectively, for the three cases throughout the treated soils. Being limited to the soils deeper than the critical depths, the comparisons in Figs. 12 and 13 show that the treatment by Case-2 resulted in significant crushing of the soil particles at the depths between 5.3 and 9.3 m, as represented by the increase in Fc (Fig. 12-b) or decrease in D50 (Fig. 13-b). However, for Cases-1 and 3, the no appreciable crushing at these depths, indicates that relative displacement of the soil particles was the dominant mechanism of deformation. In addition, compared to those by Cases-1 and 3, it is observed in Fig. 5 that the effectiveness of treatment by Case-2 is relatively the least for the soils between 6.30 and 7.30 m in depth. Very interestingly, at these depths, the corresponding particle crushing is the greatest as shown in Figs. 12-b and 13-b. These observations indicate that the intense crushing is undoubtedly the main reason for the less effective treatment by Case-2 for the soils above 8.50 m.

The main reasons for the intense crushing by Case-2 injection are most likely the following:
• Confining action: the injection sequence of Case-2 provided effective confinement for the subsequent injections of L-3 through L-7.
• Less compressibility: the soils above 8.50 m are less compressible in terms of Fc than the below ones (see Fig. 12-b).
• Crushability: the soils above 8.50 m are more amenable to crushing than the below ones, because of their larger particle size (see Fig. 13-b).

Therefore, based on the aforementioned, it is reasonable to conclude that as long as deformation is dominated by relative displacement of the soil particles, the improvement increases continuously with injection. In addition, it is recommended to avoid the condition, beyond which particle crushing becomes the dominant mechanism of deformation, so as to not lessen the gained improvement of treated soils. Evaluation of this condition is an issue that needs further investigation.

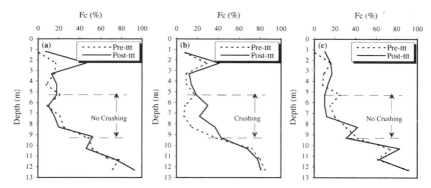

Figure 12. Comparison between pre- and post-treatment Fc: (a) Case-1; (b) Case-2; (c) Case-3.

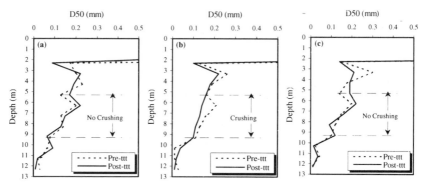

Figure 13. Comparison between pre- and post-treatment D50: (a) Case-1; (b) Case-2; (c) Case-3.

Conclusions

In this paper, analyses of the results of a field test on compaction grout piles were presented. The test consisted of eighty-seven piles injected in three cases of different spacings and pile diameters. Based on the analyses presented, the groutability and the treatment effectiveness of compaction grouting could be better understood and the following conclusions can be drawn for the soils deeper than the critical depth:

• The attained replacement ratio depends on the initial soil properties, as well as the design replacement ratio. For unconfined injections with limiting pressure of 8.0 MPa, the groutability in terms of $a_{s(Attained)}/a_{s(Design)}$ has unique relationships with the initial Fc, D50 and N-value, as shown in Fig. 3. Based on these relationships, all the design replacement ratios (10-20%) are attainable for the soils of Fc > 40%, D50 < 0.10 mm and N-value < 9. For the other soils, the attainable ratio increases with increasing Fc, decreasing D50 and decreasing N-value.

- The treatment effectiveness in terms of $\Delta N/a_s$ is highly influenced by the initial soil properties. The effectiveness increases with decreasing Fc, increasing D50 and increasing N-value. For design purposes, the treatment effectiveness in terms of $\Delta N/a_s$ may be estimated, based on the initial soil properties, using the expressions given by Eqs. 1 through 3 (for the case of unconfined injections). Then, the replacement ratio, a_s, is to be decided based on the required improvement in terms of ΔN.
- As long as deformation is dominated by relative displacement of the soil particles, the improvement by compaction grouting increases continuously with injection. However, the gained improvement may be lessened, if particle crushing becomes the dominant mechanism of deformation. Therefore, it is recommended to avoid the condition, beyond which particle crushing dominates.
- The injection sequence may be controlled to provide effective confinement and containment for the subsequent injections. The confining action has very significant influence on the treatment mechanisms and improvement. The confined injection is much more effective than the unconfined one, provided that deformation is dominated by relative displacement of the soil particles. A confined injection with replacement ratio of 15% could improve highly compressible soils much more than an unconfined injection with replacement ratio of 20%. Therefore, the confined injection may be considered as a new solution for improving the highly compressible soils, whose effective treatment is usually unexpected.

Acknlowledgement

The writers would like to acknowledge the efforts done by the officials of the Port and Airport Department, Kanto Regional Development Bureau, Ministry of Land, Infrastructure and Transport, Japan, in arranging and conducting this field test. Special thanks are extended for providing the test data of this material with all the necessary details. The interest and support of the staff of Fukken Co. Ltd., especially Mr. M. Taki, are also greatly appreciated.

References

Boulanger, R. W., and Hayden, R. F. (1995). "Aspects of compaction grouting of liquefiable soil," *J. Geotech. Engrg.*, ASCE, 121(12), 844-855.

Brown, D. R., and Warner, J. (1973). "Compaction grouting," *J. Soil Mech. and Found. Div.*, ASCE, 99(8), 589-601.

El-Kelesh, A. M. , Mossaad, M. E., and Basha, I. M. (2001). "Model of compaction grouting," *J. Geotech. and Geoenvir. Engrg.*, ASCE, 127(11), 955-964.

El-Kelesh, A. M., Matsui, T., Hayashi, K., Tsuboi, H., and Fukada, H. (2002a). "Compaction grouting and ground surface upheave," *Proc., 4th Int. Conf. on Ground Improv. Tech.*, Kuala Lumpur, Vol. 1, 323-330.

El-Kelesh, A. M., Matsui, T., Hayashi, K., Tsuboi, H., and Fukada, H. (2002b). "Slope stabilization by compaction grout piles," *Proc., 6th Int. Symp. on Envir. Geotech. and Global Sustainable Develop.*, Seoul, Korea, 535-543.

Graf, E. D. (1992). "Compaction grouting, 1992," *Grouting, Soil Improvement and Geosynthetics, Geotech. Spec. Publ. No. 30,* R. H. Borden, R. D. Holtz, and I. Juran, eds., Vol. 1, ASCE, New York, 275-287.

Meyerhof, G. G. (1957). "Discussion on research on determining the density of sands by spoon penetration testing," *Proc., 4th Int. Conf. on Soil Mech. and Found. Engrg.*, Vol. 3, 110.

Salley, J. R, Foreman, B., Baker, W., and Henry, J. F. (1987). "Compaction grouting test program Pinopolis West Dam," *Proc., Soil Improvement-A 10-Year Update,* ASCE, New York, 245-269.

Yamaguchi, S., Kozawa, D., Arata, M., Matsumoto, H., Taki, M., and Kanno, Y. (2000). "Design and construction method of compaction grouting as a ground-improving technique against liquefaction," *Proc., Int. Symp. on Coastal Geotech. Engrg. in Practice (IS-Yokohama 2000),* Vol. 1, 557-562.

Design Considerations for Inclusions by Limited Mobility Displacement Grouting

Michael J. Byle, P.E., F. ASCE

Abstract

Many factors must be considered in the design of inclusions to be created by Limited Mobility Displacement (LMD) grouting. Inclusions can serve many functions from simple displacement of the surrounding soils to cause compaction, to forming structural columns and improving continuity of rock masses. The creation of inclusions is dependent upon the grout properties, subsurface conditions, and the design objectives. Practical limitations dictate the controllable size of inclusions, while grout and soil properties, and construction procedures control the strength and shape of injected masses. This paper reviews analytical tools for design of inclusions and presents recommendations for their use.

Introduction

Limited Mobility Displacement (LMD) grouting, now popularly called "low mobility grouting" or LMG, is the injection of a stiff grout that does not mix with or penetrate the soil, displaces the substrate into which it is injected and does not travel far from the point of injection (Byle, 1997). This type of grouting injects a grout mass into the soil or soil or water filled openings in rock. This grout mass acts as an inclusion that can either be structural or nonstructural. In the effective design of LMD grouting, the nature and function of the inclusion must be clearly understood and considered in the design of a grouting program. This is especially important for applications where the grout inclusion serves a structural function, such as grout columns, underpinning, karst mitigation, or rock mass improvement.

In these applications it is critical to install the grout inclusion at the required location and size with the appropriate strength. There is always a measure of uncertainty in the prediction of grout travel, but the LMD grouting process will provide a high degree of control and a reasonable certainty that the grout will remain in the location where it is injected. There are a great many factors affecting the grout travel and properties, many of which are can only be controlled by an experienced and qualified contractor. Given an experienced contractor using good grouting practices, the measures outlined in this paper will provide a reasonable basis to develop a rational grouting program.

Design Methodologies

Limiting Injection Pressure/ Volume Relationship. There are few broadly accepted methodologies for the design of LMD grouting. Expanding cavity theory has been cited as the basis for describing the injection of grout bulbs (El-Kelesh et al., 2001, Warner, 1992). El-Kelesh et al (2001) have attempted to develop predictive equations for the behavior of grout injections including the limiting pressure and grout volume for grouting in granular soils. The procedure involves computing the pressure volume relationship developed from spherical cavity theory simultaneously with the limiting pressure volume relationship for conical shear. These approaches assume an isotropic elastic-plastic continuum; a condition that rarely exists in reality. This procedure is complex and seldom used in design. This analysis also requires fundamental soil properties such as Poisson's Ratio, Rigidity Index and Modulus of Deformation that it may be impractical to obtain for most applications. These parameters may be estimated, though the errors induced by estimation may limit the usefulness of the approach.

Most contractors rely on rules of thumb to estimate the volume of grout that can be injected in soil. Typically 0.1 to 0.2 m³ / m (1 to 2 ft³ / ft) of hole is the practical range for loose to medium dense soil within 2 meters (6 ft) of the ground surface. Higher quantities may be injected at greater depths. Typically 0.3 to 0.5 m³ / m (3 to 5 ft³ / ft) can be injected with 2 to 5 m of overburden. For very soft conditions or depths greater than 5 m, or when injecting into karst much larger quantities are safely achievable. The major limitation to injection is the heaving of the ground surface and/or excessive lateral deformations to subsurface structures. Stresses are estimated based on elastic and limit-state theory to determine affects on adjacent structures and heave of the ground surface.

A simple estimation tool is to consider the injection pressure acting on the projected area of the grout mass in the horizontal plane at the top of the injection to verify adequate overburden to resist heave (Figure 1). The resistance of the overburden is estimated as the mass of the truncated cone of soil above the injection point. A simple force balance on this free body will give a reasonable estimate of resistance. For the purposes of this evaluation, the maximum pressure anticipated should be used, since only the rate of injection is controlled in the injection process and pressure is only limited by the resistance of the soil and failure to the surface. This relation ship may be used iteratively to determine a limiting pressure-volume relationship relative to depth to be used in establishing refusal criteria.

The volume of the truncated cone is given by the following relation:

$$Vc = \left(\frac{\pi \cdot D}{3} \right) \cdot \left[D^2 \cdot (\tan(\theta))^2 + 3 \cdot r \cdot D \cdot \tan(\theta) + 3 \cdot r^2 \right] \qquad (1)$$

Where r = radius of grout bulb, D = depth to grout injection, and θ = angle of the cone. The angle, θ, varies with the soil properties following typical Mohr-Coulomb criteria.

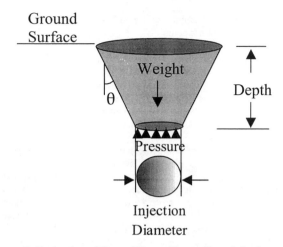

Figure 1 Evaluation of Ground Heave Due to Grout Injection at Depth

For most cases, the effect of shear on the surface of the cone may be neglected where multiple stages may cause repeated shear on this plane. A simplified force balance would be represented by the following equation:

$$p * \pi r^2 = \gamma * V_c \qquad (2)$$

Where p = grout pressure, and γ = total unit weight of the soil.

It will often be possible to inject larger masses of grout without ground heave, however, some circumstances will produce heave at lesser volumes. This simplified approach is suitable for most applications. Where the project is of a critical nature and size to warrant a more detailed study, more sophisticated testing and analyses such as finite element analysis may be appropriate.

Compaction Displacement. For compaction grouting applications, the design must define the amount of density increase required and the volume of material to be densified. The simplest approach to determining the required density increase is evaluating the soil in-situ unit weight and relative density, and then comparing that to the required relative density to provide the required performance. One then determines the change in volume that is required to achieve the required relative density by the following equation:

$$V_d = \frac{\gamma f - \gamma_o}{\gamma f} \cdot 100 \qquad (3)$$

Where: γ_o = initial in-situ total unit weight, γ_f = final desired total unit weight and V_d = percent displacement volume required.

The volume of material to be densified is determined geometrically as the volume of soil that must be improved to provide the required result. One should consider all of the ramifications of grout injection on determining this total volume. Consider that improvement of deeper layers may be required to support the added weight from the mass of grout injected. The total volume of grout should be taken as the total volume to be treated times the percent displacement volume calculated above.

Injection Volume. The volume of grout injected, as measured at the pump will be differ from the volume of displacement in the ground. This is due to volume reduction because of compression of air voids in the grout under pressure and the expulsion of water and/or paste through system leakage, pump bypass, inaccurate flow measurement methods and loss to the ground at the point of injection. The amount of loss is specific to the injection equipment and grout mixture. Proper control of limited mobility displacement grouting requires some fluid loss (Byle, 2000). The amount of this loss is typically small and not significant relative to the design, however, it is important for construction control to evaluate volume of the inclusion to verify that the required dimensions are achieved. The injected grout volume, V_g , should be determined from the measured volume using the adjustment factor, $F_{vr,}$ as follows:

$$V_g = V_d * F_{vr} \qquad (4)$$

There is very little in the way of measurement data to properly calibrate this factor. An example calibration is given by the following case. In this case, grout columns were injected into a layer of soft peat. The pump was calibrated using a 1 cubic foot box. The stroke count was recorded and volume of grout computed based on the calibration. The total volume injected was recorded as 10.3 cubic feet (0.96 m³) discounting the volume required to fill the pipe. The injected column was subsequently extracted and measured. The volume of the column was measured as 10.7 cubic feet (0.99 m³). This represents a 4% increase in the volume measured by pump strokes. In this case F_{vr} = 1.04. There was nothing in the mix design that would have produced expansion of the grout. The error in volume measurement is likely due to an automatic stroke counter that failed to count all strokes or to partial strokes of the piston not being accounted for properly. A comparison of the injected quantities to the measured column volume is presented in Figure 2. This figure demonstrates that even with low mobility grout, the grout may migrate away from the point of injection.

Where accurate volume measurements are made, factors very near but slightly less than one would commonly be expected. This is due to the fact that there will always be some amount of fluid loss from dilatent cement based grout which will produce a corresponding decrease in the volume of the grout mass as it is compressed by the injection pressure. Where injections are made into reasonably well-drained soil, it is

Injection Ratio

Difference Between Measured and Pumped Volume as a Percentage of Pumped Volume

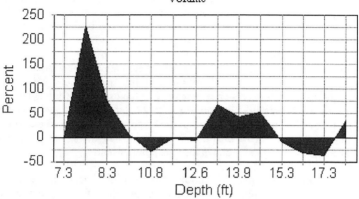

Figure 2 Comparison of Measured Grout Column Dimensions to Injection Volume Measured by Pump Strokes.

expected that the grout solids will be densely packed under the injection pressure and fluid will be lost to the surrounding ground. The volume of grout within the injection riser and hoses must be deducted from the total volume injected as this volume is removed with each pipe section as the riser is extracted.

Unless specific field measurements are made to verify this factor for specific equipment and grouting conditions a value near unity is recommended. Absent specific data, F_{vr} should be taken as 1.0. Determination of the correct ratio can be determined by excavating test injections and measuring the volume of the hardened grout mass.

Volume Measurement. As noted above, the measurement of grout flow is a critical element in the evaluation of grouting performance. Volume measurement is also an economic factor, since many contracts include a pay item for quantity of grout injected. In compaction applications, the true injected volume is an exact measure of displacement and, hence, a direct measure of densification. For the creation of inclusions, such as grout piles, the volume injected is a measure of the size of the grout body created.

Volume measurement of compaction grouting the field is typically of limited accuracy. Often the volume injected is measured by counting strokes of a piston pump. The volume is computed either by using the cylinder displacement and assuming the pump injects a full cylinder volume on each stroke, or by calibrating the pump by pumping into a barrel of known volume or a 1 cubic foot (0.28 m³) box.

There are significant sources of error in volume measurement in this fashion. These include: inadequate accounting for partial strokes, incomplete filling of the cylinders with stiff grouts, inaccurate mechanical stroke counters that either skip or add strokes, inaccurate manual stroke count and grout bypassing worn seals. Additional factors affecting the accuracy of volume measurements by this method are many and include: worn swing tube wear plate, air content of the grout, fluid losses through fittings and joints, spillage, wasted grout, and backward flow. The backpressure and rate of pumping will also affect the volume displaced per pump stroke.

Errors in measured volumes may be obtained where partial strokes are either over or under counted as full strokes, where there is leakage in the system, or where compaction of the grout occurs under pressure. One example of the latter condition would be the use of air entrainment in the grout. After grout is drawn into the pump cylinder, air bubbles entrained in the grout are compressed and forced into solution by the pumping pressure. This pressure is maintained in the ground resulting in a smaller volume injected than is drawn into the pump piston.

The typical method of pump calibration is of marginal value as the pumping is usually done at a slow rate and without backpressure and for most pumps requires estimation of partial strokes to fill the container. To have any value, the calibration container should be at least 10 times the volume displaced by each piston stroke. This would limit the error due to estimating partial strokes to less than 5 %.

Comparison of stroke counts to daily mix volumes have indicated errors of +30% to -50% from the stroke count volume depending on the method of calibration used. Where pump strokes are used for volume measurement, the pump stroke total volumes should be checked against batched volumes for each batch and for a running total on a daily basis. This means that at the end of each day's pumping, the total of stroke counts plus measured waste volume should be equal to the total batched volume. Where the grout is delivered to the site in ready-mix trucks, the volume of grout in each truck should be verified. Off site mixing adds another level of uncertainty to the equation. Delivery tickets can also be in error and the volume carried in each truck should be checked to assure a reasonable calibration of the grout injection.

New flow measurement devices based on magnetics and acoustics are becoming available that can measure the flow of very stiff grout may make this correction less of an issue in the future. Such meters, properly calibrated will eliminate much of the current uncertainty in volume measurement.

Grout Strength Design. The strength requirements for the various applications of LMD grouting vary with the application. Where the grout serves a structural purpose, the grout strength must be adequate to support the required loading; otherwise, strength may be of lesser importance.

Compaction Grouting. The grout strength required for compaction grouting is usually not a critical factor. There is usually no minimum strength requirement for compaction grouting except that the grout must be at least as strong as the surrounding soil. This in most cases will require grout strength to be relatively low. The required strength should be computed as the greater of the undrained shear strength, or the consolidated drained strength of the soil after compaction. This will require consideration of the initial ground stress and may result in higher strength being required at depth. Technically, a compaction grout need not contain cement if it has internal friction equal or greater than the compacted soil. Typically compaction grout strengths are low on the order of 50 to 300 psi (340 to 2100 kPa)

Columns. Common structural applications involve the creating columns of grout or making structural connections in a rock mass. The difficulty in these applications is determining the effective cross sectional area to compute stress. Since the grout injection diameter is a function of the resistance of the surrounding soil or rock to displacement, except in very uniform conditions, one cannot be entirely sure what the actual injection dimensions will be prior to injection. While the limit volume computation presented in equations (1) and (2) above, will provide a reasonable starting point in estimating the injection size, the actual injection size will vary due to such factors as anisotropy and other non-homogeneity.

There are two basic approaches to handling this uncertainty. The first, and most conservative, is to design the strength based on the diameter of the injection hole. There is a high degree of certainty that this diameter will be achieved and any additional injection quantity serves to increase the factor of safety. This may be a reasonable approach for critical structures where high strength grout can be used at reasonable spacings. The second approach relies on a statistical approach and assumes that an average injection diameter will be achieved and that any deviations in the grout diameter will be randomly distributed. This is a common approach and is reasonable for grouting in non-layered deposits such as random fills, colluvium, or broken rock and where there are a large number of columns such that a reduced capacity in any one column will not be critical. In this case a minimum injection per stage must be specified and stages must be small enough that a reasonable assurance of continuity is obtained.

For layered deposits, a third more complex approach may be used where the hole diameter approach is not feasible. This approach involves a detailed evaluation of the subsurface layering to estimate the probable behavior of grout injection in each layer. Additional controls, such as reduced stage lengths and/or reduced pumping rates, may be used in specific layers to alter the grout behavior in critical zones to produce the required dimensions. Consideration should be given to using top down grouting, where each stage is drilled through and injected below the previous stage, where layers of substantially different stiffness would permit grout to migrate away from the target layer to softer, less stiff layers above or below. Layered soils will require detailed grouting design and specifications that are closely monitored in the field to validate the design assumptions. Test grouting and excavation of control columns may be required.

For grouted columns, the capacity of the column is controlled by end bearing and skin friction, similar to other types of piles. End bearing may be enhanced by injection of a larger bulb of grout at the column base. For most granular soils, the skin friction may be conservatively taken as the shear strength of the improved soil using normal at-rest earth pressure. The grout injection will induce higher residual stresses that will increase the frictional resistance above the at-rest earth pressure; however, the residual stresses decrease over time to a fraction of their original value (Schmertmann and Henry, 1992). Where columns are spaced sufficiently close to produce a uniformly stressed volume of soil, the corrected residual stresses may be used in the computation of column frictional capacity. For single columns, the at-rest earth pressure should be used.

Rock Binder. Where the grout is used to form structural connections in a rock mass, the grout must have sufficient strength to distribute stresses between rock blocks. The strength required is dependent on the size of the voids or joints filled by grout and the degree of contact expected between grout and rock. Where the space in the rock contains soil, the mechanism for binding the rock may be a combination of grout strength and soil compaction. In foundation applications, tensile strength is not usually a requirement. Therefore, the grout injection should be performed in such a way as to compact soils present sufficiently so that the combination of compacted soil and cured grout will have sufficient compressive strength to support the required loading. If the soil is not displaced or compacted sufficiently, the soil will control the mass behavior and may permit undesirable movements. In these cases, the grout should be designed for a reasonably high strength.

Typically, grout strengths of 1,000 to 4,000 psi (7000 to 28,000 kPa) are specified for rock binder applications depending on specific requirements. Where the rock is being improved to permit the installation of driven point bearing piles, and the possibility exists for piles to be supported entirely or partially on grout, a higher strength grout would be required. Where the grout binds a larger mass of rock together to support a spread footing at a relatively low bearing pressure, a lower strength may be used.

Karst Mitigation. A procedure for mitigating the effect of small karst features has been developed by Schmertmann and Henry (1992). This procedure is based on soil arching between grouted columns to span over openings that may have been missed by the grouting. Based on an experimental model, analysis, and full-scale field installation, a design chart was developed to determine the effect of varying columns size and spacing on a square grid pattern (Figure 3).

An alternate approach to karst remediation is referred to as cap grouting, where a low mobility grout is injected at the top of rock to fill erosion domes in the soil and seal openings in the rock surface. The design of cap grouting is based on the determining hole spacing based on site specific, conditions. Holes must be spaced close enough

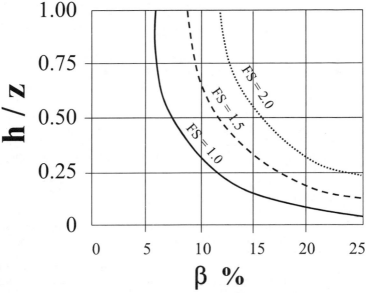

Figure 3 Design Chart for Grout columns of height, h, in soil of depth, z with a grout lateral volume strain, β, to achieve various safety factors for a grout column spacing equal to ½ of the layer thickness. (After Schmertmann and Henry, 1992)

for grout to seal the openings at the soil/rock interface, but wide enough for economy. The grout consistency and nature of the karst features will dictate the frequency of injections required. A trial program that includes split spacing may be used to verify grouting effectiveness and to optimize the injection hole spacing.

A third approach is void filling. In this approach, grout holes are drilled into the rock and voids encountered are filled with limited mobility grout. This process is completely controlled by the geologic and rock conditions. It is essential to have a good understanding of the rock structure and the probable extent of solution features to target grouting to the correct locations. Geophysical tomography or other special investigative methods may be necessary to determine the required extent and volume of grouting and determine appropriate orientation and spacing of grout holes to intercept solution features at reasonable intervals.

Composite Design

In reality, the grout inclusions in LMD applications may perform many functions. The grout inclusions may reinforce soils, provide shear resistance, distribute stresses and perform other functions in conjunction with the soil or rock. To fully model this composite behavior would require complex modeling beyond the scope of this paper.

Some of the tools described above may be used to evaluate some aspects of the design. One must not assume that any one analysis will be sufficient in and of itself. Unique aspects of each application must be appropriately considered.

Where appropriate, issues of strength, structural capacity, flexure, strain compatibility and composite behavior must all be appropriately addressed. For many applications, only one or two of these issues are significant. A reasonable approach is to evaluate the composite action of grout inclusions based on other design procedures for similar structures and evaluate the comparative behavior of grout to that of more common structural elements. Composite properties of soil and grout may be estimated based on shear strength, displacement ratio, and densification; however these should be used with caution and validated by constructing test sections and conducting in situ verification testing.

Completing the Design

In any LMD grouting program, at a minimum, the following must be determined and specified: grout strength, maximum hole spacing, and depth of grouting. Additional factors may include maximum and/or minimum grout injection per stage, stage size, bored hole size, and specific injection locations, where critical. The level of detail required will depend on the nature of the application and the tolerance of the design for deviations. Where the inclusions serve only to displace and densify a soil mass, only the target density and limits of improvement are truly necessary. Where columns are designed to provide structural support, the locations of columns may be critical to proper support and should be specified. In rock mass binding applications, the location and orientation of geologic features may dictate the requirements for grout injection orientation and spacing and it may be appropriate to specify these.

In general, the specification should always clearly state the purpose of the grouting and function of the grout inclusions. Wherever possible, the grouting contractor should be made a partner in the design. This limits misunderstandings and generally produces a superior result. Where LMD grouting is being competitively bid, except in the case of soil compaction, more design details must be specified. Wherever performance cannot be readily verified or where clear acceptance/rejection cannot be defined, a method specification is appropriate. Otherwise, a performance type specification with defined measures of performance with clear acceptance/rejection criteria is preferred. Pre-qualifying LMD grouting contractors is recommended wherever practical in competitive bid situations.

Verification methods and measures should always be used to provide some assurance that the objectives of the grouting program have been met. Tools for selection of verification methods have been developed by the ASCE Committee on Grouting (1995). The design of the grouting program must consider how it will be verified and how critical aspects of the grouting necessary to the proper performance will be controlled. Proper performance should not be left to chance.

References

ASCE Committee on Grouting (1995) *Verification of Geotechnical Grouting.* Byle, M.J. and Borden, R.H. Eds. Geotechnical Special Publication No. 57, ASCE Press, pp. 1-77.

Byle, M.J. (1997) "Limited Mobility Displacement Grouting: When 'Compaction Grout' is Not 'Compaction Grout', Proceedings of Sessions by the Grouting Committee of the Geo-Institute/ASCE in conjunction with the GeoLogan '97 Conference, *Grouting, Compaction, Remediation and Testing GSP No.66*, pp 32-42.

El-Kelesh, Adel M., Mostafa E. Mossaad, and Ismail M. Basha, (2001) "Model of Compaction Grouting" *Journal of Geotechnical and Geoenvironmental Engineering*, Vol. 127, No. 11, November 2001, pp. 955-964

Schmertmann, J.H. and Henry, J.F. (1992)" A Design Theory for Compaction Grouting." *Grouting Soil Improvement and Geosynthetics,* R.H. Borden, Ed. ASCE Geotechnical Special Publication No.30, pp 215-228.

Warner, James (1992) "Compaction Grout; Rheology vs. Effectiveness," *Grouting Soil Improvement and Geosynthetics, Geotechnical Special Publication No. 30,* Borden, R.H., Holtz, R.D., and Juran, I. Eds., ASCE, 1992. pp 229-239

Characterization of Fractured Rock for Grouting Design Using Hydrogeological Methods

Åsa Fransson[1]
Gunnar Gustafson[1]

Abstract

This paper aims at briefly presenting a methodology for characterization of fractured rock for grouting design using hydrogeological methods. The conceptual model is based on a grouting fan and is built up by fractures inferred from hydraulic tests and geological mapping. Instead of the commonly used Lugeon value, the specific capacity (Q/dh i.e., flow divided by difference in hydraulic head) is central since it has shown to be a robust parameter, which can be related to transmissivity and fracture aperture. Fracture aperture is important for grouting design due to its influence on both penetration length and grout take. The methodology described for estimation of transmissivity and aperture distributions has potential for further development for computer use, which would enable a fast analysis of data from hydraulic tests and geological mapping at a working site. Based on aperture distribution and expressions describing the spreading of grout, the choice of input parameters such as grout properties, pressure and borehole distance could be improved. Furthermore, the transmissivity and aperture distributions for probe holes give a general description of rock, which is used for the interpretation of data from individual grouting boreholes. This description of fractured rock for grouting should be a good basis for further discussions and development as well as facilitating the choice of strategy.

Introduction

The construction of a laboratory facility at Äspö, the Äspö Hard Rock Laboratory, Sweden, to investigate questions related to the deposition of nuclear waste has given rise to a need for further understanding and development of grouting technique. This is crucial, since the construction of access ramps, transport tunnels and deposition tunnels must meet demands for safe and controlled conditions that facilitate suitable sealing.

---

[1]Department of Geology, Chalmers University of Technology, SE-412 96 Göteborg, Sweden; phone +46 31 772 2057; asa.fransson@geo.chalmers.se; +46 31 772 1928; g2@geo.chalmers.se

Furthermore, information about the spreading and amount of grout injected is useful in analysing the behaviour of a deep repository when it is finally sealed. The work was initiated by the Swedish Nuclear Fuel and Waste Management Company, SKB.

The characterization for grouting is looked upon as an engineering problem and the methods used are hydraulic tests and geological mapping. Simplifications are made to develop a method and a geometrical model, which should consider the prevailing conditions, such as limitations in time and availability of data.

Conceptualization

Sealing of tunnels by grouting is a method commonly used to minimize inflow of water and to enhance the stability of the tunnel. To get an idea of what strategy to use before grouting a section of a tunnel, water loss measurements are performed in the boreholes of the grouting fan, resulting in a number of Lugeon values (Houlsby, 1990 and Kutzner, 1996). These values represent the volumes of water that are injected per unit time and per metre of borehole at a given pressure. However, they do not give any information about the discontinuities that actually transmit the water.

The grouting parameters considered for the conceptual model presented herein are mainly the penetration of grout and the grout take, which are both influenced by the aperture of the fractures, see e.g. Gustafson and Stille (1996) and Eriksson et al. (2000). The penetration of grout, I, can be expressed by the difference between the injection pressure of grout and the water pressure, ΔP, the aperture, b, and the flow limit of the grout material, τ_0, according to:

$$I = \frac{\Delta P b}{2\tau_0} \tag{1}$$

The ability of a fracture to transmit water, or its transmissivity, T, is also dependent upon the fracture aperture, which makes it a key parameter in this work. The transmissivity is obtained by hydraulic tests and the hydraulic aperture is estimated using:

$$T = \frac{\rho g b^3}{12\mu} \tag{2}$$

Besides the aperture, b, the transmissivity is also influenced by the acceleration due to gravity, g, and the density and viscosity of water, ρ and μ. This is also referred to as the cubic-law (de Marsily, 1986). Hydraulic tests can be performed either by pumping or injecting water since the same theoretical considerations apply to "wells" that extract water and those that inject water (Fetter, 1994).

The conceptual model developed for this characterization is based on one individual grouting fan where the boreholes cross fractures known to have varying ability to transmit water as well as different orientations and fracture lengths, see principal sketches in Figure 1.

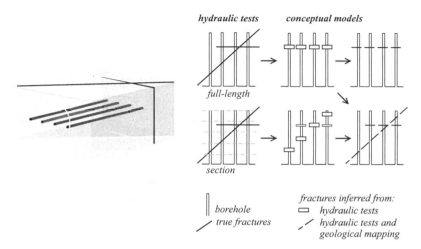

Figure 1. Principal sketches of a grouting fan where four boreholes cross two fractures. The upper part represents full-length hydraulic tests as well as the resulting conceptual models consisting of one fracture whereas the lower part from the section tests enables a more detailed description.

Hydraulic tests and geological mapping are used as main investigations and the intersected fractures are assumed to be two-dimensional (2D) features with radial flow which should provide a useful and robust description to help solve this engineering problem, since it allows two-dimensional analyses for both hydraulic tests and grouting predictions.

Fractures inferred from hydraulic tests and geological mapping should be connected to form a simplified model. The model thus consists of stacked 2D features along boreholes, which, if possible based on transmissivity and orientation, are connected to other features of adjacent boreholes. The figure shows conceptual models inferred from hydraulic full-length and section tests. Here, fracture orientation is considered important to increase the probability of intersecting the fractures, whereas the length gives an indication of the connectivity and where connection between boreholes may be expected.

Further, data from longer boreholes (probe holes) are used to calculate a probability of conductive fractures and a distribution of transmissivities, here referred to as $T(p)$. From the transmissivity distribution, an aperture distribution, $b(p)$ is estimated and, since the grout take is proportional to the transmissivity and the penetration length is proportional to the aperture these distributions are valuable.

Results and discussion

Transmissivity and based on this work, the easier obtainable specific capacity, Q/dh, are important for grouting design since they provide information about the discontinuities that actually transmit the water. Firstly, see Figure 2, transmissivity and specific capacity are important because the estimated apertures, b(T) (or b(p) if a distribution) and b(Q/dh), give guidance about the groutability or what fractures we can expect to seal and what type of grout to use. Further, the result or the transmissivity and inflow (Q ≈T·dh) that would remain after sealing these fractures can be related to a tightness criterion.

In Fransson (2001), six boreholes were used to characterize a limited volume of rock and Figure 3 presents specific capacities from 0.5-metre section tests along these boreholes. Based on fracture mapping of cores and logging of boreholes, the largest specific capacities of boreholes 2-6 originate from the same fracture which indicates that the conceptual model in Figure 1 can be used as a basis to construct a simplified model. In this case, the specific capacity of the entire borehole or full-length hydraulic tests would have been a good description of the main conductive fracture. This is seen by comparing Q/dh:1, which is the largest specific capacity of each borehole and Q/dh:All, which represents the summation of specific capacities along each borehole. Q/dh:1 for borehole 1 and Q/dh:2 and Q/dh:3 originated from shorter fractures which did not seem to be connected.

These investigations and what was presented in Fransson (1999) indicate that the specific capacity is a robust parameter and that the median specific capacity, Q/dh_{50}, of several boreholes crossing the same fracture is close to the effective or cross-fracture transmissivity. This was also the case for the boreholes presented in Figure 3. The effective or cross-fracture transmissivity is a representative value of the fracture and the differences between the median and the other specific capacities or apertures indicate the variations in aperture within the fracture.

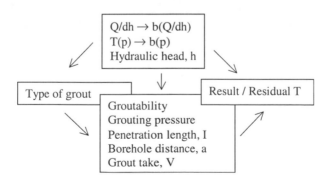

Figure 2. Sketch showing how the different parameters could be used in a grouting strategy.

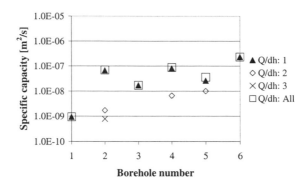

Figure 3. Specific capacities for 0.5-metre sections measured in six boreholes. Q/dh:1 is the largest specific capacity and Q/dh:All, is the summation of all specific capacities along each borehole.

Based on this conceptual model, penetration lengths could be estimated from obtained apertures and subsequently compared and adjusted to the borehole distance, a, see Figure 4. In this example, one fracture is fully sealed and the other partly sealed due to a smaller aperture. Here as well, a residual transmissivity could be estimated based on transmissivities of those fractures that are expected not to be sealed or only partly sealed (see Figure 2).

From a transmissivity distribution, $T(p)$, (estimated from hydraulic tests and geological mapping of a probe hole, see Fransson, 2002), a distribution of apertures, $b(p)$ (Eq. 2), and subsequently penetration lengths, $I(p)$ (Eq. 1), can be estimated, see Figure 5.

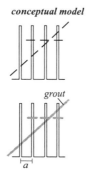

Figure 4. Model based on a grouting fan (see Figure 1) where estimated apertures and assumed pressure and grout properties give one completely sealed and one partly sealed fracture.

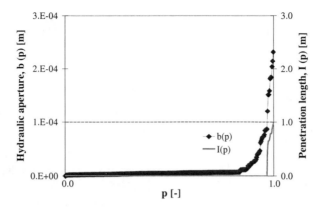

Figure 5. An example of an aperture distribution, b(p) (Eq. 2) and an estimated distribution of penetration length, I(p) (Eq. 1). Fractures with apertures below $1 \cdot 10^{-4}$ m are assumed not to be groutable.

Here, apertures smaller than $1 \cdot 10^{-4}$ m are assumed not to be groutable. Using this concept, the penetration length and borehole distance should be compared and adjusted to improve the grouting design. A residual transmissivity could be estimated based on transmissivities of those fractures that are expected not to be sealed or only partly sealed.

    A transmissivity distribution like the one in Figure 5 was estimated for a probe hole in the vicinity of the boreholes represented in Figure 3. The presence of one major conductive fracture was in accordance with what could be estimated from that distribution.

Conclusions

Based on these investigations and what was presented in Fransson (1999) it was concluded that specific capacity (Q/dh) from short duration hydraulic tests is robust enough to describe fracture aperture which influences both penetration of grout and the grout take. Using the specific capacity would be an improvement compared to the commonly used Lugeon value since it is more easily linked to the transmissivity and the aperture of a fracture. Investigations show that the median specific capacity was found to be close to the effective cross fracture transmissivity. Data from probe holes were used to calculate a probability of conductive fractures and a distribution of transmissivities. From the transmissivity distribution, an aperture distribution is estimated and, since the grout take is proportional to the transmissivity and the penetration length is proportional to the aperture, penetration length distributions could also be estimated. This could be used as guidance for choosing type of grout, grouting pressure, grouting borehole distance and give an idea of the ability of the remaining ungrouted fractures to transmit water. Furthermore, the probe holes can be used to improve the predictions for, and interpretation of data from, individual grouting boreholes. These data, obtained along individual boreholes using hydraulic tests and geological mapping, can be linked to form a simplified model. This was tested and verified during a field experiment as well as strengthened by other similar grouting fans.

The methodology described for estimation of transmissivity and aperture distributions has potential for further development for computer use, which would enable a fast analysis of data from hydraulic tests and geological mapping at a working site. Based on the aperture distribution and expressions describing the spreading of grout, input parameters such as grout properties, pressure and borehole distance could be chosen. Furthermore, the transmissivity and aperture distributions for probe holes give a general description of rock, which is used to improve the interpretation of data from individual grouting boreholes. This description of fractured rock for grouting should be a good basis for discussions and facilitate the choice of strategy.

Acknowledgements

The financial support provided by the Swedish Nuclear Fuel and Waste Management Company is highly appreciated.

References

Eriksson, M., Stille, H., and Andersson, J. (2000) Numerical calculations for prediction of grout spread with account for filtration and varying aperture. *Tunnelling and Underground Space Technology*, 15(4), 353-364.

Fetter, C.W. (1994) *Applied Hydrogeology*. Prentice-Hall Inc., New Jersey.

Fransson, Å. (1999) Grouting predictions based on hydraulic tests of short duration: Analytical, numerical and experimental approaches. Licentiate thesis, Chalmers University of Technology, Department of Geology, Göteborg.

Fransson, Å. (2001) Characterisation of a fractured rock mass for a grouting field test. *Tunnelling and Underground Space Technology*, 16(4), 331-339.

Fransson, Å. (2002) Nonparametric method for transmissivity distributions along boreholes. *Ground Water*, 40(2), 201-204.

Gustafson, G., and Stille, H. (1996) Prediction of groutability from grout properties and hydrogeological data. *Tunnelling and Underground Space Technology*, 11(3), 325-332.

Houlsby, A.C. (1990) *Construction and design of cement grouting. A guide to grouting in rock foundations,* John Wiley & Sons Inc., USA.

Kutzner, C. (1996) *Grouting of rock and soil*, A. A. Balkema, Rotterdam.

de Marsily, G. (1986) *Quantitative hydrogeology. Groundwater hydrology for engineers*, Academic Press Inc., San Diego.

Rock Mechanics Effects of Cement Grouting in Hard Rock Masses

PhD S. Swedenborg[1] and PhD L-O. Dahlström[1]

Abstract: Large effort has been put into understanding how cement grout penetrates and hydraulically seals a fractured hard rock. Whether the grout sealing will endure the stress re-distributions due to rock excavation is, however, sparsely discussed. This paper presents a research program that investigated the principal rock mechanical differences between grouted and ungrouted joints to obtain input data for modeling. From direct shear tests of grouted and ungrouted rock joint replicas, it was found that cement grout acts basically as a 'lubricant' in a joint subjected to shear stress. The 'hydraulic failure' of a grouted rock joint was found to be a function of dilation and coincides with the joint peak strength. Numerical modeling was performed to study the consequences in tunneling. Under adverse geological conditions, failure of grouted joints may propagate into the surrounding rock mass up to twice the tunnel diameter. The investigation showed that the shear resistance of a grouted hard rock joint was lower than an identical ungrouted joint. At low normal stress levels, an initial strengthening cohesive effect was noted.

1. Introduction

1.1. Pre-grouting and rock mechanics

One problem occurring while tunneling below the ground water level is the inflow of water. Pre-grouting can considerably reduce the transmissivity of the rock mass in the tunnel vicinity. Research conducted during the last four decades has revealed many of the mechanisms that govern the penetration of grout into rock joints. Work procedures and quality control systems have been developed to optimize grouting results. For pre-grouted tunnels, a limited number of control holes are commonly drilled to verify dry conditions ahead of the tunnel face. Should leakage despite this occur, it is assumed that the water bearing structure had not been intercepted by any of the grout or control holes. The research described here gives another explanation to unexpected leakage.

The excavation of a tunnel causes rock stress re-distribution. Rock mass deformations due to excavation may introduce shear stresses and subsequent dilation, capable of deforming the grouted joint and causing hydraulic failure[2] of the previously sealed

---

[1] NCC Engineering, Gothenburg, Sweden
[2] Hydraulic failure in this paper refers to when a grouted joint re-gains its ability to conduct water

joints. This process is governed by joint properties, the orientation of joint sets and by the ambient rock stress field.

1.2. State of the art

Direct shear test on cement grouted joints were conducted by Coulson (1970) on fabricated joints from cores of coarse and fine-grained granites produced by using a modified Brazilian splitting technique. For normal stress levels common in tunneling situations, these tests concluded that the shear strength of grouted joints was considerably lower than that of similar ungrouted joints. Another important observation was the transition of failure mode. At normal stresses below 0.8 MPa, a bond failure was observed in the interface between grout and rock. Little deformation was required to reach the peak value in this failure mode. At higher stress levels, the failure cut through the grout rather than at the rock/grout interface. To reach the peak strength in this mode, more deformation was needed.

In agreement with above, Barroso (1970) found a reduction of shear strength of grouted joints compared to the ungrouted joints when subjected to stresses similar to those around tunnels. He also noted a change of failure mode, depending on the normal stress level.

For both studies, the magnitude of the grout thickness in the joint was kept constant over the entire sample area, but varied in different tests. This excluded the possibility of assessing failure when both rock-to-rock contacts and grouted voids interact since the influence of asperities was excluded

Carter and Ooi (1988) reported results of direct shear tests on concrete-rock interfaces. They used samples consisting of a lower part diamond-cut sandstone surface upon which a concrete upper half was casted. The setting simulates the interface between rock and cast in-place concrete. The objective of the work was to characterize the strength and deformation behavior of a pile shaft when loaded. The shear resistance measured was separated into a cohesion, C , friction, ϕ , and dilatancy, i components. Dividing the shear deformation, u, into elastic deformations (u^e) and plastic deformations (u^p) and fitting exponential curves to the data, the model becomes:

$$\Delta u = \Delta u^e + \Delta u^P \tag{1}$$

$$C = C_0 \qquad\qquad \text{if} \qquad\qquad u^P < \delta \tag{2}$$

$$C = C_0 \cdot exp\left(-k_1 \cdot \left(u^P - \delta\right)\right) \qquad\qquad \text{if} \qquad\qquad u^P \geq \delta \tag{3}$$

,where δ is the threshold value of plastic shear displacement, at which damage of the cohesive capacity commences. In the same manner, the mobilized frictional angle ϕ is a function of the plastic shear displacement:

$$\phi = \phi_0 \cdot \left(1 - exp\left(- k_2 \cdot u^P\right)\right) \tag{4}$$

For the dilatancy i, a similar function emerges from fitting the test data:

$$i = i_0 \cdot exp\left[- k_3 \cdot \sigma \cdot \left(\left(1 - u^P / \lambda\right) \cdot q_u\right)\right] \quad \text{if} \qquad 0 \le u^P \le \lambda \tag{5}$$

$$i = 0 \qquad\qquad\qquad \text{if} \qquad u^P > \lambda \tag{6}$$

,where σ = normal stress; λ = plastic shear displacement when dilatancy ceases and q_u = uniaxial compressive strength of the weakest material. The model describes an interface between two materials, rather than a composite bond, such as a grouted rock joint. It is the author's opinion that this model can be used to give a good indication of the joint behavior.

2. Performed direct shear tests

2.1. Preparatory tests

The senior author on diamond-cut flat surfaces performed the tests. These tests had two objectives; 1) to verify that the selected replica material intended for the casted joint was suitable to accurately simulate crystalline rock joints and 2) to investigate the basic shear behavior of cement grouted joints. Tests were performed in the direct shear box at Chalmers University of Technology, Gothenburg, Sweden. The sample size was 100x100 mm. Samples were prepared and grouted submerged in water and then left for 7 days of curing.

A number of parameters were varied in the tests. The standard setting was; water cement ratio 0.8; normal stress σ_n=4MPa; joint aperture or grout thickness t =3mm; horizontal failure plane, meaning parallel with the shear loading axis.

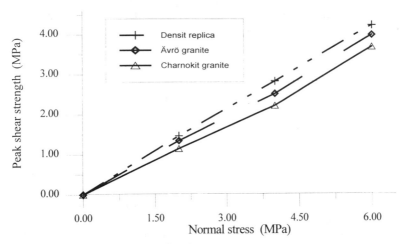

Figure 1 Measured basic frictional angles.

The replica material used was Densit T2. It is a cement based floor toping product with mechanical parameters almost identical to competent crystalline rock. Its wet flowing properties allowed for complete filling of the joint replica forms.

The basic friction angle ϕ_{bas} of the replica material, Densit T2, was slightly higher than the reference crystalline rocks used (Ävrö granite and Charnokit granite), see **Figure 1**.

But when grout was placed between the simulated joint surfaces, this difference disappears which is shown in **Figure 2** below. This demonstrates that the filling governs the joint mechanical behavior, not the original joint surfaces.

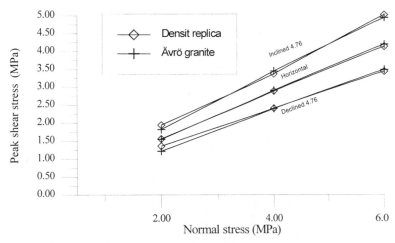

Figure 2 *Results of shear test on grouted, 3mm joints.*

From the large number of tests performed it was seen that the main failure plane occurred near the interface between rock and grout, but always slightly inside the grout. A Mohr-Colomb based failure criterion appears reasonable. However, the contribution of the cohesion diminishes after failure. As a result, a bi-linear failure is appropriate, and can be represented by the following:

$$\tau_{peak} = C_0 = \tau_{adhesion} \qquad \text{if} \qquad u^P = 0 \tag{7}$$

$$\tau_{peak} = \sigma_n \cdot tan(\phi_{grout}) \qquad \text{if} \qquad u^P > 0 \tag{8}$$

In conclusion, full cement grouting of a void will increase the peak shear capacity at low normal stress, but will decrease it at higher stress levels common to many rock engineering situations. Both $\tau_{adhesion}$ and ϕ_{grout} depend on the *wcr* of the grout used.

The failure of fully grout filled diamond-cut joints is shown in **Figure 3** along with the basic friction for Densit T2, i.e. the ungrouted case, for comparison.

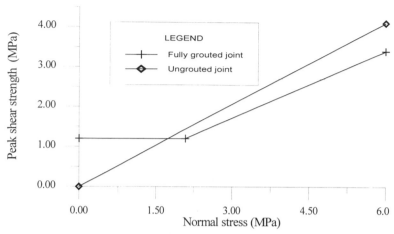

Figure 3 *Principal failure envelopes of fully grouted joints.*

The failure mode at higher normal stress levels can be separated into two categories, with or without multiple shear failures (Riedel shears[3]) in the grout layer. The Riedel shears propagated through the grout layer, allowing for local steep sliding within the grout layer. This caused considerable contraction of the grout layer. This phenomenon did not interact with the main failure plane, located close to the rock-grout interface. Riedel shears are associated with contraction of joints, which cannot appear in rock joints, therefore was this failure mode not investigated further.

It should be kept in mind that these test were on hard crystalline rocks. Low strength geological materials with low basic friction angle (i.e. like shales or mudstone etc) were not investigated.

2.2. Preparation for direct shear test of grouted rock joint replicas

A shear box at University Joseph Fourier in Grenoble, France was used for this study. Cast replicas were made of Densit T2 with a uniaxial strength of 210 MPa, which had been found feasible in the preparatory test.

The original joint sets had to be reproduced twice to prepare the molds. The *à-priori* estimated Joint Roughness Coefficient was in the range 5 to 9 for the four selected samples. Two of the joint samples originated from a fine-grained granite, the other two from a coarse-grained granite. After adjusting and fixing the samples into the removable shear box frames, a center hole were drilled in the lower sample for joint access.

---

[3] Multiple Shear failure in thin layer due to the stress situation

2.3.   Grouting of rock joint replicas

From the literature, it is known that penetration of cement suspensions depends much on the size of the void opening. For this testing the sample halves were fitted to each other and a normal stress of 0.5 to 1.0 MPa was applied and mechanically locked to resist the grouting pressure. A silicone rubber sealing was fitted in the gap between the two frames affixing the joint sample to hold the fluid grout.

Grout was injected through the center hole in the lower sample. The grout had cement particle size $d_{95}<30$ μm, a water-cement ratio wcr between 0.7 and 0.9 and contained 1 % plastizicer.

2.4.   Testing of grouted rock joint replicas

Prior to installing the samples in the shear box, the grout hole was re-drilled and a coupling was fitted to apply a water pressure in the joint plane.
Samples were loaded to the normal stress of the test in a single step to avoid possible plastic normal deformations of the grout due to cyclic loading which could affect the hydraulic seal. A water pressure of 2 MPa was then applied to the center hole. A pressure gage was connected to the system and a loss of water pressure within the system would indicate the hydraulic failure.

The shear stress was applied to the sample at a rate of 1,0 mm per minute while automatically logging deformations, stress levels and water pressures. The total shear displacement was 10 mm, thereafter shearing was reversed to its original position.

3.   Results of direct shear tests on grouted and ungrouted rock joint replicas

The main objective of the test program was to study the difference between the grouted and ungrouted joints and to establish mechanical parameters to be used for rock mechanical modeling.

3.1.   Comparison of grouted and ungrouted rock joint replicas

When comparing ungrouted with grouted samples, applied normal stress or water-cement ratios injected were varied. The evaluation of the obtained results was made with a model based on Carter and Ooi (1988), Eq. (1) – (4). When varying the normal force, the data could easily be fitted to the equations. The effect of different wcr could not be fitted into the curves so an average was used. The difference of shear resistance due to wcr indicates that grout is being grinded in the rock-to-rock shear zones and lowering the friction.

Unfortunately, when back-calculating the JRC for the samples used, the interval was small, JRC ranged from 5.6 to 6.9. The variation, that might exist, depending on the joint surface roughness could not be evaluated.

The results are exemplified in **Figure 4** below.

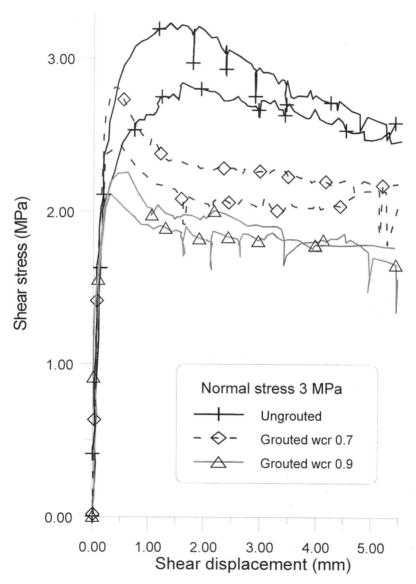

Figure 4 *Representative curves of grouted and ungrouted rock joint replicas*

The curves show that grouted joints reach their peak shear strength when the cohesive part, $\tau_{adhesion}$, reaches its peak and then fails. Un-grouted joints reach their peak shear strength when the joint dilatancy is fully mobilized. Ungrouted joints require larger relative shear displacement to reach peak strength.

In the research it was found that most of the normal stress is transferred via rock-to-rock contacts, the peak strength is a function of the cohesion and mobilized friction. The peak strength of grouted joints is reached when plastic deformations commence, i.e. when the grout cohesion fails. The friction angle between the rock-to-rock contacts is the basic frictional angle of the host rock material since the walls remain in contact as before grouting. Note in **Figure 4** that the dilatancy does not contribute to the peak strength.

Following relation is representative this concept:

$$\tau_{peak,grouted} = \sigma_n \cdot \tan(\phi_{bas,rock}) + \tau_{adhesion,grout} \tag{9}$$

When the grout cohesion fails, plastic deformations commence which cause dilation. Therefore, the peak shear strength and the hydraulic failure occur concurrently for grouted joints.

In the process of hydratization, the curing grout fills most of the void space. Therefore, as soon as plastic deformations commence, hydrated cement will interact in the friction plane between the joint sides, reducing the friction. After failure, the shear resistance will increase due to dilatancy, but reduce due to lubricating action of fine grout particles in the rock-to-rock shear zones.

Furthermore, assuming that the cohesive failure of the grouted bond is brittle, the initial part of the shear-displacement curves is mis-leading. The apparent joint shear stiffness $[\Delta\tau/\Delta u^e]$ is not a parameter of the joint, but a function of the elastic shear modulus of the samples and the test machine deformations.

3.2. Assessment of input parameters for numerical modeling

The following behavior of grouted rock joints can be assumed, based on the performed direct shear tests and be used for numerical modeling;

$$\Delta u = \Delta u^e + \Delta u^P \tag{1}$$

$$C = C_0 \qquad\qquad \text{if} \qquad u^P = 0 \tag{10}$$

$$C = C_0 \cdot \exp\left(-k_1 \cdot u^P\right) \qquad \text{if} \qquad u^P > 0 \tag{11}$$

,where u^e is the elastic threshold value of shear displacement, i.e. when damage of the cohesive bond starts. The mobilized friction angle is also a function of the shear displacement, since the joint has rock-to-rock contacts;

$$\phi = \phi_{rock} \cdot \left(1 - exp\left(-k_2 \cdot u^e\right)\right) \tag{12}$$

However, since the grout decreases the frictional shear resistance, a correctional function is needed:

$$\phi_{red} = (\phi_{rock} - \phi_{res,grout}) \cdot \left(1 - exp\left(-k_4 \cdot u^p\right)\right) \tag{13}$$

In the grouted case, the effective friction angle after cohesive failure becomes

$$\phi_{effective} = \phi_{rock} - \phi_{red} \tag{14}$$

For the dilatancy, the function developed by fitting of test data is:

$$i = i_0 \cdot \left(1 - exp\left(-k_3 \cdot u^p\right)\right) \qquad if \ 0 \leq u^p \leq \lambda \tag{15}$$

$$i = 0 \qquad\qquad\qquad\qquad if \qquad u > \lambda \tag{16}$$

The effect of asperities has been omitted. This is because;

- The effect of aspiritis is limited at large scale. Barton (1981) suggested that it could be assumed to be in the range of 1° for numerical modeling of rock masses.

- When examining the joint surfaces after shearing, the damage of the grouted joints is significantly less than of the ungrouted ones. The grout fills the inverts of the joint surface and prevents interlocking, which occurs in ungrouted joints. The effects of aspiritis can be assumed to be incorporated in the frictional and dilatancy factors.

4. Numerical modeling

A numerical model provides an opportunity to study the principal stress and deformation behavior of rock masses given certain boundary conditions. For some engineering disciplines, the method may be used for detailed design. But for rock engineering, where input data for any rock mass and the *in-situ* rock stress state are highly uncertain, study of general behavior is more appropriate.

4.1. Assessment of input parameters for numerical modeling

Typical values were evaluated from direct shear tests by fitting test results to Eq (13) – (16). The results used for numerical modeling are presented in Table 1.

Table 1 *Typical parameters for numerical modeling.*

Exponential constant	Physical values
$k_1=2000$	$C_0=\tau_{adhesion}= 0.6$ MPa
$k_2=15000$	$\phi_{rock}= 33°$
$k_3=1500$	$\phi_{grout}= 25°$
$k_4=1000$	$u^e=0.0001$ m

For the numerical modeling, the threshold value, u^e, was assumed the same for the commencement for cohesive failure and development of dilatancy. A small relative shear displacement without dilation may be possible. From the data available, 0.1 mm appears to be a reasonable average value.

4.2. Modeling results

Numerical modeling was performed with rectangular and circular tunnels with 35 m of rock cover and with an area of about 80 m². Modeled grouted joints were systematically oriented in different positions. The horizontal principal stresses were set at 0.35 and 3.5 MPa for the two cases analyzed while vertical principal stress was hydrostatic. The model boundaries were made approx. 5 times the diameter of the opening to avoid boundary effects. Normal rock support would have only been partially activated when deformation exceed failure for the grouted joints and was therefore excluded from the models (Holmberg, 1991). Figure 5 presents model results enlarged around the tunnel opening.

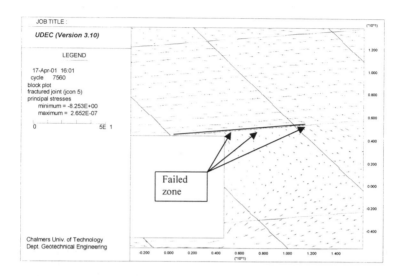

Figure 5 *Rock mechanical interaction with failing grouted joint. Primary stress field, σ_h=3,5 MPa, σ_v= 1 MPa. Secondary stress field plotted. Failed zone = re-opened joint.*

The modeling indicates that the compressive stress field around the opening suppresses shear displacements and dilation that maintain the integrity of the grouted joints water seal. The most adverse joint orientation is parallel to the principal shear stress. This shear stress will open the grouted joint until the reactive forces due to the normal stress acting on the joint equals the driving shear stress.

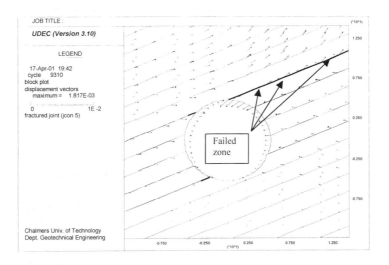

Figure 6 *Grout sealing failure due to rock mass deformations. Primary stress field, σ_h=3,5 MPa, σ_v= 1 MPa. Displacement vectors plotted. Failed zone = re-opened joint.*

Hydraulic failure may propagate far from the tunnel perimeter due to other mechanisms than shown in **Figure 5** above. In **Figure 6** it is caused by horizontal stress release due to tunnel excavation. Due to the jointing, the roof section will deform to the right while the right hand wall side will deform to the left.

5. Conclusions

The investigation on grouted rock joint replicas compared with ungrouted samples demonstrate that the grout joint is brittle and that small deformations can initiate dilation of the rock joint and break the grout seal. The direct shear tests also showed that grout gives a joint a certain cohesion but lower the frictional resistance. Cement grouting increase joint shear strength at low normal stresses but reduces the shears resistance at higher normal stress levels.

From the numerical modeling, a compressive stress field around the underground opening appears to be an essential prerequisite for successful pregrouting in tunnels. Due to the brittle failure of grouted joints, the failure of the grout seal is a deformation problem of the rock mass.

From the presented investigations, some practical conclusion may be drawn. If a leakage in a pre-grouted tunnel needs to be reduced it appears feasible to seek the origin of the seepage at distances from the perimeter of the tunnel. Outside of the compressed zone, the joint apertures have not been diminished due to increment of normal stresses, thus may still be possible to grout successfully.

6. Acknowledgement

The authors wish to thank SweBeFo, Sweden for allocating funds for the presented research, performed at Chalmers University of Technology, under the guidance of Prof. Ulf Lindblom. We also want to express our gratitude for the excellent co-operation with UJF at Grenoble and Prof. Marc Boulon. The use of the direct shear device at Laboratory 3S was quite essential for this study.

REFERENCES

Coulson J.H. (1970): The effects of surface roughness on the shear strength of joints in rock. *Ph. D thesis,* University of Illinois at Urbana-Champaign, U.S.A.

Barroso M. (1970): Cement grouts and their influence on the shear strength of fissured rock masses. *Laboratorio Nacional De Engenharia Civil*, Lisbon, Portugal.

Barton, N. 1981. Modeling Rock Joint behavior from *In-Situ* Block Tests. *ONWI Tech. Report Publ. ONWI-308*

Carter and Ooi (1988): Application of a joint model to concrete-sandstone interfaces. *Proceedings of the international conference on numerical methods in geomechanics held in Innsbruck 1988* Balkema, Rotterdam

Holmberg, M. 1991. The mechanical behavior of untensioned grouted rock bolts. *PhD Thesis, Division of Soil and Rock Mechanics*. Royal Institute of Technology, Stockholm, Sweden.

Swedenborg, S. 2001. Rock mechanical effects of cement grouting in hard rock. *PhD Thesis, Department of Geotechnical Engineering*, Chalmers University of Technology, Gothenburg, Sweden

Subsidence Mitigation Using Void Fill Grouting

Darrel V. Holmquist, P.E.[1] and Damon B. Thomas, P.E. [2] in
conjunction with Kent Simon, P.E.[3]

Abstract

Void fill grouting is a technique which has been extensively used by CTL/Thompson, Inc. to mitigate mine subsidence potential resulting from abandoned underground coal mine workings. The technique involves rotary drilling from the surface down into the mine workings and pressure injection of grout into underground mine voids and rubble zones to fill voids.

Communities of Rock Springs, Glenrock and Hanna, Wyoming were built over abandoned coal mines. Over time, the rock comprising the roof of the mine begins to fracture and collapse into the open mine. This process continues until the space is either occupied by rubble or the caving reaches the surface, threatening public health and safety. To successfully mitigate the potential of subsidence reaching the surface, a thorough understanding of the subsidence mechanism is required and the proper mitigation method selected. The paper describes typical subsidence mechanisms and mitigation methods.

The depth to mining is important when determining whether or not subsidence will reach the surface. Also important are; the thickness of the mined seam, rock bulking characteristics and strength of the overburden rock. This information can be obtained by investigative drilling, laboratory testing on material samples and careful analysis. The paper presents brief descriptions of analytical methods to determine critical mine depth. Once the critical depth to the mine is determined, a pilot drilling and grouting program is undertaken to determine the optimal hole spacing for mitigation.

During construction, holes are drilled on a grid spacing as determined in the pilot program and low strength grout is injected through the drill holes into the mine workings. Critical items during grouting are; grouting sequence, slump variation, pressures, cutoff quantities and verification drilling.

CTL/Thompson, Inc. has successfully used this technique in Rock Springs and Glenrock, Wyoming for over 13 years with construction costs exceeding 35 million dollars. Verification drilling, and the lack of insurance claims or surficial subsidence features indicate that void fill grouting is an effective method to mitigate the effects of subsidence due to abandoned underground mines

---

[1]Senior Principal Consultant, CTL/Thompson, Inc., Denver, Colorado.
[2]Project Manager, CTL/Thompson, Inc., Denver, Colorado
[3]Project Officer, State of Wyoming, Department of Environmental Quality, AML Division.

Mining Techniques

Mining in Wyoming underground coal fields typically utilized a "room and pillar" technique (Figure 1). The mines were accessed by vertical or sloped shafts called entries, with air shafts excavated parallel to the entry. Once the entry shaft penetrated the mineable coal layer, motor roads or haulways were cut parallel and perpendicular to the strike of the seam. Rooms typically 6 meters wide and 30 to 90 meters long were then cut with pillars approximately 4.5 to 12 meters wide between rooms for support. During operation of the mine, between 30 to 60 percent of the coal was extracted by room and pillar mining.

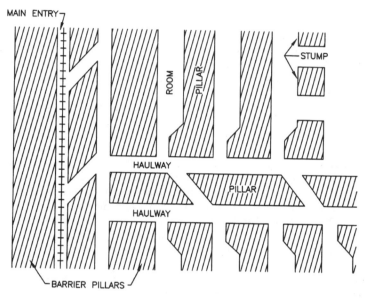

Room and Pillar Mining
Figure 1

Typically, 4.5 to 11 meter wide sections of the support pillars were removed leaving "stump" pillars. Following the retreat mining, the remaining pillars and roof coal were shaved. By the time the mines were closed, the overall coal extraction could reach 75 to 85 percent.

Closure and abandonment of the mines following the retreat mining was typically not well documented. As a result, mines were often larger and more extensively developed than shown on the mine map. Investigation has shown that many mines were closed by dumping debris, timbers, trash and coal slack back into the entry. Air shafts and hoist shafts were often closed with a relatively shallow 'plug' with soils and debris placed above the plug.

Subsidence Mechanism

Rock overlying the extraction is stressed and will subside into the extraction. The occurrence of subsidence and the mechanisms by which the overburden rock is distressed and displaced depend upon physical properties of the overburden, coal and floor materials, the size and depth of individual extractions and extraction ratios achieved. Subsidence may be caused by failure of the mine roof, coal pillars or mine floor materials. Subsidence may take the form of sinkholes or a chimney-type caving, a gentle sagging or trough-type subsidence, or settlement of backfill materials in entry or air shafts. The following paragraphs discuss subsidence mechanisms.

Caving Subsidence

Subsidence can produce sinkholes or depressions at the ground surface by caving of materials overlying comparatively shallow mine working. Caving occurs as the roof over an opening fractures and collapses into the space where the extraction has occurred. This process continues until the space is either occupied by debris or the caving reaches to the surface (Figure 2).

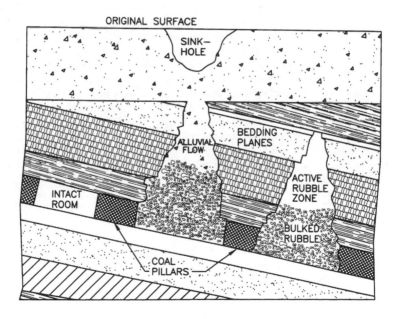

Caving Subsidence Mechanism
Figure 2

Caving is common over room and pillar operations. The depth to mining and thickness of the bedrock are critical in determining whether or not the subsidence feature will reach the surface. Also important are the thickness of the extraction and bulking and strength characteristics of the overburden rock.

The size of sinkholes caused by caving is controlled by the geometry of the mine and properties of the overburden. The depth of the sinkholes depends on the depth to mining, thickness of the extraction, and bulking properties of the overburden. The aerial extent of surface depressions is largely controlled by the size of the mine opening. Research has found that sinkholes typically are circular or elliptical in shape and not larger than the size of the extraction causing them. The experience in Rock Springs and Glenrock indicate caving is the most likely subsidence mechanism, except for two shallow mine areas in Rock Springs. The majority of the documented subsidence events in Rock Springs and Glenrock were less than 10 meters in diameter.

Trough Subsidence

Trough-type subsidence is a sagging of the overburden triggered where large extraction ratios and panel sizes are achieved, both in areas of room and pillar mining and longwall mining. This generally occurs as a caving of the immediate roof followed by sagging of overlying strata. Trough-type subsidence is the common mechanism over longwall mines in the United States and Europe. The presence of internal barriers and the low width to depth ratios helps reduce the magnitude of displacement. Experience in the Wyoming area indicates trough subsidence is very rare and occurs only in very shallow mines (at depths of approximately 15 meters) in isolated areas in Rock Springs.

Trough-type subsidence over room and pillar mining will be localized as compared to the area-wide troughs developed by longwall mining. As with sinkholes, the depth and extent of troughs will depend on the depth to mining, physical properties of the overburden, and extraction ratios achieved. The shape of depressions will be irregular due to the presence of remnant pillars. Like longwall mining, subsidence over retreat mining will develop rapidly due the high extraction ratios achieved. Additional movements have occurred from recompression of rubble or re-orientation of beds which have sagged. Two sudden and dramatic collapses occurred as a result of water injection into mined areas.

Shaft Collapse

The subsidence potential associated with entry or air shafts is high, because of the sudden and catastrophic nature of movement which can occur. Although small in area, shafts can be dangerous because of the haphazard way in which they were backfilled. A shaft in Rock Springs opened up next to City Hall and was over 9 meters in diameter and more than 6 meters deep.

Evaluation of Subsidence Potential

The depth to mining is the critical factor used in evaluating the subsidence potential at a given site. The alluvial soils found in Rock Springs and Glenrock tend to flow into open holes and provide little strength in resisting subsidence. Therefore, the thickness of the bedrock over the mine becomes the critical factor in determining the subsidence potential in these areas.

There are several methods used to determine a minimum bedrock thickness above the coal seam beyond which caving subsidence will not be expected to reach the surface. The methods vary with respect to the treatment of subsidence mechanisms, mine geometry, geology and mining method.

Mine Geometry

With this method, the critical dimension affecting subsidence are the thickness of cover or overburden height (H) and the extraction thickness (h). Piggot and Eynon (1977) suggest subsidence will not propagate to the ground surface over room and pillar workings where the overburden to extraction thickness ratio (H/h) exceeds 10.

Bulking Factor

Caving of the roof above a mine can continue until the extraction and collapse area is filled with broken and bulked rock or the caving reaches the surface. With this method, the height to which caving can occur is based on the coal seam thickness and the bulking of the collapsed rock. The increase in the volume of the collapsed rock is referred to in terms of its "Bulking Factor" (Piggot and Eynon, 1977). The Bulking Factor is calculated in terms of thickness and defined as the original extraction height minus any remaining void divided by the height of the rubble zone above the mine roof.

National Coal Board's Method

The National Coal Board of the United Kingdom published a handbook (1975) which presents descriptive models that accurately predict ground movements over longwall mines in Europe. While useful in predicting subsidence in longwall mines, the applicability of this method in Wyoming is limited due to room and pillar mine geometries. Use of this method results in significantly overestimating the depth and aerial extent of the subsidence features.

Complementary Influence Function

Complementary influence functions were developed (Sutherland and Munson, 1983) based on two elements: the mined element and an unmined element, assigning a surface response to each. The response of the unmined element is based on the elastic response of the strata overlying the structure while the response of the mined element is related to the breaking of the immediate roof and non-uniform distribution of voids in the rubble. This method is useful in both room and pillar and longwall applications and was used extensively in the Rock Springs area. The model reasonably predicts the shape of the subsidence feature, but does not predict the depth of the feature nor whether or not the subsidence will reach the surface.

RUBBLE Model

The RUBBLE computer model uses a finite element method to predict rubble heights. The RUBBLE model (Benzley and Krieg 1982) is based on the principles of continuum mechanics and analyzes the geomechanical processes of roof failure and collapse caused by underground mining. Subsidence predictions using the RUBBLE model have correlated well with field measurements in the United States and Wyoming. The RUBBLE model was extensively used in the Glenrock area. Properly calibrated, this technique has provided the most accurate prediction of subsidence potential when compared to locally documented subsidence data.

Subsidence Risk

Based on the analyses performed, subsidence risk is typically categorized as either 'low' or 'significant'. Areas around shafts or adits are classified as 'significant' due to the unreliable method of abandonment and catastrophic nature of subsidence in these areas. The remainder of the mine is classified based on the critical depth to the coal seam.

Areas designated as 'low' subsidence risk are the deeper areas and generally have no development restrictions. Areas designated as 'significant' are the shallower areas and must either be mitigated or avoided.

Mitigation

A large portion of the mines in Rock Springs and Glenrock were classified as 'significant' subsidence risk based on a critical bedrock thickness of 30.5 and 27.5 meters of bedrock cover, respectively. In both communities, areas of significant subsidence risk have been previously developed. Over the years, subsidence has reached the surface in numerous areas in the form of sinkholes. The ongoing appearance of new subsidence events prompted mitigation.

The mitigation technique selected for use by CTL/Thompson, Inc. in these communities was 'void fill' grout injection. The intent of void filling was to reduce the amount of existing void within the mine workings, thereby reducing the potential height of caving and decreasing the risk of subsidence.

The construction process involved drilling small diameter borings into existing mine voids and rubble zones followed by grout injection. A pilot program is performed prior to actual mitigation to determine the optimal hole spacing to ensure proper grout flow. The pilot programs conducted in Rock Springs and Glenrock indicated optimal hole spacing of 9 and 12 meter centers, respectively.

Grout

A single, low-strength grout mix comprised of water, cement, fly ash and aggregate was used. The properties of the injected grout play perhaps the biggest role in the successful mitigation of the mines in these communities. These properties include the flowability, aggregate type, set time and strength.

Flowability

The conditions of the mine vary from drill hole to drill hole. Some areas had large amounts of open void where other areas had tight rubble with very little void. This variation makes injection of a single grout mix difficult resulting in poor grout flowability.

The keys to flowability are water content, fly ash and rounded aggregate particles. The mix design for the projects was the responsibility of the contractor. Initially, contractors had a very difficult time developing a mix design which provided strength and flowability. Specifications called for a minimum of 3.5 MPa strength over a range of slumps from 5 to 20 centimeters to provide for adequate flow. The adjustable slump allowed rapid modification of the grout in the field which enabled the adaptation of the grout to varying conditions found within the mine workings. To promote flowability, the higher slump grout was used in areas of tight rubble, helping the grout travel through the smaller voids between the bulked rock.

Flowability was further enhanced by using fly ash in the mix. Fly ash was selected because of the rounded particles significantly aided the mixes ability to flow. Fly ash also improved the strength of the mix.

Aggregate Type

The type of aggregate used in the grout also affects flowability. The specified aggregate consists of natural, clean, hard, tough, durable, rounded particles with a maximum size of 1.5 centimeters. Manufactured sands were strictly prohibited from use.

The difference between a rounded sand and a manufactured sand not only affects flowability, but also constructability. Manufactured sands have sharp faces which tend to not only reduce the flowability of the grout in the rubble zones, but also showed a tendency to clog or lock-up in the grout pipe. The use of natural sands not only made pumping easier, but also acts as a natural lubricant within the grout to help flowability through tight rubble zones.

Set Time

The set time of the grout mix was critical in that the time it took to empty a mixing truck was longer than if pouring concrete directly down the chute. The mixer truck sends the grout into the hopper of the grout pump. The pump then injects the grout through a series of pipes placed in the drill hole. The rate of pumping is generally between 3.8 and 23 cubic meters per hour.

Some fly ash has a tendency to hot flash, and set quickly. The use of such a fly ash in the grout mix can not only restrict flowability through the mine, but may also set in the pipe. Fly ash conforming to ASTM C 618-93 was specified to help prevent flash setting.

Strength

The grout mix specified was to have a minimum compressive strength at 28 days of 3.5 MPa over the full range of slumps. This strength value was selected to provide enough support for the overlying bedrock, yet also remain removable if grout traveled shallow enough to be excavated during utility work or future construction.

In addition, the low strength allowed a significant amount of fly ash to be utilized in the mix. The fly ash was used as bulk filler material and a workability agent. This resulted in the ability to lower the amount of cement used in the grout mix which ultimately made the grout a more cost effective approach to mitigation.

Mitigation Effectiveness

Mitigation of abandoned coal mines is only practical if the effectiveness can be verified. During and shortly after mitigation, verification drilling, coring, geophysical logging and visual identification of drill cuttings were used to confirm the presence of grout. Likewise, over the past 13 years, the lack of insurance claims or surficial subsidence features in mitigated areas also lend to the effectiveness of the void filling technique.

Verification Drilling

Verification drilling and coring was used to confirm the effectiveness of void filling in both Rock Springs and Glenrock. During the course of the mitigation, areas were identified where additional evaluation of the grout flow was required. Typically, these areas were between the holes of the grid spacing and in areas of low or questionable grout injection volumes.

During verification drilling, the additional holes are drilled throughout the mitigated areas. Verification holes were geophysically logged where possible. If conditions made geophysical logging not possible or if roof contact was not demonstrated, grouting was attempted.

Coring

Coring was done in mitigated areas to also confirm the presence of grout. When coring was conducted, holes were rotary drilled to a point above the projected mine interval then cored through the grouted interval with a 6.35 to 8.5 centimeter diameter barrel. Core samples retrieved were boxed and kept as visual documentation of the effectiveness of the mitigation.

Dispute Avoidance

Contracts by state law are hard dollar bids. Given the paltry knowledge regarding the extent, depth and roof condition, it is inevitable that disputes will arise. Tightly written contract specifications and disclosure of potential disputes address the unknowns before bidding. Disclosures address issues such as utility damage, geologic unknowns, contract quantities and special timing or equipment requirements needed to complete the project. Contract provisions are included in both General Conditions and Technical Specifications to handle these unknowns when they occur. As part of the disclosure process, mandatory Pre-Bid meetings where specifications, limitations and contract ground rules are clearly and concisely explained should be required. Questions both verbal and written should be taken and meeting minutes, and answers provided in written form to all who attend.

Contractors are "pre-qualified" to bid on the contract based on past work experience. Prospective bidders are evaluated based on the successful completion of projects of similar size and value involving methods and equipment similar to the contract. Names, addresses and phone numbers for references are required and should be verified as part of the process. Bidders submit detailed history demonstrating at least five years of experience with similar projects. It is very helpful to ask references who were the superintendents and "would you hire them again?" The references evaluate the adequacy of the contractor's equipment and methods in completing the work specified in the contract.

Emphasis is placed on the contractor's personnel. The project manager or field superintendent, who will have the ultimate authority should have at least five years of previous experience with similar projects, and one year or more in his current position. Check his individual references and include contract language which does not allow the replacement of key personnel, except at the request of the owner or engineer.

The Pre-Qualification package is reviewed by both the engineer and owner. The qualification determination is made jointly, although the owner assumes ultimate responsibility. The owner reserves the right not to prequalify a contractor on the basis of experience, equipment, personnel, or poor performance on any previous project, disclosed or not. The decision of the reviewer should be final and without appeal. Sample Pre-Qualification questionnaires may be obtained from the authors.

The Partnering process should be included in the contract specifications as a requirement and a pay item. Partnering is a cooperative commitment among all stakeholders to achieve certain goals and avoid or resolve disputes. Partnering should be done before the "Notice to Proceed" and be facilitated by an independent professional. The goals of the process are to:

1)   Introduce all the players and develop the concept of a team approach, rather than an "us versus them" concept. Successful completion of the project becomes a mutual goal.

2)   Develop a mutual trustful, respectful and cooperative spirit among the players.

3)   Discover each player's concerns, fears and wishes for the project and develop workable solutions in advance.

4)   Develop a project charter which presents the project desires and goals and commits each signator to work toward making the project successful.

5)   Develop a conflict resolution chart committed to solving problems at the lowest possible level. Each level is identified with names, phone numbers, level of authority (in dollars) and time allotted to resolve the conflict. Each level should have representation of the AML, municipality, engineer, contractor, and other interested parties.

In order for the process to be effective, all the players must participate, from field technicians and laborers to the CEO's. Without full participation, the results are seldom effective. The entire group needs to know that the process has the backing of everyone including the top management. Participants should include the owner, Contractor, Engineer, Home Owners Association, Business Groups affected, Public Works, Elected Officials and the press. This usually results in 25 to 35 people at the sessions. An effective facilitator can control and stimulate the group to achieve the best outcome. On new projects or where the players are not familiar with each other, use of a quick and easy personality profile has greatly aided the participants in understanding how to motivate and deal with each other.

Our experience with Partnering has been extremely good, in only one contract out of 18 was it necessary to have a follow-up session in order to resolve a conflict. The process acts to alleviate perceived conflicts prior to initiation of the project and promotes a spirit of cooperation from the CEO down to the field level personnel.

Dispute Resolution

If project disputes cannot be avoided or resolved through the Partnering process, a claim may be filed by the contractor. There are several alternatives presented which may be incorporated in the contract specifications to allow resolution prior to taking the issue to court. Dispute resolution alternatives include arbitration, mediation, dispute review boards, mini-trials and master's court. These resolution alternatives vary in cost and implementation and may be binding or non-binding.

Summary

CTL/Thompson, Inc. in conjunction with the State of Wyoming, Department of Environmental Quality, Abandoned Mine Lands Division have used the void fill grouting technique to successfully mitigate over 96 surface acres of land. This land was already developed and not only had low property values, but posed public health and safety issues. Since mitigation, property values have improved, funding from the Federal Housing Administration has been reinstated and public confidence restored.

We feel void fill mitigation is a viable, proven mitigation alternative to reduce the risk of subsidence due to the presence of abandoned underground mines. In addition to the work already performed in Rock Springs and Glenrock, Wyoming, some municipalities in the Denver, Colorado area are considering the use of void fill grouting in undeveloped areas. Such mitigation would be unprecedented in the Denver region and would allow previously 'unusable' land to be developed.

1114 GROUTING AND GROUND TREATMENT

Reference Documents

Benzley, S. E. and Krieg. R. D. (1982). "A Continuum Finite Element Approach for rock Failure and Rubble Formation." *Int. J. Numerical and Analytical Methods in Geomechanics, Vol. 6*, pp 277-266.

CTL/Thompson, Inc., Job No. 19,592-I. (1993). "Report of Investigation", McDonald Mine, AML Project 8A-II, Glenrock, Wyoming.

Morrison, C.S. and Holmquist, D.V. (1989). "Analysis of Subsidence Potential using Complimentary Influence Functions", *Proceedings of the Evolution of Abandoned Mine Land Technologies.*

National Coal Board Mining Department. (1975). "Subsidence Engineers' Handbook", Great Britain.

Piggot, R. J. and Eynon, P. (1977). "Ground Movements Arising from the Presence of Shallow Abandoned Mine Workings" in Large Ground Movements and Structures, J.D. Geddes ed., Wiley & Sons N.Y.

Oravecz, Kalman I. (1977). "Measurement of Surface Displacements caused by Extraction of Coal Pillars", *Large Ground Movements and Structures*, J. Wiley and Sons, N.Y.

Roenfeldt, M.A. and Holmquist, D.V. (1986). "Analytical Methods of Subsidence Prediction", *Proceedings of the 1985 Conference on Coal Mine Subsidence in the Rocky Mountain Region.*

Sutherland, Herbert J. and Munson, Darrel E. (1983). "Subsidence Prediction for High Extraction Mining Using Complimentary Influence Functions. SAND82-2949, Sandia National Laboratories, Albuquerque, New Mexico.

MINING GROUTING: *a rational approach*

W F Heinz[1]

Abstract:
Mining grouting in South Africa has always been associated with deep mines. Certain techniques and equipment used are a result of the very high pressures resulting from the large depths of South African Mines.
Some of the techniques specifically developed within the South African mining environment are

1.   Precementation of deep shafts up to 2400m.
2.   Cover grouting to develop or sink under or through rock formations in safety.
3.   The successful impermeabilisation of rock masses with 'thin, unstable" cement grouts.
4.   The conveyance of cement-sand slurries over many kilometres.

"Grouting is more an art than an engineering science". This statement may be true but in essence it has always been an admission of our lack of understanding of the success of grouting. In recent years cement and chemical grouting have developed a new dynamism driven by a better understanding of the grouting process, by an improved understanding of the behaviour of grouting materials, by the development of new grouting materials (micro fine cements) and techniques (jet grouting) and, of course, by many new good publications (books and articles) and research on the subject. Other factors such as environmental concerns and computers have also contributed to this new dynamism in the grouting field.

Many of these developments have been initiated in the civil engineering field such as dam grouting, tunnelling, etc.; grouting in underground mining conditions is conspicuously absent in research, development and literature. This paper presents the development over many years and the State-of-the-Art of South African mining grouting and endeavours to present a more *rational* evaluation and appreciation of the achievements of the early grouting engineers. It also presents a more *rational* approach in grouting particularly for mining conditions, keeping in mind recent developments in grouting engineering and possible developments in future.

1. INTRODUCTION

About 50% of the world's gold has been mined in South Africa over the previous century. Gold production was first recorded in 1871 in the Northern Transvaal. The

---

[1] Chairman. Rodio South Africa Pty Ltd., P O Box 714, Halfway House, 1685, South Africa.
Email: rodio@rodio.co.za

Witwatersrand Basin, the richest gold field in the world, was discovered in 1886 near Johannesburg.

However, the South African gold mining industry has been declining steadily in recent years. In 1910 gold production was 234.25 tons. In 1970 South Africa produced 1000.4 tons of gold which was approximately 80% of the entire world production, during 2002 South Africa produced slightly less than 400 tons which was approximately 25% of the world production.

Mining was originally confined to a 70km belt along the reef outcrops of the Witwatersrand basin. Today gold is mined up to 4,000m below surface. ERPM in the East Rand passed the 3,000m mark 45 years ago and reached a world record of 3,428m below surface in 1959. At present investigations are underway to evaluate the exploitation of ultra-deep ore bodies down to 5,000m below surface.

By necessity the access shafts have been extended deeper and deeper. During the early 60's, 2,000m was thought to be the maximum depth of a shaft. In the 90's the length of the winch was extended to 3,000m; Western Areas South Deep Shaft is currently the deepest shaft in the world at 2,994m.

The problems associated with deep gold mines have resulted in extraordinary achievements. The extreme heat (rock temperature up to 60°C) and pressure encountered at depth, the hardness of the rock, the ever-present underground water, gases (methane and hydrogen sulphide) etc. all required innovative solutions.

Grouting using mainly cement slurries was introduced at a relatively early stage in the development of the gold field; without the cementation technique the enormous development of the Gold Mining Industry would probably never have taken place.

2. HISTORICAL BACKGROUND

Cementation or mining grouting developed during the previous century. Innovation and development was driven by the economic upswing of the coal and gold mining industries during the beginning of the 20th century and technical problems related to mining.

The contributions of South African mining engineers in the field of grouting in mining and construction has been greatly underestimated.

Mining cementation relates mainly to two fields: shaft sinking and underground cementation for water control and sometimes strength improvements.

Some of the first precementation work was executed by Portier in France in 1864, however, the South African cementation techniques are closely related to the name Albert Francois, a brilliant Belgian mining engineer.

Francois developed his cementation process in 1896 for the purpose of shaft sinking. However, Francois' most important contribution was the invention of a high pressure cementation pump (see Fig. 1) and the realization that cement at high pressures (up to 5,000 lb/sq inch, 350 kg/cm²) can be introduced into the minutest fissures in rock formations. The process was first brought to England in 1911 and was applied at the Hatfield Colliery in Yorkshire. Subsequently the Francois Cementation Process was applied with great success in many English collieries particularly in Yorkshire and the Midlands in the years following WWI (World War I).

Francois' involvement on the South African gold fields commenced in 1917. In 1914 Francois offered his services to a Johannesburg mining house but it was only two years later when significant water inrushes occurred (4,500,000 gal/day, 20,457m³/day) at ERPM that Francois was requested to assist. At ERPM 26th cross cut south, water was struck and in the 30 level west drive, water was encountered unexpectedly at high pressure. (Krynauw, 1918). The Francois Cementation Process was successful. The same process was used to create a grout curtain underneath an arch dam in the Mazoe Valley, Rhodesia in 1918; again the process proved successful

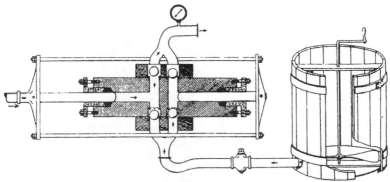

Fig. 1.--Diagrammatic Sketch of Water End of   Cementation Pump and Mixing Tub.

In 1917 the Francois Cementation Syndicate was formed; two years later this was changed to the Francois Cementation Company (Africa) Ltd as a wholly owned subsidiary of the parent company in the UK. Today the company still operates successfully as Cementation-Skanska.

By the end of 1918, the Francois Cementation Process was well established; the following projects had been completed successfully: Comet Deep (ERPM), 26th cross cut, 30th level West Angelo Deep 28th West Dam, 29th West Cross cut; Geduld Proprietary Mine Ltd: 2nd and 3rd level South, 4th level South Drive; Daggafontein Mine No 2 Shaft and No 4 Shaft, Brakpan Mines. In addition the Rand Water Board Pumping Station and the Mazoe Dam in Southern Rhodesia were successfully grouted and repaired with this process.

The Francois Cementation process was described by Krynauw (1918), who worked closely with Albert Francois during 1917, as follows:

"An essential condition in the introduction of cement into fissures and cracks is that the injection should be done under a considerable pressure, the object being, firstly, to overcome the contra pressure of water present in the fissure; secondly, for the purpose of forcing the cement as far as possible into the minute cracks, and, thirdly, for the purpose of squeezing out the superfluous water from the cavity which is being filled with cement pulp, and thus leave the cement in a condition most suitable for its rapid and efficient setting."

After the discovery of the Witwatersrand Goldfields, it was soon realized and verified that the gold bearing reefs continued below the water bearing dolomites. In 1910 the West Rand Estates Ltd located a shaft site commonly known as Pullinger near the centre of the Venterspost mining area. It was a circular shaft; sinking began in 1910, the shaft was lined with German Haniel and Lueg's cast iron tubbing and was abandoned in 1911 at a depth of 97ft (29.5m) as two of the most modern electric pumps at the time could not cope with the inrushes of 208 000 gal/hour (946m³/hour) and several mud rushes.

Subsequent applications which successfully established the Francois Cementation process for shaft sinking in South Africa were the sinking of Daggafontein Mines (no. 2), West Springs No. 1 and two circular shafts at the Brakpan Mines. At Vogelstruisbult GM in 1934 the cementation process was significantly improved by using more modern drilling methods.

During earlier cementation projects at Daggafontein and West Springs long diamond drilling holes up to 200ft (61m) were drilled. The holes were drilled inclined, eighteen holes were drilled for a 45ft by 8ft shaft. At Vogelstruisbult GM for the first time percussion equipment was used to drill the cementation holes. At Venterspost this method was further improved to drill 40ft (12m) cementation cover holes to allow a 30ft (9m) advance.

However, a milestone was achieved by sinking Shafts No. 1 and No. 2 at the Venterspost Gold Mining Company. After the failure of the Pullinger shaft in 1911 it was decided in 1934 to use cementation for sinking the new shaft in the vicinity. Cementation work at Venterspost was divided into three phases (Allen and Crawhall, 1937/38):

a) "The pre-treatment of fissured dolomite free from decomposition.
b) The stabilisation of sand and wad deposits occurring in the upper zones of the dolomite.
c) The correction of errors arising from insufficiently intensive ground preparation."

The sinking of these two shafts was remarkable for several reasons:

1. For the first time it was shown that the cementation process was successful under the most hazardous conditions where previous attempts had failed such as sinking through dolomite AND

2. For the first time a shaft was sunk through "Wad" (hydrated manganese oxide), which is low in density, highly compressible and, except for the absence of fibres, not unlike peat and often water logged. Wad occurs in large horizons in the upper strata in dolomite. In some zones it was necessary to place a concrete mat 2 to 4 ft in thickness to be able to effectively treat this material. The cementation process for this material was what would today be termed a combination of the "Soil-Frac" method and compaction grouting. Crawhall states that: "the muds have proved themselves amenable to cementation, and the effect of injections is to squeeze the soft ground near the shaft into such a compact condition that it is impervious to the vast volume of water impounded in similar ground outside the immediate area of the shaft. This produces a stable enough condition to permit excavation. More usually the cement builds up in ever thickening layers from some plane of weakness, but is also found to cut through the mud dividing it into innumerable compartments separated by films of cement."

In parallel to these important advances in the mining grouting field, civil engineering cementation advanced rapidly driven by the great dam building era in Europe, particularly in Switzerland, Italy and Spain.

Several geotechnical companies specializing in cementation works were founded during this era; timeously for the purpose of entering the great era of dam and tunnel construction.

During the 20 years leading up to WW II, many important innovations and new grouting techniques were developed. The now well established Lugeon water test was developed by Maurice Lugeon (Lugeon 1933), the tube-à-manchette was invented, the split spacing technique became common place. The Hoover Dam (Simmonds, 1953) with its specific grouting problems also contributed significantly to this field. The essential elements of present day dam grouting techniques had been developed by the end of the thirties.

By 1946 batching plants had been developed, producing large volumes of high quality cement slurries for dam grouting activities such as grout curtains and rock consolidation e.g. at Mattmark (Blatter, 1961)

In the early phases of development of the Orange Free State Goldfields (OFS Goldfields) during the fifties, civil engineering grouting techniques developed for dams, tunnels and foundations were applied successfully in the mining field.

Now for the first time, civil engineering grouting methods were applied with effort to mining engineering problems. The most notable techniques introduced were high speed, high shear mixing of cement grouts, the use of packers at large depths, improved grouting pumps and mixers, high volume automatic batching plants, more advanced control and monitoring equipment but possibly most importantly the introduction of grouts with higher densities i.e. smaller w:c ratios.

The Orange Free State Goldfields were discovered in 1938. Development was interrupted by WW II but expansion continued rapidly after the war. During some years in the fifties, more than 30 deep shafts were at various stages of development in South Africa. At present (2003) shafts and mines are being closed and the OFS goldfield has been reduced to a few shafts only.

While the underground cementation process had matured by the fifties, pregrouting of deep shafts received a new impetus during the development of the OFS goldfields. Some notable precementation work was executed at Harmony (Newman, 1956), President Brand (Mudd, 1959), FSG, Western Deep levels (1960), Hartebeestfontein and Buffelsfontein.

The precementation at Buffelsfontein in 1961 was the first precementation executed by utilizing large automated batching plants using high speed/high shear mixing, "thicker" mixes and packers at depth.

3. SHAFT SINKING
In the shaft sinking context grouting is applied to:

    a)　Pregrouting or precementation of shafts
    b)　Cover grouting during shaft sinking

Sometimes a shallow grout curtain is constructed around the shaft collar. As this would require a typical dam grouting technique, we disregard this application for the purpose of this paper.

a) Pregrouting or precementation of shafts

Pregrouting or precementation of deep shafts prior to sinking has been applied in South Africa since the fifties, with considerable success. Geological and geohydrological considerations are decisive parameters determining success or failure of a precementation, indeed its desirability, economically and otherwise. The following information and data is required:

1) Comprehensive and exhaustive collection of geological information and data relevant to the project. In particular
    a)　Geological data, logs of all boreholes at the site and in the vicinity of the project
    b)　Aerial photographs and magnetic surveys to detect possible faults and dykes, infrared photography in dolomitic terrain.
    c)　Analysis of prominent joints and fissure patterns.
    d)　Compilation of information on water: water tables (perched), quality, direction and flow of water.

2) Characteristics and values of the in situ stress filed.
    a)　Hydrofracturing data which will provide direction and value of stresses.
    b)　Determination of hydrofracturing stresses and possibly determining of similar values for existing fractures, weak joints etc.

Benefits of pregrouting of shafts can be summarised as follows:

1) It increases the safety of the sinking operation.
2) It minimizes the inflow of water and gas;
3) It minimizes the time lost due to additional grouting operations (cover grouting) during sinking and hence minimizes standing time of costly shaft sinking crews and equipment;
4) It provides improved rock strength for excavations in the immediate vicinity of the shaft area (grouted fissures have been found up to 60m from the pregrouted shaft);
5) It provides detailed information of the geology of the proposed shaft site and possibly information on ore grades in the shaft vicinity;
6) It reduces the number of intermediate underground grout stations during shaft sinking operations.
7) It allows the mine to start mining earlier in some cases several months, which should result in earlier positive cash flow from mining operations.
8) Ideally pregrouting is done outside the shaft perimeter with no interference between the sinking teams and the pregrouting teams or contractor.
9) Shaft sinking is an expensive operation, hence time allowed for supporting activities such as grouting during sinking operations though critical, are often curtailed to the detriment of the final result. Pregrouting minimizes this possibility.

Pregrouting can be done before sinking commences or during sinking. If done during shaft sinking the pregrouting holes should lead the sinking operation by several hundred metres.

Grouting Pressure

During recent years much progress has been made in determining allowable grouting pressures for shaft pregrouting projects.

The role of fluids and slurries in fracture propagation has been researched intensively in recent years.
The most significant contributions in this field result from oil and gas fields development (Cleary, 1997). The production of oil and gas fields is enhanced by hydraulic fracturing. Therefore, fracturing is well researched and most importantly its parameters can be determined in situ. Therefore, hydrofracturing can be used to determine in situ allowable pressures to be applied to pregrouting. (Cornet 1992, Smith 1987).

The first consideration concerns the water table. This will determine the minimum pressure required to move slurries into the rock formation.

Current practice in pregrouting of deep shafts in South Africa uses a sealing pressure at a certain depth of 2,5 times the in situ nominal hydrostatic pressure at that depth i.e. as determined from the in situ acting water table (See Figure 2.)

For the typical case where the water table is several hundred metres below the surface, this rule seems to be reasonable. If in addition one considers that typically underground water pressures are found to be two thirds of the nominal hydrostatic level as determined from the in situ water table – then this rule results in grouting operating pressures which are safely below the hydrofracturing pressure.

The maximum sealing pressure used at Western Deep Level no. 2 shaft (Muller, 1960), was 1,0 lb/sq in/ft to about 750 ft and 1,7 lbs/sqin/ft below 3000ft.
In oil well terminology this is termed the fracture gradient where values range from 0,45 lb/sqin/ft to 1,15 lb/sqin/ft with typical averages around 0.8 lb/sqin/ft (Smith 1987). These values refer to host rocks of oil and gas typically sedimentary rocks.

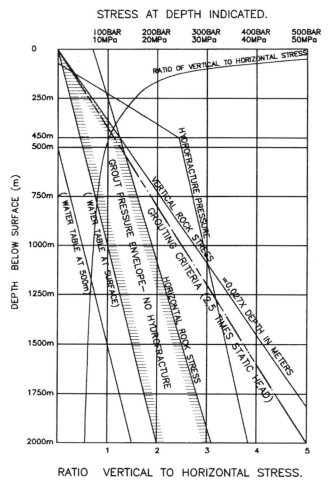

Figure 2: Grouting Hydrofracture Criteria

Figure 2 illustrates that the pregrouting pressure in the no- hydrofracture zone may be too low to achieve proper grout penetration. In the absence of other effects, fracturing quickly seeks out and attains an orientation perpendicular to the minimum stress direction, this is the path of least resistance where least work is required to propagate the fracture. Also fractures can strongly interact with each other by disturbing the stress field in which the other fracture grows. (Cleary, 1989).

Firstly it is imperative to achieve the depth of the shaft in safety. Secondly water inflow must be minimized and depending on the cost, reduced to negligible flows also with a view to operating the shaft at minimum cost. Thirdly it is advantageous to create a strengthened zone around the shaft to improve safety during sinking and later operations and maintenance. Finally, all these requirements are to be fulfilled at an acceptable cost.

In order to achieve this, pressures have to be sufficient to obtain adequate penetration at the same time controlling hydrofracturing which may open fissures and fractures that may be difficult to close again by grouting.

Current practice to grout the formation up to 2.5 times the hydrostatic head seems reasonable to 1,250m. A rough guide would indicate the use of 70% of the hydrofracturing pressure in order to remain largely within the elastic deformation of the rock formation. State-of-the-Art requires doing proper hydrofracturing tests in the formations at the site, this will furnish the stress field, direction of the minor principal stress, fracturing pressure, possibly permeability and above all fracturing parameters of pre-existing fractures and weak planes.

Financial and technical benefits of pregrouting are often questioned; although some examples shown in Table 1 of early pregrouting projects clearly show the advantages of pregrouting of shafts as opposed to only cover grouting (grouting operations during sinking).

Table I: Pregrouting/Cover Grouting

Shaft	Depth (m)	Cement Absorption			
		Pregrouting (tons)	Cover (tons)	Pregrouting (%)	Cover (%)
President Brand 2D Vent	1402	1468	576	72	28
Harmony No. 2	1687	734	61	92	8
Stilfontein Toni Shaft	1303	609	501	55	45
Kinross No. 1 & No. 1A	1681	7761	93	99	1

Table I shows that time savings can be achieved although in some cases such as Toni Shaft (Stilfontein G.M, now closed) economic advantages are not as obvious.

b) Cover drilling during shaft sinking

In earlier shaft sinking operations at the beginning of the 20th century, cover grouting utilized (Jeppe, 1946) diamond drilled holes up to 200ft. The holes were drilled at a slight angle to the vertical so that they ended some 7 to 9 ft outside the perimeter of the shaft e.g. in a 45ft by 8ft shaft 18 holes were drilled.

The rate of diamond drilling was rather slow especially in cherty dolomite e.g. at Vogelstruisbult G.M the depth of the cover holes was 126ft, drilled at an average of 3ft/day (MacWilliam, 1935). As progress was too slow it was decided to drill percussion holes initially to a depth of 15ft; the shaft was then sunk 11ft. Rapidly these percussion holes were extended to 20ft and eventually 30ft. After cementation of the 30ft holes, the shaft was advanced 27ft.

Today most shafts are circular, sometimes the stretched circular shaft is used where a short flat usually less than 10% of the shaft diameter is introduced between two semicircles. Irrespective of whether a shaft has been pregrouted, the Mines and Works Act requires that holes are drilled and grouted ahead of excavation to prevent blasting into uncontrollable quantities of water which could flood the shaft. Furthermore, it is desirable to have a dry shaft
  a) to avoid an increase in relative humidity of the incoming mine ventilation system
  b) to avoid degradation of the shaft steelwork and
  c) to reduce ventilation and pumping power requirements over the life of the shaft.

Hence the importance of proper cover grouting procedures.
State-of-the-Art practice is to drill eight to twenty-four holes (depending on shaft diameter) up to 50m but typically 36m at 75° to 85° below the horizontal and so spaced and raked that the toe of each hole overlaps the collar of the succeeding hole on the pitch circle circumference. This results in a truncated cone of spiral boreholes at least one of which will intersect randomly orientated planar fissures drawn through the cone. These rounds are repeated every 30m so that the shaft is always at least 6m inside cover.

Grouting under these conditions is still very much an artisan job, where decisions on thickening procedures are made on site at the face. Slurries with W:C ratios of 6:1 to 4:1 are used to start the grouting. Thickening the slurry is relatively fast as time is of essence during shaft sinking, grouts of W:C ratios of 1:1 to even thicker are used if no pressure is achieved.

Cover grouting has one major drawback in that cementation and sinking cannot be done at the same time. A major part of the time may have to be spent on grouting, and as grouting time is a function of geological conditions and hence is difficult to estimate and control, grouting may be rushed at the expense of the quality of sealing.

4. UNDERGROUND CEMENTATION

Underground cementation originally developed by Francois and successfully applied in many mines was at the time described as follows:

1. Diamond bore holes were drilled and on completion if cementation was found to be necessary, a casing or pipe with high pressure valves attached was inserted into the hole and cemented securely into position.
2. A cement pump was then run with a thin mix of about 3% for as long as 15 shifts unless the fissure had been closed in the meantime. The mix was then thickened to 10%, if pump pressure did not increase the mixture was again thickened to 30% with the addition of sawdust. (Cowles 1930).
3. The pump pressure was invariably increased by these means after which the mixture was thinned and pumping was continued until the hole was sealed.
4. The cement was allowed to set for a few hours, the hole was then redrilled. If the fissure was not completely sealed, the process was repeated until the fissure was sealed.

The basis for this approach was described by Voskule (1930) as follows:

"It is a well-known fact that a thin solution of cement and water, under normal conditions. sets very slowly. When high pressures are applied the cement settles out of the mixture and sets hard in a comparatively short time as an incrustation on the walls of the container or cavity; this process being repeated as more mixture is injected. In this way the cavity, the fissures and exits from the cavities are reduced in size, thus increasing the pressure required to displace with cement and drive out the excess of water. From subsequent examination of cemented strata it appears that the exits eventually become choked up and the surplus water has to be pressed out through an increasing thickness of cement filter, thus giving rise to the very high pressures required towards the completion of the injections."

In earlier days (before 1910) cavities and fissures were merely filled by gravity with the mixture. No attempt was made to increase the percentage of cement to water after the cavities had been filled. Sometimes small hand pumps were employed.

Today's standard and understanding of the grouting process would regard this procedure as technically unsound and time consuming, hence uneconomical. However, in fairness to the original pioneers in the mining grouting field, the procedure described by Voskule (1930) was successful; hence the motivation of this paper to present these early successes in order to facilitate a more rational evaluation of these techniques.

Grouting with thin, unstable slurries

Current civil engineering grouting philosophy requires thicker rather than thinner grouts or more correctly stable grouts rather than unstable grouts.

The ideal grout should behave like water and have negligible viscosity and yield stress during the dynamic phase i.e. during penetration; only thin grouts behave in this way; also the ideal grout requires instant strength once it has reached its final position and is required to perform its task. The final in situ quality of "thin" grouts as well as the danger of hydrofracturing at higher pressures during grouting are the concerns regarding thin grouts.

"Thick" or dense grouts choke fissures. Therefore, the tendency today is to use grouts as thick as possible (not to choke fissures) and as stable as possible and attempt to reduce the flow parameters such as viscosity and yield stress by adding superplastisizers.

The most important limiting factor of "thick" grout is penetration. Where, for economical reasons, penetration of many metres is essential, grouts should be as thin as empirical tests will justify. However, much thinner ratios than W: C 4:1 are not justified, particularly as rheological parameters do not change significantly for slurries thinner than W: C 2:1 for particulate suspensions.

It is important to realise that "stable" grouts are really grouts stable under gravitational forces only. Practically all cementitious grouts are unstable at high pressures. In "Ultra Deep Grout Barriers" (Heinz, 1993) the author introduced the concept of **static** and **dynamic** phase grouting.

Stable means either sedimentation is so slow that it is almost negligible or thixotropic action, hydration or other reactions and possible forces prevent sedimentation.

It is helpful, indeed necessary, to distinguish between STATIC PHASE grouting and DYNAMIC PHASE grouting of particulate suspensions.
The ideal static phase of cement grouting is the measuring cylinder where sedimentation is predominantly influenced by:
> gravity, very low particle velocity, stationary continuous phase, some particle interference.

In contrast, in the dynamic phase of cement grouting the sedimentation process is predominantly regulated by:
> High velocity resulting from high pressures, forces which change the direction and value of the resultant force on the particles in contrast to gravity only, different velocities between the suspended particles and suspending phase, selective sedimentation (pressure filtration).

Both phases require control and manipulation. It is incorrect to assume as is typically done that if the static phase is "stable" the dynamic phase is also "stable". Stable in the dynamic phase requires the properties of the grout to remain similar before and after moving through the rock mass.

High pressure mining grouting is **dynamic** phase grouting and hence is fundamentally different from **static** phase grouting.

Recent mining grouting practice has tended towards "thicker" slurries, however, the basic technique of pumping thin grout first is still utilised.

Typically water is pumped for two to three hours. This will give the artisan an indication of the transmissivity of the rock formation. This will be followed by "thin" grout approximately W:C 6:1, whereupon the slurry is thickened to attempt to obtain pressure.

5. CHEMICAL GROUTING

While cement grouting is still widely used in underground, mining applications, chemical grouting has increasingly supplemented and in some cases even replaced cement as a sealer of water and methane carrying rock formations.

Although AM9 was used successfully for water control underground at Kinross, this was soon discontinued because of its toxicity. At present mainly two types of chemical grouts are used (brand names): Polygrout and Supergrout.

Polygrout is a polyurethane that is water activated. Polygrout comes in two forms (both pure liquids). One is soluble in water as well as water activated and sets as low as 5% in solution. The other Polygrout is water activated but not water soluble.

Both set to a rubber like consistency and expand up to 3 times their volume when setting. Polygrout is in many ways the ultimate grout. It has no solids, sets in 1 minute to 60 minutes in virtually any water, and it is acid resistance. It can be used to seal small leaks to huge fissures.

The disadvantages of Polygrout are its relatively high price (80 times the price of normal OPC) and the fact that it cannot be redrilled as it is too "rubbery". Its main advantage is the reaction with water hence its effectiveness in stopping large water inrushes.

Supergrout is a "cementitious" type grout setting hard like cement. It is a modified oxychloride inorganic grout. Bentonite can be added to modify the flow properties and to help sealing. Supergrout A is a liquid and Supergrout B is a powder. The powder is mixed into the liquid and is ready to be pumped. Normal grouting equipment can be used.

Supergrout has less solids than cement and is more colloidal in size, so it exhibits superior penetration of fissures, and is used throughout the goldfields and platinum mines of SA to seal methane as well as water. Supergrout sets in about 4-6 hours at room temperature but is temperature sensitive and sets in about half that time underground.

Supergrout is reasonably priced (20 times the price of normal OPC) and is easy to use with normal grouting equipment; it can also be redrilled easily.

6. CONCLUSION

The achievements of the early mining grouting engineers in South Africa and elsewhere at the beginning of the previous century, were quite extraordinary.

A deciding factor in the development of high pressure grouting in deep mines was the invention of the high pressure cementation pump by Albert Francois over 100 years ago.

Discovery and development of the gold fields in South Africa and the particular problems encountered in deep mines were the driving forces of the developments in mining grouting.

Mining grouting techniques as developed in South Africa and other mining countries, have often been criticized, in some cases even been denounced as "black art". These critics have little understanding of the actual grouting techniques and have certainly not studied the successes achieved in this field. Indeed the successes are the most important justification for the application of these mining grouting techniques.

Grouting with thin, unstable cement slurries is possibly the most criticized and controversial element of mining grouting, nevertheless it is still being applied at present with success. Practically all particulate suspensions are unstable at high pressures.

In an endeavour to reconcile civil engineering and mining grouting, it is helpful, indeed necessary to define a new concept viz. **static** and **dynamic** phase grouting described in the paper. Somewhat akin to the characterization of hydraulic flow by the Reynolds number, this new concept attempts to redefine the fields of application and the limits of these phases and highlights the need to include pressure, rock and fissure characteristics as well as slurry properties in the parameters that determine successful grouting procedures.

7. REFERENCE LIST

Allen, W. & Crawhall, J.S. (1937). "Shaft-Sinking in Dolomite at Venterspost." Papers and Discussions, Association of Mine Managers of the Transvaal.

Biccard Jeppe, C. (1946). "Gold Mining on the Witwatersrand." Papers and Discussions, The Transvaal Chamber of Mines, South Africa.

Blatter, C. E. (1961). "Vorversuche und Ausführung des Injektionsschleiers in Mattmark." Schweizerische Bauzeitung, Heft 42, 43. Oktober.

Cleary, M. P. (1997). "Technology Transfer for Hydraulic Fracturing." www.gri.org

Cleary, M. P. (1989). "Effects of Depth Rock Fracture." Proceedings ISRM-SPE, International Symposium, Balkema.

Cornet, F.H. (1992). "The HTPF and the Integrated Stress Determination Methods." Pergamon Press. *(HTPF: Hydraulic test on preexisting fracture)*

Cowles, E.P. (1930). "Underground Cementation." Third Empire Mining and Metallurgical Congress, Johannesburg.

Heinz, W.F., (October 1993) "Extrem tiefe Injektionsschürzen/Ultra Deep Grout Barriers" Proceedings, International Conference on Grouting in Rock and Concrete, Salzburg, Austria.

Krynauw, A.H. (1918). "Cementation Process Applied to Mining – Francois System". The Journal of the Chemical, Metallurgical and Mining Society of South Africa. Johannesburg, South Africa.

Lugeon, M. (1933). "Barrages et Geologie." Lausanne. Reimpression photostatique 1979. Poligrafico Pedrazzini, Locarno.

MacWilliam, K.J. (1931 – 36)." Notes on Cementation at Vogelstruisbult Gold Mining Areas, Limited." Papers and Discussions, The Transvaal and OFS Chamber of Mines, South Africa.

Mudd, R.A. (1958 - 59). "Some Notes on Pre-grouting at President Brand Gold Mining Company, Ltd." Papers and Discussions, The Transvaal and OFS Chamber of Mines, South Africa.

Muller, T.F. & Skeen, C. (1952) "Shaft-Sinking on the Virginia and Merriespruit Mines." Papers and Discussions, Association of Mine Managers of the Transvaal.

Newman, S.C. (1956 - 57). "Pre-cementation at No. 2 Shaft, Harmony Gold Mining Company, Ltd." Papers and Discussions, The Transvaal and OFS Chamber of Mines, South Africa.

Simmonds. A.W. (1953). "Final Foundation Treatment at Hoover Dam." Vol. 118 of the Transactions of the ASCE.

Smith, D.K. (1987). "Cementing" SPE Monograph, Society of Petroleum Engineers.

Voskule, G.A. (1930). "The Cementation Process." Third Empire Mining and Metallurgical Congress, Johannesburg.

Innovative Grouting Solves Geotechnical Issues:
Five Case Histories

H. Clay Griffin & Richard M. Berry[1]

Abstract

Some grouting projects are of particular interest because grouting was not initially considered in the range of alternative solutions. And some grouting projects are interesting because they involve unusual physical or technical challenges. To be worthy of note, a grouting project has to be wildly successful against long odds. Occasionally, a grouting project meets all three criteria and is of interest to the engineering community at large. This paper presents brief case histories of five such projects, with the goal of encouraging engineers to consider grouting in the "short list" of possible solutions to geotechnical problems.

Background

In the most simplistic terms, geotechnical grouting is just the introduction of a relatively fluid material into the ground. Practiced from the days of Julius Caesar, cementitious slurry is still poured into holes using almost no equipment and even less expertise. However, at the other end of the spectrum, successful modern grouting contractors have adapted special equipment and techniques to solve an incredible breadth of daunting geotechnical problems. These specialty grouting firms routinely advance injection casings into difficult soil and rock formations, and inject a wide range of custom-compounded grout materials to achieve the specific ground modification desired.

In surprisingly many cases, engineers supporting industry and construction take on problems in the ground without even considering grouting as an alternative. Those unfamiliar with modern grouting practices may consider it a "fringe" technology or a last resort. Grouting was not among the first alternatives in any of the projects described in this presentation, and was only introduced when earlier solutions became unviable due to cost, schedule, or technical limitations.

[1] H. Clay Griffin is President of Rembco Geotechnical Contractors, PO Box 23009, Knoxville, TN 37933, Tel 865-690-6917 Fax 865-690-9135. Email: info@rembco.com
Richard M. Berry is a grouting consultant to Rembco.

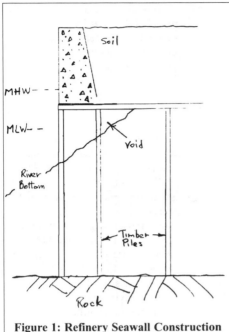

MHW

MLW

Void

River Bottom

Timber Piles

Rock

Soil

Figure 1: Refinery Seawall Construction

A seawall at a refinery. An oil refinery located along the Delaware River was expanded just before the Second World War. The increased capacity was made possible by creating new real estate on dredged fill at the riverfront. The fill was placed behind a concrete gravity wall constructed at the outer edge of a timber pile wharf. The retaining wall is 2.4m (8 feet) tall, approximately 120m (400 feet) long, and tied back with deadman anchors for lateral stability. Pipes were placed through the concrete wall for loading and unloading oil tankers. The normal low tide level of the river is about 60m (2 feet) below the bottom of the concrete seawall, rising approximately 2m (7 feet) at high tide. A sketch of the approximate construction is shown in Figure 1.

Decades of refinery operations had resulted in considerable contamination of the ground with various petroleum distillates. Over many years, the tidal rise and fall of the river gradually induced leakage of contaminants through the concrete at expansion joints, anchorage locations, and pipe penetrations. Even more disturbing was the soil erosion at the base of the seawall. At high tide, the groundwater level rose throughout the site. At low tide, the groundwater relieved beneath the retaining wall, carrying light distillates that would then float to the surface of the river in a visible sheen. The company installed absorption booms along the seawall. But as time went on, the leakage of contamination worsened until servicing of the booms became unmanageable.

The company first decided to construct a new sheet pile retaining wall in front of the existing seawall. However, the cost of the new wall was prohibitive, especially considering the permitting restrictions, disruption of terminal operations, and potential failure of the existing wall. Ultimately, a better solution was formulated through a variety of grouting techniques. Grouting with acrylamide gel solved discreet leaks at the anchorages and pipe penetrations. This water-thin grout also permeated fine cracks in the concrete. The expansion joints were sealed using a hydrophobic polyurethane grout without having to remove the old deteriorated joint material. This work was accessed from the river at low tide when the surfaces of the faults were visible.

Figure 2: Grout Placement

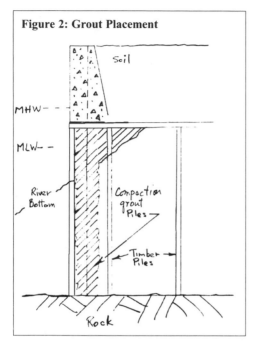

A limited probe-drilling program revealed that soil loss beneath the seawall was extensive, extending up to eight feet behind the wall in some locations. This led to concerns that the seawall would settle as the old wharf deteriorated. The solution was to support the seawall on compaction piles. 90mm (3.5 inch) diameter holes were drilled vertically through the 2.4m (8-foot) depth of the concrete, and injection casings were advanced approximately 11m (35 feet) to bedrock. Cohesive, mortar-like compaction grout was injected to compact the soil and form piles approximately 30cm (1 foot) in diameter from bedrock up to the base of the seawall. Low mobility grout (LMG) was then injected to fill remaining voids beneath the wall and around the row of piles.

Figure 2 illustrates the grout placement. The seepage stopped soon thereafter and the oil booms were removed. The entire project was completed for $70,000 with no disruption to refinery or terminal operations.

A distressed effluent pipeline. At a large paper mill on Canada's west coast, a leak appeared near the shoreline where an underground 1.1m (42 inch) fiberglass pipeline carried about 8 million liters (2 million gallons) per hour of mill effluent to diffusers in the ocean, as shown in Figure 3. The leak surfaced above a "Y" that combined the effluent from two defoaming tanks into a single line. The pipeline was buried 6m (20 feet) below sea level, under 7m (22 feet) of sand fill, and the surface flow from the leak was estimated at approximately 30L/s (500 gpm). A CCTV survey to visually assess the damage was unsuccessful due to the

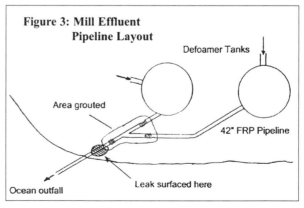

Figure 3: Mill Effluent Pipeline Layout
Defoamer Tanks
Area grouted
42" FRP Pipeline
Ocean outfall
Leak surfaced here

turbidity of the effluent. However, a sonar survey inside the pipe revealed an offset of nearly five inches at one joint of the "Y". Besides the obvious environmental impact, mill managers feared that the leak would soon wash away enough of the bedding material to allow a catastrophic failure of the mill's only effluent line. A failure of this type would require a lengthy shutdown of all production operations.

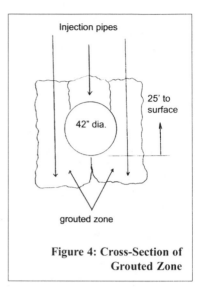

Figure 4: Cross-Section of Grouted Zone

The paper company had a spare "Y" for just this situation, and the decision was made to excavate and replace the damaged piece. However, due to it's location (only a few feet from water's edge) a cofferdam would have to be constructed before any excavation. The cost of nearly one million dollars for this conventional type of repair was accepted as inevitable. But the four to six week schedule incurred more risk and downtime than the company could tolerate. A pipeline consultant suggested the possibility of grouting to stabilize the pipeline until a cofferdam could be constructed and a shutdown scheduled. The company agreed and called in a specialty grouting contractor (Rembco).

A grouting program was quickly devised which would stabilize the bedding of the "Y" to prevent further movement. A secondary goal was to attempt to reduce the leakage to a more manageable rate until proper repairs could be made. Several constraints complicated the grouting program:

• The three joints of the "Y" were shielded by corrugated metal pipe, offering poor external access to the leak source.
• An internal injection packer could not be used due to the size and offset in the joint.
• The fiberglass pipe was less than 19mm (¾ inch) thick and could easily be accidentally penetrated by the drilled-in injection casings.
• Grouting pressures had to be kept low to avoid further damage to the fragile pipe.
• The grouting material had to be somewhat flexible to avoid creating "hard spots" around the pipe which could induce shear failures.
• The mill would not shut down. Flow could be diverted to a clarifier for only two hours each day. Repairs had to be made "on the fly".

The program was developed as a two-phased approach. First, the bedding around the joints would be grouted with an accelerated, water-catalyzed, hydrophilic urethane to fill washed-out voids and provide toughness at the joints. Then the bedding around the rest of the "Y" would be stabilized with acrylamide grout, a water-thin gel grout that offers a flash set controllable from two seconds to two hours. Any fine-grained bedding remaining around the pipe would easily be permeated by the acrylamide. Injection casings were to be drilled to a depth 0.6m (2 feet) below the pipe invert, passing within one foot of each side. Casings would also be drilled down at the centerline just to the top of the pipe. The injection pattern was designed to create a continuous envelope of grouted soil approximately two feet around the pipe. Figure 4, on the previous page, shows a cross-section of the planned grout zone and Figure 5 is a plan layout of the injection locations.

Crew, equipment, and materials were quickly mobilized and flown to the remote site. As the drilling got underway, groutability tests were run using soils collected at the site. Maximum pumping rates of 0.19L/s (3 gpm) for the urethane and 0.31L/s (5 gpm) for the acrylamide were determined to achieve maximum permeation without fracturing or displacing the formation. The void area was estimated to constitute 45% of the soil volume. At the pressures involved, the expansion of the urethane foam grout was estimated to be 6½ to 1.

Injection casings were drilled in first in the joint locations, using a specially modified open-center drag bit that would not damage the pipe on contact. The effluent flow was temporarily diverted to a clarifier, and urethane grout was injected at the joint closest to the surface leak. After the first two hours of urethane grouting, effluent flow was restored to the pipe. To the surprise and delight of all, the flow to the surface was reduced to about 20 gpm. Completing injection at the other two joints, a total of 1,325L (350 gallons) of urethane was injected through 18 casings.

Acrylamide grouting was then undertaken in the remaining locations to complete the stabilization effort. A total of 9,000L (2,375 gallons) of acrylamide was pumped through 15 injection casings, grouting the formation to complete refusal. During this grouting, all backflow to the surface stopped. The tightness of the new grout seal was confirmed by probe drilling at ten locations, all locations indicating a complete seal.

This grouting program that was devised as a temporary patch became a permanent solution. The total cost of the grouting was $125,000 and the work was performed without disruption of the mill's production. The major seven-figure repair was never needed.

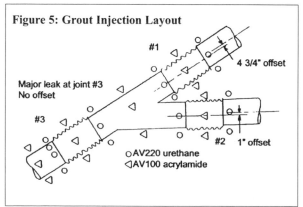

Figure 5: Grout Injection Layout

#1

4 3/4" offset

Major leak at joint #3
No offset

#3

#2   1" offset

o AV220 urethane
◁ AV100 acrylamide

An underground fire. Residents of a small town in northeastern Tennessee began to complain of smoke and noxious fumes coming from the hillside woods next to the mainline railroad track just outside of town. It seems that the woods had been set afire as a Halloween "prank", but once the underbrush had burned off, the ground itself was found to be slowly burning away. An investigation by the railroad determined that the 90m by 30m (300 by 100 foot) area had originally been a shallow valley that was filled back in the steam engine era with depleted coal "clinkers", a plentiful byproduct of railroad operations. Although not useful as a fuel, the clinkers still contained enough coal to smolder for years. The railroad responded by striking an agreement with the local fire department to extinguish the fire. After dousing the ground surface with approximately 15 million liters (4 million gallons) of water, the fire department pronounced victory. But several days later, the smoke returned . . . and the fire burned on.

Railroad project engineers, working with the grouting contractor (Rembco), designed a grouting program that would permeate the ground with a mildly cementitious material, depriving the fire of air and stabilizing the ground that was quickly turning to ash. Probe drilling across the area indicated a clinker depth of up to 3m (10 feet) with silty clay below as a relatively impermeable base. The fire had reached the tracks and was undermining the ballast as the fill was reduced to ash. Besides extinguishing the fire, stabilization of this area was essential to maintain normal operations of the railroad.

An underground fire poses several significant safety hazards to workers involved in the abatement. The fire burns away roots of the trees, allowing them to fall in the work area with little or no warning. Additionally, even as the fire consumes the clinker fill, the topsoil and root mat tends to bridge over the near-molten ground until it is loaded . . . perhaps by a human footstep. And,

Figure 6: Drilling Injection Casings

of course, the ground surface is extremely hot, with occasional small flare-ups and smoke. Before grouting operations began, the larger trees >76mm (3 inch) diameter were felled as a safety precaution. These trees were arranged as matt-like working platforms for the grouting crew. Even so, boots became a major consumable on the project.

Injection casings were advanced to the base of the fill on a grid spacing of approximately 2m (7 feet), with hand-held pneumatic rotary drills, as shown in Figure 6 on the previous page. A fluid, low-strength grout, comprised of flyash, cement, and water, was mixed on-site in high-shear colloidal mixers and injected through the casings to permeate the clinker fill,

Figure 7: Injection of Grout to Permeate Fill.

as shown in Figure 7. After completing the initial grid, any remaining "hot spots" were treated in the same fashion. Approximately 142m³ (5,000 cubic feet) of grout was injected to completely extinguish the fire.

Water at a salt mine. The Strategic Petroleum Reserve (SPR) was established in 1975 as part of a U.S. energy sufficiency initiative. The plan was to store up to one billion barrels of crude oil in deep salt caverns along the Gulf coast. In order to "jump start" the SPR operations, an existing salt mine was acquired from Morton Salt at Weeks Island, Louisiana. Crude oil was to be stored in the two mined levels at 180m (600 feet) and 275m (900 feet) below sea level. In a salt mine, water infiltration is more than a nuisance... it dissolves the roof and walls, causing structural failures and, in the worse cases, loss of the mine to flooding. For more than 10 years, a specialty grouting firm (Rembco) provided consulting, inspection, and grouting services for the SPR at Weeks Island to control water infiltration. Many interesting and challenging grouting operations were conducted in the shafts, access drifts, and storage caverns. But an area that was not even targeted for grouting became the focus of significant effort . . . the "Wet Drift".

During the last part of their mining operation, Morton had started a new drift in a direction which was to go around a notoriously wet salt zone. However, about 45m (150 feet) down the drift, a huge source of leaking brine was encountered. When the SPR took over the mine, the decision was made to isolate the wet drift with a concrete bulkhead to protect against catastrophic flooding. The potential pressure on the bulkhead was 1725 kPa (250 psi). With a drift cross-section of 6 x 9m (20 x 30 feet), the force on the bulkhead could be as much as 96 millionN (20 million pounds). The concrete bulkhead was 9m (30 feet) thick with several circumferential keys into the salt. Almost as an afterthought, a 46cm (18 inch) diameter pipe was cast into the center of the bulkhead, with a submarine-type door on each end to allow access to the flooding drift. The water level increased behind the bulkhead as expected, but then the unexpected happened. Adjacent drifts near the same level began to leak as well, effectively bypassing the bulkhead through innumerable cracks in the formation. Mine engineers and bureaucrats alike decided that grouting was the only hope for saving the mine.

If the working conditions had been configured by El Diablo himself, they could not have been more miserable. The only access for crew and equipment into the working area was through the 9m (30-foot) long pipe with a 432mm (17 inch) inside diameter. A taut cable was rigged in the top of the pipe to carry litters made from plastic pipe. We were able to strap a worker on his back on such a contrivance and pass him through the man-way if a disabling emergency were to occur. Litters also carried equipment of all types: lights, harnesses, portable breathing equipment, drills, scaffolding, tools, packers, pumps, etc. Men went in head first, hands over head in order to fit. Lighting was barely adequate, ventilation was poor, water covered the floor and poured from the roof. It was not a place for the claustrophobic.

Monitoring of inflow showed that significant water was coming through cracks in the floor or low in the walls. However, a small area of the roof held the highest potential for catastrophic failure, with discrete leaks registering up to 4.7L/s (75 gpm).

A special aluminum drill mast was built and outfitted with a fast hydraulic rotary head. The components were carefully chosen so they could be man-handled through the access pipe and into the desired position. The 48kw (65 hp) diesel-powered hydraulic unit was located outside the drift, with hoses running through the pipe to the control console inside. Probe drilling indicated that the largest water source was located above and considerably beyond the wet drift, rather than directly overhead. A grouting pattern was designed that required drilling 46m (150 foot) long holes through the 6m (20 foot) high roof in a fan-shaped array centered about 30 degrees above horizontal. First, 50mm (2 inch) diameter surface casings were set about 3m (10 feet) into competent salt to prevent structural failure at high pressures. Full-flow ball valves were affixed to the surface casing so that heavy flows could simply be turned off. Grout holes of 38mm (1.5 inch) diameter were then advanced through the surface casing to full depth, using saturated brine as the flushing fluid. The two pictures (Figures 8a and 8b on the following pages) were taken during drilling operations in the roof of the drift. Recall that everything in the picture went in through a clear opening of 432mm (17 inches).

Figure 8b: Rotary Drilling with Specialized Hydraulic Equipment

A wide range of grouting materials and admixtures were considered and tested for this work. Any water used in the grout mix had to be saturated brine (about 30% salt) to prevent further erosion of the formation during injections. The material had to permeate small cracks, have controllable set characteristics, and retain long-term stability . . . all in a salt matrix. Testing and trials led to the selection of two primary grouts for the project. Fine cracks and limited fracture zones were grouted with a 15% acrylamide solution with a modified catalyst system. This gel-type solution grout penetrates well due to its extremely low viscosity (2cps), and offers excellent control of set time over a wide range of injection conditions. Polymerized acrylamide at 15% is a firm gel and an excellent waterstop in narrow sections, but it has little structural value and virtually no adhesion. Large voids, fissures, or "rotten salt" zones encountered were grouted with microfine cement to provide significant strength in the formation. Generally, cement grout (or any colloidal suspension of particles) could not permeate a moderately fractured formation of salt. But microfine cement particles are so small (<2.5 microns), that the grout penetrates as well as most solution grouts.

The wet drift was ultimately "dried out" by this successful grouting program, protecting our government's investment of more than $3 billion, 180m (600 feet) underground.

A defective caisson and its repair.

During the construction of a six-story steel-framed building in Nashville, Tennessee, an interior column of the structure was found to have settled about 50mm (2 inches). The building had begun to suffer some stress and showed severe bending of some interior steel components. An investigation showed that the deep foundation for the column, a 1.1m (42 inch) diameter caisson, was founded on a thin layer of rock only 300mm (1 foot) thick. Immediately below this rock layer, a 1.5m (5 foot) seam of soft clay overlaid competent bedrock. Simply put, the caisson stopped about 1.5m (5 feet) short of its intended target.

Figure 8a: Drilling Injection Holes in a Salt Mine

The project engineers first decided to support the column footing on four drilled steel micropiles, removing the load from the faulty caisson. However, poor access would have made implementation of this solution costly and extremely disruptive. The general contractor called a grouting contractor (Rembco) on the "off chance" that some sort of grouting might offer a more acceptable alternative.

Compaction grouting is a ground improvement technique that compacts soil through the injection of a cohesive mortar-like grout between 0-50mm (0-2 inch) slump. The grout tends to stay in a cohesive mass, or bulb, displacing and compacting the soil around it. A plan was formulated around using compaction grout to displace the 1.5m (5-foot) clay seam between the base of the caisson and bedrock.

Three holes were cored through the length of the caisson and injection casings were advanced through the clay seam to bedrock. The thick cementitious compaction grout, designed to reach 34,450kPa (5,000 psi) in 28 days, was injected at a fixed volume per foot as the casings were retracted. The injection pressure steadily rose as the grout bulb displaced the clay, due to the increased compaction of the surrounding soil. This grout was allowed to set overnight and then two more holes were cored to bedrock to assess the results. The cores showed continuous grout from the caisson to bedrock, except for a 50mm (2 inch) mud seam that remained at the top of the grout. A slightly thinner grout, about 100mm (4 inch) slump, was injected to displace the last bit of mud. Pressures built quickly in this last stage, as the soil had already been intensely compacted, but pumping was continued somewhat beyond the theoretical volume required to displace the mud. At a pressure of nearly 6,890kPa (1,000 psi), the caisson actually began to lift back toward its original position. Calculations of the building and caisson dead loads, along with the skin friction on the caisson, indicate that the final grouting lifted approximately 354,000kg (780,000 pounds)! The grouting was completed in less than two days, without impacting any other construction activities on this congested site.

Conclusions

The case histories presented above are all non-standard applications of grouting. In fact, the field of geotechnical grouting thrives on "non-standard" applications. So, when should grouting be considered?
• If ground conditions don't meet the project requirements . . . consider grouting.
• If solutions to subgrade problems are expensive . . . consider grouting.
• If disruptions to existing operations must be avoided . . . consider grouting.
• Before you tear it out or dig it up . . . consider grouting.

Discuss your project with grouting specialists when developing the list of alternative solutions to geotechnical problems. Although the application of grouting technology may not be obvious, the case histories presented in this paper illustrate how specialists in this field may offer just the "edge" you need.

Long term performance of grouts and the effects of grout by-products.

Stephan A. Jefferis[1]

Abstract

This paper sets out a brief case history of the design of a cement-bentonite cut-off wall grout for use in very aggressive groundwater conditions and presents a basic procedure for the estimation of the design life of a grout barrier under such conditions. The paper then considers two perhaps unexpected effects of grouts, the formation of a soap-like material in silicate grouted ground which hindered slurry tunnelling operations and the release of ammonia from jet grouted ground.

Introduction

The properties of grouts, including strength and permeability, may be affected by aggressive materials in the ground. In order to predict the rate and significance of attack a basic procedure is outlined for the estimation of the time it will take for an aggressive permeant to penetrate through a cement-bentonite grout and how the permeability of the grout will vary during the process. Two more unexpected interactions of grouts and soils are then considered. These reactions have been sufficiently important to affect construction works but might not be recognised as risk factors by those unfamiliar with grout chemistries.

Permeability of cement based grouts subject to aggressive ground conditions

If a grout barrier is to be used in an aggressive environment, then there will be concerns about its effective life and in particular how the permeability of the barrier will vary over time as a result of the reaction. In the 1992 conference of this series of meetings on grouting the author presented a paper on modelling chemical attack of cement based grouts (Jefferis, 1992). The paper presented a very simple model for the overall permeability of a mass of grout in terms of the permeability of a reacted zone and an unreacted zone as shown in Figure 1a. The model assumes that as the aggressive groundwater moves through the grout a reaction front develops between

----

[1]Professor in Civil Engineering, University of Surrey, Guilford, Surrey, GU2 7XH, England; Phone +44 (0)1483 689118; s.jefferis@surrey.ac.uk

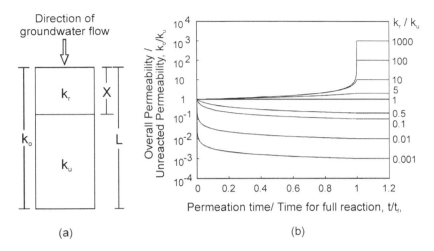

(a)   (b)

Figure 1a, b. Single reacted zone model for groundwater - grout interaction and
theoretical plots of permeability as a function of time

grout which has reacted with the groundwater and a downstream region where
reaction is yet to occur.   The reaction is assumed to cause a change of grout
permeability from an initial unreacted permeability k_u to a reacted permeability, k_r.
More complex models involving multiple reactions zones can be developed but
obtaining the necessary data to calibrate the models is difficult.

If the grout mass is the only effective hydraulic barrier, the rate of advance of
the reaction front will be controlled by the flow rate of liquid through the grout and
the nature of the reaction between the grout and chemicals in the groundwater. In the
1992 paper it was shown that the time for full reaction of a mass of grout is given by:

$$t_f = \frac{V_t L^2}{2H} \left(\frac{1}{k_u} + \frac{1}{k_r} \right)$$
(1)

Where V_t is the volume of groundwater required to fully react with one cubic metre
of grout according to the reaction which changes the permeability from k_u to k_r, H is
the driving head of water and L the flow path length in the grout mass. The overall
permeability, k_o of the grout mass at time, t (t < t_f) is given by:

$$\frac{1}{k_o^2} = \frac{1}{k_u^2} + \left(\frac{1}{k_r^2} - \frac{1}{k_u^2} \right) \frac{t_f}{t}$$
(2)

Figure 1b shows a set of theoretical curves for permeability as a function of time for a
reactive permeant following Eqs 1 and 2 for different values for the ratio of the

reacted permeability to the initial unreacted permeability (k_r/k_u). A key feature of these curves is that, for any reaction that significantly increases permeability ($k_r/k_u > 10$), prediction of the final permeability of the reacted grout mass is impossible until the reaction front has fully penetrated the grout mass. The early parts of each of the curves overlie one another. Short-term permeability tests cannot be used as an indicator of long-term permeability of the grout mass. The final reacted permeability of the grout cannot be determined until the reaction front has passed through the sample and for low permeability grouts this will require very long test times.

Eq 2 for the permeability as a function of time is the most basic equation and can be extended and generalised by considering parallels with contaminant migration in soils from which it can be shown that the retardation of the reaction front, R (the ratio of the velocity of the reaction front to the interstitial velocity of the water in the pores of the grout) in a material of porosity, n is given by:

$$R = \left(1 + \frac{V_t}{n}\right) \qquad (3)$$

In principle, this retardation can be used with standard procedures for contaminant migration in soils to predict the rate of groundwater / grout interaction including the effects of both advective and diffusive fluxes of contaminant. However, the standard procedures do not normally allow for a change of permeability on reaction and for the situation described below a particular advection / diffusion / reaction computer code had to be written to model the situation.

It should be noted that the change of permeability can be a reduction or an increase depending on the nature of the chemical reactions between groundwater and grout. The retardation is a function of the reaction volume ratio V_t and this ratio will decrease as the concentration of reactive species in the groundwater increases or the concentration of reactive material in the grout decreases (i.e. the retardation will be less for reactants at high concentration or in areas of more dilute/poor grout).

For reactions which increase the permeability of the grout ($k_r > k_u$), the effect of the greater velocity of the reaction front for lower values of V_t may be to localise a zone of failure though this will be offset by transverse diffusion from the advancing front. In the field this could lead to the formation of more permeable 'worm-hole' or fissure type features in a grout mass and it would be interesting to know whether these have been observed to occur in grout grounds.

For reactions that reduce permeability ($k_r < k_u$), the effect will be a more rapid reduction in permeability in the 'vulnerable' areas thus tending to 'repair' them. The author has observed reduction in permeability following reaction of aggressive chemicals with cement-bentonite grouts as a result of the precipitation of calcium carbonate (from bicarbonate species in landfill leachate) and iron oxyhydroxides (from dissolved iron in acid mine drainage waters) – though there can be a subsequent increase in permeability as discussed below. The effect of reaction front velocity is very important and a topic which needs further analysis. Procedures need to be developed to predict the types of reaction that will lead to local failure, especially the potentially rapid, 'wormhole' type failures.

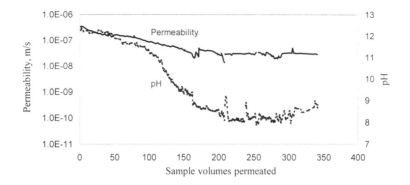

Figure 2. Effect of water permeation on the permeability of a cement-bentonite grout

The effect of water permeation. As discussed above the interaction between
cementitious grouts and chemicals in groundwater is complex.   However, even
leaching by pure water is not simple and shows a number of interaction effects.

Figure 2 shows a plot of the permeability of a sample of a cement-bentonite
grout permeated with tap water as a function of the number of sample volumes of
water passed through it (note: sample volumes is used as a convenient non-
dimensional measure of the amount of leaching to which the sample has been
subjected and it is in line with the definition of V_t as used in Eq 1. For chemical
reactions it is also theoretically more relevant than 'pore volumes', a measure much
used in the literature on contaminant-barrier interaction – but more appropriate to
physical effects such as the change of dielectric constant that occurs when an aqueous
permeant is replaced by a non-aqueous permeant). The figure also includes data on
the pH of the effluent from the sample. The data are for a test on a single sample but
have been repeated with several samples of comparable mix design.

The permeation was continued until the effluent pH approached a plateau at
about 8, a process which took nearly 3 years in the laboratory. The initial pH of the
effluent was over 12 due to the presence of sodium and potassium hydroxides and
free calcium hydroxide in the cement-bentonite mix. Once these hydroxides have
been removed continued leaching causes the degradation of the calcium silicate
hydrates, the main binding phase in Portland cements, with the release of further
calcium hydroxide and the production of hydrates of lower calcium to silicon ratio.
The reduction in pH is an indicator of this damage to the hydrate structure and not
surprisingly during the permeation there was visible shrinkage of the sample.

The permeation was interrupted on a number of occasions and on the re-
establishment of flow the initial effluent was of slightly higher pH as can be seen
from the 'spikes' on the pH plot of Figure 2. In contrast the permeability showed less
change after the interruptions. It can be seen that the cement-bentonite grout showed
a continuing reduction in permeability up to about 160 sample volumes of permeant
at which point the permeability began to level off. The reduction in the permeability
of cement-bentonite on leaching has been well established and thus the behaviour

shown in Figure 2 is as expected. However, the figure denies a thesis current in the literature that the permeability of cement-bentonite continues to decrease without a discernible limit. It also provides some information on the factors which underlie the permeability reduction. These include:

a) The deposition of calcium carbonate in the pores of the grout as a result of the reaction of bicarbonate ion in the permeating water (this was open atmospheric carbon dioxide) with calcium ions initially from free lime in the mix and once the pH dropped below about 10.5 from the calcium silicate hydrates of the cement;

b) The leaching of free lime from the cement;

c) The leaching of cement hydrate materials, and especially calcium from the calcium silicate hydrates, once all free lime had been leached. This will reduce the cementing activity of the hydrates and lead to some shrinkage of the sample if it is under a confining stress – samples immersed in water but not subject to confinement have been found to show some softening but without shrinkage;

d) The deposition of any suspended matter present in the water – as clean tap water was used this is not expected to be a significant effect. However, it is known that algae can develop in the slow moving water of permeameter pipework if this is exposed to sunlight. In the test programme there was no visible evidence of algal deposits in the pipework.

Each of the mechanisms a, b and c can be investigated in term of the V_t for reaction but this is beyond the scope of this paper. Mechanisms (a) and (d) may be expected to decrease the permeability of the material and (b) to increase it as a result of loss of solid volume. Mechanism (c) would be expected to soften the material and also increase its permeability. Softening of the sample, was indicated by consolidation (of order 10%) despite the low confining pressure (30 kPa) used in the tests. Mechanism (d) would be expected to cause a continuing reduction in permeability and although it may have had some effect it cannot explain the observed plateau in permeability after about 160 sample volumes. It follows that mechanisms (a), (b) and (c) appear to be the important controls and their combined effect is a reduction in permeability. It is instructive to use Eq 1 to calculate the time it would take for this plateau of permeability to be achieved in the field. For a 1 m thick grout mass subject to unit hydraulic gradient and of initial permeability 10^{-7} m/s reducing to 10^{-9} m/s the required time in the field is 2500 years (which will still take 5 years to model in the laboratory with a 76 mm long sample and a gradient of 40).

Water leaching thus reduces the permeability of the cement-bentonite grout and the surprising conclusion is reached that provided the material is confined, as there is some shrinkage on leaching, the effectiveness of a cement-bentonite grouted barrier will improve with water permeation for a very substantial time. Confinement is key but may not always be available, for example in the voids of a gravel where the permeability of the grout may reduce but the shrinkage within the soil pores and the resulting opening of flow channels may lead to an increase in permeability.

Effect of more complex permeants. The problem of a complex permeant and the need to predict long-term performance in a short time was a major issue at the Aznallcollar dam site near Seville in Spain (Jefferis and Fernandes, 2000). The situation was as follows:

In April 1998, the dyke impounding part of the tailings pond at the Aznalcóllar -Los Frailes mine failed, releasing 1.3 Mm^3 of tailings and 5.5 Mm^3 of acid water which affected a 45 km stretch of the Agrio and Guadiamar Rivers and covered an estimated area of 2616 hectares of riverbed, riverbanks and farm land. The owners, Boliden Apirsa, organised a comprehensive programme of work to mitigate the consequences of the spill and to ensure long-term security. This included the complete sealing of the tailings remaining within the pond and a 2.8 km long cement-bentonite-geomembrane cut-off wall to prevent seepage of the acid and sulfate rich groundwaters to the adjacent river. The work was carried out according to an emergency timetable which did not permit the extensive programme of laboratory testing that would normally precede cut-off design for such aggressive conditions. The design was therefore developed by extrapolating experience to the limits and applying Eq 3 to estimate the retardation of the reaction front in the grout. The groundwater was extremely aggressive and the worst case figures used for the design were as shown in Table 1. The concentrations in Table 1 are given in terms of mg/litre and milliequivalents per litre (meq/litre) as the latter gives an indication of the relative reactivity of the various species.

Table 1. Worst case groundwater chemistry.

Species	mg/litre	meq/litre	Species	mg/litre	meq/litre
pH	2.8	0.0016	Iron	13	366
Sulfate	9038	188	Zinc	9	278
Calcium	388	19	Manganese	8	210
Magnesium	382	31	Copper	68	2

The acid, sulfate, magnesium, iron, zinc, manganese and copper will all react with the calcium hydroxide, and the calcium silicate and aluminate hydrates present in a cement grout and the problem was to predict the effect of the reaction and the possible life of a barrier formed from a cement-bentonite grout in a slurry trench cut-off wall. From consideration of the nature of the reactions of each of the species with the grout, V_t values can be estimated for the groundwater/grout reaction. This showed that sulfate attack leading to the formation of gypsum and then thaumasite and ettringite (calcium sulfoaluminate and calcium silicate sulfate carbonate) would be the most rapid, least retarded reaction. Both the minerals thaumasite and ettringite have a needle-like habit and the growth of the crystals within a grout mass can be highly disruptive leading to an increase in permeability. In contrast, the effect of the acid at pH 2.8 (in this case a strong acid, sulfuric acid) will be trivial as there is much more acid neutralising material in cement than there is sulfate reactive material.

Magnesium, iron, copper, manganese (and zinc depending on pH) all will precipitate as hydroxides or oxyhydroxides, often of very fine particle size. The initial effect of these precipitates is likely to be a reduction in permeability as they will have a pore blocking effect but in the longer term there may be an increase in permeability as the reactions will lead to the degradation of the main cementing component in the mix, the calcium silicate hydrates. The overall effect was therefore expected to be an initial reduction in permeability followed by a longer term increase. The contractor for the works, Rodio Cimentaciones Especiales S.A., proposed a

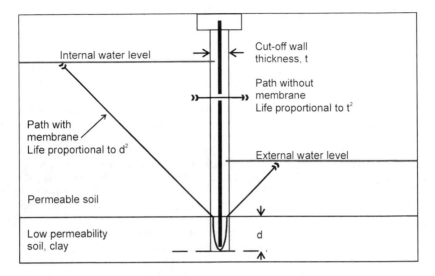

Figure 3. Schematic of geomembrane in cut-off wall showing toe-in to aquiclude

cement-bentonite-pulverised fuel ash (pfa) mix with over 350 kg/m^3 of pfa in the mix and just sufficient cement to ensure that a permeability of 10^{-9} m/s could be achieved. This was a novel mix design with a much greater quantity of solids than has been used in a cut-off wall before and calculations suggested that this could give an acceptable design life if used in conjunction with a geomembrane.

The most sensitive path with the cement-bentonite-geomembrane system was identified as being under the geomembrane at the base of the wall where it toes into the low permeability layer, see Figure 3. Because of the emergency nature of the works, laboratory tests on the mixes had to be commenced in parallel with the barrier construction. For these tests a simulated worst case groundwater was used with samples of the mix as designed and at lower cement and pfa contents. As predicted, the effect of permeation was an immediate drop in permeability by a factor of two or more due to the clogging action of the reaction precipitates. A steady decline in permeability followed and by 100 days the permeability had dropped to about 5 x 10^{-10} m/s. Thereafter there was a trend of gently increasing permeability. Breakthrough of the sulfate rich waters occurred at about 100 samples volumes and the highest recorded laboratory permeability at the end of the tests was 6 x 10^{-9} m/s.

Modelling the reaction front. As discussed above, the flux of aggressive chemical into the wall will be a combination of advection and diffusion, retarded by the chemical reaction between the ions in the water and the cement-bentonite material. In the laboratory diffusion may have only a modest effect but in the field where timescales will be much longer its effect cannot be ignored and therefore a computer code was written which considered advection, diffusion and retardation by reaction. This code was used to analyse two situations: firstly, the effective lifetime of a barrier without a geomembrane before breakthrough of the contaminated groundwater and secondly, the breakthrough time for the path underneath the geomembrane where it toed into the low permeability layer, see Figure 3.

As the wall was installed as an emergency measure, the design could not be based on field data. For the purpose of estimation of the design life, it was assumed that the unreacted part of the cement-bentonite material would follow the observed laboratory behaviour with water, an initial permeability of 10^{-7} m/s, dropping according to a power law relationship, to 10^{-10} m/s at one year (the permeability would expected to drop below 10^{-10} m/s but the preliminary testing was not regime long enough to confirm this and therefore no credit was given for it). The initial permeability of the material on contact with the aggressive water was assumed to be 10^{-8} m/s, reduced from the 10^{-7} m/s for water because of the immediate clogging effect of the aggressive groundwater. The final fully reacted permeability was also assumed to be 10^{-8} m/s (the highest observed laboratory permeability after reaction proved to be 6×10^{-9} m/s when contaminant breakthrough was achieved in the permeability tests – at a time when wall construction was nearly complete).

Using the advection-diffusion-retardation computer program and allowing for a change of permeability with both time and chemical reaction, it was shown that the time required for the permeability of the grout to increase to 10^{-9} m/s in the most sensitive pathway, under the geomembrane at the base of the wall, would be over 110 years (for a 1 m head difference across the wall and for the 1.5 m deep toe-in which had been used in the works). Without the geomembrane the most sensitive pathway would have been directly through the wall, a nominal path length of 600 mm which would give an estimated working life of 5 years.

The life without the geomembrane is very short as the high permeability of the young material allows rapid movement of the reaction front. The longer flow path under the wall allows the bulk of the material to develop a low permeability which then controls the rate of advance of the reaction front. This analysis demonstrates a very important benefit of using a geomembrane in aggressive conditions provided there is a deep toe-in.

The above assumptions on the permeability are conservative and the actual effective life of the wall should be in excess of the figures quoted above. Furthermore the 110 year time estimate is for the cement-bentonite component of the wall to reach a permeability of 10^{-9} m/s. At this time the overall permeability of the wall should still be much less than 10^{-9} m/s if the geomembrane is still effective and controls flow across the main area of the wall except at local defects and in the toe region.

Chemical effects of cement grouts on the soil

The previous example of a reaction of groundwater with a cut-off wall shows the effect of groundwater chemistry on a grout. However, grouts can also have an effect on the local soils and grouts products can themselves produce problems.

Soil pH. One of the effects of cement grouting is to raise the pH of the soil which is in contact with the grout as the hydration of cement liberates calcium hydroxide and some sodium and potassium hydroxides. The pH of saturated calcium hydroxide solution is 12.5 and this will be somewhat increased by the presence of the sodium and potassium hydroxides. pHs of over 12.5 are therefore to be expected for freshly hydrated cement (this will reduce over time if the cement is exposed to atmospheric carbon dioxide, an acid gas or if the alkalis from the cement react with soil materials). Soil grout mixes of pH>12 are not unusual after jet grouting. In general this can be expected to be beneficial as it can promote some pozzolanic reaction with clays in the soil to improve the strength of the soil-grout mix.

Soil organic matter. There is a considerable literature on organic matter in soils but in engineering works rather little attention is paid to the nature of the organics. These are likely to be dominated by carbonaceous matter but there may also be some nitrogenous matter, which in deep soils is likely to be in the reduced form. For example, inorganic nitrogen (as opposed to organic nitrogen) is likely to be in the form of ammonium species rather than nitrate. Ammonium species that may be present in a soil include: ammonium ion, ammonia gas in solution and, if the soil is not water-saturated, ammonia gas itself. These species can be related by equilibrium reactions as follows:

$$NH_4^+ + OH^- = NH_3 + H_2O = NH_{3gas}\uparrow$$

For pH values of less than about 7.5 and normal soil temperatures effectively all the ammonium species will be present as ammonium ion (NH_4^+) and as such they are likely mainly to be sorbed onto soil constituents such as clays or to form part of the organic fraction. Because of the sorption, the ammonium ion concentrations in the porewater may be low and there will be little tendency for the ammonia to be liberated as a gas. However, as the pH is increased above about 7.5 the equilibrium moves towards ammonia (NH_3) gas in solution so leading to a desorption/release from the soil (which also may be promoted by ion exchange, for example by calcium from a cement grout replacing ammonium on the clay). The ammonia gas in solution will be in turn in equilibrium with ammonia gas in air if the soil is exposed. The movement in the position of the equilibrium with changing pH is initially quite limited but accelerates rapidly above a pH of about 9.

Increasing the temperature has the same effect as increasing the pH, it drives the equilibrium towards ammonia gas in solution and thence to ammonia in air. This will be significant if heat released by hydration of cement raises the soil temperature.

Although the equilibrium moves with increasing pH to favour ammonia gas in solution, it is important to note that ammonia gas is very soluble in water and even at high pHs or moderately elevated temperatures it may remain largely in solution.

In the course of tests on estuarine soils from various sites, the author has found ammoniacal nitrogen levels of 0.2 to 2 g/kg (0.02 to 0.2% by weight). This ammoniacal nitrogen may be held on clay minerals in cation exchange sites or sorbed onto or a component of the organic matter but, if released, it could raise the concentration in the groundwater to a few tens of mg/litre. This level is high enough to be harmful to fish and aquatic life but generally if there is any tendency for migration, the ammonia will be harmlessly re-absorbed onto the surrounding soils – the zone of grouted soil and hence of ammonia in pore water is likely to be trivial when compared with the mass of surrounding soil. However, the liberation of ammonia into the porewater can have unfortunate consequences if grouted material is to be excavated in an enclosed or poorly ventilated environment, for example, tunnels or deep excavations. Scoping calculations with the chemical speciation programme PHREEQCI (Parkhurst, 1995) show that, for a groundwater ammonia concentration of 10 mg/litre in a soil at a temperature of 15°C with pHs in the range 10.5 to 13, the equilibrium concentration of ammonia in air will be in the range 20 to 23 ppm by volume. The reason for the small change in air concentration with pH is that for pH>10.5 effectively all the ammonia in the groundwater is in the form of ammonia gas in solution. Further raising the pH can have only a limited effect on the amount of ammonia gas in solution and it is this that controls the concentration in the air (change of pH does not effect the partition between ammonia gas in solution and in the air; all that it affects is the partitioning between ammonium ion in solution and ammonia gas in solution).

The quantity of air that can be contaminated with ammonia from the soil is very substantial.   For example, an ammonia concentration of 20 ppm by volume corresponds to 14 mg/m^3 of air. If a soil contains 0.2 to 2 g/kg of ammonia and all the ammonia in one cubic metre of the soil were released it could produce 20 ppm by volume, a significant and unpleasant level, in 25,000 to 250,000 cubic metres of air.

Ammonia in a tunnel. A particular example of the analyses set out above was the ammonia observed in a section of tunnel in Singapore (Bracegirdle and Jefferis, 2001). In 1984 and 1985, tunnels forming part of the Singapore Mass Rapid Transit were constructed through Marine Clay, a member of the Kallang Formation. The clay is very soft and has a significant, albeit variable, organic content. Because of concern over the possible settlement and damage to buildings along the route of the tunnel, compressed air working was mandatory; the use of versatile earth pressure balance tunnelling machines capable of dealing with the variety of conditions along the tunnel route had not become widespread at that time. The provision for a jet-grout annulus was added to the contract at the last minute, with the intention of providing additional security and the opportunity to reduce compressed air pressures.

The annulus formed around the tunnel was on average 2.0m thick, and was formed over a 500m length of the route of the twin running tunnels. The grout injected was typically a 1:1 water/cement mixture with a lignosulphonate plasticiser. The tunnels, constructed using an excavator mounted in a conventional shield, passed through the grouted annulus between 6 and 12 months after jet-grouting.   Soil temperatures of up to 45°C were recorded, which is 15°C in excess of the ambient soil temperature. In addition, ammonia gas at concentrations of up to 36ppm by volume was recorded in the tunnel. Heavy demands were placed on the supply of compressed air to meet cooling and ventilation requirements.

The excessive temperature in the ground was shown to be the result of the heat generated by the hydration of the cement in the grout, a feature that may not always be considered when rich concrete or grout mixes are used underground. Heat release may be substantial and the effects may persist for a considerable time, long enough to have impact on the pore pressures as well as chemistry.

Prevention of ammonia release in grouted grounds is very difficult as many grouts including cement grouts and silicate grouts will raise the pH of the soil and therefore could cause ammonia release problems. If promoters are planning to use alkaline grouts in nitrogenous soils (or indeed almost any soils as the levels of nitrogen required are very small) in enclosed environments, then consideration should be given to ammonia generation and the provision of appropriate ventilation.

By-products from sodium silicate grouts

Sodium silicate is a highly alkaline viscous liquid which is widely used as a grout for sandy soils, especially in temporary works. It may be regarded as silica dissolved in a sodium hydroxide solution and on setting it is the silica component of the mix which polymerises to form a coherent structure. The role of the sodium hydroxide is to keep the silica into solution and if it is neutralised for example by an acid, the silica will precipitate as a gelatinous mass. This is the basis of many silicate grout systems, though silicates also may be hardened by reaction with strong electrolyte solutions such as calcium chloride or by polymerisation with sodium aluminate.

Silicate ester grouts. If acid is added to sodium silicate, there will be an immediate flash precipitation of silica and the system would be useless as a grout unless the silicate and hardener are injected separately as in the Joosten process.

Organic esters which slowly hydrolyse to produce an acid to neutralise the sodium hydroxide can be used to give a controllable set time. The nature of the ester controls the rate of reaction while the quantity of ester per unit volume of silicate controls the degree of neutralisation (100% neutralisation is not required). Typically esters of polyhydric acids (e.g. methyl and ethyl esters of succinic, glutaric and adipic acids) are used so that the hydrolysis produces polycarboxylic acids (molecules with several acid groups) with greater alkali neutralising power per mole than monocarboxylic acids. The hydrolysis reaction is of the form:

ester + water = alcohol + acid

In the alkaline environment of a sodium silicate solution, the equilibrium moves in favour of alcohol + acid as the acid is removed by reaction with the sodium hydroxide and the alcohol remains (typically methyl or ethyl alcohol). The rate of hydrolysis of the ester controls the rate of neutralisation and hence the set time of the grout. The set time may be varied to a limited extent by varying the ester and silicate concentrations (though this will also vary the degree of neutralisation). For major variations in set time, the type of ester must be varied and manufacturers may supply a range of esters. For optimum gelation, a substantial proportion of the alkali should be neutralised according to a reaction of the type:

acid + sodium hydroxide = sodium salt of acid + water

However, as the esters tend to be more expensive than sodium silicate there can be a temptation to use as little ester as will give some set. Whilst this may produce a gelled product it will not make optimum use of the sodium silicate nor produce the most durable product. Thus when designing silicate/ester systems, it is important to recognise the underlying chemistry and to match the quantity acid which is produced per unit weight of the ester (the saponification value) to the alkali content of the silicate. Far too often in research projects the need for appropriate neutralisation has been overlooked resulting, unsurprisingly, in reports of poor durability.

Saponifcation reactions. The neutralisation of the alkali in the sodium silicate with the organic acid is sometimes referred to as a saponifcation reaction and therefore perhaps it should be no surprise that the by-products of the neutralisation of the alkali can be a soap type material. Furthermore, silicates are used as stabilisers in some detergents to keep the dirt from the wash in suspension and prevent it re-settling into the clothes. A soil stabilised with a silicate-ester grout can therefore produce a rather stable foam if mixed with water. This was found on a slurry tunnelling project in Cairo (a tunnelling process using a machine with a sealed face, pressurised by slurry and where the spoil is carried to the surface by circulating the slurry). The breakout chambers for each drive of the tunnel were grouted with a silicate-ester grout prior to tunnelling and as soon as tunnelling was started on each drive, the slurry turned to an unpumpable foam which made tunnelling almost impossible. Fortunately the author deduced the role of the grout and an appropriate antifoam agent was researched and applied. This was reasonably effective but tunnelling in the grouted zones remained difficult. The message is simple, avoid silicate grouts neutralised with esters (or comparable materials) on slurry tunnel drives. With other types of tunnelling machine, the grout actually could be beneficial as the soap effect could reduce head torque but the author is not aware of any field applications or data.

Conclusions

Cement and chemical grouts are complex chemical systems. The ground and groundwater in contact with them can affect their performance and, as shown in the brief case histories, the grouts can in turn affect conditions in the soil. It is hoped that this paper will make the reader sensitive to some of the possible chemical reactions that otherwise might not be recognised until a problem has occurred.

References

Bracegirdle, A. and Jefferis, S.A. (2001). Heat and ammonia associated with jet-grouting in marine clay, *Third British Geotechnical Association Conference on Environmental Geotechnics*, Edinburgh.

Jefferis, S.A. and Fernandez, A. (2000). Spanish dyke failure leads to developments in cut-off wall design, *International Conference on Geotechnical and Geological Engineering*, Melbourne, Australia.

Jefferis, S.A. (1992). Contaminant - grout interaction, *ASCE Specialty Conference, Grouting, Soil Improvement and Geosynthetics*, New Orleans.

Parkhurst, D. (1995). *PHREEQC a computer program for speciation, reaction-path, advective-transport, and inverse geochemical calculations*, US Geological Survey.

Mix Design and Quality Control Procedures
for High Mobility Cement Based Grouts

M. Chuaqui[1] and D.A. Bruce[2]

Abstract

Measures of success for any grouting program should include superior technical performance and cost effectiveness. These can be achieved by designing grouts with properties that are specifically tailored to the application. This requires a fundamental understanding of the fluid and set performance characteristics needed for a specific application. For high mobility cement based grouts (HMG), these properties include bleed, segregation, resistance to pressure filtration, control of particle agglomeration, anti-washout characteristics, rheology, evolution of cohesion with time, set time, matrix porosity, ultimate strength, resistance to chemical attack, and durability. A description of how each property is quantified, evaluated and optimized is provided, and related to appropriate standards. A three-step process for the design and quality control of an HMG project is outlined. The first step is a laboratory-scale testing program to determine basic formulations, optimized for performance characteristics and cost. The second step is full-scale trial batching performed on site with the materials and equipment that will be used on the project. The third step is quality control testing during production grouting to ensure that the grouts being used are being batched correctly and will perform appropriately in situ. A digest of mix HMG designs used on recent projects is provided for illustration and reference.

[1]Principal, MC Grouting, Inc., 29 Haliburton Ave., Toronto, ON   M9B 4Y5 Canada;
Tel: (416) 695-2593; Fax: (416) 695 2399; email: marcelo.chuaqui@sympatico.ca.
[2]President, Geosystems, L.P., P.O. Box 237, Venetia, PA 15367, U.S.A.;
Tel: (724) 942-0570; Fax: (724) 942-1911; email: dabruce@geosystemsbruce.com.

1.   Introduction

The term HMG refers to the family of high mobility cement based grouts. These have a rheology which is best measured by either a Marsh cone or flow cone and not by a slump cone. As opposed to low mobility grouts (LMG) (Byle, 1997), HMGs have a relatively low viscosity and therefore high mobility. A fundamental understanding of the performance characteristics required of a grout formulation is necessary to responsively design a grout for a specific application. Whereas HMGs traditionally comprised only cement and water, with the use of other materials such as sand or accelerator considered only in extreme conditions (e.g. "runaway" takes), recent years have seen major changes in grout mix formulation, especially in the United States. Routinely, projects are now employing suites of balanced, stable particulate grouts whose fluid and set properties are achieved by the use of multiple additives, as well as variations in the waster content and cement characteristics. Such HMGs are characterized by low bleed, superior resistance to pressure filtration, and controlled rheology.

The development of these HMGs on any particular project is best done in a three-phase process. During the first phase, a series of formulations, each suited for injection under the specific site conditions, is developed through a laboratory-testing program. During the second phase, on site and prior to production, the mix designs are replicated to investigate any changes in properties due to differences in materials, mixing equipment or procedures between the laboratory testing and production grouting. During this phase the baseline data for the quality control program are also established. During the third phase, during production grouting, the properties of the grouts are verified regularly to ensure that grouts are being batched correctly.

This paper provides a summary of the key performance parameters for HMGs and how to measure them. The influence of the individual component materials is discussed, and is illustrated by reference to a digest of mixes used in recent projects in the United States.

2.   Evaluation of HMG Properties

In order to responsively design a HMG, it is necessary to understand what fluid and set properties are desirable. For different applications the relative importance of the properties will change, but in general the properties listed below are usually if not always desirable. Table 1 summarizes the standard quality control tests used in the field.

2.1   Rheology and the Evolution of Apparent Viscosity and Cohesion with Time

The rheology of an HMG is characterized by apparent viscosity, cohesion, and internal friction. Figure 1 depicts the behavior of Newtonian and Binghamian fluids. Water and true solution grouts behave as Newtonian fluids, while stable HMGs behave like Binghamian fluids. The behavior of both the fluids depicted in Figure 1 reflects no internal friction. The behavior depicted in Figure 2 is characteristic of a grout with internal friction, and a shear strength that depends on fluid pressure. Unstable grouts behave unpredictably, acting alternatively as a Newtonian fluid and then as a Binghamian fluid with internal friction (Gause and Bruce, 1997). Fluids with high internal friction are not optimal for injection, as high pressures are required to pump them over significant distances.

Table 1. Standard Field Quality Control Tests for HMGs

EQUIPMENT	TEST	DESCRIPTION
Marsh Funnel	Apparent Viscosity	The Marsh time of the grout can be measured in accordance with the method described in API Recommended Practice 13B-1 with a Marsh funnel and a calibrated container. The test is performed by filling the Marsh cone to the bottom of the dump screen and then measuring the time for 0.26 gallons (1 liter) of grout to flow through the funnel.
Penetrometer/ or Shear Vane	Cohesion and Time to Initial/Final Gelation	Either a penetrometer or shear vane type test will be used to measure the amount of time required for the grout to reach initial gelation (cohesion of 100 Pa) and final gelation (cohesion of 1000 Pa).
API Filter Press	Pressure Filtration Coefficient	The pressure coefficient can be measured with an API filter press. The test is performed by pouring a 0.42-quarts (400-ml) grout sample into the top of the filter press. The sample is then pressurized to 0.7 MPA. The test is run until all the water is expelled from the sample. The value of the pressure filtration coefficient is then calculated with the following equation: $$K_{pf} = \frac{\text{volume of filtrate}}{\text{volume of sample} \times (\text{time in minutes})^{(1/2)}} \times 1$$
250-ml Graduated Cylinder – Glass	Bleed	The bleed capacity of the grout can be measured in accordance with the method ASTM C940 with a 0.26-quart (250-ml) graduate cylinder. The test is performed by pouring grout into the cylinder to the 0.21-quart (200-ml) level. The sample is then left undisturbed for two hours before the amount of bleed water is measured.
Baroid Mud Balance	Specific Gravity	The specific gravity of a grout can be measured in accordance to the method described in API Recommended Practice 13B-1 with a Baroid Mud Balance. The Baroid mud balance is a calibrated scale that is used to measure the specific gravity. Micromotion flow/density meters and hydrometers are also used in practice.
Vicat Needle	Initial and Final Set Times	The initial and final set times can be determined with the Vicat needle testing apparatus. The vicat needle is set at the surface of the grout sample and released. Initial set is reached when the needle only penetrates 1-inch (25-mm). Final set is reached when the needle does not penetrate the surface of the grout sample.

The cohesion (c) as shown in Figure 1, corresponds to the yield stress. The dynamic viscosity is shown by η, the plastic viscosity is shown by η_B and the apparent viscosity is shown by η'. The smaller the cohesion, the closer the plastic viscosity and the apparent viscosity are to each other. In the case of a Newtonian fluid, the cohesion is zero and the plastic viscosity and the apparent viscosity are equivalent and referred to as the dynamic viscosity.

The cohesion controls how far a grout will penetrate an aperture of given radius at a specific pressure, while the viscosity determines the flow rate and the therefore the grouting time for an aperture of given radius at a specific pressure (Lombardi, 1985): $L \alpha C = (p \times r)/ (2 \times C)$, where: L = length of the channel, p = the applied pressure, r = radius of the channel, C = cohesion of the grout.

Therefore a low viscosity will optimize permeation grouting in a soil with small pores or a rock mass of fine fissures. Grouts for these applications should not however have so low a viscosity that they will travel long distances without appreciable pressure drop.

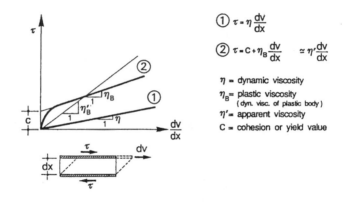

① $\tau = \eta \dfrac{dv}{dx}$

② $\tau = C + \eta_B \dfrac{dv}{dx} \quad \simeq \eta' \dfrac{dv}{dx}$

η = dynamic viscosity
η_B = plastic viscosity
    (dyn. visc. of plastic body)
η' = apparent viscosity
C = cohesion or yield value

Figure 1: Rheological Behavior 1) Newtonian Fluid, 2 Binghamian Body (De Paoli et al., 1992).

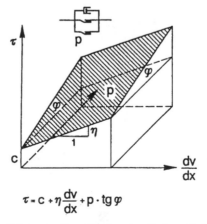

$$\tau = C + \eta \dfrac{dv}{dx} + p \cdot tg\, \varphi$$

Figure 2: Rheological Behavior for a Binghamian Body with internal friction (De Paoli et al., 1992).

Grouts to combat active flow conditions have higher viscosity and cohesion that enhance antiwashout characteristics. The high cohesion is not disadvantageous in such circumstances, since penetration of small apertures is not required in this application.

If the fluid properties of the grout change in an uncontrolled or unpredictable manner with time it is therefore not possible to properly control or analyze the injection process. With the exception of a thixotropic grout, it is typically desirable for a grout to maintain a constant viscosity for a period of time equal to the injection time and then for its viscosity to increase rapidly until initial set is reached. Thixotropic grouts have a low viscosity while being in turbulent motion (being sheared during injection) and much higher viscosity when no shear is applied. This property is beneficial when injecting grouts into open voids because the grout can be placed where it is desired without it flowing away after pumping ceases.

The viscosity of a grout at a given age is typically indexed with a Marsh Funnel for non-sanded grouts and with a flow-cone for sanded grouts. Water has a "Marsh time" of 28 seconds. The procedure for measuring sanded grouts is similar and is described in ASTM C-939. When the Marsh time is in the range of 35 to 50 seconds (low viscosity) good correlation exists between apparent viscosity and true viscosity of the fluid (Deere, 1982). For higher viscosity grouts and partially set grouts, the measurement of cohesion becomes more important. An instrument such as a shear vane, a penetrometer, a Lombardi plate or a Baroid rheometer can be used.

2.2   Resistance to Pressure Filtration

Injecting grouts into small apertures is similar to pressing the grout against a filter material. Depending on the formulation of the grout, the water can be forced out, creating a filter cake at the borehole wall. With time, the filter cake makes the formation inaccessible to further injection at that location. HMG resistance to pressure filtration is typically measured with an API filter press. Two tests are possible: measurement of the pressure filtration coefficient K_{PF} and measurement of the cake growth coefficient K_{pc} (De Paoli et al., 1992). These tests are described in Table 1.

The value of the respective coefficients is then calculated as follows:

$$K_{pf} (min^{-1/2}) = \frac{volume\ of\ filtrate}{volume\ of\ sample} \quad X \quad \frac{1}{(time\ in\ minutes)^{1/2}}$$

$$K_{pc} (mm\ x\ min^{-1/2}) = (thickness\ of\ cake\ in\ mm)\ X\ \frac{1}{(time\ in\ minutes)^{1/2}}$$

Resistance to pressure filtration (and hence ability to penetrate) is inversely reflected by the magnitude of the calculated value. The relationship between cohesion and pressure filtration coefficient is shown in Figure 3. Unstable mixes result from the high water cement ratios which were common in U.S. practice for decades (e.g. Albritton, 1982).

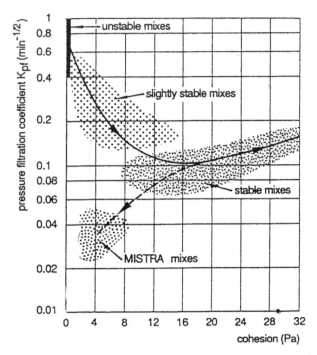

Figure 3. Relationship between resistance to pressure filtration and cohesion for different types of mixes (De Paoli et al., 1992). (Note "MISTRA" refers to a "Modified Stabilized Cement Grout.")

By utilizing combinations of cement additives and water solids ratios pressure filtration coefficients below 0.01 min$^{-1/2}$ can be achieved, while maintaining the apparent viscosity under a 60 second Marsh time: this significantly enhances the penetrability potential of HMGs.

2.3   Bleed

Bleed develops as the cement particles settle due to the effects of gravity and allow free water to develop as a discrete volume. If a grout has high bleed capacity it will not fully fill the pore space within the soil or fractures in a rock due to the bleed water which forms as it sets. Such effects have been demonstrated when using unstable microfine cement grouts in fine sands (Helal and Krizek, 1992). For stable HMGs, bleed should be as low as possible (preferably less than 2%), but in no case should be more than 5%.

2.4   Water-Repellant/Anti-Washout Characteristics

If a grout is being placed below the water table, it is undesirable for the grout to disassociate. This characteristic becomes especially important when there is the potential of HMG encountering moving water: it will be diluted, so reducing its effectiveness and potentially posing an environmental threat. Therefore, a grout with good water repellant characteristics is preferable, since it will displace water with minimal dissolution.

Conventional tests to quantify washout resistance are applicable to concretes, and not HMGs. A test for grouts has been specially designed (Gause and Bruce, 1997), wherein a clear plastic chamber allows grout to be placed either in static or dynamic water conditions. The "integrity" of the grout is measured by its efficiency in displacing a known volume of water from the chamber. The lower the volume of grout required to fill the chamber, the better the anti-washout/water repellant characteristics of the grout.

Alternatively, anti-washout characteristics can be evaluated through a series of visually demonstrative, qualitative tests, such as pouring the grout through a column of freestanding water to observe the extent of dispersion of the grout particles. Samples can also be tested by pouring water into a container half filled with grout. With either test, the sample can then be allowed to set so that the amount of laitance/bleed (if any) formed on the surface can be measured. This is a good example of a pragmatic, responsive field test.

2.5   Prevention of Particle Agglomeration

The maximum particle size of the hydrated solids in a grout is a key factor that determines the dimensions of soils pores or fractures that can be penetrated. In principle this can be resolved by reducing the particle sizes of the cement, especially the coarse portion. However, if the particles within the suspension are agglomerating during mixing and pumping then the effective maximum particle size is increased and so certain soils or fractures simply become inaccessible. The nature of the particle agglomeration within a mixed HMG can be determined by measuring the particle size distribution in a laboratory by light scattering, absorption or diffraction methods. However, it is not practical to measure it elsewhere directly. An indirect measure is provided by evolution of viscosity with time (as the particles form flocs the viscosity increases). Usually, additives such as super-plasticizer are used to reduce the amount of particle agglomeration (Section 3).

2.6   Hydration Control

The ability to accelerate or retard the set of grout is critical for certain grouting applications.   When grouting in moving water, an accelerated set time is highly desirable and when inline mixing at the bottom of the hole is conducted, set times lower than 1 minute can be achieved.   Conversely, during multiple pass soil grouting, a retarded initial set can be highly desirable, since retarded set times allow multiple passes to be conducted without having to attempt to hydrofracture cured grouts. By retarding the set time it is therefore possible to re-inject certain horizons several times with different formulations, so permitting further penetration and/or densification.

The initial and final set times can be determined with the Vicat needle testing apparatus. The vicat needle is set at the surface of the grout sample and allowed to drop: initial set is reached when the needle only penetrates 1-inch (25-mm). Final set is determined when the needle does not penetrate the surface of the grout sample.

2.7   Prevention of Filler Segregation

For sanded HMG, it is important that the sand within the grout remains suspended and evenly distributed. If the sand falls out of suspension, it does not become part of the grout matrix and furthermore it becomes very difficult to pump the grout. This property can be evaluated through qualitative tests. These tests include sawing a cured sample in half to inspect the distribution of the sand or physically feeling or observing the fluid grout sample after it has been allowed to sit undisturbed for a period of time in a clear walled container.

2.8   Matrix Porosity of Cured Grout

Set grouts with low matrix porosity are more durable since water penetration potential is correspondingly reduced.   This property is very important for environmental cut-off applications were very low permeabilities are required or when durability and resistance to chemical attack are important. This parameter is measured through triaxial permeability tests. The permeability can be reduced by reducing the water:cement ratio and/or adding materials with a very small particle size such as silica fume. Further details are provided by Littlejohn (1982).

2.9   Unconfined Compressive Strength.

It is essential for structural HMGs to achieve their target strength once they have cured. This property can be measured by performing cylinder or grout cube breaks to determine the unconfined compressive strength.   Sampling and curing procedures should reflect actual field conditions.

3. The Roles of HMG Components

3.1 Note on Execution of Tests

The mixing equipment used for laboratory-testing programs should be selected so that it mimics the intensity and effectiveness of mixing that will result from the production equipment. For example if a high shear colloidal mixer is to be used on site then a small high shear colloidal mixer, or a high shear Hobart type mixer must be used in the laboratory. If agitator trucks are to be used for mixing the grout onsite then a drum mixer is appropriate in the laboratory. The mixer should be sized so that small batches can be prepared and a large number of formulations can be prepared for all tests.

3.2 Materials

3.2.1 Cements

There are several different types of cements available for specific purposes. These include cements with different particle size gradations, and cements chemically formulated to be resistant against specific chemical attack or to provide high early strength. Selecting the appropriate cement type and range of water cement ratios is the initial step in developing a series of site specific HMG formulations. An HMG must have enough solids to be stable and reach an acceptable strength and durability. The use of grouts with high water cement ratios (i.e. greater than 1.5 by weight) is typically disadvantageous due to the resultant reduced injectability, reduced stability, increased bleed, increased matrix porosity and reduced durability. The use of grouts for ground treatment with a water cement ratio of lower then 0.45 by weight is not common due to the high costs and the limitations of mixing and pumping equipment.

Portland cements are the most common and best-known cements used world wide as the basic ingredient for HMGs:

- Type I portland cement is accepted as the general-purpose cement for the majority of grouting projects when the special properties of the other types are not required.
- Type II portland cement is manufactured to resist moderate sulfate attack and to generate a slower rate of heat of hydration than Type I. Type I/II cement is often sold.
- Type III portland cement is used when high early strength is required. It is considered for applications were fast sets are required. Also because it consists of finer particles it can be used to grout slightly smaller apertures than can be penetrated with Type I.
- Type IV portland cement generates less heat during hydration than Type II and develops strength at a slower rate than Type I. It can be used for applications were a large mass of grout will be placed and high hydration temperatures are unacceptable.

• Type V portland cement is manufactured for use in grout exposed to severe sulfate action.

• Microfine cements have been ground longer and finer to allow penetration of finer fissures or soil pores. These cements are available with a variety of different properties and may contain blast furnace slag as well as portland cement. They are typically characterized by Blaine Fineness and the maximum particle size. The Blaine fineness is a measure of the specific surface area. Typically the maximum particle size is less than 8 microns, with the bulk of particles under 4 microns with Blaine fineness in excess of 8000 cm^2/g.

3.2.2   Other Materials

Formulating a suitable HMG typically involves balancing the potential positive and negative impacts of different additives against each other. It is critical that a laboratory testing program be performed as the interactions between different additives can sometimes be unpredictable. On several occasions, the authors have overcome compatibility issues by simply switching the brand of bentonite being used, or the supplier of Type F Flyash during the lab testing phase of mix design. It is advisable to acquire all the chemical additives from one supplier, to help ensure compatibility.

The most commonly used additives include:

• Super-Plasticizer - Several different types are available including naphthalene sulphonate-, lingo sulphonate,- and melamine-based materials. These chemicals cause each cement particle to adopt a negative charge, and so they electrostatically repel each other. This reduces the viscosity of the grout by inhibiting particle agglomeration. Since the particles are then better dispersed, more surface area is available for hydration, the grout pore space is reduced, and crystals from adjacent particles can interlock more regularly and more strongly. This leads to enhanced strength and durability (Gause and Bruce, 1997). These chemicals also typically retard the initial set of the grout. Typical proportioning is between 0.5% to 2% by weight of cement. Ligno sulphonate also acts as a retarder.

• Bentonite - Bentonite stabilizes the grout, increases its resistance to pressure filtration and increases its viscosity. It will reduce the ultimate strength of the grout. Excellent data are provided in Deere (1982) and Littlejohn (1982). There is a wide variety of grades and types of bentonite. However, for most grouting operations, pure, chemically unaltered Wyoming sodium montmorillonite is optimal. The sequence and quality of mixing is critical. During mixing, the dry cement should be added to a pre-hydrated bentonite slurry, since when the cement first comes into contact with the bentonite, the viscosity of the slurry rapidly increases. However, as mixing continues, the HMG becomes less viscous. The initial increase in viscosity is caused by the mutual flocculation of the negatively charged bentonite particles with the positively charged cement particles. The reduction in viscosity occurs as the smaller bentonite particles completely coat the cement particles, masking the latter's positive charge (Jefferis, 1982). If high shear mixing is not used or the cement is added too rapidly to the mixer, then it is unlikely that the cement particles will fully disperse as they come into contact with the bentonite. Then flocs of cement particles become coated with bentonite and are difficult to break up (Jefferis, 1982). Bentonite should be hydrated for 12 hours prior to being used unless tests show that equivalent hydration can be achieved with a high shear mixer. Grouts with bentonite that has not been hydrated can be subject to durability problems due to cracking. For HMGs for ground treatment, bentonite should not be used at percentages of higher than 5% by weight of cement due to its adverse affect on strength.

- Flyash – Both Type C and Type F Flyash are pozzolanic materials, which also improve the particle size distribution of HMGs. They enhance resistance against pressure filtration and increase the durability of the cured grout. Variable amounts of cement replacement can by used for different applications. It is important to note that Type C Flyash expands and when used in dosages over 20% can cause durability problems in grouts.

- Silica Fume (Micro Silica) - Silica Fume is a microfine powder (< 1 micron) that also improves the particle size distribution of HMGs. It therefore, enhances resistance against pressure filtration, and increases durability and strength of cured grout by reducing its matrix porosity. It also makes the grout more water repellant. Typical proportioning rates are less than 10% replacement of cement (by weight).

- Welan Gum - Welan gum is a high molecular weight biopolymer. It acts as a thixotropic agent and significantly enhances resistance to pressure filtration. It also increases the cohesion of HMG, which makes the grout slightly more water-repellant. The dosage required to achieve a particular reduction in pressure filtration coefficient is dependent on the quality of mixing. The better the mixing the more effective the Welan gum is in reducing the pressure filtration coefficient. Welan gum does not reduce the pressure filtration coefficient significantly when used in agitator trucks but does enhance the thixotropy of the grout. Typical proportioning is about 0.1% to 0.2% by weight of cement.

- Anti-Washout Agents – A modified cellulose ether such as Master Builder's Rheomac UW450 material significantly enhances resistance to washout, reduces the pressure filtration coefficient, and makes the grout thixotropic. This additive is not compatible with naphthalene sulphonate plasticizer and moderately compatible with Whelan gum and bentonite due to the sharp increase in the HMG viscosity. Typical proportioning is 0.2% to 1.0% by weight of cement.

• Hydration Controls – There are three distinct concepts:

a) Accelerators: There are several different kinds of accelerators the most common of which are sodium silicate and calcium chloride. Calcium chloride must be dissolved prior to adding it to the grout. Although calcium chloride is an efficient accelerator it can have some negative impacts on performance. It may adversely affect sulphate resistance and it can corrode steel (Littlejohn, 1982). Sodium silicate reacts with the calcium ions liberated during initial hydration of the cement to form calcium silicate gel. This reaction causes a rapid increase in the viscosity of the HMG, making the grout very cohesive. This is an exothermic reaction and if high dosages of sodium silicate (>20% by weight of cement) are used, a flash set can be achieved but the resulting cured grout will have low strength and durability. At lower dosages, an initial rise in viscosity occurs but thereafter viscosity remains constant until a rapid set occurs some time later. The cohesiveness of the accelerated grout can provide excellent anti-washout characteristics.

b) Retarders: Extend the gel and set times of grouts in a controllable fashion. Set times of several days are achievable with some of more recent products.

c) Hydration Inhibitors: These are two-component systems involving the use of a stabilizer and an activator. The stabilizer forms a protective coating around the cement particles that stops the hydration process, for up to 72 hours. When the activator is introduced to the grout dissolution of the protective barrier occurs allowing the commencement of hydration and so normal crystal growth (Gause and Bruce, 1997).

4.   Digest of HMG Mix Designs

There has been a rapidly growing number of cases in the United States where multi-component HMGs have been successfully used in a wide variety of applications and conditions. Space restrictions prevent even summary descriptions of the mix design logic. Nevertheless, the authors thought it would be useful to provide examples of typical HMGs used recently in certain major projects (Table 2).

Table 2: Summary of projects for which HMG mix designs are provided.

PROJECT	PURPOSE OF GROUTING	MAIN REFERENCE
Oak Ridge National Laboratory, TN	To encapsulate radio-active waste contained in trenches via multiple injections	Berry and Narduzzo, 1997
Penn Forest Dam, PA	Rock mass curtain grouting of dam foundation	Wilson and Dreese, 1998
Tims Ford Dam, TN	Remedial curtain in karstic limestone to reduce 8000 gpm seepage	Bruce, Hamby and Henry, 1998
Limestone Quarry, WV	Curtain in karstic limestone to cut-off 40,000 gpm seepage	Bruce, Traylor and Lolcama, 2001

Details of the mix designs used in these projects are provided in Tables 3 to 6, respectively. It must be noted that in each case, the HMG design was tailored to the specific site conditions and project goals.

Table 3: Mix designations and properties for HMG mixes at ORNL.

MATERIAL	MIX DESIGNATION				
	MF	MIX W	MIX E	MIX K	MIX T
Water (lbs/Kg)	100/45.5	100/45.5	100/45.5	100/45.5	100/45.5
Bentonite (lbs/Kg)	2/0.9	2/0.9	3/1.4	5/2.3	7/3.2
Silica Fume (lbs/Kg)	5/2.3	6/2.7	6/2.7	6/2.7	6/2.7
Type F FlyAsh (lbs/Kg)	0	0	0	0	10/4.5
Pumice (lbs/Kg)	13/5.9	8/3.6	16/7.3	16/7.3	16/7.3
Ligno-Sulphonate - Retarder (lbs/Kg)	0.5/0.2	0.8/0.4	1/0.5	1.4/0.6	2/0.9
Welan Gum (lbs/Kg)	0.05/0.02	0.05/0.02	0.08/0.004	0.08/0.004	0.08/0.004
High Early Cement (lbs/Kg)	0	50/20	80/40	80/40	80/40
Microfine Cement (lbs/Kg)	65/30	0	0	0	0
Specific Gravity	1.41	1.28	1.43	1.44	1.49
Flow Cone (Sec)	34	35	45	45	51
Shear Vane (Pa)	333 @ 18 hrs	0 @ 18 hrs	1000 @ 18 hrs	360 @ 24 hrs	573 @ 24 hrs
Brookfield Viscosity Mtr (Pa)	90	550	720	350	No reading.
Initial Set @20C (hrs) Vicat	104	85	93	No Measure	186
KPF x 10^{-3} $(min)^{-1/2}$	2.75	37.7	4.2	3.6	3.8

Note that these mixes were prepared in a production scale mixer in large batches to verify the properties obtained earlier during bench scale tests.
MF – microfine, W, E, K, T – regular cement

Table 4: Mix designations and properties for HMG mixes at Penn Forest Dam, PA.

MATERIAL	MIX DESIGNATION			
	A	B	C	C+
Water (lbs/Kg)	210/95	165/75	120/55	120/55
8% Bentonite Slurry (Gal/Liters)	8/30	8/30	8/30	8/30
Portland Cement (lbs/Kg)	188/85	188/85	188/85	188/85
Type F FlyAsh (lbs/Kg)	80/36	80/36	80/36	80/36
Welan Gum (lbs/Kg)	0.22/0.1	0.22/0.1	0.22/0.1	0.22/0.1
Super Plasticizer (oz/liter)	45/1.4	45/1.4	30/0.9	30/0.9
Sodium Silicate (gal/liter)				4/20
Accelerator (gal/liter)				2-5
Yield (gal/liter)	44/168	38/148	33/128	40/150
Specific Gravity	1.49	1.55	1.64	1.58
Marsh Cone (Sec)	45-50	70-80	>120	
Flow Cone (Sec)			20	
Bleed at 3 hours (%)	1	0	0	0
Initial Set (hrs)	2-3	1-3	1-2	0.016
Final Set (hrs)	5-6	4-6	3-5	
KPF x 10^{-3} $(min)^{-1/2}$	0.05	0.05	0.05	0.05

Table 5: Mix designations and properties for HMG mixes at Tims Ford Dam, TN.

MATERIAL	MIX DESIGNATION			
	MIX A	MIX B	MIX C	MIX D
Water (lbs/kg)	141/64	141/64	94/42	94/42
Bentonite (lbs/kg)	4.7/2.1	9.4/4.2	4.7/2.1	4.7/2.1
Cement (lbs/kg)	94/64	94/64	94/64	94/64
Rheobuild 2000B (oz/liter)	15/0.5	30/0.9	20/0.6	20/0.6
Rheomac UW450 (oz/liter)	0	0	0	5/0.2
Specific Gravity	1.39	1.40	1.53	1.53
Bleed (%)	<5	<1	<1	0
KPF x 10^{-3} $(min)^{-1/2}$	<104	<42	<42	<42
Comp Strength @ 28D (PSI)	500	500	800	800
Flow Cone (Sec)	35	50	60+	100+
Stiffening Time (hh:mm)	4:30	4:30	4:00	4:00
Hardening Time (hh:mm)	10:30	8:30	8:00	8:00
Thickening and Thinning	**Mix Designation**			
Additional Bentonite (gal/liter)	8/32	16/64	8/32	8/32
Additional Water (gal/liter)	9.9/39	2.8/11	4.2/16	4.2/16

Table 6: Mix designations and properties for HMG mixes at Limestone Quarry, WV.

Material	Mix Designation				
	A0	A1	A2	A3	A4
Water (lbs/kg)	82/310	106/401	103/390	105/397	102/386
8% Bentonite Slurry (Gal/Liters)	68/257	20/76	26/98	26/98	25/95
Cement (lbs/kg)	729/331	856/389	878/399	895/406	866/393
Flyash (lbs/kg)	638/290	749/340	768/349	783/355	866/393
Rheobuild 2000B (oz/liter)	62/1833	73/2159	74/2188	76/2247	73/2159
Rheomac UW450 (oz/liter)	19/562	22/651	22/651	23/680	22/651
Specific Gravity	1.53	1.55	1.60	1.63	1.64
Bleed (%)	0	5.5	3.5	2	2
Marsh Time	55	80+	120+	N/A	N/A

5.   Final Remarks

Although balanced, stabilized HMGs have been used in Europe for far longer, it is only since the late 1990's that their efficiency and effectiveness has been exploited in the U.S. However, the success of several high profile projects when using such HMGs has driven a rapid acceptance of the logic and principles involved. Perhaps the clearest indicator of this has been the fact that one of the industry's largest "users" – the US Army Corps of Engineers – has fully embraced the concepts to the extent that the use of multicomponent HMGs is now standard. Major projects in the private sector are being similarly executed.

References

Albritton J.A., (1982). "Cement Grouting Practices US Army Corps of Engineers." Proceedings of the ASCE Specialty Conference in Grouting in Geotechnical Engineering, New Orleans, Louisiana, pp. 264-278.

Berry, R.M. and Narduzzo, L., (1997). "Radioactive Waste Trench Grouting – A Case History of Oak Ridge National Laboratory Grouting"- *Grouting – Compaction, Remediation, and Testing,* Proc. of Sessions Sponsored by the Grouting Committee of the Geo-Institute of the American Society of Civil Engineers, Logan UT, Ed. by C. Vipulanandan, Geotechnical Special Publication No. 66, July 16-18, pp. 76-89.

Bruce, D.A., Hamby, J.A. and Henry, J.F., (1998) "Tims Ford Dam, Tennessee: Remedial Grouting of Right Rim". Proc. Annual Conference, Dam Safety 1998, Association of State Dam Safety Officials, Las Vegas, NV, Oct. 11-14, 13 pp.

Bruce, D.A., R.P. Traylor, and J. Lolcama. (2001). "The Sealing of a Massive Water Flow through Karstic Limestone." *Foundations and Ground Improvement,*

GROUTING AND GROUND TREATMENT

Proceedings of a Specialty Conference, American Society of Civil Engineers, Blacksburg, VA, June 9-13, Geotechnical Special Publication No. 113, pp. 160-174.

Byle, M.J. (1997). "Limited Mobility Displacement Grouting: When "Compaction Grout" is NOT Compaction Grout", *Grouting – Compaction, Remediation, and Testing*, Proc. of Sessions Sponsored by the Grouting Committee of the Geo-Institute of the American Society of Civil Engineers, Logan UT, Ed. by C. Vipulanandan, Geotechnical Special Publication No. 66, July 16-18, pp. 32-42.

Deere, D.U. (1982), " Cement-Bentonite Grouting For Dams," Proceedings of the ASCE Specialty Conference in Grouting in Geotechnical Engineering, New Orleans, Louisiana, pp. 279-300.

De Paoli, B., Bosco, B., Granata, R. and Bruce, D.A. (1992). "Fundamental Observations on Cement Based Grouts (1): Microfine Cements and the Cemill Process." Proc. ASCE Conference, "Grouting, Soil Improvement and Geosynthetics", New Orleans, LA,. Feb. 25-28, 2 Volumes, pp. 474-485.

Gause, C.C. and D.A. Bruce. (1997). "Control of Fluid Properties of Particulate Grouts: Part 1 – General Concepts." *Grouting – Compaction, Remediation, and Testing*, Proc. of Sessions Sponsored by the Grouting Committee of the Geo-Institute of the American Society of Civil Engineers, Logan UT, Ed. by C. Vipulanandan, Geotechnical Special Publication No. 66, July 16-18, pp. 212-229.

Helal, A.M and Krizek, (1992), "Preferred Orientation of Pore Structure In Cement-Grouted Sand," Proceedings of the ASCE Specialty Conference in Grouting in Geotechnical Engineering, New Orleans, Louisiana, pp 526-540.

Jefferis, S.A., (1982), "Effects of Mixing On Bentonite Slurries and Grouts," Specialty Conference in Grouting in Geotechnical Engineering, New Orleans, Louisiana, Feb. 10 to 12, pp. 62-77.

Littlejohn, G.S. (1982), "Design of Cement Based Grouts," Proceedings of the ASCE Specialty Conference in Grouting in Geotechnical Engineering, New Orleans, Louisiana, Feb. 10 to 12, pp. 35-48.

Lombardi, G., (1985). "The Role of Cohesion in Cement Grouting of Rock," 15th ICOLD Congress. Lausanne, Switzerland, Q58, 13 p.

Wilson, D. and Dreese, T. (1998). "Grouting Technologies for Dam Foundations," Proceedings of the 1998 Annual Conference Association of State Dam Stafety Officials, October 11-14, Las Vegas, Neveda, Paper No. 68.

Fly Ash Utilization in Grouting Applications

Ayse Pekrioglu[1], Ata G. Doven[2], Mehmet T. Tumay[3]

Abstract

In addition to many environmental benefits, fly ash utilization provides end – products with superior engineering and physical qualities as well as economic benefits, considering the utilization potential in construction – related applications such as; cement production and concrete products in the form of highway pavement concrete, structural concrete, and roller compacted concrete, bricks, blocks and paving stones; artificial lightweight aggregate, structural fills or embankments, stabilization of waste materials, mineral filler in asphalt paving, flowable/structural fill and grouting mixes.

The fly ash, being a cementitious coal combustion by – product, promises high volume utilization in grouting applications by addition of other mineral admixtures when necessary. The fly ash grouts provide required engineering performance in improving ground stability by increasing strength and shearing resistance as well as reducing the permeability of soils treated with enhanced technical, rheological, durability and economic advantages over sand and cement grouts. The fly ash grout is also cost – effective when alternative suspension, emulsion or solution materials are considered.

In this study, fly ash grout composite formed of various combinations of high volume fly ash, cement, lime and high range water reducing chemical admixture has been investigated to define short – term engineering performance in terms of physicochemical (chemical compound analysis, unit weight, void ratio, specific gravity, linear shrinkage, hydraulic conductivity) and mechanical properties (unconfined compressive strength and flexural strength), excluding durability as the long term engineering performance. The overall short – term engineering performance indicates high volume fly ash utilization potential in grouting applications.

[1]Research Assistant, Civil Engineering Department,
Eastern Mediterranean University, Gazimagusa, TRNC, via Mersin 10, Turkey
phone +90 392 6302335, fax +90 392 3650710; ayse.pekrioglu@emu.edu.tr
[2]Vice Dean, Faculty of Engineering; ASCE member;
Eastern Mediterranean University, Gazimagusa, TRNC, via Mersin 10, Turkey
phone +90 392 6302027, fax +90 392 3651217; gurhan.doven@emu.edu.tr
[3]Associate Dean for Research, College of Engineering, ASCE fellow;
Louisiana State University, Baton Rouge, LA 70803, USA
phone +1 225 5789165, fax +1 225 5784845; mtumay@eng.lsu.edu

Introduction

Permeation grouting is required when inaccessible voids require filling to improve ground stability, increase the shearing resistance, strength and reduce the permeability. This is normally achieved by injecting suspensions, emulsions and solutions into the ground to improve the geotechnical properties of the soils and rocks.

Fly ash grouts are suspension compositions produced on site, normally using fly ash, Portland cement and water. Fly ash has been used for many years as an alternative to sand and cement grouts. It has important technical, rheological, durability and economic advantages over sand and cement grouts. The advantages of using fly ash grouts are (United Kingdom Quality Ash Association, UKQAA):

- Reduced bleeding
- Increased working life
- Improved pumpability and flowability
- Reduced permeability
- Increased compressive strength and durability
- Increased yield per unit weight
- Reduced water/solids ratio
- Economy

Grouts are used in rock bolting, in construction of grouted anchors, grouted piles, in solidification of ground by jet or consolidation grouting, in repairing and rehabilitation of dams and deteriorating highway structures, in industrial applications such as structural grouting of columns, bearing plates and machine foundations. There are various types of grouts classified as polymer based, cement based and clay grouts, which are used depending on the nature and location of the application, the engineering properties of the grout and material to be grouted, and the engineering performance required following the grouting process (Halow, 1982; Welsh, 1984).

The most common grouting materials used in geotechnical applications are cementitious materials such as Portland cement, fly ash, lime, silica fume, etc., in combination with or without sand and clay; most usually in the form of Portland cement or microfine cement grout; and, as increasingly promoted in time due to the environmental and economical benefits, in the forms of fly ash – lime grout, fly ash – cement – clay grout, fly ash – cement – sand grout, and specialist grouts in which water reducing, accelerating and retarding admixtures are added to conventional cement based grouts to adjust the specified rheological properties, including thixotropy, viscosity, density and setting time.

The particle size of the suspended solids within grout is quite important in geotechnical applications in terms of the grouting process efficiency and overall engineering performance of the grouted soil due to the reason that these solids may penetrate in the soil at reasonable pressure and rates if they are smaller than the size of the voids. The researches indicate that the maximum particle size must not exceed one-third to one-tenth of the size of voids (Littlejohn, 1982; Welsh, 1984).

The objective of this research work is to investigate the physicochemical, rheological and mechanical properties of high volume fly ash composite composed of mixture combinations of fly ash, cement, lime, silica fume, water reducing admixture and water for use in grouting applications, which is beneficial in terms of grouting process efficiency besides environmental and economical concerns.

Materials

Cement. The cement used was Type V with strength class 32.5 MPa, the chemical composition being presented in Table 1. The Blaine fineness of the cement is 3.184 cm^2/gr and its specific gravity is 2.95.

Fly Ash. An ASTM Class C fly ash from Soma Thermal Power Plant, Turkey was used. Its specific gravity is 1.89 and chemical compositions are presented also in Table 1.

Lime. It is hydrated calcium lime provided by a local company. The specific gravity of lime is 2.17 and its chemical composition is presented in Table 1 as well.

Admixture. A melamine based polymer dispersion water reducing admixture (WRA) Sikament FFN was used in all of the groups of mixtures.

Table 1. Chemical Analysis of Fly Ash, Cement and Lime

Oxides (%)	Fly Ash	ASTM C 618		Cement	Lime
		Type C	Type F		
SiO_2	45.20			26.03	N/A
Al_2O_3	19.77			6.82	0.38
Fe_2O_3	5.74			3.22	0.30
CaO	15.09			53.41	70.89
MgO	2.11	<5.0	<5.0	3.64	1.95
SO_3	1.36	<5.0	<5.0	2.25	N/A
Loss on Ignition	2.24	<6.0	<6.0	1.88	24.59

Mixture Proportions

The proportions of the mixtures are given in Table 2. Thirteen mix groups were studied in the laboratory research program. F, C, L indicates fly ash, cement and lime respectively and the following digits refer to the percentage of fly ash, cement and lime in the mixes as given in Table 2.

Table 2. Material Composition of Mix Groups

Group No	Mix Groups	Fly Ash (%)	Lime (%)	Cement (%)	WRA (%)
1	F100	100	0	0	0
2	F95C5	95	0	5	0
3	F90C10	90	0	10	0
4	F80C20	80	0	20	0
5	F95L5	95	5	0	0
6	F90L10	90	10	0	0
7	F80L20	80	20	0	0
8	F80C15L5	80	5	15	0
9	F80C10L10	80	10	10	0
10	F80C5L15	80	15	5	0
11	F80C20FFN0.25	80	0	20	0.25
12	F100FFN0.25	100	0	0	0.25
13	F80C15L5FFN0.25	80	5	15	0.25

Methodology Followed in the Experimental Program

All the mixtures were mixed for 2 minutes in the mixer. For each mix group, six 40x40x160 mm prisms and twelve 55x110 mm cylinders were casted and consolidated by vibration. After 3 – 4 days, specimens were extracted from the molds and the specimens were kept moist by wrapping up with two folds with foil in between nylon cellophane paper. Then they were kept in curing room at 20±1 ^{0}C and 70% relative humidity. The cylindrical specimens were used for the determination of compressive strength and hydraulic conductivity, and the prismatic specimens were used for the determination of flexural strength and linear shrinkage.

For the determination of the time of efflux of mix groups, the mix groups' ingredients were mixed in the mixer for 1 minute. The water content of all mix groups were adjusted for a slump range between 150 to 250 mm. The slump range corresponds to a flow time of 60 to 30 seconds respectively through a standard flow cone. Three to four trials were made until 60 sec flowability was obtained. By plotting the time versus water content, the necessary water content for obtaining 250 mm slump was determined.

The compressive strength tests were carried out on 55x110 mm cylinder specimens according to the requirements of ASTM C39-96. Flexural strength of mix groups were found by using simple beam with center-point loading. This test method is used to find the modulus of rupture of specimens.

Linear shrinkage test was performed according to ASTM C531-95. The readings were taken as long as the shrinkage continued (more than 14 days). Then they were dried in a 60^{0}C oven for three days and the changes in length were measured according to ASTM C531-95.

Hydraulic conductivity tests were performed according to ASTM D5084-90. Cylindrical specimens were used to measure the hydraulic conductivity of mix groups. 300 kPa backpressure and 400 kPa cell pressure were applied at each specimen.

Initial and final setting times were found from the tests that were performed according to ASTM C191-99. This method covers determination of time of setting of hydraulic cement by means of Vicat needle.

Experimental Results

Analysis of Physical Properties. The results of experiments conducted for the dry unit weight, void ratio and specific gravity experiments conducted on all mix groups are given in Table 3. The unit weights varied between 0.992-1.096 kN/m^3 with F80L20 group having the lowest and the F80C15L5 group having the highest unit weight. Addition of lime to fly ash decreased the unit weight. Similar results were also obtained in specific gravity tests, which showed that the cause of these variations is solely due to the specific gravity and grain size distribution of the ingredients.

Table 3. Physical Properties of Mix Groups

Group No	Mix Groups	Dry unit weight (kN/m^3)	Void Ratio (%)	Specific Gravity
1	F100	1.021	58.67	1.62
2	F95C5	1.025	56.10	1.60
3	F90C10	1.089	50.60	1.64
4	F80C20	1.091	50.32	1.64
5	F95L5	1.055	54.50	1.63
6	F90L10	1.010	55.45	1.57
7	F80L20	0.992	61.29	1.60
8	F80C15L5	1.096	52.37	1.67
9	F80C10L10	1.083	52.35	1.65
10	F80C5L15	1.038	57.03	1.63
11	F80C20FFN0.25	1.089	48.76	1.62
12	F100FFN0.25	1.082	49.72	1.62
13	F80C15L5FFN0.25	1.075	49.77	1.61

Analysis of Mechanical Properties. Unconfined compression, and flexural strength tests were performed to find the mechanical properties of mix groups which provides useful data in determining the strength and deformation characteristics of grout such as shear strength at various lateral pressures, angle of shearing resistance, strength in pure shear, deformation modulus, bearing capacity and creep behavior.

Unconfined Compression Test. The unconfined compressive strength (UCS) values of all groups are given in Table 4 and Figure 1.

Table 4. UCS Values of Mix Groups at 7, 28 and 90 Curing Days

Group No	Mix Groups	Average Unconfined Compressive Strength (MPa)		
		7 days	28 days	90 days
1	F100	0.03	0.64	3.18
2	F95C5	0.05	2.34	4.73
3	F90C10	0.07	4.04	7.33
4	F80C20	0.22	6.58	9.76
5	F95L5	0.10	1.29	4.09
6	F90L10	0.11	1.71	5.75
7	F80L20	0.11	1.15	6.07
8	F80C15L5	0.28	3.00	9.05
9	F80C10L10	0.25	3.39	8.09
10	F80C5L15	0.18	2.00	8.31
11	F80C20FFN0.25	0.11	6.13	11.8
12	F100FFN0.25	0.04	3.26	5.74
13	F80C15L5FFN0.25	0.08	4.38	9.14

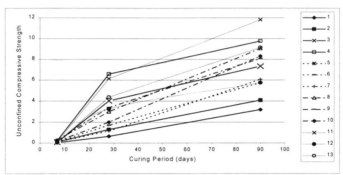

Fig 1. The Rate of UCS Gain of the Mix Groups.

Strength development in grout mixtures is directly related to cement content and water content, particularly when Class F fly ash is used. Ultimate strengths may gradually increase well beyond the 28-day strength, perhaps even beyond 90 days (Figure 1).

The rate of strength gain for fly ash-cement groups is less than fly ash-lime groups. The addition of WRA decreased the water content requirement for the same slump values, which led to the increase in the UCS values of mix groups both at 28 days and 90 days.

7, 28 and 90 days UCS values are given in Figure 2. It may be seen from Figure 3 and Figure 4 that optimum cement replacement is approximately 17 % – 20 % in the long term, however 20 % – 26% in the short term. For lime replacement, this range is 14 % – 17 % in the long term and it is 13 % – 16 % in the short term in terms of UCS results. The change in the optimum lime and cement content for 28 and 90 days curing period is solely related to the rate of hydration and the strength gain.

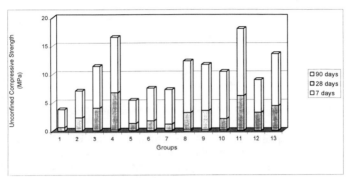

Fig 2. 7, 28 and 90 Days UCS of Mix Groups.

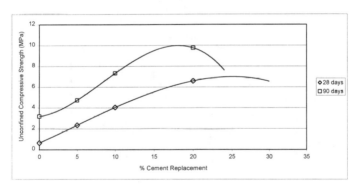

Fig.3. Optimization of Cement Replacement based on 28 and 90 Days UCS

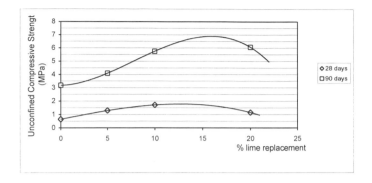

Fig.4. Optimization of Lime Replacement based on 28 and 90 Days UCS

Flexural Strength Test. The determination of flexural strength (the modulus of rupture) is essential to estimate the load at which the structural members may crack. The information is useful in the design of pavement slabs and airfield runway as flexural tension is critical in these cases. The flexural strength values of the mix groups range approximately between 20 – 30% of UCS values (Table 5).

An average value of 20 percent of the UCS is considered to be a fairly accurate estimate of the flexural strength of pozzolan stabilized base mixtures.

Table 5. Flexural Strength Values of Mix Groups

Group No	Mix Groups	Flexural Strength (MPa)
1	F100	0.199
2	F95C5	1.073
3	F90C10	1.816
4	F80C20	0.582
5	F95L5	0.628
6	F90L10	0.805
7	F80L20	0.498
8	F80C15L5	1.226
9	F80C10L10	1.242
10	F80C5L15	0.866
11	F80C20FFN0.25	1.019
12	F100FFN0.25	0.253
13	F80C15L5FFN0.25	1.242

Figure 5 shows that there is considerable difference between the UCS and flexural strength ratio of those lime modified and cement or cement and water reducing admixture modified fly ash groups, which indicate the enhanced degree of hydration due to direct interaction between the silica in fly ash and the free calcium hydroxide rather than the bound one to the other oxides in cement. The flexural strengths reported in the literature ranged from 1.0 to 2.3 MPa for the Class C fly ash slurries with 2% to 7% cement contents (Halow, 1982), as also recorded for the cement replaced mix groups in this study.

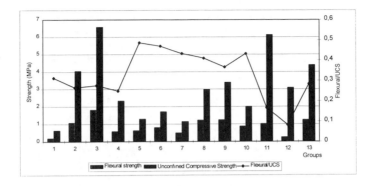

Fig. 5. Flexural Strength and UCS Values of Mix Groups

Analysis of Linear Shrinkage and Setting Time Values. Setting time is very important for grouts so that the grouting process may be performed within a relatively long period of time. Initial and final setting times and linear shrinkage values are given in Table 6. The linear shrinkage increased with increasing lime content.

In general fly ash replacement to concrete is not expected to show great expansion that will cause deterioration in the concrete. In this research, the expansions of all groups did not exceed 0.8 % of that specified by ASTM C 151-89. Addition of cement to fly ash has decreased the expansion. Also addition of WRA has decreased the expansion of latter three groups (Group No 11, 12 and 13).

Table 6. The Setting Times and Linear Shrinkage Values of Mix Groups

Group No	Mix Groups	Setting Times		Linear Shrinkage (%)
		Initial	Final	
1	F100	27 hrs. 15 min.	53 hrs. 15 min.	0.743
2	F95C5	22 hrs. 30 min.	36 hrs.	0.649
3	F90C10	20 hrs.	32 hrs. 45 min.	0.625
4	F80C20	17 hrs. 45 min.	28 hrs.15 min.	0.590
5	F95L5	15 hrs. 30 min.	46 hrs. 30 min	0.707
6	F90L10	13 hrs. 15 min.	37 hrs.	0.869
7	F80L20	10 hrs.	33 hrs. 15 min.	0.937
8	F80C15L5	17 hrs. 30 min.	29 hrs. 15 min.	0.760
9	F80C10L10	15 hrs. 45 min.	28 hrs. 15 min.	0.788
10	F80C5L15	14 hrs. 30 min.	27 hrs. 30 min.	0.821
11	F80C20FFN0.25	16 hrs. 15 min.	27 hrs. 15 min.	0.574
12	F100FFN0.25	25 hrs. 15 min.	48 hrs.	0.686
13	F80C15L5FFN0.25	15 hrs. 45 min.	24 hrs. 45 min.	0.715

Analysis of Hydraulic Conductivity Test. The hydraulic conductivity values are given in Table 7 and Figure 6. The range is between 1.28×10^{-7} cm/sec to 1.42×10^{-8} cm/sec. Addition of WRA decreased the permeability. The use of fly ash as landfill cover or liner material provides advantages due to its practicality in applications, better engineering performance than natural soils for stability, and hydraulic conductivity values less than 10^{-6} cm/sec. The Resource Conservation and Recovery

Act Subtitle D specifies the maximum hydraulic conductivity value as 1.0×10^{-7} cm/sec. Recent studies (Hwang, 1994) indicate that mixing of fly ash with native soils consisting clays decreases the hydraulic conductivity below 10^{-7} cm/sec. The coefficient of permeability values of lime added specimens are greater than that of only fly ash specimen.

Table 7. Hydraulic Conductivity Values of Mix Groups

Group No	Mix Groups	K_{av} (cm/sec)
1	F100	1.42 E-08
2	F95C5	8.36 E-07
3	F90C10	7.35 E-07
4	F80C20	4.74 E-07
5	F95L5	4.51 E-07
6	F90L10	2.52 E-07
7	F80L20	2.77 E-07
8	F80C15L5	8.88 E-07
9	F80C10L10	3.83 E-07
10	F80C5L15	3.80 E-07
11	F80C20FFN0.25	3.45 E-07
12	F100FFN0.25	3.40 E-07
13	F80C15L5FFN0.25	1.28 E-07

Fly ash reacting with available lime and alkalies generates additional cementitious compounds that act to block bleed channels, filling pore space and reducing the permeability of the hardened cement paste. The pozzolanic reaction consumes calcium hydroxide, which is leachable, replacing it with insoluble calcium silicate hydrates. The hydraulic conductivity values for high fly ash content grout mixtures have been found to decrease with increasing cement content, generally being in the range of 10^{-6} to 10^{-7} cm/sec.

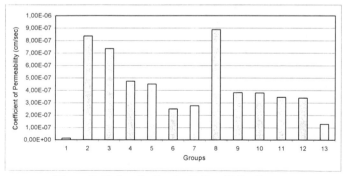

Figure 6. Hydraulic Conductivity of Mix Groups.

The hydraulic conductivity of material is affected by its density or degree of compaction, its grain size distribution and its internal pore structure. Pozzolan stabilized base materials generally have hydraulic conductivity values between 10^{-6}

and 10^{-7} cm/sec. The hydraulic conductivity of well-compacted fly ash ranges from 10^{-4} to 10^{-6} cm/sec, which is roughly equivalent to the hydraulic conductivity range of silty sand to silty clay soil.

Conclusion

Following conclusions may be drawn in terms of the experimental results:
- No segregation in the grout composite is expected during grouting process, since the particle size range within a quite narrow limit.
- The penetration capability of the grout composite is expected to be superior to the conventional cement based grouts due to enhanced suspension characteristic and low unit weight of the grout composite, which is quite close to that of water.
- The UCS and flexural strength of mix groups (other than the control group) promises satisfactory performance against lateral earthquake loads in grouting applications. The soil–grout composite mix and rock–grout interface are to be investigated to determine the performance of the grout composite in jet grouting and rock–bolting applications.
- The initial and final setting times indicate sufficient working period for jet grouting applications. The setting time may be adjusted by increasing the cement content based on the setting time of any specific job.

References

Halow, J.S., and Covey, J.N. (1982). "The Use of Fly Ash Grouts for Concrete Pavement Maintenance", *Proceedings of Sixth International Ash Utilization Symposium*, The National Ash Association, West Virginia, USA.

Hwang, J.Y. (1994), "Fly Ash Processing and Utilization", Institute of Materials Processing, Michigan Technological University, Michigan, USA.

Littlejohn, G.S. (1982). "Design of Cement Based Grouts", *Proceedings of ASCE Specialty Conference on Grouting in Geotechnical Engineering*, New Orleans, Louisiana, USA.

United Kingdom Quality Ash Association, UKQAA. "Pulverized Fuel Ash for Grouts", Technical Data Sheet 3, http://www.ukqaa.org.uk/DSheet03/ grout001.html, 12/15/2001.

Welsh, J.P. (1984). "Innovative Cement Grouting", SP-83, American Concrete Institute, 175 pp.

Additives and Admixtures in Cement-based Grouts

Alex Naudts[1], Eric Landry[2], Stephen Hooey[3], and Ward Naudts[4]

Abstract

Additives and admixtures are used in cement and non-cement-based grouts to modify their fluid and set characteristics. The ability to modify all fluid and set characteristics increases the durability, strength and penetrability of grout. Well designed formulations create balanced stable suspension grouts that reduce the cost of any grouting operation through increased grouting effectiveness by minimizing the cohesion and maximizing the penetrability of a grout via proper application of additives and admixtures.

Introduction

The use of a variety of additives and admixtures in microfine and regular cement-based grout formulations has become routine in advanced professional grouting practice. They play an important role in producing more durable grouts with improved rheological and set characteristics than could be obtained using only neat cement suspension grouts.

Neat cement grouts contain two products that are not very compatible: water and cement. Stokes' law governs the sedimentation of the particles; the finer they are, the slower the process. Segregation causes bleed paths to form unless a low water/ cement ratio is used. The resistance against pressure filtration of these grouts is poor, compared to balanced, stable suspension grouts. In rock and structural grouting, this phenomenon causes bleed water migration through the gelling grout. This creates pathways and bleed-pockets in the upper part of the cracks and crevices in which erosion and chemical attack can take place. Especially when grouting in water saturated media, laitance is formed which has no strength and is erodible. In soil grouting the use of unstable grouts results in poor penetration and an unreliable end product.

To obtain durable suspension grouts, the calcium hydroxide formed between the chemically resistant, hard ettringite in the cured cement paste, needs to be tied up

[1] President, ECO Grouting Specialists Ltd. Grand Valley, ON. L0N 1G0. Tel: (519) 928-5949; anaudts@ecogrout.com
[2] Project Manager, ECO Grouting Specialists Ltd; elandry@ecogrout.com
[3] Project Manager, ECO Grouting Specialists Ltd; shooey@ecogrout.com
[4] Grouting Engineer, ECO Grouting Specialists Ltd; wnaudts@ecogrout.com

and also turned into ettringite. The ability to modify viscosity, cohesion, bleed, gel-time, pressure filtration, and strength of cement-based grouts is essential to ensure a cost effective and high quality end product. These modifications facilitate multiple grouting passes in soil grouting applications.

The "DNA" defining a modern day balanced suspension grout should be the particle size distribution, resistance against pressure filtration, bleed, initial and final gel and set times, evolution of cohesion and strength, not merely the water/powder ratio.

The properties of cement-based suspension grouts

The properties of cement-based grouts can be classified into two categories:

- Fluid characteristics: cohesion and its evolution with time, thixotropy, viscosity, pressure filtration coefficient, bleed, and initial and final gelation.

- Set characteristics: initial and final set-time, unconfined compressive strength, durability, resistance against chemical attack, and permeability coefficient.

Properties and characteristics of suspension grouts can be significantly altered by means of additives and admixtures. Often compromises need to be made between conflicting characteristics in the search for the ideal grout for a particular project. The fluid characteristics of the grout can be evaluated to assess their effectiveness for the formation to be grouted by applying the 'amenability theory' (defined below) to assess the suitability of the selected formulation for the size of apertures and pore channels.

The grouting engineer must have a good understanding of the characteristics and the function of each component in the grout to formulate a balanced, stable, suspension grout with the desired rheology and set characteristics. It is obvious that the filling of a karst requires a different formulation than the filling of fine fissures in a rock formation or structure. Grouts placed in flowing water must have higher cohesion than grouts placed in stagnant water or in dry circumstances.

Fluid characteristics of cement-based suspension grouts

Viscosity, Cohesion and Thixotropy: For excellent background information on these matters, refer to publications by Khayat and Yahia (1997), Skaggs, Rakitsy, and Whitaker (1994), Naudts (1994) and Lombardi (1985).

The cohesion, apparent viscosity and thixotropy of a grout are related but impact differently on the rheology of a fluid. *Cohesion*, is a measure indicating how far and how easily the Binghamian fluid will flow through pore channels of a given size under the given pressure. *Thixotropy* relates to the time in which a certain cohesion value is obtained and the impact of the shear rate in obtaining this increase in cohesion. *Viscosity* is strictly a measure of flowability, and can be measured with a coaxial cylindrical viscosity meter or even with a Marsh cone. The latter two properties are closely related: the higher the cohesion, the higher the apparent viscosity. Ideally, admixtures only marginally impact on the cohesion or viscosity at moderate to high shear rates, but cause a major increase once the grout moves slowly or is at rest. These grouts are often referred to as "stable suspension grouts with shear thinning rheology".

It has been recognized that the behaviour of suspension grouts can be characterized as a Binghamian fluid, as opposed to a Newtonian fluid. In a Binghamian model, interparticle forces between the solids result in a yield stress that must be exceeded in order to initiate flow. The shear stress "τ" represents the transition between fluid-like and solid-like behaviour. The cohesion represents the resistance to flow until a minimum force is applied. The cohesion "C" - more so than the viscosity - is the property that prevents suspended particles from settling, or a grout from sagging in wide apertures.

It is clear from equation (1) that cohesion, plastic viscosity, and shear rate influence penetrability of a suspension grout. Therefore, the higher the cohesion and viscosity the higher the shear stress and hence the (grouting) pressure that will need to be applied to move the grout through the pore spaces.

(1) $\tau = C + \eta_B \, (\, dv/dx) \,$ or $\, \eta \, (dv/dx)$

Where: τ = shear stress; C = yield value or cohesion; η_B = plastic viscosity; dv/dx = shear rate; η = apparent viscosity

The ease of placement and susceptibility to wash-out of a grout are directly related to its thixotropic properties. Thixotropic agents impart a unique property to the cement-based suspension grout – as the shear rate applied to the grout falls there is a rapid increase in cohesion and viscosity (Figures 1 and 2). As the grout is being placed, the velocity and hence the shear rate drops at the fringes of the grout cylinder, and the cohesion and viscosity rise substantially. This in turn increases the resistance to washout but reduces the ability to flow. Even at very low concentrations (0.05% by weight of cement or less) pre-hydrated biopolymers used in the base mix design provide a significant thixotropic effect as shown in Figure 1. When these admixtures are properly prehydrated and neutralized, they are very potent to produce suspension grouts with shear thinning rheology.

The penetrability of a grout into a formation is determined by the pressure applied, its particle size distribution, and the cohesion of the grout (Håkansson. et al., 1992). For a cylindrical flow channel of radius (r) it has been demonstrated that the distance "L" that a grout will penetrate through a flow channel with diameter 2r, is governed by the applied pressure (p) and cohesion of the grout (C) (Lombardi, 1985).

$$L = \frac{p \times r}{2\,C}$$

Where: L = length of the cylindrical flow channel; p = applied pressure; C = cohesion of the grout; r = radius of cylindrical flow channel (or half the thickness of a seam or fissure).

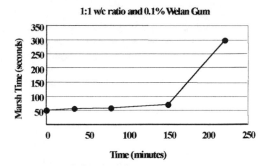

Figure 1. Evolution of Viscosity

The penetrability of grout is directly related to its cohesion. In general, the higher the cohesion of a grout the higher is the resistance to pressure filtration as described below and shown in Figure 3.

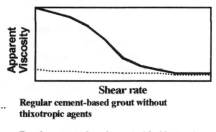

Shear rate

... **Regular cement-based grout without thixotropic agents**

— **Regular cement-based grout with thixotropic agent (biopolymer)**

Figure 2. Thixotropy

Resistance against pressure filtration: If circumstances permit grouting with high pressures the grout's penetration can be enhanced if the grout has good resistance to pressure filtration. If an unstable grout is used, the grout will pressure filtrate (dry pack) causing a considerable reduction in penetration.

The pressure filtration coefficient is defined as the volume of water lost in the pressure filtration test divided by the initial volume of grout in the cylinder of the apparatus, divided by the square root of the of the filtration time in minutes.

$$Kpf = \frac{volume\ of\ water\ ejected}{initial\ volume\ of\ grout\ x\ (\ filtration\ time\ (\text{min}))^{\frac{1}{2}}}$$

Neat cement grout mixes and unstable mixes have very poor resistance to pressure filtration. During grouting, unstable mixes will be subjected to substantial and almost immediate water loss where flow channels are narrow. The remaining grout forms a filtrate in the pores or fissures close to the injection point and refusal is reached quickly.

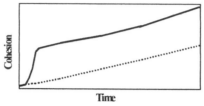

--- Balanced stable cement-based grout: gradual evolution of cohesion

— Thixotropic grout: rapid evolution of cohesion

Figure 3. Evolution of Cohesion

A balanced suspension grout refers to a grout that has a good resistance (low Kpf) to pressure filtration and a low to moderate cohesion. The relationship between pressure filtration and cohesion, based on recent in-house testing, is illustrated in Figure 4. Regular stable cement-based grouts have a low Kpf but high cohesion will limit their grout spread. By introducing high range water reducers the cohesion of the grout can be reduced without adding water, thus maintaining high resistance to pressure filtration.

The amenability coefficient (Ac) (Naudts, 1995)

The amenability coefficient (Ac) is a determined by dividing the apparent Lugeon value of the formation (Lu_{gr}) determined by using grout as the test fluid; by the initial Lugeon value of the formation determined during water testing (Lu_{wa}). A correction factor is applied to the Lugeon value determined using the grout to take into consideration the higher viscosity of the grout over water. For field applications in rock grouting, the following equation is used to calculate the apparent Lugeon value:

$$Lu_{gr} = \frac{flow(l/min) \; x \; 1m \; x \; 143 \; psi \; x \; V_{marsh\,gr}(sec)}{1 \, l/min \; x \; L_{zone}(m) \; x \; P_{effective}(psi) \; x \; 28 \sec}$$

with $V_{Marsh\,gr}$: efflux time for 1 litre of grout, via the Marsh Cone.

Amenability is a measure of the suitability for a given suspension grout to permeate fissures and apertures, accessible to water, in the grout-zone. Since Lu_{wa} is directly related to the aperture size of the fissures and open pores intersected by a borehole, it is clear that not all of these apertures are accessible to a selected grout. Only the fissures or pore channels that are wide enough will accept the grout. Hence the permeability coefficient established with grout is directly related to the aperture width of the fissures intersected by the borehole that can accept grout. By dividing Lu_{grout} by Lu_{wa} the percentage of open apertures intersected by a borehole accessible to grout is given.

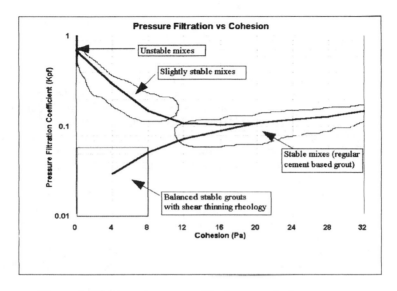

Figure 4. Resistance to pressure filtration vs. cohesion

The amenability coefficient (determined during the first few minutes of grouting) will immediately signal if the selected formulation is suitable, somewhat suitable or not suitable at all. The amenability coefficient indicates where, or if, regular cement-based grout is no longer suitable and one has to resort to a formulation with finer particles. Expensive and useless grouting operations can be avoided if the amenability of the formation is recognized to be too low.

The amenability coefficient is calculated early in the operation. The apparent Lugeon value, however, can be calculated on a regular basis as the grouting operation progresses. The apparent Lugeon value combines flow and pressure: the main grouting data. In a properly executed grouting operation, the apparent Lugeon value will gradually decrease with time. If the apparent Lugeon value remains constant or increases, it means that gouge or drill cuttings are being displaced or that the grout is sagging or is being washed out.

If on the other hand, the apparent Lugeon value is decreasing quickly, it means that "particle pick-up" is taking place or that the solids content is too high or that a grout with a lower pressure filtration coefficient is needed. The role of additives and admixtures is to optimize the effectiveness of formulations in relation to the hydrogeological conditions encountered in the formation.

Set characteristics of cement-based suspension grouts

Gel-time and set-time: The gel and set time of grout are affected by numerous factors including: the use of fillers, particle size distribution of the grout, ambient grout temperature, the use of retarders and accelerators, the solids content of the formulation and the medium being injected. The use of some retarders and accelerators also affects the durability and strength of the grout.

Bleed: The residual permeability of a rock formation injected with grout can be estimated by using the amenability coefficient as previously described, but also the bleed of the grout has to be taken into account since bleed pockets will be formed.

In order for a grout to be considered 'stable' by the classic school of grouters, the bleed should be less than 5%. Ideally, bleed should be less than 2%. The use of superplasticizers masks the ultimate bleed in a suspension grout by slowing down the settlement of the cement particles. For this reason, bleed should also be determined from samples (kept covered to prevent evaporation) that have reached final set.

Unconfined compressive strength: The strength of grout is affected by numerous factors including: solids content, type and percentage of fillers, cement content, grain size distribution, water content, as well as the presence of certain retarders, accelerators, and viscosity modifiers.

The resistance against pressure filtration and the execution of the grouting operation have a major impact on the ultimate strength of the grouted medium. During an extensive laboratory testing program using the In-situ Soil Injection System (ISIS) in 2000 (Landry, et al., 2000) using slag and non-slag based microfine cements, the authors discovered that sands grouted with only one pass had considerably lower strength than the same sands grouted in several passes allowing pressure filtration to take place. The difference could amount to a factor six!

Durability and resistance to chemical attack: Although related to strength properties the durability of a grout can be considerably enhanced through the use of various admixtures that will react with excess calcium hydroxide to form secondary ettringite. Through the use of accelerated aging tests, whereby samples of cured grout are exposed to an environment that is harsher (at a higher temperature, exposed to dynamic attack by placing the samples in a shaker bath) than the expected field conditions. Furthermore, the ability of the grout to withstand chemical attack over time can be evaluated.

Matrix Porosity and Permeability: The matrix porosity and permeability of a set grout can be an important factor. A more pervious grout is subject to faster deterioration that could, in turn, be accelerated if the environment has a low pH. Calcium hydroxide can be more readily leached out from the grout, if the permeability coefficient is high. Alternately, if a pervious grout is required in situations where strength is required but drainage is necessary, specially formulated cement foams can be used that have very high permeability and good strength characteristics.

Common additives and admixtures used in suspension grouts

The characteristics that are desirable in a suspension grout have been outlined above. How are additives and admixtures best used to produce a stable, balanced suspension grout? There is a myriad of additives and admixtures available on the market. Many products are sold under brand names. Manufacturers and distributors do not like to reveal details of the actual chemical composition. The application of a few additives and admixtures are detailed below as a guide for the reader.

Additives

Slag: Slag is a non-metallic by-product created during the smelting process of iron ore. In the smelter, layers of dolomite are placed between the ore. Slag is generally composed of silicates, aluminosilicates, calcium oxides and other base components. Slag is commonly used as an active component in the grout. It takes part in the hydraulic reaction between water and cement. It also reacts with other additives to create ettrengite. On its own, slag will cure very slowly and therefore must be used in conjunction with cement.

The low hydraulic reactivity of slag is ideal to delay initial set. This is particularly useful to control the set time of ultrafine cements. Most brands contain large amounts of slag and are blended with cement and other additives and admixtures before they are ground down to an ultra fine powder. Because of the large specific surface (Blaine fineness in excess of 8000 cm^2/g) and low matrix permeability and better chemical stability of ultrafine cement particles, the reactivity of the pure Portland cements is too high. Slag is commonly added to extend the set time of microfine cement-based grouts and to create a reaction with the calcium hydroxide. This reduces the matrix porosity and enhances the chemical resistance.

Fly-ash: Fly-ash (Type F) is a rather inexpensive filler with pozzolanic characteristics. This product reacts with calcium hydroxide (free lime) that is generated by the hydraulic reaction between the cement and water to form secondary ettringite. This creates a more durable grout, especially in environments where the free lime could be easily leached out. Fly-ash also slightly reduces segregation and enhances the water repellant characteristics of the grout, as well as, its resistance against pressure filtration.

Fly-ash slows down the hydraulic reaction and strength development. Fly-ash also reacts with lime to form a low strength grout. Grouts containing high concentrations of type F fly ash (70 – 90 % of the total solids content in the mix) cure very slowly and can be used for backfill grouting operations, in compensation grouting applications and flowable fill.

Fly-ash is a valuable component used to make ultrafine cement by either wet milling for immediate use in microfine cement-based grouts (Naudts and Yates, 2000).

It is blended with Portland cement and plays the same role as slag but reaches a lower unconfined compressive strength (at the same concentrations and mixing ratios).

There are two common types of fly-ash available: type C and type F. The difference lies in their chemical composition due to the various types of coal used in the combustion chamber. If the concentration of type C fly ash exceeds 15% by weight of cement, it could lead to rapid deterioration of the grout.

Natural Pozzolan: Natural pozzolan can be found in a natural state in different rock formations or they can be made from clay or schists burnt into cinder. Commonly known natural pozzolan such as pumice and trass have been effectively used to chemically react with the calcium hydroxide created during the hydration of cement to form secondary ettringite. By tying up the free calcium hydroxide, a more durable and competent end-product is achieved. The strength gain in the curing grout is slower and the exothermic reaction is less pronounced, which is beneficial when grout is placed in large quantities.

Trass: clay-phyllosilicate (1-1 structures) This natural pozzolan has been used for over 2,000 years. Trass-enhanced mortars have stood up for 20 centuries in marine environments. Trass is a powderised tuff stone mined in Andernach (Am Rhein) in Germany. Trass is extensively used in grouts for grouting sewers because it enhances the chemical resistance of the end product. Because of its ability to react with the free lime, the bond to concrete of grouts containing natural pozzolans is considerably better compared to those that lack it.

Silica fume: Silica fume is a by-product of the production of silicon or alloys containing at least 75% silicon. The silica fume particles collected are spherical in shape with an average diameter of approximately 0.1 micron. The small particle size makes it act as small ball bearings, keeping the larger cement particles into position in the mix, enhancing the penetrability of the grout.

Silica fume is also useful to reduce the permeability of the cured grout and as a result enhance its durability. The stability and resistance against pressure filtration characteristics of a grout can be enhanced with the introduction of silica fume. Silica fume is also known to provide water-repellant characteristics to the grout.

Hooton (1990) researched and documented the qualities of silica fume enhanced grouts. During the last 15 years the authors have tested and used silica fume in numerous grouting applications. Especially for the grouting of structures, silica fume enhanced grouts are very suitable.

Typical concentrations of silica fume in cement-based grouts vary between 4 and 10% b.w.o.c. Because of its fineness, it is a most valuable additive to ultrafine cement-based grouts. It also enhances the wet milling of cement-based suspension grouts.

Bentonite: Bentonite is one of the most common additives used in cement-based suspension grouts. It enhances resistance to pressure filtration, reduces bleed, enhances stability and penetrability, and increases the cohesion and viscosity of the grout. The barrel yield value of the bentonite is indicative of the stabilizing effect of the bentonite. A barrel yield of 110 or more is recommended for grouting applications.

Bentonite is a clay phylosilicate (2-1 structure) characterized by its great affinity for water. For grouting applications, a high yield sodium montmorillonite is recommended. They are composed of two layers of silica and one of aluminum. The weak Van der Waal link between the layers of silica makes it possible for bentonite particles to electrostatically attract other molecules of the opposite polarity. This phenomenon is particularly useful in grouting applications to curtail the segregation (i.e. minimize the bleed and reduce the pressure filtration coefficient) in cement-based suspension grouts.

Bentonite should be used as a pre-hydrated slurry. The authors found out that if the bentonite is not pre-hydrated first, it will lead to the cracking of the cured grout. Even if bentonite is introduced as a slurry, it should always be entered first in the mix. The mixing order has an impact on the rheology and even the strength of the cured grout. When bentonite is introduced as the last component, its stabilizing effects are vastly reduced.

Locally available fillers: The formulator should be familiar with the chemical composition of the locally available fillers such as waste products from industries, 'soils' and minerals, and consider incorporating them into the mix to produce inexpensive yet durable grouts. Initially, only the fluid characteristics of the mix can be assessed. The set characteristics and durability of the end product are typically not known and must be carefully considered, unless the grouting operation is only of a temporary nature.

Admixtures

Dispersants, high range water reducers and superplasticisers: These admixtures are used to reduce the viscosity of the grout and to reduce the speed of segregation of a given grout and consequentially, enhance its stability.

Deflocculators enrobe the cement particles with a film of the same negative charge. Consequentially, the particles reject each other and the formation of macro-flocs is prevented. Typically deflocculators used in grouts are naphthalene sulphonate or melamine based.

One important reason water reducers, dispersants, and superplasticisers are used is to improve the mixing of all ingredients. Consequentially, these admixtures should be introduced at the beginning of the mixing cycle and not at the end. In the case of mixes characterized by low viscosity, it is appropriate to introduce dispersants at the end of the mixing cycle. The latter saves money as the late introduction of dispersants requires smaller amounts.

Viscosity modifiers: Viscosity modifiers can be used to provide a wide range of flow properties. Viscosity modifyiers with high molecular weight which are highly cross linked, when used in a fully neutralized solution or colloidal suspension, will provide a very "short flow" rheology. Short flow rheology can be characterized by gelled, highly water repellant consistency similar to mayonnaise.   On the other hand, viscosity modifiers with a low molecular weight, more lightly cross-linked, will produce suspension grouts with "long flow" rheology, more resembling a "syrup".   One of the key attributes of viscosity modifiers is that they impart "shear thinning".   This means that when pressure is applied to the grout, it will flow.   When the driving pressure is removed, the grout will quickly "thicken up".

Minute quantities of bio-polymers such as natural gums and starches, cellulose ester derivatives and hydrophobically modified polyarcylate polymers, polyethelene oxide or polyvinyl alcohol, make it possible to change the rheology of a suspension grout dramatically.   In order to maximize the qualities of these products, they should be properly pre-hydrated and chemically neutralized.

Based on our research, viscosity modifiers can increase the resistance against pressure filtration by a factor 2 to 10, depending on the grout formulation.

Retarders, grout stabilizers: In the past, lignosulphonate - based superplasticisers were used to slow down the curing of cement-based suspension grouts.   This was done to facilitate multiple grouting passes for soil grouting or to prevent the loss of access to a given grout zone. The effect of a commercially prepared retarder on the gel and set times of microfine cement-based grout is shown in Figure 5.

Citric acid has been used for the same purpose.   A major loss in strength was noticed, in spite of citric acid's minimal impact on initial and final gelation times.

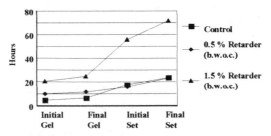

Figure 5. Gel and set time modification using a retarder

Schwarz (2000) performed very valuable work on the use of superplasticizers in conjunction with isopropyl alcohol to delay the evolution of the cohesion and extend the initial gel time. Isopropyl alcohol is an effective retarder that, when used in small quantities, will not affect the final strength of the grout. The isopropyl alcohol acts as a 'super' superplasticizer, which prevents the initial setting stages of the reaction from occurring.   During or following the final stage of grouting, sodium silicate can be injected which reverses the retardation process and causes the grout to set quickly.

Conclusions

Admixtures and additives are essential to create stable regular cement-based suspension grouts. When properly formulated, grouts will have the desired flow and set characteristics which results in a lower residual permeability of the medium to be grouted. There is a trade-off between stability and viscosity of a grout without sacrificing injection rate, penetrability and spread. The capability to modify all characteristics of grout results in a superior overall product. Consequentially, the effectiveness of a grouting operation is optimized, which results in a better end product and a reduction of the cost of the grouting program.

References

DePaoli, B., Bosco, B., Granata, R., and Bruce, D. (1992), "Fundamental Observations on Cement Based Grouts: Microfine Cements and the Cemill Process." *Grouting and Soil Improvement.* Volume 1. pp. 474-499.

Krizek R and Schwarz L. (2000). "Evolving Morphology of Early Age Microfine Cement Grout. Grouting and Ground Modification". *Advances in Grouting and Ground Modification.* ASCE. 181 – 199.

Håkansson, U., Hässler L. and Håkan, S. (1992). "Rheological Propeties of Microfine Cement Grouts with Additives." *Grouting, Soil Improvement and Geosynthetics.* ASCE. 551-563.

Hooton, R.D., and Konecny L. (1990). "Permeability of Grouted Fractures in Granite." *Concrete International.* July. pp. 48 - 56.

Khayat, K.H. and Yahia, A. (1997) "Effect of Welan Gum-High-Range-Water Reducer Combinations on Rheology of Cement Grout." *ACI Materials Journal.* V. 94, No. 5 September - October. pp. 365 - 372.

Landry, E., Lees, D, and Naudts, A. (2000). "New Developments in Rock and Soil Grouting: Design and Evaluation." *Geotechnical News.* September. 38 - 44.

Lombardi, G. (1985) "The Role of the Cohesion in Cement Grouting of Rock" 15th Congress de Grande Barrages. Lausanne. pp. 235 - 260.

Naudts, A. (1995). "Grouting to Improve Foundation Soil." *Practical Foundation Engineering Handbook.* Ed. Robert Wade Brown. McGraw-Hill. pp. 5.277-5.400.

Naudts, A and Yates, M. (2000) "New On-site Wet Milling Technology for the Preparation of Ultrafine Cement-based Grout." Canadian Geotechnical Conference. Montreal. October.

Skaggs C.B., Rakitsy W.G. and Whitaker S.F. (1994) "Applications of Rheological Modifiers and Superplasticizers in Cementitious Systems." *Special Publication 148,* ACI, Detroit. pp. 191-207.

Evaluation of Fly Ash and Clay in Soil Goutıng

S. Akbulut[1], A. Saglamer[2]

Abstract

Improvement of the mechanical and hydraulic properties of soils by grouting is one of the most widely used techniques in soil stabilization. The use of added pozzolanic materials such as silica fume, and fly ash, as well as clay to improve the physical, mechanical, and fluidity properties of cement grouts has been researched in recent years. In this study, the usage of grout additives fly ash and clay in soil grouting, and the effects of these grouts on soil strength have been researched in laboratory tests. First, grouts with added fly ash and clay in different amounts were prepared, and then these grouts were injected into soil samples. The unconfined compressive strength of grouted samples was determined for 7 and 28 days of curing time. It is shown that fly ash and clay improved fluidity of the grouts. The results of fly ash and clay grouted samples were compared with the cement grouted samples and the test results were evaluated against each other.

INTRODUCTION

Today, numerous applications in geotechnical engineering are done by sealing the pores in the ground, in order to increase bearing capacity and decrease settlement. At the same time, this processing reduces the adverse effects of earthquake motions and prevents liquefaction both in the laboratory and in field cases. Successful grouting depends on the filling of voids in the soil sufficiently and the presence of the connections between the soil particles and grout material. Although Portland cement is widely used in cement grouts, to make more successful grouting, and to decrease the cost of grouting, some industrial by-products such as silica fume (SF) and fly ash (FA) are also used in (Pierre et al, 1984; Sandra and Jejjrey, 1992). The purpose the use of these additives in grouts was to increase the stability of the grout, to save expense, and to investigate the effects of these materials on the strength of grouted samples. In addition, many chemical products are especially used in the grouting of granular soils (Krizek et al, 1992). The main factors on the properties of cement grout are additive content and water/cement ratio, which affect soil strength and plasticity after grouting (Incecik and Ceren, 1995). Grout mixtures vary widely depending on site conditions, desired results, available materials, and economics (Borden and Groome, 1984).

[1]Assistant Professor, Dept. of Civ. Engrg., Ataturk Univ., Erzurum 25240, Turkey.E-fax:1(603)687-4965, E-mail: sakbulut@atauni.edu.tr
[2]Professor, Dept. of Geotech. Engrg., Civ. Engrg. Faculty, Istanbul Technical Univ., Ayazaga, Istanbul 80626, Turkey. E-mail: asaglam@itu.edu.tr

In this laboratory study, the granular soil (sand-gravel) that was prepared with relative density of Dr = 0.30 and was grouted by fly ash (FA) and clay (CL). The grouting tests were conducted using w/s ratios of 1, 1.5, and 2 under grouting pressure of 100 kPa. After grouting, the grouted samples were left in the mold for 24 hours or more to enable some hardening to take place, and then they were removed. The samples were cured in a damp room at 7 and 28 days. Grouted samples with relative density of Dr = 0.30 were subjected to unconfined compression tests at 7 and 28 days (Akbulut, 1999). As the results of fly ash FA and clay (CL) grouted samples as compared with cement grouted samples, it seems that fly ash (FA) and clay (CL) improved the compressive strength of grouted samples, but cement grouted samples provided more strengthening than those of additive grouted samples.

2. MATERIALS USED IN THE TESTS

Soils: Alluvial soil, a sand-gravel mixture, used in the tests was classified as a poorly graded sand-gravel mixture SP with uniformity coefficient (Cu) of 4.57 and coefficient of gradation (Cc) of 0.87. The grain-size distribution of the soil used in the tests is varies between 9.5 and 0.3 mm sieves and the frictional angle, $\varnothing_{30} = 39°$, was found in soil samples with relative density of Dr= 0.30. The soil passing through a 0.3 mm sieve was not used in the test specimens because it is under grouting limits (Kutzner, 1996).

Cement (CM): Portland cement used in the tests has specific gravity of $\gamma = 31.3$ kN/m^3, specific area of $s = 3063$ cm^2/gr, and compressive strength of $\sigma = 59$ MPa at 28 days.

Fly Ash (FA): Fly ashes are classified by their chemical structures and they are named F and C type fly ash according to ASTM C 618. $SiO_2+Al_2O_3+F_2O_3$ is defined as pozzolanic activity and if this value is greater than 70%, then it can be defined as F type fly ash. If this value is greater than 50%, then it can be defined as C type fly ash. Mechanical and physical properties of the fly ash, which was obtained from Thermal power of Afsin Elbistan, used in the grouting tests depend on the coal composition, flame temperature, and fly ash filtration conditions. Fly ash (FA) used in tests are classified C type fly according to ASTM C 618 due to the fact that its pouzzolanic activity is less than 70% (Akbulut, 1999).

Clay (CL): Clay used in the tests was taken from vast clay formations in the Erzurum-Oltu Valley. It is can be defined as high plasticity clay and has a moisture content of 6% under natural conditions. The Liquid Limit (WL) of the clay is approximately 70% and it has value in use in soil grouting (Bell, 1993).

3. EXPERIMENTS

Grouts: The grout components were CM = 100% (additive proportion of FA, and CL = *0%*), FA = 5, 10% and CL = 5, 10% by total weight of solid material used in grouts. Water/ solid ratios (w/s) of 1, 1.5, and 2 were chosen for all grouts during the tests. First, water was poured into the mixture bowl and then additives were added into the water little by little. Finally, the cement was poured into the

grout mixing to prevent the flocculation of particles (Incecik and Ceren, 1995). A mixer, which has a speed of 1400 rpm, was used in the preparation of grouts. Then, the grouts were poured into the pressure tank to inject them into the prepared soil samples for grouting.

Preparation of soil samples: A flex-glass cylindrical mould was used in preparation of the soil samples to be grouted. The weight of the amount of gravel which fills the mould and gives the required relative density was calculated and this weight of material was poured into the moulds, it was then lightly compacted down with a rod as well as tapped on the sides of the mould until the required unit weight was reached. All soil samples were prepared in relative density of Dr = 0.30 for grouting tests, and then the prepared samples were successfully grouted by means of CM, FA, and CL additive grouts.

Grouting Tests: The prepared soil samples were injected with cement based grouts which have FA = 0, 5, 10% and CL = 5, 10% in different proportions by total weight of grout material, w/s = 1, 1.5, and 2, under grouting pressure of 100 kPa. After grouts had been prepared, they were poured into the pressure tank which was fed sufficient air pressure, and then samples were injected. When the valve on the injector was opened, the sample was grouted directly. After reaching enough grouting pressure and sufficient grouting, the grouting process was stopped. During the grouting tests, the grouting pressure was observed by means of manometer on the injector. After grouting, the grouted samples were left into the mould for 24 hours to enable some hardening to take place, then the samples in the mold were cured in water pond for 7 and 28 days (Akbulut, 1999). After curing, the samples were subjected to unconfined compressive tests and the strength values of grouted samples were calculated from the test results. Experiments were repeated in sufficient number for each for the soil samples.

4. RESULTS FROM EXPERIMENTS

The samples which have relative density of Dr = 0.30 were grouted by cement and fly ash and clay additive grouts with different w/s ratios under grouting pressure 100 kPa. The changes of compressive strength of grouted samples with fly ash and clay percent, curing time and w/s ratio for 7 and 28 days are given in Table 1 and Fig. 1, 2, 3, 4, 5 and 6, respectively. Initial tangent modulus calculated for 7 and 28 days samples are also given in Table 1 and Fig.6.

The compressive strength of FA/CM grouted specimens is less than that of cement grouted specimens at 7 and 28 days. 5% fly ash grouted samples gave more compressive strength than that of 10% FA grouted samples. The lowest compressive strength took place in 10% fly ash grouted samples with w/s=2 (Figure 1a and b).

The addiction clay (CL) decreased the compressive strength as compared with neat cement grouted specimens at 7 and 28 days of curing. %5 clay grouted specimens yielded more compressive strength than that of 10% clay grouted samples. The highest compressive strength was calculated from cement grouted specimens (Figure 2a and b).

The compressive strength of grouted specimens was increased by time and the highest compressive strength was obtained for 28 days. The highest compressive strength was obtained in the samples grouted by means of grout with w/s ratio = 1, and a lowest compressive strength was obtained in the samples grouted by grouts with w/s ratios = 2 (Figure 3a and b, and 4a and b).

Compressive strength of all grouted samples decreased disproportionally with the increase of the *w/s* ratio of grouts. Although the highest decrease in compression strength took place between *w/s* = 1 and 1.5, the lowest decrease in compressive strength of grouted samples was observed between w/s = 1.5 and 2. The highest compressive strength was obtained in grouted samples, in *w/s* ratio = 1, and a lower amount of compressive strength was obtained in grouted samples, with *w/s* = 2. While any increase in mixture water makes grouting easier, it also decreases the strength of the grouted samples (Figure 5a and b, and 6a and b).

Elastic modulus of grouted samples decreased with an increase in w/s ratio and the lowest elastic modulus values took place in samples with w/s ratio of 2 and samples grouted using fly ash and clay (Figure 7a and b, and 8a and b).

Table 1.
Compressive strength and elastic modulus of grouted samples in A-series tests for 7 and 28 days

Test	Additives	Additive Ratio (%)	w/s (kg)	Comp. Strength 7day MPa	Comp. Strength 28day MPa	E 7day MPa	E 28day MPa
1	CM	100	1	2.676	3.475	7042	9391
2	CM	100	1.5	1.062	2.106	3662	6194
3	CM	100	2	0.607	0.942	2890	4485
4	FA/CM	5/95	1	1.571	3.029	5343	8460
5	FA/CM	5/95	1.5	0.88	1.236	2750	4594
6	FA/CM	5/95	2	0.573	0.893	2640	3674
7	FA/CM	10/90	1	1.467	2.604	4097	6705
8	FA/CM	10/90	1.5	0.644	0.937	2800	3488
9	FA/CM	10/90	2	0.219	0.736	1068	2875
10	CL/CM	5/95	1	2.210	3.031	5755	8169
11	CL/CM	5/95	1.5	0.831	1.055	3089	4042
12	CL/CM	5/95	2	0.376	0.835	1129	3630
13	CL/CM	10/90	1	1.8	2.969	4687	8002
14	CL/CM	10/90	1.5	0.764	1.12	3144	4869
15	CL/CM	10/90	2	0.26	0.493	1198	2969

5. CONCLUSIONS

In this study, the influence of fly ash and clay additives to cement grouts used to stabilize a granular soil was researched by model grouting tests under laboratory conditions. The grouts were prepared in different water/solid (w/s) ratios, and than they were injected into the soil samples. In the tests, the compressive strength of grouted samples was examined. The results of grouted samples with additive fly ash and clay were compared with those of neat cement grouted specimens.

-Based on the tests, the effect of fly ash and clay content, curing time, and water/solid ratio in grout on the strength of grouted samples has been determined. The detailed conclusions are as follows:

-The strength of grouted samples was reduced due to an increase in additive content of fly ash and clay. While %5 addicts FA and CL decrease the compressive strength 13% and 12%, %10 addicts FA and CL decrease the compressive strength 25% and 15% for 28 days, respectively.

-The strength of grouted samples increased over time and therefore, the highest strength values for all grouted samples were observed at 28 days of curing.

-The water content in the grout mixture makes grouting easier, but it also reduces the strength of grouted samples higher water/solid ratio of grout.

-The elastic modulus of grouted samples showed different values according to additive content and a decrease in *w/s* ratio of grouts increased the elastic modulus at 7 and 28 days.

-In the tests the compressive strength of fly ash and clay grouted samples gave less values than that of cement grouted samples. However, the use of fly ash and clay in soil grouting made outstanding improvement to the mechanical properties of soil. The tests indicate that fly ash and clay can be used in the soil grouting for evaluation these materials with the added value of decreasing costs.

REFERENCES

Akbulut, S., 1999, Improvement of geotechnical properties of granular soils by grouting, PhD Thesis, The Institute of The Technical University of Istanbul, Turkey

Bell, F.G., 1993, *Engineering treatment of soils*, Published by E&FN Spon, pp. 10-160, London

Borden, R and Groome M.D., 1984, Influence of bentonite content on the pumpability of compaction grouts, *Innovative cement grouting*, ACI, Publication SP-83, pp. 115-129, Detroit

Incecik, M. and Ceren, İ., 1995, Cement Grouting Model Tests, Bulletin of The technical University of Istanbul, Volume: 48, No: 2, 305-317, Istanbul, Turkey

Krizek, R. J., Michel, D. F., Helal, M. and Borden, R. H., 1992. Engineering Properties of Acrylate Polymer Grout, Grouting, Soil Improvement and Geosynthetics, Geotechnical Engineering Division of ASCE, Volume: 2, No: 30, pp. 712-724, Lousiana

Kutzner, C., 1996. Grouting of Rock and Soil, pp. 10-195, Balkema, Netherlands

Pierre, C., Balivy, G., and Parizeau, R., 1984. The Use of Condensed Silica Fume in Goruts", Innovative Cement Grouting Publication sp-83, pp. 1-18, ACI, Detroit

Sandra, T., Jeffrey, C. E., 1992. The Effects of Fillers and Admixtures on Grout Performance, Grouting Soil Improvement and Geosynthetics, Geotechnical Engineering Division of ASCE, Volume: 1, No: 30, pp. 337-349, Louisiana

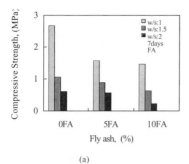

(a)

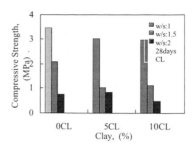

(b)

Figure 2. Compressive strength of grouted samples versus additive clay content; (a) 7 days, (b) 28 days of curing

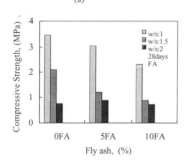

(b)

Figure 1. Compressive strength of grouted samples versus additive fly ash content; (a) 7 days, (b) 28 days of curing

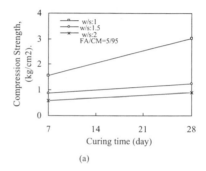

(a)

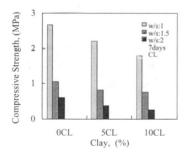

(a)

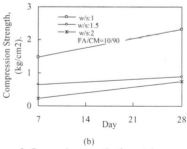

(b)

Figure 3. Compressive strength of grouted samples versus curing time; (a) 5% fly ash, (b) 10% fly ash

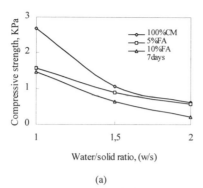

(a)

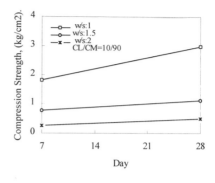

(b)

Figure 5. Compressive strength of grouted samples versus curing time; (a) 5% cement, (b) 10% cement

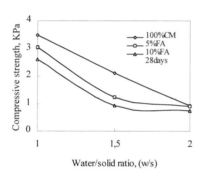

(b)

Figure 4. Compressive strength of grouted samples versus water/solid ratio of grout; (a) for 7 days, (b) for 28 days

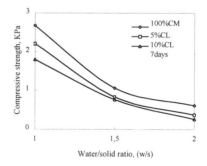

(a)

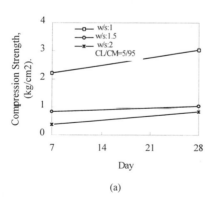

(a)

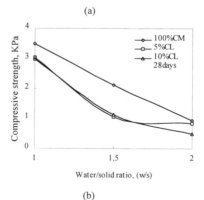

(b)

Figure 6. Compressive strength of grouted samples versus water/solid ratio of grout; (a) for 7 days, (b) for 28 days

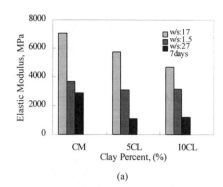

(a)

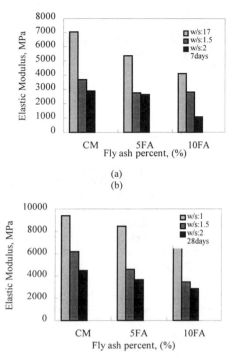

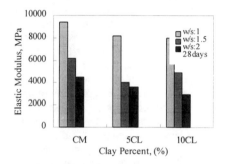

(b)

Figure 7. Elastic modulus of grouted samples versus fly ash percentage; (a) 7 days, (b) 28 days of curing

Figure 8. Elastic modulus of grouted samples versus clay percentage; (a) 7 days, (b) 28 days of curing

New On-site Wet Milling Technology for the Preparation of Ultrafine Cement-based Grouts

Alex Naudts[1], Eric Landry[2]

Abstract

The use of ultrafine cement-based grout has been gaining importance for rock and soil grouting over the last few years. One of the more dramatic technical innovations in the grouting industry that has been sought is the ability to mill, on-site, a fine or ultrafine cement-based grout using regular cement, and locally available additives (fly-ash, pumice, slag, bentonite, clay, tailings, catalysts, sand). This paper presents the results of extensive research and testing program executed during the development of a mobile Particle Size Reduction Mill (PASREM).

The search for an inexpensive and effective milling machine for on-site production of ultrafine cement-based suspension grout has been ongoing since the early 1990's. The application of on-site milling would solve the problems associated with the use of classic pre-prepared ultrafine cement. This includes the agglomeration of particles that often results in a grout with a higher average particle size than desired, reducing penetrability of the grout, and the high cost of ultrafine cement.

On-site milling would also allow for the use of many locally available products and additives that could enhance the final grout and reduce the cost of producing the grout.

PASREM is a mobile milling machine used to produce ultra fine cements from readily available portland cements and additives. This process is preferably done by injecting a balanced cement-based suspension grout through the PASREM, immediately prior to grouting. This is further referred to as the wet milling process. The PASREM process can also be used to mill the aforementioned dry powders to microfine size, to produce a classic (bagged) microfine cement.

The History of Wet Milling

The wet milling process is not a new concept. In 1992, Rodeo announced the first wet mill production process (Bosco et al., 1992) referred to as the Cemill[R] process. This process did not break through a number of problems: the coagulation of particles, rapid gelation of the grout, a sensitive and complex operation as well as some practical factors.

---

[1] President, ECO Grouting Specialists Ltd., 293199 8th Line, Grand Valley, ON, L0N 1G0; Tel. (519) 928-5949; anaudts@ecogrout.com
[2] Senior Engineer, ECO Grouting Specialists Ltd., elandry@ecogrout.com

In 1993 Sandia National Laboratories also undertook the development of a wet milling process (Ahrens, 1993). This process was slow and cement particles began to coagulate and flocculate in the machine and the end product did not have the desired rheology (high viscosity and cohesion). This process turned out to be impractical for permeation grouting.

In 1999, a wet mill in Germany met criticism because the mill hardly reduced the particle size (Steiner, personal communication).

The complications associated with earlier attempts to develop a practical wet mill were addressed during the development of a milling machine described in this paper.

Objective

The object of the program was to develop a feasible technology to produce fine and ultrafine cement-based suspension grouts via a wet milling and mixing process using cement. The main benefits of such an operation are the ability to custom mix and mill cement-based grouts to the desired particle size from regular cement, and locally available additives and admixtures.

The use of a wet milling process reduces the reliance on classic dry milled and bagged ultrafine cements that are expensive and generally available in developed countries or at a significantly increased cost in remote areas or underdeveloped countries. Regular cement is generally available everywhere. For many soil and structural permeation grouting projects applications however, regular cement is only marginally suitable for use. Especially where grout curtains for environmental applications are executed, the requirements for lower residual permeability have led to increased demand for improved mixing and fine and ultrafine cement-based grout.

In many grouting operations, the amenability of the formation to be grouted (Naudts, 1995) using regular cement-based suspension grout is not well known until the operation is well under way. It is therefore difficult to estimate the quantities of ultrafine cements that will be required. This leads to a compromise, in quality (allowing higher residual permeability in the rock grouting) when not enough ultrafine cement is ordered, or unnecessarily high costs when too much has been purchased.

Mixing and milling on-site suspension grout to the required particle size - only the quantities that are actually needed - has been the desire of those who want to democratise the grouting industry. Especially since the amenability theory quantifies the percentage of apertures accessible to water that are also accessible to a given grout (with particular rheology and particle size) it is possible to determine how fine the grout actually has to be in order to fill the fissures and pores of the medium to be treated.

For rock grouting, properly balanced, stable, cement-based suspension grout will penetrate clean apertures that are approximately two times larger than the largest particle size of the grout used. This means that properly formulated regular cement-based suspension will penetrate fractures as narrow as 90 - 100 micron (as was demonstrated in the Stripa Research Program). If ultrafine cement-based suspension grouts are used, the aperture width still accessible to these grouts is 20 - 30 micron. To put this in perspective: one fissure 80 microns wide (not injectable with regular cement-based grouts) per meter of borehole will produce a residual permeability of 3×10^{-5}

cm/s. By using ultrafine cement-based grouts the residual permeability of the of the treated rock formation can drop by a factor of 2 to 20 depending on the amenability of the formation for regular cement-based grout.

For soil grouting, on the other hand, soils with an in-situ hydraulic conductivity value of 8×10^{-2} cm/s can be permeated by regular cement-based grouts. Properly formulated ultrafine cement-based grouts (d_{95}<10 micron) will homogenously permeate soils with a permeability coefficient of 5×10^{-3} cm/s over a radius of 0.80 m.

Research and Development

The majority of problems encountered with grout prepared in wet milling machines were associated with the coagulation, flocculation and early gelation of the grout. It was felt that with the advancements in the field of grout admixtures that these problems could be overcome.

Liquid admixtures (viscosity, cohesion, and set modifiers) were introduced in the grout mix during milling using variable speed dosage pumps to each of the milling chambers. The rate of admixture injection could then be varied in each chamber to produce a grout with the desired rheology.

The second problem that plagued some early mills was the long milling time required to achieve a significant reduction in particle size. Initial research, therefore, focussed on the determining the most efficient milling process for reducing particles from an average size of 50 micron to as small as 3 micron in a short period of time. A number of field trips to were taken to observe various types of milling machines available in different industries. The machine selected for development was a mill with an upper chamber and a lower milling chamber. The chambers are filled with different types, sizes and volumes of grinding media. A diesel engine turns a shaft that causes the mill to oscillate and the grinding media to impact at the speed of sound. Adjusting the position of the shaft and the speed of rotation can vary the motion and size of oscillations of the mill.

During the R&D process, numerous modifications were tested including using different sizes of grinding media in each of the chambers, adjusting the volume grinding media in the chambers, changing the speed (rpm) of the mill, altering the shape of the oscillation curve, and the resident time of the grout in the mill. During the R&D process, specialists with extensive experience with a variety of milling machines assisted to determine the optimum parameters for particle size reduction. The information available from the mining industry was, at best, unreliable, and, at worst, conflicting. The research involved running numerous tests to establish the optimum configuration for the mill and to maximize efficiency by adjusting operating procedures.

Samples of wet milled product were retrieved for each testing operation and sent to a laboratory for particle size analysis. Two particle size analysers capable of measuring a particle range of 1 micron to 100 micron were used for this research project.

After several months of testing, a system was developed to reduce the particle size of a regular cement-based suspension grout significantly while maintaining adequate production rates and desirable rheological characteristics. The best results were obtained when the grinding media in the upper chamber were of a larger size than those in the lower chamber.   This produced a two-phase milling process.   After extensive testing to optimise operational parameters on the mill, particle size analysis revealed that there was a dramatic reduction average particle size when all the milling parameters were "right-on".   There was, however, always a faction of the particles that were deemed to be too large (12 - 18 micron range).   It was determined that the resident time required to reduce the particle size of this faction would be detrimental to production rates.

Research into the use of separators, commonly used in other industries, produced a viable option for removing the larger faction of particles.   The concept of further separating the larger faction of cement grout particles (as even regular cement-based has a fine faction) impacted on other characteristics of the grout.   Several separators systems were tested to determine the optimum type necessary to remove the larger particles.   The testing resulted in a device referred to as a 'PASREP', that uses two separators placed in series or used separately, depending on the reduction of particle size necessary for that particular application.   The next step was to integrate the milling and separation process.

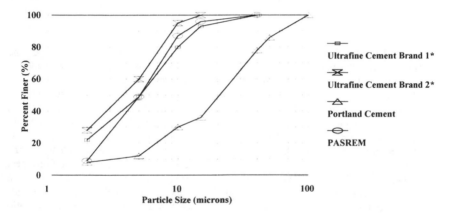

Figure 1. Particle size analysis of wet milled and classic ultrafine cement-based grouts

The Integrated Wet Milling and Separation Process

The production of fine and ultrafine cement-based suspension grout is a two-phase milling process followed by a separation process. The initial grout mix is produced in a paddle mixer, concrete truck or a high shear mixer containing water and any of a combination of regular cement, fly-ash (type F) pumice, slag, silica fume, bentonite, clay, retarders, superplasticizers and viscosity modifiers. The suspension grout is pumped into the upper chamber of the milling machine at a predefined pump rate. The suspension grout undergoes a first phase of milling in the upper chamber. The resident time of the upper chamber can be independently adjusted from the lower chamber.

Admixtures are introduced during the milling process either into the upper and/or lower chambers. Just prior to the cement grout exiting the mill more admixtures can be added to modify the milled grout prior to use. If required, the milled grout is run through one or both sets of separators to remove the majority of the larger particles. The fine fraction is then pumped in an agitator tank for immediate use. The coarse faction is pumped back to the upper chamber of the mill. In this way, the milling time required to reduce the particle size of the larger faction does not impact on the overall production rate of the mill.

Results

Figure 1 compares the particle size distribution of a regular cement-based grout, two ultrafine cement-based grouts prepared with commercially available ultrafine cements, and a wet milled cement-based grout prepared using regular cement. This illustrates the significant reduction in particle size between regular cement-based grout and regular cement-based grout after passing through the PASREM mill. A further reduction in particle size is achieved after the grout is passed through the separators. The two step wet milling process produces grout from regular cement that is comparable in particle size to grouts prepared using commercially available ultrafine cement.

By increasing the resident time of the grout in the mill the particles become finer. Since the permeability of a grout is predominantly dictated by its rheology, stability (resistance against pressure filtration) and cohesion, it is vitally important to place significant emphasis on these factors, in addition to the particle size distribution of the grout. In many instances, if the grout being injected into a formation has a favourable particle size distribution but poor resistance to pressure filtration, high bleed, and low cohesion, the end product will still be of poor quality. Freshly milled grout was injected into a series of identical sand columns (5% silt, 13 % moisture content) and the time to percolate through the column was measured. Although the test is far from ideal (Landry, et al, 2000), it provides a relative measure for the penetrability of a grout in a given medium. It was discovered that most commercially available ultrafine cement-based grouts, properly prepared in a high shear mixer, with the same water: powder ratio as the wet milled grout, did not permeate as quickly through the sand cylinders as the wet milled grouts. Although some of these commercially available ultrafine cements were verified to have a lower mean particle size and a finer d_{95} than the wet milled grouts. In the opinion of the authors, the intense high shear mixing in the mill provides the explanation for the results.

Table 1 illustrates some of the grout properties of ultrafine grout produced in the PASREM mill. The wet milled cement grout was prepared using water, regular cement, slag bentonite, naphthalene sulphonate, and a retarder.

Discussion

The demand for a workable wet milling process is best demonstrated by the fact that several others in the grouting industry have attempted or are still in the process of designing such a system.

For rock and structural grouting, the suitability of a given grout formulation is determined by its amenability coefficient. In essence, the initial permeability of the formation is determined via a water test (Lu_{water} using Caron's equation) and in turn compared with one the permeability of the formation obtained using grout as a test fluid (Lu_{grout}). If $Lu_{water} = Lu_{grout}$ it means that all accessible apertures intersected by the borehole to water are also accessible to grout. This means that the particle size is adequate to pass through all the pores (soil) or fissures (rock). If Ac (Lu_{wa}/Lu_{grout}) is less than say 0.5 then less than 50 % of the apertures accessible by water are accessible to the grout. The means that a finer or more amenable grout is required to adequately treat the medium being grouted.

The need to mill the grout to a finer particle size is therefore governed by the amenability coefficient. The production rate and the need to use the separators follow from the aforementioned considerations.

Table 1. Wet milled ultrafine cement grout product data

Property	Standard or Test Method	Results
Specific Gravity	API RP 13B-1	1.42 g/cm^3 (water/powder ratio: 1 to 1)
Bleed	ASTM C-940	0-4%
Initial Viscosity	API RP 13B-1	29-35 seconds
Pressure Filtration	API Filter Press	45 x 10^{-3} minute$^{-1/2}$
Cohesion	Lombardi Plate	1-3 Pascals – slow evolution in time until pump time has lapsed. Easy to adjust the viscosity by adding rheology modifiers during or after the milling process.
Initial Gelation	Wally Baker Shear Vane	adjustable between a few hours to 48 hours
Final Gelation	Wally Baker Shear Vane	adjustable between a few hours to 48 hours
Chemical Resistance		can be adjusted to enhance the chemical resistance against a variety of chemicals

Some drawbacks to the use of classic bagged ultrafine cement that the use of the wet milling process overcomes are:

- dry bagged ultrafine cement can cost up to 10 times more than regular cement;
- availability of ultrafine cements is limited and transportation costs from the manufacturers to more remote projects may be prohibitive;
- estimating grout volumes quantities required for a particular project can be difficult, leading to a shortage or excess of expensive utrafine cement;
- ultrafine cements have particle sizes 5 to 10 times smaller than regular cement with few options in between; and,
- dry bagged cements hydrate over time, causing coagulation of particles increasing the average particle size (i.e. they can have a short shelf life).

The cost of wet milling cement grout on-site is relative to the size of the project and the volume of milled product required. The cost of the mill would include mobilization and demobilization, operation of the engine and mill, and the cost of the cement and other additives and an operator. For small projects the costs of mobilization and demobilization typically cannot justify the use of a wet mill. As the size of the project increases, the relative cost saving of using a wet mill become apparent. The cash cost of producing, milling and mixing an ultrafine cement grout using regular cement is much lower than the cost of bagged ultrafine cement. By using the wet mill to produce fine or ultrafine cement grout from regular cement problems associated with shortages or oversupply of material are reduced since regular cement is usually readily available, and potential oversupply can easily be used for other construction applications.

Conclusion

The nine month long, intensive research and development program succeeded in developing a wet mill capable of producing a balanced, stable fine cement-based suspension grouts with a significant reduction in particle size from regular cement-based grouts. The wet milling process not only produces better-mixed grout, it also reduces the particle size of the grout considerably. The milling process has been optimised; the size and shape of the milling media in the milling chambers, oscillation cycle and resident time and rpm have all been studied to improve efficiency. The wet milled grout has a better resistance against pressure filtration, lower cohesion, and hence better penetrability than grouts with comparable particle size distributions and composition that are prepared in a high shear mixer.

The wet milling process produces grouts with higher amenability for the formation to be treated, bridges the gap between regular and ultrafine cement and regular cement and makes grouting operations more economical as a result.

Acknowledgements

The authors would like to thank Mr. Roger Willoughby of the National Research Council of Canada   (NRC) for his assistance through this research and development project; and the NRC for its generous financial support.

References

Ahrens. E. (1993). "Test Plan - Sealing of the Disturbed Rock Zone including Marker Bed 139 and the Overlying Halite below the Repository horizon at the Waste Isolation Pilot Plant." Internal publication, Sandia National Laboratories, Albuquerque, New Mexico.

Bosco, B., Bruce, D., A., De Paoli, B., and Granata, R. (1992). "Fundamental Observations on Cement-based Grouts (2): Microfine Cements and the Cemill[R] Process." Geotechnical Engineering Division, ASCE. New Orleans. pp. 486-489.

Landry, E., Lees. D, and Naudts, A. (2000). "New Developments in Rock and Soil

Grouting: Design and Evaluation." *Geotechnical News.* September. pp. 38-44.

Naudts, A. (1995) "Grouting to Improve Foundation Soil." *Practical Foundation Engineering Handbook.* Ed. Robert Wade Brown. McGraw-Hill. pp. 5.277 - 5.400.

Related Material

Heenan, D., and Naudts, A. (2000). "Advanced Grouting Program at Penn Forest Dam Results in Reduced Construction Costs and High Quality Product." *Geotechnical News.* June. pp. 43 - 48.

Characterization of a Non-Shrinkage Cement Grout Used for Water Pipe Joints

C. Vipulanandan[1], M.ASCE and Y. Mattey[2]

Abstract

    Cement-based grouts are used at the Prestressed Concrete Cylinder Pipe (PCCP) joints to seal and protect the steel cylinder from the surrounding environment during the service life of the pipes. If a proper grout-water mix was not used, the grout could lose water, harden prematurely and shrink in the joint, causing long-term maintenance problems. Hence, it is critical to develop field test methods to ensure the quality of the grouts used in the water pipe joints.

    In this study both field and laboratory samples were used to determine the properties of a non-shrinkage cement grout. The laboratory study was undertaken to evaluate the effect of water content on the working and mechanical properties of a non-shrinking grout used for interior and exterior joint sealing of PCCP pipes. By varying the water content, the behavior of various grout mixes was investigated. Over 130 field samples were tested and the results are compared to those of the laboratory samples. Field specimens exhibited a very large variation in the compressive strength as compared to the laboratory specimens. A relationship between unit weight, water content and compressive strength of the grout has been developed. The variation of compressive strength with curing time for various grout-water mixes has been developed. Based on these test results, using a grout/slurry balance to ensure quality of the grout in the field has been recommended.

[1] Chairman, Professor and Director of Center for Innovative Grouting Materials and Technology (CIGMAT), Department of Civil and Environmental Engineering, University of Houston, Houston, Texas 77204-4003 (Phone: (713)743-4278 Fax: (713)743-4260 e-mail: cvipulanandan @uh.edu).

[2] Graduate Student, Center for Innovative Grouting Materials and Technology (CIGMAT), Department of Civil and Environmental Engineering, University of Houston, Houston, Texas 77204-4003 (Phone: (713)743-4291).

Introduction

Since the early 1940's more than 28,000 miles of large diameter prestressed concrete cylinder pipes have been placed in service, principally as water transmission mains, operating at pressures as high as 250 psi. Although the pipes have generally performed well, over 500 failures due to exterior corrosion have occurred since 1955 [Price et al., 1999]. Corrosion can be caused by cathodic protection problems and/or damaged to joints resulting in exposure of the joints to corrosion. When done correctly, grouting can ensure a good joint seal to protect the steel cylinder from corrosive environment. Hence it is critical to ensure that quality grouts are used at the joints. If the grout is not correctly designed, it can lose water, harden prematurely and shrink in the joint, resulting in numerous problems [Vipulanandan et al. 1992, 1998 a & b]. Hence field test methods must be developed to control the quality of grouts in the field.

The grouts used in the exterior pipe joint construction are usually a cement-sand mixture with good flow properties. To achieve good pumping characteristics, aggregate is generally limited to sand [SP-83, 1984; Grout Guide, 1981]. Hence, methods to characterize the flow properties of the grout must be selected from the ASTM tests to better characterize the grouts. A laboratory study was undertaken to evaluate the working and mechanical properties of a non-shrinking grout used for exterior joint sealing.

Specifications on Grouts

Based on a comparative study of the AWWA M9, ASTM C 1107 and City of Houston (COH) and Texas grout specifications (nos. 02507 & 02511) the following observations are advanced [ASTM, 2001; Vipulanandan, 1999; AWWA 1995]:

(1) The required 7-day and 28-day compressive strength in the COH specification for the grout is 17 MPa (2,500 psi) and 34.5 MPa (5000 psi), respectively.

(2) Time for backfilling the trench recommended by AWWA M9 is much less than that recommended by COH. Grout should not be squeezed out of the joint by prematurely backfilling and loading the joints with construction equipment.

(3) No on-site testing methods are currently recommended in any of the specifications reviewed. In order to ensure the quality of the grout, it is important to develop a reliable on-site test for the grout.

(4) Proper placement of the grout in the plastic wrap (diaper) (9 to 12 in. (225 to 300 mm) wide) must be emphasized to ensure void free grout joint. Proper diaper material can isolate the grout from the contaminated region.

Objectives

The objectives of this study are as follows: (1) to evaluate the properties of a commercially available non-shrinkage grout used for interior and exterior joint grouting, (2) to determine the effect of water content on the setting time, flowability and compressive strength of the grout, and (3) to develop a field procedure for quality control of the grouts used in the field.

Experimental Program

In this study field samples were obtained from an ongoing water pipeline project in Houston, Texas. Laboratory samples were prepared under controlled conditions in the CIGMAT Laboratory at the University of Houston.

(i) Methods

(a) Flowability

Two instruments were used for determining the flowability of the grouts. These are known as flow table and flow cylinder methods and are generally recommended for cement mortar and slurries (Vipulanandan et al., 1996-98).

(i) Flow Table (ASTM C 230): In this test, the table is raised and dropped 10 times in 6 seconds by rotating a hand wheel. Flowability is calculated as 100(d-4)/4, where d is the diameter of the material spread. The average weight of the sample tested each time was about 500 grams.

(ii) Flow Cylinder (ASTM D 6103): This procedure consists of placing a 3 in. diameter x 6 in. long open ended cylinder vertically on a level surface and filling the cylinder to the top with grout. The cylinder is then lifted vertically to allow the material to flow out onto the level surface. Good flowability is achieved when there is no noticeable segregation and the material spread is at least 8 in. in diameter. Flowability is defined as 100 (d-3)/5, where d is the diameter of material spread. The average weight of the sample tested each time was about 1300 grams.

(b) Setting time

Different instruments were used for determine the setting time of the grout. The methods are the Vicat needle and penetration test.

(i) Needle Test (ASTM C 191): The Vicat needle was used to determine the initial and final setting times of the grout. The penetration of the 1.0 mm diameter needle was monitored with time. By definition, the initial time of set is the time corresponding to a needle penetration of 25 mm and the final time of set is the time corresponding to a needle penetration of less than 1 mm.

(ii) Penetration Test (ASTM C 403): The penetration resistance was also used to determine the initial and final setting times of the cement grouts. Initial and final setting times are defined as the elapsed time to reach a resistance of 500 and 4000 psi (3.5 and 28 MPa), respectively.

(c) Mechanical Properties

(i) Unconfined Compressive Stress-Strain Relationships (ASTM C 109): Both cubic and cylindrical specimens were used for testing. The cylindrical specimens for uniaxial compression tests were 38 mm (1.5 in.) in diameter with height varying from 65 to 90 mm. Compression tests were performed using a screw-type machine with a capacity of 10 kips. The displacement rate was kept constant at 0.031 mm/min. The specimens were trimmed and capped to ensure parallel surfaces. At least three specimens were tested under each condition.

Results and Discussion

(i) Field Study

Over 130 cubic grout samples were collected from an ongoing COH project over a period of 4 months. The tests were performed on 2-in cube samples and the variation of 7-day and 28-day strengths are shown in Fig. 1 in a probability plot and the results are also summarized in Table 1. Variation of the cube strengths can be approximated by a normal distribution.

Table 1. Summary of Strength from Field Samples

	Compressive Strength		
	7-day	28-day	Ratio 7-d/28-d
Sample size	74	58	58
Average (psi)	5535	8429	0.69
Standard Deviation (psi)	2168	3057	0.19
Covariance	39%	36%	27%
Range	2325 to 10390	5075 to 14708	0.43 to 0.98

1 MPa = 145 psi

Based on this analysis the following observations are advanced.
(1) There is a very large variation in the 7-day and 28-day grout strengths. The coefficients of variation were 39% and 36% for the 7-day and 28-day strength respectively. This large variation in the strength may be partly due to the fact that there was no quality control of the grout mix on-site.

(2) Less than 1% of the samples failed the grout strength requirement of 17 MPa (2,500 psi) after 7-days of curing. All samples pased the strength requirement of 34.5 MPa (5,000 psi) after 28-days of curing.

(3) The 7th to 28th day strength ratio was in the range of 0.43 to 0.98 with an average of 0.69.

(ii) Laboratory Study

A commercially available non-shrinkage grout (NS), a mix of cement, sand and additives, specified by the COH 02507 was selected for this study. The same grout was used in the field. Unit weight was measured using a slurry balance (manufactured by Baroid) as shown in Fig. 2. Unit weight can be measured to an accuracy of 0.1 lb/gallon (11.8 grams/L). Flowability of the grouts was measured using the flow table (ASTM C

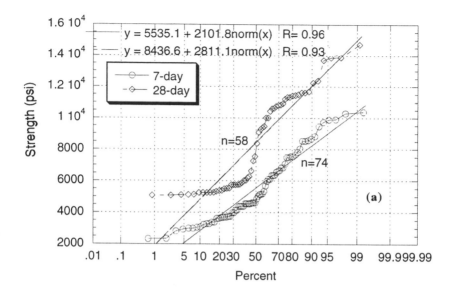

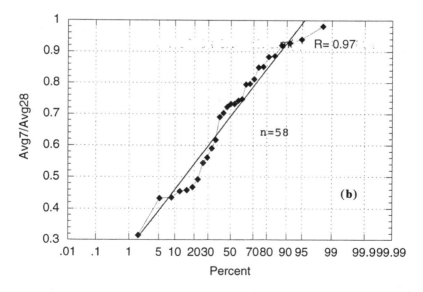

Figure 1. Variation of Field Grout Strength on a Probability Plot (a) 7-days and 28-days Strength and (b) Strength Ratio. (145 psi = 1 MPa)

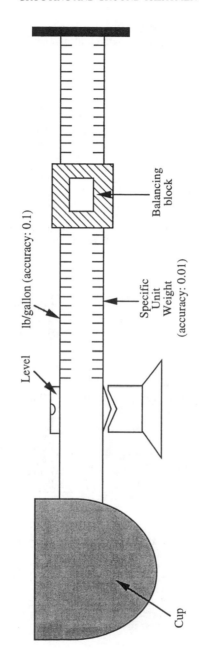

Figure 2. Grout Balance to Measure the Unit Weight of Fresh Grout.

230) and flow cylinder (ASTM C 6103). Both 2-in (50 mm) cubes and 1.5-in (38 mm) cylinders were used in determining the strength of the non-shrinkage grout.

(a) Unit Weight

The variation in the unit weight of the non-shrinkage grout with water content as measured by the slurry balance is shown in Fig. 3. The balance was sensitive in the range of water contents investigated in this study. The unit weight of the grout increases with the addition of water and reaches a maximum around 18% moisture content. At this water content, the entire void is filled with water. Further addition of water reduces the unit weight of the grout, since denser solids (cement and sand in the dry grout mix) are replaced with water. Also the variation of the 7-day and 28-day compressive strengths (cube specimens) with water content is shown in Fig. 3. By specifying the strength requirement for the grout, it will be possible to select the range for the grout unit weights that can be used in the field and the amount of water to be used in preparing the grout mixes. Selecting the unit weight based on the required 7 and 28-day strength will help in developing quality control of the grout in the field using the slurry/grout balance.

(b) Flowability

The flowability of the non-shrinkage grout with water addition is shown in Fig. 4. When the water content is increased beyond 15%, the flowability is increased rapidly. Grout with 15% and 20% water contents are characterized as plastic and fluid grouts respectively. Both flow table and flow cylinder methods gave similar results.

Table 2. Setting Times for Non-Shrinkage Grout Mixes (Vicat Needle)

Grout Mix	Setting Time (minutes)		Remarks
Water-to-Grout Ratio (%)	Initial	Final	
15 (plastic)	50	70	Fast setting. Rapid gain in strength and stiffness.
20 (Fluid)	180	300	Additional water substantially increased the setting times

(c) Setting Time

Setting time for the grout mixes were determined using the Vicat needle (ASTM C 191) and the results are summarized in Table 2 and Fig. 5. For the plastic mix no change in Vicat needle or penetration resistance reading was observed up to 50 minutes. For the plastic and fluid mixes the 1000 psi (7 MPa) penetration resistance (ASTM C 403) was achieved in 60 and 200 minutes respectively. This is very close to the results observed with the Vicat needle. The time to reach 4000 psi (28 MPa) resistance for the plastic and fluid mixes were 70 and 260 minutes respectively. Increasing the water content from 15 to 20 % resulted in more than three fold increase in the initial and final setting times.

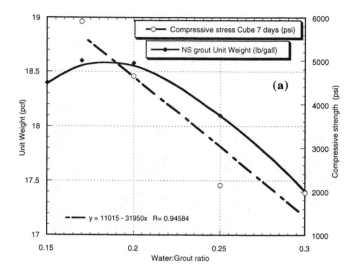

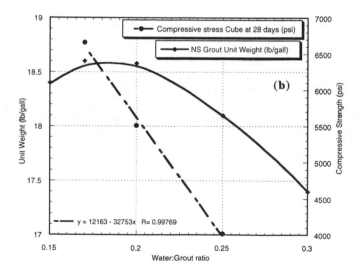

Figure 3. Variation of Unit Weight and Compressive Strength with Water Content for the Non-Shrinkage Grout Cured for (a) 7-days and (b) 28-days (10 pcf = 1.6 kN/m³; 145 psi = 1 MPa; 1 lb/gal = 11.8 g/L).

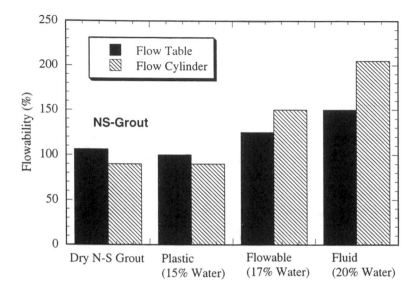

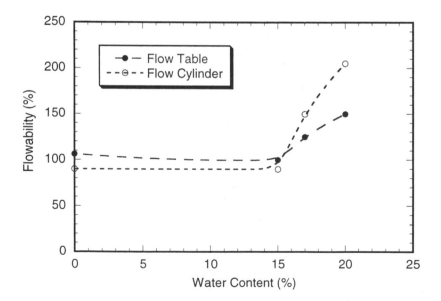

Figure 4. Variation of Flowability with Water Content for Non-Shrinkage Grouts.

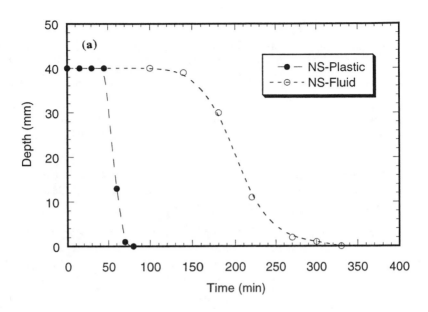

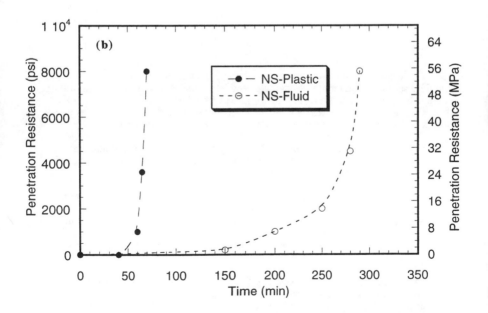

Figure 5. Setting Time for Non-Shrinkage Grouts (a) Vicat needle and (b) Penetration Resistance.

(d) Compressive Strength

A series of tests were performed to determine the effect of water content and curing time on the compressive strength of grouts. The compressive strength incresed with the curing time (Fig. 6; Table 3) and reduced linearly with the water content (Fig. 3). For the laboratory grout mixes the 7th to 28th day average strength ratio was in the range of 0.8

(water content $\leq$ 20%). When the water content was 25%, the grout didn't meet the strength requirement on the 7th and 28th days of curing. The cube strength was higher than the cylinder strength and the strength ratio (cylinder strength/cube strength) was in the range of 0.8 to 0.9 for the plastic grout and 0.6 for the fluid grout.

Table 3. Strength Results for Cement Grouts Tested in the Laboratory

Grout Mix	Compressive Strength			Remarks
Water-to-Grout ratio	7-day (psi)	28-day (psi)	Ratio 7/28	
15 (Plastic)	4231	5268	0.80	Meets COH specs.
17 (Flowable)	5909	6659	0.89	Meets COH specs.
20 (Fluid)	3818	4950	0.77	Meets COH specs.
25 (Fluid)	2144	4013	0.53	Not acceptable
30 (Fluid)	1970	Not tested	-	Not acceptable

1 MPa = 145 psi

Conclusions

The behavior of a non-shrinkage cement grout was studied using laboratory and field samples. The grout was characterized based on its flowability, setting time and mechanical properties. Property relationships have been developed to guide the designing of the grout mixes used at the pipe joints. Based on the experimental results and analysis of the test data following conclusions are advanced:

1. **Flowability**: Both flow table and flow cylinder methods had similar results for the grout mixes (dry, plastic and fluid) investigated. Flowability increased rapidly above 15% moisture content.

2. **Property Relationship**: Unit weight-water content- compressive strength relationships for the grout has been developed. Based on the design strength requirement, these relationships can be used to design the grout mixes. Unit weight measurement using the grout/slurry balance could be adopted as the field test for the grouts. The grout had the maximum wet unit weight at 18% water content.

3. **Strength**: The compressive strength of the flowable grout increased with curing time at varying rates based on the water content. The strength decreased lineraly with the increase in moisture content. Field samples had very large variation in strength as compared to laboratory mixes. For the laboratory grout mixes the 7th to 28th day strength ratio was in the range of

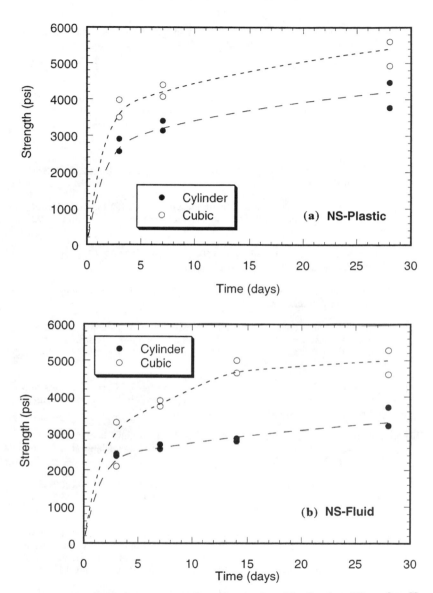

Figure 6. Variation of Compressive Strength with Curing Time for Non-Shrinkage Grouts (a) Plastic (15% Water Content) and (2) Fluid Grout (20% Water Content) (145 psi = 1 MPa).

0.8. The cube specimens had higher strength than the cylindrical specimens and the strength ratio varied with the moisture content and curing time.

Acknowledgment

This study was supported by the Center for Innovative Grouting Materials and Technology (CIGMAT) under grants from Lockwood Andrew and Newnam Inc. and the City of Houston. The support is very much appreciated.

References

[1] American Concrete Institute (1984), Innovative Cement Grouting, ACI, SP-83, , Detroit, 175 p.

[2] A Professional's Guide to Non shrinkage Grouting (1981), U.S. Grout Corporation, Engineering and Technical Center, Fairfield, Connecticut, 192 p.

[3] Annual Book of ASTM Standards (2001), Volume 04.02 Concrete and Mineral Aggregates, ASTM, Philadelphia, PA.

[4] AWWA Standards (C301 and C304) and Manual M9 (1995), American Water Works Association, Denver, Colorado.

[5] Price, R. E. and Lewis, R. A.(1999), "Evaluation of Concrete Pressure Pipelines and Failure Prevention," CIGMAT Proceedings, Construction and Rehabilitation Activities Related to Houston & Other Major Cities, Houston, Texas, pp. I7-I8.

[6] Vipulanandan, C. and Shenoy, S. (1992)" Properties of Cement Grouts and Grouted Sands with Additives," Proceedings, Grouting, Soil Improvement and Geosynthetics, ASCE, pp. 500-511.

[7] Vipulanandan, C. and Jasti, V. (1996)" Development and Characterization of Cellular Grouts for Slipling,with Additives," Proceedings, Materials for New Millennium, ASCE, pp. 829-839.

[8] Vipulanandan, C. and Jasti, V. (1997)" Behavior of Lightweight Cementitious Cellular Grouts," Proceedings, Grouting, Geotechnical Special Publication No. 66, ASCE, pp. 197-211.

[9] Vipulanandan, C. Weng, Y. and Zhang, C.(1998a)"Role of Constituents on the Behavior of Flowable Fly Ash Fill Behavior of Lightweight Cementitious Cellular Grouts," Proceedings, Recycle Materials in Geotechnical Applications, Geotechnical Special Publication No. 79, ASCE, pp. 137-152.

[10] Vipulanandan, C. (1998) (Editor) Proceedings, Grouting: Compaction, Remediation and Testing, ASCE, GSP 66, 227 p.

[11] Vipulanandan, C. (1999), Investigation of Prestressed Concrete Cylinder Pipe (PCCP), Bedding and Grout Materials, Final Report, Lockwood Andrew and Newnam Inc. and City of Houston - Surface Water Transmission Program (SWTP), Houston, Texas, 80 p.

Experimental Investigation of Factors Affecting the Injectability of Microcement Grouts

by

M.C. Santagata[1], Member ASCE and E. Santagata[2]

Abstract

The paper illustrates some of the results obtained from an extensive experimental program, conducted on grouts manufactured with a pozzolanic microcement (D_{98} = 10 μm) and an acrylic-based superplasticizer, and consisting of: a) injection tests in laboratory prepared sand columns, and b) rheological tests performed employing a high resolution rheometer.

The data presented in the paper highlights the effects of varying the water-cement ratio of the mixtures (w/c = 1 to 2.75), the admixture dosage, (0-1.6% in terms of active polymer by mass of cement) and the injection pressure (p_{inj}= 50 and 200 kPa). An attempt is made to link the results of the injection tests to the rheological properties of the grouts. It is found that, for the reference medium investigated (Ticino sand), at a given injection pressure, provided that the viscosity exceeds a threshold value, there is a unique and linear relationship between the height penetrated by the grout inside the sand column and its viscosity. A similar relationship, albeit shifted to higher values of the viscosity, applies for a lower injection pressure. Small changes in the viscosity of the grout dramatically affect the height of sand permeated by the grout.

Medium-related effects are also addressed. Based on the results of injection tests in five sands it is observed that small changes in the sand's gradation significantly affect its groutability, with increasing difficulties associated with grouting less uniform sands. Overall, the height penetrated by the grout correlates with the hydraulic conductivity of the medium.

Introduction

Microcement grouts are mixtures of an ultrafine hydraulic binder, water and admixtures, which can be employed in the treatment of porous systems to increase their stiffness, strength and durability, and reduce their permeability. These effects are desirable in many engineering applications and microcement grouts can provide an effective and economical solution in those situations in which the use of chemical grouts may be too costly or not completely environmentally safe.

For effective use in the field, grouting mixtures need to be thoroughly characterized in the laboratory with respect to their performance-related, project-specific, required properties. In particular, determination of flow properties by means

---

[1] Assistant Professor, School of Civil Engineering, Purdue University, 1284 Civil Engineering Building, West Lafayette, IN 47907-1284 – mks@ecn.purdue.edu

[2] Full Professor, Department of Hydraulics, Transportation and Civil Infrastructures, Politecnico di Torino, Corso Duca degli Abruzzi 24, 10129 Torino, Italy – santagata@polito.it

of both rheological tests and simulative injection tests may provide guidance for selection of the optimal mixture.

Injection tests in sand columns have limitations due, in particular, to the typically one-dimensional nature of the test, and one should be cautious about directly extrapolating the results obtained in the laboratory to the field. For example, Mittag and Savidis (1999) have shown that these tests lead to significantly underestimate filtration effects associated with flow of particulate grouts through soil. In addition, laboratory tests generally consist of a single injection, and thus do not accurately represent the field process, which involves multiple grouting passes.

However, these tests do provide an excellent means to perform a preliminary assessment of the groutability of a soil, to establish fundamental, albeit not always quantitative trends of behavior and compare grouts manufactured with different materials. In this capacity, they have been employed by a number of researchers (e.g. De Paoli et al. 1992, Perret et al. 1997).

In previous experimental work by the first author (Santagata et al. 1997, Santagata and Collepardi 1998) injection tests in laboratory prepared sand columns were used to highlight how the improvement of the properties of a sand, as measured by the compressive strength following grouting, is ultimately related to the water-cement ratio of the mixture. It was also shown that, through the use of superplasticizing admixtures, grouts with values of the water-cement ratio significantly lower than those typically employed can be used for the improvement of relatively fine sands. Finally it was observed that, particularly in the case of high solid content grouts, traditional groutability criteria that rely exclusively on geometrical considerations (e.g. on the relationship between the D_{10} of the sand and the D_{95} of the binder – Mitchell 1992), are not sufficient to establish if, and to what degree, a porous medium can be permeated, and that it is necessary to develop an understanding of rheology-based groutability limits. However, in these studies, the flow properties of the grout and their relationship with its "injectability" were analyzed only in a limited manner by making use of a very simple empirical laboratory test procedure (i.e. the Marsh cone test).

More can be learned on the flow properties of grouts by making use of rheological tests in which the fluid is strained in a simple way under well-established boundary conditions. Several researchers (e.g. Håkansson et al. 1992, Krizek et al. 1993, Schwarz and Krizek 1992) have followed this approach and employed one particular device (the Brookfield rotational viscometer) to study the effects of compositional parameters and mixing methods on the flow properties of microcement grouts.

However, significantly more advanced rheometrical techniques (e.g. parallel plate, cone and plate, coaxial cylinders rheometers), which have been widely used for a number of years in the field of polymers, and which have, more recently, been applied to some civil engineering materials (e.g. cement pastes, Lei and. Struble 1997), are today available. These devices are not only equipped with instrumentation that ensures highly accurate and repeatable measurements of the rheological properties, but can also be used to perform a broader range of tests (see equipment description below).

In the research presented in this paper, the Authors carried out an extensive experimental program in which the data obtained from injection tests in sand columns was integrated with the results deriving from rheological tests carried out on grouting mixtures by means of a high-resolution rheometer. The goals of the project were to highlight the effects of a number of test and mixture variables on the rheology of microcement grouts in the unhardened state, and to define and/or validate simple, performance-related testing protocols which may be used for the acceptance, design and control of grouting materials. While the detailed analysis of the rheological behavior of grout mixtures as a function of composition and curing time is presented elsewhere (Santagata and Santagata 2002), this paper focuses on the work which specifically addresses the issue of injectability.

Materials

Grouts

In the investigation one industrially-blended microcement, containing 60% Portland clinker, 36% natural pozzolan and 4% natural anhydrite as a set regulator, was considered (see Table 1 for the composition of the raw materials). The microcement is characterized by the particle size distribution shown in Figure 1, and by a characteristic diameter at which 98% of the binder is passing, equal to 10 microns.

Table 1 – Composition of the raw materials

% by mass	LOI	SiO_2	Al_2O_3	Fe_2O_3	CaO	MgO	SO_3	K_2O	Na_2O	Cl
Clinker	0.22	22.47	5.07	1.94	65.49	2.11	1.40	0.88	0.15	0.01
Pozzolan	1.2	61.68	18.10	7.91	3.37	3.09	0.00	3.01	1.24	0.03
Anhydrite	6.57	3.12	0.95	0.33	36.09	2.56	49.60	0.26	0.09	0.04

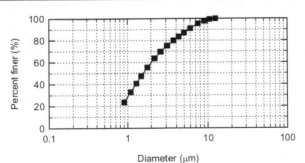

Figure 1 – Particle size distribution of the microcement

Grouts were prepared with given values of the water-cement ratio and by adding, when required, an acrylic based superplasticizer, the action of which is linked to the inhibition of flocculation of the cement particles through a steric hindrance effect. Research performed both on concrete mixtures (e.g. Collepardi et al.1993) and grouts (Santagata and Collepardi 1998) indicates that this admixture is more effective than melamine and naphthalene based superplasticizers in enhancing the fluidity of cement based mixtures, and in reducing the fluidity loss in time. In addition, unlike many other admixtures, its effectiveness appears to be independent of when it is added during the mixing process (e.g. Collepardi et al. 1993).

Granular media

The reference granular medium employed throughout the testing program was natural Ticino sand, a uniform sand ($C_u\sim2$) mainly formed by quartz crystals, which was already used in previous investigations (Santagata et al. 1997, Santagata and Collepardi 1998). In order to evaluate medium-related effects on grout injectability, two other less uniform sands were considered: a basaltic crushed sand ($C_u\sim6.5$) and a coarser crushed unbound material ($C_u\sim10.6$), formed by a mixed array of siliceous and calcareous particles. The particle size distributions of the three materials are compared to each other in Figure 2. Finally, in the last part of the investigation the basaltic and mixed-origin sand were sieved and recombined to replicate the grading of the reference Ticino sand.

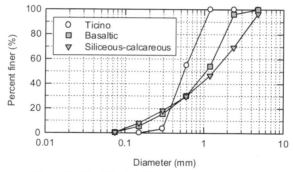

Figure 2 – Particle size distribution of the three sands

Methods

Injection tests

Injection tests were carried out on laboratory prepared sand columns as a direct means of evaluating the injectability of microcement mixtures, thereby highlighting both grout-related and medium-related effects. Figure 3 shows a schematic view of the apparatus employed for these tests, a slightly modified version of the system originally developed by De Paoli et al. (1992). Sand columns, 4 cm in diameter and approximately 46 cm tall, were prepared by following a standard protocol, which included compaction of the sand inside a PVC tube, saturation of the column with water, and measurement of the hydraulic conductivity under a constant hydraulic head. After completion of this preliminary phase, the grout was prepared, introduced in the pressure vessel and then injected in the column at a constant pressure. Time elapsed between preparation and injection was carefully controlled and kept equal to 5 minutes throughout the testing program. Pressure-related effects were also considered in the project, by performing injection tests at two different values of the injection pressure (approximately 50 and 200 kPa [7.3 and 29 psi]). Measurements carried out on the samples were limited to the evaluation of the height of the medium permeated by the grout.

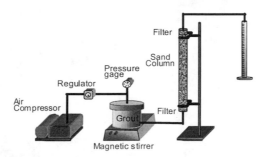

Figure 3 – Schematic view of the apparatus employed for the injection tests

Rheological tests

Rheological tests were carried out on the microcement grouts by employing a controlled stress rheometer (HAAKE Rotovisco RT20), previously used for the characterization of bituminous binders for paving applications. While the rheometer is not specifically designed for the analysis of low-viscosity fluids, it proved to be extremely useful in the evaluation of the flow properties of grouting mixtures, provided that specific test protocols were followed and that calculated viscosity values were not regarded as absolute.

As shown in Figure 4a, the rheometer, consists of three major components: a chamber which houses the specimen (here shown in the lowered position for specimen set-up); the chamber that contains the sensors, the thermoresistances and the conduits for temperature control, and which, once set-up is completed, is in contact with the specimen chamber; and the third chamber where the electrical motor used to apply the torque (with resolution of $8 \cdot 10^{-8}$ N·m [$5.9 \cdot 10^{-8}$ lb·ft]) and the sensor for measuring the angular rotation (resolution of $6 \cdot 10^{-5}$ rad) reside. Various measuring tools can be used in conjunction with this rheometer. In the investigation presented in this paper, all tests were performed making use of the corrugated parallel plate configuration (a fixed bottom plate and an upper plate connected to the rotor shaft, both 20 mm [0.79"] in diameter) employing a fixed gap of 0.5 mm [0.02"](Figure 4b).

The rheometer is computer controlled during all phases (connection of measuring tools to rotor shaft, gap and torque application, etc.). The PC is also responsible for temperature control (between 0° and 500°C, with a resolution equal to 0.01°C) performed through a system of thermoresistances and conduits through which a cooling liquid produced by a cryostat is circulated.

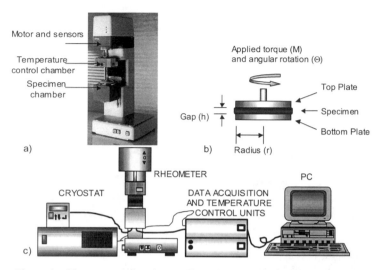

Figure 4 – Rheometer and testing configuration used for the rheological tests

Tests can be carried out both in the controlled-stress (CS) mode, typical of the equipment, and in the controlled-rate (CR) mode, which the equipment can emulate as a result of the fast closed-loop control of torque and rotation. Both steady-state single point measurements and multi-step stress (or strain rate) measurement ramps can be performed. In addition, oscillatory (or "dynamic") tests can also be carried out.

Batches of grout for the rheological tests were prepared from 50 g of microcement and manually mixed in a metallic bowl for 60 seconds to reach homogeneous conditions. After a fixed time the test samples were again mixed for 30 seconds and then poured directly upon the lower plate of the testing system. The overall experimental program involved over 500 tests, which were performed all at room temperature varying both compositional parameters (microcement, w/c, superplasticizer), curing time, and test protocol (type of test, strain rate, etc.) to highlight the effects of time and grout composition on parameters such as yield stress (τ_o) and plastic viscosity (η_p) (Santagata and Santagata 2002).

The tests that are especially significant in the context of the groutability study described in this paper are of the constant rate type. In this test a constant strain rate is imposed for a certain duration, while the resulting torque is measured. The unit shear stress (τ) and the corresponding viscosity ($\eta = \tau / \gamma_p$) can then be calculated from the torque. The test yields a time-stress (or time-viscosity) curve, which describes the flow response of the grout. From each constant rate test one point used to define the constitutive relationship between strain rate and shear stress is obtained. Figure 5 shows an example of the time-stress curve obtained from a CR test performed on a much "thicker" mixture than those considered in the injection tests. The curve is characterized by: an initial peak value of the shear stress τ_{max}, (and correspondingly of the viscosity, $\eta_{max}=\tau_{max}/\gamma_p$), which describes the initial structure of the mixture; a subsequent decrease of the shear stress (and of the viscosity), which reflects the fact that as the suspension continues to be sheared, its initial structure is destroyed; a final equilibrium condition (τ_{eq} and η_{eq}), which is for the most independent of the stress history of the specimen. The CR tests presented in this paper were all performed five minutes after mixing, imposing a constant value of the strain rate γ_p equal to 16.96 s^{-1} for a duration of 300 seconds.

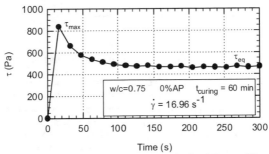

Time (s)

Figure 5 – Example of a time-stress curve obtained from a CR test

Results

Injection tests were carried out on columns of the standard Ticino sand by using mixtures in which the water/cement ratio and the admixture dosage were varied while employing a single ultrafine binder. As expected, variations in the composition of the grouting mixture significantly affected the height of penetration in the injection columns. This is shown in Figure 6, where penetration heights obtained employing an injection pressure of 200 kPa [29 psi] are plotted as a function of the water/cement ratio of the grout. The data show that the height permeated by the grout steadily increases with water-cement ratio and that two threshold values of w/c can be identified: one below which the grout cannot permeate the sand column (equal to 1.5 in the case of the microcement considered in this investigation in presence of no admixture), and one which allows complete permeation of the column (2.75 in this case).

As shown in Figure 6, in absence of a superplasticizing agent, the water cement ratios required to obtain significant permeation of the sand column are quite high (>2.5). Under these conditions grouts may, however, be characterized by poor stability, particularly under pressure. In addition, the improvement of the engineering properties (increased strength and reduced hydraulic conductivity) associated with the use of grouts with such high w/c is generally limited (e.g. Santagata and Collepardi 1998).

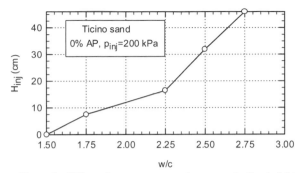

Figure 6 – Effect of water-cement ratio on penetration height

Superplasticizing agents can, however, be effectively used to allow permeation of grouts manufactured with much lower water-cement ratios. As a result of the lower w/c, for equivalent levels of permeation the use of such admixtures produces a significant improvement in the mechanical properties (e.g. Santagata and Collepardi 1998). In this experimental program, an acrylic polymer (AP) based admixture was used in increasing dosages, up to 1.6% of pure product by dry mass of cement (corresponding to a 5.3% dosage of the superplasticizer in its commercial form, which is a 30% solution of the polymer), to improve the injectability of grouts manufactured with w/c =1. As shown in solid symbols in Figure 7, an increase in the dosage of the admixture (%AP) caused an increase of the penetration height, with complete permeation obtained with an admixture dosage equal to 1.2%. Comparison with the data obtained with no superplasticizer (Figure 6) indicates that the admixture reduces the mixture's water without any loss in the fluidity by about 55-65% per percentage of pure product (i.e. ~ 17-20% reduction in mix water per percentage of superplasticizer).

Injection tests were carried out using two different values of the injection pressure (200 and 50 kPa). The results of the tests for the lower injection pressure are plotted in Figure 7 alongside the data for the 200 kPa injection pressure. As shown in the figure, both sets of data present very similar trends with the data for the 50 kPa injection pressure shifted to the right, reflecting the greater dosage of superplasticizer required to achieve a certain permeation of the sand column. Both sets of experimental data may be described by simple exponential curves of the following type:

$$h = \alpha \cdot e^{\beta(\%AP)}$$

where the exponent β is very similar for the two plotted curves and is thought to be dependent mainly upon the microcement and the sand; and the coefficient α increases with injection pressure.

The results above demonstrate that to evaluate the effects of an admixture or to compare different grouting products by means of injection tests, the pressure should be adequately selected to highlight differences. Moreover, in certain applications it may be significant to use in the laboratory an injection pressure which may be close to the one actually used in field applications.

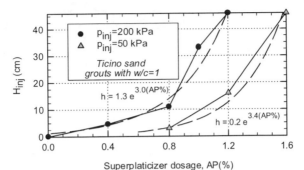

Superplaticizer dosage, AP(%)

Figure 7 – Effect of admixture dosage and injection pressure on penetration height

To develop correlations between injectability and actual flow properties of the grouts, rheological tests were performed on the grouts used in the injection tests, employing the apparatus and procedures described in the previous section (control-rate mode, with a constant flow gradient equal to 16.96 s^{-1}, all grouts cured for 5 minutes before testing). The shear stress and viscosity values at peak (τ_{max} and η_{max}) and at equilibrium (τ_{eq} and η_{eq}) conditions derived from these tests are summarized in Table 2 alongside the composition data for the grouts, and the penetration height measured in the injection tests (for pressures equal to both 50 and 200 kPa).

Table 2 – Results of rheological tests on grouts used for injection tests
(where more than one test was performed average values are reported)

w/c	%AP	τ_{max} [Pa]	η_{max} [Pa·s]	τ_{eq} [Pa]	η_{eq} [Pa·s]	H_{200kPa}(cm)	H_{50kPa} (cm)
1	0.4	108.0	6.52	105.1	6.19	4.5	0
1	0.8	93.4	5.52	90.0	5.31	11	3
1	1.0	95.6	5.66	89.2	5.27	33	_
1	1.2	93.4	5.52	88.6	5.24	46	16
1	1.6	_	_	88.2	5.20	_	46
1.5	0.0	99.3	5.87	94.5	5.59	0	_
1.75	0.0	95.1	5.62	93.1	5.50	7.5	_
2.25	0.0	94.2	5.58	90.0	5.31	16.5	_
2.50	0.0	92.0	5.44	89.5	5.28	32	_
2.75	0.0	93.2	5.51	88.9	5.24	46	_

Table 2 indicates that while, as expected, in all cases τ_{eq} is smaller than the corresponding τ_{max}, the two values are quite close. In fact, the time stress curves obtained from these tests are not characterized by the sharp decrease in shear stress observed in Figure 4. Analysis of the extensive database available for a much broader range of conditions indicates that as the w/c increases and the time from mixing decreases, not only do the measured values of the shear stress decrease, but also it becomes harder to identify the peak conditions (Santagata and Santagata

2002). Moreover, it is found that for all mixtures the parameters measured in peak conditions (τ_{max} and η_{max}) are typically very sample dependent and do not always show clear and monotonic dependencies from compositional factors. As a consequence, in order to define relationships between the rheology of the grouts and their permeation properties, it was considered more appropriate to make use of the parameters evaluated at equilibrium (τ_{eq} and η_{eq}), in free-flow conditions, which not only are less dependent on the stress history of each test sample, but also better reflect the conditions in which the grout is employed.

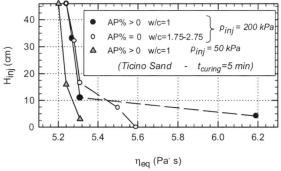

η_{eq} (Pa·s)

Figure 8 – Relationship between injectability and viscosity

Figure 8 plots the height permeated versus the viscosity of the mixture for tests conducted with an injection pressure of 200 kPa with grouts prepared both with (solid circles) and without (empty circles) the acrylic admixture. Note that both the injection tests and the rheological tests were performed 5 minutes after mixing the grout. It is observed that when significant permeation of the sand column is obtained (H_{inj} >15 cm), the data points for the two different kinds of mixtures fall on the same band and that the height permeated by the grout is essentially linearly related to the viscosity of the mixture. The range in the values of viscosity that ensure successful permeation of the column is quite limited (<5.28 Pa·s), and small changes in the viscosity of the grout significantly affect the permeation. For values of the viscosity exceeding this threshold value the two data sets diverge and the grout is found to permeate within the sand column more than what would be expected by extrapolating to higher vales of η_{eq} the linear relationship valid at lower viscosities (particularly in the case of the superplasticized grout). It is hypothesized that when the viscosity is such that the grout encounters significant resistance to its penetration inside the porous medium, the grout tends to separate. When this occurs the microcement particles fill the voids between the sand grains, and the mixture that permeates the column is much poorer in solids, and therefore characterized by lower viscosity, than the original one. As a result it penetrates further than expected. However, if the grout is unstable, as might be the case of the high w/c neat grouts, a plug is formed as the microcement particles separate from the water, and further permeation of any fluid is reduced, if not completely impeded. Note that it is also possible that in the case of the superplasticized grouts, the admixture not only enhances the flow properties of the grout, but also modifies the interaction of the grout with the soil particles, possibly facilitating the penetration of the grout in the medium, even for higher values of the viscosity.

Since the data plotted in Figure 8 demonstrate that the injection height of the columns is extremely sensitive to variations of the viscosity of the grouts, it is confirmed that in order to achieve a certain penetration of a grout inside a porous medium, its rheological properties need to be closely controlled, and their dependence on the composition of the grout clearly understood. However, even for the same grout, the rheological properties vary significantly in time, as the hydration process proceeds and the grout starts to set. Understanding the factors that determine the rate at which this process occurs, and how it may be controlled, is crucial, given that in practice some time may go by before the grout is injected. Additional rheological tests were conducted to establish the change in the viscosity as a function of time. These tests confirm that the acrylic superplasticizer is extremely effective also in retarding the hydration process and thus reducing the fluidity loss: the greater the dosage of superplasticizer, the slower the increase in viscosity with time. For example, in the case of the mixture with w/c=1 and 1.2% AP, the viscosity was found to be unchanged (i.e. equal to 5.24 Pa·s) after 60 minutes from mixing, and thus full permeation of the sand column would still have been possible at that time; even after additional 60 minutes the increase in viscosity was very modest (η_{eq} = 5.28 Pa·s) indicating that significant (~32 cm) permeation of the column could still have taken place (Figure 8). On the other hand stiffening of the neat grout was observed to progress at a much faster rate.

The relationship between groutability and viscosity is also dependent on the injection pressure. Also included in Figure 8 are the results obtained for grouts all prepared at a w/c of 1 with different dosages of the acrylic superplasticizing admixture, and injected at 50 kPa. The results show that the injection pressure significantly affects the permeation height obtained with a grout of a given viscosity, and that as the injection pressure decreases so does the range of viscosity values that allows successful injection. However, the sensitivity of the injectability of a grout to its viscosity, as measured by the slope of the H_{inj}-η_{eq} curve, does not appear to be significantly dependent on the injection pressure.

The results presented in Figure 8 are relative to one reference sand. While similar relationships between rheological properties of the grout, groutability as measured by the height of permeation and injection pressure are to be anticipated, from a quantitative point of view the observations made above have to be considered medium-specific. The following figures highlight how significant the role of the medium to be permeated can be, and how even modest changes in the characteristics of the sand can lead to significant differences in its groutability. Figure 9 compares the results of the injection tests presented above (Figure 6) for Ticino sand, to tests performed using the two other granular materials characterized by greater coefficient of uniformity and larger percentage of fines. The data, which are all relative to neat grouts, indicate that for the less uniform sands the groutability is greatly reduced. Significantly greater water-cement ratios have to be employed to ensure some penetration of the grout inside the sand column. However, regardless of the water cement ratio of the grout, for both sands the height that can be permeated does not exceed 20 cm. This is due to the high w/c, which causes the mixture to become unstable, especially under the action of the injection pressure. As a result, the microcement particles are retained by the sand "filter" and permeation cannot proceed further. These results, as well as the discussion above on the data presented in Figure 8, highlight another important property of suspension grouts that requires consideration, and that this paper does not address, i.e. the pressure filtration coefficient (e.g. see De Paoli et al. 1992).

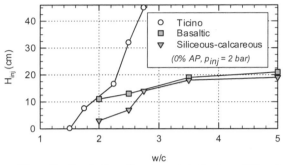

Figure 9 – Penetration of neat grouts in the three different sands

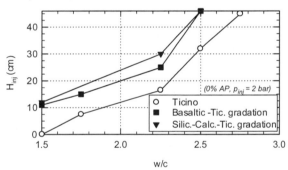

Figure 10 – Penetration of neat grouts in sands with the same gradation

Medium-related effects were further assessed by performing injection tests in columns prepared with the same two sands sieved and recombined to replicate the particle size distribution of the reference Ticino sand. The results of these tests performed using grouts with no superplasticizing admixture, are presented in Figure 10. The data emphasize the effect of particle shape, and show that even for sands characterized by identical particle size distributions, there may be some differences in the w/c required to obtain a certain penetration of the grout.

Figure 11a plots the height permeated by a reference grout (w/c=2, 0%AP) in the five different sands considered in this investigation versus the hydraulic conductivity of the medium, as measured in the constant head tests performed prior to injecting the grout. The data fall on one line indicating that the level of penetration of a grout inside a porous medium is controlled by the hydraulic conductivity of the medium, which is a direct function of its porosity (Figure 11b).

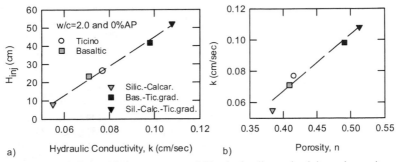

a)   Hydraulic Conductivity, k (cm/sec)   b)   Porosity, n

Figure 11 – Relationship between groutability, hydraulic conductivity and porosity

Summary and Conclusions

This paper presents the findings of a laboratory study on the rheological properties and injectability in fine sands of microcement-based grouts.   Injection tests performed inside laboratory prepared columns of a reference sand highlight how increased penetration inside a porous medium can be achieved by either increasing the w/c of a grout or, by adding a superplasticizing admixture in increasing dosages. The injection pressure also plays an important role.   In particular, the data suggest that the relationship between permeation height and w/c (or superplasticizer dosage) is unique for a given binder and sand, and that a decrease in the injection pressure (from 200 to 50 kPa) causes the curve to shift to higher values of the w/c (or admixture dosage).

Comparison of the results of the injection tests with the data obtained from controlled strain rate tests performed on the grouts making use of a high resolution rheometer indicates that there is a linear relationship between the viscosity of the grout and the distance penetrated, with small changes in the viscosity of the mixture dramatically affecting the extent of penetration. The interval of viscosity values that ensures injection of the grout is very narrow and decreases as the injection pressure is reduced.

In the case of neat grouts relatively large values (>2.5 for the microcement and the sand employed in the tests) of the water-cement ratio are necessary to obtain these viscosity values.   However grouts with equal flow properties but significantly reduced w/c can be manufactured employing superplasticizing admixtures, particularly using dosages higher than those commonly used in manufacturing concrete. In the case of the acrylic based superplasticizer used in the experimental program, the reduction in water is about 60% per percentage of active product used, and significant benefits in terms of reduced fluidity loss with time are also observed.

The paper also presents the results of injection tests in other sands characterized by different degree of uniformity, particle shape and percentage of fines. For a given grout the groutability of a sand as measured by the height permeated during the injection test correlates with its hydraulic conductivity. Reductions in the hydraulic conductivity by as little as a factor of two dramatically reduce the penetration of the grout. For two of the sands great difficulties were encountered in performing the injection, regardless of the rheology of the mixtures, highlighting how in the field the presence of lenses of finer material may critically affect the success of an injection.

References

Collepardi, M., Coppola, L., Cerulli, T., Ferrari, G., Pistolesi, C., Zaffaroni, P., and Quek, F. (1993), "Zero slump-loss superplasticized concrete," *Proc. 18th Conf. on Our World in Concrete and Structures*, Singapore , 73-79.

De Paoli, B., Bosco, B., Granata, R., and Bruce, D.A. (1992), "Fundamental observations on cement based grouts: traditional materials," *Proc. Grouting, Soil Improvement and Geosynthetics*, GSP No.30, ASCE, New Orleans, Feb.25-28, 474-485.

Håkansson, U., Hässler, L. and Stille, H. (1992), "Rheological properties of microfine cement grouts with additives," *Proc. Grouting, Soil Improvement and Geosynthetics*, GSP No. 30, ASCE, New Orleans, Feb. 25-28, 551-563.

Krizek, R.J., Schwarz, L. and Pepper, S. E. (1993), "Bleed and rheology of cement grouts," *Proc. Grouting in Rock and Concrete*, Widmann, Ed., Balkema, 55-64.

Lei, G.W. and Struble, L.J. (1997), "Microstructure and flow behavior of fresh cement paste," *J. Am. Ceram. Soc.*, 80 (8), 2021-2028.

Mitchell, J.K.(1992), "The role of soil modification in environmental engineering applications," *Proc. Grouting, Soil Improvement and Geosynthetics*, GSP No. 30, ASCE, New Orleans, Feb. 25-28, 110-143.

Mittag J. and Savidis S. A. (1999), "The use of microfine cements for the horizontal sealing of construction pits in Berlin," *Proc. International Conference on Anchoring and Grouting towards the New Century*, Zhongshan University Publisher, 264-267.

Perret, S. Ballivy, G., Khayat, K. and Mnif, T. (1997), "Injectability of fine sand with cement-based grout," *Proc. Grouting: Compaction, Remediation and Testing*, GSP No. 66, ASCE, Logan, UT, July 16-18, 289-305.

Santagata, M.C., Bonora, G., and Collepardi, M. (1997), "Superplasticized microcement grouts," *Proc. CANMET-ACI Conference on Superplasticizers and Other Admixtures in Concrete*," Rome, Italy, 177-195.

Santagata, M.C. and Collepardi, M., (1998), "Selection of cement-based grouts for soil improvement," *Proc. Grouts and Grouting*, GSP No. 80, ASCE, Boston, Oct. 18-21, 177-195.

Santagata, E. and Santagata M. (2002), "Rheology of microcement grouts", in preparation.

Schwarz, L. and Krizek, R.J. (1992), "Effects of mixing on rheological properties of microfine cement grout," *Proc. Grouting, Soil Improvement and Geosynthetics*, GSP No. 30, ASCE, New Orleans, Feb. 25-28, 512-525.

Treatment of Medium to Coarse Sands by Microcem H900 as an Alternative Grouting to Silicate-Ester Grouts

Murat MOLLAMAHMUTOGLU[1]

Abstract

The aim of this paper is to present comparative laboratory studies conducted on commercially available OPC, Microcem H900 and silicate-ester grouts. It has been found that Microcem H900 has better flow properties, and bleed characteristics than OPC. Further more, its permeation into medium to coarse sand is as effective as silicate-ester grout and the strength of the sand gained by the injection of Microcem H900 into sands is higher than that of silicate-ester grouted sand.

Introduction

Due to the inability of OPC grouts to permeate such soil formations as fine to medium and / or medium to coarse sands and the problems associated with permanence and toxicity of chemical grouts, advanced studies have revealed that microfine cement based grouts may be utilised to overcome the difficulties mentioned above and hence, an opening in the market has appeared for the manufacture of very fine cements.

Therefore, in this laboratory investigation attention is focused on the rheological properties of Microcem H900, the penetrability of this grout into medium to coarse sand and its effectiveness in terms of strength and permeability of the grouted sand in comparison with OPC and silicate-ester grouts.

Grout Properties Of Microcem H900 And OPC Suspensions

To mix grouts having different water / cement ratios, the quantities of each constituent, i.e. water and cement, were weighed in separate plastic containers. The

[1]Gazi University, Civil Engineering Department, 06570 Ankara, Turkey Phone: +90(312) 231 74 00 ext. .2241 Fax: +90 (312) 230 84 34 e-mail: molla@mmf.gazi.edu.tr

water was then transferred into a larger plastic drum and the cement was introduced gradually, whilst operating an electric drill with whisk attachment at 1300 rpm within water. Grouts were agitated for up to ten minutes which is sufficient for experimental studies (Banfill, P.F.G., 1981) in this way and then subjected to the following tests. This method was carried out to simulate the colloidal mixing equipment commonly used in industry today.

Sedimentation

When using cement grouts to permeate into joints or pores, it is not only the rheological properties that are important and influences the success of the grouting operation but also the stability to sedimentation and the size of the cement particles (Schwarz, L. G. et al, 1992). A commonly used criteria (Bleed capacity) is that the volume of clear water on top of a 1000 ml graduated cylinder divided by the original grout volume, must be less than %5 after two hours (Deere et al., 1985). To monitor the bleed capacity of each mix a sample was taken from the agitator approximately 10 minutes after mixing and placed into 1000 ml graduated cylinder. The result of sedimentation tests (up to 24 hours) conducted on OPC and Microcem H900 grouts with different water / cement ratios are given in Tables 1 and 2.

As seen from the tables OPC has a higher bleed capacity. This is attributed to differences in the grain-size distribution of the two materials. While grains of Microcem H900 range in size from 2 µm to 40 µm 80 % of OPC grain sizes are in between 10µm to 100µm . Furthermore, after two hours of sedimentation period the bleeding capacities of neat OPC and Microcem H900 suspensions with w/c ratios of 0.8 -1.2 are 19 %, 40 %, 1% and 4% respectively. This indicates that Microcem H900 is a stable cement grout.

Viscosity

Viscosity tests were performed using Haake Rotovisco RV20 viscometer at 15 to 30 minutes to determine plastic viscosity and yield stress of Microcem H900 and OPC grout suspensions. Immediately after mixing was completed, the grout was poured into the container and testing was started. Measured values (at 20 ^{0}C) in Table 3 shows that there is no noticeable difference in viscosity of OPC and Microcem H900 grout suspensions but as the particle's size becomes smaller the yield stresses of suspensions increases. This may be due to the fact that hydration rate increases with the increase in specific surface area which in turn results in resistance between particles against shearing (Littlejohn, G. S., 1982). The specific surface areas for OPC and Microcem H900 are about 4000 cm^2 and 7700 cm^2 respectively.

Set Time

Setting process may be considered as having two stages; an initial stage in which the fluidity of the grout decreases to a level at which it is no longer pumpable and a second stage in which the grout hardens and attains significant strength termed 'final set' (Schwarz, L. G. et al, 1992). For a successful grouting operation, it is necessary to determine the initial set time of suspension grouts. For this aim, Vicat needle apparatus (ASTM C191-49) was used to define the set time (the limit of pumpability) of OPC and Microcem H900 grouts and the results given in Table 4 indicate that increase in water-cement-ratio increases set time of all grouts but more remarkable the set time of Microcem H900 suspensions. Furthermore, set time of cement grouts is influenced by particle size. That is to say, as the specific surface area increases hydration rate increases thus reducing the set time.

Grout Properties Of Silicate-Ester Solution

The desired grout solutions were made by mixing the pre-measured quantities of distilled water, sodium silicate (ICI Grade M75) and ester hardener (Rhone Poulenc 600B) in a glass jar sealed with a rubber stopper. The glass jars were then fixed into a special laboratory mixer and mixed thoroughly at a speed of 1 Hz for approximately two minutes. The grout mixes adopted (frequently used in practice) in this research were 50 %, 60 % by volume of sodium silicate and 7 % by volume of Hardener 600B. In each mix the remaining component was water

Syneresis

Although most of the silicate formulations are considered permanent materials, the end product is subject to phenomena, which often tend to cause doubt about permanence. A newly made silicate-ester grout gels, preserved in airtight bottles for three years, exuded water and reduced in volume (between 3 % and 10 % of total volume of grout mix). This phenomenon is called syneresis and occurred at a decreasing rate with time (Mollamahmutoğlu, 1992)

Viscosity

As mentioned before, Haake viscometer (rotational viscometer) was used to determine the viscosity of the silicate-ester grouts at 15 to 30 minutes and the viscosity values obtained at ambient temperature of 20 ^{0}C for 50 %, 60 % by volume of sodium silicate and 7 % by volume of Hardener 600B mixes are 0.007-0.008 Pa.s. (1.015×10^{-6}-1.16×10^{-6} lb/in^2.s)and   0.009-0.013 Pa.s. (1.305×10^{-6}-1.885×10^{-6} lb/in^2.s) respectively.

Set (Gel) Time

Gel time for 50 %, 60 % by volume of sodium silicate and 7 % by volume of Hardener 600B grout mixes obtained at 20 °C are 46 and 55 minutes respectively

Sand Specimen Preparation and Grouting

A medium sand whose grain size is inbetween 0.5 mm and 1.4 mm was used throughout with a relative density of 70 % sand specimens were prepared under submerged condition using steel moulds, 38 mm in diameter and 300 mm long, which were split into two pieces longitudinally. The grout injection system consisted of an air pressure controlled grout chamber connected to the bases of grouting moulds. The sand filled moulds were injected with grout at a constant pressure to ensure permeation. Grouting pressure for chemical and cement grouts were 20 kPa (2.9 lb/in^2) and 80 kPa (11.6 lb/in^2) respectively.

Permeation of Grouts into Medium to Coarse Sand

The sand was permeated by OPC, Microcem H900, and silicate-ester grouts separately. For both cement grouts, water-cement ratio of suspensions was adopted as 1.2 since it has a very common practical use as stable mix in industry. The percentages of each component in silicate-ester grout solutions were selected for the same purpose as mentioned above. The sand in use was easily permeated by silicate-ester grouts under pumping pressure of 20 kPa (2.9 lb/in^2) but with Microcem H900 grout permeation was achieved by 80 kPa (11.6 lb/in^2). However with OPC grout permeation into the same sand was impossible with the same w/c ratio although the pumping pressure was increased up to 250 kPa (36.25 lb/in^2). Even then grouiting was refused. When specimen moulds were dismantled it was observed that only 1/5 of sample height was permeated by OPC grout but then hydraulic fracture occured.

Unconfined Compression Tests

Twenty four hours after injection the grouting cells were dismantled and samples were cut to the required length for unconfined compression tests. The grouted sand samples were then cured under sealed condition, namely sealed plastic envelopes in 100 % relative humidity curing room. The grouted samples were tested at a strain rate of 1.52 mm/min. for different time intervals. While the unconfined compressive strength of silicate-ester grouted sand specimens remained constant after 24 hrs, the unconfined compressive strength of microcem grouted sand specimens increased with time (Table 5) and the strength gained by Microcem H900 grout is much higher than that of silicate-ester grouted sand.

Permeability Tests

With a simple arrangment permeability tests were conducted on grouted specimens without removing them from their moulds. No flow was observed from micro cement as well as slicate-ester grouted sand specimens within the period of 40 days' measurement.

Conclusions

The results obtained from this experimental study are follows:

1. The bleed capacity for various w/c ratios for neat Microcem H900 is considerably lower than that of neat OPC for the same w/c ratios. Moreover, Microcem H900 grout suspensions having w/c ratios in between 0.8 and 1.2 are stable grouts since the bleed capacity at the end of two hours for the above mentioned w/c ratios are less than 5%.

2. As the specific surface area increases, yield stress and the plastic viscosity increase. For instance, while the yield stress and plastic viscosity for neat OPC grout with w/c ratio of 0.8 are 2 Pa $(2.9 \times 10^{-4}$ lb/in^2) and 0.02 Pa.s. $(2.9 \times 10^{-6}$ lb/in^2.s) they are 13 Pa (18.85×10^{-4} lb/in^2) and 0.03 Pa.s. (4.35×10^{-6} lb/in^2.s) respectively for the same w/c ratios of neat Microcem H900 grout.

3. The set time of a given type of cement is influenced by particle size. As the particle size becomes smaller the hydration rate increases thus reducing set time. The set time for OPC grout having w/c ratio of 0.8 is about 12 hrs but the set time for Microcem H900 grout with the same w/c ratio is around 7 hrs.

4. Permeation into sand having particle in between 0.5 mm and 1.4 mm and a relative density of 70% at 80 kPa (11.6 lb/in^2) grouting pressure was achieved by neat Microcem H900 with w/c ratio of 1.2. However, with neat OPC grout having the same w/c ratio permeation was impossible into the same sand although the grouting pressure was increased up to 250 kPa (36.25 lb/in^2) in which hydraulic fracture took place.

5. Silicate-ester grouts have substantially better flow properties and shorter set times than cement grouts. Furthermore, grouting of the sand with 50 and 60 percentages of silicate-ester grouts was accomplished at a low grouting pressure of 20 kPa (2.9 lb/in^2) indicating that it has better permeation capacity than that of Microcem H900 grout. However, permeation into the same sand was also possible with Microcem H900 grout where the OPC grouts failed.

6. Silicate-ester grouts are subjected to syneresis, which may affect the performance of grouting in terms of permeability and unconfined compressive strength with time and may result in pollution in the ground (Mollamahmutoglu, M., 1992).

7. Unconfined compressive strength of grouted sand with Microcem H900 is much higher than that of silicate-ester grouted sand. For example, unconfined compressive strength of microcem grouted sand with w/c ratio of 1.2 is 4425 kPa

(641.625 lb/in^2) whereas the unconfined compressive strength of 60 % silicate-ester grouted sand is 413 kPa (59.885 lb/in^2) at the end of 28 days.

8. Throughout 40 days permeability testing there was no flow through the sand specimens injected with both micro cement grout and silicate-ester grouts, which indicates that permeation of Microcem H900 into pores of medium to coarse sand was as effective as silicate-ester grouts.

References

Banfill, P.F.G., (1981) *Rheology of Fresh Cement and Concrete*, E&F.N. Spon, London.

Deere, D.U., Lombardi, G., (1985) *Grout Slurries Thick and Thin, Issues in Dam Grouting*, ASCE, N.Y., 156-164.

Littlejohn G.S., (1982) Design of Cement Based Grouts, *Proc. of ASCE Specially Conf. on Grouting in Geotechnical Engineering*, New Orleans, Louisiana, 35-48.

Mollamahmutoglu, M., (1992) *Creep Behaviour of Silicate-ester Grouted Sand, PhD Thesis*, University of Bradford, UK, p.251.

Schwarz, G.L., Krizerk, J.R., (1992) Effects of Mixing on Rheological Properties of Micro Fine Cement Grout, *ASCE Conference on Grouting, Soil improvement and Geosynthetics*, New Orleans, 512-525.

Table 1. Bleed Capacity Results for OPC

W/C	Wet density (Mg/m³)	Mix temp. (°C)	Ambient temp. (°C)	Time (hours)			
				1	2	3	24
				Bleed Capacity (%)			
0.8	1.60	20	Approx. 17 °C	17	19	19	18
1	1.51	22		32	32	32	31
1.2	1.45	22		40	40	41	40
1.4	1.40	23		45	44	45	45

Table 2. Bleed Capacity Results for Microcem H900

W/C	Wet density (Mg/m³)	Mix Temp. (°C)	Ambient temp. (°C)	Time (hours)			
				1	2	3	24
				Bleed Capacity (%)			
0.8	1.61	19	Approx. 17 °C	1	1	1	0
1	1.52	19		1	2	2	2
1.2	1.45	20		4	4	6	10
1.4	1.38	21		6	12	15	15

Table 3. Viscosity of Microcem H900 and OPC Grout Suspensions (at ambient temp. of 20 °C)

W/C	Viscosity (Pa.s.)		Yield stress (Pa)	
	Microcem H900	OPC	Microcem H900	OPC
0.8	0.03	0.02	13	2
1	0.02	0.02	4.7	1
1.2	0.02	0.02	1.6	1
1.4	0.02	0.01	1	1

Table 4. Set Time Data for OPC and Microcem H900 Grouts
(at room temperature of 17~20 °C)

Grout type	W/C	Mixing time (minutes)	Initial set time (minutes)
OPC	0.8	10	710
	1.0	10	720
	1.2	10	760
	1.4	10	770
Microcem H900	0.8	10	210
	1.0	10	420
	1.2	10	600
	1.4	10	765

Table 5. Unconfined Compressive Strength of Grouted Sand

Grout type	Unconfined compressive strength (kPa)				
	1 day	7 days	14 days	28 days	48 days
Microcem H900 (w / c =1.2)*	2000	3100	4123	4425	4428
Silicate-ester 50:7:43*	305	305	305	305	305
60:7:33*	413	413	413	413	413

* Average values of three identical samples for each time intervals

Formulation of high-performance cement grouts for the rehabilitation of heritage masonry structures

S. Perret[1], G. Ballivy[2], D. Palardy[3], and R. Laporte[4]

Abstract

Heritage masonry structures are an integral part of our landscape. Heritage structures located in Canada and the United States are relatively young compared to European heritage structures but they can show signs of deterioration due to our harsh climate.

Repairing these structures using grout injection is a very interesting technique since it involves preserving the historical character of the structures. The formulation of injection grouts for such repair must be based not only on physical properties of the deteriorated mortar, but also on chemical properties. Compatibility between the existing medium and the repair material is a major factor in avoiding new deterioration related to poor mechanical behaviour of injected grout or chemical reactions with deteriorated mortar. Other parameters such as rheology, injectability and stability of repair grouts must be considered to ensure the effectiveness of grout injection.

Many mineral additives and chemical admixtures are available for the formulation of cement-based grouts used to make repairs. Some of these products have been tested in the laboratory to optimise high-performance cement grouts used to repair heritage hydraulic masonry structures in Canada. The characteristics of these grouts and their performance during field work are presented in this paper. The effectiveness of the grouting treatment was controlled by sonic tomography conducted before and after grouting: several techniques are illustrated and their results are very sensitive to the actual grout intake in the different locations of the structure treated.

[1] Ph.D. candidate, Dept. of Civil Engineering, Université de Sherbrooke, Sherbrooke, PQ, J1K 2R1, Canada; phone 819-821-8000; sperret@hermes.usherb.ca
[2] Professor, Dept. of Civil Engineering, Université de Sherbrooke, Sherbrooke, PQ, J1K 2R1, Canada; phone 819-821-7115
[3] Materials Engineer, Labo S.M. inc., Varennes, PQ, J3X 1P7, Canada
[4] Program Manager, Heritage Canals and Engineering Works, Public Works and Government Services Canada, Hull, PQ, K1A 0M5, Canada; phone 819-997-6102

Introduction

The restoration of historic masonry structures must preserve their historic nature as well as ensure their stability in order to keep the structure operational. Cement grouting is a well-known and interesting method to repair and reinforce such structures. However, choosing appropriate grouts is often a critical phase. The condition of the structure and the degree of deterioration determine the type of repair and grout needed (Palardy, 2001).

A proper evaluation of degradation in the structure with non-destructive testing, a good characterisation of cement-based grout before doing field work and monitoring of injection grouting are necessary to avoid problems during injection and guarantee success. Numerous cement-based grouts were developed and formulated at the Université de Sherbrooke over the course of several injection grouting projects.

Origin and type of degradation

Deterioration of masonry structures can be physical, chemical or biological. Cement-based grouts used for the injection of masonry structures have to be characterized using laboratory tests in order to formulate high-performance mixes adapted to the medium that will be treated. Before making tests on grouts, an important step has to be considered: an analysis of the causes for the deterioration. Repairs to masonry structures, whether the structures are built of bricks, stone or concrete, are often carried out without first analyzing the causes for the deterioration. Consolidation and restoration of historic structures initially seem to improve the condition of masonry walls but in many cases ultimately cause further deterioration because the problem was misidentified or the wrong repair techniques and materials were used. Serious damage can result from the interaction between the cement of the grout and the deteriorated material already in the wall (mortar, bricks, stone or concrete).

In order to evaluate deterioration that may be linked to the restoration of a heritage masonry structure, both the original materials used to build the walls and any "new materials" used in making repairs have to be considered. Recognizing the type of chemical deterioration that occurred in a masonry structure is an important prerequisite to choosing repair materials. This step will help prevent the chemical reactions that caused the deterioration from occurring again between the existing materials in the structure and the products used to complete the repairs.

Chemical deterioration

With regard to the composition of joint mortars, the development of hydraulic cements made it possible to switch from lime mortars to hydraulic lime mortars (lime mixed with pozzolan) and cement mortars. These mortars are less porous and therefore better able to withstand the effects of water. However, using mortar rich in binding agents, which are harder and more watertight than stone, can create wet areas along the joints because moisture is not being transferred between the two materials. The result is a concentration of solutions and serious cracking along the edges of the joints (Jeannette, 1992).

Compared with the materials used to build walls, repair materials such as Portland cement and hydraulic lime do not really differ in terms of their mineralogical composition. Both types of materials contain compounds with good hydraulic performance (calcium aluminates and silicates). The hydration products of these compounds lead to reactions with the sulfates and the formation of ettringite and thaumasite. Using cement grout to consolidate the core of a masonry wall can also, if a very large quantity of ettringite or thaumasite forms, compromise the structure's static equilibrium.

Mechanical deterioration

Depending on location, the extent of each type of degradation will vary (Perret et al., 2000). In Canada, winter conditions cause important deterioration related to freeze-thaw action, while there is little biological deterioration because of low humidity rates in our country. The climate in areas along the St. Lawrence Valley is hard on the mortar in masonry buildings. The amount of annual precipitation is relatively high, and the many freeze-thaw cycles, especially in mid-winter after a period of rain, are particularly harmful to structures. The climate along Canada's east coast is similar but milder. Central Canada is colder but receives less precipitation. Usually, the deterioration we must deal with in this country is primarily deterioration of the mortar joints in masonry walls.

Water, whether in the form of humidity or heavy rain, seeps into masonry walls by whatever means it can: porosity of the mortar, open cracks, etc. Water erodes the components of the wall, especially the mortar in the joints. This type of deterioration is often seen in fortifications where water is able to seep into the structure from the embankments behind the ramparts.

By the same token, upward leaching through masonry walls from the ground, the foundations or adjacent embankments also causes deterioration. Even more importantly, these solutions often carry salts through the structure (Jeannette, 1992). These salts, which vary depending on the composition and location of the soil, can have a natural or an artificial source. For example, the salts in de-icing agents are very harmful to masonry.

The presence of water that has travelled through a masonry wall to the interior of a structure can lead to biological deterioration and, particularly in buildings in Canada, deterioration attributable to freeze-thaw cycles. Frost splitting occurs when ice expands beyond the space occupied by the solutions with which the stone or mortar is saturated. Not all solutions freeze when ice forms; the result is that solids and liquids occur simultaneously. When the ice expands, it squeezes the solutions and causes serious cracking of the materials. Saturation is therefore a crucial factor: when a wall is mostly unsaturated, ice expands into the empty pores; in a saturated wall, however, it affects the water (Jeannette, 1992).

When mortar becomes damp, freezing can create significant hydraulic pressure that causes the mortar to crack. Cracking can be reduced by adding air pockets to the mortar: when the water freezes, the pockets act as expansion chambers and dissipate the hydraulic pressure.

In the early 1990s, a study of the deterioration of mortar joints conducted at a number of national historic sites in Canada confirmed that many types of mortar have been used to repair masonry joints (Fontaine et al., 1998). Cement mortars were commonly used in the 1950s and 1960s, and some are still in place and doing their job. Standard practice, however, is for them to be systematically removed from the joints because their high density does not help ensure that masonry structures last a long time. During the 1970s and 1980s, conventional lime putty mixtures were used; however, they disintegrated five to ten years after they were installed. The teams that maintain Canada's heritage buildings are now going back to bagged mortars even though their adhesion often diminishes within a few years.

Characterization methods for cement-based grouts

According to Ullrich and Maus (1989), no universal standard can be established for heritage structures; rules and recommendations can be made and applied, however, adjusting them to specific conditions as needed. Grouting can be used to:
- rebond to the stone loose mortar that needs reinforcing;
- fill cracks and cavities inside walls;
- increase the bearing capacity of the original masonry;
- create areas that can withstand heavier forces (new construction);
- tie anchors to the masonry or prevent anchors from corroding.

The properties of grout can be divided into two categories (Beall, 1993): properties of the material (workability, water retention, initial slump, slump after water suction, etc.) and properties of the hardened material (bond, durability, deformability and compressive strength).

The first step in identifying the properties of fresh grout is determining the formulation. The ingredients and the final product must be compatible with the old materials in the masonry structure being repaired. There is no test available for this parameter; the chemical and mineralogical properties of the components have to be identified, and an effort made to prevent any negative interaction.

The workability of grout is determined by a series of rheological tests: cohesion, density, outflow, plasticity and viscosity (Beall, 1993), currently used by the research group of the Université de Sherbrooke. Injection grout is evaluated not only in terms of its fluid consistency and homogeneity, but also its injectability and penetration. Table 1 lists the tests conducted to determine the properties of grout used to repair masonry. Table 2 describes the tests that can be done on hardened grouts.

Binding agents in grouts

Physically, cement materials must be chosen on the basis of the properties of the materials used in building the masonry structure, the amount of deterioration and, more specifically, the particle-size distribution of the deteriorated mortar (Binda et al., 1997). In extreme cases, where the deterioration has produced a fine to medium, relatively dense material with few cavities, the quality of grouting will be determined by the size of the particles in the cement that is used. Microfine cements may be required in those situations. The chosen grouts must also be very fluid so that they properly penetrate the structure, and durable when hard so that there is no new deterioration. The mixtures must therefore be stable so that there is no sedimentation or filtration under pressure. For all these reasons, the cement grouts and mortars used in recent years to repair structures have contained chemical and/or mineral additives that improve their fluidity, penetrability and durability (Perret et al, 2000).

Table 1. Fresh grout tests

Parameters	Standards	Description
Temperature	No standard	---
Density	Recommendation from the American Petroleum Institute	Baroid mud scale
Fluidity	**Derived** from ASTM D4016	Rotary viscosimeter (measurement of viscosity, shear threshold)
Outflow time	**Derived** from ASTM C939	Flow through the tip of a Marsh cone of given dimensions
Spread	**Derived** from ASTM C143	Spread on a plate of a given volume of grout put in a mini-cone
Cohesion	No standard: Lombardi plate test (Lombardi, 1985)	Measurement of the mass of the grout that adheres to a plate immersed in the grout
Static bleeding	**Derived** from ASTM C940	Measurement of the quantity of water that bleeds onto the surface of a given volume of grout
Forced bleeding	**Derived** from ASTM D5891-95 and the API (1991)	Measurement of the quantity of water that bleeds out when a given amount of pressure is applied
Washout	No standard	Washout of a volume of grout by an equal volume of water
Injectability	NF P 18-891	Ability of grout to pass through a column of sand of given particle size
Setting time	Derived from ASTM C953	Measurement of the penetration of a Vicat needle

Table 2. Hardened grout tests

Parameters	Standards	Description
Adhesion	**Derived** from ASTM C952	Grout-masonry unit bond test
Compressive strength	**Derived** from ASTM C942	Compressive strength tests of 50 mm cubes
Durability	ASTM C267	Evaluation of the effect of site conditions on grout properties (mass of sample, appearance, compressive strength)
Absorption	ASTM C1403	Evaluation of the percentage mass absorption of water by a grout sample. Deduction of an approximate value of the porosity of the sample

In most grouting projects and laboratory tests conducted by the Université de Sherbrooke, Type I, Type III and blast furnace slag-based microfine cements were used. Grouts containing semi-hydraulic lime were also considered and characterized in the laboratory because their properties are similar to the original mortar: solid yet flexible, and capable of absorbing moisture but also quick-drying (Perret et al., 2000). No lime-based grouts have been injected during field work yet. The properties of the different binding agents are presented in Table 3.

Characterization and optimization of grout formulations have shown that well-formulated mixtures contribute to quality restoration work (Palardy et al., 2000; Perret et al., 2002). Such restoration work and laboratory studies of lime grouts are presented in the next pages.

Table 3. Characteristics of binding agents (according to manufacturers)

Agent	Base	Particle size	Specific surface (m^2/kg)
Type I	Portland	$D_{max} > 100$ μm	225 to 320
Type III	Portland	$D_{max} = 100$ μm	325 to 420
Microfine	Slag	$D_{max} = 100$ μm	1,000
Semi-hydraulic lime	Lime		1,100

Case study no. 1: Laboratory formulation of lime grout

Rheological tests were conducted at the Université de Sherbrooke on grouts made from Type I, Type III and microfine slag cements. In each case, a given proportion of cement was replaced with semi-hydraulic lime. The primary objective was to gauge the impact of the added lime on the fluidity and stability of the mixtures. The three cements have different specific surfaces, making it possible to evaluate their penetrability into relatively permeable areas or fairly wide cracks. The impact of the added lime on the injectability of the mixtures was also studied.

In successive series of tests, 20%, 40% and 60% of each type of cement was replaced with lime. As expected, adding lime reduced the compressive strength of the hardened grouts and increased their tendency to absorb water. However, the added lime also reduced the fluidity of the mixtures and therefore their ability to penetrate into fine soils or fine cracks. Table 4 presents results obtained with grout containing only 20% lime. Descriptions of tests are provided in table 1.

Table 4. Properties of fresh grouts

Binding type	I	I + SHL	III	III + SHL	MF	MF + SHL
Cement (% mass)	100	80	100	80	100	80
Semi-hydraulic lime (% mass)	0	20	0	20	0	20
W/B (water/binding agent ratio)	0.8	0.8	0.8	0.8	0.8	0.8
Superplasticizer (%, mass of solids relative to binding agent)	1.2	1.2	1.2	1.2	1.2	1.4
Marsh cone outflow (700 ml) (s)	23	24	33	23	25	27
Mini-cone slump (mm)	200	200	ND	186	187	185
Washout mass loss (%)	9.0	18	9.5	17	14	14
Static bleeding after 2 h (%)	28	2	0	0	0	0
Forced bleeding after 10 min. (%)	17	23	18	37	20	37
Apparent viscosity at 300 rpm (Pa.s)	0.009	0.017	0.011	0.018	0.016	0.022
Shear threshold (Pa)	0.5	2	0.5	2.0	0.5	1.0
Injected soil height (mm)	360	270	360	270	360	180

A comparison is made between reference grouts such as type I (I), type III (III), microfine cement (MF) and a mixture of these cements and semi-hydraulic lime (SHL). The static stability of grouts was reduced by the introduction of lime but a problem appeared with forced bleeding that increased with the introduction of semi-hydraulic lime. An important increase of washout mass loss was obtained with the introduction of lime in type I and type III cement grouts. This increase did not appear with the addition of lime in microfine cement grout; washout mass loss was already high for the reference microfine grout. A study concerning the addition of anti-washout admixture in grouts containing lime can be of interest. However, it is important to consider that lime grouts should not be injected in saturated media: the setting of such grout must be done in the presence of air.

The injectability of grouts was tested using columns 40 mm in diameter and 360 mm long. The columns were filled with a granular soil with a granulometric distribution ranging from 0.008 μm to 10 mm. This soil simulated a deteriorated mortar with an hydraulic conductivity about 10^{-3} m/s. Adding lime tend to decrease the penetration of the mixtures into the granular soil. This tendency is directly related to the increase in viscosity attributable to the addition of lime. The lime grouts showed good performances even if they did not go through the total height of the columns. These are the first formulations tested in our laboratory and other studies will be done to evaluate other types of superplasticizer and other chemical admixtures.

Case study no. 2: Restoration of a masonry bridge pier

The acoustic tomography method and cement-based grouts resistant to washout were used to repair a pier of a railway bridge built in 1887 in Montreal. Deterioration on the outer section of the pier consisted mainly of:
-   the erosion of jointing mortar on the upstream side in the tidal range,
-   the scaling of repointing mortar caused by repetitive freezing and thawing cycles,
-   the presence of cracks in the masonry stone of approximately 0.5 mm in width.

Acoustic tomography of the internal mass of the pier revealed the presence of weak zones.

Portland cement and a microfine cement-based grout were optimised in the laboratory and injected into the pier core and the outer surfaces of the pier (Perret et al., 2002). Large volumes of grout were injected locally; fluidity and stability performance criteria were reached owing to the addition of antiwashout agent. For the different grouts tested, water-to-cement ratios were between 0.4 and 0.7. Superplasticizer content was about 1.0 to 1.7% (percentage of dry solids relative to the mass of cement) and only 0.015 to 0.03% (percentage of dry solids relative to the mass of cement) of antiwashout agent was used.

The post-injection quality assessment was performed using destructive and non-destructive tests. Drilling tests and tensile bond strength tests were done on cored samples of jointed rocks; good tensile strengths ranging from 0.05 to 0.5 MPa were obtained with the new bonding of blocks.

Acoustic tomography was also used as a non-destructive technique to evaluate the quality of treatment. Four sections were evaluated by acoustic tomography within the pier. Injection grouting improved the overall condition of the structure: average elastic wave velocities increased from 2,000 to 4,000 m/s (Rhazi et al., 2000). The correlation between quality improvement and injected volumes of grout are presented in Figure 1.

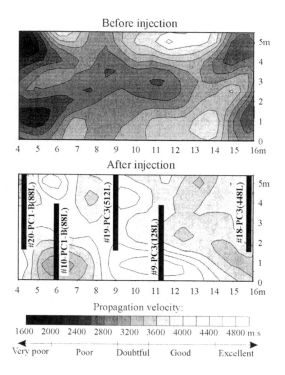

Figure 1. Acoustic tomography survey on the masonry bridge pier: (a) before and (b) after grouting (Perret et al., 2002)

Case study no. 3: Grouting and dismantling of the breast wall of a lock

Over the last 10 years, the research group at the Université de Sherbrooke carried out several lock rehabilitation projects on Rideau Canal, Ontario (Ballivy et al., 1997). Simulations of cement-based grout penetration into deteriorated historic hydraulic masonry structures were performed. During restoration work, the breast wall of the lock located in Upper Beveridge, Ontario, was injected and dismantled to observe grout propagation.

Microfine cement-based grouts were formulated with superplasticizer for the grouting of six vertical boreholes. One half of the wall was grouted with these mixtures containing an antiwashout agent, while the other part of the wall did not contain such an agent (Palardy et al., 2000). Coloured powder was added to the different grouts for easier identification after dismantling of the breast wall.

Fluid and stable grouts were optimised and the formulations are presented in Table 5. The high penetrability of these mixtures allowed the grouting of cracks 5 to 6 m in long and 1 mm thick.

The dismantling of the breast wall provided a great opportunity to observe the propagation of grouts and their effectiveness. Even if grouting pressure and injected volumes of grout are recorded during the injection process, a part of the injected material can be lost through preferential route out of the zone treated. As this problem is impossible to detect, acoustic tomography is an interesting technique which must be applied before grouting. This technique can give information about the presence of large voids and, consequently an estimation of volumes of grout needed (Perret et al, 2001).

Table 5. Formulation of microfine cement grouts – Upper Beveridge lock, Ontario (Palardy et al., 2000)

Formulation[1]	Field characterisation[2]	Grouting of
W/C : 1.2 SP : 0.7% AW : 0.1%	MC: 33 s Density: 1.4 T: 12° C	Rock foundation
W/C : 0.8 AW : 1.2%	MC: 29 s Density: 1.52 T: 18° C	Breast wall
W/C : 1.2 SP : 1.0% AW : 0.1%	MC: 28 s Density: 1.43 T: NA	Breast wall

[1] W/C : water-to-cement ratio   SP : superplasticizer   AW : antiwashout agent
[2] MC: Marsh cone flow time for 700 mL

Conclusion

Rehabilitation of heritage masonry structures involves a good evaluation of the degradation existing in the structure in order to understand the origin of deterioration and optimise a grout formulation adapted to the problems encountered. Non-destructive techniques, such as acoustic tomography survey before grouting, can be used to obtain more information about the degradation of structures. Disintegrated mortars and cracks in masonry structures have to be grouted with special cement-based grouts such as microfine cement-based grouts and mixtures containing mineral or chemical agents to guarantee injectability and stability of mixes. After grouting, drilling tests and water tests can be substituted by a second acoustic tomography survey. This non-destructive method gives good results for post-injection quality assessment.

1252   GROUTING AND GROUND TREATMENT

Acknowledgements

The authors gratefully acknowledge the Natural Sciences and Engineering Research Council of Canada (NSERC) and the Quebec Provincial FCAR fund (Fonds pour la formation de chercheurs et l'aide à la recherche). They also appreciate the collaboration of governmental agencies and private industry participating in field work: Rideau Canal, Parks Canada, Canadian National Railway. The technical assistance of Martin Lizotte and Danick Charbonneau is highly appreciated.

References

American Petroleum Institute (API). (1991). *Drilling fluid materials, Specification 13A, Section 4*, American Petroleum Institute.

Ballivy, G., Mnif, T. & Laporte, R. (1997). "Use of high-performance grout to restore masonry structures." *Proc. of the 1st Conf. of the ASCE Geo-Institute "Grouting: Compaction, Remediation and Testing"*, GSP No. 66, ASCE, New York, 289-305.

Ballivy, G., Perret, S., Rhazi, J., Palardy, D., Laporte, R., & Gagnon, E. (2001). "Rehabilitation of hydraulic masonry heritage structures: injection of special cement-based grouts and tomographic control." *7th International Conference STREMAH 2001*, Italie, Bologne, 527-536.

Beall, C. (1993). *Masonry design and detailing for architects, engineers and contractors*, McGraw-Hill, New York.

Binda, L., Modena, C., Bariono, G., & Abbaneo, S. (1997). "Repair and investigation techniques for stone masonry walls." *Construction and Building Materials*, 11(3), 133-142.

Fontaine, L., Thomson, M.L., & Suter, G.T. (1998). "Masonry in Northern Climates: A Prologue." *APT Bulletin*, 29(2), 39-40.

Jeannette, D. (1992). "L'altération des monuments." *Terroirs et monuments de France, Itinéraires de découvertes*, BRGM, France, 355-364.

Lombardi, G. (1985). "The role of cohesion in cement grouting of rock." *15th Congress on Large Dams*, International Commission on Large Dams, Switzerland, III, 235-262.

Palardy, D. (in progress). *Élaboration d'une approche méthodique pour l'injection de coulis en milieux fissurés*. Ph.D. Thesis in Civil Engineering, Université de Sherbrooke, Canada.

Palardy, D., Perret, S., Ballivy, G., & Laporte, R. (2000). "Étude de la pénétrabilité du coulis de ciment dans les structures de maçonnerie dégradées à caractère historique : effet d'un agent colloïdal." *Can. J. Civ. Eng.*, 27, 642-654.

Perret, S. Khayat, K.H., Gagnon, E., & Rhazi, J. (2002). "Use of injection grouts for the repair of a 130-year old masonry structure." *Journal of Bridge Engineering*, ASCE, 7(1), 31-38.

Perret, S., Palardy, D. & Ballivy, G. (2000). *Technical evaluation of cement grouting methods in heritage structures,* internal report, Université de Sherbrooke for Public Works and Government Services Canada.

Rhazi, J., Ballivy, G., Saleh, K., Tremblay, S., Gagnon, E. & Goupil, F. (2000). "Techniques de contrôle des travaux de réparation par injection." *Proc. of the 53rd Canadian Geotechnical Conf.,* 985-992.

Ullrich, M., & Maus H. (1989). "Engineering examination of existing buildings previously repaired." *International Technical Conference*, ICCROM, Italy, 423-430.

Sealing of Dilatation Joints with Polyurethane Resins

Jozef Hulla[1], Peter Slastan[2], Drahomir Janicek[3]

Abstract

Shortly after putting into operation of the Gabčíkovo Danube Project, the navigation locks began to leak. It was necessary to complete and modernise the monitoring system so that this would enable a more precise location of the leaking areas and also monitoring of efficiency of the performed rehabilitation activities. Additional sealing was done in several stages. Joints were sealed with polyurethane resins and the degraded gravel bedrock was sealed with traditional grouting.

Introduction

Sometimes high quality sealing of the civil engineering constructions do not fulfil their main role.

This paper presents an example from the Slovak Republic (Figure 1) describing leaking dilatation joints between the blocks of the navigation locks in Gabčíkovo and successful application of polyurethane resins to control the leaks (Hulla and Hummel 2000).

----

[1]Member of Slovak Society of Civil Engineers, Prof. Eng. DrSc., Slovak University of Technology, Department for Geotechnics, Radlinskeho 11, 81368 Bratislava, Slovak Republic; hulla@svf.stuba.sk; phone 00421259274666; fax 00421252925642.
[2]Member of Slovak Society of Civil Engineers, Eng., CarboTech Slovakia, M.R.Stefanika 32, 01001 Zilina, Slovak Republic; carbotech@za.internet.sk; phone and fax 00421415623281.
[3]Member of Czech Society of Civil Engineers, Eng., CarboTech Bohemia, Lihovarska 10, 71603 Ostrava - Radvanice, Czech Republic; carbotech@carbotech.cz; phone 00420696232801; fax 00420696232994.

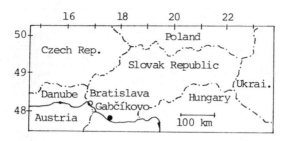

Figure 1. Location of Gabčíkovo navigation locks.

Design of navigation locks and the subsoil

There are two navigation locks in Gabčíkovo (Figure 2). Each of them has a width of 34 m and a length of 275 m and consists of six sections, other two sections form upstream and downstream lock heads (Figure 3). Dilatation joints between the sections are sealed in one layer with shaped rubber. Polystyrene boards were inserted in the joints under and above the rubber sealing to provide the desired width of the joints (25 to 60 mm) that was meant to enable deformation of the rubber sealing (Binder 1999).

   Foundation works on the navigation locks were carried out in a sealed excavation; vertical walls were made of a self-hardening suspension (Hydrostav Company Bratislava), Váhostav Company Žilina was in charge of grouting sealing of the bottom. All sealing elements were of a very high quality; only 0.080 $m^3 s^{-1}$ of water were pumped from the drilled wells in a long time period while the project estimated up to 1.0 $m^3 s^{-1}$ (Hulla et al. 1986).

   The sealing walls and the grouted bottom remained in the ground, removed were only the upper parts of the walls under the downstream lock head and at the emptying channels. This has created exactly defined boundary conditions in the subsoil and in the area around the navigation locks during their operation.

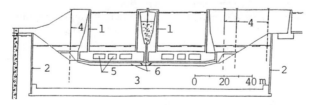

Figure 2. Sectional view through the navigation locks: 1- rubber sealing of dilatation joints, 2 - sealing walls, 3 - grouted bottom, 4 - observation boreholes, 5 - channels, 6 - strip foundations under the dilatation joints.

   Subsoil of the navigation locks is formed by gravel layers ranging down to the a depth of over 300 m, beneath which it is neogene clays. Permeability coefficients of the gravel soils were in the range a 10^{-4} to 10^{-1} m s^{-1} between the

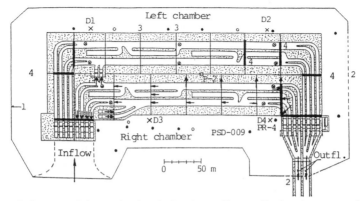

Figure 3. Layout of the navigation locks: 1 - sealing walls, 2 - the removed upper part of the walls, 3 - dilatation joints, 4 - leaking joints positions from the modelling, o - observation boreholes permeable above the foundation joint of the locks, • - observation boreholes permeable under the foundation joint of the locks, D1 through D4 - deformation measuring boreholes, ⊗ - entrance openings to the channels, ↑ - leaking positions after emptying of right lock.

foundation joint and surface of the grouted bottom encompass, without any regularity, and thus they form a very permeable non-homogeneous medium.

Monitoring system

Monitoring system of the water level and flow velocity regimes in the area around the navigation locks was created of boreholes and of several preserved pumping wells that have perforated tubes in various depths above or under the foundations of the navigation locks. Locations of some observation boreholes are shown in Figure 3; the tubes are equipped with sensors enabling continuous monitoring and storing of information on water levels in computer memories.

The levels in the observation boreholes used to raise by 6 m, in extreme cases by 14 m, during the filling of the leaking locks, in dependence on time (Figure 4). Filtration velocities ranged from 10^{-6} up to 10^{-3} m s^{-1} (Figure 5). After the additional sealing of dilatation joints would be fully accomplished the water levels in the observation boreholes should not raise during filling at all. The values of filtration velocities should decrease.

Each section was equipped with at least four control height spots for monitoring of navigation locks settlement using very precise levelling. Maximum values reached 160 mm, of which approximately 130 mm was reached already before putting into operation of the dam. The settlement is of a consolidated nature, it was not settled finally yet, but increases are already insignificant in the presence. In case the leaking areas would not have been repaired, destabilisation processes causing intensive washing out of fine grain particles could develop in the subsoil what would lead to increased settlement.

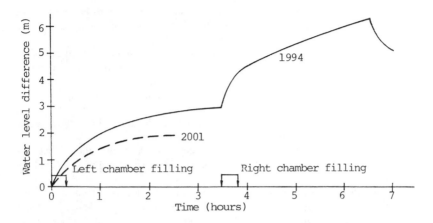

Figure 4. Development of water levels in the observation borehole PDS-009 in the year 1994 (damaged joints) and in the year 2000 (after the first stage of additional sealing).

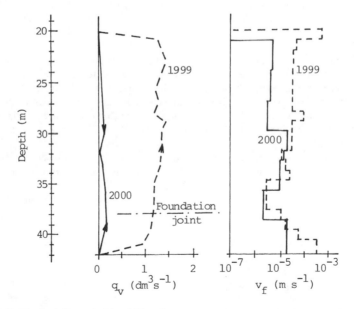

Figure 5. Vertical flows (q_v) and filtration velocities (v_f) in the observation borehole PSD-009 in 1999 (damaged joints) and in the year 2000 (after the first stage of additional sealing).

The need for information on the behaviour of navigation locks during operation is needed to design a comprehensive automated system. Data on shift in longitudinal, transversal, and vertical direction have been obtained using two automated tachoemeters and a net of sensing spots.

Shifts of the navigation locks are negligibly small in the longitudinal direction. Vertical walls move away by 15 mm due to water pressure in the transverse direction when the locks are full. The individual sections are pressed into the gravel subsoil during filling and while the locks are full, and the individual blocks rise up by 7 mm, in some cases, during emptying and while the locks are empty. Water load is constant, however, the quality of the individual concrete blocks and resistance of the gravel soils vary thus causing different vertical deformations.

All sealing and additional sealing systems must allow for differential movement in dilatation joints in transverse as well as in vertical directions without being damaged.

Special deformation measuring boreholes were built in the area around the locks that enable monitoring of height changes in the gravel subsoil. These boreholes are very important for evaluation of stability of the navigation locks. The respective layers are compressed in the course of water level changes and during movement of fine grain particles in the area around the locks and the surface of the site sinks. Information from four deformation measuring objects (D1 to D4 in Figure 3) showed, that height changes in the area under the foundation joint of the navigation locks are small in the present. After perfect removal of leaks, increments of height changes in the subsoil of the navigation locks would be precluded and such processes would be stopped even in higher levels above the foundation joint in the area around the locks.

The level of groundwater changes in the area around the leaking navigation locks and inside the area sealed by cut-off walls and grouted bottom depending on water level in the navigation locks during filling, operation and during emptying of the locks. Water flows through the leaking areas, increases the water level in the vicinity above the water level in the discharge channel, flow velocities increase, and sand particles move in the area around the water discharge from the construction into the gravel soil.

Measures were adopted in order to the minimise the destabilisation processes in the subsoil and in the area in the vicinity immediately after this situation had been discovered. This measures required operation of only one navigation lock at one time and it permitted both navigation locks to be filled at the same time only in exceptional cases.

Requirements for sealing materials and the additional sealing process

The basic requirement for implementation of rehabilitation measures was to enable a regular, only in extreme cases partially limited, operation of the navigation locks for international navigation. Rehabilitation system assumed several stages of sealing of the damaged dilatation joints and grouting of degraded gravel bedrock of the navigation locks as necessary only in a limited extent (Hulla et al. 1998, Mucha and Bansky 1998).

Basic requirements for the sealing materials were as follows: long-term performance, reliable connection with concrete sections, curing and aging under water, the materials may not be fragile, they must enable elastic deformations which commonly occur during operation of navigation locks, they may not harm the water quality.

Polyurethane resins were chosen as the most suitable materials on the basis of various sources of information on solution of similar problems abroad (Cornely, 1988, 2002). The resins were produced by CarboTech and they were imported to Slovakia by CarboTech-Slovakia.

A two-component polyurethane resin Bevedan - Bevedol WFA was used by the Ingmat Company (Faix, 1999) for sealing of the damaged dilatation joints in the deepest areas, under the bottom of the navigation locks, as well as in canals, approximately 15 m under the water level in empty locks. Bevedan is a polyisocyanate - diphenylmethanediisocyanate, Bevedol WFA is a mixture of polyoles and admixtures. In the ratio of 1:1 it reacts and creates a hard, arduous resin. After contacting water, foaming begins within 25 to 65 seconds, the end of foaming is within 45 to 95 seconds (the data apply to temperatures within 10 to 15°C, the periods of foaming are longer under lower temperatures). The degree of foaming is 3 to 10, limiting period with the temperature of 30°C is shorter than 15 minutes. Limiting period is measured from the beginning of mixing of the components, after this period adhesion to the surface is bigger than 1 MPa. Also this period takes longer under lower temperatures. However, polyurethane resins enable to obtain a strength of joints of up to 6 MPa.

Components Bevedan and Bevedol WFA are transported using a special pump to a previously prepared opening after thorough mixing. They form a foam in contact with water but the foamed material is pushed out by subsequently transported grouting mixture, which does not come into contact with water anymore and thus it creates without foaming a hard, impermeable material without pores. This leads to sealing and re-enforcing of the environment filled by the resin.

The Bevedan component is a harmful substance and when working with this product it is necessary to strictly comply with the specified procedures. However the foamed and cured resins do not pollute water environment and some of them have even hygienic certificates for use even in contact with drinking water.

The aforementioned properties have been obtained in laboratory conditions which may significantly vary from the actual conditions, mainly in temperature. There admixtures are known that enable to modify qualities and attributes of polyurethane resins directly on the construction site. However, their use must be supervised by experts.

Sealing work began of the vertical dilatation joints from the surface of side walls up to the water level in the empty locks within the pilot stage already in 1994, shortly after the leaks in the navigation locks were found. There were obvious leaks in almost each dilatation joint of the right wall of the left lock approximately 1 m above the water level in the empty lock, which at that time was at the level of 112 m above the sea level. These works were executed by several companies using various systems. However, most joints were sealed with polyurethane resins. Parts of the dilatation joints between the rubber sealing and wall surfaces were sealed after removal of the filling.

A special fender with semi-circular shape with rubber sealing on the contact surfaces was fabricated in Hydrostav to be implemented in additional sealing works on the vertical joints under the water level in the empty locks. Water had been pumped out from its inner part without major problems and the joints could be repaired in a wet environment.

Efficiency of additional sealing work, carried out within the pilot stage on the vertical dilatation joints, did not bring any major positive changes in the water level and flow velocity regimes in the area around the navigation locks.

Thus the first important stage of additional sealing works in 1999 was focused on the most permeable dilatation joints in the foundation slab, in the area of the intake and oftake canals under the bottom of the locks mainly. Additional sealing works were implemented also in vertical joints between the lower water level and the bottom of the locks on this occasion.

The most permeable locations in dilatation joints were located more precisely by numerical modelling (Mucha and Bansky 1998). Flow modelling have been employed to search for such leaking joint positions that would resembled as closely as possible the real water flow in the area around locks that had been obtained from analysis of water level regime in the observation boreholes. The most permeable joints are marked by thicker lines in Figure 3.

The damaged dilatation joints were characterised by washed out polystyrene and heraclite distance filling. It was necessary to clean them with pressure water in any case. Plates with openings for introduction of tubes through which dilatation joints above and under the damaged rubber sealing were filled by two-component polyurethane resins Bevedan - Bevedol WFA using special pumps under pressure of 15 to 20 MPa were fastened to the surface of the joints.

The pumps, sealing materials, oxygen cylinders, process equipment, TV monitor for control, monitoring and supervising of the diving activities with video-cameras and for continuous audio connection with divers were located on a craft near to the respective access opening. All important information, such as condition of the joint before additional sealing, process of additional sealing, and condition of the joint after additional sealing works were recorded on videotapes.

The performed work was extremely demanding not only because of the fact that the work was carried out 15 m under the water but also by time limitations due to the ship navigation through the neighboring lock. Most of the activities were carried out in winter months at the end of 1999 under low temperatures which also negatively affected the course of reaction processes of sealing materials.

Regular inspection procedures in the course of implementation of works were supplemented by monitoring of water levels in the navigation locks and in the selected observation boreholes during normal operation of the locks. Final evaluation of the efficiency of the additional sealing works in the first stage was carried out by analyses of water level and flow velocity regimes in observation objects, which are built in the area around the locks in the course of special operation of the navigation locks.

Changes in the water level regime

Differences in water levels of the observation spots increased during filling of the navigation locks due to presence of leaking parts and their particular values depend on the period during which the lock is full, on the cycles of filling and emptying of the locks during normal navigation operation, on differences in water levels of the intake and oftake canals, on the technical conditions of the locks, as well as on other factors.

The Figure 4 presents level development of the PSD-009 borehole in case of a full left navigation lock. The right navigation lock was also filled after 3.5 hours in 1994. The level increased by 6.2 m after 6.5 hours. Due to the additional sealing of the left lock was the borehole level lower by 0.8 m after 2.5 hours in the year 2000 than by the unsealed lock in 1994.

A special operation system of the locks was introduced in order to eliminate at least a part of the aforementioned factors. Both locks were empty for ten 10 hours at first, one lock was full then two hours, and after it has been discharged normal operation of the locks followed. Both locks were empty for 10 hours afterwards, and then the same lock was full two hours. The same procedure was applied also for the second lock. Changes in water levels were monitored in observation boreholes before implementation of additional sealing works and after completion of the first stage of additional sealing works using this system.

Tail water effects were difficult to be excluded, respectively the significant influence of the level difference in the inlet and outlet canals. Final evaluation could be made only after correction of the obtained results.

Figure 6 shows average values of differences in water levels in observation boreholes after correction of the data to a unified 20.6 m difference in water levels in the inlet and outlet channel for the right and left lock before, during and after implementation of additional sealing works.

We may conclude that the average values of increased water levels in the observation objects decreased after the first stage from 4.45 to 2.65 m when the left lock was full for two hours, compared to the condition before implementation of additional sealing works

The average value of increased water levels before implementation of additional sealing works (1.9 m) increased after the first stage (to 2.45 m) when the right lock was full for two hours. The situation in the right lock seemed to have worsened at the first sight; however, this was not the case because, due to additional sealing of the left lock, the drainage effect decreased for the right lock, with free water discharge by open gates and emptying canals. This led consequently to an increased water level in gravel soil in the area around the navigation locks during special operation.

Changes in the velocity regime

Gravel soils in the navigation locks subsoil are very heterogeneous. Permeability coefficients range from 10^{-4} to 10^{-1} m s^{-1}. Furthermore, natural

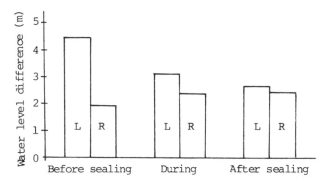

Figure 6. Efficiency of the sealing in the changes of average level differences in the observation boreholes at the 2 hours full left (L) or right (R) lock.

heterogeneity was increased also artificially by spillage of grout slurry into the most permeable locations of gravel soils during grouting of the construction excavation bottom. Water level regime does not necessarily represent an objective mean for characterisation of groundwater flow in such an environment. It was found out in many cases that high water levels appeared in boreholes with slow flow velocities and on the other hand in the boreholes with low water levels there were found high water flow velocities.

Measurements of water levels in the observation boreholes were, therefore, supplemented by measurements of filtration velocities. Filtration velocities were calculated on the basis of changes of parameters of vertical motion of water in the permeable parts of observation boreholes, which resulted from interconnection of different pressure horizons (Halevy et al. 1967).

The measured flow velocities allowed also to evaluate stability of sand particles in the gravel skeleton by comparison with critical velocities during which sand particles begin to move. The existence of such processes was tested also by washing in of sand particles into observation boreholes; the depth decreased in some boreholes almost by 10 m during three years.

The efficiency of additional sealing works was manifested by changes of directions and by decreased filtration velocities. Statistics has shown medians of filtration velocities (Figure 7) for the left lock being full before implementation of additional sealing works (3×10^{-5} m s^{-1}) decreased after implementation of additional sealing works (to 1.7×10^{-5} m s^{-1}), in case of the full right lock the values after implementation of additional sealing works remained the same (1.5×10^{-5} m s^{-1}).

Further research and additional sealing processes

In spite of the positive results brought about by the pilot and the first stages, a better efficiency of the additional sealing works was generally expected. It was, therefore, important to check the quality of the provided work first of all. It was carried out

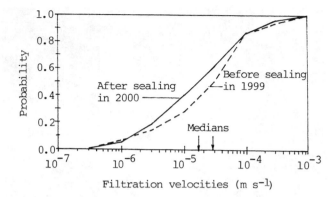

Filtration velocities (m s⁻¹)

Figure 7. Empirical distribution functions - efficiency of sealing works in the changes of velocity regime by a full left lock.

under water in exceptionally difficult winter conditions and many of the workers were not enough experienced.

The additional sealing within the pilot and the first stages, as well as the enlarged filling capacity enabled to empty the right navigation lock almost completely, thus directly examine the dilatation and operating joints and the concrete construction locks by the end of November 2001 (Stoličný and Cábel 2001).

The water level was by 13 m lower on the bottom of the intake and oftake canals compared to the level outlet canal. It was needed to pump approximately 0.7 m^3s^{-1} to keep up this difference in course of normal operation of the unsealed left lock. A part of this amount flew in around the technological locks of the intake and oftake canals. The working joints and concrete constructions enabled practically no water seepage.

It was possible to locate the water inflows through unsealed areas in the dilatation joints precisely (as outlined in the Figure 3). The majority of the concentrated inflows occurred at the side walls of the edge canals, one of them on photograph presented in the Figure 8. Concentrated water inflows through horizontal dilatation joint were less intensive what is a very important information for the stability of the navigation lock subsoil. Hydrodynamic destabilisation processes of the gravel soils have shown especially on the side walls of the navigation locks.

Three dilatation joints, that have been additionally sealed within the first stage, were able to be checked after the water outflow from the right navigation lock. It has shown that the extremely difficult working conditions in a constrained surrounding 15 m under the water level did not allow a final seal up. The partial additional sealing has to be also appreciated because it created conditions for a complete water discharge. Water will be able to be pumped out from both locks gradually and the polyurethane resins would finally seal the dilatation joints damaged rubber sealing in dry conditions in the second stage. State of the gravel subsoil under the navigation locks can be examined at the same time, and if needed additionally

Figure 8. Water inflow into the canal through the damaged dilatation joint sealing in the left wall after the right lock was pumped out (Stoličný and Cábel, 2001).

grouted by a cement mixture. The water levels should not change their positions in the observation boreholes in the surroundings after the second stage and the surface settlement should be also stopped in the area close to the locks.

Conclusions

Large quantities of water flowed into the vicinity gravel environment during the navigational locks operation through the dilatation joints. Intensive hydrodynamic effects washed out sand particles thus endangering the lock stability.

Polyurethane resins were used for additional sealing works in several stages. Their effectiveness was controlled through changes in the water level and velocity regimes in the observation boreholes in the area around.

Vertical joints were sealed down to the water level in the outlet canal without any evident changes in the water level and velocity regimes in the pilot stage; water from the intake and oftake canals under the bottom of the locks could not have been pump.

Vertical joints that had occurred between the level in the outlet canal and bottom of the navigation locks were sealed in the second stage, as well as selected joints in the canals. These demanding job was done by divers under the water level. Effectiveness of the provided work has shown in the positive changes of water level and velocity regimes, and enabled a complete water pumping from the right navigation lock, examine the dilatation joints and locate the leakage areas precisely. A similar situation would be found probably also in the left navigation lock.

Gradual discharge and full sealing of the damaged joints in both locks would take place along with additional inspecting and eventual grouting of the gravel soils under the bottoms of the locks, if needed, in the third stage. Effectiveness of the work would easily be checked and monitored in the long term by the water level regime in the observation boreholes that should not react to the operation of the navigation locks.

The authors of this paper would like to express their thanks to the responsible staff of the Water Management Constructions Enterprise Bratislava and River Danube Basin Gabčíkovo and acknowledge them for providing information, as well as to the Scientific Grant Agency of the Slovak Republic for the support in the 1/9066/02 Grant Project.

References

Binder, J. (1999): "Untightnesses of Gabčíkovo navigation locks." *Inžinierske stavby,* SSCE, 47(4), 114-117.

Cornely, W. (1988): "Polyurethane grouts for waterstop in tunnels and tunnelling." *International Congress on Tunnels and Water,* Madrid, 413-418.

Cornely, W. (2002): "Sealing with Carbotech Fosroc synthetic resins." *Reinforcement, sealing and anchoring of rock massif and building structures 2002,* VSB-TU, Ostrava, 6-15.

Faix, D. (1999): "The Gabčíkovo dam - rehabilitation of the foundation slab of navigation locks." *Eurostav,* 5(8), 47-48.

Halevy, E. et al. (1967): "Borehole dilution techniques: a critical review." *Isotopes in Hydrology,* IAEA, Vienna, 531-564.

Hulla, J. et al. (1986): "Deep foundation pits for the Gabčíkovo hydropower plant." *8th Danube-European Conference on Soil Mechanics and Foundation Engineering,* DGEG, Essen, 113-118.

Hulla, J. et al. (1998): *Proposal of additional sealing works in Gabčíkovo,* Slovak University of Technology, Report, Bratislava.

Hulla, J. and Hummel, J. (2000): "Original state and additional sealing of dilatation joints of navigation locks with polyurethane resins." *Inžinierske stavby,* SSCE, 48 (7-8), 218-225.

Mucha, I. and Banský, L. (1998): *Modeling of water flows in the foundation zone of the Gabčíkovo navigation locks,* Consulting group for groundwater, Report, Bratislava.

Stoličný, J. and Cábel, J. (2001): *Evaluation of the Pumping of Water from the Right Navigation Lock Gabčíkovo in the Days from 19th till 24th November 2001,* Watereconomy Construction, Report, Bratislava.

Irreversible Changes in the Grouting Industry Caused by Polyurethane Grouting: An overview of 30 years of polyurethane grouting

Alex Naudts[1]

Abstract

Water reactive polyurethane grouts were introduced into the grouting industry during the late sixties by the Takenaka company in Japan under the trade name TACSS. It became possible to inject "one component" grouts without potlife that do not easily wash out and react with the ground water. Because of environmental scrutiny, the first series of TACCS were replaced by solvent-free, hydrophobic, MDI based polyurethane prepolymers. Whilst remarkable successes were booked in mining and geotechnical engineering projects, more and more these products were used for permanent seepage control for sealing concrete structures. Hydrophilic polyurethanes were also introduced in Japan predominantly for the latter application. They contained solvents and were TDI based. Their high reactivity and high dilution ratio with water made them attractive to practitioners.

In 1980, the N.V. DeNeef Chemie obtained the exclusive rights for TACSS for most places on earth and the successes in stopping major leaks in tunnels changed classic grouting (sodium silicate cement combinations) and seepage control grouting (acrylamide grouting) because of practical and environmental considerations. After the N.V. Denys brought similar products to the industry in 1980, more manufacturers jumped on the bandwagon. By the mid-eighties there were more than 10 manufacturers of polyurethane grouts. Several new and improved hydrophobic water reactive urethanes were developed during the eighties as a result of this new trend. A few manufacturers created closed cell, water reactive hydrophobic polyurethanes. The era of custom-made formulations, tailored to the project, started.

Water reactive hydrophilic polyurethanes came under close scrutiny because of longevity problems. The classic two-component polyurethane foams, used in mining were gradually introduced in geotechnical engineering. For permanent seepage control, in concrete structures two-component polyurethane elastomers became popular. The introduction of hydro-block in France for major inflow control was another remarkable development. Extensive research was performed, especially in Scandinavia to establish life time expectancy of hydrophobic water reactive polyurethane. Pioneering research was done to establish mathematical models to understand the flow of P.U. through fine fissures. This paper focuses on the engineering aspects of polyurethane grouting with in the background the history of these fascinating products. It elaborates on the various types of applications illustrated with case histories for each type.

[1] President, ECO Grouting Specialists Ltd., Grand Valley, ON. L0N 1G0. Tel. (519) 928-5949; anaudts@ecogrout.com

Background

Polyurethane grouts are probably the most popular type of solution grouts. They have been used for more than 35 years and have contributed to the democratization of the grouting industry.

The term "solution grouts" pertains to grouts that behave like Newtonian fluids. Contrary to suspensions grouts, which behave like Binghamian fluids, solution grouts do not contain particles. The term "chemical grouts" is misleading, since cement-based suspension grouts are complex chemical grouts too. Solution grouts are injectable into very fine apertures, not accessible to (even microfine) suspension grouts.

Scientific analysis of polyurethane grouting has been under-represented. For too long polyurethane grouting has been considered the business of "small time operators" only and substantially ignored by the larger practitioners. Polyurethane grouting was once the trade of the "finesse grouters" (predominantly operating in seepage control work). They in turn, have become a major force in the grouting industry.

Polyurethane grouts encompass a large family of solution grouts. It is therefore very difficult to make generic statements for the entire group. Most polyurethane grouts are non-evolutive or true solution grouts. A true solution grout is characterised by a flat viscosity curve, followed by a sudden increase in viscosity, immediately prior to gelation or curing. Acrylamide, acrylate and most polyurethane based grouts are typical examples of true solution grouts.

The following categories of polyurethane grouts are, in general, recognised by the industry:
- Water reactive polyurethane grouts
- Two component foaming grouts (polyol-isocyanides combination)
- Two component polyurethane elastomers

Water reactive polyurethane grouts in turn have two main sub-categories:
- hydrophobic polyurethane prepolymer grouts
- hydrophilic polyurethane prepolymer grouts

The two aforementioned sub-categories and the two other categories are further divided in sub-families of polyurethane grouts.

The sensitivity (i.e. the ability to control factors influencing the reaction or curing pattern during the reaction) of most polyurethane grouts is typically lower than the sensitivity of cement-based suspension grouts.

The longevity of the end product and its chemical resistance are typically far superior than cement-based suspension grouts (Naudts, 1990). The toxicity of most of the polyurethane grouts is usually not an issue and approval for use in potable water applications has been obtained for several polyurethane categories of grouts.

Solution grouts have made it possible to treat media with a measurable permeability, whether wood, glass, masonry, concrete, rock or soils. Because of some irresponsible use of some solution grouts (especially from an environmental standpoint) and their selection for ill-conceived applications, solution grouts have generated some bad publicity in the past. Environmental issues are usually very complex, and often quite controversial, hence solution grouts, unjustifiably have received a bad reputation. Water reactive hydrophobic polyurethane grouts are now commonly used. They are successfully used for seepage control in concrete structures, for soil and rock grouting under flow conditions in mining and for environmental applications.

Polyurethane Grouts: Overview

Water Reactive Polyurethanes: Polyurethane prepolymer grouts have one thing in common: they react with the in-situ available (ground) water to create a foam or gel that is either hydrophobic or hydrophilic. They are "one component" products using "the enemy", the water, as a reaction partner to create the end- product. The catalyst (a tertiary amine) is not considered a component since it only affects the rate and the direction of the polymer forming process (Hepburn, 1992). Adding more catalyst, only speeds up the gelation process. Surfactants are added to the resin to prevent collapse of the foam and to create small uniform cells.

Water reactive polyurethane resins are classified into two sub-categories:
- Hydrophobic polyurethane resins: they react with water but repel it after the final (cured) product has been formed.
- Hydrophilic polyurethane resins react with water but continue to physically absorb it after the chemical reaction has been completed.

During the exothermic reaction, the hydrophobic polyurethanes expand and penetrate pervious media: fine cracks (as narrow as 8 micron) and soils (with a permeability coefficient as low as 10^{-4} cm/s. The penetration is greatly enhanced by the formation of CO2, independent from the grouting pressure. Hence the name: "active" grouts. Their penetrability is determined by their viscosity and reaction time. Credible research work on the penetration rate of hydrophobic water reactive polyurethane into fine fissures has been conducted by Andersen (1998). Andersen' s work reveals that the penetration of fine apertures is a very slow process, which requires multiple hole grouting to obtain a economically justifiable and technically sound grouting operation. The gel time is linked with a rapid increase in the viscosity. Prior to gelation, the viscosity of the grout decreases because of the formation of CO_2. During gelation, the viscosity increases substantially and prevents further penetration.

Hydrophilic polyurethanes form either a gel or foam during the reaction, depending on the type and/or the amount of water they are being mixed with. Not all hydrophilic polyurethane grouts expand during their initial reaction. The hydrophilic polyurethanes, especially in North America, often have been utilised to fill the gap in the market left by AM9, when production of this product was "officially" terminated (?) in 1978. There is, however, a serious problem with rapid deterioration and water absorption by the hydrophilic polyurethanes after reaction.

Water-reactive prepolymers are high molecular grouting materials. They are primarily produced by mixing a polyol with an excessive amount of poly-isocyanides to form a low prepolymeric compound, containing some free OCN groups. The injection resin is composed of this prepolymer, plasticizer, diluting agents, surfactants and the amine catalyst.

The mechanism of reaction among the isocyanides, polyol and other components is rather complicated. In simple terms the following happens:

1) The reaction between the isocyanides and the polyol yields a pre-polyurethane.
2) The reaction of poly-isocyanides with water liberates carbon dioxide and urea derivatives.
3) The reaction of poly-isocyanides with ureido develops molecular links and high molecular formation.

These reactions occur because of the existence of the free OCN groups in the grout, which can react with the compounds containing active hydrogen atoms, such as hydroxy, water, amino and ureido. The hydrogen atoms move to link up with the nitrogen atoms of the poly-isocyanides and form high molecular polymers.

Subcategory 1: Hydrophobic "semi-rigid" and "rigid" Hydrophobic Water Reactive Prepolymers: They are mainly used for blocking water inflows and soil grouting in the presence of moving ground water or when high strengths are required, when excellent chemical resistance is required in environmentally sensitive areas or when durability is an issue.

At the turn of the century there were approximately 20 manufacturers involved in the manufacturing of these grouts. The oldest products on the market were the TACSS series, manufactured by Takenaka in Japan. The DeNeef company deserves credit for making polyurethane grouting popular in a great deal of countries.

These products react with the in-situ water and expand during the exothermic chemical reaction, releasing carbon dioxide. They are totally stable after reaction, but have limited flexibility. The reaction time can typically be adjusted between 45 seconds up to one hour by adding a tertiary amine-based catalyst.

Figure 1 illustrates a typical reaction curve for water reactive hydrophobic polyurethane prepolymers. First there is an induction time during which the viscosity of the water/P.U mix remains constant followed by the reaction time, during which the foaming starts. The latter is associated with a decrease in viscosity followed by a very rapid increase in viscosity prior to final gelation. The reaction pattern of hydrophobic polyurethanes displays a shotgun reaction pattern.

Heating the grout (max. 60°C - and never directly applying the heat to the grout-containers) speeds up the reaction. If shorter set times are required, a resident tube is used: water or cement-based suspension grout is injected via separate lines into a resident tube, in conjunction with the prepolymer as indicated in Figure 2.

This way, the grout has already passed the "induction phase" of the chemical reaction, when introduced into the formation or the water-flow. By providing a grouting set-up, in which the "water introduction point", can be continually altered (mobile feeder-pipe through stuffing box connected to the resident tube), the resident time can be continually changed, and hence the reaction time. Grout wash-out can be eliminated under most circumstances.

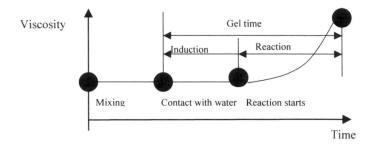

Figure 1. Typical reaction curve for water reactive hydrophobic polyurethane prepolymers

The water cut-off series of prepolymers has been developed for stopping major inflows and seepage control. They are most suitable for the rather crude applications when water has to be stopped, but not necessarily "to the last drop". As a result, they are ideal trouble shooting products for stopping or controlling major inflows in geotechnical applications. Grouted soil or cured grout, has a residual permeability in the 10^{-4} to 10^{-6} cm/sec range, which will allow minor seepage to take place.

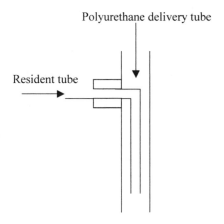

Figure 2. Resident tube

Products within the same family of hydrophobic polyurethane differ due to: viscosity, miscibility with water, purity of the MDI, reactivity at elevated pressures, toxicity (solvent content and type of isocyanides), cellular structure of the freely foamed PU. and residual hydraulic conductivity of grouted medium.

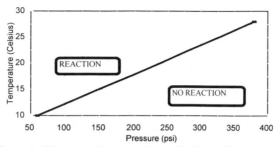

Figure 3. Water reactive prepolymer (for Tacss 20)

Several manufacturers have obtained potable water approvals, for their water-reactive hydrophobic polyurethane prepolymers.

In order to start the reaction, there is a minimum enthalpy required, which is higher, as the pressure is higher. This means that, at a given pressure, if the enthalpy is too low (temperature too low) the reaction does not start unless the products are mixed thoroughly with the water. Naudts (1986) established a reaction-diagram for some of the hydrophobic polyurethane grouts (Figure 3.)

Above the "reaction" line the reaction takes place, and below this line reaction does not occur unless mechanical agitation of "in situ water" and the polyurethane grout is provided. The (turbulent) flow of the urethane-water mixture through cracks and channels often provides for enough in-situ mixing. Some of the more recently developed prepolymers react regardless of pressure and groundwater temperature.

During grouting, the carbon dioxide, generated during the chemical reaction, will generate additional pressure, as the grout flows through the cracks and pore channels, pushing the grout into very fine cracks and crevices. The reaction pressure in totally confined circumstances is over 2.5 MPa (350 psi) as was determined in the "bomb test" by Naudts (1986). Other researchers (Tjugum 1991; Burgstaller 1989) indicate lower values, mainly because polyurethane was given the chance to expand during their experiments, compressing small quantities of air in the test chamber. The formation is, in reality, never completely sealed and the expansion pressures are in the order of 0.5 to 1 MPa (70 -150 psi).

Where it is absolutely necessary to control the reaction times it can be achieved by injecting the prepolymer as a two component grout, with water or cement-based suspension grout being the second component, separately introduced into the manifold. By using the header-pipe as resident pipe, and selecting an induction period, a little longer than the resident time, it is possible to control an inflow. This is a fairly sensitive operation as the flow pattern in the structure to be sealed is continually changing during the grouting operation.

Only MDI-based prepolymers should be used. The vapour pressure (at ambient temperature) of the solvent free MDI based products is low. However, ventilation is still required when working in enclosed spaces in order not to exceed the maximum T.L.V. limits of 100 ppb (8 hour period), as determined by the International Institute for

Isocyanides.
The long term durability of grouts has been researched via accelerated ageing tests. Credible work has been done by the Swedish National Testing Institute (Jakubowicz, 1992). Accelerated ageing tests by Naudts (1990) revealed that in highly acidic environments; polyurethane is anticipated not to break down more than 10% over a 100 year time period. In an alkalic environment (such as in concrete or in limestone formations), a weight loss of 35% has been projected over a 45 year period because of hydrolytic degradation of the cured polyurethane. Soils grouted with polyurethane, according to Takenaka (1976) and research work by Oshita, et al. (1991), do not display signs of long-term degradation.

Please note that even properly formulated cement-based suspension grouts do not perform nearly as well as water reactive polyurethane grouts in an accelerated ageing test, even under "normal" conditions.

The unconfined compressive strengths (UCS) of soils grouted with polyurethane prepolymers strongly vary from type to type. The "rigid" types produce a much higher strength than the semi-rigid types. Years ago, solvents were used to reduce the viscosity of the "rigid" polyurethanes to enhance penetrability; this is no longer acceptable for environmental reasons. The grain size distribution of the soil has also a significant impact on the UCS of the grouted soils. The grouting pressures also influence the UCS of the grouted soil and its residual permeability. The water content and the permeability coefficient of the soil play an important role too. In order to avoid unpleasant surprises, lab simulation injection tests such as the ISIS tests as described by Landry, et al. (2000), provides reliable results in predicting the injectability and the characteristics of the grouted soils.

Examples of Field Applications:

Soil Grouting: Containment of toxic spill in overburden after train derailment in Kinkempois (Belgium) 1990.

In order to prevent toxic chemicals from flowing into the river a grout curtain was installed within hours after the spill occurred due to a train derailment. Steel sleeve pipes were installed at close spacing through the silty sand overburden overlaying the bedrock. It is important to use sleeve pipes of good quality to prevent backflush through the sleeves. A fast curing, chemical resistant, "rigid" water reactive polyurethane was selected for this project.

Rock Grouting: Sugai Piah Hydroelectric Project (Malaysia) 1993.

A grout curtain through fractured granodiorite and a gouge filled fault zone was installed under flow conditions (20 litres/second) and high pressure (1- 3 MPa). Water reactive polyurethane was injected in conjunction with cement-based grout to form a grout curtain with a residual permeability of less than 0.1 Lugeon. The grout curtain extended radially 20 meters around a concrete plug, near the bottom of a drop shaft. Multiple hole grouting via multiple port sleeve pipes using secondary and tertiary and quaternary grout holes were utilized. Balanced, stable cement-based grouts were used in upstream holes at the same time as the polyurethane grouts were injected in the downstream holes. The permeability of the formation was systematically reduced from over 300 Lugeon to less than 0.1 Lugeon without suffering washout of grout.

Sealing soil piping conditions through joints in deep tunnels, and coffer dams below the water-table

Most subways in Western Europe consist of multiple level tunnels, as deep as 30 metres (100') below the groundwater table in fine granular soils. Failures in joints have devastating consequences.

Failure in joint between slurry-wall panels near the "Midi Tower" in Brussels (Central station) - 1982.

Several tunnels in the Brussels Subway were constructed using the slurry wall system. During the excavation of the second underground level, a leaking joint suddenly gave away during the night, causing a major inflow of water and fine sand into the tunnel. A massive sinkhole was formed at the Midi Square immediately beside a 45 story building. Bentonite inclusions formed in the joints between panels of the slurry walls suddenly blew out.

Soil drifting into the tunnel disturbed a large area, endangering the foundation system of the Midi-Tower. Within hours an emergency grouting program was launched. It involved prepolymer drenched pads and wedges blocking the soil-inflow, within the hour. The water inflow was stopped within eight hours via horizontal sleeve pipes installed near the failed joint. A soil grouting program (using vertical sleeve-pipes) involved cement-based suspension grouts, sodium silicates and polyurethane prepolymers, to create a soil conglomerate on the outside of the joint, to full depth, successfully protecting the tunnel joint against future failures, and allowing a safe structural rehabilitation program from inside the tunnel.

Pre-excavation grouting to control or prevent inflows in rock-tunnels and shafts
Francoeur Mine Shaft sinking. Val D'or, Quebec.

In spite of systematically drilling pilot holes, a steeply dipping, water bearing fault zone went undiscovered during a shaft sinking operation at the Francoeur Mine. After blasting a round, a major in-rush of water occurred on one side of the shaft under high pressure (4.5 MPa (630psi)). An emergency bulkhead was created at the bottom of the shaft to channel the flow through relief pipes. Hydrophobic polyurethane in conjunction with stable cement based suspension grout was injected to stop the inflow. The water-bearing feature was systematically grouted with a combination of polyurethane and cement grouts. This approach made it possible to blast within one hour after completing a grouting operation. The shaft sinking was only interrupted for 5 days. This system has been used numerous times to eliminate down-time when driving tunnels through unstable water bearing rock formations.

Tunnel Grouting in Hardrock Tunnels: Emergency situations

In 1987, during the installation of a deep tunnel in Geneva (Switzerland) for CERN (Centre European de Recherche Nucleaire), major infiltration problems were encountered. The 27 km long rock tunnel is located 160 m below the water-table. The tunnel was excavated with a four meter diameter WIRT TBM. At one point, a chalk horizon with interbedded clay seams was intersected. The clay extruded from the seams resulting in a major inflow. A cement grouting program was launched. After injecting 600 metric tons of neat cement grout and cement with sodium silicate in combination, interrupting tunnelling operations for more than 60 days, no reduction in inflow was measurable. The total inflow (aquifer limited) remained steady at 400 litres/second (6,300 gpm). A decision was made to proceed with water reactive hydrophobic polyurethane prepolymers. The total inflow was reduced to 37 litres/second in only two days of grouting to meet design criteria.

Dam and dike grouting to prevent seepage through the abutment or the embankment

An extensive rock grouting program was performed in Montezic (French Alps) in the Digue de Monnes (owned by EDF) in 1988 to stop further leaching of an old grout curtain made using sodium silicates and cementitious grouts. A sleeve pipe grouting program was launched, to restore the integrity of the old dam. The casing grout, surrounding the sleeve pipes clogged some of the minor fractures, which reduced access of the grout to the formation. It is poor grouting practice to use casing grout around sleeve pipes for rock-grouting. Towel packers (or MPSP barriers) would have been much more appropriate (in lieu of the sleeve pipes) to create a grout curtain along the centre-line of the dike. Rock hydro-fracturing was required to access and grout the open cracks (with flowing water) and to eliminate seepage through this dike. The set time of the water reactive grout was selected at 75% of the "flow through" time.

Renovation of Man-Accessible Sewers

This is one of the most common applications for this type of water reactive prepolymer. One of the first rehabilitation projects on man-accessible sewers subject to internal erosion (even soil piping conditions) occurred in Mechelen (Belgium) in 1977. The technique, now applied all over the world, only slowly found its way to North-America. One of the first jobs of this kind in North America has been executed at Niagara Falls (Canada).

Sewer Rehabilitation on Dorchester Road, Niagara Falls, Ontario. (1988/89)

During 1988 and 1989 the pavement of this very busy road collapsed a few times causing some spectacular accidents (vehicles driving into substantial sink holes. A 25-year-old sewer, located at a depth of about 7 metres below the pavement appeared to be the cause of the problem.

Loss of fine soil particles through leaking joints triggered an "internal erosion process" which in turn created voids above the watertable, eventually turning into sinkholes. Remedial measures that were undertaken consisted of the following:

• Installation of a proper ventilation system in the sewer-line.
• Installation of bulkheads and pumping the incoming sewage to the next manhole.
• Cleaning of the section to be repaired.
• Drilling of at least four injection packers in each joint as pressure relief holes and sealing the zone between the packers with hydraulic cement or with pads drenched in prepolymer.
• Grouting of the joints (from packer to packer) with a water reactive prepolymer until the grout travels through the entire joint sealing it completely.
• Installation of sleeve-pipes (anchored in place, using prepolymer collar pads) from the inside of the sewer-line followed by a cement suspension grouting program, embedding the sewer-line in cement based suspension grout and providing the required lateral uniform support to the sewer line to prevent ovalisation and cracking.

More than 1.5 km of sewer grouting has been carried out in Niagara Falls since September 1989, successfully rehabilitating the old sewer for a fraction of the cost building a new sewer.

Subcategory 2: Hydrophobic Water Reactive Polyurethane Grout: Hydrophobic "Flexible" Prepolymers: These products remain flexible over time and do not undergo volume changes after curing. The foam contains a certain percentage of open cells, when curing under atmospheric pressure. As a result, grout has to cure under pressure to form an effective elastic seal.

Flexible prepolymers should only be used for joint and crack-injection and seepage control, but not for soil grouting. The high viscosity of the prepolymer is an advantage when inflows have to be sealed through wider apertures. The high viscosity increases the resident time of the prepolymer in the formation. Flexible hydrophobic prepolymers are less reactive than most other water reactive prepolymers.

Sault Ste. Marie Locks rehabilitation. Michigan, U.S.A.

The U.S. Army corps of Engineers (USACE) Sault Ste. Marie Division embarked on a substantial crack injection program to stop the seepage after specialist contractors had only limited success. The USACE executed the work with in-house staff that were trained by outside consultants in proper crack injection techniques: angled drilling towards the mid-point of the cracks; acid flushing with phosphoric acid; and multiple hole grouting until grout travels to adjacent packers.

It is noteworthy that a crew without any previous experience in grouting successfully and permanently sealed more than 10000 lineal feet of damp and badly leaking cracks, regardless of the brand of flexible prepolymer used. It was confirmed that the travel times through fine cracks, as derived from the mathematical model based on Anderson's research work (Anderson in 1998), was predictable. It takes more than

15 minutes for polyurethane to travel 1 meter through a crack of 20 micron under a pressure of 500 kpa (70psi). USACE staff experimented with the use of acid flushing and treated several sections without acid flushing. These tests revealed irrefutable evidence that acid flushing is mandatory for permanent sealing of cracks.

Hydrophilic Water Reactive Polyurethane Grouts

These products are grouted in conjunction with water and form a hydrophilic gel or a hydrophilic foam (depending on type and mixing ratio). They are often more efficient than the hydrophobic prepolymers, to stop major inflows, because they are more reactive (i.e. they react much faster and the duration of the reaction is much shorter).

Hydrophilic gels appear not to be stable in time. The rate of deterioration differs from product to product. The cured gels tend to physically absorb water, thereby increasing their porosity, losing their bond and some or all of their waterproofing and mechanical characteristics. Some products do not survive accelerated aging tests. Since these products absorb water after gelation, they exert a swell pressure that sometimes cannot be sustained by the structure, resulting in structural damage.

Utmost care is recommended when selecting these products for an application. When the products are mechanically confined, the post-swelling is rather a positive characteristic since a tighter "gasket" is created. It is the author's experience however, that these products are questionable for long-term solutions.

Polyurethane Elastomers (2 Component Grouts)

These products consist of 2 components: the polyol (usually a poly-ether polyol) on which a catalyst is added to select the gel time; and, the isocyanate: preferably an M.D.I. (Diphenylmethanediisocyanate) type, since T.D.I. (Toluene Diisocyanate) types pose severe health hazards at ambient temperatures.

The first generation of polyurethane elastomers has been used in Germany since the early sixties under the name "polytixon". The polytixon products are T.D.I. based; oils were used to lower the viscosity, and to dilute the polyol.

Since 1984, a new generation of polyurethane elastomers has been introduced to the grouting industry resulting in a considerable improvement to the rehabilitation and seepage control grouting.

In cured form these grouts are totally inert and hydrophobic, and most of the above products remain flexible in time, even at low temperatures. These products have an excellent penetrability in cracks and are more suitable than the classic epoxy grouts for structural repair work in concrete, because of their flexibility. An engineered approach towards continuous monitoring methods of cracks has been long overdue. Wiechmann (1990) developed a continuous monitoring device which allows to evaluate movements of joints and cracks, pinpoint the cause and determine in a meaningful and engineered fashion the appropriate remedial measures (methodology and product selection).

The tensile strength and the bond of the urethane must exceed the tensile strength of the concrete, to be suitable to seal active cracks in concrete structures. It is our experience that cracks and joints fluctuating more than 40% in width are very difficult to seal, unless a "cap" on the upstream side of the joint or crack can be created.

Applications of Hydrophobic Urethane Elastomer

Sealing cracks and joints in concrete and masonry.

Whenever practically possible, polyurethane elastomers should go selected over water reactive polyurethanes. The appropriate type has to be selected for the application at hand.

Injectable tube applications (contact grouting)

Grouting, via pre-placed injectable tubes, is gradually becoming an integral part of the design to create water-tight construction joints, or to "glue" old concrete to new concrete, or as an alternative to water-stops.

Air and water leaks often occur through cold joints in structures. To prevent these types of problems, injectable tubes were invented in Germany in 1982 and are now used world-wide. This is one of the most genuine applications where grouting is used as a construction tool, rather than as a remedial measure.

P.U. Elastomers Subcategory1: Water Compatible Hydrophobic Elastomers

These products displace water in cracks and cure without foaming or reacting with the water, to form an hydrophobic elastomer, with acceptable adhesion to the medium. The elastomer is not affected by wet/dry cycles and is stable in time. These products are suitable for grouting into water-bearing formations, cracks or joints (not running), and for grouting into injectable tubes filled with water to create an elastic "gasket" between two members. They are also very suitable for repair work of wet cracks in concrete structures.

Pedestrian Tunnel Rehabilitation, Ottawa, Canada, 1989.

Persistent seepage had been a serious nuisance in the pedestrian tunnel between the central block and west block on Parliament Hill in Ottawa. Water bypassed several waterproofing membranes (copper, rubber) and entered the tunnel via the expansion joints. During the rehabilitation program, carried out in December 1989, a water repellant, two-component polyurethane grout was injected, forcing the water out of the joint and resulting in a flexible completely inert two-component polyurethane elastomer.

P.U. Elastomers: Subcategory 2: Hydrophilic Polyurethane Elastomers

These grouts swell out after they have cured in contact with water. They are predominantly used for crack-injection in dry or damp joints or for use in injectable tubes. Once cured, they swell out in a predictable way when in contact with water, to form a tight gasket.

The major difference with the pure one component hydrophillic prepolymers is that the elastomers have a hydrophobic matrix, with "active antennas", attracting water. The percentage of swelling is predictable.   These elastomers do not lose their mechanical characteristics, because of their hydrophobic matrix. Typical applications include: pickotage ring in shafts; sealing around flood bulkheads; prefabrication of "swell-seals" of any size or dimension for joint sealing or prefab applications; injectable tube grouting (tunnels, parkades, swimming pools, etc.); and cast-in-place expansion joints.

Fjellinjen Tunnel in Oslo (Norway)

In the design of this tunnel the installation of injectable tubes in every cold, construction and expansion joint was specified. The injectable tubes were all grouted, after the structure had settled, with a two component polyurethane elastomer, Polycast EXP. This grout cures to form an elastic rubber. When water penetrates the joint, the grout swells out to a maximum pre-determined amount. Similar applications took place in the Liefkenshoektunnel (Antwerp Belgium 1990) and the Piet Heyn tunnel (The Netherlands 1993)

Hydroblock

This is a two-component "auxiliary grout" (cement-urethane) grout developed by Verstraeten in the late 1980's. A-component consists of a partly cementitious based suspension grout, on which a particular type of amine has been added.   The B component consists of a "cocktail" of polyol. The mixing ratio (A/B) is approximately 10-15/1. Both products are injected via concentric pipes, whereby the B component is only introduced at the exit point of the grout-line. No static mixer is required. The reaction time (in spite of the water flow) of the grout a less than a second. The cement grout gels instantly, forming spaghetti-like stringers, clogging up the seepage paths. The "soft grout spaghetti" continues to cure like a regular cement-based suspension grout, providing a permanent cut-off.

Two-component Polyurethane Foam Grouts

The result of the reaction between a polyol (R-OH) and an isocyanate (R1-NCO) is the creation of a polyurethane. Depending on the type of polyol, blowing agents, catalysts, a wide variety of foams with different characteristics can be formed, only differing in: density, cellular structure, compressive strength, reaction pattern (cream time - tack free time), water absorption, fire resistance

The isocyanate has a high affinity for water and thus has a tendency to "steal" the isocyanate, leaving not enough isocyanate for the polyol to form a complete reaction. Typical applications include: stabilization of unstable rock in mines; sealing previous formations in front of flood bulkheads; sealing gaps and joints around ventilation doors; filling damaged "air-caissons" (locks); and, pipeline applications such as temporary floating blocks for river crossing, and erosion cushions).

Conclusions

Although polyurethane grouts have been extensively used worldwide for almost four decades, some large grouting firms still have not caught on to this trend. Polyurethane grouts provide solutions to problem situations especially when major inflows are present or when long term performance is required. Although more durable than cement-based grouts, the water reactive hydrophobic polyurethane grouts do not last "for ever" in highly alkaline environments but are very durably in neutral and acidic environments. Sound research work has been performed confirming what practitioners experienced pertaining to penetrability of polyurethane grouts in cracks: acid flushing and multiple hole grouting are absolutely necessary prior to grouting cracks in concrete and is advantageous when grouting limestone formations.

With increased awareness of environmental matters, most polyurethanes offer a solution of grouting in potable water environments.

It is important to be aware that most hydrophilic, water reactive polyurethanes continue to absorb water while degrading. Two component polyurethane foams have branched out from their classic applications in mining to geotechnical applications.

Understanding the characteristics of the large variety of polyurethanes is necessary for proper product selection. Most of the research work on polyurethane grouts has been performed in Scandinavia. There is still more to be learned about this fascinating family of products that democratised the grouting industry.

References

Andersen, H. (1998) *Chemical Rock Grouting. An experimental study on polyurethane Foams.* Chalmers University of Technology, Goteborg. S-41296.

Burgstaller, C. (1989) *Numerische Simulation von Schauminjetionen auf der Basis von experimentellen Untersuchungen.* Nontan Universitat Leoben, Germany.

Hepburn, C. (1992) *Polyurethane elastomers.* Elsevier Applied Science. London, New York.

Jakubowicz, I. *Determination of Long Term Properties for Polymer Based Grouts* conference proceedings. Nordiskt Symposium I berg injektering. Ed. Lundblom Chalmers University of Technology, Goteborg. pp 50 – 56.

Landry, E, Lees, D, and Naudts A. (2000): "New Developments in Rock and Soil Grouting: Design and Evaluation." *Geotechnical News.* September 2000. pp 38 – 44.

Naudts, A. (1986). "Research & lab testing program on the use of acrylamide and polyurethane grouts under high pressure." IMC. Esterhazy. Unpublished.

Naudts, A. (1990) "Research Project Fort Cady Minerals: Chemical resistance of modified cement-based grout and polyurethane grouts Accelerated ageing test." Unpublished.

Naudts A., (1994) "Soil Grouting to Improve Foundation Soils." *Practical Foundation Engineering Handbook.* ed. R.W. Wade. McGraw Hill. pp 5.340 – 5.363

Oshita, T., Kitano, M., and Terashima, K. (1991) "Long Term Durability of Sails Solidified with Hydrophobic Polyisocyanate type grout." *Society of Materials Science.* Japan. December. pp. 1552 – 1557

Takenaka. (1976) Tacss know-how books. Internal Publication

Tjugum A (1991). *Utrikling av testnetade for injeksjonsmidler I beveglige riss.* Norwegian Institute of Technology. Trondheim

Verstraeten, P. personal communications

Weichmann, M. (1990) "The need for continuous monitoring by accurate instrumentation in analysing structural movements." Second International Grouting Conference in Canada, Toronto.

Wet-Dry Cyclic Behavior of a Hydrophilic Polyurethane Grout

C. Vipulanandan[1] M.ASCE, Y. Mattey[2], David Magill[3] and
Steve Hennings[4]

Abstract

Polyurethane grouts are used for controlling leaks in civil infrastructure facilities. Due to the wide range of environmental conditions in which the polyurethane grouts are used, there is increased interest in better characterizing the behavior of hydrophilic polyurethane grouts. Polyurethane grout selected for this study had a free volume expansion of over 800% for water-to-grout (W/G) mix ratio of 0.5. In this study the effects of W/G ratio and volume change (ΔV) on a solidified hydrophilic grout properties was investigated. Unit weights of cured polyurethane so prepared varied from 1.1 kN/m^3 (7 pcf) to 10 kN/m^3 (63 pcf). The behavior of cured polyurethane grout was studied under wet and dry cycles. Each cycle had a week of immersion in water followed by a week in air. During the test, weight change and volume change were measured up to forty wet-dry cycles. Parameters influencing the swelling and shrinkage of the grout have been identified. Using multiple regression analysis a relationship has been developed to represent the maximum swelling in terms of cured grout parameters.

[1] Chairman, Professor and Director of CIGMAT, [2]Graduate Student, Center for Innovative Grouting Materials and Technology (CIGMAT), Dept. of Civil and Environ. Eng., University of Houston, Houston, Texas 77204-4003 (Phone: (713) 743-4278; Fax: (713) 743-4260 e-mail: cvipulanandan @uh.edu)

[3] President and [4] Technical Director, Avanti International, Webster, Texas 77598-1528. (Phone: (281) 486-5600   Fax: (281) 486-7300)

Introduction

When faced with leaking problems wastewater systems that are structurally sound, grouting is an effective method of rehabilitation and polyurethane based grouts are popularly used. The ideal grout for these applications should have low viscosity to minimize pumping pressure, good gel time, and the ability to make the soil around the leak impermeable [Karol 1990; Bodocsi et al. 1991; CIGMAT News 1995; Vipulanandan et al. 1996 a & b; Concrete Construction, 1998]. Hydrophilic polyurethane grouts when mixed with water will expand and seal the leaks. But the grout behavior during and after curing is not well understood.

Polyurethane chemistry is extremely versatile and is under development all the time [Goods, 1982; Aronld, 1995; Arefmanesh, 1995; Yasunaga et al. 1997]. Polyurethane forming is a continuous process and the cell formation is considered to be an important stage since it determines the durability of the material. By varying the proportions of the components in the system it allows foams to be produced which have a range of densities of less than 1 pcf up to 70 pcf and with extremely useful chemical and mechanical properties [Vipulanandan et al., 2000; Mattey, 2001]. The porosity and unit weight of the polyurethane foam could influence the performance of the materials. Only limited information is available on polyurethane grout performance for civil infrastructure related applications.

There has been an extensive growth in the use of polyurethane grouts in repairing and maintaining underground structures such as sanitary sewers and storm water collection systems and disposal pipeline systems throughout the world in the past four decades [Karol, 1990; Vipulanandan and Jasti, 1996b]. It should be noted that there are no ASTM standard methods for testing polyurethane grouts and there is only limited data reported in the literature. Lack of standard testing procedures for polyurethane foams for use in civil infrastructure applications makes it difficult for the design engineer to select this material. It is not clear how pressure, temperature, and microstructure of the grout are affected if the free volume expansion is restricted, which is the case in most leakage control applications. Durability of these foam grouts under wet-dry conditions is important for selecting grouting materials for these applications.

Objectives

The objective of this study was to investigate the performance of polyurethane grout under wet-dry cycles. The specific objectives are as follows:

(1) Develop   pressure-temperature-time   relationships   for   preparing polyurethane grout specimens under controlled volume change.

(2)   Quantify the changes in the polyurethane foam grouts under wet-dry cycles after selecting the testing conditions.

(3)   Determine the relationship between maximum swelling and selected polyurethane grout properties.

Materials and Testing Program

(a) Net Grout:

A commercially available AV-202 multigrout (Avanti International, Webster, Texas) was selected for this investigation since it reacts with water in any proportion to form a foam or gel. The grout resin was dark brown in color with a viscosity of 2500 cps (at $30^{\circ}C$) and a specific gravity of 1.15. Depending on the proportioning of the resin to water, a range of products from very porous foam to gel can be obtained.

(b) Specimen Preparation

It was of interest to prepare specimens with controlled volume change with uniaxial expansion under K_0 condition [Vipulanandan et al., 2000; Mattey, 2001]. Total volume of the mold was 100 mL. Using specially instrumented 100 mL molds and various grout and water mixes, specimens were prepared with controlled volume change. Changes in the pressure and temperature were monitored for a length of time till readings remained almost unchanged. Water to grout ratio was varied from 0.5 to 8 [Vipulanandan et al., 2000]. The volume change during specimen preparation was limited to between 0 and 150% of the initial volume of grout and water mix. The specimens were removed from the molds and stored in moist airtight zip lock bags till the time of testing to prevent moisture loss and minimize changes in the specimens.

Specimens were prepared with controlled volume change, based on water-to-grout ratios. The amount of resin and water to be added was determined by the allowable volume change needed. Resin and water were mixed at room condition (temperature (23 ± 2 °C) and at relative humidity (50 ± 5 %)). Using the thermocouple, temperature increments were monitored during the curing process. These measurements were taken until temperature returned to its initial temperature (or temperature increment of zero). The change in load was measured with time to an accuracy of 2 lb. [Vipulanandan et al., 2000]. The pressure was monitored for a period of 3 to 6 hours, or as needed.

(c) Unit Weight

For each cured specimen length and diameter of specimen was measured with a Vernier caliper (least square of 0.01mm). When measuring the specimen, special care was taken to not apply pressure on the foam specimen. The specimens were weighted to an accuracy of 0.1 g. The results reported are average of three readings.

(d) Volume Expansion

In both free expansion tests and controlled volume tests, polyurethane grout was allowed to expand in one direction. The change in volume of an expanding material may be expressed as

$$\Delta V = (\frac{V_f - V_i}{V_i}) * 100$$ [1]

where V_i = Initial volume of unreacted material (resin + water),
         V_f = Final volume of reacted material,
         ΔV = Percentage Volume Change due to foaming.

The volume expansion ΔV was thus defined as the ratio of the change in volume of reacted (hardened) with respect to unreacted materials (resin + water). Volume expansion describes how many times the material has expanded with respect to original volume of the material mixed and is equal to the void ratio.

(e) Wet-Dry Cycle Test

Water absorption during wet and dry cycles for the foam grouts was determined by measuring the change in both weight and volume when the specimen was immersed in tap water (drinking water at room temperature, pH of 8) for a week and the following week placed at room condition (temperature of 23 ± 2 °C and relative humidity of 50 ± 5 %). The room condition was selected to represent the dry condition in this study, since these grouts are used for controlling leaks in structures at or near room condition. Volume and weight were recorded on a daily basis during the test period. At the end of the immersion period, the difference between the initial weight and final weight was the weight gain in the specimen. Similarly change in volume was determined. At the end of the dry period, the difference between the initial and the final weight was due to the loss of water. Results recorded were the mean of a minimum of three measurements. Tests were conducted continuously for 40 cycles. The detailed procedure used for this test is described in CIGMAT Standard GR 3-00.

Test Results and Discussion

In this study, a hydrophilic polyurethane resin was mixed with varying the amount of water up to 8 times the volume of the resin [Vipulanandan et al., 2000]. During curing under controlled volume change, temperature and pressure were monitored. The unit weights of the specimens varied from 1.1 kN/m³ (7 pcf) to 10 kN/m³ (63 pcf).

(a) Curing Process

Pressure (P)-Temperature (T)-Time (t) Relationships: The changes in pressure and temperature with curing time for a grout mix with 150% volume change is shown in Fig. 1. In all the cases investigated the temperature rise was faster than the pressure. For gout mixes with water-to-grout mix ratios of 1 and 6 the peak temperature were reached in 2 and 8 minutes respectively. While the temperature continued to decrease rapidly the pressure remained almost unchanged after reaching the peak for longer period of time. According to the grout supplier's brochure (Avanti International) the foam-time for the water-to-grout ratios of 1 and 6 are 1.5 and 8 minutes respectively. These foam-times appear to correspond better to the time to reach maximum temperature than maximum pressure.

In this study pressure as high as 1.3 MPa (180 psi) has been measured. Increasing the water-to-grout ratio and allowable volume change during curing will result in reduced pressure increase. Increasing the water-to-grout ratio also reduced the maximum temperature raise in the mix. For the variables investigated the maximum temperature rise (above ambient) was about 30°C.

(b) Unit Weight versus Water-to-Grout Ratio (W/G)

The relationship between the polyurethane foam unit weight and the W/G ratio was determined through a series of tests and the variation is shown in Fig. 2. The relationship between the unit weight (γ in pcf) and water-to-grout ratio (W/G) in terms of volume (α_v) can be represented as follows:

$$\gamma = 13.3 + 5.15\,\alpha_v \qquad [2]$$

The coefficient of correlation was 0.83 and α_v varied from 0.5 to 8. It must be noted that 10 pcf is equal to 1.6 kN/m³.

(c) Wet-Dry Cycles

When the W/G ratio was 0.5 and ΔV=150% (unit weight of 34 pcf/5.4 kN/m³) the maximum weight gain during the wetting (swelling) was 190% (Fig 3(a)). During drying, the minimum weight gain was 0%. The grout repeated its performance over 40 cycles without any degradation (disintegration/cracking of the specimens). The maximum and minimum volume changes were 100% and 20% respectively. Although weight returned to the original value the volume did not and had residual swelling. Performance of another grout mix with W/G

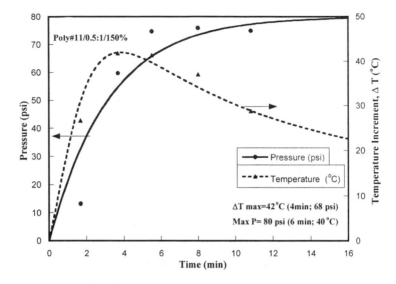

Figure 1. Pressure-Temperature-Time relationships for curing polyurethane grout mix with water-to-gout of 0.5:1 and volume change of 150%.

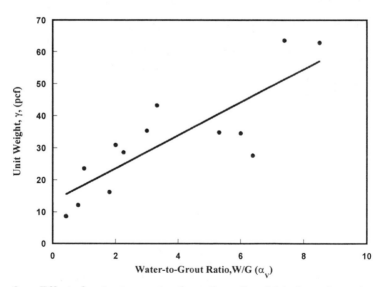

Figure 2. Effect of water-to-grout ratio on the unit weight of cured grouts.

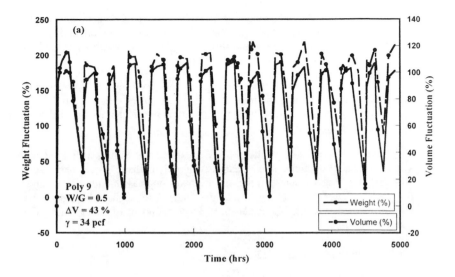

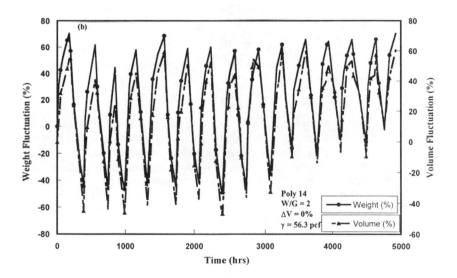

Figure 3. Wet-dry cycle test with swelling and shrinkage (a) water-to-gout of 0.5:1 and volume change of 43%; (b) water-to-gout of 2:1 and volume change of 0%.

ratio of 2 and $\Delta V = 0\%$ (unit weight of 56 pcf / 9 kN/m^3) the maximum weight gain during the wetting (swelling) was 60% (Fig 3(b)). During drying, the maximum weight loss was -45%. Additional weight loss is possibly due the loss of water from the grout.

The grout repeated its performance over 40 cycles without any degradation. The maximum and minimum volume changes were 50% and -30% respectively. In this case neither weight nor volume returned to the original state and there was residual shrinkage.

Inspecting the test results showed that when W/G was increased the swelling capacity (weight based quantity) of polyurethane grouts decreased. Also the swelling capacity of the grout mixes increased with the volume expansion (ΔV) of the grout (Fig. 4). Higher the ΔV, the higher the voids in the specimens and hence the grout can hold greater amount of water during the water immersion.

(d) Multiple Regression Analysis

When analyzing the maximum swelling capacity (by weight) of the polyurethane grout, the following linear equation was assumed to predict the behavior of the grout [Mattey, 2001].

$$Swell_{max} = A_{swell} + B_{swell} * \frac{W}{G} + C_{swell} * \Delta V .\qquad [3]$$

Using the test data and performing the multiple regression analysis leads to the following relationship

$$Swell_{max} = 114.3 - 13.2 * \frac{W}{G} + 1.3 * \Delta V \qquad [4]$$

Model parameters B_{swell} and C_{swell} suggest that the swelling capacity of the grout is more dependent on the water-to-grout ratio than it is on the volume expansion ΔV. The maximum swelling capacity decreased with increase in W/G ratio. On the other hand, the maximum swelling capacity increased with ΔV. This correlation can be seen in Fig. 5 where the model prediction is compared to the experimental results. Of the data considered, 25% had higher than predicted swelling capacity and the measure-to-predicted swelling ratio was 1 with a correlation factor of 0.99.

Conclusions

This study focused on developing a method to prepare polyurethane grouts under controlled volume change and monitor the pressure and temperature during curing. The water-to-grout ratio was varied from 0.5 to 8 and the volume change (controlled volume test) during curing varied from 0 to 150% (based on initial liquid volume). The behavior of cured polyurethane grout specimens was studied under wet and dry cycles. Each cycle had a week

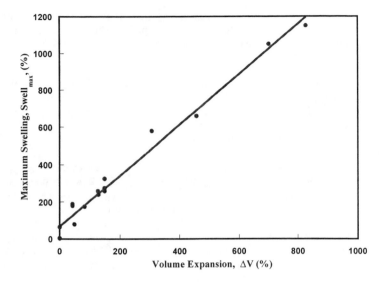

Figure 4. Relationship between maximum swelling and void ratio (processing volume change (ΔV)).

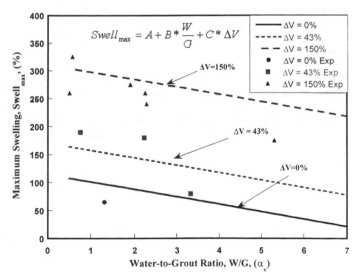

Figure 5. Predicted and measured maximum swelling in polyurethane grout.

of immersion in water and another week in air. The unit weight of cured grouts varied from 1.1 kN/m^3 (20 pcf) to 10 kN/m^3 (63 pcf). During the tests, weight change and volume change were measured up to forty wet-dry cycles. Based on the test results, the following observations can be advanced:

(1) Pressure-Temperature-Time Relationships: This is affected by the water-to-grout ratio and the volume change allowed for the curing grout mix. Maximum temperature and pressure are out of phase and the temperature peaked first. Increasing the water-to-grout ratio reduced the maximum pressure and temperature.

(2) Swelling depends on the water-to-grout ratio and ΔV. Swelling (weight based) decreased with increased W/G and decreased with ΔV. When the specimens were immersed in water, they reached the maximum swelling, which decreased with increased W/G ratio.

(3) Shrinking was related to water-to-grout ratio. In general, the shrinking capacity of the grout decreased with increased W/G ratio. When W/G ratio was 2 or higher, there was negative residual shrinkage during the dry cycles.

Acknowledgment

This work was supported by the Center for Innovative Grouting Materials and Technology (CIGMAT) under grants from various industries and the National Science Foundation (NSF) (CMS-9526094).

References

[1]   Annual Book of ASTM Standards (2001), Section 4 (Construction) and Section 8 (Plastics), ASTM, Philadelphia, PA.

[2]   Arefmanesh, A. and Advani, G. (1995), "Nonisothermal Bubble Growth in Polymeric Foams", Polymer Engineering and Science, Vol. 35, No. 3, pp. 252-259.

[3]   Arnold J. C. (1995), "The Effect of Physical Aging on Brittle Fracture of Polymers", Polymer Engineering and Science, Vol. 35, No. 2, pp. 165-169.

[4]   Bodocsi, A. and Bowers, M. T. (1991), "Permeability and Acrylate, Urethane and Silicate Grouted Sands with Chemicals, Journal of Geotechnical Engineering, Vol. 117, No. 8, pp.

[5]   CIGMAT News and Literature Review, Vol. 1, No. 3 (1995), Center for Innovative Grouting Materials and Technology (CIGMAT), University of Houston, November 1995 (htttp://gem1.uh.cive.edu)

[6]   CIGMAT Standard GR 3-00 (2000) "Standard Test Method for Wet and Dry Cycle Resistance of Grouts and Grouted Sands," Center for Innovative Grouting Materials and Technology (CIGMAT), University of Houston, Texas. 4 p.

[7]   Concrete Construction (Oct. 1998), "Repair, Protection and Rehabilitation, pp. 898-890.

[8]   Goods, G. (1982), "Flexible Polyurethane Foams, Chemistry and Technology," Applied Science Publishers, London, England.

[9]   Karol, R. H. (1990), Chemical Grouting, Marcel Dekker Inc., New York, NY, 465 p.

[10] Klempner, D. and Frisch, K. C. (1991), "Handbook of Polymeric Foams and Foam Technology," Hanser Publisher, New York, New York.

[11] Lowther, J. and Gabr, M. A. (1997), "Permeability and Strength Characteristic of Urethane-Grouted Sand," Proceedings, Grouting, Geotechnical Special Publication No. 66, ASCE , pp. 197-211.

[12] Malyshev, L. I., Korolev, V. M. and Skokov, V. G. (1995), "Use of Synthetic Resin for Grouting When Repairing Structures and Performing Antiseepage and Strengthening Works," Hydrotechnical Construction, Vol. 29, No. 12, pp. 690-693.

[13] Mattey, Y. (2001), Mechanical Properties and Microstructure of Polyurethane Grouts, M.S. Thesis, Department of Civil and Environmental Engineering, University of Houston, Houston, Texas, 172 p.

[14] Tonyan, T. D., and Gibson, L.J. (1992), "Structure and Mechanics of Cement Foams, " Journal of Materials Science, Vol. 27, pp. 6272- 6378.

[15] Vipulanandan, C. Jasti, V., Magill, D. and Mack, D. (1996a), "Shrinkage Control in Acrylamide Grouts and Grouted Sands," Proceedings, Materials for the New Millennium, ASCE, Washington D.C., pp.840-850.

[16] Vipulanandan, C. and Jasti, V. (1996b) "Development and Characterization of Cellular Grouts for Sliplining," Proceedings, Materials for New Millennium, ASCE, pp. 829-839.

[17] Vipulanandan, C. and Jasti, V.(1997) "Behavior of Lightweight Cementitious Cellular Grouts," Proceedings, Grouting, Geotechnical Special Publication No. 66, ASCE , pp. 197-211.

[18] Vipulanandan, C., Mattey, Y., Magil, D. and Mack, D.(2000) "Characterizing the Behavior of Hydrophilic Polyurethane Grout, Geotechnical Special Publication No. 104, ASCE , pp. 235-245.

[19] Yasunaga, K., Zhang, X. D., and Macosko, C. W. (1997), "Skin Development in Free Rise, Flexible Structure and Mechanics of Cement Foams, " Journal of Celluar Plastics, Vol. 33, pp. 528- 544.

Hot Bitumen Grouting: The antidote for catastrophic inflows

Alex Naudts[1] and Stephen Hooey[2]

Abstract

Hot bitumen grouting technology has continually evolved since its early applications almost a century ago in France, Germany and the USA to seal persistent leaks in tunnels, below dams and for erosion protection along canals. Advancements in the industry especially in the field of monitoring and grouting equipment has made the injection of hot bitumen in conjunction with cement based suspension grout, the most economical, practical and sure solution to stop major inflows through, below or around structures. These applications proved the effectiveness of the hot bitumen grouting technique to stop major water inflows and stabilize water bearing, cohesionless soils, in a fast, predictable and economical way. This paper elaborates on a few remarkable field applications, one of which was likely the largest grouting effort ever undertaken.

History of hot bitumen use in grouting

One of the oldest references to bitumen for use in bitumen grouting was for the construction of the Tower of Babel recounted in *The Antiquities of the Jews* (IV:3), by Flavius Josephus in the first century A.D. Bitumen has a long history in antiquity and through the Middle Ages for use in waterproofing applications and warfare.

By the end of the 19th century grouting with hot bitumen for remedial repair work on dams and seepage control in rock tunnels was introduced. Bitumen was first used at European dams in Switzerland and France and later at dam sites in North America. There is documented evidence in the records of Puget Sound Power and Light that hot bitumen was used during the 1920's at Lower Baker Dam near Seattle, and on some projects for the Tennessee Valley Authority.

There is documented evidence that virtually all deep shafts in the Dutch, Belgian, French and German coal mines sunk near the end of the 19th century and the beginning of the 20th century were enveloped in bitumen poured or injected as a hot melt in the annular space between the liner and the formation.

From the first two decades after the Second World War, there are a few more documented cases of the use of bitumen for stopping major water inflows. The use of

---

[1] President, ECO Grouting Specialists Ltd., 293199 8th Line, RR #1 Grand Valley (Toronto), ON, Canada. Tel: (519) 928-5949; anaudts@ecogrout.com
[2] Project Manager, ECO Grouting Specialists Ltd; shooey@ecogrout.com

emulsions, and their associated failures, had given bitumen a very bad reputation. Both from a technical and environmental perspective the emulsions were sensitive and often did not provide the desired results.

The selection of the type of bitumen to use during the first applications was often controversial. Residual seepage was often caused by extrusion of the bitumen from wide seams or cracks because of a lack of strength at ambient temperature. The selection of inappropriate grades of bitumen, such as road bitumen with low solidification point, as well as, the problems discussed above, made the use of hot bitumen for grouting applications almost extinct.

The development of different types of environmentally friendly bitumen expanded its use in the grouting industry. Oxidized blown bitumen replaced the softer bitumen typically used in road paving applications and emulsions in grouting, and played an important role in marine, civil and mining applications, mainly for seepage control.

During the late 1970's and early 1980's bitumen was used as a permanent grout for lining and sealing purposes on nuclear waste disposal sites in Germany and France, including the Manche nuclear waste deposit. It was also used in tunnel grouting and for contact grouting around concrete plugs in abandoned salt and potash mines in Europe.

The use of hot bitumen was 'rediscovered' during the early 1980's with the success of the Lower Baker Dam and the Stewartville Dam grouting projects. Unfortunately the selection of the wrong type of bitumen by an inexperienced contractor at hydroelectric power projects near Waterton (New York) and Sault Ste. Marie during the mid-1980's temporarily dampened the revival of bitumen grouting.

Hot bitumen grouting made a remarkable comeback during the late 1990's. Projects in Asia, New Brunswick, West Virginia, and Wisconsin demonstrated that the application of bitumen technology is an efficient, economical and powerful tool to prevent or stop seepage and major leaks.

The nature of most bitumen grouting projects involved emergency situations in which very serious water inflow problems needed to be solved. This has actually hampered the exposure of bitumen grouting in the mining and civil engineering world, since clients often do not wish that detailed information be disseminated on their misfortune or problem situation.

How hot bitumen grouting works

Dr. Erich Schönian, one of the first scientists to study the penetrability and behavior of hot bitumen, documented remarkable findings regarding bitumen penetration in cracks (Schönian, 1999). As the bitumen grout comes in contact with water, the viscosity of the grout increases rapidly resulting in a lava-like flow. A hard insulating crust is formed at the interface between water and bitumen and shelters the low viscosity, hot bitumen behind it. The "crust" or "skin" is remelted from within when hot bitumen continues to be injected.

When hot bitumen is injected into a medium saturated with water, it cools quickly at the interface with water. Steam is created at that point, decreasing the viscosity of the bitumen. The steam acts as an "air lift" drawing the bitumen into its pathway through small and large fissures or pore channels.   The center of the bitumen mass remains hot, and continuously breaks through (remelting) the skin formed at the interface of the bitumen and water. There is absolutely no wash-out. The faster the water flows, the faster the bitumen cools off. The skin prevents wash-out while the "sheltered" hot bitumen behind the skin behaves as a Newtonian fluid, penetrating in a similar fashion as solution grouts

Because bitumen has good insulating characteristics, it can be injected for a very long time (days – even weeks) into the same grout hole without the risk of either premature blockage or wash-out. The width of the fissures accessible to hot bitumen depends on the duration of the grouting operation. The longer the grouting operation, the finer the apertures the bitumen will penetrate.   Hot bitumen will penetrate fractures as small as 0.1mm as demonstrated during the Kraghammer Project in 1963, described below.

When hot bitumen cools it is subject to significant thermal shrinkage. This phenomenon is partially overcome in smaller fractures if pressure continues to be applied and warmer bitumen pushes the cooling bitumen into the shrinkage gaps. Cement-based suspension grout is often injected in conjunction with hot bitumen to compensate for the thermal shrinkage of the bitumen; to make the bitumen less susceptible to creep; and to increase the mechanical strength of the end product.

Holes that are being grouted with hot bitumen rarely come to refusal (zero flow at highest allowable pressure) when the initial pump rate is kept high (to establish a heat sink and prevent rapid cooling of the bitumen causing premature refusal).

The advantage of bitumen over other grouting systems to stop or control water flow, especially under high pressure and at high flow rates, is that blown bitumen will never wash-out. Modifying the mix design of cement-based grouts (used as balanced stable grout with thixotropic viscosity modifiers) or solution grouts (especially water reactive hydrophobic polyurethane) to prevent the grout from washing out while obtaining a large grout spread can be very difficult and often impossible. The project may require numerous boreholes, large volumes of grout, and several attempts, if it is at all successful. In order for the grout not to wash out, it needs to possess adequate cohesion to form a conglomerate larger than the particle size that corresponds with the "critical particle size" under the given flow conditions. The critical particle size refers to the size of the particle that will just be moved by a flow of a given velocity. As the velocity increases, the critical particle size increases.   With bitumen, the skin has a very high cohesion, while the "interior" has a low cohesion and viscosity. With classic suspension grouts, on the other hand, the grout has virtually the same rheological characteristics (including cohesion) throughout the grout cylinder. If a high cohesion is required to prevent wash-out, it will limit grout spread and penetrability. Bitumen combines the best of both worlds: a skin prevents wash-out while the low viscosity bitumen penetrates fine pathways. Hence the need for many grout holes if cement-based suspension grout or solution grout are used; and the need for fewer grout holes if hot bitumen grout is used. If the cohesion of the grout is too low, the grout will wash-out. It should be noted that, as more apertures and pathways are plugged, water travels faster, increasing the critical particle size and hence the required cohesion of the grout to prevent wash-out. This makes the requisite rheological characteristics of 'closure' grouts even more onerous.

Special considerations for using hot bitumen grouting

Based on the analysis of the various bitumen projects conducted during the last twenty years it is safe to conclude that hot bitumen always meets its objectives in the short-term. For long-term success, experience, knowledge and a sound engineering design must be applied.

The equipment and set-up are generally more complex for bitumen grouting than for the applications involving regular cement based grouts or solution grouts. The operating temperature of the surface pipe system need to be in the range of 180 – 225 °Celsius (356 - 437 °Fahrenheit). Moreover, a supply of hot bitumen needs to be obtained and maintained at the requisite temperature. The bitumen should, ideally, be delivered to the site in heated and insulated bulk tankers with the potential to boost or adjust the temperature on site in a custom built grout plant.

The piping system used during grouting to deliver the hot bitumen from the bitumen pumps to the sleeve pipe "stinger" located at the end of the bitumen grout hole, must either be pre-heated with hot oil, heat trace, or steam, potentially through a re-circulation system. Additionally, the grout pipes must be insulated and equipped with temperature sensors and pressure gauges. The flow rate, total volume of grout injected, and grouting pressure must be monitored and recorded in real-time. This allows informed decisions to be made while the operation is in progress. Additional safety measures for dealing with hot materials need to be respected, following general and specific site procedures. With proper safety procedures in place, the use of appropriate equipment and execution by well-trained and attentive crews under the supervision of an experienced engineer, hot bitumen grouting can be conducted safely and effectively.

It is required to continuously monitor the subsurface conditions for signs of cement grout wash-out by measuring the pH of the water, evaluate temperature changes as recorded via down hole sensors to assess the spread of the bitumen, and to interpret changes in apparent Lugeon value (i.e. changes in flow rates and injection pressures). The apparent Lugeon value is the permeability coefficient of the formation using grout as a test fluid (Landry, et al., 2000). It is noteworthy that during the execution of a hot bitumen grouting program the apparent Lugeon value typically decreases with time (contrary to cement grouting operations) due to the excellent penetration properties of the bitumen into the formation.

Environmental issues

The injection of hydrocarbons into soil, rock or structures immediately raises environmental concerns. There are, however, many types of bitumen available with a wide range of characteristics. The desirable type for use in grouting is a "hard" oxidized environmentally friendly type of bitumen with a high solidification point.

Oxidized blown bitumen has a long history of successful use for lining (potable) water reservoirs in California (over 40 years) and in 1987 Washington and Oregon State wildlife authorities have used it for lining fish hatcheries ponds.

Oxidized bitumen has proven to be in compliance with American Water Works Association (AWWA) standards for leachate resistance of materials for use in potable water applications. Indeed, it is now routinely used for water pipeline lining applications. It could be considered the most environmentally friendly grout presently available on the market

Examples of field applications

Lower Baker Dam, Washington, U.S.A., 1920's, 1950's, 1964 and 1982

One of the oldest documented bitumen grouting operations took place more than 75 years ago to stop major leaks through the limestone foundation below the abutments of the Lower Baker Dam, operated by Puget Sound Power and Light.

Slotted steel pipes, pre-heated via a central steel cable were used. The operation was reported to be very successful in the short-term. Creep of the bitumen, however, opened secondary flow channels and resulted in further erosion of clay filled seams in the limestone foundation.

Another bitumen grouting operation was performed some thirty years later at the same dam, using virtually the same techniques with similar results. This was repeated again during the 1960's. It was concluded that bitumen grouting was not efficient since leaks reoccurred within a year.

In 1982, the owner attempted to use water reactive polyurethane prepolymers. It was injected via a "residence pipe system" which enabled the set time to be adjusted within a second. After weeks of attempts, all parties came to the conclusion that the polyurethane grouts either set in the immediate vicinity of the grout pipe or that it washed out rendering it impossible to make polyurethane grouts work under the prevailing flow conditions (many fine fissure with a flow through time less than 20 seconds).

Hot bitumen grouting was again successfully used to curtail the 2,200 litres per second inflow. At the end of two weeks of grouting, the leak was reduced to 2% of its original flow. The owner could not be convinced to use bitumen in conjunction with cement-based grout. The leak gradually again increased over time and has stabilized at a rate of approximately 1,000 litres per second.

Stewartville Dam, Ontario, Canada – 1980's (Deans, et al. 1985)

The Stewartville Dam, located on the Madawaska River, measuring 63 metres high, and 248 metres long, was constructed in 1948. Hot bitumen in conjunction with cement-based grout was successfully used to seal a 22,000 litres per minute inflow beneath the dam foundation under a full reservoir head of 46 metres. The foundation is composed of predominantly massive competent limestone. Zones of weathered micaceous limestone susceptible to erosion occur on some bedding planes and joints. Leakage and wash-out gradually enlarged these zones and eventually water began to enter the foundation drainage gallery. This application illustrates the high degree of control to perform a surgical strike using hot bitumen in conjunction with cement-based suspension grout since the leaks had to be stopped while not plugging the adjacent foundation drains.

Cement-based suspension grouting programs carried out between in the 1970's and early 1980's were ineffective. Grout clogged one drain with no measurable reduction in seepage. The grout typically washed out in spite of the use of rheology enhancing additives such as straw, sawdust, and other "innovative" additives.

Continuing on the same course of action was clearly not productive. Following additional geotechnical investigations, extensive water and dye testing, and a review of available grouting technologies, it was decided in 1983 to proceed with the injection of hot bitumen in conjunction with regular cement-based grout to compensate for the thermal shrinkage of the grout and produce a durable end product.

The main challenge in the grouting operation was to grout under conditions of high flow, in fractures up to 0.2 metres wide, and within 7.5 metre of the foundation drain, which had to be kept open. Dye tests indicated less than 5 seconds "flow-through" time below the dam.

Once the bitumen grouting began, reduction in seepage flow was immediately noted in the inspection tunnel connected to the foundation drain. After a few minutes of grouting the leakage from the treated portion of the dam was reduced to 10 % of the original inflow and was completely stopped after six hours of grouting. Cement-based grout was injected in the same seams upstream of the bitumen grout holes.

A similar grouting program conducted in 1984 beneath the northern portion of the dam reduced seepage to almost nothing from over 9,000 litres per minute.

The visco-plastic properties of the hot, blown bitumen combined with controlled injection rates prevented excessive travel of bitumen from interfering with the foundation drains. Combined with the injection of cement based suspension grout, a durable end product was created. Post-grouting drilling revealed a good bond between the bitumen, cement-based grout and limestone.

Kraghammer Sattel, Germany, 1963(Schönian, 1999)

The Bigge reservoir in Germany, located in the Kraghammer Saddle is situated on highly fractured permeable alternating strata of greywacke slate and sandy partly calcareous clay slate. Bitumen grouting was used to reduce the permeability of the formation beneath the reservoir dam.

The unique feature of this project was that the success of the bitumen grouting was checked by the excavation of two inspection tunnels through the grouted formation. Of significance is that a portion of the formation was grouted with regular cement based grout, and the volumes injected and final permeabilities were compared to the areas of the formation grouted with hot bitumen.

The results of the testing yielded three interesting findings:

• Grout takes per metre of borehole were considerably less over zones injected with hot bitumen than in zones injected with cement based-grout.

- Bitumen was found to have filled seams in the rock up to 3.5 metres from the boreholes and penetrated seams as narrow as 0.1 mm
- Zones with initial permeability in the 1-10 Lugeon range were successfully grouted with hot bitumen proving that hot bitumen is very suitable to treat formations with low initial permeability values.

Drainage Plug in Abandoned Open Pit Mine Tunnel, Asia

In the mid 1990's the plug inside an old access tunnel, connecting a very large open pit mine used for tailing impoundment with the river failed. The reservoir was filled with millions of cubic meters of liquid waste and tailings. The slurry flowed out of the impoundment reservoir into a river causing major environmental problems. The flow reached a peak of 7 m^3/second. The hydrostatic pressure in the tunnel (2.5 m wide, 2.5 m high and 3 km long), was in excess of 1 MPa.

In order to provide mechanical support to withstand the considerable forces resulting from the hydrostatic pressures behind the future plug, a number of large geotextile barrier bags were inflated with cement-based suspension grout in the tunnel. The geotextile bags were strapped onto steel sleeve pipes and lowered into the 175 metre deep drill holes intersecting the tunnel. The bags were inflated in stages with cement-based suspension grout to a diameter of two meters. After inflation, additional reinforcing steel was lowered into each sleeve pipe that straddled the tunnel. As a result, the slurry was forced to flow through a fence of reinforced "concrete piles".

A sophisticated grouting operation involving the local mine forces was undertaken. The grout hole was pre-heated with hot oil. Bitumen grouting started when down-hole thermocouples indicated that the temperature was adequate. Cement-based suspension grout was injected upstream of the bitumen injection point.

Some of the cement grout holes were used to inject a mortar with cohesion in excess of 500 Pascal (injected with a concrete pump using mortar supplied by transit mixers). Further upstream a low viscosity, polymer enhanced, stable suspension grout was "jet grouted" into the tailings flow. Within the hour, the flow of water and slurry through the tunnel was stopped. At that point, water started to flow only through the fissures and joints in highly sheared bedrock (highly pervious k > 500 Lu) surrounding the tunnel. Bitumen was originally traveling upstream and downstream of the injection point. Once the flow was stopped, bitumen was traveling "against the flow", drawn into that direction by steam. Soon, the first cement grout hole reached refusal (no flow at 70 bar). Within four hours, bitumen had traveled more than 40 metres upstream of the injection point and sealed, one at a time, all six cement grout holes, as far as 100 metres upstream of the bitumen grout holes.

Large pockets of tailings remained encapsulated in the cement/bitumen plug after the flow was stopped. The tailings were systematically removed via cross-hole flushing between newly drilled grout holes. The voids were filled with stable, balanced cement based suspension grout, while the formation surrounding the tunnel was grouted with microfine cement based suspension grout. The piezometers in the area slowly recovered, eventually reaching the reservoir level, some 100 metres above the crest of the tunnel.

Potash Mine, Canada - 1997

In 1997 inflow of fresh water into a potash mine in Canada had increased to a point threatening the mine's continued operation. A slow leak of fresh water had gradually dissolved the salt layer between overlying shale formation and lower basalt rock forming a large cavern. Eventually the overhanging mudstone and limestone collapsed. The dewatering system was overwhelmed and an emergency grouting program was designed and implemented to attempt to save the mine. Inflows of fresh water ranged from 10 – 15,000 cubic metres per day.

The proposed method was the injection of hot bitumen in conjunction with regular cement-based suspension grout to fill the cavern and stop the inflow. The initial "gas testing" indicated that the volume of the underground cavern above the rubble pile was 19,000 cubic meters.

The cavern was located approximately 700 metres below a large brine pond. Two 1,600 metre long drill holes, one for injection of hot bitumen and one for injection of cement-based suspension grout, were installed from surface using directional drilling.

Conceptually the grouting would encompass two phases: initial filling of the cavern with bitumen and water repellant cement grout to cut-off the leak to the mine; and, grouting of the aquifer feeding the cavern.

There was the danger that once the leak was stopped, there was the possibility that the hydrostatic pressures would rise in the cavern and the deteriorated formation could collapse sending a tidal wave of water through the mine.

Once the bitumen grouting operation was successfully launched, cement grouting was initiated. Crews of 50 people per shift performed the grouting operation around the clock. Tanker trucks filled with, hot oxidized blown bitumen were brought in from as far as 800 km away. There were as many as twenty-six insulated tanker trucks involved in what might have been the largest production grouting operation ever undertaken. The site facilities were capable of boosting the temperature of the bitumen to the required temperature. Flow, accumulated flow, pressures, hole temperature, temperature of the bitumen, and temperatures in various exploration holes near or above the cavern were all displayed in real time and monitored at several locations during the project.

Bitumen was injected at an average flow rate of approximately 25 m^3/hour for more than two weeks of continuous operation. Cement-based suspension grout formulations were injected at a rate of approximately 45 m^3/hour.

After 24 hours of bitumen grouting and cement-based grout injection into the cavern, inflow rates began to decrease and hydrostatic pressures in the formation started to rise. After three days of around the clock grouting the inflow completely stopped and formation pressures continued to rise. After five days a major collapse of the "cavern floor" occurred. Immediately the hydrostatic pressure in the formation dropped. A tidal wave rolled through the mine as millions of litres of water rushed in over a few hours. The grouting operation continued without interruption. The inflows were again substantially reduced and formation pressures rose once again after an additional five days of around the clock grouting at the aforementioned injection rates.

The inflow was completely stopped again. The spirits of the team were high – victory appeared to have been accomplished. On the thirteenth day of grouting, before the cavity was filled with grout, the formation collapsed again. The second tidal wave again flushed millions of litres of water into the mine.

A sophisticated "gas test' was again conducted, concluding that the "void" above the rubble had increased to more than 100,000 cubic meters. A final effort of injecting bitumen at a rate of 40 m³/hour and cement-based suspension grout in conjunction with sodium silicate (via concentric pipes) at approximately 60 m³/hour was launched. The grout did not wash out (as was verified via pH tests) but the hydrostatic pressure in the formation did not recover. The collapsed zone covered the size of several football fields. The rock formation had lost its structural integrity. After 15 days of continuous, around the clock grouting, the operation was terminated. More than 23 million litres of grout had been placed. It was concluded that the salt horizon had been too severely undermined to be recovered.

The fact that injection of hot bitumen and cement-based grout temporarily completely sealed an inflow of such magnitude at such a great depth is a testament to the robust nature of the technique. The failure of the plug was not a failure of the bitumen grouting technique but was a consequence of the undermining of the salt horizon surrounding the newly formed plug that had occurred over the two months prior to grouting.

Quarry in Eastern United States, 1998.

During routine mining operations in an old limestone quarry, the floor of which was located some 70 metres below the level of an adjacent river, a major water in rush occurred. Piping through clay filled karsts caused a hydraulic connection to the river that led to inflows into the quarry of over 3,000 liters/ second.

It was determined that karsts filled with erodible clay and gouge, some as high as 40 metres were acting as flow conduits. A grout curtain was installed using the following techniques:

1) In zones where small fissures and interconnected vugs governed the hydraulic conductivity cement based suspension grouts were injected.

2) In other areas where large vugs and karsts were encountered in the absence of significant water flow, a classic, low mobility grout was used.

3) Cement grouting was used to channel the flow to 'windows' where these classic grouts were no longer suitable to stay in the formation under the governing water flow. Grouting was typically conducted with the two aforementioned systems until wash-out of the grout became too severe. These 'windows' were successfully grouted using hot bitumen in conjunction with regular cement-based grout.

Milwaukee Tunnel, Wisconsin, U.S.A. —March, 2001

Jay-Dee Contractors, Inc., an established American tunneling contractor, excavated a 1.2 meter diameter tunnel in soft saturated alluvium containing coarse sand and gravel in Oak Creek, near Milwaukee, Wisconsin.

During the installation of a vertical feeder pipe into the unlined tunnel, a collapse associated with heavy inflows of ground and water occurred. Several hundred cubic meters of sand and gravel washed into the tunnel. The tunnel was approximately 35 metres below surface. The water table was 5 metres above the crown of the tunnel. No significant dewatering of the soil was permitted.

Extensive grouting from surface via open-ended pipes and sleeve pipes had been conducted for several weeks and resulted in temporary sealing of the leak. This allowed the contractor to substantially excavate the tunnel leaving a 13 m long zone straddling the breach in place. At this point the leak reoccurred at a rate of 100 litres per second.

Finally, hot bitumen was injected via sleeve pipes into the area where the collapse occurred from within the tunnel. Two temporary bulkheads were used to contain the sand and gravel inside the tunnel. Hot bitumen and regular cement-based grout were injected behind the bulkhead. The hot bitumen saturated the loose sand and gravel inside the tunnel and traveled to the collapsed area and into the surrounding soils stabilizing the entire area around the tunnel. The grouting operation only lasted a few hours. The tunneling contractor was able to excavate the collapsed tunnel and install the concrete liner without any further difficulties.

New Yung Chung Tunnel, Ilan, Taiwan, 2002

In 1998 the New Yung Chung railway tunnel was excavated into a water bearing marble formation that was bounded on both sides by weak greenschist rock with abundant gouge material present. The high ambient water pressures of approximately 50 bar (750 psi) and inflows of up to 4 cubic meters/second caused a major collapse in the tunnel.

The initial tunnel trajectory was abandoned and two new tunnels (7 meter and 5 meter diameters) trajectories were planned through the water-bearing zone. An extensive hot bitumen grouting program was implemented from within the new tunnels to enhance the stability of the rock mass surrounding the planned tunnel excavation and decrease the permeability of water bearing marble zone.

Over the course of four grouting phases, 3084 cubic meters of hot bitumen was injected into the formation via 16 bitumen delivery pipes at pressures as high as 125 bar (1800 psi). During the grouting operation hot bitumen was injected into individual bitumen grout holes. The longest operation lasted for nine days with approximately 1450 cubic meters of bitumen injected into a single hole. Grouting regimes noted during the grouting included permeation grouting, permeation grouting in conjunction with hydrofracturing, and at extreme pressures, hydrofracture grouting alone.

The hot bitumen grouting program conducted in the New Yung Chung Tunnel resulted in a reduction in the hydraulic conductivity of the formation surrounding the main tunnel of approximately 95 %. This is likely the first bitumen grouting program ever conducted entirely from within a tunnel. At the time of writing this paper, one tunnel has been excavated through the marble zone and the second tunnel is approximately 10 meters from completion. There is reported to be almost no water entering the face of the excavation, extensive bitumen has been encountered, and there have been no problems with ground stability.

Conclusions

When installed in accordance with a sound engineered design by a competent contractor with suitable equipment using the appropriate type of bitumen the "hot bitumen in conjunction with cement grouting" technique has never failed to stop inflows.

Bitumen grouting techniques can be applied safely, economically, and extremely effectively, even under adverse conditions. Last but not least: blown oxidized bitumen is probably the most environmentally friendly grout available at present.

Acknowledgements

Special thanks to Mr. Stephen Hooey, P.Eng, for his assistance in writing this paper.

The author acknowledges the trust, cooperation and assistance of numerous individuals who played a role in some of the bitumen projects described. With the risk of having left some out they are: C. Brawner, Don Bruce, J. Bruce, M. Fallet, G. Grimmig, D. Haas, L. Jaillard, Dr. K. Kading, D. Krizek, E. Landry, B. Lukajic, T. Moody, G. Moore, J. Pordon, G. Rorison, E. Schönian, W. Thrytall, M. Yates

Many thanks also to Whitman Benn, Jay Dee Contractors, and U.S.L.

References

Deans, G., Lukajic, B., and Smith, G. (1985) "Use of asphalt in treatment of dam foundation leakage: Stewartville Dam." ASCE Spring Convention. Denver. April 1985.

Landry, E, Lees, D., and Naudts, A. (2000) "New Developments in Rock and Soil Grouting: Design and Evaluation." *Geotechnical News*. September. pp. 38-44.

Schönian, Erich. (1999) *The Shell Bitumen Handbook.* Shell International Petroleum Company Ltd.

Related Material

Bruce, D., Naudts, A., and Smoak, G. (1998) "High Flow Reduction in Major Structures: Materials, Principals, and Case Histories." *Grouts and Grouting. Proceedings: Geo-Congress 98.* Boston, MA. pp. 156-175.

Naudts, A. (1995) "Grouting to Improve Foundation Soil." *Practical Foundation Engineering Handbook.* Chapter 5B. Ed. Robert Wade Brown. McGraw-Hill. pp. 5.277-5.400.

Van Asbeck, W.F. and E. Schönian. (1968) *Bitumen im Wasserbau. Band 2.* (Bitumen in Hydraulic Engineering, Vol. 2) Hũthig und Dreyer, Heidelberg, and Deutche Shell AG, Hamburg.

Liquefaction Resistance of a Colloid Silica Grouted Sand

H. J. Liao[1], C. C. Huang[2] and B. S. Chao[2]

Abstract

The Chi-Chi earthquake attacked Taiwan in September 21, 1999 caused liquefaction in some alluvial deposits inland and hydraulic fills along the coastline. In the areas with obvious liquefaction, sand boils were found on the ground surface and many buildings suffered severe settlement or tilting. To prevent buildings from future liquefaction damage, subsoil underlying the settled or tilted buildings can be improved by grouting method. Considering the groutability and durability of grouting material, a colloid silica grout was chosen as a potential grouting material to improve the liquefaction resistance of in-situ sandy soil. The liquefaction resistance of this colloid silica grouted sand was studied by a cyclic triaxial test apparatus. Test results showed that up to 4 ~ 7 folds increase in liquefaction resistance of grouted sand compared with that of ungrouted sand was observed despite the strength of this colloid silica gel was very low. Higher stress ratio and more number of loading cycles were needed to initiate liquefaction in grouted sand specimens. But a lesser liquefaction induced strain was observed. Under cyclic loading, the deformation of grouted sand specimens increased gradually with loading cycles until initial liquefaction occurred compared to a sudden increase in axial deformation of ungrouted sand specimen when initial liquefaction happened. Cyclic mobility was observed in the grouted sand specimens.

Introduction

The Chi-Chi earthquake (M = 7.3) attacked Taiwan at 1:47 am September 21, 1999. It caused more than twenty five hundred casualties and destroyed tens of thousand buildings and infrastructures. Among the damages caused by this earthquake, the liquefaction induced damages to the lifelines, buildings, and harbor facilities in alluvial deposits inland and hydraulic fills along the coastline were of interest to the geotechnical engineers. In general, the occurrence of liquefaction was widespread. But the areas with obvious liquefaction damage were only concentrated in certain

----

[1]Professor and Chairman, Department of Construction Engineering, National Taiwan University of Science and Technology, 43, Sec.4, Keelung Rd., Taipei, Taiwan, R.O.C.; phone 886-2-2737-6566; hjliao@mail.ntust.edu.tw
[2]Formerly, graduate student, Department of Construction Engineering, National Taiwan University of Science and Technology

Figure 1. Tilting of building due to earthquake induced liquefaction and lateral spread

areas. Many buildings were found suffered severe settlement or tilting (Figure 1). Among them, tilting was found to happen most frequently in 3 to 7 stories buildings founded either on spread footing or mat foundation. However, most of the tilted or seriously settled buildings did not show signs of structure damage. So, many building owners had tried several methods to level the tilted buildings after the earthquake. Among the methods used, underpinning method and grouting method were most used. The former was used to level the building by jacking up the settled foundation. But nothing was done to the subsoil. The latter leveled the tilted buildings by injecting cement grout or cement-sodium silicate grout to the soil underlying the foundation. But bringing the tilted buildings back to level was the primary objective of the grouting operation. Little attention was paid to the degree of improvement made to the in-situ soil.

So far, grouting is probably the most effective and feasible way to improve the liquefaction resistance of in-situ soil under existing buildings. As indicated by Ishihara (1985), once an unliquefiable layer with sufficient thickness is formed by grouting, the building founded on this layer will be unlikely to be damaged by earthquake induced liquefaction (Figure 2). This paper will present the laboratory results of a ground improvement research project which aimed to improve the liquefaction resistance of in-situ soil by grouting method. Considering the groutability and long-term durability of grouting material, a colloid silica grout developed by the Permanent Grouting Association of Japan was chosen here to prepare grouted sand specimens for testing. The liquefaction resistance of sand grouted with this grout was studied with a cyclic triaxial test apparatus.

Test Program And Materials

The test program of this study was divided into two parts: one was to determine the compressive strength and cyclic behavior of grouted sand; the other was to determine the strength of colloid silica grout. Since the strength of neat grout was low, it was impossible to prepare cylindrical shaped specimens for testing. Laboratory vane

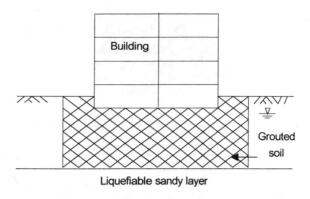

Figure 2. Grouting used to prevent building from liquefaction induced damage

shear apparatus was used instead to determine the shear strength of grout gel. In comparison, the strength of grouted sand was significantly higher than the neat grout. So it was possible to test the strength and cyclic resistance of grouted sand in cylindrical shaped specimens.

Grout

The grout used in this study was a solution type colloidal silica grout which was commercially available under the product name of PERMA ROCK grout. Colloidal silica is made by extracting alkali from sodium silicate with ion-exchange resin in the factory. This process forms colloids with a diameter of 5 - 20 nm. By breaking the electrical double layer around the colloid surface with inorganic salt reactant, siloxane radicals (Si-OH) on the surface of colloids begin to condensate and polymerize forming a structure like a pile of spherical colloids connected to each other (Figure 3). Since there is no sodium ion remained in the gel structure, this gel structure can remain stable over long time (Yonekura, 1996). In comparison, there are some sodium ions remained in the gel structure formed from the mixture of sodium silicate grout and some organic or inorganic reactants. The existence of these sodium ions will reverse the gelling process and cause gel structure to decompose sometime later. As a result, the gel structure formed by sodium silicate and some organic or inorganic reactants is only good for temporary use. But the gel structure developed from colloid silica can be used for permanent purpose. The term of "permanent" is defined here as low resolution of silicate and low volume change with time under submerged condition.

To prepare the neat grout specimen for testing, colloid silica grout was prepared following the mixing proportions shown in Table 1. A typical gelling curve of colloid silica grout with elapsed time is shown in Figure. 4. In general, the initial viscosity of colloid silica grout remains unchanged until the gelling point is reached. The gelling point may varies with the mixing proportions. But the shape of the curve does not change. Due to the low initial viscosity of the colloid silica grout, its groutability to

Figure 3. Gel structure of colloidal silica (Yonekura, 1996)

sandy soil is good. The grout strengths determined from laboratory vane shear test are shown in Figure 5. Generally, the strength of this colloid silica grout is low and its gel strength develops slowly from 6 kPa (7 days curing) to 18 kPa (28 days curing).

Test Sand

The sand tested in this study was taken from a hydraulic fill on the west coast of Taiwan near Mai-liao. Its particle size distribution curve is shown in Figure 6. Mai-liao sand has the following properties: specific gravity of particle $G_s = 2.719$, uniformity coefficient $C_u = 4.13$, coefficient of gradation $C_c = 1.238$, maximum void ratio $e_{max} = 0.996$, and minimum void ratio $e_{min} = 0.485$. It is classified as SP. To obtain a sand specimen with relative density of 50 ± 3 %, sand was weighted first and poured into a split-mold in five layers. Dry-tamped method was adopted to tamp the sand at each layer.

Grouted Sand

In preparing the grout mixture, PR ACTOR NS powder was mixed with predetermined amount of water first following the mixing proportions shown in Table 1. Then it was mixed with PERMA ROCK AT-30 solution to form the grout. To keep the gel time within workable limit, the temperature of grout was kept at 10°C. Under such a temperature, this grout gelled in about 8 minutes which were long enough to inject the grout into sand specimens. The device used for injecting grout into sand specimen was similar to Krizek et al.'s device (1992). Grout was

Table 1. Mixing proportions of grout

Solution A (500 ml)		Solution B (500 ml)	
PERMA ROCK AT-30	500 ml	PR ACTOR NS	20 ~ 30 g
		Water	485 ~ 490 ml
Grout volume = 1000 ml			

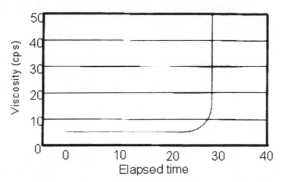

Figure 4. Typical gelling curve of colloid silica grout (Yonekura, 2002)

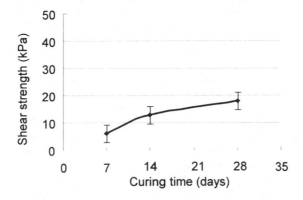

Figure 5. Shear strength of colloid silica gel at different curing times

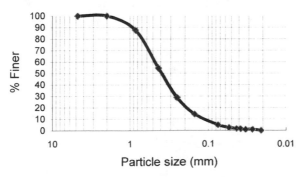

Figure 6. Particle size distribution of Mai-liao sand

allowed to permeate through the specimen from the bottom of the mold and overflow from the top. About three void volumes of grout were injected through the sand specimen. The grouted specimen was left in the split mold for a day or two before the mold was disassembled. Grouted sand specimens were put into plastic bags and cured under water.

The unconfined compressive strengths of grouted sand at different curing times are shown in Figure 7. It increased from 113 kPa (7 days curing) to 141 kPa (28 days curing). A failure plane inclined about 60 degrees from horizon was observed on the failed specimens.

Cyclic Behavior Of Grouted Sand

A SEIKEN DTC-262 type cyclic triaxial test apparatus was used to carry out the undrained cyclic triaxial test on colloid silica grouted sand. To facilitate the specimen saturation process, CO_2 was circulated through the specimen for 60 minutes followed by de-aired water. But only little water was able to flow into the specimen due to the low permeability of grouted sand specimen. During saturation, a back pressure of 190 kPa was applied to the specimen until a B value greater than 0.95 was obtained. Since the voids of grouted sand specimens were almost entirely filled with colloid silica gel, it only took about 4 hours to reach a B value of 0.95 or higher. Specimen was then consolidated at a pressure of 98 kPa. After consolidation, specimen was tested under undrained cyclic loading condition until the initial liquefaction occurred. As indicated by Lee & Vernese (1978), the cyclic loading frequency does not affect the test results of liquefaction resistance if it falls in the range between 0.17 Hz and 4.0 Hz. So, the frequency of this cyclic triaxial test was set at 0.5 Hz. The load applied to the specimen at each cycle is set according to the predetermined stress ratio. During testing, the applied cyclic load, pore pressure change, and axial deformation of specimen were recorded.

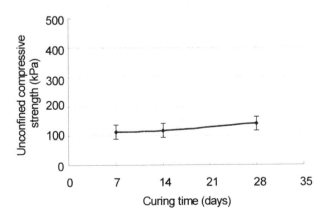

Figure 7. Unconfined compressive strength of grouted sand at different curing times

The cyclic stress ratio (SR) used to describe the liquefaction resistance of grouted sand specimen is defined as

$$SR = \frac{\sigma_d}{2\sigma_o'} \tag{1}$$

where σ_d is the cyclic deviatoric stress and σ_o' is the average effective confining pressure.

The cyclic stress ratios (SR) and the number of cycles needed to generate initial liquefaction in grouted sand specimens with different curing times are plotted in Figure 8. As shown in Figure 8, there is a clear difference in liquefaction resistance between colloid silica grouted sand and ungrouted sand. Up to 4 ~ 7 folds increase in liquefaction resistance of grouted sand is observed and more number of loading cycles are needed to initiate liquefaction. When the curing time increases from 7 days to 14 and 28 days, the liquefaction resistance of grouted sand increases 8% and 24% respectively. It is about equal to the amount of unconfined compressive strength increase over the same period of curing time.

The relationship between normalized pore water pressure ($\Delta u/\sigma_{3c}'$) and normalized time (T/T_{IL}, T_{IL} is the time needed to reach initial liquefaction) of grouted sand is plotted in Figure 9. Since the voids of grouted sand specimen were filled with colloid silica gel, cyclic load with low stress ratio was unlikely to generate sufficient void volume change and pore water pressure increase. Instead, a larger stress ratio is needed to cause sufficient void volume change within the grouted sand specimen. Such a large stress ratio tended to cause a sudden increase in pore water pressure at the initial stage (Figure 9). However, grouted sand specimen managed to withstand many more loading cycles before it reached initial liquefaction (Figure 8). For

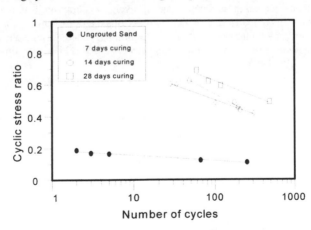

Figure 8. Liquefaction resistance of grouted and ungrouted sand specimens

example, more than 50 loading cycles (N) are needed to initiate liquefaction in grouted sand specimen at the stress ratio of 0.7 and cured for 28 days.

The double axial strain amplitudes measured from ungrouted sand and grouted sand specimens during testing are as follows: 16% (ungrouted sand), 10.67% (7 days curing), 9.33% (14 days curing), and 10% (28 days curing). It indicated that the liquefaction induced strain for ungrouted sand was about 1.6 times larger than that of grouted sand. Curing time had little effect on the liquefaction induced strain. In addition, sudden increase in specimen deformation was observed when ungrouted sand reached its initial liquefaction. In comparison, the deformation of grouted sand specimens increased gradually with cyclic loading. Cyclic mobility was observed in the grouted sand.

Conclusions

The following conclusions can be advanced based on the findings from the experimental study on liquefaction resistance of colloid silica grouted sand:

1. As long as the voids of sandy soil can be filled with grout, the liquefaction resistance of soil can be significantly increased. Even the strength of the grout is low.
2. Up to 4 ~ 7 folds increase in stress ratio was needed to initiate liquefaction in grouted sands (28 days curing) compared to ungrouted sand. Similarly, the loading cycles needed to initiate liquefaction were increased to more than 50 cycles at the stress ratio of 0.7 for grouted sand compared to 5 cycles at the stress ratio of 0.15 ~ 0.2 for ungrouted sand.
3. The liquefaction mechanism of sand had been significantly changed due to the existence of colloid silica gel in the voids. The axial deformation of grouted sand specimens increased gradually during cyclic loading until initial liquefaction occurred. In comparison, a sudden increase in axial deformation of ungrouted sand specimen was found when initial liquefaction happened. Test results indicated that the liquefaction induced strain of ungrouted sand was about 1.6 times larger than that of colloid silica grouted sand.

Acknowledgements

The Authors wish to thank Prof. Ryo Yonekura for providing information on Perma Rock grout, the National Science Council of R.O.C. for providing financial support (NSC 89-2921-2-319-005-14) for this study, and the Permanent Grouting Association of Japan for providing colloid silica grout for testing.

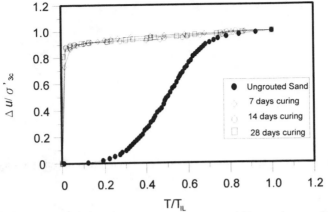

Figure 9. Change of pore water pressure response within specimens with time

References

Ishihara, K. (1985). "Stability of natural deposits during earthquake", *Proc. 11th International Conference on Soil Mechanics and Foundation Engineering*, Vol. 1, 321-376.

Yonekura, R. (1996). "The developing process and the new concepts of chemical grouting in Japan", *Proc. of IS-TOKYO'96, 2nd International Conference on Ground Improvement Geosystems*, Vol. 2, 889-901.

Yonekura, R. (2002). Personal Communication.

Lee, K. L., and Vernese, F. J. (1978). "End restraint effects on cyclic triaxial strength of sand", *Journal of the Geotechnical Engineering Division*, ASCE, 104(GT6), 705-719.

Krizek, R. J., Liao, H. J., and Borden, R. H. (1992). "Mechanical properties of microfine cement/sodium silicate grouted sand", *Proc. of Conference on Grouting, Soil Improvement and Geosynthetics*, New Orleans, 688-699.

A Study on the Optimal Mixture Ratio for Stabilization of Surface Layer on Ultra-soft Marine Clay

Byung-Sik Chun, Jin-Chun Kim

Abstract

Recently, as large constructions on the coast increase, application of a surface layer stabilization method, which is one of the improvement methods for dredged soft clay, have increased. However, there are few studies about this method. The purpose of this study is to clarify characteristics of ultra-soft marine clay and hardening agent, and to evaluate an optimal mixture ratio of hardening agent through the laboratory tests according to design of experiments. Laboratory test results were verified by statistical analysis and pilot tests. Laboratory tests were performed with hardening agents and test soils by the design of experiments, and regression equations according to relation of hardening agents materials and unconfined compressive strength were derived from the tests. Regression equations also verified its applicability to field by pilot tests.

From the results of tests, it was found that hardening agent materials such as cement, slag, fly-ash, inorganic salts, häuyne, gypsum, etc. affect compressive strength. It was defined that optimal mixture ratio which satisfies the required compressive strength from the statistical analysis. Also, it was compared that the effect of ground improvement by cements and hardening agents through the pilot tests. This study will serve data for design or construction criteria of stabilization of surface layer on ultra-soft marine clay.

----

[1]Professor, Department of Civil Engineering Hanyang Univ. 17 Haengdang-dong Seongdong-ku, Seoul, R.O.Korea; phone 822-2290-0326; hengdang@unitel.co.kr

[2] President, Ph. D, Korea Institute of Geo Technology Inc., #90-18 Songpa-dong, Songpa-gu, Seoul, R.O.Korea;kig-2000@hanmail.net

Introduction

Recently, Seoul-Pusan high speed railway, In-Chon international airport in Yongjong Island, subway of metropolises and shore structure of west and south offshore are constructed to very large scale in Korea. Structures are being constructed under poor conditions more and more. Because nation territory is small in area, extremely soft ground improvement and stabilization are necessary to territory expansion by offshore development and reclamation (KICT, 1998).

As large constructions increase in offshore, new stabilization methods for dredged soft soil are introduced. Drain and piling are preferred stabilization methods of dredged soil, but recently, stabilization methods using hardening agents are many executed for instant construction (KNHC, 1998; B.S. Chun, 1996). Therefore, design and specification of soil stabilization methods are needed for surface ground stabilization of extremely soft dredged ground.

In this study, preliminary tests were performed to get optimal mixture ratio of stabilizer ingredient, and Jin-Hae marine clay was used to get physical and chemical properties. Laboratory tests using 50 stabilized soils were performed to get optimal mixture ratio for 6 types of 16-stabilizer material, and standard mixing tables of hardening agent were determined according to ground conditions through statistical analysis. Also, it was compared the effect of ground improvement by cements and hardening agents through the pilot tests.

Laboratory Tests

The test soils are extremely soft dredged clays in Jin-Hae (at Kyung-Sang south province in Korea). These soils were dredged by using dredge pump from the bottom of the sea to the inside of the temporary revetment and water content was about 100%.

Properties of test soils. Laboratory physical and mechanical test results on subject soil are shown in Table 1. Table 2 and Figure 1 show the results of X-Ray fluorescence analysis and X-Ray Diffration analysis.

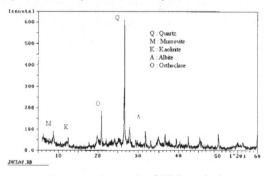

Figure 1. The result of XRD analysis

Table 1. Laboratory physical and mechanical test results

Selected test		Test results	
Water content (%)		98.0 ~ 101.0	
Specific gravity		2.65	
Liquid limit (%)		55.8	
Plasticity index		26.4	
Group symbol		CH	
Undrained shear strength (kg/cm²)		0.17	
Compaction Test	γ_{dmax} (t/m³)	1.303 (1.515)	Standard Proctor Test
	O.M.C (%)	24.88 (20.00)	(Modified Proctor Test)
Consolidation Test	C_C	0.607	
	P_C (kg/cm²)	0.27	Mikkasa method
	C_V (cm²/sec)	$5.29 \sim 8.96 \times 10^{-4}$	Log-time method

Table 2. The result of XRF analysis

Item	Chemical component (%)										
Subject soil	SiO_2	Al_2O	Fe_2O_3	CaO	MgO	Na_2O	SO_3	K_2O	TiO_2	P_2O_5	Ig-loss
	54.21	17.94	6.80	1.27	1.78	2.53	0.28	2.93	0.70	0.12	11.44

Chemical Properties of hardening agents. In this study, Hardening agent materials are selected as follows.

Table 3. Hardening agent materials

Material of hardening Agent	Level 1	Level 2	Level 3
Portland cement (P)	Type 1	Type 3	-
Slag powder (S)	4000 cm²/g (30 %, 50 %, 70 %)	6000 cm²/g (30 %, 50 %, 70 %)	8000 cm²/g (30 %, 50 %, 70 %)
Fly ash powder (F)	4000 cm²/g (10 %, 20 %, 30 %)	6000 cm²/g (10 %, 20 %, 30 %)	-
Gypsum (G)	5 %	10 %	15 %
Häuyne (A)	5 %	10 %	15 %
*Inorganic saline (M)	1 %	2 %	3 %

*Inorganic saline (M) comprises KCl, NaCl and $MgCl_2$ (KCl : NaCl : $MgCl_2$ = 5 : 3 : 2).

Table 4. Chemical components of hardening agent

Item	Chemical component (%)										Sum
	SiO_2	Al_2O_3	Fe_2O_3	CaO	MgO	Na_2O	SO_3	CaO	Ig-Loss	K_2O	
Type1 cement	20.86	5.67	2.74	62.52	3.36	0.14	2.34	0.51	1.32	1.05	100
Type3 cement	19.74	5.80	3.29	62.70	2.30	0.11	3.98	0.80	1.18	0.89	100
Slag	33.33	15.34	0.44	42.12	5.70	0.26	2.08	0.00	0.27	0.45	100
Häuyne	6.51	36.57	1.67	41.72	1.11	0.10	10.50	0.30	1.19	0.63	100
Gypsum	2.38	0.32	0.04	39.56	-	0.01	55.57	-	2.08	0.04	100

Preliminary tests. The unconfined compression tests were carried out for the soils treated by the hardening agent mixture (48 cases) and cement (Type 1 and 3).

Table 5. Hardening agent mixture

Base Cement	Mixture 1	Mixture 2	Mixture 3
P1 (Type 1 Portland cement)	P1S4_3	P1S4_5	P1S4_7
	P1S6_3	P1S6_5	P1S6_7
	P1S8_3	P1S8_5	P1S8_7
	P1F4_1	P1F4_2	P1F4_3
	P1F6_1	P1F6_2	P1F6_3
	P1G5	P1G10	P1G15
	P1A5	P1A10	P1A15
	P1M1	P1M2	P1M3
P3 (Type 3 Portland cement)	P3S4_3	P3S4_5	P3S4_7
	P3S6_3	P3S6_5	P3S6_7
	P3S8_3	P3S8_5	P3S8_7
	P3F4_1	P3F4_2	P3F4_3
	P3F6_1	P3F6_2	P3F6_3
	P3G5	P3G10	P3G15
	P3A5	P3A10	P3A15
	P3M1	P3M2	P3M3

*P : Portland cement; S : slag powder; F : fly ash powder; G : gypsum; A : häuyne; M : inorganic saline
*In this table, the term of P3S4_3 means that the amount of cement and slag powder is 70 % and 30 % of the amount of hardening agent (cement+ slag powder) of the total amount of hardening agent, and S4 means slag powder having specific surface of 4000 cm^2/g, and so on.

Preparation of test specimens. The specimens were prepared in accordance with KS F 2329 and unconfined compression test method in Japan. The procedures for preparation of specimens are as follows:

1. After water content determination by weight of oven-drying soil only, the specimens are prepared by mixing soil and hardening agents.
2. The amount of cement that is main hardening agent is 100 kg per specimen volume of 1 m^3.
3. Each additive is mixed in soil by the ratio of additive to cement.
4. The specimens are molded with water content of 100 %, increasing the amount of water content continued till specimens could be molded.
5. The specimen is prepared by mixing cement, water, additive and subject soil in mold.
6. For each condition, three specimens are prepared of size ø5 cm×H10 cm.

Preliminary test results. The unconfined compressive strength of subject soil was 0.34 kg/cm^2, that of treated soil by Type 1 Portland cement was 0.72 kg/cm^2 and that of treated soil by Type 3 Portland cement was 0.79 kg/cm^2. From test results, as water content increased, the unconfined compressive strength of specimen decreased. In cases inorganic saline or fly ash were added, the unconfined compressive strength was less than 0.8 kg/cm^2, which was lower than different additives. In cases slag powder and fly ash were added, the application was proportional to the unconfined compressive strength.

Main tests. The hardening agent materials used in the main tests were Portland cement (Type 1, 3), slag powder (specific surface of 6000 cm^2/g) and the mixture of gypsum and arwin. The unconfined compressive strength was selected as the response value. Water content (ω), the amount of hardening agent (HA) and mixing ratio of slag powder (S) were selected as Factors. Factors and their levels are shown in Table 6.

Table 6. Factors and their levels chosen in the experiments

Factors	-1.216	-1	0	1	1.216
Water content (%)	89.2	100	150	200	210.8
Amount of hardening agent HA (kg/m^3)	89.2	100	150	200	210.8
Mixing ratio of slag S (%)	25.68	30	50	70	74.32

Table 7. Design matrix

Test No.	Water content (%)	Hardening agent (kg/m³)	Slag powder (%)	Admixture J
1	100	100	30	
2	100	100	70	
3	100	200	30	
4	100	200	70	
5	200	100	30	
6	200	100	70	
7	200	200	30	Total amount of
8	200	200	70	hardening agent ×
9	150	150	50	10 %
10	89.2	150	50	
11	210.8	150	50	
12	150	89.2	50	
13	150	210.8	50	
14	150	150	25.68	
15	150	150	74.32	

*Admixture J comprises gypsum and häuyne (gypsum : häuyne = 2 : 1).

The result of main test. The unconfined compression tests using triaxial compression apparatus by strain-controlled method were performed for three specimens in same condition. The mean value of three specimens represents unconfined compressive strength, and the test results are presented in Table 8.

Analysis of main test results. Based on the main test results, second order regression analysis and response surface analysis were performed using quality management software "Quality Plus (Q+)".

Response surface analysis. Response surface analysis was performed to know the relationship between each independent variables (water content, amount of hardening agent, slag powder) and a dependent variable (unconfined compressive strength). Response surface curves based on test result and response surface analysis are presented in Figure 2.

Table 8. The result of unconfined compression tests

Test No.	Water content (%)	Hardening agent (kg/m³)	Slag Powder (%)	Soil weight (g)	Water weight (g)	Cement weight (g)	Slag weight (g)	(gypsum + arwin) (g)	Total (g)	Type 1 Portland cement 7-day	Type 1 Portland cement 28-day	Type 3 Portland cement 7-day	Type 3 Portland cement 28-day
1	100	100	30	160	160	12.3	5.3	1.96	19.6	207.78	313.72	235.05	355.32
2	100	100	70	160	160	5.3	12.3	1.96	19.6	158.63	285.57	183.35	333.15
3	100	200	30	160	160	24.7	10.6	3.92	39.2	330.89	594.39	368.17	647.95
4	100	200	70	160	160	10.6	24.7	3.92	39.2	393.97	602.63	431.93	668.26
5	200	100	30	106.7	213.3	12.3	5.3	1.96	19.6	99.38	174.81	110.95	189.92
6	200	100	70	106.7	213.3	5.3	12.3	1.96	19.6	93.69	168.54	100.26	180.50
7	200	200	30	106.7	213.3	24.7	10.6	3.92	39.2	177.17	274.39	193.06	304.31
8	200	200	70	106.7	213.3	10.6	24.7	3.92	39.2	132.14	256.53	155.78	282.33
9	150	150	50	128	192	13.23	13.23	2.94	29.4	206.40	302.15	234.36	353.06
10	89.2	150	50	169	151	13.23	13.23	2.94	29.4	354.83	588.99	392.60	648.44
11	210.8	150	50	103	217	13.23	13.23	2.94	29.4	143.52	216.80	155.29	242.11
12	150	89.2	50	128	192	7.88	7.88	1.75	17.5	129.79	214.45	136.85	224.75
13	150	210.8	50	128	192	18.59	18.59	4.13	41.3	447.24	671.00	497.47	753.90
14	150	150	25.68	128	192	19.66	6.8	2.94	29.4	129.49	246.43	139.60	265.65
15	150	150	74.32	128	192	6.8	19.66	2.94	29.4	140.38	260.46	151.07	276.35

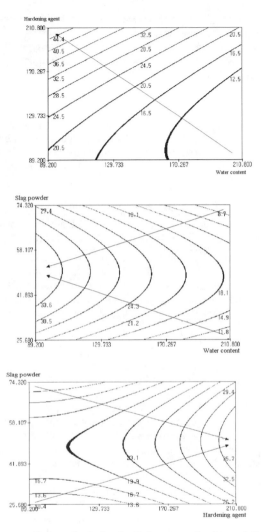

Figure 2. Response surface analysis for the 7-day strength of type 1 Portland cement

In Figure 2, the arrow indicates the direction of strength increase and the values in graphs are unconfined compressive strengths. From the Figure, the compressive strength increases as water content decreases and the hardening agent increases. The compressive strength is maximum value when the amount of slag powder is 50% of the hardening agent and the compressive strengths have little differences when the amount of slag powder is 30% and 70% of that of hardening agent, the unconfined compressive strength of treated soil is three-forth of the strength of 50 % slag ratio.

The relationship between each independent variables and a dependent variable, is easily determined from response surface curve.

In the result of response surface analysis, the strength of type III portland cement showed average 1.1 times as large as type I portland cement and 7-day strength was 55~65 %, about 60 % of 28-day strength.

In-situ Tests

In this study, surface layer stabilization method was performed in the field to verify its effectiveness on improvement of ground surface. The testing site in the temporary revetment is located in Jin-Hae (at Kyung-sang south province in Korea) and was reclaimed with soft marine clay. The properties of subject soil of this site were shown in Table 9.

Table 9. The properties of subject soil

Specific gravity	Natural water content (%)	Liquid limit (%)	Plasticity index (%)	Undrained shear strength (kPa)	Group symbol
2.65	98-101	55.75	26.43	16.68	CH

The two 7 m square testing sites (A/B site) were constructed with same condition. Type 1 Portland cement and hardening agents (200 kg/m^3) were used at A and B site, respectively, to compare the test results each other and laboratory test results.

Unconfined Compression Tests. After construction, to obtain the soil sample from test site, the sampling tube that is made of cylindrical steel pipe 500 mm long and 50 mm in diameter was inserted into the improved site. In 24 hours, soil samples were extruded from the tubes and were put in curing room. The curing time was 7 and 28 days to compare with laboratory test results. The test results were shown in Table 8.

From the test results, in the case of A site (the site improved by cement only), the unconfined compressive strength of 7 days curing was 166.6kPa and that of 28 days curing was 285.2kPa. In the case of B site (the site improved by hardening agents), the unconfined compressive strength of 7 and 28 days curing was 307.7kPa and 510.6kPa, respectively. The test results also show that the strength of 7 days curing was about 60 % of that of 28 days curing, which coincide with that of laboratory tests, and the unconfined compressive strength of soil samples obtained from A site was about 55% of that of B site.

With the same mixture ratio, the unconfined compressive strength of soil samples obtained from B site was about two of third of that in the laboratory.

Table 10. The result of unconfined compression tests

	Site A		Site B	
	7 days curing (kPa)	28 days curing (kPa)	7 days curing (kPa)	28 days curing (kPa)
No.1	184.13	280.76	268.40	-
No.2	-	-	304.60	512.08
No.3	147.44	342.57	-	470.49
No.4	158.24	300.77	293.32	478.92
No.5	-	240.05	315.10	502.37
No.6	148.92	258.10	350.32	560.35
No.7	195.71	289.89	318.24	543.08
Ave.	166.77	285.47	308.03	511.10

Plate-Loading Test. Plate–load tests were performed to estimate the bearing capacity of improved surface layers in 14 and 28 days curing.

According to "Plate-Load Test (KS F 2310)", A 30 cm circular steel plate was used as bearing plates, and H-beam and concrete block as dead load. After the application of each load increments, the settlements were measured at elapsed times of 1, 2, 4, 8, 15, and 30 min. The load increments were more than 5 stages, which was determined by dividing the predicted bearing capacity.

Generally, a half of yield strength or a third of ultimate strength is taken as the allowable bearing capacity. The allowable bearing capacity, in this paper, was estimated from the yield load that was obtained the load-settlement curve.

From the test results, the yield strength of 14 and 28 days curing was 725.94kPa and 774.99kPa at A site, and 92 t/m^2 and 108 t/m^2 at B site, respectively.

In addition, the coefficient of subgrade reaction(κ) is usually the value of q/δ at δ=1.25 mm in the load-settlement curve and it is influenced by the size and shape of bearing plate.

The test results show that the yield strength of 14 days curing was 90 % of that of 28 days curing and that the bearing capacity of site B was 1.3 to1.4 times more than that of site A.

Table 11. The result of plate load tests

	Yield strength (kPa)	Allowable bearing Capacity (kPa)	Settlement at yield strength (mm)	The coefficient of subgrade reaction (MN/m³)
A site (14 days curing)	725.94	362.97	7	133.42
A site (28 days curing)	774.99	387.50	5.5	156.96
B site (14 days curing)	902.52	451.26	7	164.81
B site (28 days curing)	1059.48	529.74	6	2432.88

Conclusions

The analysis of improvement effect of the treated soil by hardening agents by response surface analysis and the result of the in-situ tests are as follows.

1. The test soils, very soft marine clay in Jin-Hae, was classified CH by Group Symbol (USCS). And the unconfined compressive strength of subject soil was 33 kPa, that of treated soil using Type 1 Portland cement was 0.70 kPa and that of using Type 1 Portland cement was 77 kPA.

2. From the result of preliminary tests, inorganic saline and fly ash were ignored because the strength of specimens lower than different additives. So it is decided to use cement (Type 1 and 3), slag powder, Arwin and gypsum in this test by experiments design.

3. According to response surface analysis, the strength of Type 3 cement was 1.05 - 1.2 times as strong as that of Type 1 cement in the same condition, the average was 1.1 times. The strength for 7 curing days was 55 - 65% of that of 28 curing days, the average was 60%.

4. The unconfined compressive strength of the treated soil by hardening agents was maximal when the slag was 50% of the total amount of hardening agents and was not that different between 40% and 60%. Also there was almost no difference of unconfined compressive strength of the treated soil when the slag ratio was 70% and 30% of the total amount of hardening agents. And the strength was three forth of maximum strength when slag ratio was 50% of the total amount of hardening agents. From these results, it showed that slag is less influential than cement for strength as the more ratio of slag in hardening agent increase, the more the ratio of cement decrease in it.

5. According to in-situ testing on soft marine clay with water content of 100 %, the unconfined compressive strength of soil samples obtained from A site improved with cement was 55 % of that of soil samples obtained from A site improved with hardening agent, and the unconfined compressive strength of field was two third of that laboratory. Also, the strength of 7 curing day was about 60 % of that of 28 curing day and these result was coincide with that of laboratory test.

6. From the result of plate–load test, the allowable bearing capacity of 14 and 28 days curing was 362.6kPa and 387.1kPa at A site where the surface layer was improved with cement only, and 450.8kPa and 529.2kPa at B site where the surface layer was improved by using hardening agents, respectively. Also, it is shown that the yield strength of 14 days curing was 90 % of that of 28 days curing, and that the bearing capacity of site B was 1.3 - 1.4 times more than that of site A.

Acknowledgements

This paper is based on the research supported by Korea's Ministry of Construction & Transportation (MOCT), and the authors acknowledge the financial and kind support provided by the MOCT.

References

Bergado D.T., Anderson L.R., Miura N., and Balasubramaniam A.S. (1996). *Soft Ground Improvement in Lowland and other environments*, ASCE press, 1-9., 234-304.

Chun, B. S. & Choi G. S. (1996). "A Study on Soil Improvement Effects under Poor Ground Conditions", *Journal of the Korean geotechnical society.*, Vol. 12, No.2, 115-130.

Housing Research Institute, Korea national housing corp. (1998). *A study on the applying standard of soft ground improvement method*, 2-84

Im, J. S. etc. (1996). *The dictionary of geotechnical engineering term*, Engineers Book Company: 94, 367.

Japan Institute of Construction Technology Education. (1999). *Standard soil concrete test handbook*, 76-77.

Korea Institute of Construction Technology. (1988). *A study on the soft ground with shallow stabilization method* , 33-65

Kuno. (1994). *Ground improvement manual by hardening agent passed of cement type 2nd edition*, Japan Cement Institute, 232-257.

Moctsuo. (1992). *Soil stabilization method handbook*, Nikkan Kogyo, 1-14., 175-242.

Park, S. H. (1998). *The modern experiments design*, Minyeongsa Book Company, 347-385.

Park, S. H. (1987). *The regression analysis*, Daeyeongsa Book Company, 411-471.

A method for measuring and evaluating the penetrability of grouts

Magnus Eriksson[1]
Håkan Stille[2]

Abstract

This paper suggests a method for measuring the penetrability of grouts. The proposed method evaluates the penetrability of the grout based on measurements with a newly developed device where the grouts ability to pass filters of different widths is measured. After evaluation two especially descriptive parameters are obtained, a minimum and a critical aperture. These two parameters describes the penetrability of the grout and can be used for comparing grouts and in modelling grout propagation.

In the paper some measurements and evaluated results are presented. This is to study some governing factors for the penetrability of grouts but the main objective is to demonstrate the proposed method. The results confirms that the main factor governing the penetrability is the type of cement used and that variations in w/c ratio and addition of superplasticizers only have limited influence. Measurements on the filter cake shows that the thickness and the density are influenced by pressure and w/c ratio.

Introduction

Grouting in rock is a practical method to use for sealing rock from the flow of water. Common applications are to seal the foundation of a dam or to seal tunnels from inflow of water.

To facilitate development of grouting technique and predictions of grouting result research aims to clarify fundamental issues involved. One issue is for instance the flowability of cement based grouts. Flow properties of suspensions differ from those of Newtonian fluids and the flow equations are most commonly based on the Bingham rheological model (Wallner, 1976, Hässler et al, 1992; Amadei, 2001). A second issue related to cement based grouts is the limited penetration ability. This issue is of relevance for describing which fractures can be grouted and to what extent. Some authors use a limit on aperture where the grout can not penetrate (Moon & Song, 1997 and Amadei,

---

[1] Corresponding author. Division of Soil and Rock Mechanics, Royal Institute of Technology, S-100 44 Stockholm.

[2] Professor at the Division of Soil and Rock Mechanics, Royal Institute of Technology, S-100 44 Stockholm.

2000). However, research in recent years has shown that the difference in penetrability among grouts can be wide (Hansson, 1995; Schwarz, 1997; Eriksson et al, 1999; Eriksson et al, 2000) and to know their penetrability measurements are required. It has also been shown that the limited penetration ability leads to filtration of the grout and that this filtration process is different for different apertures (Feder 1993; Hanson, 1995; Eriksson et al, 1999). Which factors governs the penetrability is not fully understood and research is continuing in this field.

The work presented in this paper suggests a method for measuring and evaluating the penetrability of cement based grouts. The limited penetrability of cement based grouts has in recent years been a focus of grouting research and development to find means to increase the value of predictions and to improve the grouting result. The method presented in this paper is based on developed equipment, called "penetrability meter". Measurements with the penetrability meter results after evaluation in two main parameters, a critical and minimum aperture, which gives a span where filtration occurs in the grout and describes the penetrability. The equipment also facilitates studies of the filter cake that is formed where filtration occurs.

The suggested method gives both possibilities to examine the penetrability of grout and to obtain input parameters for modelling of grout propagation. A model to include the penetrability of the grouts in numerical calculations has been developed and presented in Eriksson et al (2000). Laboratory experiments (Eriksson, 2001) and a small field test have verified this model (Eriksson, 2002).

Some about penetrability of cement based grouts and measuring methods

The penetrability of cement based grouts is limited compared to water. This is due to the presence of particles in the suspension. Schwarz (1997) showed that the limited penetrability ability can cause both a mechanical and chemical filtration of the grout.

Mechanical filtration arises due to the presence of particles in the grout suspension. In Martinet (1998) it was shown that a valve of particles is stable if it is formed out of not more than three particles. This explains the empirical finding by Hansson (1995) where based on laboratory testing it was found that grouts with a d_{95} of less than 3 times the opening size had good penetrability. To increase the penetrability of grouts the development recent years have been towards more fined grained cements, and so called micro cement products have become commonly used. Cements with a maximum particle size (d_{95}) not larger than 15-20 μm are commonly referred to as micro fine or super fine cements.

The properties of micro cements differ from more coarse-grained cements since a larger portion have colloidal behaviour. Physio-chemical aspects on the penetrability of micro fine cements was studied in Schwarz (1997) and in respect of dispersion, grouts exhibiting strong flocculent behaviour can show low penetrability. Another aspect is attachment of particles to rock surfaces due to steric forces in between particle and rock surface.

Filtration can be observed as either as filter cakes forming or traces of chemical filtration. Filter cakes were observed in Eriksson et al (2000) and in Eriksson (2001). The filter cake is characterised by a higher density than the initial grout density and has a low permeability. Chemical filtration, where more soluble components have travelled further than the particles was for instance observed in Lagerblad (1998) and in Eriksson (2002).

Some authors suggests that the penetrability of grouts depends on the mix. Feder (1993) and Hansson (1995) for instance discusses that the penetrability depends on w/c ratio and on amount and type of additive. This was investigated in Eriksson et al (1999) and it was concluded that the influence on penetrability due to variations in w/c ratio and type and amount additive is limited and that the main influence arises from the cement type.

Another aspect is how the penetrability changes over time. This was studied in Eriksson et al (1999) where grouts based on two different kinds of cements with different particle size distributions were compared at different w/c ratios. It was shown in that study that both grouts showed time dependent behaviour in respect of penetrability. Especially strong time dependency was noticed in the grout based on a micro cement ($d_{95} < 12$ μm).

Different methods for investigating the penetrability of grout are available. In Eriksson et al (1999) the results of three different methods were compared. One method was the commonly used sand-column test (see eg. Viseur & Barrioulet, 1998). In this test the grout is injected in sand of a specified fraction. The other two methods were a column test (Feder, 1993 and Brantberger & Nelson, 1998) and the filter pump test (Hansson, 1995). In the column test, grout is injected through narrow column of specified aperture. In the filter pump test grout is sucked up through a filter of specified width.

All these methods have advantages and disadvantages. One disadvantage with all three methods is that they only give relative measures of the penetrability and that they do not give necessary input data for modelling. Therefor the motivation for developing the penetrability meter was to obtain a method for measuring the penetrability which also results in parameters usable in calculations of grout spread.

Proposed method

The proposed method for measuring and evaluating the penetrability of grouts is developed to give parameters for grouting predictions. The principle behind the method is to evaluate a critical and minimum aperture of the grout. The minimum aperture should represent an aperture limit under which no grout can enter an opening. The critical aperture should represent an upper limit over which an infinite amount of grout can pass. Between these two values a finite volume of grout passes. The volume is restricted since a filter cake forms and blocks further flow. Figure 1 illustrates this behaviour.

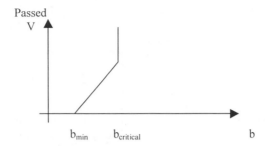

Figure 1. Illustration of the minimum and critical aperture.

The method uses an injection equipment called a penetrability meter. This equipment is shown in Figure 2. The grout is stored in a pressurised container where an outlet pipe is attached. The outlet pipe has a valve and a cap holder. In the cap holder a cap with a filter is positioned. The grout is pressed through the filter with a defined aperture and by using a number of caps with a range of filters the flocculation of particles and the passed volumes at different apertures is simulated.

Figure 2. Picture of the developed measuring device. Filter caps and measuring cylinders are seen in the picture as well as a stirring device used between measurements

The circular filters (radius 15mm) that are used consist of a mesh of thin threads that gives a number of rectangular openings. The opening have size $x \cdot x$ where x is assumed to be the equivalent aperture opening. Used widths of the filters for examining the penetrability of grouts are most often between 35 μm and 250 μm.

In the evaluation of the results the passed volume is plotted against the filter widths and the critical and minimum aperture is registered as illustrated in Figure 1 above. Over the interval where filtration is noted, a linear fit function is given to the measured values. This eliminates some of the randomness in the singular measurement. The equation of this linear function is used to define the minimum and critical aperture. As will be shown in the next section a maximum injected volume to measure must be determined. When the measured volume is larger than the maximum volume and no filter cake is formed in the filter no larger filter widths needs to be measured.

To obtain the time dependency in the penetrability several series of measurements at distinct time intervals are made. The two parameters, $b_{critical}$ and b_{min}, are evaluated for each series of measurement and plotted against time. This procedure is illustrated in Figure 3.

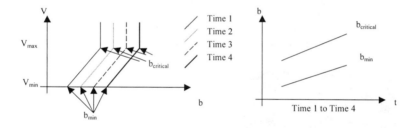

Figure 3. Illustration of how measured values are evaluated in respect of $b_{critical}$ and b_{min} and plotted against time

An example of a measurement and how to obtain the time dependency of the penetrability is given below. The measurements were made in connection with a field experiment (Eriksson, 2002). Three different series of measurements were made and the grout volume that passed through filters ranging between 35 μm and 120 μm was measured. The grout used a micro cement (UF12) with a w/c ratio of 2.5 and no additives. Based on the measurements, the b_{min} value is determined where the passed amount starts to increase and the $b_{critical}$ value where 100% of the maximum volume passed the filter. Each evaluated value is plotted against time. In Figure 4 the individual measurements on the grout at time intervals of 10, 20 and 30 minutes are shown and in Figure 5 the evaluated time dependency is shown.

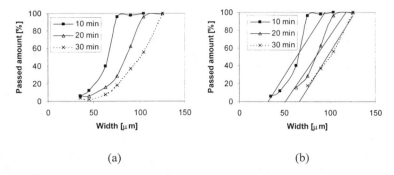

(a)   (b)

Figure 4. In (a) the measurements made at different time intervals after mixing are shown. In (b) the regression lines based on the used values are shown. The grout is based on micro cement (UF12) with a w/c ratio of 2.5 and no additives.

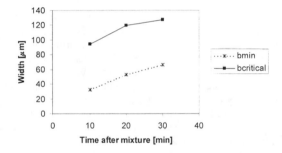

Figure 5. Penetrability of a grout as a function of time from Figure 4.

There are several measured parameters from the method that influence the result when modelling. Most obvious are the critical and minimum apertures that are obtained directly based on the curve showing passed volume against filter widths. Other parameters are the length and density of the filter cake and the density of the filtered grout. In the proposed method the filter cake that forms can be studied in respect of thickness and density as well as the filtrate. Results of this is presented in the next section.

Measurements

Introduction

In this section measurements are presented. The objectives are to demonstrate the method and to investigate some governing factors related to the penetrability of grouts.

The measurements are made to investigate how differences in w/c ratio and the addition of superplasticizer influence the penetrability of grouts. It will also be examined if the results are pressure sensitive. The different mixes that are tested all use a cement with $d_{95}=30\mu m$. The cement is a Portland clinker, developed for grouting in rock. Two different w/c ratios are tested, 0.7 and 1.0 and in some mixes 0.7 % weight superplasticizer is added. Two different pressure levels, 1 bar and 2 bar, are examined. Table 1 presents the six grouts that are tested in all together twelve tests.

Table 1. Grout mixes tested in the experiments

Grout	W/C	Additive [%]	Pressure [Bar]
A	1	0	1
B	1	0.7	1
C	1	0.7	2
D	0.7	0	1
E	0.7	0.7	1
F	0.7	0.7	2

All mixes are tested at two different time intervals, 5 minutes after mixing and 30 minutes after mixing. After the 5 minutes test the cap and the filter was cleaned and the grout in the container was kept in motion between the tests. All tests were made indoor, at a temperature of $20°$ C with a cement stored at $20°$ C mixed with tap water with an temperature of around $13°$ C.

Each test is presented with the evaluated parameters b_{min} and $b_{critical}$ and the thickness and density of the filter cake formed. In one test (Test B, 30 minute measurement) the filtrate is examined in respect of density. The filter cake thickness was measured with a slide gauge with an instrument accuracy of 1/100 mm. The estimated practical accuracy is 1/10 mm. The density of the filter cake and the filtrate was measured with a small cylinder cup with a defined volume. The cup was weighed empty and filled. The practical accuracy of the density estimation was found to be ±1%.

Results

To find the maximum volume necessary to measure a first measurement was made where as large volumes as possible with the equipment was measured. Figure 6 shows the full measurement of grout A in the penetrability meter. At a filter width of 310 mm all 20 litres in the container passed and no filter cake formed. At all filter widths smaller that 310 mm a finite volume was obtained and filter cakes formed. This indicates that the critical aperture of that grout should be in the range of 310 mm. It is however seen in Figure 6 that two distinctly different trends are seen when the results are plotted in a lin-log scale. At filter widths smaller than approximately 125 µm one trend is seen and at filter widths larger that 125 µm a second trend is seen. This is in conflict with the expected behaviour presented in Figure 1.

This result is explained by the presence of oversized particles. In the trend below the 125 µm filter it is the behaviour of the grout that is seen and in the trend over the 125 µm filter the effect of oversized particles is seen. These oversized particles could visually be observed on the filters. The oversized particles could emanate from the cement or from the mixing as well from poorly cleaned equipment. Based on these results, one could conclude that for this grout the maximum volume necessary to measure was 1 litre.

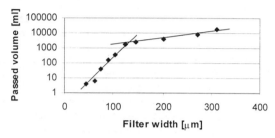

Figure 6. Passed volume at different filter widths in test on grout A.

The results concerning the evaluated parameters of critical and minimum aperture size for each grout are presented in Table 2. These were evaluated based on a maximum volume of 1 litre.

Table 2. Evaluated critical and minimum apertures in the different grouts.

Test	5 minutes after mix		30 minutes after mix	
	$b_{critical}$ [µm]	b_{min} [µm]	$b_{critical}$ [µm]	b_{min} [µm]
A	134	73	134	71
B	127	71	137	72
C	131	71	138	73
D	140	74	140	74
E	140	74	141	74
F	138	73	139	73

The results presented in Table 2 represents a limited number of test results and statistically significant results can not be concluded based on this material. However, some indications can be noted in the presented material. It is seen that all results are found in a reasonable small range, indicating that the investigated grouts are more or less the same from a practical point of view. This agrees with the findings presented in Eriksson et al (1998) where it was noted that the main factor governing the penetrability is the cement type used. Comparing the results of these grouts to the earlier presented one (see Figure 5) show differences that can be obtained when different cements are used.

In Table 2 the results A compared to D, B compared to E and C compared to F should indicate the difference in penetrability between the two choices of w/c ratio. Since all values are lower in the higher w/c ratio cases (A-C) than in the lower w/c ratio cases (D-F), this gives an indication that penetrability to some extent is w/c dependent. It is seen that larger differences occurs in the value $b_{critical}$ than in the b_{min} value which shows that $b_{critical}$ is more sensitive to variations in w/c ratio than b_{min}.

The influence of superplasticizer is based on the presented measurements not found to result in any systematic difference. The grouts without superplasticizer show the poorest penetrability in the 5 minute measurement but not in the 30 minute measurement. In the dilute grout a pronounced difference is seen with and without superplasticizer but in the thicker grout no effect is noticed. Based on these results, no conclusion concerning any positive or negative effect on penetrability from addition of superplasticizers can be made. However, one other observation can be made. The results show that the grout containing superplasticizers have more pronounced time dependency that the grouts without superplasticizer. It is noted that the results only represents one kind of superplasticizer.

From comparing the results for different pressures it is seen that no systematic difference is obtained. In the dilute grout the penetrability is lower for the high pressure and in the thicker grout a higher penetrability is noticed. The differences are noted to be small. It is therefor concluded that in the studied range of pressures, this does not influence the evaluation of the critical and minimum aperture.

Measurements of the filter cake length and density were also made. The results from these measurements are presented in Table 3 and Table 4.

Table 3. Measured filter cake thickness and density in the measurements 5 minutes after mixture.

Test	5 minutes after mix									
	Thickness [mm]					Density [kg/m^3]				
	45	63	75	90	104	45	63	75	90	104
A	9.4	9.0	11.4	9.7	7.0	1.80	1.78	1.85	1.80	1.80
B	8.9	9.7	13.2	11.3	7.6	1.79	1.89	1.79	1.68	1.89
C	11.6	13.8	13.3	13.2	13.7	1.80	1.80	1.80	1.90	1.80
D	10.9	10.3	12.5	10.7	10.5	2.00	1.89	2.00	2.00	2.00
E	9.4	9.5	10.6	13.2	11.2	1.89	2.00	2.00	2.11	2.00
F	15.4	14.0	14.2	13.0	14.1	2.00	1.89	2.00	2.00	2.00

Table 4. Measured filter cake thickness and density in the measurements 30 minutes after mixture.

Test	30 minutes after mix									
	Thickness [mm]					Density [kg/m³]				
	45	63	75	90	104	45	63	75	90	104
A	10.0	9.0	10.5	9.2	8.7	1.82	1.87	1.87	1.89	1.82
B	12.7	10.4	9.8	10.0	9.0	1.89	1.89	1.79	1.79	1.79
C	14.8	14.8	14.0	12.8	15.9	2.00	1.89	1.79	1.89	1.79
D	10.2	10.8	12.0	10.2	7.0	2.00	2.00	2.00	2.00	2.00
E	10.2	10.8	12.0	10.2	7.0	2.00	1.89	2.00	1.89	1.89
F	14.7	13.8	13.8	12.2	12.2	2.00	1.89	2.00	2.00	2.00

The presented results show systematic differences in filter cake thickness and density. The results show that the filter cake thickness is pressure sensitive while the filter cake density is w/c sensitive.

In one test (Grout B) also the density of the filtrate was studied. The result from this is shown in Table 5. The results show that the filtrate has a reduced density in comparison to the initial, 1520 kg/m³. The density in the 45 μm filter is most interesting due to the low volume obtained in that test. In the other measurements, the densities that are measured are influenced by the density of the grout that first passes the filter, i.e by grout that has more or less initial density. For filters larger than 90 μm, no density reduction can be found based on the whole sample.

Table 5. Measured filtrate densities in test B

Filter	45	63	75	90
Density [kg/m³]	1164	1403	1404	1484

Discussion

This paper has presented a method for measuring and evaluating the penetrability of grouts. The method uses a device developed for measuring, called a penetrability meter, and from the evaluation of two parameters that are obtained which represents the penetrability of the grout.

The evaluation method aims to find two parameters in the grout that define the interval over which filtration occurs. If the aperture is larger that a critical aperture no filtration should occur and an infinite volume ought to be possible to grout. If the aperture is smaller then a minimum aperture a filter cake immediately forms and no actual amount of grout can enter. Between the critical and the minimum aperture filtration occurs and only a finite volume can be injected. Based on the results presented in Eriksson (2001) and Eriksson (2002) the description of the penetrability in accordance to this method gives good possibilities to predictions of grout spread in geometries where filtration can occur.

The tests showed that it is difficult to practically verify the critical aperture. This was found to depend on oversized particles in the grout which sooner of later blocks the flow. From analysing the test results it was however concluded that two separated curve trends are seen. The critical aperture that the grout in itself inhibits was evaluated from the first part of the curve, overlooking the influence of the oversized particles.

The results showed that evaluated critical and minimum aperture was found in a rather narrow range. This is natural since only one kind of cement was used. However, comparing to the earlier measurements on other kinds of cements that also was presented, reveals that considerable differences in evaluated parameters can be obtained.

Some presented tests tried to indicate if any difference in penetrability was noticed due to variation in w/c ratio, addition of superplasticizer and for different pressure. The most clear difference was found for different w/c ratios. Concerning how the penetrability was influenced by the superplasticizer and the pressure was less clear. Since a certain stochastic behaviour must be expected and since only one test of each mix was made only indications are possible.

The filter cakes were studied in respect of thickness and densities. As with the critical and minimum aperture the ranges in results were small. The tests indicate that the length of the filter cake is pressure sensitive and the density is w/c ratio sensitive.

References

Amadei, B., (2000). A Mathematical Model for Flow of Bingham Material is Fractures. Proc.4th NARMS Conf. : Pacific Rocks 2000, ''Rock around the Rim'', (J. Girard, M. Liebman, Ch. Breeds & T. Doe eds), Balkema, 2000.

Amadei, B., Savage, W.Z., (2001). An Analytical Solution for Transient Flow of Bingham Viscoplastic Materials in Rock Fractures. *Int. J. Rock Mech, & Min. Sci.* 38, pp 285-296.

Brantberger, M., Nelson, M., (1998). Hallandsås Ridge Railway Tunnel – Development of Pre-grouting Methodology. Proc. to Underground Constructions in Modern Infrastructure, (Franzén, Bergdahl & Nordmark eds), 1998 A.A. Balkema Rotterdam. ISBN 90 5410 964 5, pp 405-409.

Eriksson, M., Dalmalm, T., Brantberger, M., Stille, H., (1999). Bleed and Filtration Stability of Cement Based Grouts – A literature and Laboratory Study. *Proc. to the National Group ISRM, Rock Mechanics Meeting 2000, Swedish Rock Engineering Research*, Stockholm, Sweden, pp 203-225 (In Swedish, English abstract)

Eriksson, M., Stille, H., Andersson, J. (2000). Numerical Calculations for Prediction of Grout Spread with Account for Filtration and Varying Aperture, *Tunnelling and Underground Space Technology*, Vol 15, No. 4, pp 353-364.

Eriksson, M., (2001). Numerical Calculations of Grout Propagation Subjected to Filtration – Comparison to Laboratory Experiments, *Proc. Rock Mechanics – a Challenge for Society, (Särkkä & Eloranta, eds.), 2001* Swets & Zeitlinger Lisse, ISBN 90 2651 821 8, pp 567-572.

Eriksson, M. (2002). Grouting Field Experiment at Äspö Hard Rock Laboratory, *Tunnelling and Underground Space Technology*, Vol 17, No. 3, pp 287-293.

Feder, G., (1993). The Pressure Necessary to Start the Grouting Procedure. *Proc. of the International Conference on Grouting in Rock and Concrete*, Salzburg, A.A Balkema, ISBN 90 5410 350 7, pp 399-402.

Hansson, P., (1995). Filtration Stability of Cement Grout for Injection of Concrete Structures, *IABSE Symposiun, San Fransisco*, pp 1199-1204.

Lagerblad, B., (1998). Undersökning av cementbaserade injekteringsmedel i sprickor – Exempel från Äspölaboratoriet. *SKB, Work report AR D-98-18*, Stockholm. (in Swedish)

Moon, H.K., Song, M.K., (1997). Numerical and Experimental Analysis of Penetration Grouting in Jointed Rock Masses. *Int. J. Rock Mech, & Min. Sci.* 34:3-4, paper No. 206.

Martinet, P., (1998). Flow and Clogging Mechanisms in Porous Media with Applications to Dams. Doctoral Thesis, Division of Hydraulic Engineering, Dep. of Civil and Environmental Engineering, Royal Institute of Technology, Stockholm, Sweden.

Schwarz, L.G., (1997).Roles of Rheology and Chemical Filtration on Injectability of Microfine Cement Grouts, Dissertation Thesis, North Western University, Evanstone, Illinois, UMI number 9814310

Viseur, V., Barrioulet, M. (1998). Critéres d'inectabilité de coulis de ciments ultrafins (Criteria of injectability of very fine cement grouts). *Materials and Structures*, Vol. 31, July 1998, 393-399.

Selection criteria of polyurethane resins to seal concrete joints in underwater road tunnels in the Montreal area

J.P.Vrignaud[1], G. Ballivy[2], S. Perret[3], and E. Fernagu[4]

Abstract

Polyurethane resins are widely used in infrastructure repair technologies, particularly to stop water inflow in underground tunnels. However, the longevity of these chemical grouts is not well documented, as their application is relatively new. This is particularly true in Canada, and especially in the Montréal area, where tunnel walls are subject to many freezing and thawing cycles in one year, initiating some debonding and water inflow followed by ice formation.

This presentation is dedicated to the characterization of various polyurethane resins under such environmental conditions. Special laboratory tests have been developed to measure the bonding strength and shrinkage of these resins injected into cracked concrete or in improperly sealed expansion joints. These traction tests and cyclic expansion/contraction tests illustrated the major role of environmental conditions (humidity and temperature) on the effectiveness of these resins.

Several well-documented applications have been conducted during the rehabilitation of two major road tunnels in Montréal. The most effective imperviousness was obtained with a semi-rigid hydrophobic polyurethane resin, but it is also very important to specify how to prepare the fissures and how to inject these resins.

Introduction

For an injection project, the choice of adequate material can be made from a large number of products. The professionals who will be involved in such a project must understand the properties of each type of product in order to make the right choice. Nowadays, polyurethane resins are widely used in repairing the imperviousness of

---

[1] M.A.Sc, Dept. of Civil Engineering, Université de Sherbrooke, Sherbrooke, PQ, J1K 2R1, Canada
[2] Professor, Dept. of Civil Engineering, Université de Sherbrooke, Sherbrooke, PQ, J1K 2R1, Canada; phone 819-821-7115
[3] Ph.D. candidate, Dept. of Civil Engineering, Université de Sherbrooke, Sherbrooke, PQ, J1K 2R1, Canada; phone 819-821-8000; sperret@hermes.usherb.ca
[4] Trainee, Ecole des Mines de Nancy, France

concrete structures. Also used are such materials as silicate gels and acrylamide resins or acrylates (Naudts, 1995).

Silicate gels are very advantageous economically, but their longevity is generally rather low. Also, silicate gels possess a relatively high viscosity and can deteriorate considerably under important water inflow or during their chemical transformation. Silicate gels are mainly used to consolidate foundation soils that cannot be injected with a cement grout.

Acrylamide resins present the advantage of having a very low viscosity and an easily controllable setting time; however, there are problems with processing, toxicity and very low mechanical strengths.

Over the last few years, polyurethane resins have revolutionized the field of chemical grouts and widened their range of application, due to their suitability to a wide variety of uses. They are used to seal cracks or water inflows and seal construction or expansion joints. They are also used to consolidate or stabilize foundations (soil or rock), prevent movements in structures or fill cavities behind tunnel walls, masonry walls or under slabs.

Properties of polyurethane resins

Joyce (1992) defines three modes of action of polyurethane foam injected in a crack :
♦ a chemical action, through bonding of the foam to concrete;
♦ a physical action, through anchoring of the foam to pores and voids in concrete;
♦ a mechanical action, through pressure of the expansive foam in the crack.

The main properties of polyurethane resins injected into cracks can be explained through an understanding of how they react to water.

Polyurethane foams can be classified as two types according to their behaviour with water : hydrophilic foams and hydrophobic foams. Hydrophobic foams do not absorb water, while hydrophilic foams are capable of absorbing it, like a sponge. It is the type of isocyanate used in the reaction that gives the polyurethane foam the property of being hydrophobic or hydrophilic. Hydrophilic foam is the result of the reaction between a TDI (TolueneDiIsocyanate) type isocyanate, while water repellent foam is the result of the reaction of an MDI type (Diphenylmethane DiIsocyanate). Both types of resin have different behaviours, as listed in Table 1.

Hydrophilic resins present the advantage of bonding well to wet concrete, but the disadvantages of having poor durability and a poor volumetric stability to freeze-thaw and wetting-drying cycles. Another disadvantage of hydrophilic foam is that its structure depends very much on the quantity of water used during the reaction. Injection of a hydrophilic resin in the field must be coupled with an injection of water to ensure the resin has enough water to react. Excess water remains in the cells of the foam and if the material dries, there is a risk of shrinkage. The adjustment can be difficult in the case of cracks where there is a strong water inflow, since it is not really possible to determine the amount of water that will be used for the reaction.

On the contrary, hydrophobic polyurethanes present the advantage of being very stable to freeze-thaw and wetting-drying cycles, but the disadvantage of having a relatively low bond to wet concrete. To conduct the injection of a hydrophobic resin, as the quantity of water used for the reaction is very low, the technique consists in injecting water in the crack before the injection of resin. Then, the low quantity of water that will be present in the crack will be sufficient to make the resin react. Therefore, the method is easy and the foam produced is nearly uniform, whatever the quantity of water present in the environment injected.

Table 1. Comparison of properties of hydrophobic and hydrophilic resins

Properties	Hydrophobic	Hydrophilic
Bonding to wet surfaces	Poor	Good
Structure of the final product	Independant of the quantity of water	Dependant on the quantity of water
Speed of reaction	Very quick	Quick
Expansion (free)	10 to 30	5 to 8
Volumetric stability to freeze-thaw and wetting-drying cycles	Good	Poor
Volumetric stability after reaction	Good	Poor
Injection technique	Simple (resin only)	Double (resin + water)

Polyurethane foams can be classified as three general types depending on their deformability: flexible foam (or soft foam), semi-rigid foam (or semi-flexible) and rigid foam.

The use of polyurethane injection resins is relatively new and the selection criteria to pick the appropriate resin for differing applications are not well established. Some authors and researchers have determined selection criteria for applications in injection of polyurethane resins but, these criteria are mostly interested in the expansion reaction of the foam, the production of CO_2 and the final structure of the foam obtained, and not in its mechanical properties.

Thus, the mechanical characteristics of the foam are not very well detailed. Some authors (Karol, 1990; Anderson, 1998) indicate only that the foam must possess a "high strength" without referring to either deformability, nor bonding capability of the foam. Yet, these parameters are very important in repair projects because they determine the durability of the foam and its resistance to cycles of opening and closing of cracks.

Traction and expansion/contraction tests on samples injected with polyurethane foam

A testing program has been developed at the Université de Sherbrooke to simulate failure of imperviousness during the opening of a crack. Research carried out shows that no normalized test has been developed to evaluate injected polyurethane foams (Vrignaud, 2000).

Two types of stress can bring about failure of imperviousness in a crack injected with polyurethane: either the pressure of the water is too strong upstream, or the foam can not prevent the opening of a crack such as occurs during the freezing period. If the injection is carried out uniformly, without joints and the depth of penetration of the grout is sufficient, the foam, which is water resistant, can be considered strong enough to resist the pressure of upstream water. We present in the following pages test methods to determine the durability of polyurethane foams under traction tests and cyclic expansion and contraction of joints.

Preparation of samples

The samples tested are made of concrete prepared in the shape of a parallelepiped and have a larger width at both extremities for attaching the traction apparatus. The samples are 206 mm in height and have a section measuring 127 mm × 127 mm, except for the two ends where the section is 127 mm × 177 mm (Figure 1).

Preliminary tests carried out on small cracked concrete cylinders have shown that when the sample is injected, it is very fragile. Therefore, it is important to design the set-up in such a way that the sample is not manipulated between the moment of injection and the time that the traction test is performed. That is why two holes are drilled before making the set-up in press. Preparation of the samples is divided into five phases:

a) Casting of concrete cubes:
Concrete samples are cast in wooden formwork. The concrete mix is designed in such a way that slitting of the sample produces only one clear and regular crack. The concrete mix is made of Type III cement and reaches a compression strenght of 60 MPa , after 28 days

b) Drilling of a 45° angle hole
For practical reasons, an inclined drilling is carried out before fracturing of the sample. Its location is presented on Figure 2.

c) Fracturing of the samples by slitting:
The samples are fractured by slitting, with the action of a hydraulic press applying a continuous compression load on the sample. Failure of the sample is sudden: the failure surfaces obtained on both parts of the sample bypass the aggregate and are in general very regular, which ensures good uniformity of the samples.

d) Drilling of a vertical injection hole:
For each sample, a hole 25 mm in diameter is bored vertically in each cylinder. This vertical hole intercepts the inclined one. During injection, the closing of the hole is ensured by a mechanical plug.

e) Conditioning of samples:
The samples are placed in the desired humidity and temperature conditions.

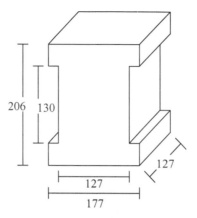

Figure 1. Concrete sample

Injection of polyurethane foam
First, the sample is secured to the press with four aluminium bars attached to the main frame using a threaded rod. Then, the opening of the crack is precisely set at the desired value (2 mm or 5 mm) by registering the displacement of the actuating jack.

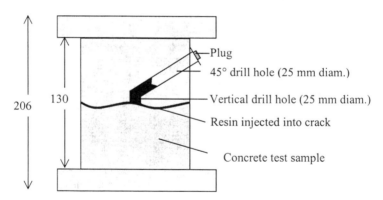

Figure 2. Drilling of an injection hole and distribution of the resin during injection

The injection mould is then placed around the sample. This is the main difficulty in the injection of polyurethane resins in the laboratory since the pressure developed by the reaction of the foam is so high that it is difficult to contain the foam in the crack. Tests carried out on other samples have shown that the loss of grout is impossible to prevent unless a perfectly sealed and extremely strong injection mould is built. However, in that case, the crack would be closed and there would be no air outlet: the proper distribution of grout would be hindered. In fact, in the case of *in-*

situ injection, the loss of grout is inevitable (La Penta, 1992). Therefore, such losses do not present any difficulty if they are kept at a minimum. On the laboratory sample, this loss will be more important than in the field due to the scale factor. To ensure the confinement of the foam, a mould made up of four wooden planks and a rubber membrane can be used. he set-up is held in place with the help of two form clamps during injection (Figure 3).

Injection of the grout can then be carried out. The acquisition system registering load and displacement is activated. Displacement is also measured to make sure it is maintained at zero during injection and that there is no vibration of the press. The grout is injected by gravity, thanks to a plastic tube and a funnel set-up in the injection hole of the sample. When all the grout has been injected, a mechanical plug is inserted and tightened in the tube to prevent reflux of the foam.

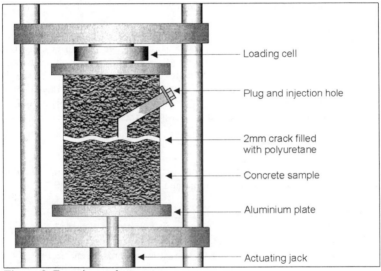

Figure 3. Experimental set-up

Methodology and tests parameters
Several parameters are recorded during the both tests:
- *during reaction of the injected resin*: speed of reaction and pressure due to foam expansion during injection are studied;
- *after reaction of the injected resin*: strength and movements during traction tests or strength of resins under expansion and contraction tests are measured.

The traction test is carried out under controlled displacement by imposing a traction displacement at a given speed. The resulting load on the sample is measured. The speed of displacement of 0.1 mm /minute is chosen in order to impose the stress as slowly as possible. The test is completed when resin debonds from the concrete.

To find out the durability of the polyurethane foam when subjected to seasonal variations in temperature, a cyclic test was implemented. The frequency and amplitude of displacements were then chosen arbitrarily :

- The amplitude of displacements was chosen in order to simulate the stress imposed on the foam during major temperature variations. First, the amplitude is fixed to be 2 mm for crack openings of 5 mm and, secondly, at 1 mm for crack openings of 2 mm. These values correspond to deformations ($\varepsilon = \Delta l / l$) of 40 and 50 %, respectively.
- Since preliminary tests have determined that the behaviour of foam stabilizes after 20 to 25 cycles, 30 displacement cycles are applied to the sample.
- Finally, since the cyclic displacements are imposed on the foam *in-situ* very slowly the chosen cycle frequency is as low as possible. For practical reasons, the duration of the test is, however, limited: 30 cycles are carried out in 45 minutes, for a period of approximately 88 seconds per cycle.

It is considered possible to control the temperature of the grout in the field, whatever the season during which it is injected. For this reason, tests are carried out with the resin being maintained at a temperature of 20°C before injection. However, the temperature of concrete is not controllable in practice. In order to see if temperature influences the injection of resin and the characteristics of the injected foam, different temperature conditions can be chosen for concrete samples. Finally, two saturation conditions can be considered: dry or humid conditions.

Analysis of some results obtained

The tests presented in this paper have been done with an hydrophilic resin, a flexible hydrophobic resin and a semi-rigid hydrophobic one.

Evolution of pressure during injection
Immediately after injection of polyurethane foam, pressure recorded increases because of the reaction of foam. Then, at the end of this reaction, when there is no more expansion of foam, pressure decreases for hydrophobic resins and stabilises for hydrophilic ones. The saturation of hydrophilic resins with water could explain this phenomenon.

Traction tests
Traction tests were performed with wet joint in concrete. For all the tests, failure occurred at the contact between foam and concrete. For hydrophilic resin, bonding was about 20 to 40 kPa and deformation at failure reached 20 to 50%. For semi-rigid hydrophobic resin, bonding was reduced to 5 to 10 kPa and deformation at failure was about 2 to 5%. Then for flexible hydrophobic resin, bonding to wet surfaces and deformation were nil.

For hydrophobic resins, bonding due to chemical action of foam on concrete is almost nil. For hydrophilic resins, physical and chemical actions are important: the foam is able to anchor in the roughness of the joint.

Cyclic expansion/contraction tests

Two different tests were done during this study. For hydrophobic resin, the foam was submitted to compressive cyclic displacements because of its low resistance to traction. For hydrophilic resin, the foam was submitted to compression (75%) and traction (25%).

Cyclic tests showed different behaviours for semi-rigid hydrophobic resins and hydrophilic ones. On the one hand, hydrophilic resins behave as elastic foam. On the other hand, a tensile strength develops during expansion applied to semi-rigid hydrophobic resins and these ones take a long time to be back to their initial shape.

During expansion/contraction tests, the 30 cycles decrease the bonding between semi-rigid hydrophobic foam and concrete. But this test, has no effect on the bonding between hydrophilic foam and concrete. If contraction/traction cyclic tests are applied , then this bonding is decreased too.

Conclusion

An experimental set-up has thus been developed to test the mechanical behaviour of foam injected in a crack. It was possible to study the durability of injection from the point of view of the seasonal opening and closing movements of cracks. Based on these laboratory tests, it was easier to understand the behaviour of various types of semi-rigid, flexible, hydrophobic and hydrophilic resins.

The results obtained from the experimental set-up have shown that polyurethane foam is a difficult material to test in the laboratory, which explains why there is no normalized test to help in the determination of which are the most effective products in the industry. The set-up developed presented the advantage of being simple and it helped determine the tensile strength and elongation to failure of the foam injected. It also presented the disadvantage of not taking into account all the factors that can influence foam. It is difficult, in particular, to confine foam and reproduce a similar test from one sample to another.

Hydrophobic semi-rigid resins react very quickly, expand greatly and have a very small viscosity (100 cP). The obtained foam bonds, but poorly, to wet surfaces and has a good volumetric stability under freeze-thaw and wetting-drying cycles.

Hydrophobic flexible resins react slower, expand less and have a higher viscosity (600 cP) than semi-rigid resins. The foam obtained after reaction of the resin has a very good volumetric stability under freeze-thaw and wetting-drying cycles. The disadvantage of these resins is a nil bonding with wet surfaces.

Hydrophilic semi-rigid resins react slower, expand less and have a higher viscosity (500 cP) than hydrophobic resins. The obtained foam has poor volumetric stability, but bonds well to wet surfaces. During the drying of such resins, significant shrinkage occurs. As a consequence, hydrophilic resins are usually used in saturated medium such as swimming-pools and hydroelectric structures.

Based on the results obtained during traction and cyclic tests, single-component, hydrophobic and semi-rigid resin has been therefore recommended for injection work on road tunnels in Montréal. These products have been chosen for their quick reaction and for their good volumetric stability under freeze-thaw and wetting-drying cycles.

Acknowledgements

The authors gratefully acknowledge the collaboration of Ministère des Transports du Québec (Department of Transportation, Province of Quebec) participating in field work. They also appreciate the financial support of Natural Sciences and Engineering Research Council of Canada (NSERC) and the Quebec Provincial FCAR fund (Fonds pour la formation de chercheurs et l'aide à la recherche). The technical assistance of Martin Lizotte, Georges Lalonde and Danick Charbonneau is greatly appreciated.

References

Anderson, H. (1998). *Chemical rock grouting: an experimental study on polyurethane foams.* Thesis, School of Civil Engineering, Department of Geotechnical Engineering, Chalmers Univ. of Technology, Goteborg (Sweden).

Joyce, J.T. (1992). "Polyurethane grouts used to stop water leakage through cracks and joints." *Concrete Construction*, 551-555.

Karol, R.H. (1990). *Chemical grouting*, Ed. Marcel Dekker, New York.

La Penta, B.A. (1992). "Tunnel seepage control by the interior grouting method." *Grouting, Soil Improvement and Geosynthetics*, 1, 436-448.

Naudts, A. (1995). "Grouting to improve foundation soil." *Practical Foundation Engineering Handbook*, Ed. Robert Wade Brown, New York, 5277-5400.

Vrignaud, J.P. (2000). *Contribution à l'étude des résines polyuréthanes d'injection utilisées pour le traitement des infiltrations d'eau dans les tunnels en service.* M.A.Sc, Department of Civil Engineering, Université de Sherbrooke, Sherbrooke, Québec, Canada.

Soil Grouting: Means, Methods and Design

Daniel Lees[1], Marcelo Chuaqui[2]

Abstract

Permeation grouting and controlled hydrofracture grouting are two soil grouting techniques used for the purpose of improving soil properties. Although there are many other techniques, the focus of this paper will be on permeation and hydrofracture grouting and will be referred to as soil grouting throughout the paper for simplicity. Soil improvements include stabilization, strengthening (bearing or bond) and permeability reduction. This paper will discuss the means and methods available for performing permeation grouting. The purpose of this discussion is to provide information for developing practical designs to meet project specific goals. The discussion will include topics such as grouting methods, mix designs, equipment, quality control, monitoring, injectability constraints and determinations.

Soil grouting is an often misunderstood and overly simplified technology. This misunderstanding can result in inadequate design and construction methods. The ideas conveyed in this paper are based on the authors' experiences in a number of soil grouting projects as grouting consultants and contractors.

1. Introduction

Permeation grouting is the process of injecting flowable grout into the pore spaces of soils for the purpose of improving the soil properties. The grout must be injected at a controlled rate so that it can flow into the pore space before excessive pressure builds causing hydrofracture or displacement of soils. Hydrofracture grouting is performed when the properties of the grout and soils are such that permeation of the soil is not possible and the grouting pressure causes the soil to fracture creating 'lenses' of grout and moderate densification of the surrounding soils. Controlled hydrofracture grouting occurs when the individual overseeing the grouting operation is aware that hydrofracturing is occurring. The volume of grout injected must be carefully

[1]Project Manager, Geo-Foundations Contractors Inc., 135 Commercial Drive Bolton Ontario Canada; tel (905) 857-6962; fax (905) 857-0175; email drlees@usl-1983.com
[2]Principal, MC Grouting, Inc., 12 Northglen Ave., Toronto, ON M9B 4R6 Canada; tel (416) 695-2593; fax (416) 695 2399; email: marcelo.chuaqui@sympatico.ca;

GROUTING AND GROUND TREATMENT

controlled and any adjacent structures or underground utilities should be monitored. Some degree of hydrofracture grouting can be necessary when zones of soil are encountered with high silt content preventing the same extent of permeation achieved in surrounding soils. Minor hydrofracturing can provide some soil improvement and provide access to soils, which are permeable with the grout being used.

Soil grouting is performed for both strengthening and permeability reduction applications. Some common structural applications include soil modification for:

- ❑ temporary support for tunnels and excavations,
- ❑ improved bearing capacity,
- ❑ improved bond characteristics and bond area for anchors and piles.

Some permeability reduction applications include:

- ❑ environmental cut-off walls
- ❑ grout curtains for dams and tailings ponds,
- ❑ seepage control during tunneling and excavating
- ❑ seepage control into structures.

Each of the applications mentioned above, requires different degrees of grouting and the complexity of the grouting program will be a function of the soil conditions and the application. That is to say that each grouting program will be different depending on the goals of the project and a standard set of procedures should not be followed for all soil grouting programs.

2. Construction Methods

A number of different aspects of a soil grouting program will be discussed in detail in this section. This discussion will summarize the conditions for which soil grouting is practicable and the limits of soil grouting capabilities. After this discussion, conclusions will be drawn based on the topics discussed and on the author's experience on soil grouting projects. These projects involve soil grouting for:

- ❑ soil stabilization for tunneling,
- ❑ soil stabilization prior to shaft excavation,
- ❑ construction of environmental cut-off walls,
- ❑ preventing water seepage through an earth/limestone rubble dam into a paper plant,
- ❑ encapsulation of hazardous and low-level solid radioactive waste.

2.1 Grout Injection

The most reliable and widely accepted means of performing soil grouting for most applications is by utilizing 'tube a manchete' (TAM) or also referred to as a 'multiple port sleeve pipe' (MPSP). Both MPSP and TAMS will be referred to as sleeve pipes throughout the remainder of this paper. A sleeve pipe is a pipe that has ports drilled out in equal increments along the length of the pipe. The ports are covered with a rubber sleeve that acts as a one way check valve. When fluid pressure is applied from inside the pipe the rubber sleeve deforms allowing the grout to flow out of the pipe.

When the pressure is released, the rubber sleeves close preventing grout to flow back into the sleeve pipe. The rubber sleeves also prevent water and soil from entering the pipe, when there is external pressure against the sleeve pipe. Double packers are used to isolate one or more sleeves to provide grout injection into discrete soil zones.

Methods such as 'end-of-casing' grouting should be reserved for void filling, soil preconditioning, karst grouting, compaction grouting and low mobility grouting. This method, when used with high mobility grouts, does not permit control over where the grout is placed and does not allow for multiple passes of grout injection. Its use can only be justified where the performance criteria are very low.

The means of installing the sleeve pipes is best left to the contractor with some simple restrictions. Usually a cased or augured hole is required to provide temporary support prior to installing the sleeve pipe. Water and air can be used to aid in the drilling operation however drilling fluids, whenever possible, should be avoided as they can cause a permeability reduction surrounding the drill hole, which will prevent adequate permeation during the grouting operation. The important aspects of installing the sleeve pipe are that the pipe and sleeves are not damaged, the pipe is not bent and the casing grout is an appropriate formulation (see section 2.3.4) and installed to completely encapsulate the sleeve pipe. The casing grout is typically a very weak, bentonite-cement mix designed to provide multiple use of the sleeve ports and maintain isolation of each sleeve port zone. The casing grout should be stable (0% bleed) and of very high viscosity to prevent permeation into the soil, filling the annular space only. The casing grout should be installed by tremmie grouting or by installing a packer in the sleeve pipe and injecting the casing grout from the furthest sleeve from surface.

The combination of casing grout, sleeve pipes and double packers make it possible to inject grout into discrete zones, several times. These are critical conditions for a successful and predictable soil grouting program.

2.2 Soil Injectability

The injectability of a soil depends on the properties of the soil and the properties of the grout being injected. The relevant soil properties include silt content, particle size distribution, relative density (soil compaction) and water content of the soil. The relevant grout properties are viscosity, gel time, stability and particle size, if a particulate grout is being used.

The soil properties can be estimated by performing split spoon sampling, which provides 'N' values and samples that can be used to determine the particle size distribution and water content of the soil. Alternatively, soil probing can be performed either with standard cone penetrometer tests (CPT) or by the calculating the specific energy required to penetrate the soils when drilling holes. However, this method will not provide any details on the grain size of the soil which is very important in determining the permeability of the soil. Figure 1 (Baker, 1982) provides an illustration of soil injectability relative to particle size distribution of soils.

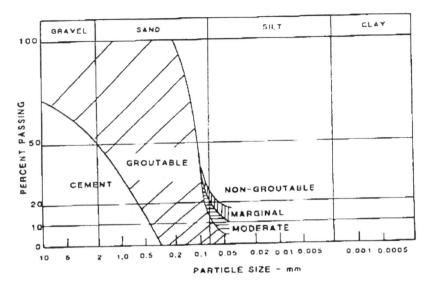

Figure 1: Grain Size Ranges for Chemically Groutable Soil

Obtaining the d_{10} (cm) of the soil and applying it to Hazen's equation can generate another approximation of injectability of soils.

Hazen's Equation $k(cm/s) = C(d_{10})^2$

with C generally ranging from 100 to 150 depending on the gradation of the soil.

Hazen's equation provides an estimate of the hydraulic conductivity of the soil and has demonstrated to be relatively accurate in undisturbed sandy soils. With this information, an estimate of the soil injectability can be provided as follows (Landry et al, 2000):

$k > 1 \times 10^{-1}$ cm/s, injectable with regular cement grouts;
$k > 5 \times 10^{-3}$ cm/s, injectable with microfine cement grouts;
$k > 1 \times 10^{-4}$ cm/s, injectable with solution grouts.

This information is only an estimate and provides a means of generating a starting point. Whenever possible, a small scale field test in the same ground conditions as expected for production grouting work should be performed to determine the injectability of the soil. Water pressure testing should be performed to generate a base line and for comparison with the assumptions made (see section 2.10.2).

Permeation tests should also be performed using the anticipated grout formulations. Often, there is not enough time or budget to perform a separate field test. In these cases, it is advantageous to start the grout operation in the least critical areas. In this way, testing can be performed and changes to the grout mix and grout types can be made if necessary.

2.3 Mix Designs

Based on the material components, permeation grouts can be divided into two basic categories; solution (chemical) grouts and particulate grouts (usually cementitious). Within these two categories exist a wide variety of grouts particularly with respect to microfine cements and solution grouts. The mix designs in this section will focus on grouts with which the authors are most familiar, that is regular cement based grouts, microfine cement based grouts and sodium silicate solution grout. Acrylate based grouts can be used when a more durable true solution grout is required.

2.3.1 Regular Cement Based Grout

Regular cement based grouts are the least expensive of the permeability grouts. Regular cement based grouts are used for permeating soils of relatively high permeability (greater than 10^{-1} cm/s). In general, gravels, sandy gravels and coarse sand with low silt content can be permeated with regular cement grout. Although the soil conditions may suggest that permeation with a cement grout is possible, it is still very important to design the grout to be stable and of an appropriate viscosity to allow control of the flow and grout spread. Of equal importance is the equipment used to prepare the grout and the people involved in the grouting operation.

Type 3 cement is a high-early strength cement. The reason that this cement causes high-early strength is due to the particle size of the cement. The cement has been milled finer that type 1 creating smaller particles and thus more surface area for reaction. The smaller particle size is naturally advantageous for permeation grouting and due to its relatively inexpensive cost, makes it a practical material choice for a regular cement based grout. The following list provides a basic range of grout components in developing a regular cement based grout.

- ❑ Water:cement   0.8 - 2.0 by weight;
- ❑ Bentonite   1 – 2 % by weight of water;
- ❑ Plasticizer   0.5 – 2 % by weight of cement;
- ❑ Retarder   0.5 – 3 % by weight of cement.

With mixes that have a water to cement ratio greater than 2.0, it is difficult to control bleed and maintain stability in the mix. Mixes that are not stable (more than 5% bleed) will result in residual permeability in the soil caused by the presence of bleed water separating out of the grout after injection. Mixes that are highly unstable are hard to pump as the particles can settle out of the mix at low flow rates. Unstable grouts are also susceptible to pressure filtrating. This can cause blockages in the grout lines as well as premature refusal due to pressure filtration of the grout close to the injection point. Although the amount of bentonite can be increased to stabilize the mix, this will decrease the grouts strength and increase the viscosity of the mix. The recommended grout components given above can be used to produce stable grout mixes with varying levels of viscosity.

In general, bentonite should be introduced as a prehydrated slurry and the water in the slurry must be accounted for in the total water content of the mix. Typical bentonite slurry consists of 6-8% bentonite. When batching the grout, each component should be introduced in a consistent order. Normally the order is as follows; water, bentonite slurry, cement, fillers, additives.

There are many additives that can be used to enhance the fluid and set properties of a grout. In this paper we will only discuss the properties that are most relevant to standard soil grouting projects.

Retarding the set-time of a grout is important when performing multiple passes or phases in the same sleeve-pipe. Retarded set times allow multiple passes without having to use special high-pressure pumps to hydrofracture cured grouts.

Injecting grouts into soil is similar to pressing the grout against a filter. The water can be literally forced out of the grout. When this occurs, water is injected into the soil while a filter cake forms at the injection point. Eventually, the filter cake makes the formation inaccessible. This is another reason why it is so important to use a stable grout mix which is resistant to pressure filtration. This can be accomplished by increasing the solids content, using bentonite, welan gum and other additives.

There also are a variety of 'fillers' that can be used in place of some of the cement in order to decrease or increase strength, reduce cost, improve stability, reduce matrix porosity, enhance durability, and improve penetrability. Such products include; flyash, slag, silica fume, pumice and lime among others.

The project goals and planned methodology will dictate the properties that are required from the grout. Grout mix design testing should be performed prior to performing the grouting program in the field. The purpose of this is to develop grouts with the properties required for successful execution of project. An understanding of how each component in the grout mix will affect the rheology and characteristics of the grout should be obtained during the test program. This will permit the logical and educated decisions to be made in the field during the grouting program.

2.3.2 Microfine Cement Grouts

Microfine cement mixes are designed in a similar manner and with many of the same admixtures and additives as for regular cement grouts. Many of the same material ratios can be used in regular cement grout as for microfine cement grouts. Again, it is important to perform some mix design tests prior to production grouting for the same reasons given above as well as to determine how the microfine product will perform as a suspension grout.

The most important aspects to consider in a microfine grout are cost, particle size, particle size distribution and strength. Microfine cements are typically characterized by the specific surface area or Blaine fineness. The various microfine products are produced from different materials and thus result in different rheology and set properties. Of utmost importance in selecting a microfine cement is the consistency or quality control of the product. Microfine cements are used because of the ability of the reduced particle size to permeate smaller pore spaces. If the microfine product is not of a consistent particle size the microfine grout will only perform as well as the larger particles permit.

2.3.3 – Sodium Silicate Grouts

Sodium silicate is the most commonly used solution grout for soil permeation. Sodium silicate is a two component system consisting of sodium silicate and reactant. The various chemicals used as neutralizing agents differ greatly in composition and in the chemical reaction that takes place resulting in varying gel times and gel strengths. Temperature can also have a significant effect on the gel time and may require adjustments to the concentrations of the chemical components.

Often the sodium silicate component is diluted with water to reduce the viscosity of the solution grout. Sodium silicate grouts with similar viscosity to that of water are achievable.

The following are two samples of mixes used on a project in Milwaukee, Wisconsin and in York, Pennsylvania.

Table1: Sodium Silicate Solution Grout Mix Designs

Component A (volume)	Component B (volume)
70% type N sodium silicate	93% water
30% water	7% Diacetin
Component A	Component B
80% type N sodium silicate	80 – 85 % water
20% water	15 – 20 % Glyoxyl
	Calcium Chloride*

*Calcium chloride was used to adjust the gel time as necessary.

The first mix design provided a low viscosity solution grout with a gel time of approximately 20 minutes and the second had a gel time of approximately 30 minutes. Both mix designs were used successfully in conjunction with cement based grouting to provide temporary support for a tunnel excavation.

The two components are prepared in separate holding tanks and kept separate. A double piston pump is used to pump the separate components and mix them together in-line with the use of a static mixer. When developing the mix design, the time in which the grout has to travel from the static mixer to the injection point should be considered for control of the gel time. By placing the grout head close to the injection points and performing multiple hole grouting (discussed later) the effect of this time lapse can be reduced. It is important to carry out a mix design test program when using products where historical data is not available.

2.3.4 Casing Grouts

The purpose of casing grout is to encapsulate the sleeve pipe with a weak grout, which can be easily fractured to permit grouting of discrete zones of soil. The grout should be of very high viscosity to prevent permeation of the soil and of low enough strength to allow fracturing with the same equipment used in the grouting program.

The grout mix should also be designed to provide an initial set time within 12 – 18 hours. The following is a sample mix design used on several grouting projects.

Table 2: Casing Grout Mix Design

Water	130 liters
Cement (type I/II)	43 kg (1 bag)
Bentonite	5% by weight of water

2.4 Equipment

The mixing equipment for particulate grouts is an important aspect of the quality of the mix produced. A brief description will be provided here on some of the more important components and aspects to consider when setting up a grout plant. An excellent source with more detail on grout mixing and equipment can be found in Houlsby's book on foundation grouting (Houlsby, 1990). This book is an excellent source for describing the various components of grouting equipment.

Batching of particulate grouts should be performed using a high-shear (colloidal) mixer. "A high-speed rotor produces violent turbulence and high shearing action. This gives a rapid and thorough mixing" (Houlsby, 1990). This high shearing action is important to ensure that each particle is held in suspension separately. If the proper mixer is used, the grout will be transferred to an agitator as a suspension of individual particles, free of clumps or flocs. The mix design will determine if the mix will remain in this condition for injection into the soils. The shearing action is accomplished by housing the high speed rotor in a very closely fitting housing. Colcrete and Chemgrout are two manufacturers of this type of mixer. These mixers are capable of producing a grout mix in as little as 15 seconds of mixing. As the cement (and other particulates) are emptied into the mixer, the water and cement are pulled through the rotor and ejected back into the mixer. The mode in which the water-cement mix is ejected is also an important factor in the efficiency of the mixer. The circulation process should create a vortex in the mixer that allows the dry ingredients to mix with the water prior to shearing with the rotor. If this vortex does not form, large clumps of dry ingredient can form and may plug up the flow to the rotor. Some testing needs to be performed on the mixer to determine the optimum mixing volume of the mixer. If the volume is too high it may be difficult to create a vortex. If the volume is to small, the vortex may be so violent that the dry ingredients can fall down the middle of the vortex and plug the path to the rotor. The optimum mixing volume produces a high quality grout at an efficient rate.

Once thorough mixing is complete the grout is transferred to an agitator holding tank. This is usually performed by simultaneously closing the valve to the mixer and opening a valve for transfer to the holding tank. The purpose of the holding tank is to provide a steady supply of grout, which is being constantly agitated to aid in holding the particles in suspension. A grout mix, which is properly formulated and stable, will require very little agitation. The agitator paddles should be designed to keep the perimeter and bottom of the agitator free of grout build up. A hole in the bottom of the agitator allows the grout to flow to the grout pump by gravity feed.

The grout pump should be of the progressive cavity type, which is easy to control the flow and pressure and provides a steady flow of grout. The rotor and stator on these pumps need to be checked for wear from time to time. The grout is pumped to a 'header' where the flow is split into various lines. The number of lines depends on how many holes are being grouted at once. There should always be one extra grout line (return line) that brings the grout back to the agitator. Valves are located on all grout lines with the valve on the return line being used to control the pressure. Flow meters are installed after the grout header on each line. Pressure gauges are installed as close to the hole on each line as possible and a gauge should be located near the pump. This allows for a check of line losses and may be used as an indicator for determining when the grout lines are building up with grout (increasing line losses over time will indicate grout build up or increasing viscosity of grout).

Sodium silicate grouting is performed utilizing a double piston pump. The two components are held is separate containers. It is advantageous to install a mechanical stirrer in the two holding tanks to ensure that the water-silicate and water-hardener remain adequately mixed. The authors' experience are with double piston pumps which are air powered and operated off one cylinder providing a 1:1 mix ratio. This system has worked well for the mix designs provided earlier in this paper. The two components are mixed together using an in-line static mixer (hose with baffles over a sufficient length, to provide complete mixing). The static mixer is usually located as close to the grouting source as possible to delay the start of gelling and viscosity increase.

2.5 Real-Time Monitoring

As mentioned in the introduction, permeation grouting requires the injection of grout at a controlled rate so that it can flow into the pore space before excessive pressure builds causing hydrofracture or displacement of soils. In order, to determine the mode of a grouting operation real-time monitoring of grout pressure and flow is required.

The degree of sophistication required from the real-time monitoring systems depends on the project and the experience of the people performing the work. The authors have been on projects ranging from dip sticks and manual pressure gages systems to projects where automated data acquisition with computer control of pumps is implemented through magnetic flowmeters, pressure transducers and hydraulic flow-control valves.

At a minimum, there must be a means to accurately determine the pressure and flow at any instance. Furthermore, a time based graph of the pressure and flow is required to determine the mode of the grouting operation.

Different theories or methods of assessing a grouting operation exist, such as Amenability Theory and the Grout Intensity Number Principle. These systems can be implemented into the real-time monitoring system and simplify the role of assessing the grouting operation.

It is the authors experience that the use of a real-time monitoring system, that utilizes flowmeters, pressure transducers and real-time display of the grouting data yields better control of the grouting operation. These systems can be purchased commercially or custom built with analog to digital converters, dataloggers and a PC type computer.

2.6 Multiple Hole Grouting

Multiple hole grouting is the most efficient way to perform soil grouting, particularly in the latter stages of a grouting program when grout takes at individual locations can be low. The number of holes that can be grouted simultaneously will depend on a number of factors.

- ❑ the number of pressure transducers and flow meters available,
- ❑ the homogeneity of the soil,
- ❑ the grout flow vs. the grout batching capacity,
- ❑ type of computer system (if any) being used to monitor or control pressure, flow, and pumps,
- ❑ the experience and abilities of the grouting technician.

In general, grouting of four holes simultaneously is productive and manageable. It is possible to perform grouting on more holes but more than 6 at a time can become unmanageable for an individual. If the grout takes are high (usually during the first pass of a grouting program) the number of holes which can be grouted at once will be dependant on the rate at which a consistent, high quality grout can be produced. In the later stages of a grouting program, when grout takes can be very low, it is a necessity to perform multiple hole grouting. This allows for the total grout production to remain high and therefore the operation remains economical. Furthermore, multiple hole grouting allows for consistent grout rheology, even when grout take per hole is low. If the total grout take is too low, it is likely that the grout will begin to increase in viscosity and may cause premature refusal.

When performing a multiple hole grouting program it is important to grout soil zones of similar permeability (as determined from a water test or data from previous grout passes). Grouting zones of different permeability can be difficult to manage, as the grout will tend to flow to the more permeable zones.

2.7 Multiple Passes

The purpose of soil grouting is to saturate soils with grout within a defined radius of the grout injection point. The extent of the saturation depends on the specific goals of the grouting program. The most reliable way to achieve saturation of the soil in close proximity to the injection point is by performing multiple passes. Grout will always take the path of least resistance and unless the soils pore spaces are completely uniform it is not possible to sufficiently permeate the soil with one grout injection pass. Performing multiple passes controls the grout spread and systematically reduces the permeability of the soil. While grout refusal may not be achieved in every grout pass, it is essential to achieve refusal prior to changing grout types (i.e.:

going from cement grout to sodium silicate). As many as six grout passes may be necessary to completely saturate the soil and achieve refusal with each grout type. However, it may only be necessary to perform 1 or 2 grout passes to satisfy the goals of the grouting program.

It can be argued that the soil can be saturated in one pass if the grout mix design is adjusted in response to the grounds acceptance of the grout. While this may be theoretically possible, it is rarely practical. This process would limit the grouting operation to focusing on one hole at a time and the use of small grout batches. It is far more efficient to perform multiple hole grouting and multiple passes.

2.8 Multiple Phases

A grouting phase refers to injection with a specific grout type. Depending on the project goals, it may be necessary to have multiple phases in addition to multiple passes. For example, a sleeve may be injected two times with regular cement based grout, and two times with sodium silicate, during a grouting project. This sleeve therefore, has a total of four passes and two phases.

Multiple phases are particularly important when trying to achieve low residual hydraulic conductivities for long term environmental cut-off walls. The goal of this type of grouting program is to place the most economical and durable grout during the first passes and later place smaller quantities of the less durable and less economical solution grouts.

The process of injecting grout in multiple passes through subsequent phases allows for the final product to be progressively built. The status of the process can be checked along the way by performing water pressure tests.

2.9 Controlled Hydrofracture Grouting

Controlled hydrofracture grouting may be necessary at times, during a permeation grouting program, when attempting to permeate soils with high silt contents. Soil zones may be encountered with a much lower permeability than the surrounding soils. Performing controlled hydrofracture grouting can treat these zones. The process involves increasing the grout injection pressure until grout flow is observed. The injection pressure should then be reduced to that being used for permeation grouting. This will permit some permeation of soils surrounding the 'impermeable' zone and leave a grout lens within the impermeable zone. Performing multiple passes in this zone is important to create a system of lens' which will densify the soil and provide strength to the soil similar to the way in which a trees roots can provide stabilization. The grouting pressures required to hydrofracture the soil should not be maintained while allowing grout flow. This is not controlled hydrofracture grouting and can be dangerous particularly when working adjacent to structures.

2.10 Quality Assurance and Quality Control

The QA/QC is an important aspect of any soil grouting program and can be separated into three phases.

2.10.1 Pre-Production

During this phase the grout mixes are developed. The gel and set times, specific gravity, viscosity, bleed, pressure filtration coefficient and compressive strengths are determined and adjusted to meet the project goals. During this phase, typically bench scale tests are performed prior to mobilization to site. This phase of the QA/QC program will help set base lines for performing quality control during production grouting. The information gathered in the mix design tests will be of use in developing a plan for the grouting program (how to increase/decrease the viscosity of the mix etc.).

2.10.2 Production

During this phase water testing and quality control testing of the mixes is performed. The first step in the production phase of the QA/QC program is performing water flow tests in enough of the sleeve pipes (minimum 25%) to get a good idea of the permeability of the various soil horizons. Water flow tests are performed using the same equipment as for the grouting program. Water is injected as the test fluid and allowed to flow until the flow stabilizes (2 – 3 minutes). The pressure and flow are recorded and the hydraulic conductivity is determined. The hydraulic conductivity can be compared to that estimated from equation 1. This information can be used to adjust the initial mix designs and to map out the treatment zones for multiple hole grouting (i.e.: that is group the zones of similar hydraulic conductivity).

During the production grouting, real-time monitoring of the flows and pressures should be performed as discussed in section 2.5. Assessment of the grouting operation is performed in real-time. In addition to real time monitoring, records should be kept (manually) of zones grouted, pressures, flows and grout takes. This information should be recorded, as a minimum, at the start and end of each grout zone and at 5-minute intervals (much more frequently if a real-time monitoring system is not being used).

Grout batches should be tested regularly for consistency. Records should be kept indicating mix design, batch number, viscosity (marsh funnel) and specific gravity. Samples should be taken to verify gel and set times.

2.10.3 Post-production

Typical post grouting confirmation testing includes performing water testing through sleeve pipes that have not been grouted, performing SPT's or CPT's. The testing performed after grouting should be similar to the pre-grouting tests so that a comparison can be made. Analysis of the data obtained during grouting should be performed to determine the extent of the grouting performed.

3.0 Conclusions

The information provided throughout this paper demonstrates the potential complexity and diversity among soil grouting programs. It is important that the individuals involved in designing, constructing and inspecting a grouting program have experience in the field of soil grouting.

The grouting program must be designed within the limits of permeation grouting capabilities and to a degree of complexity that matches the goals, budget and schedule of the project. That is to say that a building requiring upgrading of the foundation capacity would be designed differently than a low risk temporary excavation or tunnel support. The former would require improved soil strength and durability while the latter may only require temporary cohesion of the soils.

The design and specification should also be developed so that adjustments can be made as more information is gained during the grouting program. The exact nature of the soils usually becomes best understood as the grouting program progresses. If there is already a plan in place to deal with changes as they occur, there exists a better chance that the project goals will be met.

Once a permeation grouting application is identified, the selection of a competent and experienced contractor is important. The contractor needs to have the capability to perform careful planning, and analysis of drilling and grouting data. Furthermore, the contractor should be capable of soil grouting using any and all of the means and methods laid out in this paper. That is not to say that each and every aspect of these methods will be required on an individual project. However, just as the designer must develop a specification which provides room for modification, the contractor must have the capabilities to change his methods when necessary.

References

Baker, W (1982). "Planning and Performing Structural Chemical Grouting." Proceedings of the Conference on Grouting in Geotechnical Engineering, ASCE, New Orleans, pp 515-539.

Houlsby, A. C. (1990). "Construction and Design of Cement Grouting." John Wiley & Sons, Inc., New York.

Landry, E., Lees, D., and Naudts, A. (2000). "New Developments in Rock and Soil Grouting: Design and Evaluation." Geotechnical News, Volume 18, Number 3, 38-44.

SOIL SOLIDIFICATION WITH ULTRAFINE CEMENT GROUT

By James Warner[1], F. ASCE

Abstract

Since they first appeared in Japan about 30 years ago, ultrafine cements have become commonly available. Grouts containing these cements can be formulated to permeate most any granular soil including fine sand. In this application, they have substantial advantage over the more traditionally used chemical solution grouts, in that significantly higher strengths can be obtained, and at a lesser cost. Additionally, cement grouts are non-toxic and do not suffer strength regression with time, as do most chemical grouts. Ultrafine cement can be of any one of three different origins, including common portland cement, slag based cement, and a combination portland-pozzolan blend. However, there are fundamental differences in the penetrability of grout derived from the cement of these different origins. Extensive research and experience has shown the grain size of the cement alone does not control the penetrability of the resulting grout into soil. Both grain shape and surface condition of the cement are important, as is the polar intensity. The results of an ongoing evaluation program, as well as actual case histories of projects utilizing ultrafine cement grout are presented.

Introduction

Grouts based on ultrafine cement have been used on many projects involving the solidification of granular soils; however, not all performance has been satisfactory. Investigation of many less than successful applications, as well as subsequent research, has continually shown the performance of these grouts to be strongly influenced by the derivation of the cement. Ultrafine cements of three different fundamental origins and methods of manufacture are commonly available. These include products based on ordinary portland cement, granulated blast furnace slag cement, and blended cement which is a combination of portland cement and a pozzolanic material. There are fundamental differences in the chemistry of the final product derived from these different basic ingredients however, and this can have a dramatic effect upon the penetrability of any resulting grout.

---

[1]Consulting Engineer, Mariposa, California

Portland cement consists of a mixture of calcareous and argillaceous minerals that are combined, crushed, pulverized, and burned in a rotary kiln. This produces a material known as clinker, which comprises pellets of about ½ inch (12 mm) in diameter. The clinker will contain four principal chemical components, tricalcium silicate, dicalcium silicate, tricalcium aluminate, and tetracalcium aluminoferrite. The proportions of these primary constituents will vary however, due to differences in the beginning raw minerals. The hardening rate of compositions made with the resulting cement is largely dependent upon the exact proportions of these compounds, especially the tricalcium aluminate and it is virtually always too rapid. In order to provide the proper setting time, an appropriate proportion of gypsum is thus added in the final manufacturing operation. The gypsum is the last ingredient introduced into the cement and its proportion which is variable, is dictated by the exact chemistry of the other ingredients, which comprise well over 90% of the total product.

Relative to grain size, ASTM C 150 "Standard Specification for Portland Cement" specifies only the required Blaine fineness, which is a measure of the specific surface area of all of the cement grains in a specific volume of cement. The Blaine specific surface area is usually expressed either as cm^2/g or m^2/kg. The typical Blaine fineness of cements expressed in cm^2/g is as follows:

- Ordinary portland cements – Type I   3000 – 5000
- High early strength cement – Type III   4000-6000
- Ultrafine cements - > 8000

The specific grain size distribution of available cement varies enormously. Although Type III high early strength portland cement, is required to have a higher Blain fineness than either Type I or II, neither the maximum particle size, nor the specific grain size distribution, are mandated. This is due to the varying properties of the original clinker from which the cement is ground, the minerals of which that clinker is composed, and the particular manufacturing process. Even though, on average Type III cement is finer than Type I, the author has encountered Type I cements of which the largest particles were actually finer than those of some Type III he has experienced.

Ultrafine Cement Origin

Some ultrafine cements are made by further grinding of an already produced ordinary portland cement in which the gypsum has previously been combined, whereas other ultrafine products are ground to their final fineness, after which the proper amount gypsum is added. In the former case, setting time is less predictable, as the gypsum, which is softer than the other constituents, grinds more readily and thus provides excessive surface area for reaction (Taylor 1990). The proportionate increase in surface area will always be excessive and the setting time thus accelerated. For this reason, retarders such as citric acid are sometimes interground during the processing. Portland cement-citric acid combinations are extremely sensitive to temperature changes, so the proportions of the additive must be matched to the temperature of both the working environment and the medium being injected. Problems of flash set and difficulties with both the setting time and rate of strength gain, have been reported with cements of this origin.

In products wherein the portland cement clinker is ground to its final fineness prior to the gypsum addition, care can be taken so as to match the gypsum quantity precisely to the specific reactivity and actual surface area of the particular cement. Uncontrollable setting time is thus not as great a problem with this manufacturing process, as the reactive surface area of the gypsum can be accurately controlled. When selecting ultrafine cement based on Portland cement only, it is thus important to understand the method of manufacture and its influence on setting time.

Pozzolanic materials such as slag and pumice, contribute little if any early hydration activity. Consequently, cements that contain appreciable amounts of these compounds, set more slowly and behave similarly. Such pozzolanic materials are available separately and are sometimes included as a distinct component of a grout mixture, however they can also be interground into special cements during the manufacturing process. Whereas ultrafine cements founded on portland cement can have very fast set times, especially in hot environments, such is usually not the case with cements containing appreciable amounts of a pozzolanic material. Long setting times are usually an advantage when grouting in fine voids such as the pore system of soil, so retardation is seldom a problem. Well established accelerating admixtures are available however, and their behavior is quite predictable, so faster setting times for grouts based on these products can be readily achieved, if desired.

Several different manufacturers market ultrafine cement based on granulated blast furnace slag which is a pozzolanic material. As with any pozzolanic material, it must be combined with a discrete proportion of calcium, which is required to initiate hydration and setting. This can be from any source, but is typically provided by inclusion of a minor fraction of finely ground ordinary portland cement. The reaction and set time are dependent upon the amount of calcium included, and the setting rate is usually controlled so as to be much slower than that of ordinary portland cement. Slag based cements thus typically promote less early surface hydration activity than those which are based on portland cement alone. Pumice is another pozzolanic material which is the basis for at least one readily available ultrafine product. As with the slag cements, it contains a minor amount of portland cement to initiate the reaction.

Grain Size

Whereas conventional thinking would expect the penetrability to be closely related to the grain size distribution and especially the maximum grain size of the cement particles, such has repeatedly proven to not be the case. The author has consistently found that some ultrafine cements do not penetrate readily into granular soils even though some of their coarser cousins do. Similar performance has also been reported by several contractors who have used a variety of different ultrafine products, and this has been further confirmed by a large scale test program that will be detailed shortly. Figure 3, provides the grain size distribution of several ultrafine products, as well as typical Type I and III cement.

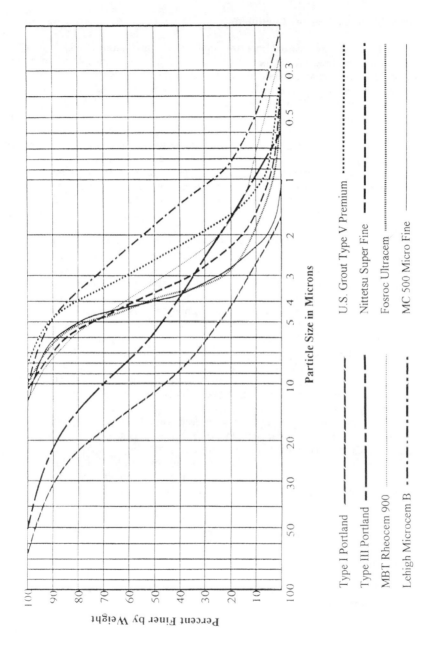

Percent Finer by Weight

Particle Size in Microns

Type I Portland
Type III Portland
MBT Rheocem 900
Lehigh Microcem B

U.S. Grout Type V Premium
Nittetsu Super Fine
Fosroc Ultracem
MC 500 Micro Fine

Grain Shape and Surface Condition

It is well recognized that the grain shape and surface condition of a cementitious material effects its pumpability as well as the penetrability of grouts made therewith. As an example, the inclusion of fly ash, the particles of which are perfect spheres, will dramatically improve the pumpability and penetration of a common cementitious suspension grout. The same can be said for entrained air which is disbursed as extremely fine spherical bubbles and greatly improves the pumpability of all cementitious compositions. Rough odd shaped particles are far more likely to become separated and hang up in the formation pore structure than are smooth spherical ones. Because the individual particles of ultrafine cement are extremely small, sophisticated equipment is required for their study. Whereas grain size has been amply discussed in the literature, relatively little investigation of the chemistry or surface condition has occurred and a literature search failed to locate any such data published in English.

In addition to loss of the larger cement particles, the advancing grout tends to pick up soil fines which might exist in the formation being grouted. As the distance from the point of injection increases, these accumulate to an ever greater extent and in combination with the larger cement particles from the grout, tend to form a filter. This is especially so in soils which contain abundant fines. Interestingly however, the grout will penetrate for some distance before a significant blockage in the pore system occurs. In such cases, the loss of cement with increasing distance from the injection probe has been confirmed on many jobsites, through strength testing of core specimens obtained at varying distances from the point of injection. Ultrafine cement grouts thus usually require a closer spacing of the grout probes, than would be the case for a chemical solution grout. While this results in greater cost for the probes, such will usually be more than offset by the lower cost of the grout material.

A further and apparently very significant contributor to the penetrability of cementitious grout is the polar strength of the cement grains. Common cements are strongly polar and the individual grains are thus powerfully attracted to each other. This adversely effects penetrability, and is the reason that high shear mixing is of such great importance. The attraction to gather remains strong however, even after good initial mixing. Slag base cements, on the other hand, are weakly polar, so they tend to remain as separate elements in the mixed grout. This explains the reason why slag base cement grouts tend to permeate readily even when high shear mixing is absent, which has been witnessed by the author on several actual projects.

Evaluation of Various Cements

The variable properties of different ultrafine cements, and especially those resulting from the different manufacturing methods, have a dramatic affect on the penetrability of the resulting grout. This has been clearly demonstrated through a series of large scale tests, conducted to evaluate the penetrability of ultrafine cement grouts into the pore structure of sand. These have been conducted as part of the field demonstration at the Annual Short Course on Grouting offered by the University of Florida, and held in Denver, Colorado, each year from 1997 through 2000. A fifth demonstration/test was held during the American Society of Civil Engineers Geo-Institute, GeoDenver conference, which was held in August, 2000, also in Denver, Colorado.

In these tests, grouts prepared from a variety of different ultrafine cements were injected into identical transparent columns, carefully filled with specifically graded sand, as illustrated in Figure 2. The columns consisted of transparent plastic tubes with an inside diameter of 7 ½ inches (188 mm) and a height of 5 feet (1.5 m). They were equipped with top and bottom plates that were bolted to flanges on each end. A ¾ inch (19 mm) internal diameter threaded fitting for the grout inlet was attached to the bottom plate.

The tubes were filled from the top with a kiln dried, commercially processed, quartz sand. The sand was graded such that 100% passed a No. 30 mesh sieve with 40% passing a No. 40 sieve. A specific amount of sand was placed into the top of the tubes while they were subjected to vibration by way of a pneumatic vibrator at their base. The removable top plate was then secured in place to restrain any upward movement of the sand that might occur. Several small holes penetrated the top plate to allow venting of air and grout, should it penetrate that far.

Figure 2. Grout injection into sand columns during demonstration.

Although the scope and control of the tests has expanded and improved through the years, all operations from the very beginning have been performed by well experienced grouting professionals, and the level of performance is the highest that could reasonably be expected on a production grouting job. Brand new, high quality full scale mixing and pumping equipment has been used which has included units from three of the major manufacturers of grouting equipment, Chemgrout, Colcrete, and Hany. The grout mixers have all been of high shear design and both plunger and progressing cavitation (Moyno) pumps have been used.

Grouts composed of a typical Type I cement, as well as a variety of different ultrafine cements, have been mixed and pumped into the columns. Ultrafine cements from as many as six different manufacturers have been used at least once, and several of them have been employed for each of the five years the event has taken place. The grout was injected into the bottom connection at a uniform pressure of 10 psi (69 KPa), for a period of not more than 20 minutes. A standard circulating grout injection system was used, with the pressure of the grout going to the individual columns being controlled by manipulation of suitable valves at the grout header, as can be observed in Figure 3.

Figure 3. Controlling grout flow with a standard circulating injection system.

Quality control tests, which might conceivably be employed in normal grouting, have been used in the monitoring and control of the tests. These include grout temperature, density, Marsh funnel and ASTM C 939 flow cone efflux times, and obviously control of the grout pressure. Whereas the temperature and other environmental conditions have varied over the period of time the tests have been conducted, trends to date pointing to the most penetrable grout compositions are surprisingly clear, and the slag based ultrafines have shown superior performance.

This is interesting because they are not necessarily the finest ground product, as can be readily observed in Figure 1.

The relative performance of grouts containing the different ultrafine cements is shown in Table 1. Therein, the penetration rates shown, if less than 60 inches (1524 mm), was for the 10 psi pressure being held for a full 20 minutes. While it must be understood that these data appropriately apply only to injection of these grouts into similar sand materials, the results are consistent with the experience of the author on several varied field applications involving solidification of a variety of different sandy materials.

Table 1 - Penetration of various Ultrafine Cement Grouts into experimental sand columns.

Cement Type	1999 Penetration - inches (mm)	2000 Penetration - inches (mm)
Type II Portland	4 (102)	11 (279)
Nittetsu Super Fine	60 (1524) 7 minutes	60 (1524) 4 minutes
Fosroc Ultracem		60 (1584) 2 minutes
MC 500 Micro Fine		60 (1524) 8 minutes
U.S. Grout Type V	57 (1448)	10 (254)
MBT Rheocem 900	9 (229)	24 (610)
Lehigh Microcem B		32 (813)

Simply because a particular grout performed relatively poorly in these tests does not mean that it is an inferior product for all applications. The best grout properties for one particular use are not necessarily the best for all, and a particular mixture that does very well in one case may perform quite poorly in another. Interestingly however, quite consistently, the grouts made with slag based ultrafine cement have exhibited the best performance for injection into sand, even though they are not the most finely ground. Such performance is consistent with the author's real-world experience on many projects where ultrafine cement grout has been injected into granular soil.

Case Histories

Seismic retrofit of the 70 year old San Francisco Civic Center required undermining of existing foundations by as much as nine feet. The foundation soils are basically, round grained, beach sands, with virtually no cohesion. The ground, which was to remain under the foundation elements, was solidified with a slag base ultra-fine cement grout. As can be observed in Figures 4 and 5, this allowed for otherwise unsupported excavation for the new extended footings. The specifications called for the grouted mass to obtain a minimum unconfined compressive strength of 100 psi (69 kPa) and required that "*the grouted sands will extend from the bearing area beneath the footings to be stabilized, extending on a 2 to 1 H:V line, to 2 feet below the bottom of the adjacent underpinning excavations.*" The means and methods to accomplish this were otherwise left to the contractor.

The grouting contractor's original plan was to place rows of primary grout holes at a spacing of six feet (1.8 m). After these were injected, the spacing was to be split to three feet (0.9 m) and secondary injections performed. Based on a calculated soil porosity of 27%, an injection quantity of grout sufficient to form a four foot (1.2 m) diameter column would be made. To obtain the required strength,

Fig. 4 – Vertical cut against foundation.   **Fig. 5 – Foundation undermined.**

ultrafine cement grout would be mixed with a water:cement ratio of six, by weight of the cement. No other ingredients were to be included except for a high range water reducer supplied by the manufacturer of the cement as an essential part of any mix for which it is used.

A test section was grouted according to the above parameters and then excavated to expose and evaluate the propriety of these initial injection criteria. Insufficient overlapping of the grout columns was found to occur. By comparing the obtained diameter of those columns with the quantity of grout injected, the actual soil porosity was established to be about 35%. Moreover, the resulting unconfined compressive strength of the solidified mass was found to be substantially greater than required and in fact, many of the specimens tested broke at well over 1100 psi (7.5 MPa). Although this was an added benefit from the standpoint of support, it was a huge problem for the general contractor who had to perform the excavation. And of some significance, the strength of the grouted mass was much greater near the injection hole than at some distance therefrom, indicating the ultrafine particles were filtering out as the distance of travel increased.

The hole spacing was then reduced to two feet (0.6 m) and the grout quantity injected was calculated to form a three foot (0.9 m) column, based on a porosity of 33%. The water:cement ratio was also raised to eight. This was found to result in complete interlocking of the individual injections. Further, the strength was reduced to about 300 psi (2 MPa), which was still much greater than required, providing a good safety factor for the grouting contractor while reducing the effort required for the excavation. Even this "reduced" strength was far greater than what would be possible to attain from the originally contemplated solution grout.

The grout was mixed in a high speed vertical paddle mixer. It was then transferred to one of two adjacent agitators. The injection was from the bottom up in preplaced ¾ inch (19 mm) injection pipes. These had been driven into place with a temporary plug of a rivet in the leading end. The pipe was raised a foot and the rivet easily knocked off immediately prior to injection. In all, more than 120,000 gallons (454,000 L) of grout was used. The use of ultrafine cement grout provided another significant advantage on the very congested construction site. Being supplied in paper bags and requiring only water to prepare, space for storage tanks or tanker trucks which would normally be required if a chemical solution grout were used, was not needed.

When the terminal building at the Kansas City International Airport was originally constructed, the excavated space behind the basement retaining walls was backfilled with very permeable, cohesionless granular fill sand. A contractor building four access tunnels connecting to the building in 1997, encountered this sand fill. It caved into his excavation, exposing utilities and undermining adjacent pavement. To mitigate the problem, ultra-fine cement grout was used to solidify the cut face of the required excavations. The grout was injected from the top down, through jet pipes that were marked at one foot intervals. The pipes were positioned at a 15 degree angle, toward the proposed excavation. They were jetted with grout into the formation, with a pause at each foot interval as required to place the predetermined amount of grout, as shown in Figure 6.

For most of the work, the probes were placed on three-foot centers and the calculated amount of grout to form a four-foot column was injected. Where the excavation was greater than about 10 feet deep, a second row of holes was provided, split spaced, and three feet behind the initial row. Only the bottom portions of these holes were grouted. The grout consisted of a slag base ultrafine cement and sufficient water to provide a water:cement ratio of five. A high range water reducer supplied with the cement was also included. It was

Fig. 6 – Jetting of Ultrafine Cement Grout.

mixed in a high shear colloidal mixer and transferred into an agitator tank which supplied a 2L8 Moyno pump, as can be seen in Figure 7. The pump was set for an injection rate of 2.6 cubic feet (75 L) of grout per minute.

Upon excavation, the sand fell freely from the injected columnar shaped masses, Figure 8. They withstood the elements, including several cycles of freeze and thaw, for a period of several months. Both vibration and impact forces were created during demolition of the adjacent concrete wall, but these had no apparent effect on the solidified mass. The spacing used for the injections would have been excessive had overlap been required so as to allow trimming back to a smooth surface plane. However, it was entirely satisfactory as constructed, and effective in prevented sloughing of the retained sand.

Fig. 7 – Portable grout plant fed from truck.   Fig. 8 – Sand fell off solidified columns.

In another example, ultrafine cement grout was used to solidify cohesionless sand to facilitate the excavation of a basement addition to a large Victorian house in San Francisco. Therein, injection probes were installed in rows on a two foot (0.8 m) spacing. The number of rows of holes was dependent upon the height of the resulting otherwise unsupported vertical cut, as illustrated in Figure 9. A single row was used for cuts less than about five feet (1.5 m), two rows for cuts up to about seven feet (2.1 m), and three rows for depths to about ten feet (3 m). All of the holes extended a depth of at least one foot (0.3 m) deeper than the bottom of the new excavation. The face row of holes extended to the original ground surface, whereas those in the second row were only injected to within about three feet (1 m) thereof. Where a third row was required, it terminated five feet (1.5 m) below the ground surface.

The injection probes were completed using a primary-secondary split spacing sequence. The face row was completed first, followed by the second, and finally the third. In all, some 40,000 gallons (151,400 L) of slag based ultrafine cement grout with a water:cement ratio of 0.6 were used. The amount of grout injected at any one stage varied according to its position and grouting sequence. The benefit of the work can be readily observed on the right side of Figure 10, wherein ungrouted sand is freely running into the excavation.

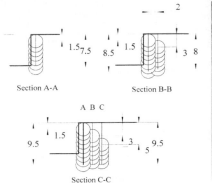

Section A-A   Section B-B

Section C-C

Fig. 9 – Typical grouted Sections for different heights of cut.

Fig. 10 – Final excavation. Note running sand in ungrouted area on right.

Interestingly, the individual columns can be seen upon careful examination of Figure 10. Therein, the smooth surfaces are where the solidified material has been trimmed back. Likewise, the cohesionless nature of the sand can be observed by the running conditions on the far right of the Figure, which is beyond the grouted zone.

Conclusions

Although not all applications have been successful, ultrafine cement has been advantageously used in numerous instances. Wide spread experience backed by large scale test evaluations, has shown that the work can be done reliably. Clearly, the use of a cement based on granulated blast furnace slag is required if maximum injectability of the grout is to occur. Because fines from the soil, as well as some of the larger size particles of the cement will tend to form a filter with increasing distance from the point of injection, the probe spacing must be somewhat closer than is required for solution grouts. This might be as near as two feet (0.6 m) for finer sands and those that contain fines, whereas three feet (0.9 m) is widely used for clean open formations. The resulting strength of a mass injected with cement grout, can be significantly higher than that obtainable with traditional solution grouts. Further, the cementitious grouts do not deteriorate with time as do most solution grouts. And of significance, a cost savings is usually experienced through use of ultrafine cement grout.

References

Taylor, H. F. W., (1990) "Cement Chemistry", Section 4.1.4, pp 99-100, Academic Press, London.

The Groutability of Sands - Results from One-Dimensional and Spherical Tests

Jens Mittag[1] and Stavros A. Savidis[2]

Abstract

In addition to the application of empirical injection criteria also one-dimensional injection tests have become a suitable technique for the evaluation of the groutability of sands. Both procedures have been applied to the injection of microfine cement suspensions for the sealing of construction pits in Berlin, Germany. The mostly unsatisfying results clearly indicate the limitations of the above mentioned evaluation practice.

Starting from this background an extensive investigation program has been carried out at the Geotechnical Institute at the Technical University of Berlin, Germany. Different combinations of sand and microfine cement grout have been used in one-dimensional and spherical injections, accompanied by rheological and grain-size analyses.

This paper focuses on the filtration effects which represent the main difference between the penetration of chemical grouts and that of suspensions of microfine cements. The deposition of particles within the soil skeleton leads to a dynamic reduction of the hydraulically effective pore-space and to a change in the rheological properties. Although basically known, these effects are not considered in the formulation of injection criteria.

---

[1]Project Engineer, GuD Consult GmbH, Dudenstraße 78, 10965 Berlin, Germany, mittag@gudconsult.de

[2]Professor; Geotechnical Institute, Technical University of Berlin, Sekr. TIB1-B7, Gustav-Meyer-Allee 25, 13355 Berlin, Germany, savidis@tu-berlin.de

Introduction

In the last decade a number of large building projects started in the centre of the formerly divided city of Berlin. To avoid lowering of the high ground water table, only trough-type excavations have been used. The quantity of water that was allowed to flow into the open excavation pits through the walls or the bases was limited to 1.5 liter/s per 1000 m² of watered area.

The bases of the open excavation pits have been sealed with back anchored underwater concrete slabs, jetted floors or injected floors (Fig. 1). After 1995, when the widespread use of silicate-grout systems was stopped by the authorities due to possible groundwater pollution, microfine cements were proposed as an alternative. More than ten construction projects with areas of up to 5000 m² have been carried out in Berlin using microfine cements. However, the sealing layers produced often could not meet the strict requirements with respect to the allowable leakage rate.

Motivated by these results a series of laboratory test has been carried out at the Geotechnical Institute of the Technical University of Berlin. In the following sections the tests conducted are described and main findings are summarized. A detailed description of the investigation program can be found in MITTAG (2000).

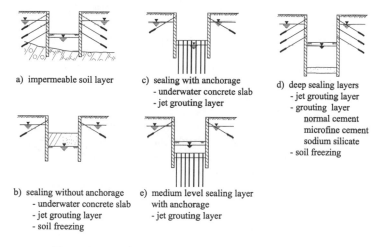

a) impermeable soil layer

c) sealing with anchorage
  - underwater concrete slab
  - jet grouting layer

d) deep sealing layers
  - jet grouting layer
  - grouting layer
    normal cement
    microfine cement
    sodium silicate
  - soil freezing

b) sealing without anchorage
  - underwater concrete slab
  - jet grouting layer
  - soil freezing

e) medium level sealing layer
  with anchorage
  - jet grouting layer

Figure 1: Construction methods for trough-type excavations

Evaluation of Groutability - Empirical Criteria and Injection Tests

During the design process of grouting work the consultant or the site engineer has to deal with some uncertainties. Even if he knows that the soil can be penetrated in principle by a specific grouting material, he has only little information about the maximum injection depth, appropriate grout mixture or allowable injection pressure. In particular, specifying an optimum raster space requires reliable data about the possible penetration radius.

The evaluation of the capability of a specific injection material depends on a large number of parameters, such as soil conditions, grout mixture, injection parameters and type of ground treatment. If only one of these conditions changes an elsewhere successfully used product becomes inappropriate. However, the currently used criteria consider only part of these important parameters.

It can be stated, that the evaluation of the groutability on the basis of empirical criteria can only give a rough idea about the suitability of an injection material. This applies in particular to the injection of microfine cement suspensions into sands and even more to their use for deep sealing layers in open excavation pits.

To overcome these difficulties and to get a better understanding of the penetration behaviour of microfine cement suspensions special one-dimensional injection tests have been developed. The criterion used in Germany (SCHULZE et al., 2002) can be described in short as follows:

1. Prepare a soil column which length is at least 4 times the diameter and saturate it with water.
2. Inject the microfine cement suspension with constant rate into the column from bottom to top.
3. The test result is positive if three times the pore volume could be injected with a pressure below 600 kPa.

Although this kind of test is already much more meaningful than the comparison of particle-size distributions there are still some important shortcomings: The penetration velocity is not clearly prescribed. Moreover, there is no relation given between the laboratory and the in-situ conditions.

But which injection rate or velocity in one-dimensional tests is appropriate for the simulation of spherical conditions? This question is difficult to answer, because in the three-dimensional case due to the geometry the velocity of the suspension is decreasing along the radius. In addition it has to be taken into account that the center of a spherical injection will by penetrated much more intensively than the outer areas. As will be shown later these fundamental differences between one-dimensional and spherical conditions are of great importance for the filtration behavior and the rheological properties of the suspensions.

Investigation Program: One-Dimensional and Spherical Injections

To avoid the above mentioned problems spherical injections have been planned and carried out in laboratory scale. Fig. 2 shows the experimental setup. The equipment has been originally developed for testing soil columns and then extended to spherical injections. The injection funnel (Fig. 3) represents 1/33 of a sphere and enables penetration depths of approx. 75 cm. An important requirement within the test program was the prevention of unintentional cracks. Therefore an overburden pressure of up to 350 kPa could be applied on the top of the funnel using a rubber membrane.

Several pressure transducers have been attached to the injection funnel. Thus, it became possible to measure the pressure drop in different funnel sections as a function of injection progress.

To allow a comparison of the results of both one-dimensional and spherical injections the test conditions have been specified as similar as possible. For all tests presented in this paper the microfine cement FINOSOL U (manufactured by Dyckerhoff Baustoffsysteme GmbH, Germany) with a water-to-cement-ratio of 5:1 has been used. Rheological investigations of the suspension with a shear-stress controlled rheometer showed a shear-thinning behavior which can be described by the power-law flow model.

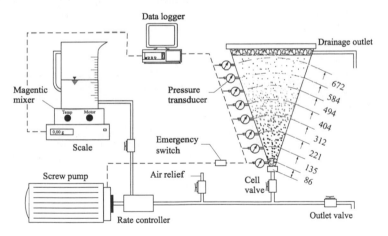

Figure 2: Experimental setup for spherical injections

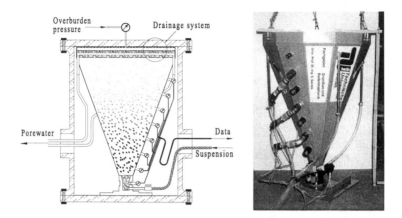

Figure 3: Representation of the injection funnel

The sand F34 (from a gravel pit in Frechen, Germany) was placed at a relative density of 50% and saturated from the bottom prior to grout injection. Fig. 4 shows the particle size distribution for different FINOSOL products, for the test sand F34 and for sands from Berlin construction sites. Applying the well-known groutability ratio $N = D_{15(soil)}/D_{85(grout)} = 24$ it can be concluded, that FINOSOL U should be able to penetrate the sand F34 used in the tests.

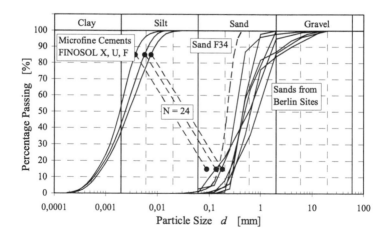

Figure 4: Particle size distributions of sands and microfine cements

Test Results

Fig. 5 represents the results of ten one-dimensional injections in PVC tubes of 10 cm in diameter and 100 cm long. The injections have been carried out with five different injection rates. The figure shows the great influence of penetration velocity on the injection pressure.

Fig. 6 shows the injection pressures from four of total nine spherical injections. The injection rate used for this tests was 0.24 liter/min which is equivalent to 8 liter/min for the whole sphere and a common value in practice.

While the one-dimensional injection showed no difficulties for the penetration progress, the spherical injections clearly indicate a deviation from the theoretically assumed sublinear pressure increase. For a higher equivalent rate of 13 liter/min the maximum radius reached was 55 cm. This limited groutability was not expected from the one-dimensional tests and the application of empirical grouting criteria.

Therefore, the subsequent investigations have been focused on the filtration process which leads to a decrease of the hydraulically effective pore-space and a change of the rheological properties of the suspension.

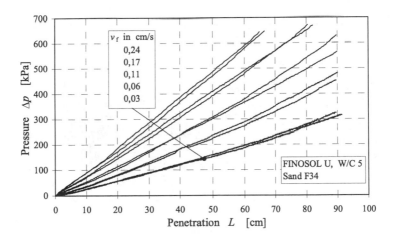

Figure 5: Results of one-dimensional injections with different filtration velocities v_f

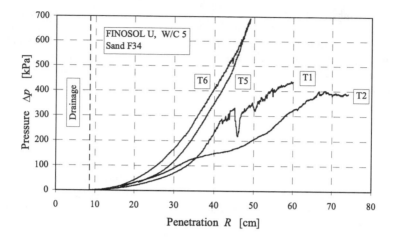

Figure 6: Results of spherical injections with an equivalent injection rate
of 8 liter/min

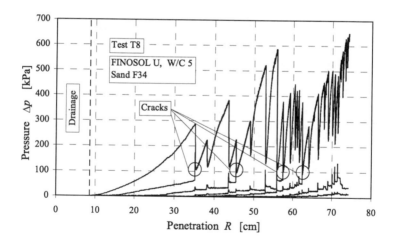

Figure 7: Development of cracks during spherical injection with an equivalent
injection rate of 13 liter/min

Fig. 7 and 8 are good representations of the events close to the injection center. After removing the grouted soil from the funnel a long surface crack was found. This crack was located at the position where the pressure transducers have been attached to the funnel and probably caused by some small soil inhomogeneities. The decrease of pore-space first leads to a strong increase of the injection pressure and the subsequent crack development yields to a pressure drop. Then the pressure increases again and the crack reaches the next transducer. When the test was finally stopped after reaching a radius of almost 75 cm the first five transducers displayed the same pressure.

Accordingly, all tests in which no cracks were monitored during the injection showed a penetration of less than 55 cm and no surface or inner cracks.

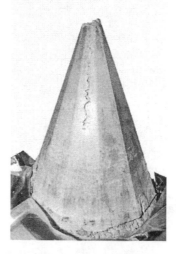

Figure 8:   Grouted soil body from test T8 with long surface crack

For closer investigations of filtration effects the cement content, the permeability and the compressive strength have been determined after 28 days on the grouted soil bodies from both one-dimensional and spherical tests.

The determination of ignition loss has been proven as very stable and most suitable method to represent the change in cement content. The ignition loss was determined after heating samples from 105°C to 1050°C. Fig. 9 and 10 show the results for some one-dimensional and spherical injections. The distribution of ignition loss along the column is rather similar and almost linear with distance for all one-dimensional tests.

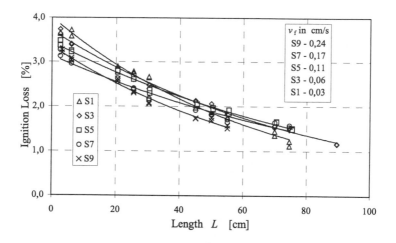

Figure 9: Ignition loss versus penetration length for one-dimensional injections with different filtration velocities v_f

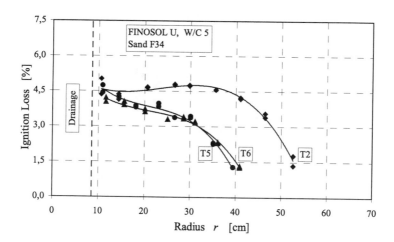

Figure 10: Ignition loss versus penetration radius for spherical injections with an equivalent injection rate of 8 liter/min

Fig. 10 shows a generally higher ignition loss near the injection center and a clearly different distribution for the test T2 with crack.

The ignition loss describes both the content of already deposited cement particles and the content of particles still in suspension at the moment the test has been stopped. The higher ignition loss near the injection center indicates a greater amount of already deposited particles and a reduced hydraulically effective pore-space which is finally responsible for the strong increase of injection pressure.

As already stated the intensity of penetration or filtration near the injection source is much higher for spherical conditions than for the one-dimensional case. That is why column injections with comparable penetration lengths show considerably less filtration effects.

Modeling of Filtration

In the research project the attempt was made to simulate the thinning of the suspension and the reduction of the hydraulically effective pore-space by application of modified filter models. In a first step the thinning of the suspension due to particle deposition was taken into account using a relation between the decreasing cement content and the corresponding rheological characteristics. In a next step it was possible to back-calculate the hydraulically equivalent pore radii at various penetration depths and sections from the measured pore pressure differences.

All this calculations were based on condition that the theory of deep-bed filtration is applicable and the change in particle concentration C over a filter length x can be described by $\partial C/\partial x = -\lambda \cdot C$. The filtration factor λ represents the hydraulic conditions and the properties of the particles, the fluid and the filter and is usually assumed to be a constant as long as the filter is still relatively clean.

While this assumption might be acceptable for many technical processes its validity is only limited for the injection of microfine cement suspensions. Since the concentration of particles is very high in this kind of suspension the deposition is rather intensive and the filtration process will be influenced not only by particle-grain interactions but also soon after the start of the test by interactions between already deposited particles and particles still in suspension.

Due to the complexity of these processes and the rapid changes of the hydraulic conditions within the filter system the usual assumptions for the theory of deep-bed filtration cannot be applied to the simulation of spherical microfine cement injections.

Although the results of one-dimensional injections could be back-calculated with acceptable accuracy, additional tests with considerable longer penetration depths and a accordingly more intensive filtration showed effects similar to those described above for spherical injections.

Conclusions

Within the investigation program the shortcomings and limitations of one-dimensional injection tests for the evaluation of the spherical groutability could be demonstrated. Due to different geometrical conditions the results from one-dimensional injections can hardly be transferred to the spherical case. This applies in particular to the estimation of the possible penetration depth, which is a basic information for the design of grouting projects for sealing purposes.

Although a lower injection velocity leads to a lower pressure or a higher penetration depth in the one-dimensional case we must recommend to use a (mostly higher) velocity which is comparable to the conditions near the injection center of the spherical case. Otherwise the one-dimensional injection can easily lead to an overestimation of the extent of groutability. A practical alternative consists in conducting spherical tests as described above.

References

Mittag, J. (2000). *Investigations on the Filtration Behaviour of Microfine Cement Suspensions* (in German). Publication of the Geotechnical Institute at the Technical University of Berlin, Vol. 27.

Schulze, B. et al. (2002). Guideline for grouting works with microfine cements in sand (in German). *Bautechnik*, vol. 79, 499-508 (part 1), 589-597 (part 2).

GROUTED COFFERDAM FOR AN INTAKE STRUCTURE IN MIXED ROCK AND GRAVEL ENVIRONMENT

BY

K. Ramachandra[1], A. Wern[2], R. J. Kapadia[3] and S. K. Shim[4]

ABSTRACT

The Ilijan Power Plant is located in the island of Luzon in Philippines. Washington Group[5] is the Engineering, Procure, and Construct (EPC) Contractor with Daelim as the Civil Sub-contractor. The cofferdam for the intake structure had dimensions of 36m x 30m and extended 9.0m below the ground water level. The subsurface conditions at the site are very complex and highly variable. Three alternating layers of dense cemented conglomerate and highly permeable open structured gravel/cobbles were encountered at the intake structure location. An order of magnitude permeability of 1×10^{-1} cm/sec was estimated by simple pump tests from a large test pit at the site. The intake structure is about fifty meters (50m) from the Pacific Ocean thus excluding de-watering as a solution for construction.

The paper describes details of subsurface conditions, the intake structure and various alternates considered for construction of a cofferdam. Schedule, reliability of the proposed solution and cost were the driving factors in the selection process with the first two being the higher priority. The selected solution had redundancy and recovery schemes planned to address unanticipated subsurface conditions and to assure timely completion of the intake structure as planned. The selected solution utilized steel sheet piles, jet grouted columns, soil anchors, and cement- sodium silicate grout mixture as a bottom seal plug placed at a significant distance below the bottom of excavation. On completion of these activities, the intake structure was built on schedule in a relatively dry condition with only two small sump pumps to handle minor seepage.

BACKGROUND:

The Ilijan power project is located in Batangas Province in the Southern portion of the island of Luzon in Philippines. The plant site is about 25 Km south of Batangas city. The site is located on the side of a hill and bounded on two sides by the ocean.

[1]Consulting Geotechnical Engineer, [2]Consulting Civil Engineer, [3]Supervising Civil Engineer, Washington Group International, Princeton, NJ, USA [4]Vice President Construction, Daelim Industrial Co. Ltd., Seoul, Korea.
[5] A successor company to the former Raytheon Engineers & Constructors, Raytheon Ebasco Overseas Ltd., and United Engineers International.

Washington Group International Inc is the Engineering, Procure, Construct (EPC) Contractor with Daelim as the Civil Subcontractor. The plant has been designed as a 1200 MW combined cycle power plant with a once through seawater cooling system requiring 28.4 cubic meter per second water supply. The cofferdam for construction of the intake structure had dimensions of 36m x 30m. The bottom of excavation was subjected to a head of 8.5m hydrostatic pressures in highly pervious gravel and cobbles traversed by layers of weak to well cemented conglomerate. The project site is covered mostly by poorly consolidated tuff and clastic sedimentary rocks. Three basic geologic units were identified in the subsurface investigation program. The first unit was a coral reef limestone in the upland northern portion of the site. The second unit was tuff that was identified in the East and Northeast portions of the site. The third was a sand and gravel conglomerate that was identified in the main plant area below the reef limestone and tuff. Conglomerate, sands and gravel were expected in the intake area. The subsurface investigation report classified the conglomerate as moderately weathered and weakly cemented in the upper portion. The composition was described as pea to cobble size rock fragments of andesite, limestone, and basalt and metavolcanic rock with cementation by calcium carbonate. Very coarse conglomerate beds with rock boulders were also occasionally encountered during the investigations. Water pressure tests with packers were performed in four bore holes. The hydraulic conductivity was estimated as 3.25 x 10^{-5} cm /sec. This low value might have resulted from the weakly cemented nature of conglomerate tested with packers in the drilled hole. Due to the complex and variable nature of geology at the site and the need for excavating below the sea level, it was proposed to dig a deep test pit. The pit would be excavated below the water table and rates of water ingress including effects of tide observed. Unfortunately this investigation was cancelled due to difficulty in getting access from the villagers during the design phase.

DEWATERING CONCERNS

The power plant required a 70m X 56m X 11 to 14 m deep excavation for the basement. The cooling water box culverts required an extended excavation of 6.5 meters below the sea level. During construction, it was observed that the water in the partially excavated basement was connected to the sea. The water level lagged the tide by about one hour. The distance to the sea was about 190 meters. The fluctuation range was about 0.6 m compared to the tidal range of 1.5m. It was also observed that the conglomerate was not cemented and highly pervious. It was clear that the previous estimated permeability value was highly misleading and the actual permeability was higher by several orders of magnitude. De-watering was not technically feasible, as pumps would not be able to keep pace with continuous recharge from the ocean. Also de-watering would adversely impact the relatively "sweet" water overlying the salt water, which was being used by the villagers for drinking. The design was altered, by raising the basement bottom elevation to avoid de-watering below the sea level.

SITE INVESTIGATIONS AND RESULTS

Four (4) new borings were drilled in the intake area. Due to problems of drilling in the conglomerate formation, reverse circulation air rotary drilling was performed. Samples were deposited every meter in plastic bags. Penetration rates were observed to differentiate cemented layers from non-cemented layers. It was concluded that the subsurface consisted predominantly of non-cemented layers of gravel and cobbles of size up to 0.3 m in diameter. Some cemented layers were also noted in depths of interest but continuity was suspect. A crude pump test in the main plant basement area had revealed that the hydraulic conductivity could be in the order of 1×10^{-1} cm/sec. Test wells in the intake area were abandoned due to breakdown of equipment. Daelim performed a permeability test and estimated the permeability as 2.6×10^{-1} cm /sec. The stratigraphy at the intake structure was determined as:

```
———————————————————————  El +1.0m
              - - - - - - - - - -  El +0.5m Ground Water level
Sands and Cobbles
———————————————————————  El –9.1m
Conglomerate  1
———————————————————————  El-10.3m
Sands and Cobbles
———————————————————————  El –14.3m
Conglomerate 2
———————————————————————  El -15.6m
Sands and Cobbles
———————————————————————  El –16.8m
Conglomerate 3
———————————————————————  El.-18.0m
Sands and Cobbles
```

The conglomerate layers identified above had weak to high cementation. A typical core of cemented Conglomerate is shown below:

The non-cemented conglomerate has been classified as sand and cobbles above.

A finite element analysis estimated that as much as 190 cubic meters/minute of water need to be de-watered in a non- grouted cofferdam for the intake structure. It was decided to study various alternatives internally and with Daelim, the civil subcontractor.

ALTERNATIVES

The three (3) major alternatives considered are described below. Several other alternatives were discarded as unreliable and costly.

a. Rectangular caisson sinking :

The concrete caisson had dimensions of 32m x 28m with cutting edges. It was to be sunk from E.L +1.5 to −13.5m, excavated to a depth sufficient to place a five-(5) feet thick tremied concrete plug to prevent blow-in. The caisson would be de-watered and the permanent structure built. About 8250 cubic meters of structural concrete and 3810 cubic meters of tremie concrete were estimated. The estimated duration of caisson sinking was about fifteen (15) months and costs were high. Uncertainties included non-uniform sinking, misalignment of screen guides due to potential out of plumb installation, refusal on top of cemented conglomerate and possibility of edges pulling apart. It was concluded that risks were high and not acceptable on an EPC project. This solution was abandoned.

b. Diaphragm wall with bottom seal :

A 50m internal diameter concrete diaphragm wall built from E.L. +1.5 to −14.5m with concrete ring beams was considered. A three-meter (3m) thick sodium silicate plug deep below the excavation invert would then be grouted to prevent blow-in of the invert. The excavation was up to E.L −8.5. The projected construction period of the diaphragm wall was about seven (7) months with unknown time required for import of specialized equipment. A circular wall would also take more site space in a very tight area. Construction of slurry wall through cemented conglomerate would require chiseling requiring more time. Construction of slurry wall near to an ocean in a saline environment would require a highly experienced specialty subcontractor. This alternative was rejected due to uncertainties of time, requirement of importing specialized equipment and personnel, costs and no redundancy.

c. Sheet pile, jet grouted columns and bottom seal:

This alternative evolved by mutual discussions between Washington Group and Daelim as team members with a common goal of meeting the overall project schedule. Since twelve-meter (12m) long sheet piles were confirmed to be available on time, it was decided to use them. Earth anchors were added for stability reasons, as the twelve meters (12m) length were considered too short. The three-meter (3m) thick bottom Sodium silicate plug was retained to prevent bottom blow up of the interior. Jet grouted columns were added for sheet pile stability and to minimize seepage through and below the sheet piles. This alternative was accepted as it was:

*   economical
*   flexible

- meeting the overall project schedule
- able to adjust to subsurface uncertainties

DETAILS OF THE CHOSEN ALTERNATIVE

A typical section of the chosen alternative is presented below:

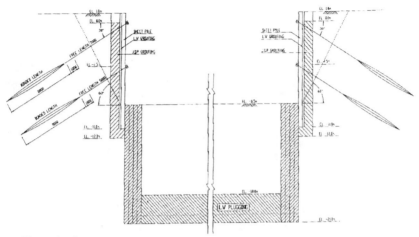

The main elements were:

- Installation of a 36m x 30m sheet piled cofferdam: The sheet pile chosen was Larssen steel sheet pile LX 32 with a section modulus of 3201 cm^3 /m. The sheet piles were driven by a five- (5) ton vibro hammer. Drilling was necessary through hard conglomerate layer number 3 in south and south west areas of excavation. Earth anchors were added for stability, as the sheet piles available were only 12m in length.

- The two rows of multi strand earth anchors were replaced by one row of earth anchors and internal rakes due to excessive water flow in the holes drilled for the second row of anchors. The 40 tons capacity earth anchors were spaced 2.4 meters apart. The anchors were 15 meters long, inclined 30° from the horizontal with a bonded length of 8 m. The anchor grout strength was 240 kg/cm^2.

- A bottom seal silicate plug 3m thick was installed between el.-18 and −21m to prevent bottom blow up before excavation to El -8.5m. Hole spacing was 0.8m and 960 holes were required. The revised grout components were the equal mixture of Liquid A (250 liters of sodium silicate and 250 kg of water) and Liquid B (200 kg of cement and 420 kg of water). The proposed Gel time was 15 seconds. The unit weight of hardened grout sample was 2.21 ton / m3. The grout pressure was 15 kg/cm2.

The uplift force was checked against the surcharge as follows:

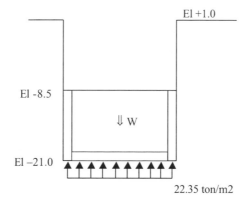

Uplift force = 21.7m x 1.03 ton/m3 = 22.35 ton/m2
Surcharge   = 12.5m x 1.03 ton/m3 (γ_{sw}) + 12.5 m x 0.9 ton / m3 (γ_{sub})
          = 24.13 ton/m2

$$SF = \frac{Surcharge}{Uplift} = 1.08$$

In order to increase the Safety Factor, about 1,500 tons of cement was added to the grout to increase the surcharge weight.

The main elements of the bottom plug were:

    a. Insert 100mm steel casing
    b. Wash inside the casing
    c. Install grout tubes in casing
    d. Place sealing material between the tube and the casing
    e. Extract casing
    f. Inject the bottom seal grout

Jet grouting was performed adjacent to the sheet piles on both outer and inner sides. The purpose of jet grouting was to provide lateral support to insufficiently embedded sheet piles and to act as a water barrier.

The exterior jet grouting consisted of one column about 1.0m diameter spaced 0.8m apart center to center from El 0 to El –12. About 113 holes were required for the exterior JSP grouting. The exterior jet grout column's function was to essentially act as a water barrier and enable placement of earth anchors. The compression strength of the grout was 100 kg/cm^2 .

The interior jet grouting consisted of three (3) columns about 1.0m diameter spaced 0.8m apart and the center row staggered to get an overlap. The jet grouted columns were installed from El -8.5 (invert of excavation) to El – 21.0. About 309 holes were required for this purpose. The interior jet grout column's function was not only to act as a water barrier but also to resist earth pressures. This load was estimated as 24 tons/meter and designed to act on three jet grouted columns net section of 1.0m x 2.5m. The compressive strength of the grout was 100 kg/cm^2. Bending and shear stresses were checked and found to be satisfactory.

During the design, there was a concern that the exterior jet grouted column adjacent to the sheet-pile (from El 0 to El − 12) may not be continuous. Since water tightness was of paramount importance, it was decided to grout the space between the jet grouted column and sheet pile by sodium silicate − cement grout used for the bottom plug.

It can be seen that the above elements assured success by enveloping all potential subsurface surprises during construction. This solution provided flexibility to make changes as necessary. It gave confidence to the Project Management that the intake structure would be built as per the project schedule.

DESIGN CHANGES DURING CONSTRUCTION

The main changes were:

- The originally envisaged two rows of multi strand earth anchors were replaced by one row of earth anchors and internal rakes due to excessive water flow in the holes drilled for the second row of anchors.
- In the course of the drilling, the drill bits were jammed in the gravel/sand layers and resulted in broken, missing tools and therefore, the drilling speed did not reach the target schedule. In order to recover the delayed progress, the idea of decreasing number of grout holes was conceived by enlarging the effective grout diameter from 0.8m to 1.5m.

  Two methods for the enlargement of effective grout area were used. One method was to delay the grout gel time by decreasing the unit quantity of sodium silicate (250 liters to 161 liters) in Liquid A to penetrate into the void of soil and increasing the unit quantity of cement (200kg to 380kg) in Liquid B. The other method was to increase the grout injection pressure to more than 60 kg/cm2 from the original 15 kg/cm2.

  The result was found satisfactory by test of injection rate, gel time and the unit weight of cored sample of grouted area. By applying this design change, the required grout holes were reduced to 440 holes from the original 960 holes. The unit weight was 2.21 tons/m^3.
- As the work progressed and the subsoil conditions were better understood, the bottom seal silicate plug elevation and the grout thickness were changed from the original elevation of -21.0m (-18.0 to -21.0m) to −17m (-14.5m to -17.0m).

- As the gel time was delayed, the method of injection was modified as follows:

  a. Drill the hole with 100mm steel casing

    b. Inject the bottom seal grout
    c. Continue injection of the seal grout while extracting the steel casing.

- The original design required seal grouting between the sheet pile and the jet grouted columns due to concern about continuity between jet grouted columns. It was found that there was no need for seal grouting as jet grouted columns effectively sealed any leakage across the sheet pile interlocks.

FIELD DATA

The efficacy of grouting was verified by excavation of a 3.0m deep test pit within the cofferdam prior to the full excavation. The free water within the soil voids was pumped out and the inflow of water observed. The inflow was found to be small which could be easily handled by a small pump. The field quantity of the grout was compared to the estimated theoretical quantity. The grouted area was mapped by using the field data such as location, hole diameter, and injected quantity. The map provided the assurance that all the required area was fully covered by grouting.

The criterion to commence excavation below the ground water level was that the inflow should be controllable by a few small capacity pumps. The measured quantity of inflow at the excavation level of –6.0m (sea water level about +0.5m) was found to be about 33 gallons per minute in an excavation area of about 36m X 30m. At the final excavation level of –8.5m, the inflow was found to be 103 gallons per minute. The ground water inflow was managed easily by a few pumps. The base of the excavation was found to be stable assuring the integrity of the bottom seal plug. No other ground water measurements were taken as construction of the intake structure was commenced immediately on reaching the full excavation depth.

Three inclinometers were installed to measure the deformations of the sheet piled wall during the excavation. The deformation varied generally between 1 cm and 5.5 cm which was tolerable for the temporary construction condition as the bottom slab for the intake was constructed immediately thereafter.

CONCLUSION

The performance of EPC projects for power plants requires that the construction of deep foundations be performed so that that schedule is never jeopardized. The construction of the cofferdam for the Ilijan cooling water intake, in an environment of highly permeable and variable gravels and rock on the edge of the ocean, required the development of a design that was capable of performing its function while being flexible enough to meet any changes to the environment as it was being constructed. Once under construction, major changes in cofferdam design or, worse yet, failure of the cofferdam to meet its objective of a dry hole would jeopardize completion of the cooling water system and ultimately plant start-up. All of these criteria needed to be met on a confined site, relatively isolated, where importation of sophisticated materials or equipment was time consuming. The cofferdam design chosen and executed was a choice of using readily available materials and adaptability to any unexpected conditions met as it was being constructed without affecting schedule. The use of steel sheet piling, while standard cofferdam practice, was limited by the availability of sections. The use of jet grouting to offset the short length of steel piling and to improve the water tightness of the steel piling was an adaptation of normal practices. The use of a bottom grout mat, substantially below the excavation level, was a further adaptation of normal practices to prevent bottom blow out. The extensive use of grouting assured that failure would not occur and that leakage into the excavation could be controlled. It was completely successful in providing, on time, a dry environment for construction of the intake structure.

A photograph showing excavation in the dry is presented below:

Humboldt Bay Nuclear Power Plant Repair Project
Monica M. Rourke[1]

Humboldt Bay Power Plant was one of the first nuclear power plants built in the United States over 30 years ago. The reactor was constructed utilizing a 60'ft. diameter caisson with internal divider walls, which was, sunk approximately 70'ft. below sea level, mucked out, plugged with a 6'ft. thick underwater tremie concrete placement and then pumped out. By design, the tremie concrete was covered by a 6" layer of drain rock, a 6" concrete slab, and in some areas a 3/16" steel liner plate which functioned as a tank for holding liquid (water) within the structure. The drain rock and a series of drainpipes through the divider walls routed any leakage to a sump where it could be pumped out.

The joint between the caisson, a carbon steel shoe referred to as the "cookie cutter" and the tremie concrete has always been subject to some leakage. Since the power plant was shutdown during the late 1970's, the caisson inleakage has been monitored and documented by the engineering staff.

The inleakage had dramatically increased following the 1989 Loma Prieta California Earthquake. At one point in June of 1997 the flow rate increased to almost 10,000 gallons per day from the 1996 average of 6,000 gallons per day, which seriously concerned the owners as the power plant, sits on an active earthquake fault. This increase in flow rate and the possibility of future earthquakes resulted in the owner's determination to seriously reduce or stop the leak.

The Owner reviewed reports from the geotechnical engineering community, which indicated several different repair options. The Owner paid for a series of mock-up test panels to be injected on-site to determine if the repair was feasible via chemical grout injection. Several different grouts were tested and rejected. To complicate matters, radioactive material was discovered in the drain rock layer of the west suppression chamber, following a review of the gravel layer probe investigation

A revised grout injection plan was approved. This plan divided the west suppression chamber into eight 5'-0"wide cells. It was agreed that a water diversion system was to be installed to divert water from each cell during the injection operations. This process would eliminate excess flow of grout material into the caisson sump and minimize possible contamination through secondary leakage.

It was also agreed that injection operations would continue by injecting grout into the gravel drain rock layer of each cell number 1 through 8, west to east. By injecting those cells in the pre-engineered pattern from west to east a contiguous waterproofing barrier in the gravel layer would be formed. Once it was established that a solid and impermeable dam had been achieved through injection of the gravel drain rock layer, preparations

---

[1] BBZ USA, Inc., Director of Technical Services

would be made to begin injection of an acrylate ester resin directly into the wall/floor joint tremie concrete interface.

The grout injection operations were successful.   The plant was approved for decommissioning. It has been five years since the injection and the plant has not experienced any secondary leakage.

Historical Background

Humboldt Bay Power Plant was one of the first nuclear power plants built in the United States having begun operations in 1963.   The Power Plant's Unit 3 Reactor sits on the shoreline of Humboldt Bay in Eureka, California. The reactor building was constructed utilizing a 60'ft. (18.3m) diameter caisson with internal divider walls, which was sunk approximately 70'ft.(21.3m) below sea level, mucked out, plugged with a 6'ft. (1.8m) thick underwater tremie concrete placement and then pumped out.

The joint between the caisson, a carbon steel shoe referred to as the "cookie cutter" and the tremie concrete has always been subject to some leakage. Since the power plant was shutdown during the late 1970's, the caisson inleakage has been monitored and documented by the engineering staff of Pacific Gas and Electric, the owners of the plant.

By design, the tremie concrete was covered by a 6" (15 cm) layer of drain rock, a 6"(15 cm) concrete slab, and in some areas a 3/16" (4.7mm) steel liner plate which functioned as a tank for holding liquid (water) within the structure. The drain rock and a series of drainpipes through the divider walls routed any leakage to a sump where it could be pumped out. The caisson inleakage had dramatically increased following the 1989 Loma Prieta California Earthquake. The increase trend continued through 1993 and by 1995 the trend began to look exponential. At one point in June of 1997 the flow rate increased to almost 10,000 gallons per day from the 1996 average of 6,000 gallons per day, which seriously concerned the owners as the power plant. The increase in flow rate and the possibility of catastrophic flooding resulted in the owner's determination to seriously reduce or stop the leak.

The problem of ground water leakage had been investigated in the past, the cause of the problem had never been clearly identified.   The consulting engineers hired by the owner scheduled a meeting with the original structural engineer and general contractor. Thirty years worth of memory, plus current test information and investigative reports yielded no new information relative to the actual cause of the problem, but focused on the suspected inleakage paths.   The group's general knowledge of the caisson construction provided meaningful evaluation as to the possible sources of the leakage.

In view of the construction history and the recent observations of leakage activity, the group concluded that the cutting edge was the most suspicious path for water travel into the structure.

Plan View of the Unit # 3 Reactor

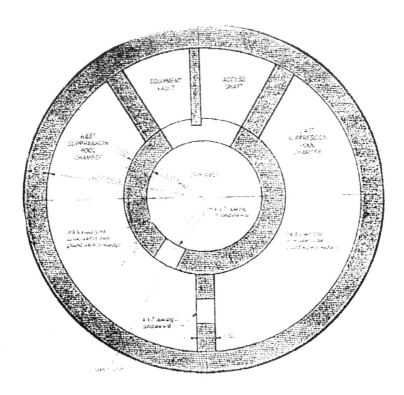

PLAN VIEW OF CAISSON
AT ELEVATION (-) 66'-0"
HUMBOLDT BAY POWER PLANT
UNIT 3

Detail of Caisson Cutting Edge-Tremie Concrete Interface

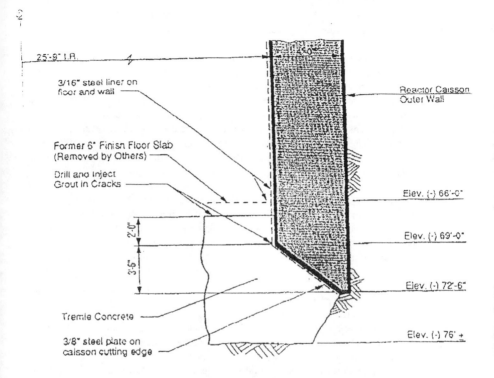

25'-9" I.R.

3/16" steel liner on
floor and wall

Reactor Caisson
Outer Wall

Former 6" Finish Floor Slab
(Removed by Others)

Drill and Inject
Grout in Cracks

Elev. (-) 66'-0"

Elev. (-) 69'-0"

2'-0"

3'-6"

Elev. (-) 72'-6"

Tremie Concrete

3/8" steel plate on
caisson cutting edge

Elev. (-) 76' ±

DETAIL OF CAISSON CUTTING EDGE-
TREMIE CONCRETE INTERFACE
HUMBOLDT BAY POWER PLANT
UNIT 3

Suspected Source of Inleakage
(1) Cookie Cutter to Tremie Concrete Joint
(2) Crack in the Tremie Concrete
(3) Crack in the Caisson and Corrosion Breach in Cookie Cutter/Liner Steel
(4) Communication through Drain Rock, Slab, and under Steel Liner Plate

Consulting Engineer's Recommendation for Repair Options
In June of 1997, the consulting engineers completed their investigation. The conclusion of this investigative report indicated that the leakage centered mainly in the West Suppression Chamber of the reactor. Several repair options were reviewed in subsequent reports presented to the owner covering possible points of water intrusion, i.e.; water leakage through cold joints or laitance between concrete pours, water leakage through voids on the sloping edges of the steel cutting edge, damaged caused by past earthquakes, leakage through the caisson wall, leakage through the jet holes inside the caisson walls and leakage through the Reactor Equipment Drain Tank

Alternative Repair Option to Grouting with Polymers
Another option that was discussed with the owner was an alternative method to place a concrete overlay over the entire base floor, including both the East and West Suppression Chambers and the Central Circular Section. While a possibly viable remedy, premium costs associated with radioactive working environments made this alternative option too costly for direct consideration.

Owners Decision to Repair
Due to the limitations in the control of leaking water, it was not feasible to expose the source of the leak and to inject grout into the primary fault. It appeared that the most favorable method of stopping the leak was to block off the flow of water through the drain rock and then, if determined necessary, to drill through the steel liner, slab drain rock, and tremie concrete or to uncover the Tremie concrete and drill into the tremie concrete to inject grout into the tremie concrete to the cookie cutter joint.

Grouting Options
(a) Fill the leaking joint;
(b) Block the drain rock;
(c) Combination of items a and b.

Mock Up Testing
To assist in product and contractor selection, the Owner planned to prepare a mock-up test panel, which would simulate the leak and the conditions of the drain rock. The simulation assumed that a "point leak is the worst condition versus a longer linear crack". The mock-up was designed to allow the demonstration of three applications with different products or different product/accelerator combinations. The mock-up was constructed of plywood with an clear plastic top and sides for viewing in the dimensions of 4'ft.x 8'ft x 6"inches deep (1.2m x 2.4m x 15 cm deep). It was filled with drain rock similar to that found in the caisson sub-floor.

Aside from the contractor successfully demonstrating the injection of a suitable polymer into the mock-up, the selection process included numerous site requirements, which focused primarily on security, work in a radiological environment, and special site rules governing material handling and training in a controlled environment. Written criteria was released and several specialty leak repair contractors made arrangements to travel to Eureka, California to perform their demonstrations. Polyurethane was immediately ruled out because of the sensitive nature of its curing characteristics. The uncontrollable foaming and gas by-products had a disastrous affect on the mock-panel.

On June 19, 1997, BBZ USA along with a joint venture team of Jean's Waterproofing, Inc. and Marathon Coatings Company injected Duroseal Inject 2000, a two-component acrylate-ester gel into the mock-up panel. While this product successfully penetrated the gravel layer, the owner and engineers were concerned that before the material had time to cure, it would be displaced by the flow of water through the gravel layer and eventually lead to plugging up the caisson sump pump.

A meeting was held following the injection of the mock up panel to discuss what possibilities there might be in controlling the cure times of the grouts. BBZ USA introduced their injection grouts belonging to the family of hydrogels and discussed their advantages and limitations relative to the criteria of the project. It was decided that if the BBZ products were successful in all test simulations and mock-up injections that a design change notice would be issued and approved to start the work.

Nondestructive Testing
On July 8, 1997, an independent Testing Laboratory performed a nondestructive examination on the west suppression chamber of the unit 3 reactor to ascertain where the leak was coming from. The exam method utilized was an infrared thermography raster scanner with 1°F resolution. The survey was performed on panel nos. 3 - 24 in the west suppression chamber. The survey showed an area from panel 3 to panel 11 appeared to be 15°F degrees cooler in temperature than surrounding areas in the chamber. This was an indication maybe not of the water under the steel floor but an indication of the steel wall thinning and air in back of it. No water spots or void areas were detected with the infrared thermography raster scanner. A total of 22 baffle panels were surveyed.

Grouting Techniques

The requirements for the grouting operation process were based precisely on the confines of the specialized jobsite nuclear power criteria, the capabilities of the chemical grouts, and the extremely sensitive safety issues surrounding the potential catastrophic complications, which could result in the repair area environment.

The selection of the chemical grouts necessitated that certain characteristics be absolute; (1) insensitivity to water in the uncured state; (2) ability to allow the injection process to take place without uncontrolled expansion or subsequent formation of negative by-products such as; gases (CO_2), bubbles, or foam layers, and (3); the ability of the chemical grout to cure into a solid mass which would be able to withstand constant high hydrostatic pressure for a long period of time.

July, 1997 - Creating a Working Plan to Implement Repair

The general description of grouting procedures was modified following test results and examination of reports on diagnostics through the concrete to determine pressure in the gravel layer, including a complete video probe of the west chamber hole F-7.

TES Inspection Services – Non-Destructive Examination Data
West Chamber Final Floor Decon & Baffle Status Chart 07/08/97

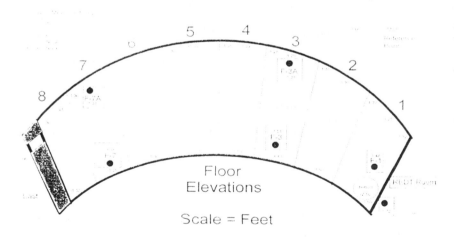

A *"Stop Work Order"* was issued by the owner and project engineering and safety teams following a review of the gravel layer probe investigation. Radioactive material was discovered in the gravel drain rock layer of in the west suppression chamber.

A Modification of the Grouting Operations 3-Step Procedure

An emergency meeting convened on August 21, 1997. With the discovery of chrome, cobalt and radioactive contamination in the drain rock layer of the west suppression chamber there were a number of issues that would have to be addressed by the general contractor, plant management staff, safety coordinators and grout manufacturer engineers. Review of the previously approved work plan outlines and procedures yielded necessary changes to the scope of work which would be modified to allow rock gravel layer sampling and/or depressurization.

Core Drilling through the Liner Plate and Concrete Floor

An immediate plan was approved to drill approximately 1-1/2"(3.8 cm) through the liner plate and concrete floor. This operation would allow a sample quantity of drain rock to be removed from the holes until a one-liter sample was gathered. This sample was to be turned over to the radiation protection division for testing.

The Revised Work Plan

The revised plan divided the west suppression chamber into eight (8) cells, 5'-0" wide or 1.5 meter wide cells. A Water Diversion System was to be installed to divert water from each cell during the injection operations. The Water Diversion System would be monitored while injection of the grout was taking place in each of the eight (8) cells. This process would eliminate excess flow of grout material into the caisson sump and minimize possible contamination through secondary leakage.

Westinghouse-HBPP Unit 3 Work Package
Fig. 1. _Suppression Chamber Grout Injection Plan_

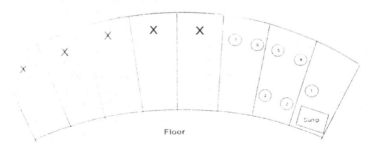

Floor

Stay 1' to 2' Away from the Wall

Modification to the Work Package was Approved

Major changes to the previously approved procedures included the elimination of the original plan of first injecting BBZ's one component acrylate ester grout, Duroseal Inject, into the void space beneath the steel liner. This void space was deemed not as critical as the voids detected under the slab in the drain rock layer. It was agreed that Duro Rapid, a non-foaming quick setting polyurethane resin would be used to fill varying void spaces beneath the slab prior to the injection of gel to consolidate and solidify a dam in the gravel drain rock layer. The leak repair team based on the diagnostics, contamination restrictions, radiation protection requirements, safety and plant licensing requirements prepared a revised work package.

Water Diversion System (WDS)

A diversion port plan was engineered locating the areas to be drilled. Water diversion ports were located approximately 6" (15.24 cm) off the outer caisson wall, between embedded T-bars and outside of the diagonal braces. The water diversion system was designed to be able to withstand a pressure of 70 psig. A pump, flow meter, and hose assembly was attached to a finished floor diagnostic port, or a previously installed diversion tap, so that this assembly would be able pump or drain water to reduce pressure above the drain rock layer.

Westinghouse-HBPP Unit 3 Work Package
Fig. 2. _Water Diversion System Installation Plan_

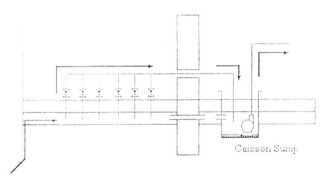

Caisson Sump

Water Diversion System (WDS) Installation

The WDS was installed by inserting the diversion port assembly, with valves closed into gravel layer as far as possible and secured by tightening the port assembly. These diversion ports were implemented to act as a method to relieve pressure from under the floor slab, and as a path of least resistance for water and grout, if any, to migrate during the injection process. Diversion flow could be stopped by allowing the gravel area to pressurize so the integrity of the packer seal could be checked sporadically during the injection process. The system was completed by connecting the tubing to the diversion ports, tying them together into a common header, and routing them into the caisson sump/reactor equipment drain tank.

The outlet of the diversion header was positioned below the elevation of the caisson sump drainpipe holes (-67'ft –20m) elevation to maximize the siphoning effect and gravity draining. The diversion valves were designed so that they could be opened individually or in varying combinations to observe flow and attempt to determine the leak location(s). Maintaining the floor depressurization was completed by siphoning or pumping into caisson sump, the reactor equipment drain tank, or a suitable container.

Westinghouse-HBPP Unit 3 Work Package

Fig. 3. *Logistics of the Water Diversion System to the Caisson Pump*

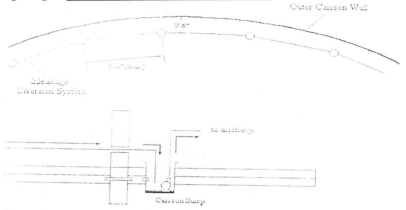

Floor Liner Plate Hold-Down Support Installation

The engineering team identified the wall support locations. The identified areas on the floors and wall were cleaned and marked for identification. A radiation protection release was obtained prior to any welding and/or any commencement of work operations for installation of the floor supports.

Description of Drilling Operations

A radiation protection release was issued identifying all floor area(s) to be drilled. Prior to drilling, a verification test of the water diversion system was completed and a floor depressurization process was conducted.

3/4" inch-(20mm) diameter holes were drilled through the liner plate and through the finished concrete slab. The depth of the concrete and gravel layers were measured and recorded, then compression-fitting packers were inserted into slab and tightened. Each packer was doubled checked and secured. The floor area or cell to be injected was strengthened by the completion of the floor hold-down supports. The installation of the floor supports and all welding and welded sections were inspected and verified to be secure prior to the start of any drilling operations.

The HBPP Caisson Inleakage Repair Approved "Typical Grout Plan":

Step 1 - Create a water barrier via grout injection into the gravel drain rock layer. Mobilize for work in the west suppression chamber. Drill a minimum of two (2) holes each, ¼" inch (6mm) in diameter in each of the eight (8) floor liner plates, between the floor beams and the wall. Insert the injection packers per the grout injection packer layout sketch provided by the grout manufacturer.

BBZ USA, Inc. HBNPP-Caisson Repair – Method Statement – Sketch 2-A

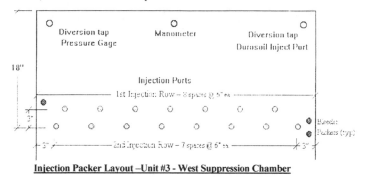

Injection Packer Layout –Unit #3 - West Suppression Chamber

Step 2 -

Open the diversion port valves and drain or pump water, as necessary, to divert and depressurize. Verify proper diversion by observing diversion flow and drain flow into the caisson pump.

Caution: If grout begins to migrate up through the diversion ports, stop grout injection, close off diversion port valve, and clean out tubing as necessary to maintain flow. Install another diversion port, as required.

Note: Mock-ups have indicated that a solid barrier can be formed in most of the gravel layer, with leakage restricted to one or more small leak paths at the interface between the top cover and the gravel.

Duro Rapid, a faster setting and higher viscosity grout, will be injected first to fill the void spaces between the finished concrete and the gravel. Drainseal will then be injected to penetrate and solidify the gravel.

Grout Set Times: **Duro Rapid - 10 seconds**
                          Drainseal - 20 seconds

Inject Duro Rapid into the gravel drain rock layer at 15 second intervals, through the first row of packers, working from one side to the other until grout is seen coming out of the bleeder packer(s), to fill the voids at the top of the drain rock. Record outflow of the cell.

Note: As pressure begins to increase upstream of the barrier, check diversion system   for any restrictions. Relieve pressure by pumping or allowing the water to free flow.
Inject Drainseal into gravel drain rock layer at ~20 second intervals, through the second row of packers, working from one side to the other until grout is seen coming out of the bleeder packer(s), to create a barrier within the drain rock. Record outflow of the cell.

Step 3 –
Reduce caisson flow to the minimum level by pumping from F1 and/or F3 to the Reactor Equipment Drain Tank and pumping from diversion ports as necessary (minimum number) to the temporary storage tank. Pump Duro Soil behind Beam A. Inject material at 20 GPM (gallons per minute)

Grout Set Time: Duro Soil - 2 Minutes

Per Grout Plan, vent from Beam A vent port. When solid, undiluted grout (Duro Soil) is seen then shut vent valve off. Maintain pressure behind dam at 25 psi for 30 seconds with injection pump. Shut Beam A injection valve, stop injection, and flush pump.

Static Pressure Test
Close off valves to the two (2) diversion ports and measure static head.

Injection of Grout - September, 1997
The leak repair applicator mobilized inside the Unit #3 reactor caisson at -66'ft.(-20 m) to begin the injection of cell 1 and cell 2 in the west suppression chamber per the outline in the Grout Plan. Injection operations commenced utilizing a dual injection operation comprised of Duro Rapid and Drainseal and began at the west end of the suppression chamber (where the leak was suspected). The building of dam in the drain rock layer was successful. Once the drain rock layer was consolidated. Water appeared more predominantly at the east end of the suppression chamber near cell 8.

The Owner wanted to change directions and begin injecting where it was evident that the leakage had increased. The leak repair contractor and grout manufacturer site engineer explained that, "water will always seek out the least path of resistance" and informed the owner that chasing the leak was not the best way to solve the problem.

A discussion followed and it was agreed that injection operations would continue by injecting grout into the gravel drain rock layer of each cell 1 through 8, west to east. By injecting those cells in the pre-engineered pattern from west to east a contiguous waterproofing barrier in the gravel layer would be formed. Once it was established that a solid and impermeable dam had been achieved through injection of the gravel drain rock layer, preparations were made to begin the injection of Duro Soil directly into the wall/floor joint at the cookie cutter and tremie concrete interface.

A large roter-stater pump had been modified to meet the requirements of size limitation and was lowered into the reactor. This specialized pump acts as a two-component static mixer and can deliver large quantities of chemical grout at one time. The pump was assembled and connected to the first injection port. The grout was poured into the vats and the water diversion system was stopped. With a restriction to 25-psi (1.7 bar) injection pressure, the contractor injected 20 Gallons Per Minute of Duro Soil into the injection packer and continued the procedure until the each cell was completed. The principle of utilizing the chemical grouting process primarily consists of a systematic method of drilling and injection for the purpose of filling large cracks and voids in concrete structures requiring high compressive finish strengths.

Total Chemical Grout Usage:

95 Gallons of Duro Rapid was injected into the drain rock layer to fill the voids between the gravel layer and concrete floor.

144 Gallons of Drainseal was injected into the drain rock layer to consolidate the gravel into a solid waterproofing barrier.

132 Gallons of Duro Soil was delivered directly into the tremie wall joint area to seal the voids and to stop the leak.

Summary

The caisson inleakage was drastically reduced immediately following the procedure - far surpassing original estimates. The structure was originally designed to handle an inleakage of approximately a 1,000gpd. Over one year later, the plant engineers who monitored the caisson inleakage sump reported that the inleakage was "barely measurable", estimated at times to be approx. 9 gallons per day. The successful results of the repair have allowed the power plant to be approved for decommissioning. As of the date of the date of this paper, five years after the repair, the plant was successfully decommissioned and secondary leakage did not appear in any of the suspected areas. The owner remains totally satisfied with the results.

References

BBZ USA, Inc. – Sketch #1 – Grout Injection Packer Installation Plan-Humboldt Bay Nuclear Power Plant – Method Statement – July, 1998 – M. Rourke, Project Engineer BBZ Bauchemie, BBZ AG - Switzerland Technical Information Data Sheets – Duro Rapid, Duroseal Inject, Duro Soil, and Duroseal Inject 2000.

PG&E Drawing 554-27-8, Section E entitled: Plan View of Caisson at Elevation -66'-0" Humboldt Bay Power Plant Unit 3

PG&E Drawing 55429-7, Detail 1 entitled: Detail of Caisson Cutting Edge-Tremie Concrete Interface Humboldt Bay Power Plant Unit 3

TES Inspection Services Non-Destructive Examination Data, Humboldt Bay Unit-Job- 2173-00 dated 07/08/97- West Chamber Final Floor Decon & Baffle Status Chart

Westinghouse Team - Work Package Humboldt Bay Power Plant Caisson Inleakage Repair Project - Work Package Number: WP-001G entitled: Suppression Chamber Grout Injection dated 08/08/97

Grout Curtain Effectiveness in Fractured Rock
by the Discrete Feature Network Approach

D.A. Shuttle[1] and E. Glynn[2]

Abstract

This paper presents a discrete fracture approach for analysis of foundation grouting in fractured rock. In this approach the grout injection boreholes and the fractures intersecting them are modeled explicitly. Grout is injected into the fractures directly and indirectly connected to the grout injection boreholes. This approach is more realistic than conventional analyses which rely on a reduction in "effective" conductivity, without considering the geometry of fracture pathways through the rock mass. The approach is applied to determine the relative effectiveness of 2 or 3 rows of grout holes for the same grouted length.

Introduction

Flow through fractured rock foundations typically occurs through networks of conductive fractures and other discrete features such as shears. Foundation grouting injects cement grout into specific discrete features connected to the grout injection boreholes. Discrete fractures not directly or indirectly connected to these boreholes are not grouted, and provide potential flow pathways through the foundation. Despite this, conventional grout analysis assumes that grouting achieves a desired average reduction in effective conductivity, without any input regarding the geometry or properties of the ungrouted fractures.

The discrete fracture network (DFN) approach (Swaby and Rawnsley, 1996; Ivanova, 1979; Long, 1984) directly models the geometry and hydraulic properties of discrete fractures in foundation rocks. The DFN method is also able to simulate the flow of water through the fracture network and the flow of grout into discrete fractures. As a result, the DFN approach is well suited for quantitative analysis of foundation grouting.

The efficiency of a grout curtain within a DFN is related to the properties of the fracture network. The spacing, size, orientation and transmissivity of fractures

---

[1] University of British Columbia; formerly Associate, Golder Associates Inc., Seattle
[2] Research Civil Engineer, US Army Waterways Experiment Station, Vicksburg MS

relative to the position of the boreholes is critical in determining what proportion of hydraulically significant fractures are grouted. Hence, by modeling the fractures and grout injection explicitly the DFN approach may be used in the pre-grouting stage to optimize the grout hole configuration. By refining the model during construction based on measured grout takes, more accurate estimates of the grout curtain efficiency may be obtained.

In this paper two DFN approaches for grouting design are described. One of these approaches is then used to determine the relative efficiency of using two closely spaced, or three more widely spaced, rows of grout boreholes. To model this efficiency a DFN network is generated, boreholes are inserted into the DFN network, and the fracture network connected to these grout hole intervals grouted numerically. The flow across the grout curtain before and after grouting, for both borehole configurations, is used to measure the effectiveness of the grouting.

Simplified Approach to the Modeling of Grout Injection

Grouting requires considerable skill and experience. The site engineer must determine the correct grout type, borehole layout, and verification approach. Although industry practice has been established from years of experience (e.g. Weaver, 1991; Houlsby, 1990), grouting techniques vary from site to site. This section describes a DFN approach for preliminary evaluation of grouting programs, before field grouting has begun, based on site characterization data.

Having completed a site investigation, data analysis and preliminary design, foundation grouting is typically performed as follows. First, a number of the boreholes are drilled. At each borehole the incremental bottom section of the hole is hydraulically tested. These hydraulic tests are used to estimate the hydraulic conductivity and volume of grout required for each borehole section. It is often assumed that the fracture aperture is correlated to the fracture transmissivity (e.g. theoretically based on the Cubic law, or experimentally Hakami, 1995). Therefore more transmissive borehole sections would be expected to take larger volumes of grout, and possibly require thicker grouts. Grout injection then proceeds in each borehole: starting with the primary holes, with secondary and subsequent tertiary holes progressively reducing the hole spacing. If grout take is much lower than expected in a given hole, the grout is thinned until a grout take is achieved. Grouting then proceeds until refusal. Once grouting has been completed in a given area, neighboring confirmation holes are hydraulically tested. It is generally assumed that if these confirmation holes show significantly reduced permeability, then grouting has been successful. If not, additional holes are installed and grouted in the vicinity.

Discrete fracture network models can be used to model this process. First, the DFN model is derived and implemented based on site characterization data (e.g. using the approach described in Dershowitz, 1995). DFN models are defined based on a combination of deterministic fractures (i.e., fractures which are known completely), conditioned fractures (i.e., fractures for which some properties such as location are known, but other properties must be defined based on population statistics), and purely stochastic fractures.

Boreholes are incorporated into the DFN model using the same geometric pattern and orientation as used in the field. Simulated hydraulic testing is then carried out in each simulated borehole to derive an effective transmissivity for each grouting interval. The procedure used to simulate the hydraulic testing is described in Dershowitz et al. (1999). As a simple approximation, the transmissivity of a borehole interval can be estimated as the sum of the transmissivities of the fractures intersecting the interval. More complex estimates take into consideration the connectivity of these fractures, and the properties of the connected networks (LaPointe et al., 1997). This transmissivity can be used to determine the initial grout mix anticipated for the interval. Borehole sections with a hydraulic conductivity of less than the target value (generally of the order of 10^{-7} m/s) are not typically grouted.

The total volume of grout for each simulated borehole interval is estimated by carrying out a network analysis of the fractures connected to each borehole (Figure 1). In this analysis, a graph theory search is used to search back into the fracture network from each borehole, until either a certain distance from the borehole is achieved or the available fracture aperture is below a specified threshold. Using this approach the total volume of grout can be estimated.

Detailed Approach to Modeling of Grout Injection

The previous section assumed a simple grouting model in which the volume of grout assigned to a borehole interval is based on interval transmissivity and connected network volume. Figure 2 illustrates a more advanced approach that has been implemented for practical applications (e.g. Shuttle et al., 2000) and which was used for the efficiency analyses described in the following sections.

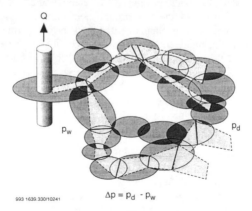

993 1639.330/10241 $\Delta p = p_d - p_w$

Figure 1 Network Analysis

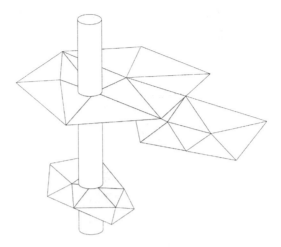

Figure 2 Grouting Simulation

In this approach, the maximum grout available for a borehole is estimated as proportional to the transmissivity of the fractures intersecting the section. The actual grout injection process is simulated in greater detail to obtain a more accurate picture of the grout pattern in situ.

First, the fractures are discretized into triangular finite elements. Grout is then injected to every finite element directly intersecting the borehole, most transmissive element first. The volume of grout taken by each element is based on the element volume, which is calculated as the product of the element aperture and element area. Following grout injection into the fracture elements immediately adjacent to the borehole, the remaining grout is then injected into the most transmissive fracture attached to the borehole, one element at a time. Grout is injected into successive elements until (a) all the available grout is used up, (b) the fracture is completely grouted, (c) a more transmissive fracture is intersected which is preferentially grouted, or (d) a specified maximum depth of grout penetration is reached. At the end of each grout injection stage, the transmissivity of each grouted element is set to a value representing concrete, typically many orders of magnitude smaller than median transmissivity of the ungrouted fracture network.

Figure 3 illustrates the resulting three-dimensional grout pattern from a single borehole. This grouting process is carried out in sequence, using the same sequence of grout sections as used in the field. As a result, if grout from a previous borehole section has filled elements connected to the current borehole section, those elements already have a low transmissivity and will not be further grouted. Thus, the grouting for many boreholes will be less than the maximum grout volume calculated initially.

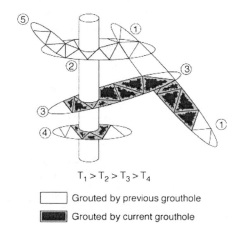

$$T_1 > T_2 > T_3 > T_4$$

[ ] Grouted by previous grouthole

[■] Grouted by current grouthole

Figure 3 Grout pattern for a single borehole section

By modeling the sequential filling of fractures with grout, this approach provides a model for grout consumption, as well as a model for the fractures that are not reached by grout, even for very close borehole spacings. As a result, the effective post-grouting permeability calculated using the grouted DFN model has the potential to be more realistic than the estimate obtained by assuming that the few confirmation boreholes provide a representative picture of the grouted rock mass at larger scales.

The detailed grouting model described above is inherently stochastic, since the majority of fractures in the DFN model are generated from population statistics, and are not measured directly. As a result, the approach does not predict the actual performance of each individual grout hole, but rather the expected behavior of boreholes overall. Thus, the approach can predict the percentage of borehole sections that will take little or no grout, the percentage of holes that will take large amounts of grout, and the mean and standard deviation of grout take per hole.

The stochastic approach also allows a calculation of grout curtain reliability in terms of the probability that significant conductive pathways through the grout curtain will remain following grouting. This is particularly valuable when grout curtains are designed for contaminant containment or prevention of piping. This is done by generating multiple realizations of the stochastic DFN model and calculating the frequency of occurrence of significant ungrouted pathways.

Grout Performance Evaluation

In engineering practice whether a grouting program is effective is defined by the objectives of the project. Typically this will require that the seepage through the grout curtain be reduced below a threshold value, or that the specified head differential be maintained.

Once field grouting has been completed, the engineer knows the quantity of grout injected into each borehole section, and the results of hydraulic conductivity tests in confirmation holes. However, the effectiveness of the grout curtain is still unknown. Figure 4 illustrates some of the potential issues:

a) high transmissivity grout sections often require a thick (larger particle size) grout. Thicker grout can preferentially enter the largest fractures in a borehole, but leave the less transmissive and smaller aperture, but still important, fractures ungrouted.

b) grout can completely miss fractures that form a portion of conductive fracture pathways not connected to the grouted boreholes.

c) grout can block fractures in confirmation boreholes without adequately sealing the neighboring fractures, giving a false indication of adequate grouting. Transmissivity in confirmation holes can be unchanged (or even increase due to changes in stress), despite successful grouting in the boreholes.

d) entire fracture sets can be missed by both grouting and injection boreholes due to relative orientations or periodic spacings.

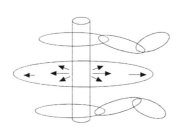

a) Coarse grout uptake dominated by a single large fracture leaving smaller fractures ungrouted.

b) Grout holes miss subvertical fractures not connected to grouthole network.

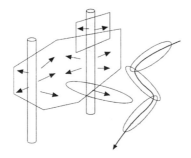

c) Confirmation holes indicate grouting success but flow continues through ungrouted fracture networks.

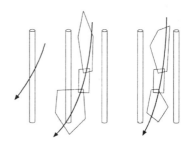

d) Periodicity of fracture locations and fracture sets leads to ungrouted fractures.

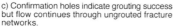

Figure 4 Issues for grouting fractured rock

Conventional grouting analysis relying on an "effective" reduction in conductance based on a limited number of confirmation borehole hydraulic tests cannot fully address the above concerns. However, they can be addressed by the discrete fracture network approach.

a) DFN grouting simulation can relate the fracture transmissivity cutoff for grouting to the type of grout used and the amount of grout take. As a result, the simulated post-grout fracture network can contain a population of ungrouted fractures that correspond to those in situ.

b) Simulated DFN grouting only fills fractures connected to boreholes. As a result, fractures not connected to boreholes do not receive grout. The post grouting flow calculated in the DFN model is based on flow through the network of remaining ungrouted fractures, rather than an effective post-grout hydraulic conductivity.

c) Confirmation boreholes can be evaluated by simulated hydraulic testing in the post-grouting DFN to evaluate the extent to which confirmation boreholes are affected by grout injection, and to what extent their behavior is controlled by local heterogeneity in the rock mass.

Simulated boreholes have the same location, orientation, and length as actual boreholes. As a result, on a statistical basis they intersect and grout the same fractures and fracture sets accessed by grouting in situ. Fractures and fracture sets that are not intersected due to orientation biases in situ are also likely to be missed by the simulated boreholes. As a result, the simulated post-grouting network stochastically reflects the conductance and connectivity of these ungrouted fractures.

Model Implementation

The purpose of the following simulations is to demonstrate a design approach to optimize grouting efficiency by explicitly accounting for the properties of the fractures. The geometry and boundary conditions of the model are shown in Figure 5, together with the location of the boreholes. The geometry of the boundaries and boreholes is intentionally very simplified to focus on fracture properties. The head gradient across the model is 100m upstream to downstream, with all other sides of the model being no-flow boundaries. In practice the true geometry and boundary conditions would be modeled, and the orientations of the boreholes optimized (e.g. Shuttle et. al., 2000).

Two grout hole configurations were considered, each having the same total grouted length of almost 4000m. The configurations are given in Table 1. Each of the boreholes is 40m in length, and was numerically grouted in four 10m sections, grouting from the top down.

Table 1 Spacing of Grouting boreholes

Number of Rows (-)	Spacing along rows (m)	Spacing between rows (m)
2	2	2.5
3	3	2.5

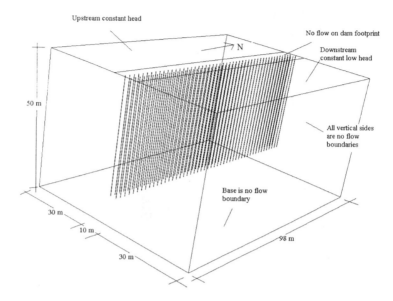

Figure 5 Boundaries of the Numerical Model

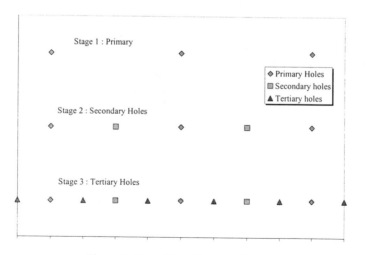

Figure 6 Three Stage Grouting Procedure

The numerical model accounts for the order in which the boreholes are grouted. The primary holes are grouted first, followed by the secondary holes, with tertiary holes completing the grouting. This is illustrated in Figure 6.

The fracture model used is summarized in Table 2. The model is based on two subvertical orthogonal sets of background fractures orientated close to E-W (set 1) and N-S (set 2). The third set of fractures represents larger "site scale" features. At a real site these would likely be mapped and explicitly represented in any hydraulic model of the site. In this analysis these features are also stochastic. Intensity is set as the fracture area per unit volume (P_{32}) as this measure is direction independent.

Table 2 Fracture Model Definition

Property	Property Definition	Set 1	Set 2	Set 3
Model	Spatial	Baecher	Baecher	Baecher
Orientation	Pole Trend (degrees)	15	110	0
	Pole Plunge (degrees)	10	10	10
	Fisher k	8	8	5
Size	Distribution	Exponential	Exponential	Exponential
	Mean Radius (m)	4	6	30
	Standard Deviation (m)	n/a	n/a	n/a
	Minimum (m)	4	4	25
	Maximum (m)	15	20	50
Intensity	P32 (1/m)	0.07	0.1	0.05
Transmissivity	Distribution	LogNormal	LogNormal	LogNormal
	Mean (m²/s)	7.00E-07	3.00E-06	8.00E-06
	Standard Deviation (m²/s)	5.00E-07	2.50E-06	5.00E-06
Aperture	Distribution	Power	Power	Power
	a	10	10	10
	b	0.65	0.65	0.65

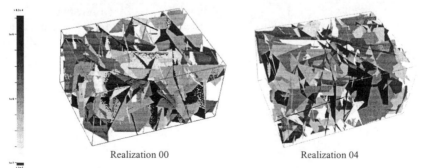

Realization 00   Realization 04

Figure 7 Fracture Network (colored by log transmissivity)

Five stochastic realizations of the fractures were generated (on a real project more realizations would be used). Each fracture realization was numerically grouted, with the maximum volume of grout pumped at each section being based on the transmissivity of each 10m borehole section. Two different volumes of grout, 40 m^3 (equivalent to an average volume of 0.4 m^3/hole) and 60 m^3 (equivalent to an average volume of 0.6 m^3/hole) were injected into the 98m wide section. The fracture networks for realizations 00 and 04 prior to grouting are shown in Figure 7.

Results

The effectiveness of grouting in these simulations is measured by the reduction in flow beneath the dam. For the 2-row and 3-row configurations this flow reduction is shown in Table 3 and Table 4 respectively. These results show that while for both grout curtain configurations injecting 50% more grout increases the grout curtain efficiency, the 2-row configuration has both a higher reduction factor and a greater increase in the reduction factor with additional grout.

Table 3 Results from 2 Row Simulations

Real	Flux (m3/s)	Grouted 40 m^3		Grouted 60 m^3	
	No grout	Flux (m3/s)	Reduction Factor	Flux (m3/s)	Reduction Factor
00	2.49E-03	6.24E-04	3.99	4.58E-04	5.43
01	4.03E-03	1.05E-03	3.85	3.33E-04	12.10
02	5.31E-03	1.71E-03	3.11	9.84E-04	5.40
03	1.89E-03	5.60E-04	3.38	4.69E-04	4.04
04	4.04E-03	1.35E-03	3.00	1.06E-03	3.80
Average	3.55E-03	1.06E-03	**3.46**	6.62E-04	**6.15**
St.Dev	1.36E-03	4.85E-04	0.44	3.36E-04	3.41

Table 4 Results from 3 Row Simulations

Real	Flux (m3/s)	Grouted 40 m^3		Grouted 60 m^3	
	No grout	Flux (m3/s)	Reduction Factor	Flux (m3/s)	Reduction Factor
00	2.49E-03	6.15E-04	4.05	4.90E-04	5.08
01	4.02E-03	1.62E-03	2.49	7.69E-04	5.24
02	5.31E-03	3.29E-03	1.61	2.65E-03	2.00
03	1.89E-03	5.62E-04	3.36	3.90E-04	4.84
04	4.04E-03	1.38E-03	2.93	9.55E-04	4.23
Average	3.55E-03	1.49E-03	**2.89**	1.05E-03	**4.28**
St.Dev	1.36E-03	1.11E-03	0.92	9.21E-04	1.33

Equally important, the 2-row configuration provides a more reliable grout curtain. At the lower volume of grout injection the reduction factor for the 2-row configuration has the smallest spread for the lower volume injection. Correspondingly, at 40 m^3 the

3-row configuration has the smallest reduction factor (1.61 for realization 02) and the largest reduction factor (4.05 for realization 00).

The reason for the poor performance of the 3-row configuration is complex. First, due to the boundary conditions the reduction factor will be very sensitive to the grouting effectiveness in the upper 10m of the foundation. In practice greater efficiency for both the 2 and 3 row configurations could be achieved by additional grouting near to surface.

For this fracture network and intensity, the larger fracture spacing results in a more dispersed grout curtain. In this instance the sub-vertical fracturing is more efficiently grouted by more closely spaced boreholes. This is most critical for the site scale high transmissivity fractures which, as shown in Figure 8, are less well grouted by the more widely spaced boreholes.

2 row Realization 01   3 row Realization 01
Figure 8 Plan view of network with grout deleted (40 m^3 of injection)

Conclusions

The presented methodology explicitly represents the discrete fractures within the foundation, and characterizes how the connectivity of the fracture network affects grouting effectiveness. When combined with a sound field grouting program, this approach has the potential to provide better pre-construction grout designs. By updating the model with measured grout takes during production, this approach also has the potential to better predict the true curtain efficiency than the conventional approach of verification holes which are only able to sample transmissivities at a limited number of locations.

One limitation of this methodology is that it requires site-specific fracture information, or estimates of properties based on similar geology. Typically much of the required raw fracture information is already gathered as part of the site investigation, but not analyzed with construction of a discrete fracture network in mind.

The presented DFN model assumed that the properties of each fracture were constant over the fracture area, e.g. only one fracture aperture and transmissivity was applied per fracture. For real projects channeling and variation in fracture roughness across each fracture may be implemented. However, site-specific data on aperture distributions within a single fracture is rare, and channeling properties will likely need to be estimated for a pre-grouting analysis. Better estimates of the channeling effects may be obtained by calibrating the DFN model against the grout takes recorded during the grouting program.

In this paper the properties of the injected grout are considered constant and unchanging with time. The maximum grout radius is used to represent time dependent changes in viscosity, and the increase in injection pressures as the grout interface moves further from the borehole. In the future more realistic grout behaviors should be incorporated to improve the prediction capabilities of the model.

References

Dershowitz, W., 1995. Interpretation and Synthesis of Discrete Fracture Orientation, Size, Shape, Spatial Structure and Hydrologic Data by Forward Modeling in Fractured and Jointed Rock Masses. Proceedings of the Conference on Fractured and Jointed Rock Masses, Lake Tahoe, California, USA, June 3-5, 1992, eds. L.R. Myer and C.F. Tsang, pp 579-86. A.A. Balkema, Rotterdam.

Dershowitz, W., G. Lee, and J. Geier, 1999. FracMan: Discrete Feature Data Analysis, Simulation, and Analysis. Golder Associates Inc., Seattle.

Hakami, E., 1995. Aperture Distribution of Rock Fractures, Doctoral Thesis, Division of Engineering Geology, Department of civil and Environmental Engineering, Royal Institute of Technology, Stockholm, Sweden.

Houlsby, A.A.C., 1990. Construction and Design of Cement Grouting: A Guide to Grouting in Rock Foundations. John Wiley & Sons.

Ivanova, V., 1998. Geologic and Stochastic Modeling of Fracture Systems in Rocks. Ph.D. Dissertation, Massachusetts Institute of Technology, Cambridge, MA

La Pointe, P., T. Eiben, W. Dershowitz, and E. Wadleigh, 1997 Compartmentalization Analysis Using Discrete Fracture Network Methods. Proceedings, Third International Reservoir Characterization Conference, Houston. NIPER, Bartlesville.

Long, J., 1984. Investigation of Equivalent Porous Medium Permeability in Networks of Discontinuous Fractures. Ph.D. Dissertation, Univeristy of California, Berkeley.

Shuttle D.A., W. Dershowitz, E. Glynn, S. Burch, and T. Novak, 2000. Discrete Fracture Network Analysis of Foundation Grouting. In Girard, Liebman, Breeds and Doe (eds) Proceedings of the Fourth North American Rock Mechanics Symposium, Pacific Rocks 2000. 31 July-3 August. Balkema, Rotterdam, pp 1369-1376.

Weaver, K., 1991. Dam Foundation Grouting. American Society of Civil Engineers.

Swaby, P. A. and K. D. Rawnsley, 1996. An interactive 3D fracture modeling environment. Society of Petroleum Engineers, Petroleum Computer Conference, Dallas, Texas, 2-5 June 1996. SPE 36004.

Performance Monitoring of Grout Curtains in Slovakian Flysh and Volcanic Rocks

Jozef Hulla[1], Dusan Chlapik[2], Robert Hok[3]

Abstract

All Slovak dams that are included in the ICOLD register were constructed in geological conditions of the Carpathian flysh except of one that was built in a volcanic rock. Their subsoils were treated by grouting courtains. This paper describes water pressure tests as unsuitable tests for determining of grout curtain effectiveness, construction conditions of dams, testing problems and results of monitoring. Determining the effectiveness of the grout courtain based on the groundwater level changes, uplift conditions, filtration velocities and developments of seepage during the operations of the reservoirs.

Introduction

The Slovak companies dealing with dams construction faced some difficulties at the beginning of their activities that were later overcome with grout curtains into the flysh rocks with variable ratio of sandstone and claystone layers.

The most challenging problems were involved with the construction of the grout curtain in volcanic rocks. The engineering-geological and hydrogeological survey had emphasised this aspect. There had been different opinions concerning depth of the grout curtain, in the design stages, thus calling for additional investigations and expert opinions.

---

[1]Prof. Eng. DrSc., Slovak University of Technology, Department for Geotechnics, Radlinskeho 11, 81368 Bratislava, Slovak Republic; hulla@svf.stuba.sk; phone 00421259274666, fax 00421252925642.
[2]Eng. PhD., Povodie Vahu, Nabrezie Ivana Krasku 834/3, 92180 Piestany, Slovak Republic; phone 00421337724620, fax 00421337625746.
[3]Eng., Povodie Vahu, Ul. Partizanov 36, 03400 Ruzomberok, Slovak Republic, phone 0042144432874, fax 0042144323319.

Many technological problems had to be resolved also during the construction. They arose after the grouting holes, mixtures, pressures and test procedures have been done. During the filling of the reservoir, all data characterising effectiveness of the treat elements and grout curtain have been carefully monitored and analysed.

Special measurements were applied using single-borehole tracer methods. Vertical water flow was measured in the boreholes. Vertical flow occurs due to the interconnection of various pressure horizons. Special boreholes were executed in grout curtain and in ungrouted medium under the curtain.

The paper describes grout curtains used under two Slovakian dams. The first is the dam Liptovska Mara in Carpathian flysh bedrock, the second is the dam Turcek in neovulcanite andesite bedrock (Figure 1).

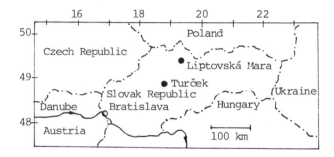

Figure 1. Location of the Slovak dams Liptovska Mara (flysh) and Turcek (andesite).

Water pressure tests

Grouting criteria based on water pressure tests are depicted in Figure 2 as a function of water discharge (Q) and corresponding pressures (p).

The older Jähde's and Lugeon's criteria were actually impossible to be met in our conditions. This is why Verfel's criteria have been used since 1983. They were initially set for pressures of 0.3 MPa. They are considered to be very progressive even now, since they allow higher losses in deeper layers (h) in which the more permeable medium cannot endanger the dam stability. Figure 2 shows criteria functions according to Kutzner (1985) used in America and Russia. In accordance with their tendencies Verfel's criteria can be simply transformed to different pressures at which water pressure tests are held.

The results of the water pressure tests have been thoroughly analysed by a wide group of our experts involved in our dam construction. In the our opinion water pressure tests (WPT) are not very reliable for curtain grouting control. At the WPT water flows out from a limited part of the boreholes undoubtedly through the most permeable position but the layer does not have to be continuous throughout the whole curtain width. Water can leak for example only to the air side but the water part of the curtain can well be convenient. The unreliability of the WPT gets even more distinct when grout curtain effectiveness is monitored during operation of the water reservoir.

Water pressure tests help to gain valuable information on permeability of natural rock medium before grouting. In case the water losses (Q) would be smaller than the criteria values (we recommend the Verfel's criteria – Figure 2) by respective pressures (p) and in respective locations, such locations need not be grouted then.

In the our opinion it is enough to provide the process' quality control on bases of grouting pressures and consumption of mixtures. Grout curtain efficiency can be determined the most reliable way in the process of reservoir filling and operation, by monitoring of the seepage regime below the dam downstream face and by comparing the results with the project assumptions. Classical calculations based on the Darcy's law and the continuity formula had been used in Slovakia in the past (for instance for the Liptovská Mara Dam). We have changed to the method of finite elements that was applied also for the Turček Dam.

Monitoring of seepage regime

The seepage regime below the dam downstream face is usually monitored by development of water levels in the observation boreholes, by development of uplifts in the surroundings of the grout curtain and by development of water outflow discharges from individual components of the drainage system. Special situations, that have occurred in our dams, required monitoring of water flow velocity in observation boreholes also by tracer single-borehole methods that have been analysed excellently by Halevy et.al.(1967).

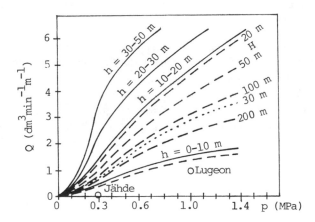

Figure 2. Grouting criteria review elaborated according to Verfel (1983) and Kutzner (1985): Q - water discharge, p - water pressure, h - depth under the injection gallery bottom, H - height of the dam, —— Verfel's criteria, – – – USA, ······ Russia.

Filtration velocities of groundwater flow, as well as, permeability coefficients of porous or fissure media can be determined on the base of electrolyte solution vertical motion or dilution process monitoring. A perforated tube with an inner diameter of 60 to 150 mm with a filter is placed in the required depth of a borehole for the single-borehole method. Results of the measurements are representative for a small area in the vicinity of the borehole given by several multiplications of its diameter. The borehole interconnects various pressure horizons almost always, and vertical water flow takes place. Less intensive vertical water flow emerges due to heterogeneous temperature distribution or because the borehole does not follow an equipotential line.

In order to measure vertical water flow in a borehole a set of equipment can be used. An immersion probe is put into the borehole; is connected to a battery powered measurement equipment placed together with tracer jet control (NaCl solution) on the surface.

Concentration dependency can be watched directly on the computer screen and the time in which the maximum concentration takes place (t_{max}) can be determined. Estimation of the vertical velocity average value requires a laboratory calibration to set the computational time and vertical discharge shall be estimated from the continuity equation:

$$q_v = v_v A = \frac{l_v \pi \left(d^2 - d_s^2\right)}{4 \times 0.266 \, t_{max}^{1.474}} \tag{1}$$

where v_v is the vertical velocity, A - cross section test tube area, l_v - vertical distance, t_{max} - peak time, d - inner diameter of the tube, d_s - outer diameter of the probe.

The measurements are repeated in appropriate depth intervals so that the whole watered part of the borehole was equally covered and the vertical water flow as a depth function could be graphically depicted.

Filtration velocity in the surrounding medium (approximately in the horizontal direction) can be calculated from the vertical water flow measurement in a borehole based on the following equation:

$$v_f = \frac{\Delta q_v}{\alpha \, d \, \Delta h} \tag{2}$$

where Δq_v is the increase or the decrease of the water discharge in the part of the borehole with the height of Δh, $\overline{\alpha}$ - borehole drainage influence coefficient for vertical flow (approximately $\overline{\alpha} \cong 20$), d - tube inner diameter.

Filtration velocity calculations according to the formula (2) are made with personal computers, results being graphically interpreted as depth dependencies. The average filtration velocity value of each borehole is given by the formula:

$$\overline{v}_f = \frac{\sum v_f \, \Delta h}{\sum \Delta h} \tag{3}$$

Permeability coefficient is given by the Darcy's law.

More intensive vertical flow in a borehole can be measured easier by borehole flowmeter.

Dilution method is used in observation boreholes with a very low water column. The tracer is usually (NaCl) introduced to water as powder. An immersion electrode probe together with simple battery conductometric equipment is used to monitor the dilution process. Filtration velocity is calculated by the formula:

$$v_f = \frac{\pi d}{4 \alpha t} \ln \frac{c_o - c_p}{c - c_p} \tag{4}$$

where d is the observation tube inner diameter, α - borehole drainage influence coefficient for the dilution method ($\alpha \cong 2$), c_o - initial concentration, c - concentration at time t, c_p - the natural concentration. The average filtration velocity values are again calculated by the formula (3).

Formula (4) assumes that solution dilution is caused by water flowing perpendicularly to the borehole axis. If there is some water flow in the direction of the borehole, the basic assumptions of the evaluation formula validity are not met and such results cannot be used. In order to eliminate the vertical flow there are devices which protect with inflatable sealings the measured area in the borehole against interference vertical flow influence (Drost, 1970) available at some working places.

Liptovska Mara Dam

Earthfill heterogenous dam on the river Váh is 52 metres high, 1225 metres long and its subsoil is formed by Carpathian flysh (paleogenous claystones and sandstones). These are treated by an very short grout curtain reaching to a depth of 10 m in the left-side bound, 20 m in the valley plain (Figure 3), 53 m in the right-side bound.

The reservoir with a total volume of 360 mil. m³ (the biggest reservoir in Slovakia) was began to fill at the time when the grout curtain had not been completed. Not even after several stages of grouting works the prescribed Verfel´s criteria were not achieved, whereas the worst results took place right under the base of the grouting gallery at the depth of 5 m. Compared with the natural medium the permeability of the curtain was five times lower, but the Verfel´s criterium was not met at 40 % of the levels checked.

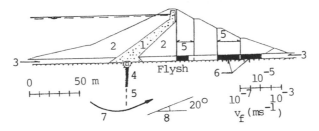

Figure 3. The Liptovská Mara dam cross-section: 1 - silt, 2 - stabilisation prisms, 3 – natural alluvium, 4 - grout curtain, 5 - observation boreholes, 6 - average filtration velocity values, 7 - area of the main water flow to gravelly subsoil at the air side, 8 - paleogenous layers slope.

Due to seepage during the reservoir filling the directions of water flow changed to the direction perpendicular to the lateral dam axis, under the dam the groundwater levels increased on average by 0.8 m and did not reach the level of the bottom drainage. Seepage can only be monitored by water flow measurement in the observation boreholes, being about 0.020 m^{3}s^{-1} for currently full reservoir. Due to the initial calculations of the project engineers a value of 0.050 m^{3}s^{-1} had been expected.

With respect to the assumed problems had the measurements of filtration velocities by single bore methods been performed regularly. The first ones took place before reservoir filling in the year 1970, observation boreholes were gradually complemented and during the last measurement in 1998 there were 106 boreholes available. The Figure 4 illustrates results from the borehole No. 33 with the most intensive flow, the maximum value of the filtration velocity of 1.4×10^{-4} m s^{-1} had been determined by the dilution method in the depth of 28.5 m, the average filtration velocity of the whole watered part of the borehole is 1.1×10^{-4} m s^{-1}.

In the valley part of the dam the average filtration velocity values in the gravel subsoil increase with the direction of water flow (Figure 3). Such phenomenon in the area can be interpreted as an underflow of a short grout curtain. These amounts are negligibly small from the water loss point of view. The most important are the hydrodynamic effects in the dam subsoil.

The measured maximum filtration velocities values are for these purposes compared with the critical values for washing out of sand particles from gravel soil pores (Hulla and Cábel, 1997) or of fine-grained particles from the rock fissures. (Ronzhin, 1974). Development of maximum filtration velocities in quarternary gravel soils and paleogenous flysch rocks under the dam is presented together with the critical velocities in the Figure 5. The results show that no situation that would be dangerous for the stability of the subsoil, as well as the whole dam had occurred during the filling of the dam, and it's operation up till now, even in spite of the fact that the curtain tightness criteria judged on the base on unsuitable water pressure tests had not been met.

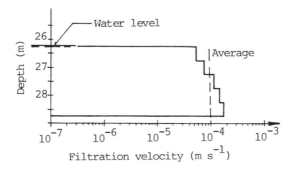

Figure 4. Dependency of filtration velocities with respect to the depth, measured by dilution method in the observation borehole No. 33, in the year 1998.

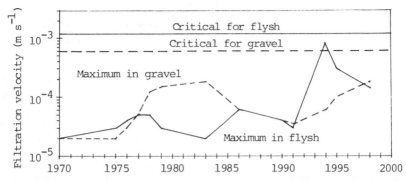

Figure 5. Develoment of the maximum filtration velocities in subsoils of the Liptovská Mara Dam, and critical velocities for the stability of the sand particles in the gravel soil pores, and for fine-grained particles in rock rifts.

Turcek dam

Rockfill dam at the junction of rivers Turiec and Ružová is 61 m high, 228 m long, with a skin bitumen-concrete sealing (Figure 6). Its neogenous neovulcanite subsoil is formed by pyroxenic-amphibolic andesites and their tuff aglomerates. The grou curtain reaches into the depths of 30 to 50 m. Reservoir volume is 10.6 mil. m^3 only.

During the investigation grouting borehole various discharges of water flew into the grout gallery from various depths. They clearly showed significantly more permeable layers in deeper positions than 30 m, maximum inflow sometimes reaching values of 0.020 m^3 s^{-1} m^{-1}. The fissures in the neovulcanite bedrock are directed mainly vertically what requires to curtain grouting in angling boreholes. In the pyroclastic rocks permeability did not go down even having done several groutings stages. However, layers with even increased permeability occurred. Water pressure test criteria fulfilment caused serious problems in some layers.

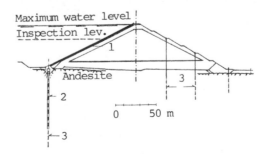

Figure 6. Cross section of the Turcek dam: 1 - bitumen-concrete skin sealing, 2 – grout curtain, 3 – observation boreholes.

A result summary is given in Figure 7, different lines depict characteristics for the natural non-grouted medium and the grouted one as well as the characteristics expressing criterial requirements. About 27 % of the checked layers did not meet the upper limit of the recommended criteria.

Despite of the knowledge mentioned a careful reservoir filling was started accompanied with thorough going measurements and analysis of the results obtained. A complex evaluation was made for the inspection water level of 757 m.a.s.l. (depicted in Figure 6). This level when compared to the empty reservoir represents an increase of 30 m, when compared to the maximum operation level, this level was 20 m lower. Another set of measurements was analysed when the maximum operational level was first reached in 1998, and during normal operation in 2001.

When compared to their status before filling the reservoir, the levels under the dam increased due to seepage from the full reservoir, as shown by the isolines from 2001 in Figure 8. The increase under the dam's exposed side is not higher than 3 m, near the grout curtain reach 10 m. It is possible to see an anomaly observed in the P-22 borehole, where the water level increased by 10.86 m, what is significantly higher than in nearby boreholes, where the observed increases were less than 1 m. The vertical water flow measurements in this borehole have shown that the site of the most intensive inflow is at the greatest depth where the borehole had loosened the horizon due to higher pressure. Thus, the increase in water pressure principally occurs in a more permeable medium at higher depths which are dependent on the increase in water level in the reservoir. The increase in water level is stabilised at the time of full operation of the reservoir.

Problems at the grouting works and apprehensions about the effectiveness of the curtain led to the construction of a high-quality uplift measuring borehole system. Short, medium, and long boreholes permit following the uplift conditions on the water side, the exposed side, and in the grout curtain directly.

The uplift values increased proportionally with the increase in the reservoir's water level. The results from measurements during the full operation of the reservoir shows the relatively high hydraulic gradients from the short and medium uplift bore-

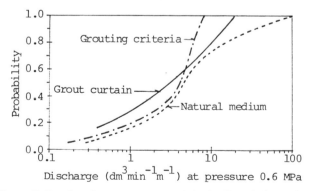

Figure 7. Results of water pressure tests in the Turcek dam subsoil.

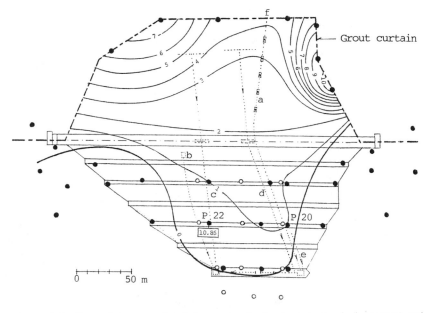

Figure 8. Isolines of the water levels increasing in the dam's bedrock (in meters) and drainage system (from **a** to **f**): o - boreholes permeable in dam body, • - boreholes permeable in bedrock.

holes document the very good effectiveness of the grout curtain. The effect of the grout curtain underflow is evident in the results from the long borehole.

Filtration velocities were evaluated in the dam subsoil due to filling the reservoir to the maximum operational level. The flow directions have declined by 30° on the exposed side in some boreholes, and the filtration velocities have increased as well.

The most significant change was observed in the P-20 borehole, where the maximum velocity was measured while the reservoir was full. The results from this borehole (Figure 9) show, that the water flows an upward direction; the main inflow area from the hydraulically most active layer was at a depth of 43 to 47 m; the maximum vertical discharge was 2.0 dm^3s^{-1} in 2001 and 14.2 dm^3s^{-1} at the first reservoir filling in 1998. Before the reservoir filling, the maximum vertical discharge being only 70 $cm^3 s^{-1}$. The maximum filtration velocity is 4×10^{-4} m s^{-1} - an absolutely maximum value in the bedrock of the dam in 2001 (3.2×10^{-3} m s^{-1} in 1998).

The changes in filtration velocity can most readily be judged by the empirical distribution functions shown in Figure 10. For the empty reservoir the median was 3.5×10^{-6} m s^{-1} (from 460 readings); for the full reservoir in 1998 it was 1.4×10^{-5} m s^{-1} (from 423 readings), for the full reservoir in 2001 it was 6×10^{-6} m s^{-1} (from 591 rea-

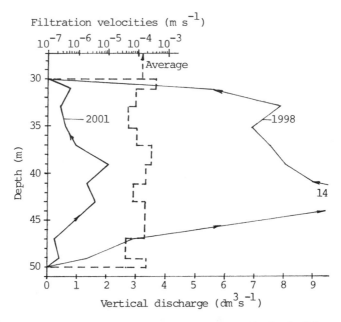

Figure 9. Monitoring the P-20 borehole's measurement results for the full reservoir.

dings). The arithmetical average value (2.2×10^{-5} m s^{-1}) was exceeded at a probability of 21 %. The critical velocity for fine silt particles moving in rock fissures 3 mm wide (0.0013 m s^{-1} - Ronzhin 1974) can be exceeded at a probability of 1 % only at the first reservoir filling.

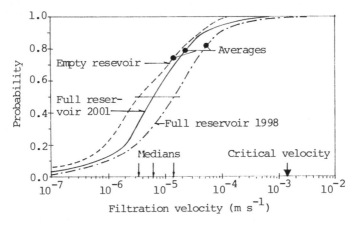

Figure 10. Filtration velocity distributions for empty and full reservoir.

The critical value of the filtration velocity was exceeded only in the P-20 borehole for the full reservoir in 1998. These are the layers in the area of the water effluent into the borehole at a depth of 44-45 m. Movable silty particles can be washed into the borehole from the surroundings. However, after some time the particle motion in the borehole's surroundings stops because the filtration velocity at a greater distance from the borehole falls below the critical value.

Despite the fact that we do not consider exceeding critical velocity in the P-20 borehole to be dangerous, the further development of the water flow in this borehole will have to be monitored thoroughly. The situation of exceeding critical velocity in other boreholes cannot be excluded.

If the critical values are not significantly exceeded, the permeability of the dam bedrock should not increase, and the seepage amounts should not go up either. Due to colmatation processes, the seepage amounts can also go down gradually.

The time development of the seepage is shown in Figure 11, where the increase procedure for the reservoir water level is also shown (thick full line). Partial seepage volumes were measured directly at different drain apertures (see Figure 8), draining water away from a certain part of the dam.

Total seepage amounts were estimated with the finite element method for the full reservoir during the dam's construction (Bednarova and Gramblickova 1995). The estimates were made alternatively for various input data; the seepage varied from 10 to 70 dm^3 s^{-1}, and the most important effect was the anisotrophy.

Direct measurements of the water amounts seeping from various drainage elements and their integration enabled us to get the maximal total value of 28 dm^{3}s^{-1} (in 1998) and 18 dm^{3}s^{-1} presently. Objective decline in seepages is the consequence of colmatation processes in the dam's bedrock.

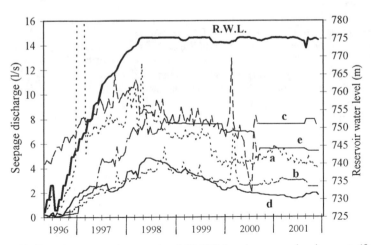

Figure 11. Turcek - reservoir water level (R.W.L.) and seepage development (from **a** to **e** - drains - see Figure 8).

The seepage amount flowing away into the effluent bed is being put to use effectively. About 70 dm^3 s^{-1} is sanitary discharge. So the loss is not one that could negatively affect the reservoir's financial operation. Their further development will be monitored thoroughly. Special attention will be paid to hydrodynamic effects which could negative affect the dam's stability.

Conclusions

According to experiences from the Slovak dams several different criteria can be used for water pressure tests (Figure 2) only by evaluating of permeability of natural rock medium and in decision making processes on the need of a curtain grouting. During curtain grouting it is enough to monitor the quality of the work on the bases of pressures and mixture consumption. In the our opinion at the high water presure the grout curtain can be faulted during tests and results are not correspond to the reality. Effectiveness of the completed curtain can be tested during the filling of the dam, and during its operation, the best way by monitoring of the characteristics of the seepage regime. Maximum filtration velocities, and critical velocities for washing out of the fine-grained particles from the gravel soil pores and from the rock fissures are important for the stability of the subsoil and bedrock, as well as the whole dam.

The authors would like to express their thanks to the managerial staff of the Slovak Water Management Enterprise in Banska Stiavnica and the Water Management Construction in Bratislava, and acknowledge them for their support in special measurement and providing of some information, as well as to the Scientific Grant Agency of the Slovak Republic for financing of the 1/9066/02 Project.

References

Bednarova, E. and. Gramblickova, D. (1995): *Model verification of the efficiency of the grout courtain under Turcek dam* (in Slovak). STU, Bratislava.
Drost, W. (1970): "Groundwater measurements at the site of the Sylvenstein dam in Bavarian Alps." *Isotope Hydrology*, IAEA Vienna, 421- 437.
Halevy, E. (1967): "Borehole dilution techniques: a critical review." *Isotopes in Hydrology*, IAEA, Vienna, 531-564.
Hulla, J. and Cabel, J. (1997): "Analysis of criteria for filtration stability" (in Slovak). *Inzinierske stavby*, SSCE, 45 (4-5), 145-149.
Kutzner, C. (1985): "Considerations on rock permeability and grouting criteria." *Proc of ICOLD*, Lausanne, 315-328.
Ronzhin, I., S. (1974): "Some criteria for evaluation of filtration stability of water structure subsoil" (in Russian). *Gidrotechničeskoe stroiteľstvo*, 24, 7, 24-27.
Verfel, J. (1983): *Grouting of rocks and diaphragm wall construction* (in Czech). SNTL, Prag. (English published by Elsevier in 1989).

Seepage Control by Grouting under an Existing Earthen Dike

Hasan Abedi, Ph.D., P.E.[1], Gary Simard, P.E. [2], and David P. Lohman[3]

Abstract

The Merimere Reservoir is a surface water supply source located approximately 3 kilometers northwest of the city of Meriden, Connecticut. The reservoir, constructed circa 1870, is impounded by an earthen dam to the north and by an earthen dike to the south. The earthen dike, which is the subject of this paper, has a concrete core wall and consists of fill resting on a low saddle of natural ground. The dike is approximately 45 m (150 ft) long and 6.6 m (22 ft) high. Evaluations of the reservoir revealed that the earthen dike was leaking at a rate of 950,000 to 1,135,000 liters per day (250,000 to 300,000 gallons per day). Several investigations were performed to determine the cause of leakage and provide remedial measures to reduce the seepage. The field investigations consisted of geologic mapping for outcrop joint trends, acoustic-emission monitoring, a seismic refraction survey, subsurface investigation and an investigation to locate the concrete core wall at the center of the dike. Bedrock outcropping at the west side of the dike is sandstone and at the east side is basalt. Based on the field investigations and engineering studies, it was concluded that the most probable location of the seepage was the highly jointed and faulted basalt along the eastern side of the dike. In order to reduce seepage through the basalt fractures, a pressure-grouting program was implemented. Cement grout formed a cutoff wall below and around the interface of the core wall and basalt bedrock. Grouting was performed in a triangular pattern in three rows with a spacing of approximately 1.5 m (5-ft). In order to evaluate the effectiveness of the grouting program, a "V-notch" weir was constructed immediately downstream of a seepage boil prior to grouting. This weir was used to measure the quantity of the seepage flow before and after grouting in order to evaluate the reduction in the seepage and the effectiveness of the grouting program. The grouting program completed at the Merimere Dike markedly reduced the seepage around the dike by over 85 percent to approximately 80,000 liters per day (21,000 gallons per day), a level to which the city of Meriden has deemed acceptable.

---

[1] Senior Geotechnical Engineer, Metcalf & Eddy, Inc., 30 Harvard Mill Square, Wakefield, MA 01880, USA; Phone 781-224-6324; hasan.abedi@m-e.com

[2] Senior Project Engineer, Metcalf & Eddy, Inc., Wallingford, CT, USA

[3] Assistant Director, Meriden Water Division, Meriden, CT, USA

Introduction

The Meriden Water Department (MWD) operates and maintains a reservoir system on the Meriden/Berlin town line consisting of the Merimere, Hallmere, Kenmere and Elmere Reservoirs. The largest of these reservoirs, Merimere Reservoir, with a capacity of approximately 1,348,000 cubic meters (356,000,000 gallons) and constructed circa 1870, is bordered on the north by an earthen dam. On the south, the reservoir is contained by an earthen dike (hereafter referred to as the dike). The dike is approximately 45 m (150 ft) long and reported to be 6.6 m (22 ft) high. Records state that the dike is constructed with a concrete core wall, although no construction information is available on the length, height or construction of the core wall. The core is supported laterally by earthen fill and is situated on a low saddle of natural ground with bedrock outcropping on each side. The MWD operates a water treatment plant at the southern end of the Merimere Reservoir that receives water directly from the reservoir via pipes that run through the dike (See Figure 1).

The MWD had estimated the leakage from the dike on the south end of the Merimere Reservoir at 950,000 to 1,135,000 liters per day (250,000 to 300,000 gallons per day) (Flaherty Giavara Associates, 1983; Orsine, Cotter and Carson, Inc., 1993). The majority of this seepage appears as a boil approximately 75 m (250 ft) downstream of the toe of the dike that quickly transitioned into a flowing stream.

Investigation of Seepage Source

The MWD retained Metcalf & Eddy to perform a geotechnical evaluation of the Merimere Dike to determine the most plausible route of the water losses and provide design and construction services to decrease the seepage rate. At the time of the study, there was no apparent seepage noted along the abutments of the dike, nor on the downstream slope. The evaluation began with a survey of the city of Meriden's records to review previous work performed on or around the dike since construction. The historical data combined with visual observations formed the basis for the geotechnical studies and led to design of the seepage cutoff. The geotechnical evaluation consisted of seven parts:

1. Historical records review and reconnaissance
2. Bedrock outcrop mapping
3. Acoustic emissions monitoring program
4. Seismic refraction survey
5. Test pit investigation
6. Geotechnical subsurface investigation
7. Concrete core wall investigation

The purpose of this paper is to report the various investigations performed to evaluate the source of the seepage and the grouting procedure utilized to substantially reduce the seepage.

Historical Records Review and Reconnaissance. The reservoir and dike were originally constructed in 1870 and numerous improvements involving the dike and water treatment plant have taken place over the last 130 years. Initial efforts to determine current conditions involved performing a review of existing documentation about the construction of the dike. Unfortunately, historical records are incomplete, resulting in a reliance on field observations to supplement historical records of the structure and appurtenances of the Merimere Dike. The only pertinent historical record was an inventory of city reservoirs. This record noted the construction date as 1870, the length of the dike as 45 m (150 ft) and the height as 6.6 m (22 ft).

Prior to approximately 1985, the City withdrew water from the Reservoir through a gatehouse located approximately 15 m (50 ft) offshore from the dike. Prior to its demolition, this gatehouse contained a 400-mm (16-in.) cast iron main intake pipe (Figure 1), a 300-mm (12-in.) auxiliary intake pipe, and associated screening. The raw water lines ran from the intake gatehouse through the dike, into a lower gatehouse located at the base of the dike, and into the water treatment plant. During a plant upgrade in 1985, both gatehouses were demolished.

Currently, the 400-mm (16-in.) pipe traverses the dike, cuts through or under the core wall and is visible in a manhole located on the outside edge of the concrete core wall. The pipe then continues downgrade to valve at the site of the former lower gatehouse, and eventually terminates at the raw water storage tank located adjacent to the water treatment plant.

Bedrock Outcrop Mapping. The site area is dominated by northeasterly trending basalt flows that form the resistant ridges of the area. The Connecticut Valley is dominated by the high ridges of basalt and the red sandstone of Triassic-Jurassic Periods. This period was geologically very dynamic.

Extensional tectonics developed as the continents separated. This tectonic period resulted in high angle normal faulting and the deposition of thousands of feet of sandstone infilling the faulted basins. The northeasterly trending high angle faults are common throughout New England.

The United States Geological Survey Bedrock Geologic Quadrangle Map indicates that the lower section of the valley is underlain by the New Haven Arkose. New Haven Arkose consists of red arkosic conglomerate and sandstone. The more resistant rocks on either side of the valley near the dike consist of the Talcott Basalt. A fault is shown on the west contact between the Talcott Basalt and the New Haven Arkose. Sandstone outcrops were observed on the southwest corner of the dike.

The Merimere Reservoir lies in a valley between two high ridges of basalt. To the East of the dike is a high ridge consisting mainly of the Talcott Basalt. The west edge of the dike blends into the natural glacial soils; no bedrock is exposed on the west side of the dike. The access road to the reservoir is along the east side of the valley.

The basalt is exposed along the east side of the access road to the reservoir. The outcrops are somewhat weathered with a coating of moss and lichens. The jointing is well developed along the entire exposure. The major joint trend parallels the regional trend, approximately N20E. These joints are consistent with the extensional tectonics and have high angle dips in either east or west directions. Two other joint trends were observed; a northerly trending set and an east-west set. For both sets, the dips are variable, dipping at high angles in either direction.

The combination of the jointing and the thickness of the flow bands create variations in the quality of the rock mass. Where the flow bands are thin, the rock quality is poor. Conversely, where the flows are more massive, the quality is better. In the outcrops adjacent to the dike, the quality of the rock was poor. Thin flows and closely spaced joints have resulted in a blocky rock mass. South of the dike the flow bands are more massive and the quality of the rock improves.

Acoustic Emission Monitoring. An Acoustic Emission Monitoring investigation was performed to identify the source of the seepage under the dike by monitoring the acoustic signals in the vicinity of the dike. Acoustic signals monitoring involves the use of pressure transducers to monitor ambient sounds in an aquatic environment. Significant amounts of water traveling through formations (such as the dike) may produce sounds that the transducers can detect. For this study, transducers were placed in the reservoir, close to the dike. Transducers were arranged so that measurements were made on a 1.5-meter (5-foot) grid for an area of approximately 30 x 45 m (100 x 150 ft). An additional transducer was placed in a small sinkhole depression located immediately upstream of the dike along Reservoir Road. A complete set of reservoir measurements was collected during the two days of the testing, and the sinkhole was monitored continuously.

The investigation revealed no large anomalies or group of anomalies that would indicate areas of concentrated leakage near the dike. It was concluded that the seepage is due to a large number of small discontinuities that were not detected.

The original plan was to perform a salt tracing study at any location that the acoustic survey suggested a major leak existed. However, since the acoustical survey did not reveal any major leakage locations, a concentrated area to perform the salt tracing could not be identified and therefore was abandoned.

Seismic Refraction Survey. A seismic refraction survey was performed to identify any layers of heavily weathered or loose material that may be the source of leakage under the dike. The survey was conducted by placing a string of geophones, along a line of interest. The ground was then struck in selected locations and the geophones recorded the initial sound waves and the time for that sound wave to bounce back from the underlying materials. Four lines totaling 375 m (1,250 ft) of coverage were selected for seismic refraction. The geophysical survey would identify any lower velocity zones indicative of highly jointed, poorer quality bedrock. The results of survey were inconclusive with regard to distinct velocity differences in the rock mass.

Test Pit Investigation. A test pit subsurface investigation was developed to identify certain key physical aspects of the dike. The first test pit program was performed along the midline of the dike to locate the actual core wall (Figure 1, TP#1). Upon location of the core wall, the test pit was extended along the backside of the wall in an attempt to determine groundwater levels. It was observed that the groundwater level on the reservoir side of the core was at approximately the same elevation as the reservoir. Groundwater was not encountered in the test pit behind the core wall to a depth of approximately 1.8 m (6 ft). The exposed core wall was approximately 0.6 meter (2 ft) below ground surface, 600 mm (24 in.) wide, and constructed of concrete. The small section of the wall that was exposed was in sound condition.

A second test pit was performed along the east side of Reservoir Road (Figure 1, TP#2). The core wall was again located along the midline. Observations of the area lead to speculation that the reservoir side of the manhole atop the dike may actually be the core wall. Although measurement of the depth of the core wall could not be performed, the bottom of the manhole was measured at 7.2 m (24 ft) below grade

Additional test pits were dug downstream of the dike approximately 9 m (30 ft) north of the groundwater boil (TP#3 & TP#4 in Figure 1). A pit was dug on each side of the 400-mm (16-in.) intake pipe. These test pits were performed not only to characterize the soil conditions but also to catalog the groundwater depths and flow volumes.

The test pit west of the 400-mm (16-in.) pipe (TP#3) was excavated to a depth of 2.4 m (8 ft). The soil consisted primarily of loose sand and gravel and was saturated approximately 0.6 meters (2 ft) below grade. The rate of infiltration was estimated at approximately 95 liters per minute (25 gallons per minute).

The second test pit (TP#4) was excavated further east along the Reservoir Road embankment. Water filled the test pit in a short time and the water level stabilized at approximately a foot below grade. Attempts to dewater the test pit were unsuccessful even with the use of several high capacity dewatering pumps. A flow of 1325 liters per minute (350 gallons per minute) was estimated with the pumps and the water level could not be lowered more than 1.5 m (5 ft) below grade. The soil was a high permeability, loose sand and gravel with little or no fine material. These flows suggested that a significant source of groundwater is located on the east side of the dike, under Reservoir Road.

Geotechnical Subsurface Investigation. The information collected from the test pits suggested that the flow was progressing through the dike in the overburden materials as opposed to the bedrock formations as previously considered. Based on these findings a subsurface drilling program was designed to investigate the overburden soils, depth to bedrock, and groundwater level. The subsurface drilling investigation consisted of 10 test borings as shown in Figure 1.

Based on the seismic refraction survey, it was anticipated that an overburden layer of approximately 6-9 m (20-30 ft) existed over denser, fractured bedrock, underlain by solid bedrock. It was the intent of the investigation to bore and sample the soil materials and core the underlying bedrock to determine its competency. However, the first boring extended to 12 m (40 ft) below grade and was eventually terminated in a very dense glacial till soil that could not be further penetrated by standard auger or split spoon-sampling techniques.

Based on the findings of the 10 borings, a generalized subsurface profile was established as shown in Figure 2. The dike appears to be constructed with loose sand and gravel fill to a depth of approximately 6 m (20 ft). Because of the exposed sandstone bedrock to the West and basalt bedrock to the East, this layer of sand and gravel was expected to extend to bedrock boundaries. Some borings encountered a medium dense to dense layer of glacial till type material at approximately 4.5 m (15 ft) below grade. Some borings encountered a layer approximately 1.2-3.6 m (4-12 ft) in thickness of weathered basalt chips and cobbles. This layer was extremely difficult to sample due to the fractured, but relatively dense nature. This layer is suspected to be highly permeable and correlates well with the level of the groundwater table.

Core Wall Investigation. An additional drilling program was performed to determine the integrity, depth and underlying material of the central core wall. This was an imperative procedure because this core wall acts as the impermeable boundary of the dike reducing the downward migration of the reservoir water.

The core wall investigation consisted of drilling three boreholes into and through the core wall. The borings are referred to as 'core holes.' Based on the samples obtained from these cores through the concrete core wall, information was provided on the construction and integrity of the concrete wall. The samples suggested that the wall was constructed solely of concrete, with no steel reinforcement noted. The integrity of the concrete was highly variable as can be expected from construction of this age. In general, the concrete wall appeared to be more deteriorated further toward the East. It is suspected that the areas with significant core wall deterioration may be contributing to the dike seepage, although this is not quantifiable.

The other objectives of the coring program were to determine the core wall depth and underlying material. Based on the core holes, the depth of the core wall was observed to range between 10.5 to 11.4 m (35 to 38 ft). The core wall appears to be constructed on the sandstone formation toward the West and transitions onto a dense glacial till moving toward the East. After penetrating the base of the concrete wall at the east side of the dike, approximately 2.1 m (7 ft) of glacial till was encountered prior to transitioning into the sandstone formation. Because basalt rock was not encountered, it appears that the embankment dam is founded mostly on the sandstone and till. The geologic contact with the basalt is at a high angle, potentially at fault near the basalt outcrops along Reservoir Road.

Seepage Source and Control Method

The leakage at the downstream end of the dike did not appear to affect the overall stability and integrity of the dike at this time. The MWD was concerned with reducing the leakage to minimize the water losses.

Based on the investigations, the suspected source or sources of seepage was narrowed to a 9-m (30-ft) wide area of the eastern abutment of the dike (i.e., under the Reservoir Road). There appeared to be a fault zone within this area, which serves as a conduit to convey the water from the reservoir to the downstream side. It was not known if there was any treatment of this fault zone during dike/abutment construction in 1870. The reservoir water could also seep through the joints in the basalt rock at the eastern abutment similar to the closely spaced northerly joints observed in the basalt outcropping along Reservoir Road.

To reduce seepage through the fault zone and the bedrock fractures, a grouting program was designed. Cement grout would form a cutoff wall below and around the interface of the core wall and bedrock. A schematic cross section of this method is shown in Figure 3. Grouting would be performed in a triangular pattern in three rows with spacing of 1.5 m (5 ft).

In order to evaluate the effectiveness of the grouting program, a "V-notch" weir was constructed immediately downstream of the present seepage boil prior to grouting. This weir was used to measure the quantity of the seepage flow prior to grouting to establish a baseline. During and after grouting, the seepage was measured to evaluate the reduction in the seepage and the effectiveness of the grouting program.

A grouting program for this application has several limitations as discussed below. MDW were apprised of these limitations prior to implementation of the program.

a.   If the source of seepage is limited to a narrow width of fault or fracture zone below the concrete core wall, then the seepage velocity may be high. When the seepage velocity is high, the cement grout may not have time to set and seal the fracture zone before it is washed away. In this scenario, a fast setting chemical grout may be selected.

b.   Fractures in the bedrock or the fault zone may extend beyond the recommended 30 ft of grouting zone. Consequently, a deeper grouting zone may have to be implemented. Also, the grout intake of some boreholes may be very high without any knowledge of grout path or destination. This has been observed in other bedrock grouting projects.

c.   Success of the program will depend on the injection of sufficient grout into intersecting joints to seal the bedrock joints. It is not realistic to assume that the initial grouting program will stop the leakage entirely since minor leakage may be occurring at points distant to the grouting program. If after completion of grouting, the seepage does not decrease sufficiently, supplemental grouting programs may have to be implemented in an attempt to reduce seepage to an acceptable level, say ten to fifteen percent of the flow prior to grouting. This second stage grouting would be assessed after original grouting is completed.

Grouting Procedure

Grouting of the bedrock was accomplished by first drilling and installing 100-mm (4-in.) PVC grout tubes to the top of bedrock. Drilling for the installation of grout tubes was accomplished with duplex rotary methods utilizing 175-mm (7-in.) steel casing. The holes were flushed clean of drill cuttings and then a cement grout (a mix ratio of 1:1) was tremie pumped into the hole. The PVC grout tube was then set inside the steel casing through the grout. The steel casing was then extracted from the drill hole.

After the necessary setup time for the grout around the PVC casing, the grout holes were re-drilled to the bottom of planned grouting zone by rotary percussion techniques. The hole diameter in the grout zone was 88 mm (3.5 in.). In order to grout the basalt rock at the abutment, angled holes were drilled. Figure 3 shows the schematics of grout holes drilled.

Three rows of grouting were performed. One row was drilled on the downstream side and two rows on the upstream side of the core wall. The grout holes were approximately 1.5 m (5 ft) apart. Grouting was conducted with a Colcrete HighShear grout plant. The main features of the grout plant are its ability to colloidally mix grout at greater than 2000 rpm and pump the grout with a vertical helical progressive cavity (Moyno) pump.

After the completion of grout hole drilling, a packer was seated at the top of grout zone inside the PVC tube. The cement grout was pumped into the grout hole through an electronic flow meter between the grout plant and the grout hole collar. The flow meter measured the flow rate and the total grout volume injected into the hole. Grouting was started with a 3:1 water to cement grout mix and thickened during grouting to 2:1, then 1:1 and finally 0.75:1 until refusal. The grout mix was thickened sequentially after mixing and injecting of 15 sacks of cement. Refusal was defined as when the rate of grout injection, at the maximum pressure, was less than one-half cubic foot of liquid grout in 20 minutes. Connection between the grout plant and the packer assembly was by 25-mm (1-in.) diameter high-pressure hoses and piping. The grouting of the bedrock was performed in a single stage.

The packer remained inflated in the grout hole until the pressure observed at the collar was dissipated, typically about one hour after grout refusal. After pressure grouting was completed in each hole, the remaining portion of the PVC grout tube was tremie grouted with a cement grout of 1:1 water to cement grout mix.

Discussions and Conclusions

Grout holes were drilled 6 m (20 ft) into the bedrock. Sandstone and basalt were encountered during drilling of the 6 m (20 ft) zone adjacent to the east abutment. This revealed presence of a fault zone. Grout-take varied from a total of 43 bags of cement to a minimum of less than one bag per hole. Some grout holes showed communication with adjacent holes. During grouting phase, the seepage water downstream of the dike was monitored to observe if any grout would wash out. No grout washout was observed. Grouting was performed in two weeks during July of 2001. A total of 280 sacks of cement were mixed and pumped into the grout zone.

Prior to and during grouting, the seepage flow rate was monitored by a 'V-notch' weir which was installed downstream of the seepage boil. Table 1 shows the seepage rates and flow reduction rates as measured at the 'V-notch' weir.

Table 1. Seepage Rates and Flow Reduction

Timeline of Measurements	Seepage Rate (Liters per day)	Percent Flow Reduction
Estimated in 1983/1993	950,000 to 1,135,000	-
One week before grouting (measured)	568,000	-
One month after grouting (measured)	117,000 to 98,000	79% to 83%
Five months after grouting (measured)	80,000	86%

As can be observed from the Table 1, grouting substantially reduced the seepage downstream of the dike. The remaining seepage may be through deeper rock fractures of the eastern abutment, the western abutment, and the sandstone underlying the dam or through other groundwater sources beyond the reservoir.

Acknowledgements

The authors wish to thank William Bent, Jose Ramos, Leo Martin, Richard Sherman, John Risitano, and Danielle Wight of Metcalf & Eddy, Inc. for their contribution during design and construction of the project. The authors wish to acknowledge GZA GeoEnvironmental of Norwood, Massachusetts that performed pressure grouting.

References

Flaherty Giavara Associates (1983), *The Merimere Reservoir Dam, Berlin Connecticut*, prepared for City of Meriden, Connecticut.

Orsine, Cotter and Carson, Inc. (1993), *Safe Yield Analysis: Merimere, Hallmere, and Kenmere Reservoir"*, prepared for City of Meriden, Connecticut.

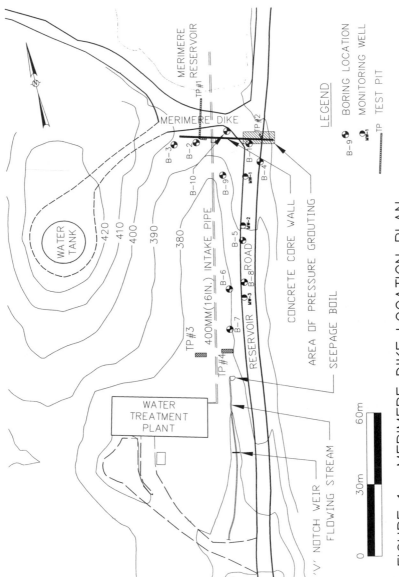

FIGURE 1. MERIMERE DIKE LOCATION PLAN

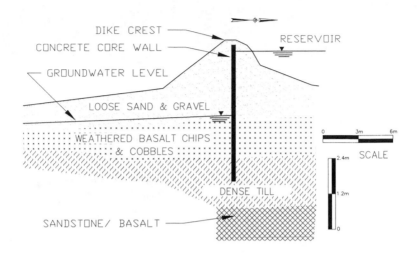

FIGURE 2. GENERALIZED SUBSURFACE PROFILE

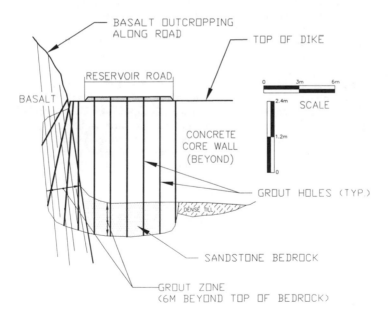

FIGURE 3. GROUTING ZONE

STATE OF THE ART IN
COMPUTER MONITORING AND ANALYSIS OF GROUTING

Trent L. Dreese, M. ASCE[1], David B. Wilson, M. ASCE[2], Douglas M. Heenan, M. ASCE[3], and James Cockburn[4]

ABSTRACT

Computer monitoring and analysis of grouting has come of age as a reliable and effective tool for better, faster, and less expensive grouting. This paper traces the development of computer monitoring and control systems, summarizes the benefits that can be realized from use of the systems, and presents the latest developments in specialized technology developed specifically for permeation grouting.

Several case history applications of successful use of computer monitoring and analysis are presented. The Penn Forest Project, a new dam construction project in Pennsylvania, was a landmark project in the utilization of computer monitoring technology and is the only large project where full scale, side by side comparison of conventional monitoring techniques and computer monitoring systems is available This project provided clear, quantitative illustration of the many technical and cost benefits that result from computer monitoring. The Patoka Lake Seepage Remediation Project, located in southwestern Indiana, was another vital project in that it was the first grouting project requiring the use of computer monitoring undertaken by the U.S. Army Corps of Engineers (USACE), and in that it has helped set the standards for contracting methods and field application. The Hunting Run Dam

---

[1] Senior Project Manager, Geotechnical Section, Gannett Fleming, Inc., P.O. Box 67100, Harrisburg, PA 17106-7100, tdreese@gfnet.com

[2] Vice President and Asst. Practice Leader, Earth Science and Hydraulics Practice, Gannett Fleming, Inc., P.O. Box 67100, Harrisburg, PA 17106-7100, dwilson@gfnet.com

[3] Vice President, Advanced Construction Techniques, Ltd., 10495 Keele Street, Maple, Ontario L6A 1R7, dheenag@agtgroup.com

[4] President, Advanced Construction Techniques Ltd., 10495 Keele Street, Maple, Ontario L6A 1R7, jcockburn@agtgroup.com

Project, recently completed for Spotsylvania County, VA, utilized new, state-of-the-art technology that advances computer monitoring and analysis of grouting to an unprecedented level of automation. This new system completely automates operations and allows real-time, comprehensive onsite display and simultaneous real-time onsite or remote analyses of grouting results.

History of Computer Monitoring

Recognition of the potential benefits and experimentation with "automated" monitoring or data recording systems for grouting started in the 1960's (Weaver 1991). Use of electronic measurement devices mated with computers was recognized as having significant potential almost as soon as desktop computers came into being in the early 1980's (Jeffries et. al. 1982) (Mueller et.al. 1982). The U.S. Bureau of Reclamation (USBR) was the first federal agency in the U.S. to experiment with the use of computers for monitoring of grouting. Utilized at Ridgeway Dam in 1982, the problems with the first system were numerous. However, this first experiment resulted in the USBR developing a comprehensive hardware and software system that would provide, generate, and record all the information that was needed for monitoring, control, and analysis of grouting (Demming et.al. 1985). The USBR implemented its use at Stillwater Dam in 1985. Although relatively little is written about the experience on Stillwater Dam, significant problems were experienced with consistently maintaining signals to the equipment. During the same time period, the USACE began using portable site recorders to obtain real-time grouting data, but severe field reliability problems were experienced (Houlsby 1990).

Beginning in the mid-1990's, there have been dramatic improvements in both the number and type of flow and pressure measuring devices, computer hardware, data acquisition software, and data management and display software. It has been proven that the use of computer monitoring systems clearly allows permeation grouting to be more technically effective, performed at a lower cost, and in less time. The systems are now sufficiently robust and user friendly to make it wise to consider their use on all future projects.

The Science of Grouting

Grouting is an especially unique type of construction. Grouting involves managing and performing dozens of simultaneous operations, each of which requires an extraordinary degree of care and any of which, if not performed properly, will result in ineffective grouting and loss of value in the project. Performing as much as 95% of the work properly can still result in nearly complete failure of a grouting program and almost total loss in program value. Studies of imperfect seepage barriers show that if only 5% of an otherwise impervious barrier is defective, the barrier efficiency in terms of seepage reduction can be as low as 10%.

Grouting is further complicated by the fact that all operations are performed underground. The fact that we cannot see the formation to be grouted or see grout permeating the voids or fractures has led to permeation grouting commonly being described as an "art". However, if proper investigation, design, contracting mechanisms, materials, and field techniques are employed combined with real-time data collection and analyses by a competent grouting engineer or geologist, engineered grout curtains can be constructed with dependable, predictable, performance and with virtually the same confidence in quality as visible "above ground" construction. Further, these measures will assure that grouting will be performed faster, better, and at lower overall cost. Grouting performed at this level is a science and is no longer merely an art.

Advantages of Computer Monitoring

The specific quality, cost, and time benefits that can be attributed to computer monitoring and analysis technology are summarized below:
1. Real time data is obtained at much smaller time intervals (5-15 sec. frequency vs. 5-15 min. frequency).
2. Eliminates missing critical events such as pressure spikes.
3. Data obtained is more accurate than data from "conventional" methods.
4. Higher grouting pressures can be used with confidence.
5. Formation response to procedure changes (mix or pressure) are known instantly.
6. Facilitates multiple hole grouting.
7. Damage or no damage to a formation due to over-pressuring can be easily detected or determined.
8. Significant acceleration of pressure testing and grouting operations.
9. More consistent grouting procedures due to central control location.
10. Reduction in inspection manpower requirements.
11. Provides detailed, permanent graphic records showing the entire time history for each operation on each stage.
12. Permits reallocation of resources to analysis of program results rather than on process management and data collection.

Many of these advantages can only be realized if plots of the apparent lugeon value (Naudts 1995) are automatically displayed in real-time by the monitoring system. The apparent Lugeon value is simply a way of expressing the permeability of the formation using grout as the permeant, and is determined using the standard equation for determining the Lugeon value with water times a correction factor. This correction factor is equal to the ratio of the apparent viscosity (marsh funnel flow time) of grout to the apparent viscosity of water. The real time graphical display of the apparent Lugeon value is extremely useful to the operator. Although the actual apparent lugeon value may not be a true measure of the formation permeability due to the assumption made in determing the correction factor, it is a measure of the relative permeability change during the grout application and does relate the injection pressure to the magnitude of the take. The display of this value allows the operator to instantly interpret the response of the formation to any changes in the grouting program such as changes to the injection pressure or grout mix. A plot of flow rate divided by effective injection pressure can be substituted for the apparent Lugeon value. Constant monitoring of the formation response to increases in pressure allows the practitioner to confidently move away from the conservative rules of thumb for grouting pressures that have been traditionally utilized. The rate of grout take and the radius of grout penetration are directly proportional to the injection pressure, and therefore, one should utilize the highest safe grouting pressure on every stage of every hole.

Dilation of rock fractures is easily ascertained from review of a plot of the apparent lugeon value versus time. As an example, Figure 1 is a grout record from Penn Forest Dam for a grouting stage from 0 to 20 feet. As indicated, a nominal grouting pressure of 20 psi was being used for this surface stage, which exceeds the U.S. rule of thumb of 1 psi per foot. The record also indicates a significant pressure spike at a time of 25 minutes. As expected the flow rate spiked at this same time and without a plot of the apparent lugeon value the impacts would have been unknown. The plot of apparent lugeon value shows a very slight increase in the formation permeability at this time (fracture dilation), but upon reduction of the pressure to previous levels the formation permeability returned to previous levels, indicating no permanent damage to the foundation. Continuous review during each grouting stage permits confirmation that the pressures being used are in fact safe.

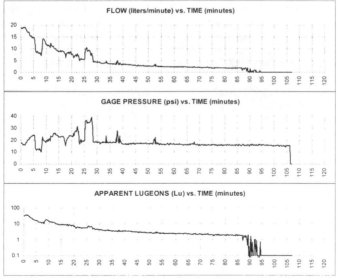

Figure 1: Grout Record indicating Fracture Dilation

Available Levels of Monitoring Technologies

Three levels of permeation grouting monitoring technologies are currently in use (Wilson and Dreese, 2002). A brief description of each level of technology and commentary on that levels applicability and considerations for use is provided below.

Level 1 Technology: Dipstick & Gage – Utilization of this technology represents the general state of practice prior to 1997. Most agency guide specifications that are publicly available are still based on this technology and many U.S. Contractors are still set up to operate exclusively with this type of monitoring. This method of monitoring is based on using a dipstick for measuring grout take, a pressure gage for observing water or grout injection pressures, and a nutating disk water meter for measuring water take. Wilson and Dreese (1998) provided a summary of the accuracy of various measuring devices for measuring flow rates and pressures. Due to the inherent inaccuracies of the measuring devices used with this technology, readings are commonly taken on 5 to 15 minute intervals to allow sufficient time for a reading change to be observed. If best practice is being followed, the engineer in charge of grouting of the stage will do limited manual calculations as the data is being obtained and make a plot of average grout take per time interval for the stage as it is being grouted. After completion of a stage, the data is later manually entered onto a large master wall chart that usually shows the hole location, and the grout take in bags of cement. Water pressure test results might or might not be shown on the chart.

This method of monitoring and control is rapidly disappearing and it is the authors' opinion that this level of technology should no longer be considered for grouting projects of any significant size or in a critical application. The minimum level of technology that is recommended for consideration based on the current state of technology and practice is the use of Level 2 Technology as described below.

Level 2 Technology: Real-Time Data Collection, Display, and Storage – Any system that uses electronic devices for measurement of flow and pressure and sends those signals to one or more other devices where the measurements are automatically displayed and recorded falls within this level of technology. At the low end of this technology, current readings of flow and pressure are displayed without any corrections for elevation, head loss, mix properties, or other factors. An XY-Recorder records the data or a data file is created by a software program for storing the data throughout the operation. This type of software functions as an automated electronic record book. At the high end of the technology, the collected data and correction factors are used in automatic calculations to produce and display corrected or calculated parameters of interest for the stage being grouted in real-time. A widely recognized software system in use at the high end of this level of technology is CAGES, which stands for Computer Aided Grout Evaluation System. CAGES software is a proprietary product developed and owned by ECO Grouting Specialists, Ltd. Two other systems that clearly operate at Level 2 Technology are a system by PARTNERmb of the Czech Republic and one from Soletanche of France. Information on these systems can be found on the internet at each companies web site. The authors are not familiar with these systems and the operation level may approach Level 3 Technology.

Level 2 Technology is a huge improvement, but it falls well short of the full potential envisioned for an automated grouting system. Data is more readily available for analysis than at Level 1, and onsite PC's enable the grouting engineer or geologist to perform better, but still very limited, analyses of the overall program as it progresses. However, onsite personnel are still faced with a mountain of numbers in which they are looking for patterns and anomalies that might or might not exist. The single paper wall chart used to plot the water testing and grouting results contains so much information that it is nearly impossible to comprehend in a timely manner, if at all. Level 2 Technology is considered applicable for any grouting project and should be the minimum level of technology considered for projects with a construction value exceeding $250,000 (in consideration of system acquisition and setup costs) or any grouting project in a critical application.

Level 3 Technology: Advanced Integrated Analytical (AIA) Systems – AIA systems are fundamentally different in nature and represent an enormous leap forward in comparison to Level 2 Technology. An AIA system provides integration of data collection, real-time data display, database functions, real time analytical and query capabilities, and CAD. The first AIA System for grouting, the IntelliGrout™ System (U.S. Patent Pending), was introduced in 2001. IntelliGrout was jointly developed and is owned by a Contractor & Engineer Team, Advanced Construction Techniques Ltd. and Gannett Fleming, Inc. Development of IntelliGrout was completed in 2000, and the system was used successfully on the Hunting Run Dam project in 2001.

IntelliGrout is a totally integrated system for data collection, monitoring, record keeping, and, most importantly, real-time onsite and offsite analyses. It not only contains all of the features of Level 2 Technology, but also includes real time graphical display of geologic features and stratigraphy, hole geometry, water test data, and grouting data, which is provided through customized programming developed within AutoCAD. AutoCAD, like the real time monitoring software, reads data from and writes data to a relational database. This relational database is the power of the system. The database allows for the generation of standard and custom reports and also allows queries to the database to search for relevant information. In addition, the proprietary AutoCAD programming is also directly linked to this real-time database, which permits real-time graphical display of grouting results on a profile. Utilizing the relational database, the system is able to perform practically unlimited, complex real-time grouting program analyses and can display the grouting results on a simple to understand and interpret, visual color display on a profile. Patterns, anomalies, compliance or non-compliance, and areas of special interest are immediately evident. The system is equipped with two monitoring stations, each with three monitoring screens (see Figure 2) to allow the operator to observe or perform multiple operations.

Figure 2: IntelliGrout Monitoring Station

A partial listing of some of its input and output capabilities and the end products produced by the system are:

- o Project parameters include choices of data units and permits identification of project specific refusal parameters. Refusal alarms indicate meeting of refusal parameters.
- o Topographic surveys are input and converted into a 3-D digital terrain model.
- o Geologic structure surfaces, consisting of units and orientations, is input in a manner similar to topographic data.
- o The curtain lines are defined through CADD and a profile of each curtain line is generated indicating the ground surface and geologic structure.
- o Holes can be defined using CADD functions and can be automatically distributed according to any criteria for depth, length, "x" feet into a geologic unit, spacing, and inclination or holes can be entered via a Hole Definition Screen within the monitoring program. The hole identification, station, elevation, and inclination are automatically determined and recorded within the database.
- o Identification and definition of stages can be done either automatically or manually.
- o Intersections of stages with geologic units are automatically calculated and the predominant geologic unit within a stage is recorded by the system.
- o Real-time data displays versus time include gage pressure, effective pressure, flow rate, current mix, and Apparent Lugeon Value. These displays are automatically printed as permanent hole records as shown on Figure 3.
- o Proposed holes appear on the CADD profile as a dashed line and color coded by hole series. After drilling, the hole is time stamped and the hole is displayed as a solid line providing real-time visual display of progress.

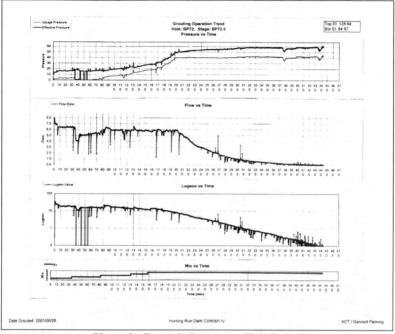

Figure 3: Example Permanent Hole Record

o   Immediately upon completion, water test and grouting results appear on the CADD drawing.

o   Display options includes the numerical Lugeon value for the stage, color coding of stages by Lugeon value, and diameter scaling of the stage by Lugeon value or radius of grout take (Figure 4).

o   CADD layering permits instant generation and viewing of any combination of information desired. Grout holes of different series (i.e. Primary, Secondary, etc.) and the drilling, grouting, and water test results for that hole series are on discrete levels within AutoCad. Therefore, the profile view of water testing and/or grouting results can be viewed on the profile for one or multiple series of holes.

o   Both water test results and grouting results can be displayed in 3-D CADD displays as cylinders scaled to magnitude of values (Figure 5).

o   The system is equipped with color plotter to produce working drawings for review or analysis or record drawings. It is also equipped with a computer projector to display grouting profiles at a scale suitable for easy viewing by multiple persons.

o   All database tables and AutoCad drawings can be electronically exported to a
    remote location for detailed review and analysis by peers.

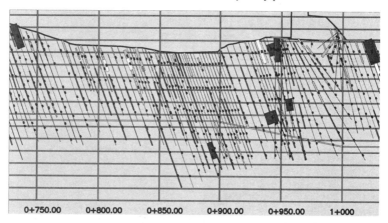

Figure 4: Water Test Results displayed in 2-D

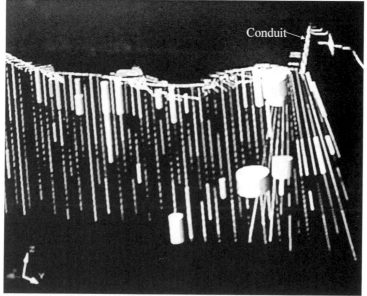

Figure 5: Water Test Results displayed in 3-D

The relational database permits the user to write queries to quickly obtain the desired information. Custom queries can be written at any time. Standard queries within the system include the following:

1. Available work listings including holes available for drilling, water testing, washing, or grouting.
2. Completed hole records (flow, gage pressure and effective pressure, Apparent Lugeon, and grout mix all displayed graphically as a function of time)
3. Searches of holes or stages for particular characteristics of interest. A useful query under this category is determination of all stages for the final hole series that have a lugeon value greater than the project performance requirement (Figure 6).

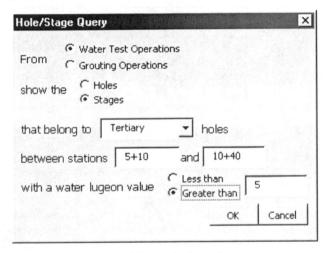

Figure 6: Example Stage Query

4. Closure Plots can be generated for both water pressure tests and grouting according to the criteria defined by the operator for inclusion in the analysis. For example, data to be included in a particular analysis can be selected based on stage depth, elevation, or geologic unit of interest and can be limited to a length of the profile using stations or primary grout hole identifications to define the area of interest (Figure 7).

AIA Technology is an enormous step forward in comparison to Level 2 Technology and is a major development in Computer Monitoring and Analysis of Grouting. AIA systems further reduce the onsite inspection staff time and optimize and increase the quality of reviews and interpretations of the results by decision makers through the data display options and the remote access capabilities. The systems are appropriate for consideration on Projects equal to or exceeding $750,000

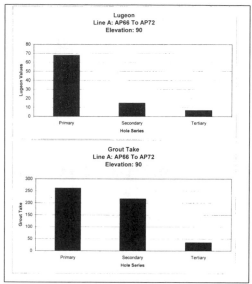

Figure 7: Example Closure Plot

in construction cost (in consideration of system acquisition and setup costs) or any project with high consequences of poor performance such as an environmental containment project. Projects of this value are of a size where the economic and technical advantages of these systems can be fully realized. Level 2 Technology can be used for these projects, but as the value and complexity of a project increases, the ability to adequately consider and interpret all of the data being collected becomes increasingly more difficult.

Case Histories

The Penn Forest Dam is critical to the water supply for the City of Bethlehem. The replacement project was constructed on a fast track schedule to avoid potential water shortages during construction. The new dam is approximately 180 feet high and 2,000 feet long and has a 3-Line grout curtain to an average depth of 140 feet. An accelerated schedule resulted in the foundation grouting being split into two separate contracts. The A-Line (first line) grouting contract specified Dipstick & Gage Technology. However, sufficient time existed before the issuance of the second grouting contract for the authors to adequately evaluate the applicability of Data Collector & Display Technology. Therefore, the B and C Line (second and third lines) grouting was specified and performed using Level 2 Technology. (Wilson and Dreese 1998). Penn Forest is the only project known to exist where a large scale, side-by-side comparison of the two technologies has been performed and where the project cost savings and time savings attributable to technology improvements are well-documented.

A detailed comparison was made of project costs for the first line, which was performed using Dipstick & Gage Technology, and the second line, which utilized Data Collector & Display Technology. The interior line was not used in the comparison, because this line was used as the closure line and was impacted by the grouting of the outside lines. For a 1,000-foot length of work in the valley, the estimated construction cost savings attributed to using the system was approximately 10%, and the savings in inspection costs were on the order of approximately 25%. The combined cost savings for this comparative length was about 20%. The estimated savings in construction time was on the order of 25%.

The Patoka Lake Seepage Remediation Project, completed by the Louisville District USACE in 2001, involved grouting a limestone ridge between the left abutment of the dam and the emergency spillway. This project was a milestone project for the USACE and reestablished the USACE as one of the industry leaders in grouting technology. Selection of the Contractor was based on Best Value Selection instead of traditional low bid contracting and the project represented the first time the USACE successfully used balanced, stable grouts and computer monitoring. The work was accomplished by a Contractor & Engineer team who performed the work, furnished and operated the computer monitoring system, and performed onsite analyses of the results. A full-time USACE geologist provided oversight of the program.

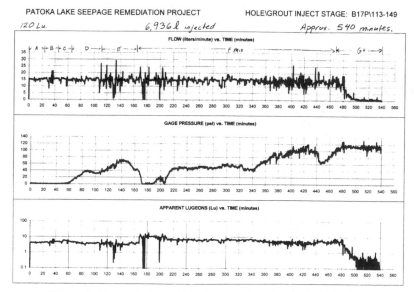

Figure 8: Example Grout Stage Record from Patoka Lake Project

In addition to the clear operational time savings and reduced inspection costs resulting from Level 2 Technology, quality benefits attributed to the system were the ability to confidently use higher pressures and the generation of superior contract records and documentation. The example grout record from the Patoka Lake Project (Figure 8) is typical of a grouting project in karst geology. Early in a grouting stage it is difficult to build pressure and the apparent viscosity of the grout mixture is increased quickly. Once pressure starts to build it is not uncommon to break through a clay seam resulting in loss of pressure. Having this information on a real-time basis and available for review is important in understanding the formation being grouted. Upon completion of the project, permeabilities in the grouted area were reduced by three orders of magnitude. Based on the results of the Patoka Lake Project, the USACE concluded that use of balanced stable grouts and computer monitoring provides a high degree of confidence that grouting can be technically effective and that grouting can be a cost effective alternative to concrete cut-off wall methods that have often been used in limestone (Flaherty 2001).

Grouting at Hunting Run Dam in Spotsylvania County, Virginia was completed in 2001. Hunting Run Dam is a water supply dam that creates an off stream storage reservoir. The grout curtain was a quantitatively designed curtain with a defined performance criterion of less than 5 Lugeons. Achieving the seepage reduction provided by a 5 Lugeon curtain was required due to the high cost of water in a pump-storage reservoir. The basic grout curtain was a single-line curtain 1,100-feet in length constructed to depths of up to 125 feet. The design included provisions to add additional curtain lines of variable depth in reaches wherever geologic conditions were found to require additional grouting to achieve the defined performance. The construction value of the grout curtain was approximately $1.1 million.

The work was accomplished using the IntelliGrout System, which was its first full-scale field trial. The ability to visually display the water testing results by hole series combined with the ability to query the database for all Tertiary hole stages with a water Lugeon value greater than 5 before grouting permitted a rapid assessment of the performance as the grouting proceeded. Locations requiring additional quaternary holes for confirmation or additional grout lines to achieve the required performance were easily identified. The analytical capabilities and analysis display features of IntelliGrout were found to be of tremendous value in locating and isolating specific geologic features requiring additional treatment and allowed highly effective program modifications to target these features to be developed in a time frame compatible with the rapid construction pace. Figures 4 and 5 show the water testing results from Hunting Run Dam in both 2-D and 3-D. The high permeability zone starting to the right of the conduit and dipping to the left and the high permeability weathered zone near the center of the valley are easily identified from either view. The weathered zone near the center of the valley was known to exist from the design subsurface investigation. However, the high permeability feature under the conduit, which followed a weathered intrusive dike, was unknown during design and the design curtain depth was short of the depth required to cut off this permeable zone. The systems operators were able to rapidly identify this zone and deepen planned holes or add additional holes to ensure that the project performance requirements were achieved. On a conventional wall chart, it is highly likely that this feature would have gone unnoticed and an area of concentrated residual leakage after construction could have easily been the result.

The IntelliGrout System also provided substantial value to the Owner. This included the economic advantages of the reduced inspection force and reduced time for peer reviews of the grouting results, as well as the owner being able to visualize and clearly understand the geologic conditions and grouting results that were the basis for program changes being made to accommodate those conditions. Output from the system provided understandable information on assurance of effective, quality construction.

References:

Demming, M. et al. (1985). "Computer Applications in Grouting." Issues in Dam Grouting. ASCE, Denver, 123-131.

Flaherty, T. "New Methods Revolutionize Grouting, Aug 2001 Engineer Update.

Houlsby, A. C. (1990). "Construction and Design of Cement Grouting: A Guide to Grouting in Rock Foundations". John Wiley & Sons, Inc., New York.

Jeffries M. G. et al. (1982). "Electronic Monitoring of Grouting." Grouting in Geotechnical Engineering. ASCE, New Orleans, 769-780.

Mueller, R. E. (1982). "Multiple Hole Grouting Method." Grouting in Geotechnical Engineering. ASCE, New Orleans, 792-808.

Naudts, A.A. (1995). "Grouting to Improve Foundation Soil", Section 5B of Practical Foundation Engineering Handbook. McGraw-Hill, New York.
Weaver K. D. (1991). "Dam Foundation Grouting." American Society of Civil Engineers. New York.

Wilson, D.B., and Dreese, T.L. (1998) "Grouting Technologies for Dam Foundations", 1998 ASDSO Annual Conference, Las Vegas.

Wilson, D.B., and Dreese, T.L. (2002) "Advances in Computer Monitoring and Analysis for Grouting of Dams". United States Society on Dams 2002 Annual Conference Proceedings. San Diego.

Numerical Simulation of Chemical Grouting in Heterogeneous Porous Media

Tirupati Bolisetti[1]* Life Member ASCE-India Section and Stanley Reitsma[1]

Abstract

A mathematical model to simulate chemical grout injection, grout curtain formation in aiding rational design of chemical grout systems in saturated porous media is proposed. Three-dimensional modular groundwater flow simulation model (MODFLOW) and three-dimensional multi-species reactive transport (RT3D) model are combined and modules for the gelling process are incorporated to simulate the grouting process. The paper investigates the influence of varying degrees of soil heterogeneity and layering on grout barrier formation through numerically generated hydraulic conductivity fields. Layer persistence and range of conductivity can be specified during generation of conductivity fields. Grout barrier performance is assessed by simulation of grout injection in a three-dimensional domain followed by determination of post-grouted conductivity field and calculation of overall grout curtain hydraulic conductivity using a flow model. The simulation results show that spatial variability control the effectiveness of grout curtain performance. About 80-90% of the 25 hydraulic conductivity fields with high variability failed to reach the desired effective hydraulic conductivity.

Introduction

Chemical grouting is the process of injecting a chemically reactive solution that behaves like a fluid and reacts after a predetermined time to form a solid, semisolid, or gel. The solid or gel reduces the hydraulic conductivity and contains the flow through the pores. It is self-evident that the grouted barriers should be continuous and should not have any preferential flow channels. Currently, no placement technology can guarantee the completeness of the engineered barrier (Sullivan *et al.* 1998). Pearlman (1999) raised two concerns with regard to the subsurface barriers: uncertainty about hydraulic conductivity of the installed barrier and barrier continuity. Heterogeneity is the main reason for discontinuous and non-uniform formation of the barrier. Hence there is a strong need to study the grouting process and grout curtain formation in heterogeneous porous media.

The main objectives of the present study are: (i) to develop a numerical model to simulate the chemical grouting process in soil media, and (ii) to numerically investigate the performance of grouted barriers in varying degrees of soil layering resulting from heterogeneity. In order to investigate the effect of soil variability,

---

[1] Civil and Environmental Engineering, University of Windsor, Windsor, Ontario, N9B 3P4, Canada
* On study leave from Regional Research Laboratory(CSIR), Bhopal, India

random hydraulic conductivity fields are generated. The varying degrees of soil layering are statistically described in terms of correlation length and variance of hydraulic conductivity. The performance of grout barrier is assessed in terms of the post-grouted hydraulic conductivity field.

Grouted Barriers: Review

Modeling of chemical grouting has triggered the interest of a few researchers in the recent years. Noll *et al.* (1992) modeled the gel barrier formation by injecting CS solution into saturated sandbox and compared with their experimental results. They used MODFLOW (McDonald and Harbaugh 1996) and numerical transport model MT3D (Zheng 1991). They did not specify the details of the gelation model used. Finsterle *et al.* (1997) presented modeling strategy of grouting and proposed equations for gelling. Their model is based on TOUGH2 (Pruess 1991). Moridis *et al.* (1999) studied the problem of grouting and compared standard engineering design with optimization design for their performance in arresting contaminant release. More recently Kim and Corapcioglu (2002) presented a two-phase approach to model the CS solution injection in unsaturated soil. Finsterle *et al.* (1997) measured increase in viscosity with time and fitted an equation to the temporal viscosity data to develop gelation model. The same model is used in this study. Tachavises and Benson (1997) numerically studied the effect of defects of predefined sizes in cutoff walls and reported that a very small but porous defect can cause flow rates three orders of higher magnitude. Sullivan *et al.* (1998) monitored a soil/cement barrier emplaced around a buried drum at Hanford geotechnical test facility using perfluorocarbon tracer (PFT). They analyzed the tracer data through numerical modeling to determine the integrity of the barrier. They observed small scale breaches of few centimeters size. As observed in the last two investigations, there is uncertainty about the barrier continuity and small breaches could affect the performance of the barrier. Monte-Carlo analysis can provide some insight into the problem and the uncertainty. This approach has been used in the analysis of seepage beneath retaining structures (Griffiths and Fenton 1993). In Monte-Carlo simulation approach, several independent equi-probable numerically simulated hydraulic conductivity fields (random conductivity fields), called as realizations, are generated and the grouting problem for each of these individual realizations is simulated.

Mathematical Model

Chemical grouting of colloidal silica (CS) solution into porous media is simulated by coupling MODFLOW and RT3D (Clement 1997) and adding modules for gelling and grout age. MODFLOW is a three-dimensional groundwater flow simulator used to determine the flow field and RT3D is a modular computer code for Reactive multi-species Transport in 3-Dimensional groundwater aquifers is used to simulate the CS injection process. Gel viscosity is indirectly incorporated in MODFLOW by changing the effective hydraulic conductivity in each cell based on gel age and concentration. A schematic diagram of the grout model and data processing is presented in a flow chart (Figure 1). MODFLOW is called from RT3D periodically (in this case it is called once every 270 seconds) to update the flow field using the changing K-field. Using velocity fields generated by MODFLOW, injection of grout is simulated using RT3D. The velocity field is then updated using MODFLOW and transport computation is repeated. When the grout is injected, the viscosity changes with concentration of grout and its age. Two reaction modules are added. One module (GROUT_AGE) keeps track of grout age. The other (GROUT_GEL) calculates grout viscosity and modifies the hydraulic conductivity. Grout age is the time since the grout reaction initiated. After the completion of grouting simulation, the performance of the barrier is assessed by testing a MODFLOW problem of the same domain with different boundary conditions to obtain effective hydraulic conductivity $K_{effective}$ of the grout curtain.

Monte-Carlo simulations, with statistically equivalent heterogeneous hydraulic conductivity (K) fields, called realizations, are conducted to assess the reliability of the grout performance. Log-normal K-fields with different correlation lengths and variability are generated using sequential gaussian simulation (SGSim) software (Deutsh and Journel, 1998). The process of grout simulation and determination of effective hydraulic conductivity is continued for each of the realizations of conductivity fields. The details on governing equations and solution techniques are presented in MODFLOW (McDonald and Harbaugh 1996) and RT3D (Clement 1997; Clement *et al.* 1998) manuals. The following paragraphs, however, present the modules developed to simulate grout age and gelling as a function of time since grout is injected and concentration of grout.

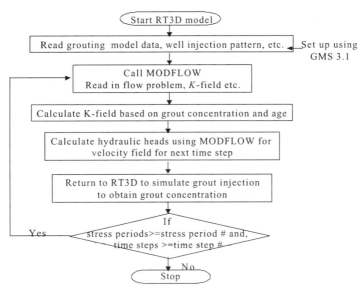

Figure 1. Flow chart of grouting model process

Gelling Module (GROUT_GEL) Gel time is the interval between initial mixing of the grout components and formation of the gel. Gel time can be adjusted by varying the composition of grout. As the grout gels, viscosity increases. For some grouts, like silicates, the viscosity gradually changes, whereas for acrylamides the viscosity remains fairly constant until the gel time is reached and then increases rapidly. Typical gel time curves may be found in Karol (1990) and Finsterle *et al* (1997).

The viscosity increase with time and concentration for CS solution is described by gel time curve (Finsterle *et al*. 1997, Moridis *et al*. 1999). The gel viscosity (Finsterle *et al*. 1997, Moridis *et al*. 1999) is given as

$$\mu_{gel} = \gamma_1 + \gamma_2 \exp(\gamma_3 t) \qquad (1)$$

where μ_{gel} is gel viscosity; t is the time (s) since the reaction started, assumed to be equal to the time since it is injected and, γ_1, γ_2 and γ_3 are fitting parameters. The parameters, γ_1, γ_2 and γ_3, of 2.0, 1.0 and 0.002197 respectively are used here and are representative of a colloidal silica gel with gel time of 45 minutes. Finsterle *et al.* (1997) proposed the above gel viscosity equation obtained based on their laboratory measurements of increase in viscosity with time and fitted the above form of equation to describe gelation. It is worth mentioning that the gelation model (Eq. 1) could also be applied to sodium silicate or acrylamide grouting as well. However, the values for parameters γ_1, γ_2 and γ_3 need to be determined in the laboratory. Due to its modular structure of the grout model, it is very easy to incorporate a different gelation model for a different chemical grout solution, if necessary. During injection grout mixes with water. The mixing process reduces the viscosity of the solution and may be described by linear and power law mixing rules (Finsterle *et al*. 1997, Moridis *et al*. 1999). The linear mixing rule is given by

$$\mu_l = X_l^{gel} \mu_{gel} + (1 - X_l^{gel}) \mu_w \qquad (2)$$

where μ_l is the mixture viscosity; μ_w is water viscosity; and X_l^{gel} is gel concentration. In this study, it is assumed that linear mixing rule is appropriate. Based on laboratory experiments of the colloidal silica composition appropriate mixing rule may be determined. In the present study the Eq. 2 is modified to prevent gelling below a certain minimum concentration level and is given as

$$\mu_l = \frac{\mu_{gel}(X_l^{gel} - X_{min})}{(1 - X_{min})} + (1 - X_l^{gel} + X_{min}) \mu_w \qquad (3)$$

where X_{min} is minimum concentration below which gelling does not occur. We use X_{min} equal to 5%. Noll *et al*. (1992) investigated the effect of diluted CS solution as low concentration as 5% and reported that the solution gels, though the resulting gel was weaker. Hence we adopted 5% as the minimum concentration to initiate gelling. Minimum gelling concentration of 5% is meant to say that gelling reaction does not get initiated and viscosity does not change below such low concentration. However, this needs to be determined for the grout composition being used. Having calculated the grout viscosity using Eq. 3 in each grid, the effective hydraulic conductivity, $K_{grouted}$, is calculated as

$$K_{grouted} = \frac{K_{original}}{\mu_l} \tag{4}$$

Grout Age (GROUT_AGE) Grout viscosity increases with time as per Eq.1. Hence the age of the grout (CS) solution needs to be tracked. This is implemented as a second reaction module in RT3D. Conceptually the grout age can be described as

$$\frac{\partial(grout \quad age)}{\partial t} = 1 \tag{5}$$

However there is a problem of premature gelling based on this equation. Initially, soil is considered as saturated and hence filled with water at time 0.0. For example, if we inject grout at some point 2.0 hrs after we started injecting at the first point, grout injected at the current point has an age of 0.0. The water that is existing from the beginning has an age of 2.0 hrs. The model based on Eq. 5 considers the age of the grout and water mixture at the current injection point as 2.0 hrs rather than 0.0 hrs. This results in premature gelling. In order to circumvent this problem, we incorporated the concept of minimum concentration, considered as 5% in this case, to initiate reaction. To incorporate minimum grout concentration, the above equation for the age of gel is modified as

$$\frac{\partial(grout \quad age)}{\partial t} = \exp\left[\frac{-10^{-9}}{\left(\max\left(\left(X_l^{gel} - X_{min}*0.9\right), \quad 10^{-4}\right)\right)^3}\right] \tag{6}$$

Thus this equation allows increase in gel age only when the concentration is higher than X_{min}.

Numerical Case Study

A three-dimensional test problem having dimensions of 420 cm x 240 cm x 100 cm is chosen as a representative unit of the large grout barrier. The domain is discretized into 5 cm grid blocks, resulting in 84 x 42 x 20 grid blocks. The domain used for the present modeling study is shown in Figure 2. The top and bottom boundaries are considered to be no flow boundaries. Left and right boundaries are assigned as constant head boundaries with a constant head of 100 cm. Front (face ABCD) and backsides (face EFGH) are considered as extending further in the same way, and thus considered as symmetric boundaries.

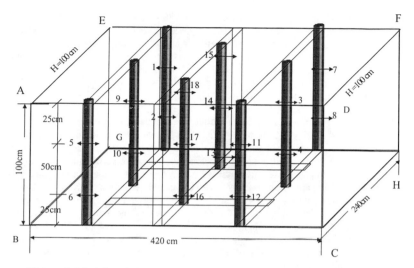

Figure 2: Modeling domain showing grout injection wells and their sequence

Figure 2 also shows the grout injection wells. A staggered injection pattern as shown in Figure 2 is chosen. The injection wells are shown in dark columns and arrows indicate the points of injection. The numbers near the injection points indicate the sequence of injection. The grout is injected through two points at each well i.e., at 25 cm from top and 25 cm from bottom, leaving injection point separation of 50 cm. The injection wells are implemented as a general head boundary, under a pressure head of 900 cm. Colloidal silica injected is assumed to have concentration of 30 %, a typical value for material as delivered. Typically diameter of grout injection wells for Tube-a-Manchette system is about 2 inches. Each injection well in this study is assumed to have an effective area of 2x2 grids which is 10 cm x 10 cm (16 sq. inches). Area of injection is reduced to half for the wells on the front (face ABCD) and back (face EFGH) boundaries to maintain symmetry. Individual wells are injected for 45 minutes at each injection point following the respective time sequence shown in Figure 2. Maximum grout mass that can be injected at each point is calculated based on the grout bulb volume of ellipsoids. For example, injection points of the middle well on the outer wall, such as 3,4,9 and 10, allowable grout mass is calculated as the grout mass that would fill in the pore volume of spherical or ellipsoidal grout bulb. For the outer (symmetric) injection points such as 1,2,5,6,... grout mass is calculated as half that of the internal grout bulbs. Hydraulic conductivity of the grout barrier is determined at end of the grout simulation using MODFLOW. In this study total simulation time is 97200 seconds. CS solution is injected for is 2700 seconds at each point. After injecting for 48600 seconds in all 18 points, the model is run for another 48600 seconds without any injection to allow grout to gel. Flow field is updated once every 270 seconds. Fully implicit upstream finite-difference scheme option of RT3D is used to solve the governing equations. MODFLOW run needed more iterations for pre-conditioned conjugate gradient scheme. Head residual and flux residual convergence criteria are set to 10^{-8} cm/s and 10^{-8} cm^3/s. Longitudinal dispersivity of 1 cm is used and the horizontal and vertical transverse dispersivities are set to $1/10^{th}$ and $1/100^{th}$ of longitudinal dispersivity. The porosity of the medium is assumed to be 0.3.

Performance Assessment A separate model is set up using the post-grouted K-fields to obtain flux through the grout walls for each of 25 realizations. The model has same domain dimensions and grid pattern but without injection wells. The constant head values on the left and right boundaries are set to 520 cm and 100 cm, respectively. During the performance assessment run we obtain flux through the barrier. Since it is difficult to compare flux values, we calculated effective barrier hydraulic conductivity. The entire head is assumed have been lost within the barrier. In order to calculate the gradient, the barrier thickness is used as the length of travel. In a heterogeneous medium determination of wall thickness, required for calculating the gradient, is difficult. Hence we calculated effective hydraulic conductivity based on an arbitrarily assumed thickness of 1.2 m which is the center to center distance of outer grout injection wells.

Monte-Carlo simulations Effect of heterogeneity is investigated using two parameters: (i) correlation scale describing extent of layering and, (ii) variability describing the range of hydraulic conductivity values. Average distance over which the spatial variation in soil properties is correlated may be called the correlation scale. Correlation refers to the layering pattern of soil formation. High correlation scale corresponds to long stratified zones and on the other hand lower correlation scale refers to lenses of soil formation stretched over smaller lengths. A correlation length of zero corresponds to completely random. Hydraulic conductivity (or any other soil property) at any point is not related to that of any other point. The correlation scale may be estimated from the correlogram or semi- variogram, which in turn may be estimated from the field data. The details on estimation of correlation scale may be found in Russo and Jury (1987) and, Woodbury and Sudicky (1991). Log-normal hydraulic conductivity scenarios are generated for a given correlation scale and variability. Twenty-five K-fields for each category are generated using SGSim program for Monte-Carlo simulations. Mean hydraulic conductivity of 10^{-3} cm/s, which corresponds to medium to fine sand, is adopted for the generated K-fields.

Effect of correlation length In the present investigation, effect of correlation length on grout performance is investigated on the 84 x 48 x 20 grid domain. For this study, conductivity fields are generated using two different correlation lengths (λ), one with λ of 1.0 m, and the other with a λ of 50.0 m. Correlation length of 1.0 m corresponds to small lenses of each category of soil spread over stretches of approximately 1.0 m. Similarly correlation length of 50.0 m corresponds to large soil feature extended over a distance of 50 m, which means total stratification in the size of the model studied. In order to study the effect of correlation length only and minimize other effects, conductivity fields with a very low variability of 0.1 are generated. Thus the simulated soil hydraulic conductivities correspond to medium sand. Statistics of $K_{effective}$ from grouted K-fields over 25 realizations are presented in Table 1. These statistical properties are the average statistics over all possible realizations used in the study. These numbers indicate the spread of performance.

Effective conductivity of the grouted barrier is considered as a metric of performance. The statistics presented in Table 1 compare the $K_{effective}$ of the grouted barriers obtained for homogenous and heterogeneous K-fields. In the first set of results, although mean $K_{effective}$ seems to be different for both the cases, the median and minimum $K_{effective}$ remains within 10%. The difference in mean is due to ineffective grouting or presence of preferential channels in the three K-field scenarios out of the 25 realizations. Thus for a given variability, $K_{effective}$ may be considered as same order for both correlation lengths. The post-grouted K-fields are classified into three categories based on grouted wall conductivity. Table 2 presents the number of realizations (K-fields) in each of the $K_{effective}$ range. For the low correlation and low variability case, all the K-fields grouted well and resulted in the lowest $K_{effective}$. For the high correlation length case, $K_{effective}$ values for three out of 25 K-fields are greater than 1.0×10^{-7} cm/s. Based on limited numerical experiments, when the variability is low, correlation length does not seem to have significant effect. The effect of change in correlation length at higher variabilities needs to be investigated.

Table 1: Statistics of $K_{effective}$ obtained over 25 realizations of K-fields

	Homogeneous	Standard Deviation (σ) of K : 0.1	
	$\sigma = 0.0$	$\lambda = 1$ m	$\lambda = 50$ m
Mean	2.823E-09	2.852E-09	2.158E-07
S. D.	-	5.288E-10	5.948E-07
Minimum	-	2.091E-09	1.908E-09
Maximum	-	3.956E-09	2.113E-06
Median	-	2.979E-09	2.798E-09

S.D. Standard Deviation

Table 2: Number of realizations in each $K_{effective}$ range

Wall conductivity cm/s	Homogeneous	Standard Deviation of K : 0.1	
	$\sigma = 0.0$	$\lambda = 1$ m	$\lambda = 50$ m
$<10^{-8}$	1	23	21
$10^{-8} < K_{effective} < 10^{-7}$	-	0	1
$>1. \times 10^{-7}$	-	0	3

Effect of variability The second parameter that is used to represent the heterogenity in soil properties is variability. In order to study the effect of range of hydraulic conductivity and degrees of layering, conductivity fields with different variability i.e., 0.1, 0.5, and 1.0 are generated. These three values represent: (i) very low range of conductivity values, (ii) medium variability corresponding to average layering effect, and (iii) very high variability media corresponding to strong layering effect and wide range of conductivity values (such as mixture of sands and silts or alternate layers of sands and silts) respectively. These conductivity fields are generated for a correlation length of 1.0 m. A smaller correlation length is selected here in order to investigate the effect of variability only. Monte-Carlo simulations using 25 realizations for each of the variabilities are carried out to assess the uncertainty in grout wall performance. Figure 3 presents the distribution of grout solution concentration under different soil conditions. Figure 3a corresponds to homogenous K field and, Figure 3b and 3c corresponds to heterogeneous K-fields with a variability of 0.5 and 1.0 respectively.

Table 3 presents the statistics of $K_{effective}$ obtained over 25 realizations of hydraulic conductivity fields with different variability, for a given correlation length. Effect of variability seems to be more severe than that of correlation length at lower variabilities. In this case both mean and median of $K_{effective}$ increased by several orders with increase in variability. For the extreme case of variability 1.0, $K_{effective}$ values obtained for all 25 realizations are greater than 1.0×10^{-7} cm/s which means these K-fields did not form the barrier properly. The same phenomena may also be observed in Figure 3, where the grout bulbs formed in the homogeneous K-field are near spheres as expected. However for the other two cases with variability of 0.5 and 1.0, the bulbs are of irregular shaped. The increase in $K_{effective}$ indicate the presence of holes causing preferential flow in barrier. Hence the design parameters such as injection well spacing, gel time may be changed to obtain the desired $K_{effective}$. Table 4 presents the number of realizations falling in each category of $K_{effective}$. For the case of low variability almost 90% of realizations resulted in target $K_{effective}$, where as, 80-90 % of hydraulic conductivity fields with higher variabilties did not reach the target $K_{effective}$.

Table 3: Statistics of $K_{effective}$ obtained over 25 realizations of K-fields

	Homoge- neous	Correlation Length (λ) of K: 1.0 m		
	$\sigma = 0.0$	$\sigma = 0.1$	$\sigma = 0.5$	$\sigma = 1.0$
Mean	2.823E-09	2.852E-09	5.098E-06	1.341E-05
S. D.	-	5.288E-10	4.627E-06	8.876E-06
Minimum	-	2.091E-09	4.170E-09	2.679E-06
Maximum	-	3.956E-09	1.776E-05	3.551E-05
Median	-	2.979E-09	3.444E-06	1.167E-05

S.D. Standard Deviation

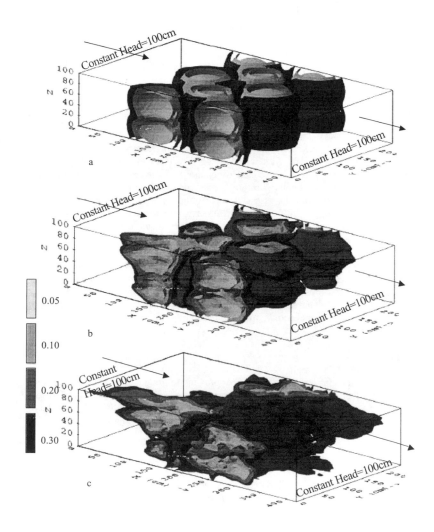

Figure 3: Grout concentration distribution in different K-fields. (a) homogeneous; (b) correlation length of 1 m and variability=0.5 and (c) correlation length of 1 m and variability=1.0. Concentration is in % by volume.

Table 4: Number of realizations in each $K_{effective}$ range

Wall conductivity cm/s	Homo-geneous	Correlation Length of K: 1.0 m		
	$\sigma = 0.0$	$\sigma = 0.1$	$\sigma = 0.5$	$\sigma = 1.0$
$<10^{-8}$	1	23	2	0
$10^{-8} <K_{effective}<10^{-7}$	-	1	2	0
$>1.x10^{-7}$	-	1	21	25

Conclusions

A modeling approach for simulating chemical grouting, developed by combining public domain codes like, RT3D and MODFLOW, is presented. Reaction modules, GROUT_AGE and GROUT_GEL, are developed to track the age of the grout and determine the reduced hydraulic conductivity due to increase in gel viscosity. The present study demonstrates the possibility of modeling the grout curtain formation and presents methodology to assess the performance of the curtain. The model provides insights into the influence of different parameters such as soil properties, grout properties on grout curtain formation and performance. The model may help in minimizing some of the expensive field tests and may be used as a rational tool in designing chemical grout systems. Conductivity fields generated using different correlation lengths, given low variability of 0.1, yielded same order of $K_{effective}$. However, for a given correlation length, increasing variability resulted in increased $K_{effective}$ values and 80-90% of the conductivity fields did not met the desire requirement of $1x10^{-7}$ cm/s. Hence the grout system has to be designed based on heterogeneity. These conclusions from limited number of realizations may be used with caution.

Acknowledgements

The funding for this research has been provided through National Science and Engineering Research Council grant, Canada Research Chair Program, and Ontario Graduate Scholarship for Science and Technology and is gratefully acknowledged. The first author thanks the Director, RRL and CSIR for sanctioning study leave.

References

Clement, T.P. (1997). *RT3D, A modular computer code for simulating Reactive multi-species Transport in 3-Dimensional groundwater aquifers*, Pacific Northwest National Laboratory, Richland, WA, USA.PNNL-11720.

Deutsh, C. and Journel, A. (1998) *GSLIB: Geostatistical software library and users guide*, 2nd Edition, Oxford Press, NY.

Finsterle, S., Oldenburg, C.M., James, A.L., Pruess, K. and Moridis, G.J. (1997). "Mathematical modeling of permeation grouting and subsurface barrier performance", *Intl Containment Tech. Conf. and Exhibition*, Feb 9-12, 1997, pp.438.

Griffiths, D.V. and Fenton, G.A. (1993) "Seepage beneath water retaining structures founded on spatially random soil," *Geotechnique* 43(4), 577-587.

Karol, R.H. (1990). *Chemical grouting*, Marcel and Dekker Inc., NY, 1990.

Kim, M and Corapcioglou, M.Y. (2002) "Gel barrier formation in unsaturated porous media," *J of Contaminant Hydrol.*, 56(1-2), 75-98.

McDonald M.D. and Harbaugh, A.W. (1996) *A modular three-dimensional finite difference flow model, Techniques in Water Resources Investigations of the U.S. Geological Survey*, Book 6., 586.

Moridis, G.J., Finsterle, S. and Heiser, J. (1999). "Evaluation of alternate designs for an injectable subsurface barrier at the Brookhaven National Laboratory site. Long Island, New York," *Water Resour. Res.*, 35(10), 2937-2953.

Noll, M.R., Bartlett C., Dochat T.M. (1992) "In situ permeability reduction and chemical fixation using colloidal silica", *Presented at the 6th Natl. Outdoor Action Conference on Aquifer Restoration*, Las Vegas, NV, May 11–13.

Pearlman, L. (1999). *Subsurface containment and monitoring systems: Barriers and beyond*, OSWER, Technology Innovation of Office, USEPA, Washington, DC. http://www.epa.gov/tio/download/remed/pearlman.pdf > April 1, 2002.

Pruess, K., (1991) TOUGH2 - *A general purpose numerical simulator for multiphase fluid and heat flow*, Rep LBL-29400 Lawrence Berkeley Natl. Lab., Berkeley, Calif.

Sullivan,T.M., Heiser, J., Gard, A. and Senum, G., (1998) "Monitoring subsurface barrier integrity using perfluorocarbon tracer," *J. of Environ. Engg.*, 124(6), 490-497.

Russo, D. and Jury W.A. (1987) "A theoretical study of the estimation of the correlation scale in spatially variable fields 1. Stationary fields," *Water Resour. Res.*, 23(7), 1257-1268.

Tachavises, C. and Benson, C.H. (1997) "Hydraulic importance of defects in vertical groundwater cut-off walls," *In Situ Remediation of the Geoenvironment, ASCE Geotechnical Special Publication no. 71*, Reston, VA (1997) 168-180.

Woodbury, A.D. and Sudicky, E.A. (1991) "The geostatiscal characteristics of Borden aquifer," *Water Resour. Res.*, 27(4), 533-546.

Zheng, C (1991) *MT3D User's Manual.* S.S.Papadapulos and Assoc., Rockville, MD.

Some Aspects on Grout Time Modelling

Thomas Dalmalm[1]

Håkan Stille[2]

ABSTRACT

Grouting has been used for many years as a sealing method for hard jointed rock and the method has been improved in a number of ways, but still there is a lack of parameters relating the sealing effort to the sealing efficiency. The grouting time is therefore here suggested as a parameter, which could be considered to improve the sealing result. It is also suggested that the grouting time is correlated to the rock mass joint system situation, where a long or short grouting time will for different rock mass joint situations mean different design. Both the grouted volume and the grouting time has an influence on the filling of the joint system in the rock mass. The time needed for sealing should be considered especially when the demand on sealing efficiency is high. The effect of grouting time is presented and discussed in some examples.

INTRODUCTION AND BACKGROUND

Time is a most important parameter in many areas, but time as a grouting parameter is not given much emphasis during the grouting process. As discussed in this article, there could be much information available by using time as a parameter, which could be used to optimise the grouting process.

Grouting as a sealing method for hard jointed rock has been used for many years and the method has been improved in a number of steps. Grouting is here referred to as injection of a cement-based grout into a hard jointed rock mass.

When grouting in hard jointed rock it is not the rock itself that is sealed, but the conductive joint system of the rock mass. The joint system has to and is therefore considered during the design of the grout works.

---

[1] Corresponding author, NCC AB, S-17080 Solna, Sweden, Phone +46 8 790 60 83, dalmalm@kth.se

[2] Professor at the Division of Soil and Rock Mechanic. Royal Institute of Technology, S-100 44 Stockholm, Sweden.

Grout work design is commonly set up by a number of requirements, such as grout proportions-, grout volume- and grout pressure- requirements. Still after many years of development there is a lack between grouting effort and achieved sealing result, and it has happened that grouting continues for days without receiving any response from the joint system of the rock mass. The grouting time is therefore here suggested as a parameter, which should be regarded during the design of grout works. The grouting time is a parameter itself, which in some occasions could be used for optimising the grouting process. To further optimise, it is suggested that the grouting time is correlated to the rock mass joint system situation, which will extend the understanding of the grouting situation.

THE ROCK MASS AND IT'S JOINT SYSTEM

The rock mass that should be sealed consists of joints. The joints may be few or many, filled or unfilled with different types of filling material. The individual joints may also vary in aperture and they may be smooth or rough. All these different aspects and others need to be regarded during grout optimisation or prediction. A full documented protocol of all joints and their parameters is not possible to produce before or after grouting because of the complexity.

The aperture variation of all joints can only be estimated, as for example by hydraulic measuring in combination with adding the number of joints cooperating. Therefore grouting optimisation and prediction is today based on both field- interpretation- and experimental data, which in combination gives a view of the present rock mass joint situation. The set of data can then be updated during progress of grouting with an active design philosophy.

GROUT TIME AND JOINT SITUATION

A long or short grouting time will for different rock mass joint situations mean different decisions. The examples given below are for to give ideas of grout spread and grouting time that could help in the decision-making process for different rock joint situations. There will always be an extra rock joint property, which could alter the proposed actions, so examples given here don't claim to be a complete list and are only suggestions that could be helpful for further work at each location. It is therefore up to the reader/grouter to decide what action is most appropriate for the present situation.

Blocky rock mass with wide joints (joint spacing 0.3 – 1 m)

If the joint situation is described as blocky, with wide (aperture >0.3 mm) unfilled joints (Figure A1), the nearby joint system will at an early stage be filled and further grout time will mainly result in filling of joints far away from the grouting point. As previously shown in Brantberger 1998 and Dalmalm 1999, the sealed zone should be as dense as possible and as close to the tunnel as possible. Therefore the sealing result will for this situation not gain much from long grout times or higher grout pressures, instead other methods as for example more grout points, high viscosity grouts and fast hardening start grout could improve the sealing.

If the rock mass contains both wide and narrow joints, a second grout fan with a grout suitable for penetrating fine joints could improve the sealing substantially. If the rock mass is very sparsely jointed or if the amount of water conductive joints are very few, it could be difficult to reach the water bearing joints. Time for sealing could for this situation be reduced by increasing the number of grout points.

Fractured rock mass with wide joints (joint spacing 0.05 – 0.3 m)

If the rock mass is described as fractured with wide (aperture > 0.3 mm) unfilled joints (Figure A2), there is a better chance to reach the water bearing structures than for the blocky rock mass. Sealing and grouting time will not gain much on an increased number of grout holes, because most of the water bearing structures are probably already connected to the grout holes. If the rock mass is fractured and the joints are narrow (aperture < 0.15 mm), sealing in the near field could gain of an increased grout pressure.

Crushed rock mass, narrow joints or filled joints (joint spacing < 0.05 m)

If the rock mass is described as crushed or disintegrated, with narrow joints (aperture <0.15 mm) or filled joints (Figure A3), the risk of sealing far away from the grouting point is less and sealing may gain on increased grout time even for low viscosity grouts. Sealing in the near zone may as well gain of an increased grout pressure, which increases the spreading speed from the grouting point.

Folded rock mass (joint spacing < 0.05 - 1)

If the rock mass is folded (Figure B1-3), the direction of the layers must be considered. Spreading in a layer may be easy, but spreading between layers could be more difficult. The grout holes should be directed so that the different layers are hit, which then will facilitate grout spread between layers. The folded rock mass could as well be divided into blocky, fractured and crushed rock mass, which not further is developed here.

Joint persistence

The joint persistence (length of joints) is an important parameter for determining the appropriate grout time. It could be difficult to decide the persistence of a joint, but even a rough indication is better than no indication. Long joints could preferably be sealed by a fast hardening grout (blocker), which stops filling of joints far away from grouting point. The fast hardening grout could then be followed by a grout, which is adapted for sealing of fine joints in the near field zone.

Joint direction

Joint direction has a large influence on different sealing strategies and is therefore considered during grout design. The joint direction could probably be used more than what is common today when selecting grout methodology. Joint groups that are directed similarly will not gain much from increased grout time as much as joint group that are intersecting each other.

Nr.		Description	Joint spacing [m]	Joint Aperture [mm]
Figure A1		Blocky rock mass. with wide unfilled joints.	0.3-1	> 0.3
Figure A2		Fractured rock mass. with wide unfilled joints.	0.05-0.3	> 0.3
Figure A3		Crushed. rock mass with narrow apertures.	< 0.05	< 0.15
Figure B 1-3		Folded rock mass. (The folded rock mass could as well be divided into blocky, fractured and crushed as above.)	< 0.05 - 1	< 0.15 - > 0.3

Porosity

The porosity is an expression for the amount of joints and cavities in the rock mass. The porosity in the rock mass could be divided into primary and secondary porosity, were the primary is the "total" porosity and the secondary regards the porosity which is part of the water transport route in the rock mass. The secondary porosity is as well called conductive porosity. The conductive porosity for a rock mass should, according to Gustafson (1986) be in the order of 0.01-1 $^0/_{00}$.

The water transport is for a rock mass directed to the conductive part of the joint plane, which in for example Janson (1998), with measurements from Äspö Hard Rock Laboratory was estimated to between 5 % and 20 %.

The correlation between porosity and conductivity has been studied by for example Brotzen (1990). From tracer tests a correlation was noted as shown in Equation 1.

$$\log p = 0.17 \cdot \log K - 1.7 \pm 0.3 \qquad \text{(Equation 1)}$$

With porosity [p] and conductivity [K]. Some authors have studied the relation between conductive porosity and permeability, which is shown in Figure 4.

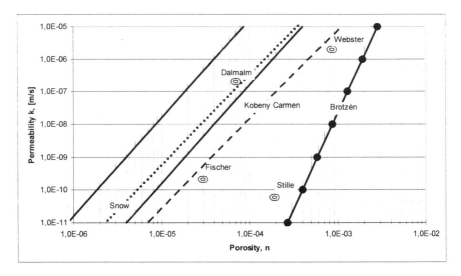

Figure 4   Correlation between conductive porosity and permeability. The area between the two smooth lines is the possible secondary porosity calculated with Equation 8. The line with dots is according to Brotzen. Snow presented the dotted line and Kobeny Carmen the broken line.

As seen in Figure 4 there is a large difference in permeability-porosity relationship. The difference is in the range of 10^{-2}, which could be explained by that, two different conceptual models for calculating the porosity has been used, see Figure 5. The left grey area in Figure 4 has been calculated by assuming a network of plane parallel joints (shown left in Figure 5). The variability (shown right in Figure 5) of the joint aperture is not regarded; instead the joint aperture is set to an average value. The average joint aperture overestimates the penetrability and underestimates the filtration of the grout i.e. the calculated sealing effect is better than in reality. The model may be used for calculating filtration and penetrability of grout, but mentioned effects has to be regarded. The Brotzén and Stille values have been calculated from tracer tests (Lundström & Stille, 1978). The tracer test gives a more correct estimate and a higher conductive porosity than the plane parallel model. The porosity's measured from tracer tests are to prefer if for example the grout volume should be calculated.

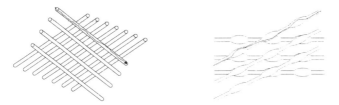

Figure 5   The calculated porosity varies with the assumed conceptual model. Left, a plane parallel model and right a model with variable aperture.

GROUTING TIME AND VOLUME

During the construction of the South Link road tunnels in Sweden extensive full-scale grout trials were carried out (Dalmalm, 2001). The trials were aimed to develop the pre-grouting concept in order to improve the sealing efficiency and to minimise potential post grouting.

The rock mass was tested with for example water pressure tests in drill holes, both parallel and perpendicular to the tunnel, see Figure 6.

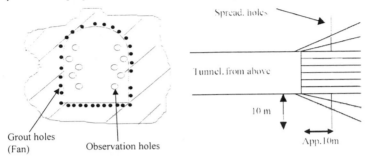

Grout holes
(Fan)   Observation holes

Figure 6   Left: Fan layout with 32 grout holes and 10 observation holes.
             Right: WPT were performed to predict the grouting and to measure the
             sealing efficiency in holes both parallel (grout- and observation holes)
             and perpendicular (spreading holes) to the tunnel.

The main bedrock types, within the trial area, were sediment gneiss and grey – redgrey gneiss-granite, which heritages from about 2000 million years. The dominating minerals are quarts, feldspar and glimmer. The geology within the trial area varies and a general judgement was that the rock mass from a grouting point of view could be described as:

- A low hydraulic conductivity, approximately $1 \cdot 10^{-7}$ m/s (app.1 Lugeon) or lower.

- Single joints, between 0.1-0.5 joints per meter, with varying conductivity.

- Narrow apertures, < 0.5 mm, with varying joint filling.

The measured average Lugeon values were very low, equivalent to a hydraulic conductivity between approximately $3 \cdot 10^{-7}$ and $2 \cdot 10^{-8}$ m/s.

A number of parameters were studied during the trials, as for example grouting time and volume. In Figure 7 typical pressure and flow rate curves are shown that leads to figure 8, with constantly increasing accumulated grout volume during the propagation of grouting. The used grout was based on the cement Cementa Injektering 30 (d_{95}= 30 μm) with an addition of 0.65 % HPM superplastisizer and specification as; Marsh cone time 38 sec., density 1550 kg/m^3, yield value ~0.2 [Pa], viscosity ~0.06 [Pas].

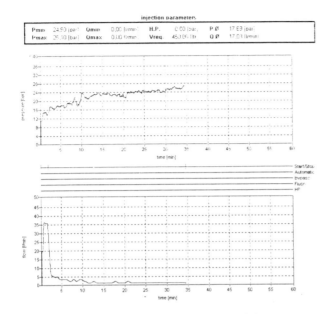

Figure 7   Pressure and flow from one grout hole.

By integrating the flow curve, the time dependency for filling the joint system is shown in Figure 8. From the figure it is noted that the volume increase does not stop by the end of the grouting, instead some kind of stop criteria was used. The stop was not strictly related to rheology and equations that assume a stop from rheology cannot here therefore be used. If the volume of the grout hole is excluded, a trend equation could be approximated, as seen in Figure 9.

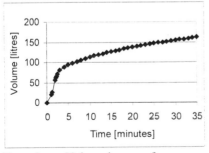

Figure 8   Volume increase from the studied grouted hole.

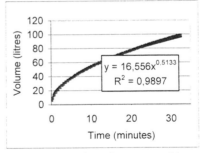

Figure 9   Volume increases, with filling of grout holes excluded.

The grout volume from spreading of grout in joint planes of the rock mass could geometrically be calculated according to Figure 10 and Equation 2. The expression is similar to what was presented in Gustafson & Stille (1996).

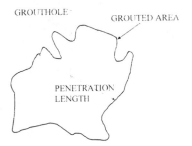

Figure 10   Grout spread in a joint plane.

$$V = I^2 \cdot \pi \cdot n \cdot L \qquad\qquad \text{(Equation 2)}$$

Where: I = penetration length   [m]
          n = porosity   [‰]
          L = effective length of grout hole   [m]

The penetration length could then be expressed as:

$$I = \sqrt{\frac{V}{k}} \qquad\qquad \text{(Equation 3)}$$

Where k is a constant, $k = \pi \cdot n \cdot L$.

If the grout volume is expressed by the relation, based on the regression in Figure 9 ($V = 16.566 \cdot 10^{-3} \cdot t^{0.5133}$ [m³]) the difference in penetration length could after integration be noted as:

$$\frac{dI}{dt} = \frac{3.30 \cdot 10^{-2}}{\sqrt{k}} \cdot t^{-0.74335} \qquad\qquad \text{(Equation 4)}$$

If the purpose is to optimise the grouting process, the increase in penetration length could be compared to the initial increase in penetration (after hole filling) as is shown in Equation 5:

$$\frac{dI/dt}{dI/dt_0(t=1)} = t^{-0.7434} \qquad\qquad \text{(Equation 5)}$$

which is illustrated in Figure 11.

From Figure 11 it could be noted that the relative penetration speed has considerable been decreased after about 15 minutes of grouting. To calculate the absolute penetration the porosity and the grout hole length as well needs to be known.

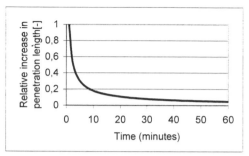

Figure 11   Relative increase in penetration length, dI/dt.

Janson (1998) simulated the time dependency for grout spread in a model tube network. It was shown based on rheology that filling of a grout hole in general could be illustrated as: 90 percent of the volume takes 10 percent of the time and remaining 10 percent of the volume takes 90 percent of the time. In the example shown here, the hole was not grouted until stop, therefore 80 percent of the volume took 10 percent of the time and the remaining 20 percent of the volume took 90 percent of the time.

The grout hole described above belongs to a fan, which consists of 32 holes. If instead grouting was stopped at half time for all grout holes and exchanged by drilling and grouting of 10 new holes, the total time for receiving the same amount of grout into the joint system of the rock mass could be decreased. An example is here given as: Half time grouting means 9 hours and 20 minutes. Drilling of 10 holes takes about 3 hours, grouting of the 10 holes up to the corresponding volume takes about 3 hours, which even includes time for changing equipment. If grouting was stopped at half time and new grout holes were drilled and grouted, more than 3 hours could be saved for this fan. The two methods can result in different sealing efficiency, which as well has to be considered together with the present joint system situation.

If the porosity of the rock mass as well is known, the absolute penetration length increase could be calculated according to Equation 4. For the rock mass at the trial site a hydraulic aperture distribution has been calculated based on water pressure tests, WPT, for all individual grout holes, joint intensity from core drilling and an assumption that about 10 % of the joints are conductive (Janson, 1998). If a drill hole is sealed by a packer and injected with water the transmissivity (T) could be calculated according to Gustafson (1986) as:

$$T = \frac{Q_W \cdot \rho_W \cdot g \cdot P_{DW}}{\Delta P_W \cdot 2 \cdot \pi} \qquad \text{(Equation 6)}$$

Where Q_W is the flow from the WPT [m³/s], ΔP_W the difference between applied water pressure and ground water pressure [Pa], g is the acceleration of gravity [m/s²], ρ_W is the density of water [997 kg/m³] and P_{DW} is a function of the length (L_b) and radius (r_W) of the drill hole, i.e. the non-dimensional expression (Gustafson, 1986):

$$P_{DW} = \ln\left(\frac{L_b}{r_W}\right) \qquad \text{(Equation 7)}$$

If in addition all joint planes along the drill hole are assumed to have the same aperture and laminar flow is assumed an expression for the hydraulic aperture (b_{hyd}) [m] could be noted as (Janson, 1998):

$$b_{hyd} = \sqrt[3]{\frac{12 \cdot T \cdot \mu_w}{\rho_w \cdot g \cdot N_w}}$$ (Equation 8)

Where μ_w is the viscosity of water ($1.3*10^{-3}$ Pas), N_w the number of conductive joints along the drill hole [No]. The following hydraulic joint aperture distribution could then be calculated, as shown in Figure 12.

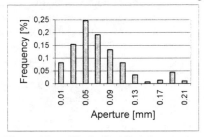

Figure 12   The hydraulic joint aperture distribution for the trial area at South Link.

Based on the distribution, the mean aperture is calculated to 0.07 mm. With an intensity of 2.3 joints per meter and an assumed conductive intensity of 0.23 joints per meter, the conductive porosity could be calculated to 0.015 $^0/_{00}$, which is low but within the range of what was presented in Gustafson (1986), that the conductive porosity should be in the interval of 0.01 – 1 $^0/_{00}$. As a comparison the Brotzén equation (Equation 1), here gives a secondary porosity of 1.09 $^0/_{00}$, which is just above the given interval.

The absolute penetration length increase could with Equation 4 be calculated, with 10 meters hole length, a conductive porosity of 0.015 $^0/_{00}$ as shown in Figure 13.

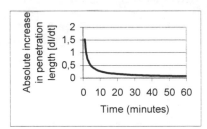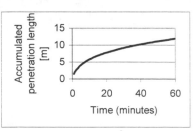

Figure 13   The absolute penetration length increase, with 0.015 $^0/_{00}$ cond. porosity.

As noted from Figure 11 and Figure 13 it is obvious that both the grouted volume and the grouting time has an influence on the filling of the joint system in the rock mass. A reduction of the water inflow to a tunnel with approximately 90 % can be done in short time, but to reach the last filling of the joint system usually takes long time.

If we have a joint system with both wide and fine joints, the wide joints will first be filled followed by the fine joints. High sealing requirements often demand that very fine joint will be sealed. The time needed for sealing should therefore be considered especially when the requirement on sealing efficiency is high.

JOINT SYSTEM AND GROUT PROPERTIES

Different joint systems need to be handled with different grout concepts in order to optimise the grout time for a given sealing effect. It has been shown by numerical calculations that a proper grout not only improves the sealing effect, it affects the grouting time as well (Eriksson, 2002). In Eriksson 2002, a numerical simulation of two cases of rock masses (A and B) was presented. The following is a summary of the presentation by Eriksson 2002.

Two different rock masses denoted A and B described as: Rock mass A is a low permeable rock with a low in situ water head (~20m). The joints in A have a mean aperture between 0.1 and 0.15 mm and appear on average every fourth meter. Rock mass B is a highly permeable rock mass with joints of significant hydraulic aperture, a mean aperture between 0.1 and 0.5 mm and a high in situ water head (~100 m), with in general one conductive joint per meter of tunnel.

For each rock mass a grout was suggested. For rock mass A, a grout (a) was suggested where the penetrability of the grout was chosen as the most important property due to that the identified joints have small apertures (0.10mm-0.15mm). For rock mass B, a grout (b) was suggested where low bleed was chosen as the most important property, because of the large apertures (0.10mm-0.50mm) in combination with the high water pressures. The different rock masses were then described in a numerical model and calculated with the two different grouts, see Table 1.

Table 1   Two different rock masses grouted with different grout concepts

	Sealing effect (%)	Average grouting time (min)
Rock mass A grout (a)	89.1	147
Rock mass A grout (b)	9.9	16
Rock mass B grout (b)	99.4	361
Rock mass B grout (a)	98.4	825

As expected, the proposed proper designs A with (a) and B with (b), gives the best sealing effect. In the rock mass with narrow joints (A), it takes long time to reach a sealing effect, but there will be almost no sealing effect without the long grouting time. For the wider joint system (B), a proper grout mix results in best sealing effect as well as shortest grouting time. The numerical calculation shows the importance of adjusting the grout for the present rock mass joint situation and the importance of observing the time during grouting. A short time was here equal to that the grout was not able to penetrate and resulted in very poor sealing effect [A (b)].

A very long time was here equal to that a lot of grout was used, pumped away to places where it won't improve the sealing [B (a)].

CONCLUSIONS AND DISCUSSION

The grouting time could in different situations be a very valuable parameter to regard and especially if it is regarded in combination with the present rock mass joint system situations. Time could be saved and higher sealing efficiency could be achieved, by observing the grouting time in combination with the rock joint system situation.

Numerical calculations have shown the importance of adjusting the grout for the different joint system situations and the importance of observing the grouting time.

As shown, a model for grout time is very important to be able to predict the grouting process. The model is also important for the understanding of the possibility to seal fine joints, which often is necessary when having high demands on sealing efficiency.

REFERENCES

Brantberger M., Dalmalm T., Eriksson M., Stille H. (1998). "Styrande faktorer för tätheten kring en förinjekterad tunnel." Division of Soil- and Rock Mechanics, Royal Institute of Technology, Stockholm, Sweden

Brotzen, O. (1990). "The study of relevant and essential flowpaths." Symposium on Validation of Geosphere Flow and Transport Models, Stockholm.

Dalmalm, T., et al. (1999). "Factors influencing the sealing effect around a pre-grouted tunnel." SAREC Conference, Johannesburg, South Africa

Dalmalm, T. (2001). "Grouting prediction systems for hard rock – Based on active design." Division of Soil- and Rock Mechanics, Royal Institute of Technology, Stockholm, Sweden

Eriksson, M. (2002). "Prediction of grout spread and sealing effect – A probabilistic approach." Division of Soil- and Rock Mechanics, Royal Institute of Technology, Stockholm, Sweden

Gustafson, G. (1986). "Geohydrologiska Förundersökningar i Berg." BeFo 84:1/86, Stockholm, Sweden.

Gustafson G. & Stille H. (1996). "Prediction of Groutability from Grout Properties and Hydrogeological data." Tunnelling and Underground Space Technology, Vol. 11, No. 3, pp 325-332, Elsevier Science Ltd, Great Britain.

Janson, T. (1998). "Calculation Models for Estimation of Grout Take in Hard Jointed Rock." Department of Soil and Rock Mechanics, Royal Institute of Technology, Stockholm, Sweden.

Lundström L. & Stille, H. (1978). "Large Scale Permeability Test of the Granite in the Stripa Mine and Thermal Conductivity Test." Swedish Nuclear Fuel Supply Co., Stockholm, Sweden.

Model Testing of Passive Site Stabilization: A New Grouting Technique

Patricia M. Gallagher[1] and Alyssa J. Koch[2]

Abstract

Passive site stabilization is a new technology proposed for non-disruptive mitigation of liquefaction risk at developed sites susceptible to liquefaction. It is based on the concept of slow injection of stabilizing materials at the edge of a site and delivery of the stabilizer to the target location using the natural or augmented groundwater flow. In this research, a box model was used to investigate the ability to uniformly deliver colloidal silica stabilizer to loose sands using low-head injection wells. Five injection wells were used to deliver stabilizer in a fairly uniform pattern to the loose sand formation. The results will be used to design centrifuge model tests in which the stabilizer will be delivered in-flight using a robot.

Introduction

Liquefaction is a phenomenon marked by a rapid and dramatic loss of soil strength, which can occur in loose, saturated sand deposits subjected to earthquake motions. Certain types of sand deposits, hydraulic fills, and mine tailings dams are particularly susceptible to liquefaction. The onset of liquefaction is usually sudden and dramatic and can result in large deformations and settlements, floating of buried structures, or loss of foundation support. Lateral spreading is a related phenomenon characterized by lateral movement of intact soil blocks over shallow liquefied deposits. Displacements caused by lateral spreading can range from minor to quite large. Gently sloping areas along waterfronts are most susceptible to lateral spreading. As a consequence, bridges can be damaged significantly due to lateral spreading

At sites susceptible to liquefaction, the simplest way to mitigate the liquefaction risk is to densify the soil. For large, open and undeveloped sites, the easiest and cheapest method of ground improvement is densification by "traditional" methods of ground improvement such as deep dynamic compaction, explosive compaction, or vibrocompaction. At constrained or developed sites, ground improvement by densification may not be possible due to the presence of structures sensitive to deformation or vibration. Additionally, access to the site could be limited and normal site use activities could interfere with mitigation activities. At these sites, the most common methods of ground improvement are underpinning or grouting. In the case of grouting, the typical method is to inject grout under pressure through closely-spaced boreholes. Typical grout materials include cement or various chemicals such as sodium silicate that are formulated with short set or gel times.

---

1 – Assistant Professor, Dept. of Civil & Arch. Engineering, Drexel University, Philadelphia, PA; pmg@drexel.edu
2 – Graduate Research Asst., Dept. of Civil & Arch. Engineering, Drexel University.

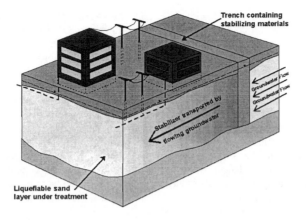

Figure 1. Passive site stabilization for mitigation of liquefaction risk.

Passive site stabilization is a new concept proposed for non-disruptive mitigation of liquefaction risk at developed sites susceptible to liquefaction. It is based on the concept of slow injection of stabilizing materials at the up gradient edge of a site and delivery of the stabilizer to the target location using the groundwater flow (Figure 1). This differs from traditional grouting methods in that low-head injection wells are used in conjunction with the groundwater to deliver the stabilizer to the entire site. Therefore, the stabilizer must have a low initial viscosity, a long induction period in which the viscosity stays fairly low, and long gel times (on the order of 50-100 days). Numerous materials were evaluated as potential stabilizers, including colloidal silica, various chemical grouts and microfine cement grouts (Gallagher and Mitchell 2002). Dilute colloidal silica (5 weight percent) was selected as the stabilizer. Colloidal silica has a low initial viscosity (about 1½ to 2 cP; water = 1 cP), a long induction period during which the viscosity remains fairly low, and long controllable gel times of up to a few months. The material cost of colloidal silica is expected to be about $75 per cubic meter of treated soil, assuming a 5 weight percent solution.

Loose sands treated with colloidal silica stabilizer had significantly higher deformation resistance to cyclic loading than untreated sands (Gallagher and Mitchell 2002). Gallagher and Mitchell (2002) did cyclic triaxial shear tests on Monterey sand samples stabilized with colloidal silica at concentrations between 5 and 20 percent. The cyclic deformation resistance of the treated sand was measured in terms of double amplitude (DA) axial strain, which is the largest difference in strain that develops during an entire cycle of compression and extension. Samples stabilized with concentrations of 20 weight percent colloidal silica showed less than one-half percent DA axial strain in 1000 cycles of loading when tested at a cyclic stress ratio (CSR) of 0.4. The CSR is defined as the ratio of the maximum cyclic shear stress to the initial effective confining stress. Sands stabilized with lower concentrations tolerated cyclic loading well, but experienced more DA axial strain. Samples treated with 5 percent colloidal silica experienced up to eleven percent strain after 100 cycles at cyclic stress ratios between 0.15 and 0.29, but remained intact during and after loading. For comparison, a magnitude 7.5 earthquake would be expected to generate about 15 uniform stress cycles (Seed and Idriss 1982). Thus, treatment with colloidal silica stabilizer significantly increased the deformation resistance of loose sand to cyclic loading.

Centrifuge modeling has been used to investigate the effect of colloidal silica treatment on the liquefaction and deformation resistance of loose, liquefiable sands during centrifuge in-flight shaking (Gallagher et al. 2002). Loose Nevada No. 120 sand was saturated with 6 weight percent colloidal silica stabilizer and subsequently subjected to two shaking events that simulated earthquake motions with uniform peak accelerations of 0.2g. The treated sand layer did not liquefy during either shaking event. In addition, significantly lower levels of strains (1/2 to 1 percent) were measured for the treated centrifuge models compared to the strains (3 to 6 percent) recorded in similar centrifuge tests done on untreated soil models.

The primary feasibility issue remaining is uniform delivery of the stabilizer to the target location. Preliminary groundwater and stabilizer transport modeling studies showed that passive site remediation could be feasible at sites with a hydraulic conductivity of 0.05 cm/s and hydraulic gradient of 0.005 and above, using colloidal silica as the stabilizer (Gallagher, 2000). The preliminary study assumed a 200-foot by 200-foot treatment area, a stabilizer travel time of 100 days and a single line of injection wells. Extraction wells would decrease the travel time required and control the down gradient extent of stabilizer delivery.

For the present study, a small box model (30.5 by 76 cm in plan and 26.5 cm high) is being used to evaluate possible delivery techniques to obtain adequate coverage of colloidal silica stabilizer, including low-head injection wells and extraction wells. This paper includes results of two box model tests (Models 6 and 9) in which loose sand was treated with colloidal silica stabilizer via low-head injection wells. The results of these and subsequent box model tests will be used to design additional centrifuge tests that will model the injection process under high-gravity levels.

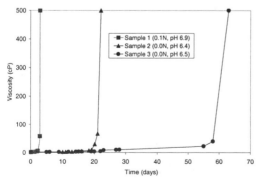

Figure 2. Typical colloidal silica gel time curves for various ionic strengths and pH values.

Colloidal Silica

Colloidal silica is an aqueous dispersion of microscopic silica particles (7-22 nm) produced from saturated solutions of silicic acid. When diluted to 5 weight percent, colloidal silica solutions have a density and viscosity similar to water (about 1½ to 2 centipoise (cP); water = 1 cP). Colloidal silica solutions can have long induction periods during which the viscosity remains fairly low and long controllable gel times of up to a few months. Colloidal silica is also nontoxic, biologically and chemically inert, and has excellent durability characteristics, making it an excellent candidate for a stabilizer. Grace Davison of Columbia, Maryland supplied the colloidal silica used for this research. It is supplied as a 30 weight percent silica solution with a viscosity of 5.5 cP and pH of 10.

The factors controlling the gel time of colloidal silica include the silica solids content, the pH, and the ionic strength of the diluted colloidal silica solution. For a given silica content, the gel time can be altered by lowering the pH and changing the ionic strength of the dilute colloidal silica solution. Example gel time curves are shown in Figure 2 for 5 weight percent solutions at different ionic strengths and pH values.

For this study, the travel time is defined as the time between mixing and a tenfold increase in the initial viscosity. For example the travel times of Samples 1 and 2 are about 2.2 and 19 days, respectively, as shown in Figure 2. The gel time is the time from the end of mixing until the colloidal silica forms a firm, resonating gel. It is approximately 4 times the travel time. The curing time is the time between the formation of a firm, resonating gel and strength testing.

Testing

Box Model Preparation

This paper presents in detail the results of two box model tests in which loose Nevada No. 120 sand ($D_r = 40\%$) was treated with colloidal silica stabilizer at a concentration of 5 weight percent. A picture of the 3-dimensional box model in presented in Figure 3. The model is constructed of 10-mm-thick (3/8") plexiglass with internal dimensions of 30.5 cm by 76 cm in plan, and a height of 26.5 cm (12" x 10.5" x 30" respectively). The box has three compartments, including a central chamber for sand placement and two outer reservoirs for groundwater control. The flow length through the sand is 46 cm (18") and each water reservoir is 15 cm (6") in length. The size of the sand compartment was selected to correspond to the interior dimensions of the model box used for centrifuge test-

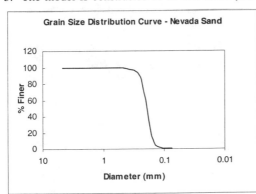

Figure 4. Gradation of Nevada No. 120 sand.

ing. One side of the sand compartment can be removed so the treated sand can be excavated for testing and visual inspection. Sampling ports in the sand compartment are used to extract samples throughout the soil profile to measure the changes in the pore fluid chemistry. Screens with a No. 200 mesh size are used between the water and soil compartments to prevent soil loss from the central chamber into the reservoirs.

Figure 3. Box model with overall groundwater flow at beginning of test.

The model was filled by pluviating the sand to a height of 20 cm (7.9") at a relative density (D_r) of 40%. A gradation curve of the sand is shown in Figure 4. After the model was filled, the up gradient reservoir was filled with water to saturate the sand. After filling, the overall groundwater flow gradient of 0.04 was established using the constant head overflow ports in each reservoir chamber. A pressure-equalizing valve was used to control pressure fluctuations in the building water supply and deliver a constant flow of water to the reservoir chamber. In Figure 3, the left and right sides of the tank are the up gradient and down gradient chambers, respectively.

Delivery wells

After the overall groundwater flow gradient was established, the stabilizer was introduced into the sand formation via delivery wells. The wells were constructed of 19 mm (¾") PVC pipe with 9, 6-mm-diameter (¼") injection ports (Figure 5) screened with nylon. The ports were arranged in one vertical row. Since the sand was loose, the wells were placed by gently pushing them into the sand deposit at equally spaced intervals as show in Figure 6.

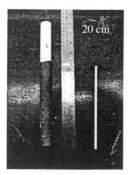

Figure 5. PVC well with holes and

Figure 6. Distribution bay on wells and start of plume progression (outlined in black).

During colloidal silica delivery, a constant head of 20 cm (7.9") from the bottom of the tank was maintained in the delivery wells. For Models 6 and 9, the heads in the injection wells were 0.8 cm and 1.1 cm, respectively, above the top of the overall groundwater level. This excess head resulted in stabilizer movement in both the up gradient and down gradient directions. A distribution bay (30.4 cm by 8 cm in plan by 5.2 cm high, 12" by 3.2" by 2") was placed on top of the wells to maintain a constant supply of colloidal silica to the wells (Figure 6). The colloidal silica solution was supplied to the distribution bay at a rate of 11.7 mL/minute using a peristaltic pump.

Colloidal Silica Delivery

In Box Model 6, colloidal silica at 5 weight percent (0.1 N, pH 6.3) was injected into the liquefiable formation via injection wells (Figure 7). The overall water flow was supplied at a rate of 0.04 m/day, the stabilizer was supplied to the distribution bay at a rate of 11.7 mL/min, and the head in the injection wells was maintained at 0.8 cm above the overall flow level. The colloidal silica solution was colored with green food dye so the investigators could visually determine the colloidal silica advancement on the top and sides of the box. Two pore volumes of colloidal silica solution were delivered to the sand formation through 5 wells spaced 5 cm (2 in) apart and located 8 cm from the up gradient water reservoir. A pore volume is the amount of stabilizer it takes to fill the void spaces in the sand. Each well had one vertical row of nine holes. The colloidal silica had a gel time of 4 days and a travel time of about 1¼ days. It was delivered to the formation over a period of 26 hours.

The progression of colloidal silica along the top and sides of the model can be seen in Figure 7. Three hours after treatment began, distinct plumes could be seen on the surface and sides of the model. The stabilizer moved both up and down gradient due to the head of stabilizer in the injection wells. The smaller plume on one side of the model (Figure 7 (c)) is likely due to restricted flow from the injection well in that area. The individual plumes had merged after 7 hours and the front progressed down gradient; however, as the front progressed, delivery remained sluggish on one side of the model (Figures 7 (g) and (h)). Due to the density difference between the colloidal silica solution and water, there was a tendency for the stabilizer to treat the bottom half of the formation first, which was noted by a time lag between the appearance of dyed stabilizer in the upper and lower parts of the formation (Figures 7 (e) and (f)).

Since the colloidal silica solution contains sodium chloride, a chloride probe was used to determine the progression of the chloride concentration during treatment. The initial chloride concentration of the stabilizer supply was measured as a benchmark. As the grout progressed through the formation, the relative amount of colloidal silica in the formation was estimated using the chloride concentration in individual ports and comparing it to the chloride concentration at the inlet of the wells. Figure 8 shows the pore water extraction process. The ports were labeled using the number of the row first and the number of the column second. Rows were labeled from top (1) to bottom (3) while columns were labeled from right to left along the removable side.

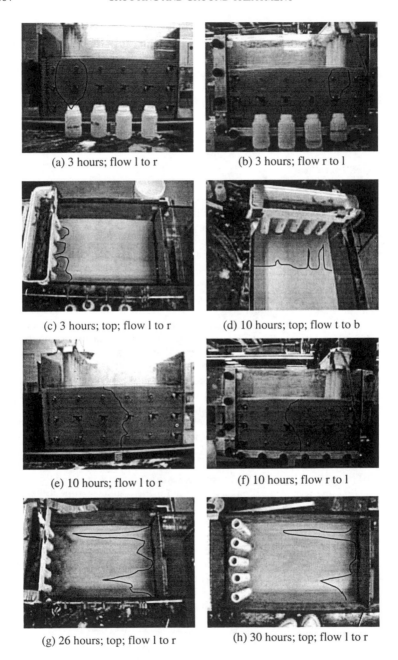

(a) 3 hours; flow l to r

(b) 3 hours; flow r to l

(c) 3 hours; top; flow l to r

(d) 10 hours; top; flow t to b

(e) 10 hours; flow l to r

(f) 10 hours; flow r to l

(g) 26 hours; top; flow l to r

(h) 30 hours; top; flow l to r

Figure 7. Model 6 colloidal silica progression.

Figure 8. Extracting a pore fluid sample via
syringe to measure chloride concentration.

Therefore, Port 2-1 is in the second row at the end closest to the up gradient edge on
the removable side. Chloride concentrations were monitored along the center row of
ports and along the last column at the down gradient edge. As expected, the chloride
concentrations increased in the down gradient pore water as the test progressed (Figure 9). After 26 hours of stabilizer delivery, of the samples from the down gradient
ports had chloride concentrations that matched the influent chloride concentration.

The colloidal silica was a firm resonating gel 4 days after mixing. The formation was
allowed to cure for 3 days after mixing and was then excavated and visually inspected
for overall stabilizer coverage. The coverage appeared to be uniform throughout the
formation.

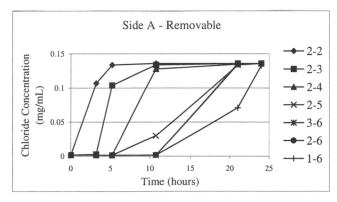

Figure 9. Model 6 chloride concentration progression during stabilizer delivery.

Model 9 was done to determine if adequate coverage of the formation could be obtained using 1½ pore volumes of colloidal silica stabilizer. Passive site stabilization will be more economical if fewer pore volumes are needed to adequately treat the formation. A 5 weight percent colloidal silica solution was used (0.15 N, pH 6.3). The travel time was about 11 hours with a gel time of 2 days. The solution was dyed with green food coloring. One and one-half pore volumes were delivered to the injection wells over the course of 7 hours at a rate of 11.7 mL/min with an overall flow rate of 0.04 m/day. The wells were placed 15 cm from the upgradient edge to permit the up gradient stabilizer flow to treat a larger area. The change in location increased the head of stabilizer in the injection wells to 1.1 cm above the overall flow.

The colloidal silica progression can be seen in Figure 10. Three individual plumes were observed 3 hours after stabilizer delivery began. These plumes merged after 4 hours of treatment. One pore volume had been delivered to the formation 3½ hours after the test began. Grout delivery was complete at 7 hours into the test. The plume continued to travel through the formation after all the stabilizer had been delivered. Once the excess head was removed from the injection wells, the overall water flow continued to propel the stabilizer down gradient. Some of the grout at the up gradient edge traveled down gradient towards the injection wells. A 2.5-cm strip of soil at the up gradient edge showed no dye by the time the stabilizer began its rapid increase in viscosity (12 hours). By 12 hours into the test, the down gradient plume had attained good coverage. The model was excavated after 2 days and appeared to be thoroughly treated.

Chloride concentrations of the formation increased on both sides of the box as the test progressed and eventually equaled the supply colloidal silica's chloride concentration. The chloride progression can be seen in Figure 11. About 10 hours after the test began, the samples from the down gradient ports had chloride concentrations that matched the influent chloride concentration.

Discussion

The model tests were generally successful in demonstrating that colloidal silica can be delivered to a loose sand formation using low-head injection wells. The results also highlight some of the feasibility issues that still need to be resolved. For example, the density difference between the stabilizer and the overall flow causes gravity flow, which assists in delivering the stabilizer to the bottom of the treatment zone. In future model tests, the injection wells will be screened in the upper half of the soil to determine the extent of settling that can be expected due to the density difference. Down gradient extraction wells will also be employed to control the extent of gravity flow.

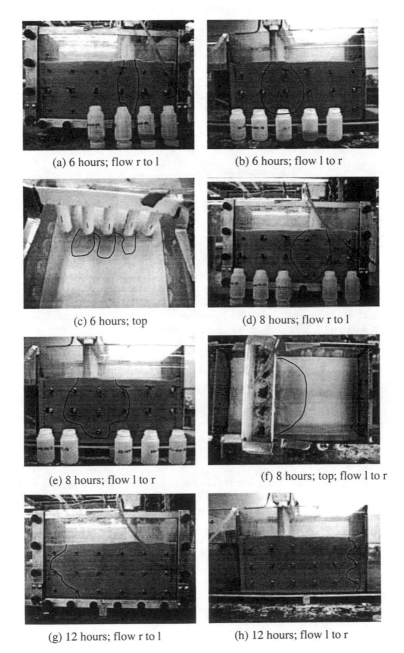

(a) 6 hours; flow r to l

(b) 6 hours; flow l to r

(c) 6 hours; top

(d) 8 hours; flow r to l

(e) 8 hours; flow l to r

(f) 8 hours; top; flow l to r

(g) 12 hours; flow r to l

(h) 12 hours; flow l to r

Figure 10. Model 9 colloidal silica progression.

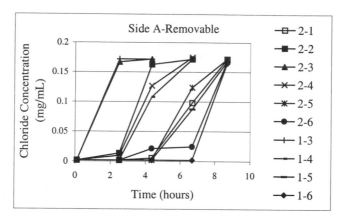

Figure 11. Model 9 chloride concentration profile.

If low-head injection wells are used, the stabilizer will travel in both the up and down gradient directions. To minimize the amount of stabilizer required, the placement of the low-head injection wells can be optimized so that full coverage can be attained in both directions by the end of the travel time. Additionally, clogged wells can prevent uniform stabilizer delivery, which was apparent in Model 6.

Monitoring the chloride concentration aided in tracking the progress of the front moving through the formation. The addition of dye to the grout was also useful in monitoring grout movement. Future work will include box model tests to determine the grout coverage that can be achieved using one pore volume of grout. Additional tests will be done using extraction wells to control the downgradient extent of grout movement.

Conclusions

The primary objective of this research was to evaluate the use of low-head injection wells as a possible delivery technique for passive site stabilization. The paper presented results from 2 model tests in which colloidal silica stabilizer was delivered to a loose sand formation via low-head injection wells. Based on the results presented herein, it appears that adequate coverage can be obtained when 1½ to 2 pore volumes of grout are delivered to the liquefiable formation. Further model tests are planned to evaluate the use of low-head injection wells coupled with extraction wells to control the down gradient extent of stabilizer delivery. Cyclic triaxial testing is planned to measure the deformation resistance of treated samples. The results of these and subsequent box model tests will be used to design additional centrifuge tests that will model the injection process under high-gravity levels.

References

Gallagher, P.M. (2000). *Passive Site Remediation for Mitigation of Liquefaction Risk.* Ph.D. Dissertation, Virginia Tech, Blacksburg, VA, 238 p.

Gallagher, P.M. and Mitchell, J.K. (2002). "Influence of Colloidal Silica Grout on Liquefaction Potential and Cyclic Undrained Behavior of Loose Sand." *Soil Dynamics and Earthquake Engineering*, in press.

Gallagher, P.M., Pamuk, A., Koch, A.J. and Abdoun, T.H. (2002). "Centrifuge Modeling Of Passive Site Remediation." *Proceedings of the 7th U.S. National Conference on Earthquake Engineering (7NCEE): Urban Earthquake Risk*, Boston, MA, July 2002.

Freeze, R.A. and Cherry, J.A. (1979). *Groundwater.* Prentice-Hall, Inc., Englewood Cliffs, NJ 604 p.

GEOPHYSICAL INVESTIGATIONS TO ASSESS THE OUTCOME OF SOIL MODIFICATION WORK.

Measuring percentile variations of soil resistivity to assess the successful modification of foundation soil by jet grouting.

Dr. Bruno Gemmi*, Dr. Gianfranco Morelli** and Dr. F.A. Bares*

Abstract

Jet grouting as a "soil-modification" procedure is one of the most controllable injection systems, capable of modifying pre-designed volumes of soil with pre-designed volumes of grout. However, the degree of soil modification achieved is a function of the ratio of jetted grout to original soil and is controlled by many variables like: adjusting jetting pressure and duration, jetted volumes and power of penetration, soil volumes injected per minute, number of nozzles used, nozzle diameters, nozzle positioning and orientation, grout composition, grout viscosity and more. One of the most important if not the most vital check to be performed upon completing any jet grouting operation is the verification that the soil has been actually modified as specified and required by the engineering design.

Electrotomography investigations (one of many types of geophysical investigations) coupled with advanced computer software procedures were used to assess the effectiveness, success, extent and quality of jet grouting work performed to provide foundations for new structures. The assessment of the extents of jetting dispersion and of grout permanence capable of effectively increase the load bearing capacity of large pockets of compressible peat and of soft silts were of particular concern.

Electrotomography was performed to:
- Ascertain and confirm in advance the overall geotechnical characteristics of the soil to be modified by jet grouting.
- Check the physical formation, the final location, the shapes, the interactions and the overall integrity and continuity of a set of jet-grouted columns laid on an equilateral triangular grid pattern as foundations of a new harbor terminal building.
- Assess the final dimensions and the yield at various depths of a number of jet-grouted test columns.
- Provide tri-dimensional models (constructed by computer data elaboration) visually showing the percentile variations of the resistivity finally achieved by the soil modified by jet grouting.

---

*Parma, Italy – Winchester MA, US, ** SOING srl, Livorno, Italy - March 2002

1. INTRODUCTION

1.0   Foreword

At a site for a new terminal building inside the Leghorn harbor in Italy, geotechnical soil characteristics for the design of new building foundations were first gathered by conventional soil exploration methods as part of standard procedures. The nature and parameters of such geotechnical information were subsequently confirmed and integrated, prior to performing jet grouting work, by surface and cross-hole electrotomography investigations. Both types of explorations were used to gather specific information about the soil conditions underlying a 1,200 m^2 wide area selected for a preliminary jet grouting test and to choose a most representative soil area within which to construct a set of jet grouting test columns.

Electrotomography investigations to assess the success of jet grouting modification work, performed upon completing 5 jet grouting test columns, allowed the production of a 3D modeling of the ground resistivity modified by the jetting. The evaluation and interpretation of the information gathered by electrotomography lead to improved soil modification specifications that replaced a planned binary jet grouting treatment ("Jet-2" using compressed air as an aid to jetting) with a cement-grout only jet grouting treatment ("Jet-1" using no auxiliary air), performed at a lower jetting pressure than originally specified.

Column coring and video camera shots taken inside the coring shafts provided physical support and confirmation of the data supplied by the electrotomography readings.

All electrotomography investigations were performed using two sets of data-collection arrays:

1.   Strings of electrodes (metallic stakes) placed on the ground surface.
2.   Strings of electrodes (stainless steel cylinders fixed to 63 mm diameter plastic pipes at pre-designed intervals and connected by multi-strand cables) installed within PVC-cased holes vertically drilled in the ground.

Final electrotomography investigations (performed from existing grade) were programmed to assess the overall results of the soil-modification operations upon completing the whole jet grouting work program (80 columns).

1.1   Technologies and methods

Geophysical investigations performed by measuring the percentile variation of the   Electrical   Resistivity   of   soil   using   multi-electrode   arrays

(Electrotomography) represent a rather innovative evolution of current soil exploration methods. The system allows computer elaboration of traditionally measured geo-Electrical data to obtain two and three-dimensional models representing, with great visual impact, actual percentile variations of resistivity in the ground. The closer the electrodes are, the more precise and reliable is the resulting model of the soil mass. Measuring electrodes are placed on the ground surface or in boreholes, according to optimized patterns allowing the acquisition of the investigative data. Normally, strings of electrodes are connected by an array of multi-strand cables conveying signals to a mobile acquisition unit.

Electrotomography investigations are performed by energizing a pair of electrodes at the time and by measuring voltage differences existing between all possible electrode-pair combinations. To minimize interference to and from utilities (underground as well as above ground), energizing by low-frequency electrical current is preferred. The numerous readings are organized along lines of electrodes or strings in boreholes. By means of computer aided analysis (using a proprietary software program capable of processing up to 5,000 readings for each electrode configuration), cross-sections and three-dimensional models of soil resistivity distribution are obtained. Readings from vertically placed electrodes display a higher degree of precision because the electrodes are physically closer to the soil layers investigated. And for consistency, such readings can be easily repeated time and time again.

A typical acquisition unit is energized with continuous current from a small 3 kw generator at 800 V max voltage and 2.5 Amp max current. Such unit may record signals from multiple electrodes up to a maximum of 256, with software control. The precision of such a device is approximately 0.25%

1.2 Technologies and methods used for the test investigations

For the test investigations here described, an IRIS Syscal R2 multi-electrode resistivity meter was used as typical data acquisition unit.

12 pipes, each 10.5 m long, carrying 16 electrodes at 700 mm intervals, each pipe placed inside a 127 mm diameter pre-drilled, cased hole were used as vertical electrodes.

Five jet grouting test columns constructed by "Jet-2" jet grouting procedures were initially scheduled to test the results of the soil modification work. The Engineer, upon a preliminary assessment of the results evidenced by a first stage of electrotomography investigations requested the construction of two additional jet grouting test columns by "Jet-1" jet grouting procedures. Fig. 1 shows the location and positioning of the seven test jet grouting columns.

A summary description of the program of electrotomography investigations used to assess the effectiveness of the jet grouting soil modification process is the following:

A. Prior to performing any jet grouting work:

I.   Acquire preliminary generic geotechnical data using parallel lines of electrical cables placed on the ground surface at 2 m intervals;
II.   Build a three dimensional model of the geotechnical characteristics of the ground extending to a depth of 12 m over a 20x40 m area;
III.   Acquire geotechnical data from cross readings of electrode-pairs (cross-hole measurements) from four vertical electrode-instrumented shafts.

B. After performing the jet grouting tests:

IV.   Acquire data from cross-hole measurements from four vertical electrode-instrumented shafts approximately 48 hrs after the completion of the first jet grouting column (C1) construction.
V.   Acquire data from cross-hole measurements approximately 48 hrs after the completion of all remaining jet grouting columns (C2 to C5) construction.
VI.   Acquire data from readings collected from four newly installed vertical electrode-instrumented holes (m1 through m4) prior to performing any additional jet grouting tests.
VII.   Acquire data from cross-hole measurements (m1 through m4) approximately 48 hrs after completing the additional Jet-1 jet grouting test column (single fluid) at 400 bar jetting pressure.
VIII.   Acquire data from cross-hole measurements (e3, e10, e12, e6 and e7) approximately 48 hrs after completing the two additional Jet-1 jet grouting test columns at 200 and 300 bar jetting pressures within the area of previously constructed columns C2 through C5.

ANALYSIS OF GEOPHYSICAL DATA

2.0   Electrical characteristics of the soil

Generally speaking, the resistivity of soil is a factor of:

- Amount of water pore saturation
- Overall soil porosity
- Degree of salinity of the fluid filling the pores
- Temperature
- Presence of organic matters like peat, hydrocarbons, solvents, etc.
- Clay content

The principal factors affecting the electrical characteristics of soil are:

- Salinity: high salinity fluids change naturally high resistivity soils like sands and gravels into conductive soils. For instance, a water body in close proximity to the seashore, i.e., affected by saltwater contamination, would exhibit lower than 5 Ohm*m resistivity.
- Grain size: fills (including demolition materials and other randomly dumped materials) generally show higher resistivity than undisturbed overburden soils due to larger and normally not uniform granular composition as well as to less effective porosity (presence of lots of discontinuities and voids of any size).
- Organic matters: resistivity lower than 2-3 Ohm*m is generally observed in soils affected by the presence of salts and acids deriving from organic decomposition of, for instance, layers of peat.
- Clays: the presence of silts and clays reduces soil resistivity to values of less than 20-30 Ohm*m.
- Insulating matters: the injection into the ground of fluids with high insulating characteristics like water/cement grouts causes a noticeable increase of resistivity due to the elimination of the pore spaces needed for the flow of electric current. However, the addition to the grouts of a 4-5% of bentonite reverses the phenomenon.
- Hydration and curing of cement-based grouts: the process of hydration and curing of injected cementitious masses does, in time and to a limited degree, affect the overall resistivity of the soil injected. This is due to interconnections between soil grains established by the grouting and to the draining off of water except of that strictly needed for hydration. The result is the production of a solidified soil matrix called "Soilcrete", initially liquid, then plastic and finally solid. The remarkable increase of resistivity of the "Soilcrete" mass during the initial phases of curing is followed by a gradual, asymptotic return to close to its starting values.

With specific reference to jet grouting operations, using mixes that have been fully hydrated prior to injection and with a good control of grout segregation and dispersion upon jetting, *the curing of the (generally cylindrical) "Soilcrete" mass is a process occurring within the treated volume,* unaffected by dry or wet soil conditions (provided ground water flows move at less than a 10^{-1}-10^{-2} velocity to avoid shear and erosion of the grouted mass).

The capability of electrotomography of recording percentile soil resistivity variations makes it an excellent instrument to investigate the failure or the success and the lateral expansion of soil improvement procedures like, for instance, jet grouting that relies on the injection of grout to modify the geotechnical characteristics of a soil.

In fact:

- Changes of percentile soil resistivity of a supposedly injected soil mass will indicate if total or partial soil modification (jet grouting column formation in our case) has been achieved. For instance, the consequence of using inadequate jetting procedures may be a significant (and useless) dispersion of grout (often due to excessive jetting pressure) that tomography would evidence as low resistivity readings in zones of otherwise expected high cement concentration and scattered high resistivity readings in zones of cement dispersion. Similarly, the use of jetting flows moving at higher than $10^{-1} - 10^{-2}$ cm/sec velocity may severely affect or even prevent the formation of grout columns due to erosion or shear of the grouted mass. An outcome that tomography would reveal as absence or scattering of higher soil resistivity values where cement-grout concentrations were expected.
- Portions of not thoroughly hydrated mixes tend to absorb surrounding water. This would be detected as an alteration of both the final volume and of the electrical characteristics of the treated soil mass.
- A large dispersion of jetted grout within the soil would significantly increase the volume of untreated soil incorporated in the "Soilcrete" mass, a constant danger when performing jet grouting. This phenomenon will be detected as an alteration of both the final volume and the electrical characteristics of the soil mass even at distances considerably larger than expected.

2.1 Electrotomography measurements (surface and cross-hole)

The test investigations extended to an approximately 1,200 m^2 wide area diagonally traversed by the old course of a filled-in canal within soils considered to closely represent the locally prevailing geotechnical conditions.

The data recorded from surface electrodes set along 10 parallel lines to obtain vertical tomography sections of the ground, closely matched soil data previously obtained by conventional soil exploration methods and identified the presence of four main geotechnical layers:

1. Fill (both canal fill and demolition fill), mostly silty, 2 to 4 m thick.
2. Clay and peat, often indistinguishable from the fill, but highly conductive (with a resistivity <2 Ohm*m due to peat presence), never thicker than 2 m.
3. Alternating silt, sand and gravel on both sides of the filled canal from 6-7 m to about 10 m of depth, characterized by noticeable but typical resistivity value changes corresponding to gradually decreasing saline contamination at increasing distances from the old canal bed. This layer included also zones of higher resistivity (up to 15-20 Ohm*m where no peat intrusions

existed) originating from larger-grained soil pockets and from resistance to marine intrusions by finer-grained soil pockets.

4. Sand and gravel below 10 m, characterized by higher permeability (thereby very open to saline intrusions) and saturated by a continuous aquifer with a 4-6 Ohm*m resistivity.

The differentiation and the transition between different layers appeared very sharp and continuous except in areas intruded by seawater, namely in or around the filled-in canal. The old canal area constituted a preferred path for seawater's intrusions. These conditions are evidenced in fig. 1 presenting plan view plots of soil resistivity recorded at different depths. Within the old canal footprint both the fill and the peat layer were thicker, while deeper down within the more permeable layers, it appeared that the saline intrusion found a preferred path. Boring logs of the soil in correspondence of the old canal confirmed that thin but very pervious layers of granular soil existed beneath the peat layers, thus confirming the information supplied by the electrotomography investigations.

An analysis of the above geotechnical information suggested performing cross-hole measurements from drill-holes set just outside the canal footprint where soil conditions closer to those prevailing in the rest of the area could be assumed to exist.

2.2 Measurements to check column C1

C1, the first jet grouting test column, was constructed to a depth of 12 m. Soil modification results were investigated by electrotomography approximately 48 hrs after its completion and again after 8 days. 17 days after completion, a core was removed from its center. Fig. 2 & 3 present a plan view of the test and a 2-D plot of data acquired at each curing date along the e1-e3 cross-hole electrotomography section cutting through the column axis.

The color scale used to identify the varying percentages of resistivity was set so that an increase of resistivity (in this case a function of the presence of cement and of its curing process) after completing the jetting would generate a positive value, i.e., a change from green to red. The blue to purple colors indicates the opposite: a reduction of resistivity. Therefore, this scale would identify areas where no soil modification occurred as well as where the degree of pore water salinity increased.

The 48 hrs readings evidenced the presence of cementitious matter only between 2.5 and 4.5 m and beyond 10 m of depth. Within the intermediate soil zone, namely across layers 2 & 3 characterized by a higher silt and peat presence, the cement grout appeared to have migrated horizontally for a distance of well over 1 m away from the axis of the column.

The 9 days readings, while confirming the above data indicated a homogeneous soil cementation to 5.50 m of depth with a small discontinuity at about 3 m of depth. It is interesting to note that the results of coring matched the electrotomography readings. In fact, while thoroughly cement-modified soil was encountered between 3.2 and 5.3 m of depth, loose, uncemented cobbles or even totally untreated soil were encountered between 5.3 and 6.5 m of depth. This zone is identified by lower resistivity colors (blue to purple). It is also interesting to note that the core sampling evidenced the presence of many large boulders and of other massive hard blocks (probable remnants of demolition), occasionally cemented by grout.

A 3-D model of the 48 hrs readings was obtained by cross-hole resistivity measurements between instrumented shafts e1, e2, e3 and e4. The models confirmed the near absence of cementitious matter in a zone around 6 m of depth and evidenced that within the first 4-5 m and below 10 m of depth the expanse of soil penetrated by jetting was considerably wider than desirable.

2.3   Investigations of columns C2, C3, C4 and C5

Column C2, constructed 9 days after column C1 to a depth of 10.5 m was investigated 6 days later and axially cored two weeks later. Columns C3, C4 and C5 were jetted four days after completing C2 and investigated 48 hrs later.

The percentile variations of soil resistivity measured by cross-hole e3-e6 and e6-e8 sections across the columns' axis performed 48 hrs after completion, evidenced a significant presence of cementitious matter within the initial 5 m of depth of both columns C2 and C3 while scarce or no cementitious matter was detected in a zone within approximately 5.5 and 7 m of depth. Notable also the fact that the diameter of column C3 below 7 m of depth mushroomed to well over 1.5 m.

Column C2, even after an additional 4 days of curing, showed very poor soil cementation between 5 and approximately 10 m of depth, a result confirmed by coring. The extensive lateral migration of grout around this column into surrounding soil layers with larger peat content and lower compressive strength was also noticeable. Clearly, such a thin and laterally scattered diffusion of grout would not contribute to effectively increase, where needed, the load bearing capacity of such a large volume of soil while constituting a waste of resources.

Similar conditions were found around columns C4 and C5. No cementitious matter appeared to exist below 5 m of depth at column C4, while, again, higher soil resistivity readings at a considerable distance from the column axis would indicate unacceptably too far lateral migration of grout.

All data collected 48 hrs after column completions were used for 3-D processing, with results showing the interaction of the jetting between the columns and the volume distribution of the grout injected. Fig. 4 presents a model based on considering only percentile variations of resistivity values higher than +10% (i.e., eliminating all colors indicating percentile soil resistivity lower than +10%). It evidences the zones where soil modification was successful as those where a substantial presence of cementitious matter was recorded (high resistivity value readings). The same model, looking from both a Northerly and a Southerly direction, shows lack of soil cementation across a band extending between 5 and 7.5 m of depth. This is in contrasts with a higher concentration of cementitious matter within the initial 4-5 m of depth and the bottom 2-3 m. In the latter portion, the lateral migration of grout appears to be too wide.

Horizontal sections made at different jetting depths starting from -11.5 m evidenced that only column C1 was reliably formed at a depth of 12 m. Higher up, a lateral migration of cementitious matter was evident up to about 8 m of depth. In correspondence with columns C4 and C5, from the same depth and up to approximately 6 m of depth, ever decreasing percentile soil resistivity variations were noticeable together with a remarkable lateral migration of grout away from the columns' axis. As expected, all horizontal sections supplied a confirmation of the grouting results evidenced by the vertical cross section readings.

Above the 4-5 m of depth the soil modification appeared to have been successful even if the lateral migration of the grout was still unacceptably large.

2.4 Investigations of a "Jet-1" column

Six weeks after the completion of column C1, cross-hole investigations were performed from a new set of instrumented holes (m1 to m4) placed inside the zone of influence of the old canal (Fig. 5) to determine if Jet-1 jetting procedures would be more effective than Jet-2 type. Fig. 6 presents a three-dimensional modeling of the soil before and after Jet-1 modification performed in the center of the area delimited by the electrotomography shafts.

Above soil representation evidenced the failure of Jet-2 to create an homogeneous and continuos cemented soil mass (most probably due to the powerful dispersing action of the high-pressure air) and the achievement of encouragingly better results with Jet-1, mostly within the peat layers. However, the still powerful penetrating action of Jet-1 @ 400 bar in correspondence with the more permeable soil layers, produced patterns of jetted cement travelling up to 2-3 m away from the axis of the column, hence, beyond the perimeter of the area of investigation. A confirmation of the latter is supplied by readings from very pervious soil layers between 6.5 & 7.5 m and 9 & 10 m of depth showing tell-tale reductions of soil resistivity values due to high scattering and migration of injected cement. The presence of a high resistivity cementitious mass within both the upper fill and the peat layers with evidence of large lateral grout migrations confirmed that an excessive jetting pressure was being used. The latter was confirmed also by a 340 mm heaving of shaft m1, located at well over 2 m from the center of the jet grouting test column.

The evidencing that a satisfactory soil modification could be achieved by a substantial and continuous deposit of cementitious matter into peat and peat-including soil but not into two very pervious layers, lead to experimenting with lower pressure Jet-1 grouting procedures. The achievement of successful soil modification using Jet-1 grouting at lower jet pressures (200 & 300 bar) between columns C1 to C5 was confirmed by readings from 6 new cross sections. This modified Jet-1 procedure was eventually and successfully specified for the completion of the improvement of the foundation soil across the whole site.

3. Conclusions

The geophysical data collected and all checks performed permit the formulation of the following conclusions:

- The soil to be modified by jet grouting did show significant variations horizontally as well as vertically. It was however possible to identify four distinct layers. The frequent and often thin intrusions of layers of silt, peat, sand and gravel between 5 and 10 m of depth made it difficult to achieve a detailed and well-scaled tri-dimensional modeling.
- The test columns (5 constructed by Jet-2 and 3 by Jet-1) did evidence the difficulty of achieving homogeneous and isodimensional soil grouting within pre-determined distances from the axis of the columns. In particular within the −5.5 to −7.0 m zone.
- Supplemental ground explorations performed onto the first two jet grouting columns confirmed the precision and reliability of the Electrotomography data readings collected along the axis of the columns to investigate the results of jetting. The readings did clearly indicate the existence of well-consolidated soil layers alternating to layers lacking any presence of cementitious matter.
- The three-dimensional modeling evidenced an excessive lateral migration of grout. The migration occurred mostly at peat layer levels. However, significant grout migration was detected at other layer elevations where it assumed the shape of large channels establishing intercommunications between columns (a phenomena known as "claquage"). Excessive grout migration could be corrected by Jet-1 remedial jetting that, however, within

very pervious soil layers, did still show occurrences of excessive lateral grout migration.

- The successful completion of Jet-1 columns at lower jetting pressures demonstrated the possibility of improving the effectiveness of jet grouting in poor soil conditions (as those existing at this site). This translated into modifying original jet grouting specifications and in specifying the performance of fill-in, remedial jetting as suggested by electrotomography readings.

- The electrotomography investigations provided the Engineer and the Owner with a simple, low cost and above all quick means to assess the notoriously often doubtful results obtained from soil improvement methods based on injecting modifying substances into difficult ground.

- The assessment of the failure of the Jet-2 type jetting and the evaluation of the reasons for the failure, lead to the tuning of jetting specifications to the parameters of the difficult ground to be treated. A clear case of dedicating additional preliminary attention to the often neglected investigation of foundation ground and of spending a very limited amount of dollars to prevent a possibly disastrous and expensive failure and claim.

- Electrotomography constitutes an additional investigative tool affording the designing of soil modification based on results of quick and relatively inexpensive test work. It does not supply "underground X-Ray" vision, but it removes a lot of the blindness that characterizes underground soil improvement procedures. A blindness that is often offset by precautionary over-design and/or by monumental failures.

- Electrotomography effectively replaces much more expensive and time consuming pot-construction investigations.

- The best conclusion for the case presented, is that timely executed electrotomography investigations could turn a potential jet grouting disaster into a successful job.

4.   APPENDIX

Figures 1 through 6

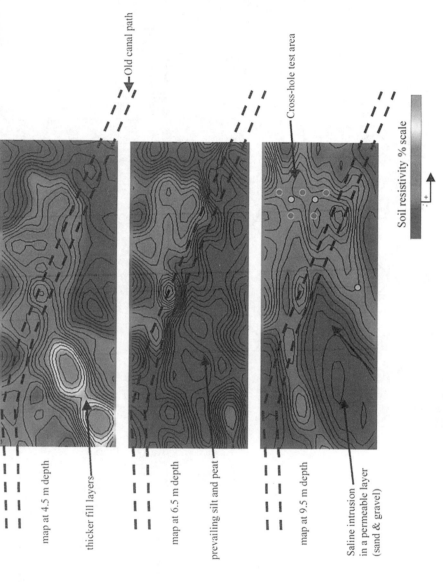

Fig. 1 : Resistivity maps generated from surface measurements

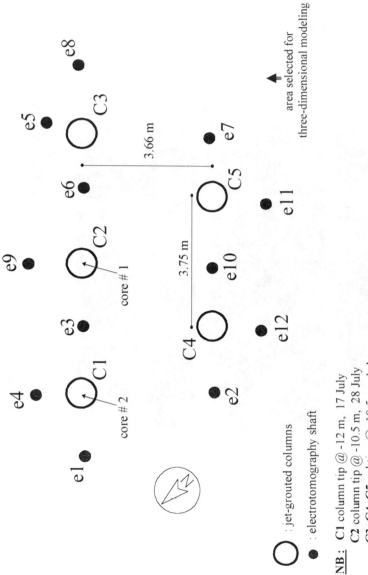

NB : C1 column tip @ -12 m, 17 July
C2 column tip @ -10.5 m, 28 July
C3, C4, C5 col.tips @ -10.5 m, 1 Aug.

○ : jet-grouted columns

● : electrotomography shaft

Fig. 2: Area investigated by cross-hole electrical tomography

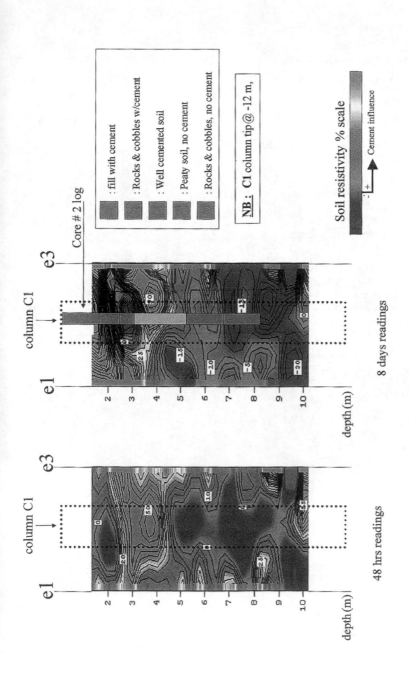

Fig. 3 : 2-D sections: % of resistivity variation (Readings taken after completing column C1)

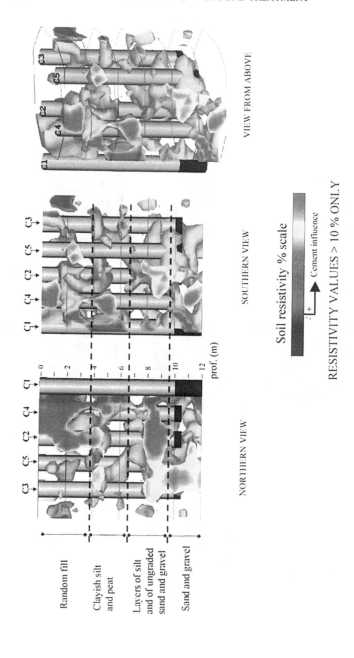

Fig. 4 : 3-D views of soil bodies with different resistivity values (48 hrs readings)

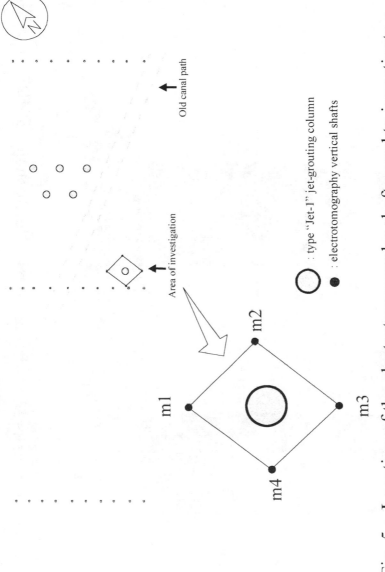

Fig. 5 : Location of the electrotomography shafts used to investigate a monofluid (Jet-1) column jetted @ 400 bar pressure.

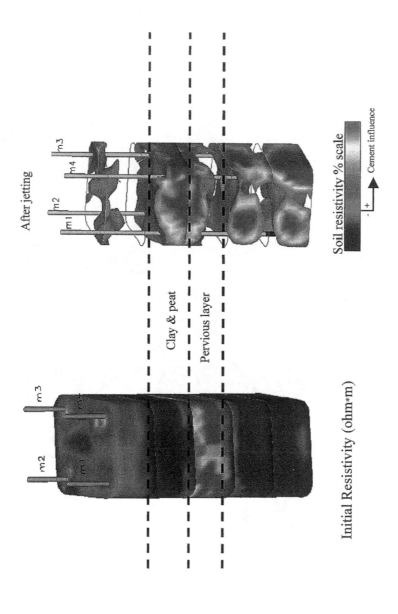

Fig. 6 : 3-D processing, before and after completing Jet-1column @ 400 bar pressure

Electro-Osmotic Grouting for Liquefaction Mitigation in Silty Soils

S. Thevanayagam[1] and W. Jia[2]

Abstract

An electro-kinetic permeation technique for injection of grouting materials into silty soils is introduced in this paper. Preliminary experimental data indicates that it is feasible to inject colloidal silica and silicate grouts into silty soils by means of a dc current. 1-D column experiments and 3-D model tests show that the soil grouted by silicate and colloidal silica by means of electro-osmosis showed significant improvement in strength. For 1-D condition, the grout penetration rate by electro-osmotic injection is about 10^{-5} cm/s under an electric gradient of 1v/cm. Typical permeation grout penetration rate by hydraulic means in silty soils is about 10^{-5} cm/s or less under a unit hydraulic gradient. Traditional permeation grouting in silty soils would require very high pumping capacity making it cost prohibitive, whereas preliminary bench-scale model tests indicate that electro-osmotic grouting is cost-effective.

Introduction

Loose saturated sands and non-plastic silty sands/sandy silts subjected to seismic excitations experience an increase in pore pressure that may lead to liquefaction, temporary loss of strength, lateral ground spreading, and vertical ground settlements. This in turn may cause foundation distress and instability of buildings and other superstructure supported on such soils (Figs. 1 and 2). Critical facilities such as hospitals, lifelines, and transportation systems founded on such soils need to be protected from such failures. Current soil improvement techniques (Table 1) to increase resistance to liquefaction can be categorized into three groups: densification (and/or drainage), reinforcement, and cementation/solidification. The choice of an appropriate technique depends on the soil type, site accessibility (accessibility beneath buildings and allowable site disturbance), and cost-benefit considerations. For clean sands and silty sands containing less than about 15% silt content, all of the above techniques are applicable. At silt contents above 10 to 15%, most of the traditional techniques are ineffective except for jet grouting and soil mixing techniques. Low permeability of silty soils renders the densification/drainage and permeation grouting techniques ineffective. Dynamic compaction and vibro-stone columns techniques may still be implemented to densify silty soils with the use of supplementary wick drains (Dise et al. 1994, Luehring et al. 2001, Thevanayagam et al. 2002). Where site/foundation soil accessibility or site disturbance is a concern, permeation grouting technique remains the most viable option, while all others become less attractive. If the soil to be improved contains high silt

[1] Associate Professor, Dept. of Civil, Struct. and Env. Eng., University at Buffalo, SUNY, Buffalo, NY 14260, theva@eng.buffalo.edu.
[2] Graduate Student, Dept. of Civil, Struct. and Env. Eng., University at Buffalo, SUNY, Buffalo, NY 14260, wjia2@acsu.buffalo.edu.

contents leading to low hydraulic conductivity, permeation grouting is also unsuitable. This paper presents a method to inject grout materials into low permeable silty soils using direct electric current.

Traditional Permeation Grouting

The purpose of permeation grouting is to fill voids as well as facilitate cementation and bonding of particles. This increases the density of the soil as well as the strength and resistance to liquefaction. During permeation grouting, a grout and a reactant is pumped at high pressures into liquefiable zones via drill holes, typically spaced at 1 to 2m apart, center-to-center (Graf et al. 1979, Andrus et al. 1995). The grout and the hardener react to make a gel or other agents that fill the voids and facilitate the formation soil particle bonds. The time required for the hardener and the grout to fully react varies from minutes to several hours depending on the grout concentrations and grout chemistry (Army Corps of Engineers 1995). The gelling time and hydraulic conductivity of the soil control the extent of grout penetration. Highly permeable sands with a hydraulic conductivity range of about 10^{-1} to 10^{-3} cm/s, depending on the grout and gel characteristics, are groutable. Silty soils have a hydraulic conductivity of the order of 10^{-5} cm/s or less. Pumping capacity and hydraulic gradients required to implement permeation grouting in silty soils are cost prohibitive. Such soils cannot be grouted successfully by means of permeation.

(a) Adapazari Area   (b) Liquefied soils

Fig.1 Tilted Buildings - Kocaeli, Turkey, Earthquake of August 17, 1999
(Soil beneath them liquefied and weakened the foundation; U.S. Geological Survey 2000)

Fig.2 Tilted Buildings - Niigata Earthquake, Japan, 1964
(M=7.5, 310 of the 1500 reinforced concrete buildings in Niigata City were damaged; 200 buildings on shallow Foundations or Friction Piles in loose sand/silty sand settled without significant structural damage due to liquefaction. Same type of buildings built on firm strata at 20 m deep did not suffer damage; http://www.ce.washington.edu/~liquefaction/html/what/what2.html)

Table 1. Soil Improvement Methods for Liquefaction Mitigation

Technique		Soil Type			Accessibility (Beneath Existing Structures)
		Sand	Silty sand	Silt	
Densification	Dynamic Compaction	Yes	*	*	No
	Vibro-Densification	Yes	*	*	No
Densification/ Drainage/ Reinforcement	Vibro-Stone Column	Yes	*	*	No
Solidification	Permeation Grout	Yes (Fines ≤5%)	**	No	Yes
	Compaction Grout	Yes	Yes	Marginal	Yes
	Soil Mixing	Yes	Yes	Yes	No
	Jet Grout	Yes	Yes	Yes	No
	Electro- Kinetic Injection (This study)	**	Yes	Yes	Yes
	Passive Grouting (Gallagher et al. 2001)	Yes	**	**	Yes

* = feasible with supplementary wick drains; ** = uncertain

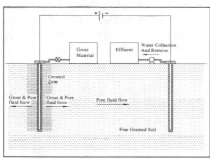

(a) Concept of Electro-kinetic Injection

(b) Vertical Grouting

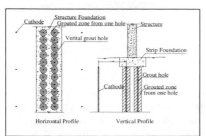

(c) Vertical Grouting

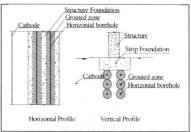

(d) Horizontal Grouting

Fig.3 Concept and strategies of Electro-kinetic Injection

Electro-kinetic Grouting

Electro-kinetic injection of grout components into silty soils is an innovative idea that may be implemented in the field in a way similar to permeation grouting, but using dc current as the means of introducing the grout/hardeners into the ground (Fig.3a). Figs.3b-d show a few typical layouts of the field application of this technique. When a d.c voltage gradient is applied across a saturated soil, in the absence of external hydraulic gradient, the following phenomena occur under the influence of the voltage gradient (Mitchell 1993): (i) pore fluid flows from anode to cathode (called electro-osmosis), (ii) positively charged dissolved ions flow from anode to cathode (called electro-migration), and (iii) negatively charged dissolved ions flow from cathode to anode (called electro-migration). Hydrolysis and electrode reactions at the electrodes may also induce changes in the pH and ion concentrations, and hence affect the soil properties and flow of ions and pore fluid. Past experience with clayey soils indicate that the electro-osmotic pore fluid flow velocity, due to a unit voltage gradient, is of the order of 10^{-5} cm/s/V/cm or larger (see Table 12.7 in Mitchell 1993). Experience in electro-kinetic treatment of contaminated soils (Acar et al. 1992, Acar et al. 1994, Acar et al. 1996, Chen et al. 1999) indicates that pore fluid and electrode chemistry significantly affect the pore fluid chemistry and flow behavior. With due consideration of these factors, it may be possible to use dc current to introduce water-soluble grouts and hardening agents into low permeable silty soils by judiciously introducing the various grout components near the electrodes. By controlling the concentrations, sequence of introduction the various components, choice of electrode material, and electrode reactions within favorable pH limits the hardening time may also be controlled. The following sections present the results from an experimental feasibility study to inject silicate and colloidal silica grouts into silty soils.

1-D Grout Injection Feasibility Tests

Fig.4 shows the experimental setup for 1-D column tests. The soil was prepared by mixing 40% sand (Ottawa sand F-55, U.S. Silica Company) and 60% silt (sil-co-sil #40 silt, U.S. Silica Company) by weight and was tamped into a Plexiglas tube (typically 5.1 cm dia., 24 cm long) at a void ratio of approximately 0.55. The specimen was typically compacted in 5 layers. After compaction of each layer, water was allowed to flow through the specimen for saturation. The compaction procedure took two days. Following the compaction, the tube was placed horizontally and water was allowed to flow through the specimen for an additional day.

A total of three tests (Test-7, Test-10, Control) were done (Table 2). Sodium silicate, colloidal silica, and water were used as the grout materials, introduced at the anode inlet tube, in tests 7, 10, and control test, respectively. The cathode outlet tube contained water. The fluid levels were nearly at the same level. The anode was made of aluminum and cathode was made of graphite. Electrolysis reactions at anode and cathode would induce changes in pH environment in the anode and cathode regions, and subsequently in the soil. If the pH environment in the anode region is highly acidic, sodium silicate and colloidal silica would polymerize and form a gel quickly, prohibiting grout penetration into the soil. For this reason, aluminum anode was used so that the pH environment would not be excessively acidic.

All specimens were subjected to an external dc voltage gradient of 1.5V/cm, measured across the electrodes, for three days. Preliminary feasibility calculations showed that this combination would produce sufficient grout penetration into the soil over a reasonable distance within a few days. A slight voltage drop is expected across the electrodes, and hence, the effective voltage gradient applied across the soil is expected to be slightly smaller than 1.5V/cm. The fluid levels at the anode and cathode were open to atmospheric pressure and the fluid levels at anode inlet and cathode outlet tubes were refilled regularly to maintain them at nearly the same level. The amount of fluid added and the electric current were recorded versus time. No noticeable rise in temperature was observed, and therefore no temperature measurements were made.

Following the electric treatment, soil samples were recovered from various locations from anode to cathode and were tested for pH and conductivity. For comparison purposes, pH and conductivity measurements were also made on two samples of the sand/silt mix prepared by mechanically mixing it with 50% silicate and colloidal silica, respectively. One cylindrical specimen was also recovered from each test (Table 2) and air-dried and tested for unconfined compressive strength. Figs.5 and 6 show the results for Tests-7 and Test-10, respectively.

Figs.5a-b show the initial and post-treatment pH and electric conductivity data for Test-7. Also shown in these figures are the pH and conductivity data for the specimen prepared by mechanically mixing soil with sodium silicate solution. The post-treatment pH and conductivity are higher than the initial values. The increase in pH, compared to the initial value of nearly 7, is due to intrusion of silicate into the soil by electro-osmosis. The increase in pH near cathode is not as high as the pH from anode to the middle of the specimen. This is because the grout has not penetrated fully across the specimen within the short treatment period. The intruded silicate also increases the conductivity of the soil. The increase in pH and conductivity is not as high as the values obtained for specimens prepared by external mixing. As shown in Fig.5c, the electric current also increased indicating an increase in soil conductivity during injection of silicate. In contrast, the post-treatment pH and conductivity of the soil subjected to control test decreased slightly. Fig.5d shows the electro-kinetic permeability (k_e) versus time, obtained from the flow volume data. The initial fluid intrusion rate is of the order of 10^{-5} cm/s/v/cm for both Test-7 and control tests. The rate decreased somewhat for Test-7 involving sodium silicate. Fig.5e shows the unconfined compression stress-strain data. The specimen injected with silicate showed a significantly high air-dried compressive strength, whereas the specimen from the control-test crumbled at a very low axial stress. This comparison with the control test also indicates that the increase in strength in Test-7 is primarily due to grout penetration rather than aluminum hydroxide precipitates. It is possible that Al^{3+} ions produced at the anode may also have reacted with the silicate material and contributed to the increase in strength in Test-7.

Table 2. Test Series 1 – Feasibility of Grout Injection

Test No.	Anode fluid	Cathode fluid	Initial pH	Initial soil σ (siemens/m)
7	50% Sodium Silicate	Water	6.56	0.024
10	Colloidal Silica	Water	6.96	0.018
Control	Water	Water	6.96	0.018

Notes: Anode = Aluminum; Cathode = Graphite; Sodium Silicate – Product N® (PQ corporation, weight ratio SiO_2/Na_2O=3.22, percentage of SiO_2 by weight = 28.7%); 50% Silicate = 50% water/50% product N® by weight; Colloidal Silica – Nyacol 215 (Obtained from Akzo Nobel, percentage of Silica by weight = 15%), the original colloidal silica solution was injected without dilution; Voltage gradient = 1.5 V/cm applied for 3 days.

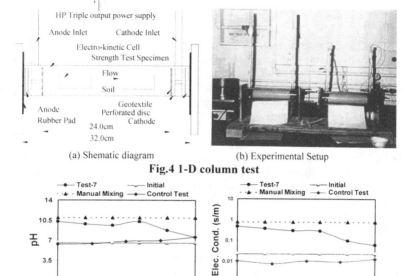

(a) Shematic diagram   (b) Experimental Setup

Fig.4 1-D column test

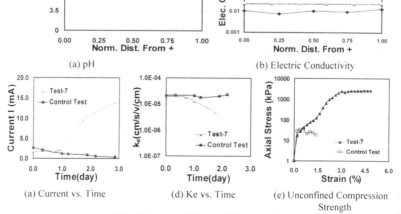

(a) pH   (b) Electric Conductivity

(a) Current vs. Time   (d) Ke vs. Time   (e) Unconfined Compression Strength

Fig.5 Groutability -- Silicate: Test-7

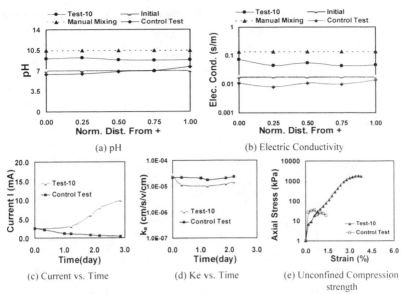

Fig.6 Groutability -- Colloidal Silica: Test-10

The results for Test-10 that involved colloidal silica grout are similar to Test-7. These above observations indicate that it is feasible to inject silicate and colloidal silica into silty soils using dc current.

Following the initial feasibility studies, additional tests were conducted to study the improvements in undrained strength of the grouted specimens in triaxial compression. Specimens were prepared in a manner similar to Test Series 1. Table 3 shows a summary of tests conducted. All specimens were prepared at a void ratio in the vicinity of about 0.55. The first two tests were control tests where the cathode and anode fluid was water. In the third test (Tri-03), 50% silicate solution was made available at the anode. In the final three tests, 50% silicate solution was introduced at the anode for about 1.5 days followed by removal of silicate solution from the anode region and replacement by (a second shot) of calcium chloride solution for an additional 1.5 days. In each test, the voltage gradient was maintained at 4.5 V/cm. After injection, a cylindrical soil specimen was removed from the electro-osmosis test cell and transferred to a triaxial test device (GDS Instruments Limited) and subjected to consolidation followed by undrained triaxial compression tests.

Fig.7a shows the deviatoric stress q ($=\sigma_1'-\sigma_3'$) versus axial strain data. Fig.7b shows the mean effective confining stress p' ($=(\sigma_1'+2\sigma_3')/3$) at the end of shear tests versus void ratio. The specimens grouted with silicate and $CaCl_2$ show high strength compared to ungrouted specimens (Fig.7b). The specimen grouted with high concentration of $CaCl_2$ tends to be brittle.

Table 3. Test Series 2: Improvement in undrained strength

Test #	Current	Anode Fluid (1st shot)	Anode Fluid (2nd shot)	Cathode Fluid	Final e
Tri-01	No	water	N/A	water	0.468
Tri-02	Yes	water	N/A	water	0.469
Tri-03	Yes	50% Silicate	N/A	water	0.435
Tri-04	Yes	50% Silicate	15% $CaCl_2$	water	0.504
Tri-05	Yes	50% Silicate	20% $CaCl_2$	water	0.566
Tri-06	Yes	50% Silicate	10% $CaCl_2$	water	0.568

Notes: Anode = Aluminum; Cathode = Graphite; 50% Silicate = 50% water + 50% sodium silicate Product N® (PQ Corporation) by weight; 10% CaCl2 solution was made by mixing 1 unit of $CaCl_2$ powder (by weight) with 9 units of water. Final e = void ratio based on water content after the undrained triaxial compression test.

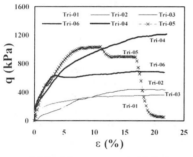

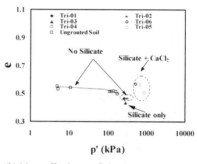

(a) Deviation stress vs. Axial Strain   (b) Mean effective confining stress vs. e

Fig.7 Undrained triaxial Compression Strength

3-D Grout Injection Feasibility Tests

Figs.8a-b show the experimental setup for a bench scale experiment conducted to evaluate the feasibility of electro-osmotic injection of grout under 3-D conditions. The soil used in this test was prepared by moist tamping a soil mix of 40% sand (Ottawa sand F-55, U.S. Silica Company) and 60% silt (sil-co-sil #90, U.S. Silica Company) by weight and saturated by water. A 2.5 cm diameter perforated Plexiglas tube containing an aluminum anode was used as the injection hole. A steel rod was used as the cathode.

The anode tube was connected to an inlet tank filled with 50% Colloidal Silica solution (50% water + 50% Ludox SM30 by weight, from Aldrich Chem. Co. Inc.). The water level in the soil was maintained at the top surface of the soil. The free fluid surface in the grout inlet tank was maintained at the same level as the top surface of the soil. The electrodes were connected to a 25v dc power supply. The grout density is 1.10g/cm^3 compared to water at 1g/cm^3. This introduces a negligibly small hydraulic head differential of 0.9 cm, compared to the voltage differential of 25V. The hydraulic conductivity of the soil is about 2×10^{-6}cm/s. The resulting hydraulic gradient is negligibly small compared to voltage gradient. The grout penetration due to this hydraulic gradient would be less than 0.1cm per day, and was considered negligible. The inflow volume of grout and electric current were monitored versus time. After 1-day the electric supply was terminated and the soil was air-dried. The energy spent during this period was 3.5 w-hr. The soil around the anode and cathode was excavated by hand trimming using a spatula. The soil around the cathode crumbled easily whereas the soil around the anode for a radial distance of about 12 cm was very hard (Fig.10). This indicates grout penetration up to that distance from the anode. Under 1-D condition, the grout penetration distance would be about 6~7 cm (=$k_e i_e t/n$, k_e= 2×10^{-5}cm/s/V/cm, i_e=1.25V/cm, n=0.33, t=1 day). But, under 3-D conditions, the voltage gradient distribution is nonlinear. 3-D numerical electro-osmotic flow simulations conducted as a part of this study indicate an average radial grout penetration of about 9 cm after one day. The predicted zone was nearly symmetric around the injection hole for short injection periods and became asymmetric towards the cathode for longer injection periods (Jia 2002).

Figs.9a-b show the current vs. time and cumulative grout inflow volume vs. time data. Based on the intruded grout volume and void ratio of the soil, the grout volume is sufficient to fully displace the water in the soil up to a radial distance of about 8 cm.

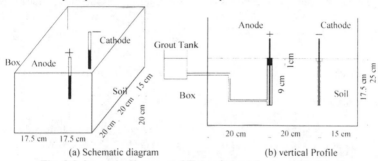

| (a) Schematic diagram | (b) vertical Profile |

Fig. 8 Schematic diagram of Experimental setup of 3-D test

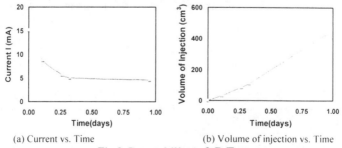

| (a) Current vs. Time | (b) Volume of injection vs. Time |

Fig.9 Groutability – 3-D Test

Fig.10 Grouted soil column around the anode

Conclusions

Results from an experimental study to assess feasibility of electro-osmotic injection of silicate and colloidal silica grouts are presented. The results indicate that it is feasible to inject silicate and colloidal silica grouts into silty soils by means of a dc current. 1-D column experiments indicate that 1-D grout penetration rate is about 10^{-5} cm/s per 1V/cm voltage gradient. Bench-scale 3-D model test results indicate a grout penetration of about 12cm for a 25V applied across electrodes spaced at 15 cm apart. Further research is ongoing to further develop this technique.

Acknowledgments

Financial support for this study was provided by MCEER, University at Buffalo, NY.

References:

Acar, Y.B., Li, H., and Gale, R., J. (1992). "Phenol removal from kaolinite by electro-kinetics." *Journal of Geotechnical Engineering*, 118(11), 1837-1852.

Acar, Y.B., Hamed, J.T., Alshawabkeh, A.N., Gale, R.J. (1994). "Removal of cadmium (ii) from saturated kaolinite by the application of electrical current." *Geotechnique* 44, No.2, 239-254.

Acar, Y.B., Alshawabkeh, A.N. (1996). "Electrokinetic remediation. I: Pilot-scale tests with lead-spiked kaolinite." *Journal of Geotechnical Engineering*, 122(3), 173-185.

Andrus, R.D. and Chung, R.M. (1995). *Ground Improvement Techniques for Liquefaction Remediation Near Existing Lifelines*, NISTIR 5714, National Institute of Standards and Technology (U.S.), Gaithersburg, MD.

Chen, J.L., Murdoch, L. (1999). "Effects of electroosmosis on natural soil: field test." *Journal of Geotechnical And Geoenvironmental Engineering*, 125(12), 1090-1098.

Dise, K., M.G. Stevens, and J.L. Von Thun. (1994). "Dynamic compaction to remediate liquefiable embankment foundation soils." GSP No.45, ASCE, Reston, VA.

Gallagher, P. and Mitchell, J.K. (2001). "Passive site remediation for mitigation of liquefaction risk." *Proc. MCEER Workshop MEDAT-2*, MCEER-01-0002, 29-32, University at Buffalo, Buffalo, NY.

Graf, E.D., Zacher, E.G. (1979). "Sand to Sandstone: Foundation Strengthening with Chemical Grout." *Civil Engineering*, Jan. 67-69, ASCE, New York.

Jia. W. (2002). *Electro-Osmotic grouting technique for soil liquefaction mitigation*, PhD Dissertation, in preparation, University at Buffalo, Buffalo, NY.

Luehring, R., Snorteland, N., Stevens, M., and Mejia, L. (2001) "Liquefaction mitigation of a silty dam foundation using vibro-stone columns and drainage wicks: A case history at Salman Lake Dam." *Proc. Dam Safety Conference*, Sept. 11-15, Colorado.

Mitchell, J.K. (1993). *Fundamentals of Soil Behavior, Second Edition*, John Wiely and Sons, Inc., New York.

Thevanayagam, S., Martin, G.R., and T. Shenthan (2002). "Ground Remediation for Silty Soils Using Composite Stone Columns (Task 150 C104E)." *Seismic Vulnerability of the Highway System*, FHWA Contract # DTFH61-98-C-00094.

U.S. Army Corps of Engineers (1997). *Chemical grouting*, ASCE Press, New York.

U.S. Geological Survey (2000). *Implications for earthquake risk reduction in the united states from the kocaeli, Turkey, earthquake from August 17, 1999*, U.S. Geological Survey Circular 1193.

GROUND TREATMENT FOR TUNNEL CONSTRUCTION ON THE MADRID METRO

Pedro R. Sola[1], A. Sarah Monroe[2],
Lucas Martin[3], Miguel Angel Blanco[3], Raúl San Juan[3]

ABSTRACT: A brief description along with a number of case histories of each of the different types of ground treatment procedures used during the construction works associated with the extensions to the Madrid Metro over the last eight years are described in this paper. An account is given of how different adverse conditions and situations can be overcome by means of these grouting techniques, ensuring at all times the safety of the excavation processes, be they using tunnelling machines or when excavating by hand, be they under buildings, roadways or pipelines.

1 INTRODUCTION

The quality of the service to the customer of an underground railway system can be measured by how accessible the infrastructure is and by the areas that it interconnects. Therefore tunnels often have to be built through densely populated and urbanised areas and stations at shallow depths for ease of access. This means that tunnel construction in urban areas for these uses have to comply with strict constraints in alignment both in plan and in section. Also, with growing demand for higher capacity, faster and safer facilities, the tunnels to be excavated are growing larger in diameter.

Shallow tunnels imply excavating though materials which are often weaker and less compacted and generally less suitable and can give rise to instability problems at the face of the excavation and cause subsidence at ground surface. This is also particularly true at entrance and exit portals at the stations. And when new underground lines link outer areas of cities, badly compacted fill materials are likely to be encountered which can be another adverse condition to affect the construction of tunnels.

For all of the above stated reasons, tunnel construction is becoming more complicated from a technical point of view and also has a greater impact on the surrounding areas. In order to respond adequately to the possible problems of excavation face instability, subsidence control and protection of structures, certain ground treatment techniques and applications have been developed for dealing with these situations.

In the following chapters, various different grouting techniques will be described along with the specific situations for which they have been applied on many parts of the new 111 km of tunnels that will have been built over 8 years during the two last extensions of the Madrid Metro (1995 – 1999 and 1999 – 2003). The figures stated correspond to the finished construction of all the lines, the last of which is scheduled

[1]Development Director, GEOCISA, Geotecnia y Cimientos, S.A., Madrid, Spain
[2]Geotechnical Engineer, GEOCISA, Geotecnia y Cimientos, S.A., Madrid, Spain
[3]Civil Engineer, GEOCISA, Geotecnia y Cimientos, S.A., Madrid, Spain

to open in 2003, and of the latest extension, approximately 98 % of the tunnels have already been completed. Along with these 9.38 m diameter tunnels constructed using both EBP machines and the hand-excavated so-called Madrid Method, there will be 26 new stations to add to the 37 built in the previous extension.

2 GROUND TREATMENT PROCEDURE USED

The different types of ground treatment which have been used extensively to deal with the various sorts of instability problems arising from the construction of the tunnels for the new metro lines in Madrid and which are to be described in this paper are based on different types of grouting techniques: jet grouting, consolidation and permeation grouting and compensation grouting.

Jet grouting is a well-known technique used for underpinning but that has now also been shown to be effective as a protective measure against subsidence in certain situations. Consolidation and permeation grouting have had many applications where loose sandy materials and water were encountered and caused serious difficulties in the excavation of sections of the lines constructed using the traditional hand excavated Madrid Method. And compensation grouting which has now become a widely used method for controlling movements of structures affected by settlement due to excavations.

3 JET GROUTING

3.1 Jet grouting design

In cases in which this type of treatment was carried out, the overburden above the tunnel was fundamentally formed of alluvial soils or sandy soils with a low fines content, with a relatively insignificant thickness of preconsolidated Tertiary substrata made up of clays and sandy clays. The deformability conditions of these types of soil, along with the risks of instabilities (chimneys), made it necessary to adopt corrective measures (ground treatment) in order to minimise the effect on nearby elements (structures, buildings, communication routes, services, etc.)

The technique used in these cases was type I jet grouting (single fluid), with the grouting parameters being adapted to the particular conditions of each site, both during the design phase and during the execution. In the design phase, the initial parameters were set, based on the available data experience, and during the execution, in accordance with the material detected during the drilling, the volume and composition of the overflow material obtained during the grouting and the results of the monitoring of movements, they were adapted to the real conditions.

The types of treatment carried out can basically be divided into three major groups:

-   Reinforcement treatment in portals. This consists of strengthening the ground by means of a "mesh" of jet grouting columns. (Figure 1)

-   "Wall" type treatment. This consists of the construction of a "wall" of grouted soil between the tunnel and the element to protect. (Figure 2)

Inverted "V" type treatment. This consists of carrying out "wall" type treatment from both sides of the tunnel. (Figure 3)

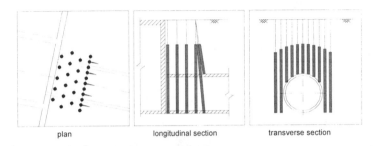

plan   longitudinal section   transverse section

FIGURE N°1. Reinforcement treatment at portals

3.1.1   Wall type treatment

This type of treatment consists of the interposition between the element to protect and the future tunnel of a discontinuous wall made up of columns of jet grouting. In this way, the deformations originated by the excavation of the tunnel as it propagates towards the element to protect are to a large degree absorbed by a structure of greater rigidity than the ground – the jet grouting wall – thus diminishing the subsidence in the protected zone. In the same way, in the event that instability phenomena were to occur in the excavation, in its propagation towards the surface, at least towards the protected zone, these are temporarily halted, permitting complementary measures to be adopted if necessary, for example consolidation grouting and/or fills.

3.1.2   Inverted "V" treatment

This type of treatment is similar to the above, with a double wall being provided, one on each side of the tunnel section. The difference compared to a single wall is basically that the injection zone is smaller, and it is not necessary to create a more superficial column of jet grouting at the point of intersection of the two walls, apart from a small length (safety margin) in order to guarantee the correct functioning of the whole.

The fundamental advantage of this type of treatment is that it deals with the entire settlement basin, thus appreciably reducing the movements along the whole length of the tunnel being treated. Furthermore, this treatment has shown to be highly efficient for minimising excavation instability problems since it constitutes a certain pre-sustaining structure above the tunnel crown. In some zones the treatment was carried out after detecting significantly higher excavation volumes than the theoretical values during the execution of the tunnel. After carrying out the treatment, the volumes of over-excavation were drastically reduced.

3.1.3   Reinforcement treatment at portals

This consists of strengthening the material above the tunnel using jet grouting. This type of treatment was done at various portals. At these locations, the ground may be decompressed due to the construction of shafts or stations, as well as not benefiting from the three-dimensional "dome" effect, thus giving rise to possible over-excavations originating and greater deformations on the surface, or even instabilities leading to cave-ins. At these points, moreover, the tunnel is usually more superficial which means that the overburden is usually made up of weaker material.

3.2 Case histories

Table 1 below shows different cases of jet grouting treatment, of the above stated types, carried out by GEOCISA during the 1999 – 2003 extension of the Madrid Metro

TABLE Nº 1: Jet grouting works undertaken on the Madrid Metro

Date	Job		Treatment type	Treatment zone (m)	Pressure (bar)	Cement take (kg/m)	metres of drilling	metres of treatment
September 2000	Metrosur Section X	Station 1 entrance and exit portals	Reinforcement	18	100-400	250-300	500	300
December 2000	Line 10 Section 2	Crossing under line C-5 of RENFE suburban service	Wall	140	400-500	150-350	2800	2200
January 2001	Metrosur Section X	Station 2 entrance and exit portals	Reinforcement	19	250-300	200-250	1500	450
August 2001	Metrosur Section X	Crossing under line C-5 of RENFE suburban service	Wall	70	200-250	200-250	850	550
August 2001	Metrosur Section X	Entrance portal to the extraction shaft	Reinforcement	6	300-350	200-250	610	530
February 2001	Metrosur Section X	Protection of buildings between chainage points 3+660 to 3+815	Inverted V	155	300-400	250-300	2600	1950
May 2001	Line 10 Section 2	Crossing under access to M-40	Wall	45	250-400	200-300	5500	2200
June 2001	Line 10 Section 2	Crossing under pipe φ 1600 CYII	Inverted V	24	100-400	100-250	2450	1610
June 2001	Metrosur Section X	Station 5 entrance and exit portals	Reinforcement	18	350-400	250-300	1900	1425
July 2001	Line 10 Section 2	Crossing under M-406 road	Inverted V	30	200-300	150-250	4200	3000
July 2001	Metrosur Section X	Protection of building at chainage 6+920	Wall	35	200-250	200-250	550	390
July 2001	Metrosur Section X	Crossing of structure (bridge) chainage 6+965	Inverted V	28	200-250	200-250	275	200
September 2001	Line 10 Section 2	Crossing under Mimbreras houses	Wall	110	200-300	200-250	2800	2350
September 2001	Line 10 Section 2	Exit of telescope No. 2	Reinforcement	8	300-400	200-250	470	220
October 2001	Line 10 Section 2	Crossing under high tension pylon	Wall	20	300-400	250-300	350	300

3.2.1 Metrosur section X (Leganés). Jet grouting wall for protection of the building around chainage point 6+920

In the section X of Metrosur around chainage point 6+920 a five-storey building was located very close to the tunnel and parallel to it, where the tunnel ran relatively superficial. The tunnel overburden was made up of soil of alluvial origin, and there was a risk both of considerable movements and of instability in the excavation face (cave-in). In order to minimise these risks a jet grouting wall was built.

During the drilling, the presence of not very dense alluvial soils down to a depth of between 10 and 12 m was indeed confirmed. Moreover, a particularly soft zone with voids was detected, which required special treatment involving first filling in the voids and then reinforcing the zone with columns of jet grouting.

The initial grouting parameters were 200 bar of pressure and cement take of 200 kg of cement per metre of jet grouting column. During the execution of the treatment the pressure was increased to 250 bar at some points, as it was found that the spoil material that was obtained was relatively high and contained only a little mixed soil.

The control of movements of the building was carried out from the start of the execution phase of the treatment. The movements registered while undertaking the treatment were not very significant (less than 2 mm), with a maximum settlement being registered after construction of the tunnel of less than around 2 mm, compared to the 10-12 mm registered in nearby untreated zones.

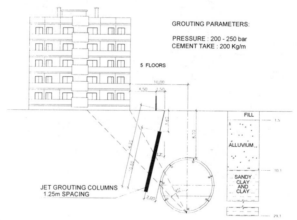

FIGURE Nº2. METROSUR SECTION X. P.K. 6+920: Wall type treatment.

3.2.2 Line 10 section 2 (Alcorcón). Inverted "V" jet grouting treatment at the crossing under accesses to the M-40 highway from Alcorcón.

The two tunnels of Line 10 section 2 around station 1+980 cross under the accesses to the M-40 highway from Alcorcón, with this access being a strategic communication route for this suburb of Madrid.

The thickness of overburden in this section was around 17 m, made up of clayey silty sands with a fines content of between 10% - 15%. In the excavation of the first tunnel about 100 m from this crossing, there were stability problems in the excavation face and major over-excavations, and it was therefore decided to carry out inverted "V" jet grouting treatment from about 80 m before this point.

The design of the treatment under the access to the M-40 highway was conditioned by the impossibility of occupying the roadway. Given this conditioning factor, the injection holes were undertaken from the hard shoulders. The drilled length of the columns varied between 25 and 55 m and their inclinations was between 43° and 65°. Due to these large magnitudes, it was necessary to use bentonite in the drilling fluid in order to be able to carry out the drilling.

The initial grouting parameters were a pressure of 250 bar and a cement take of 200 kg/m. During the execution the pressure was varied between 200 bar and 350 bar and the cement take between 150 kg/m and 300 kg/m, in accordance with both the drilling and overflow parameters obtained.

As with the previous case, control of the movements was carried out from the start of the ground treatment works on site. During the excavation of both tunnels, there were no instability problems at the face of the excavation and the movements that were registered were between 5 mm and 12 mm, values considered perfectly admissible.

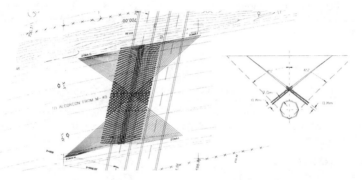

FIGURE Nº3. Line 10 section 2. Jet grouting in inverted "V" crossing under the access to the M-40

4 CONSOLIDATION AND PERMEATION GROUTING

4.1 Design of grouting using tube à manchettes

Throughout the construction works of the extension of the Madrid Metro, the use of tube à manchettes for grouting has been applied principally in the following three cases:

- Consolidation and permeation of the ground to be excavated
- Consolidation and strengthening of structures within the influence zone of the tunnel excavation
- Consolidation and strengthening of the ground at tunnel portals

4.1.1   Consolidation and permeation of the ground to be excavated

For this type of grouting a cement bentonite mix is used in a first phase of the process with the objective of consolidating the ground filling any possible voids. Subsequently a chemical grout using water, sodium silicate and a hardener whin sets once inside the voids in the ground is used in a second phase given the greater access of this grout into the smaller voids not filled by the cement grout, and expelling the water from within it. This treatment has been applied in areas where the presence of water has been detected during the excavation process.

4.1.2   Consolidation and strengthening of structures within the influence zone of the excavation

In cases where a structure could be found within the influence zone of the excavation, some kind of element was required between the structure and the tunnel in order to minimise movement induced by its excavation. This treatment has mainly been used of the protection of existing pipes located above the crown of the tunnel under construction.

4.1.3   Consolidation and strengthening of the ground at tunnel portals

In this case, the ground above the portal areas is reinforced using this grouting procedure. This ground treatment has been undertaken from the surface, generally though vertical injection holes or with slight off vertical inclinations to avoid interference nearby services, with the treatment covering areas up to 2 m above the tunnel crown.

4.2 Case histories

The following table indicates some of the works undertaken by GEOCISA during the extension of the Madrid Metro between 1999 and 2002.

TABLE Nº2. Consolidation grouting and permeation grouting works undertaken on the Madrid Metro

Date	Zone		Treatment Type	Treatment Zone (m²)	Injection hole Spacing (m)	Meters of drilling	Meters of Treatment
January 2001	Metrosur Section X	Portal at Station no. 2	Consolidation	240	1.5	1600	530
January 2001	Metrosur Section X	Entrance portal at Station no. 4	Strengthening	513	2.6	351	325
January 2001	Metrosur Section X	Consolidation of pipes. Station no. 4	Consolidation	176	1.5	250	96
March – June 2001	Line 10 to Metrosur	Adit between shaft and main tunnel	Permeation from shaft	402	1.35-2.0	4532	2800
April – June 2001	Line 8	Crossing with Line 10 – Castellana Avenue	Consolidation	2244	2.0	4600	2700
April – June 2001	Line 8.	Raimundo Fdez Villaverde St	Consolidation	565	2.0	1650	360
April – September 2001	Line 8	Consolidation pipes under M-30	Consolidation	425	1.0-1.5	575	230
September 2001	Line 10	P.K.27+122	Consolidation and Permeation	182	1.0-2.0	944	370
January 2002	Line 10	P.K.27+200	Permeation from shaft	607	2.0	840	758
February – March 2002	Line 10	P.K.27+376	Permeation from Tunnel	150	1.0	721	331
March 2002	Line 10	P.K.27+220	Permeation from Tunnel	480	2.0	460	420
March – April 2002	Line 10	P.K.27+364	Permeation from Tunnel	540	1.0	1250	1170
April 2002	Line 10	P.K.27+253	Permeation from Tunnel	460	1.5	545	489

4.2.1 Consolidation grouting near the Olympic Committee building: Line 8 of the Madrid Metro

As part of the works executed on account of the extension of line 8 of the Madrid Metro, in November 1997 the excavation was completed of one of the two parallel tunnels (the North tunnel) linking Glorieta de Mar de Cristal with the Recintos Feriales.

Following the passage of the 9.38 m North tunnel excavated using a EPB through the area near the Olympic Committee building (station 0+300 to 0+360) settlements of up to 275 mm were detected, implying a volume loss of 3.9% for an estimated settlement trough width of 12 m.

The ground in this zone is made up of quaternary fills and/or alluvial soils (NSPT<20) of variable thickness between 6 and 8.5 m, and of materials of medium to coarse type sand (NSPT<60) down to between 8.0 and 18.0 m. There is a lateral variation in what would be the top of the non-weathered soil level, with greater thicknesses of weathered soil being recorded in the area nearest the Olympic Committee building.

The treatment that was decided on as being the most suitable for the consolidation of the upper crown of the South tunnel consisted of fracturing grouting with cement grout though tubes à manchettes, and it was were carried out between January and February 1998. 28 sections were drilled at 2.0 m spacing, with the injection holes being located between the North tunnel and the South tunnel. Figure 4 shows a typical transverse section.

During the fracturing with the cement mix, grouting without pressure limit (with the maximum safety pressure of the system) was carried out in four sessions with prefixed volumes according to the grouting programme, achieving average porosities (volume fitted / apparent volume to treat) of between 12 and 13 % (figure 5), a value that lies within the magnitude of that forecasted by the execution procedure. In general, an increase in pressure was observed as volumes were accumulated from the first stage up to completion of the treatment.

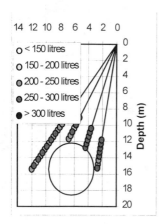
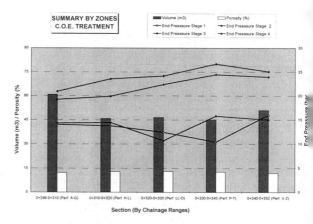

FIGURE N°4: Typical transverse section

FIGURE N°5 Distribution by zones of the grouting volumes and pressures

Together with the treatment, monitoring work was carried out in order to follow at all times the possible movements deriving both from the execution of the treatment as well as from the excavation of the South tunnel.

Following the execution of the grouting works, the levelling markers in general showed a certain amount of heave, the maximum value (around 25 mm) being recorded in the central part of the treated zone, while near the edges the heave recorded was between 2 and 10 mm. The recorded accumulated horizontal displacements, mostly perpendicular to the axis of the South tunnel and in the opposite direction to the treatment zone, was around 8 mm.

Following the execution of the South tunnel, the levelling showed an increase of maximum settlement of 10 mm, for an estimated settlement trough width of 34 mm, which is more than seven times less than the volume loss recorded in the same control section following the passage of the North tunnel.

Finally, no significant movements were observed in the actual building of the Olympic Committee, nor in the markers close to the building.

4.2.2 Permeation grouting: Stations 27+365 - 27+380: Line 10 of the Madrid Metro

In order to increase safety during excavation and reduce instability problems, ground treatment was considered necessary after having observed a large inflow of water accompanied by a localised collapse of sandy material at the tunnel face during the excavation works, which in this case was undertaken using the Madrid Method (hand excavated method, very similar to belgium method).

For the execution of the treatment, an umbrella was created from the inside the tunnel, 1.6 m thick at the crown, and reinforced with a 3 m transverse closure section. The grouting was executed in two phases, first one half of the umbrella and then the other.

The umbrella designed was 15 m long, with maximum injection hole spacing of 1.0 m and sleeves every 0.5 m along the length of the tube à manchettes installed. The maximum volume of each grouting session was set at 150 litres per sleeve.

The grouting works were carried out during February and March 2002. Prior to the grouting via the tube à manchettes, a water-reactive resin was grouted in the main entrance routes for water into the tunnel. Although cement bentonite grout was injected in fourteen injection holes in which the loss of material was detected during the drilling, the grouting was mainly done with silicates. The average volumes grouted per sleeve and the average pressures for each phase are summarised in Table 3 below:

TABLE N°3: Summary of grouting parameters

Phase	Average No. of grouting episodes	Volume/sleeve (l)	End pressure (bars)
Phase One	1.3	56.1	16.0
Phase Two (transv. closure)	1.3	160.8	22.2
Phase Two (umbrella)	1.5	197.4	17.2

As can be seen in the table 3, the average volumes were much lower in phase one than in phase two, where pressure limit was increased to 30 bars in order to enable the grout to reach the sandier lenses.   The total volumes grouted in each of the phases varied from 36.1 to 85.3 litres, with the porosities (volume filled / theoretical volume to treat) varying from 12 % to 41 %. In the second phase a volume greater that that initially predicted was grouted. Moreover, end pressures greater than 15 bars were achieved in 70% of the sleeves.

Once the treatment was completed, as the excavation face progressed it was seen that the upper sands were permeated with the grouted silicate, while in the ground below them, which is more clayey, near horizontal slabs of silicate had been created .

In this way, excavation of the tunnel was possible along the 15 m of umbrella with a controlled inflow of water.

5 COMPENSATION GROUTING

5.1 General

Compensation grouting has and is being used extensively as an effective method of protecting structures against settlement in both the last and the present extensions to the Madrid Metro system. This method, broadly speaking, consists of minimising settlement at ground level by injecting carefully calculated and controlled quantities of grout into the ground between the source of the settlement (the tunnel) and the elements to be protected (foundations of the buildings, services etc.), at specific times and locations relative to the excavation of the tunnel itself.

The design of the treatment in order to obtain grouting volumes and sequences were undertaken taking into account the predicted settlement for the area considered or assuming a percentage of ground loss taking into account previous experiences in the area. The main principle involve is the assumption, based on observed data in green field sites and in previous compensation grouting works, that the settlement produced by the excavation of the tunnels follows a near gaussian distribution and that the grout injected into the ground produces heave which mirror images this settlement. The difficulties lie in predicting the volume loss induced by the excavation and the relationship between the volume of grout injected into the ground and the volume that effectively produces heave. The factor relating these values of volumes is known as the efficiency factor. It should be noted that this is in fact a three dimensional effect.

A very important part of the design procedure, continuously adjusted during the grouting treatment, is the result of the instrumentation installed to monitor the surface, subsurface and structure movements.

The grouting procedure used in all the cases described, and which has now become standard for this type of ground treatment, is that of the three phases: pre-treatment, concurrent and observational where required. The pre-treatment prepares the ground for immediate response in the concurrent phase and the observational phase is used to produce heave once the excavation has passed but where the residual settlement are still unacceptable.

5.2 Case histories

The table below summarises the works that have been undertaken during the last six years from both of the above extension projects and also includes the most recent project, currently nearing completion to concluded by May of the present year. Accounts of the special features of a number of these examples are given in the following paragraphs

TABLE Nº4. Compensation grouting works undertaken on the Madrid Metro

LOCATION	DATE	INJECTION HOLES				TUNNEL		GROUTING
		Direction	Area (m²)	Maximum length (m)	Diameter (m)	Procedure	Volume loss (%)	Procedure
Connection of Lines 8 and 10	11/96 – 05/97	Horizontal from shafts	2880	55.1	10.5	Madrid Method	0.5	Concurrent
Line 7 section IV Ramón Gómez de la Serna St.	06/97 – 09/97	Inclined from surface	322	30	9.38	EPB	0.25	Concurrent
Line 7 section IV Santiago de Compostela St.	06/98 – 09/98	Horizontal from shaft	1050	50	9.38	EPB	0.5	Concurrent
Line 7 section III Guzmán el Bueno station	01/98 – 04/98	Near horizontal from gallery	2632	60.8	9.38	EPB	-	Observational
Line 4 section II Santa Susana St.	01/98 – 10/98	Horizontal from shaft	504	40	9.38	EPB	0.25	Concurrent
Line 1 Extension to Vallecas: Pº Federico García Lorca 36, 38	04/98 – 10/98	Near vertical	700	22	10.5	Madrid Method	-	Observational
Line 10 Tribunal station	03/00 – 11/00	Near horizontal from shaft	2571	52.5	10.5	Madrid Method	-	Observational
Line 10 to Metrosur section 1A Sanchidrián St.	09/00 – 01/01	Horizontal from shaft	875	29	9.38	EPB	0.08	Concurrent
Line 8 – section Mar de Cristal to Nuevos Ministerios	01/01 – 05/01	Horizontal, near horizontal from shaft and inclined from surface	4187	63.5	9.38	EPB	0.2 – 0.6	Concurrent
Metrosur – sections II & III Móstoles	06/01 – 12/01	Horizontal, near horizontal from shaft	4757	52	9.38	EPB	0.1 – 0.4	Concurrent

5.2.1 Tribunal Station (Line 10)

The existing Tribunal Station on Line 10 was to be extended as part of the latest metro extension project. The work involved a station cavern 21 m long by 24 m wide to be constructed using the so-called German method around the existing tunnel. Three pilot tunnels were excavated, one in the crown and one at either side. Subsequently the crown and sides of the tunnel were excavated in stages and concreted. The old station was then demolished, the remaining ground excavated and finally the invert excavated and concreted. This work was undertaken from June 2000 to October 2000.

The compensation grouting injection holes were installed from within a shaft with a slight inclination, 10° from the horizontal. Pre-treatment grouting was undertaken and a total of 84,525 l were grouted, giving grouting densities of between 25 and 62.5 l/m^2. The maximum values of heave achieved during this phase of the treatment were between 1.05 and 1.89 mm. Following this pre-treatment grouting, it was decided that given the slow speed of advance of the excavation method being used, observational grouting would be used. Should movements exceed the 2.5 mm settlement limit established as a trigger, grouting would commence. However, from the finish of the pre-treatment grouting to the end of the construction of the invert of the new station, maximum settlements observed were in the range 0.86 – 1.33 mm and thus no additional measures were required. In this particular case the works can better be described as a contingency measure rather than mitigation measures. However should larger settlements have occurred, all the required means were in place to act immediately and avoid damage to the buildings.

5.2.2 Metrosur, Sections II and III in Móstoles

The part of the latest extension of the Madrid Metro in the south west of Madrid known as Metrosur, is aimed at integrating the outer areas of Madrid with the centre of the city as well as creating easier access between these outer areas. In the suburb of Móstoles the tunnel crossed the centre of the town and five areas were identified as requiring special measures for the protection of the buildings under which it was to pass. As well as residential buildings, a school and kindergarden and a large Health Centre were also affected. The table below summarises these structures with details of the area in plan covered with the injections holes and the total lengths of sleeve port pipes installed for the compensation grouting treatment to be carried out.

TABLE Nº5. Summary of areas of compensation grouting in Móstoles

NAME	CHAINAGE	AREA (m^2)	LENGTH (m)
KINDERGARDEN	3+060 - 3+082	263	450
HEALTH CENTRE	2+780 – 2+952	2170	4170.5
ARROYOMOLINOS STREET	2+494 – 2+534	548	308
GOYA STREET	1+250 – 1+288	858	506
ALONSO CANO SCHOOL	0+447 – 0+489	918	644
TOTAL		4757	3378.5

The geology in this area is made up of a very thin layer of non-compacted fill and under which lies a succession of mainly stiff sandy clay layers with some clayey sand layer in between.

The drilling, pre-treatment, concurrent and observational grouting works were carried out between June and December 2001, with a total of 278.5 m³ of grout being injected. The case history described blow corresponds to work undertaken on the third of the areas, under the building on Arroyomolinos Street.

The 15 injection holes were drilled horizontally from a 5 m circular purpose built shaft with 3 m maximum spacing and the end of the injection holes. The level of compensation grouting was both 4 m from the base of the foundations of the building and also from the crown of the tunnel to be constructed. Throughout the grouting works the building itself was monitored by means of precise levelling of 12 points installed on the outside of the structure and two inside the building itself.

During the pre-treatment phase, 25,248 litres were injected giving an average grouting density of 46 l/m². Concurrent and observational grouting phases were undertaken and the volumes grouted were 4,620 l and 1,630 l.

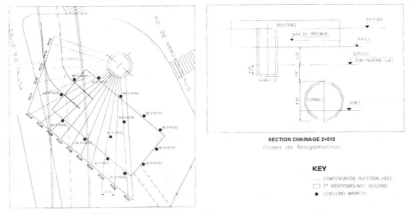

FIGURE Nº5: Layout in plan and in section of the compensation grouting injection holes

The instrumentation data indicated the end of the pre-treatment phase when an average heave of 2.8 mm was reached.

The concurrent phase of the works were undertaken while 65 m of tunnel were excavated under the building being protected. The first grouting episodes were injected when settlements of between 2.5 and 3.2 mm were observed. The volumes originally calculated were doubled as the observed settlement was already larger than the predicted ones and the tunnel had not yet crossed the whole of the area. The maximum settlements reached were 1.8 mm at a distance of 21 m from the centre line of the new tunnel and 6.3 mm at a location just above the tunnel crown. These are the values observed having undertaken a certain amount of grouting and could have been more severe should no protective measures have been taken.

An observational phase was required as following the passing of the tunnelling machine, certain control points were still below the admissible limits and additional grouting stages were executed. Volumes of 50 to 80 l were grouted at specific points and heave of between 1.2 and 1.5 mm were recorded, leaving the 15 control points at values of between 1 and 3.3 mm with an average value of 2 mm.

6 CONCLUSIONS

Given that safely is one of the prime objectives in all construction work, any additional measure proposed to ensure an adequate level of safety in difficult ground conditions or where special account has to be taken for structures, roadways and services due to their proximity to the works, are welcomed. Different types of grouting have been shown to have many applications in these situations and have been successfully used and adapted to each particular case within the works associated with the construction of the tunnels forming the new underground system in the centre and outskirts of Madrid.

ACKNOWLEDGEMENTS

The authors would like to thank the Madrid Regional Authority along with the Professors Carlos Oteo and José María Rodríguez Ortiz for their collaboration. Also particular thanks to Álvaro López Ruiz for his participation in the design of grouting mixes.

REFERENCES

Warner, J. Compaction grouting. The first thirty years, *ASCE Specialty Conference on Grouting in Geotechnical Engineering*. New Orleans. February, 1982, pp. 694-707.

Stilley, A.N., Compaction grouting for foundation stabilisation. *ASCE Specialty Conference on Grouting in Geotechnical Engineering*. New Orleans, February, 1982, pp. 923-937.

López Ruiz, A., Nuevas inyecciones químicas estructurales de base silicato en la Ingeniería Geotécnica. *Revista Informes de la Construcción del I.E.T. de la C y del C.* # 341, June 1982, Spain.

Baker, W.H., Cording, E.J. and Mac Pherson, N.H. 1983. Compaction grouting to control movement during tunnelling Underground Space Vol. 7, Pergamon Press Ltd.

Chen, X.L., Liu, Y.H., Cao,W.H. and He, Z.F. 1998. Protection for the former observatory during construction of the Yan An Dong Lu tunnel. Proc. World Tunnel Congress. Sao Paulo, ed. Balkema, Vol. 2, 1083-1088.

Moreira, J. and Flor,A. 1998. The Lisbon Metro-Strengthening of building standing above the tunnels in the city centre. Proc. World Tunnel Congress. Sao Paulo, ed. Balkema, Vol. 2, 1065-1070.

Sola, P.R., Monroe, A.S. and López Ruiz, A. 2000.Compensation grouting in the London, Lisbon and Madrid Subways. GEO-DENVER 2000. ASCE. August 5-8. Colorado USA.

González, C., Deformations around a tunnel in soft ground. PhD Thesis, Cantabria University, Spain, 2002.

GROUTING TECHNIQUES AS PART OF MODERN URBAN TUNNELLING IN EUROPE

Eduard Falk[1], Keller Fondazioni
George Burke[2], Member, Hayward Baker

ABSTRACT

Grouting techniques are more widely accepted and used in urban tunnelling because the impact on existing structures is minimal or non–existent. The compensation of settlements due to tunnel excavations became a part of underground projects within the last twenty years as grouting techniques were adapted to the specific needs of controlled soil treatment. Not only permeation grouting is being used, but fracture grouting and compaction grouting are also being used in various geological and geometrical conditions. This paper deals with the elements of compensation grouting and the organizational aspects of grouting measures in combination with different tunnelling methods. Special emphasis is placed on the contractual approach to integrate compensation grouting in a design-and-built concept. The main objective is to give a general overview about the logical sequences of decisions to make regarding settlement control during tunnelling operations.

1. INTRODUCTION

Although excavation methods are accepted and used in permanent development, urban tunnelling excavation methods are still considered a threat to existing structures. The approval of new alignments can only be obtained when a clear concept detailing the controlled handling of unavoidable settlements is presented by the designer. Because economical issues are a consideration determining whether or not technical projects move forward, it is important to provide a design concept with

---

[1] Keller Fondazioni S.r.l., Via della Siderurgia 10, 37139 Verona, Italy,
e.falk@kellergrundbau.at
[2] Hayward Baker Inc, 1130 Annapolis Road, Odenton, Maryland 21113,
gkburke@haywardbaker.com

a detailed risk analysis for the buildings or services to be protected. The area to be treated by grouting and the type of additional measures planned to ensure careful tunnelling are usually selected according to the space available for working platforms and the expected development of settlements over time. Therefore, a wide range of possible applications of grouting measures is shown in the following sections of this paper. According to the construction phase when measures are taken they can be divided into different types of approaches:

- stabilization and / or relevelling of already settled and eventually damaged structures = "repair"
- soil improvement in order to reduce total settlements = "preconditioning", "preconsolidation"
- simultaneous correction of settlements supported by sophisticated real – time monitoring = "compensation grouting"

2. SETTLEMENT PREDICTION

The prediction of the shape of a settlement trough is directly related to the soil conditions and the excavation method. It is common practice to use the volume loss as a percentage of the excavated volume to analyze the settlement contour in two- or three-dimensional models. Thus the choice of the most realistic percentage of volume loss (usually from 0.5% up to 2.0% of the excavated volume) turns out to be the most critical parameter. Practical examples show that the most important influence to this parameter comes from the controllability of the tunneling process and its ability to react on unforeseen situations. While shield-machines and TBM-operations ("mechanized tunneling") provide a limited range of possible reactions to unfavorable soil conditions, NATM excavations (small face excavations in different phases) are more flexible regarding additional measures to reduce convergence. It can be stated generally that unknown obstacles in the tunnel face are much more critical to excavation processes than slight variations within a soil type. Old water wells or remnants of manmade underground spaces can provide very dangerous situations. In such cases previous grouting operations provide additional safety because of the information provided from boring required to access the tunnel zone for grouting (Fig.1).

3. POTENTIAL DAMAGE ASSESSMENT

When considering possible damages it is important to understand that only structural damages can be evaluated easily in commercial terms. Interruptions in the use of a structure can exceed by far the value of the building itself. In addition, there are times when the preservation of the national heritage determines particular emphasis on the protection of historical buildings.

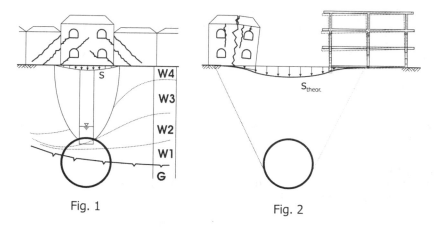

Fig. 1   Fig. 2

FIG. 1. Manmade shafts in soft soil conditions can provoke fast volume loss

**FIG. 2. The damage assessment combines the contour of the settlement trough
and the rigidity of the structure in an interactive deformation model**

Various classification systems exist to evaluate potential damages on services and structures on the basis of assumed settlement contours. More advanced models also deal with the interaction between the deformed soil and the stiffness of stiff structures (Fig.2). It is important to mention, that the real load bearing conditions of possibly modified buildings are better to be verified in a direct site visit rather than by examining of drawings. The rigidity of structures has a large impact on the development of cracks as a consequence of differential settlements. Historical buildings are usually not very rigid even when they have thick walls (Paper by Mair and Taylor 1997).

Beside the structural value of a building, the amount of a possible damage is related to the function or the idealistic worth. Typically commercially-oriented risk assessments have been overruled in the past in cases when cracked buildings where publically discussed in the media and the tunnelling operation was stopped.

4. SETTLEMENT REDUCTION OR DAMAGE REDUCTION?

The possibility to integrate grouting measures in an underground project depends on the contractual situation. If compensation grouting was not foreseen in the original project it can be difficult to mobilize the necessary equipment for a short-term intervention. It usually takes time to identify the parameters in technical and economical terms. Expensive interruptions and delays of the whole project are a common scenario.

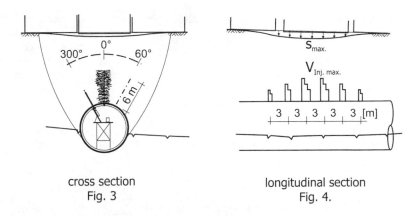

cross section
Fig. 3

longitudinal section
Fig. 4.

FIG. 3. Radial drillings were used for reconsolidation with fracture grouting: only 3 drillings with 6m length equipped with 10 grouting valves

FIG. 4. Grout quantities corresponded with settlements at the surface

When excessive settlements occur only the safety aspects are addressed immediately. Remedial work usually take place afterwards, and sometimes only after the distribution of the responsibility is clarified.

4.1 Remedial grouting after a tunnel collapse

The first example shows two situations where the mostly passive control of settlements through the TBM-operation turned out to be not sufficient. A 9m-diameter underground tunnel should be excavated in very heterogeneous soil conditions. In general the soil was wheathered granite in different stages of degradation (according to a scale from W1–slightly weathered – to W6–wheatered to clay-water bearing). While the rock outcropped at the surface in some spots the clay-like or coarse sand layers or lenses (W4) could reach 30 m depth (Fig.1).

4.1.1 Remedial grouting from the tunnel

After approximately 100 m of excavation, an ancient water well was found under an old building (Fig.1). Part of the building collapsed fortunately without personal injured. Although post-injection was intensified after this incident, long-term settlements were observed in a few places along the tunnel. This led to the conclusion that eventually more unknown voids in the ground as well as problems in the limitation of excavation volumes, could provide substantial risk for further collapses. In this case, consolidation of the loose zones and stabilization of the settling areas were obtained by radial grouting from the tunnel (Fig.3). The drillholes were sealed through the concrete lining with double preventers to protect drilling against groundwater. 2" sleeve pipes were installed in fans of 3m distance with 6 m length and soft grout mix was injected in a first phase. The grouting work was continued with injecting a water-cement-filler-bentonite mix with a density of 15,5 kN/m³, using accelerators. Because the soil found was mainly silt and fine sand, fracture grouting was used to get a wider radius of influence (in a range of 2,0 m). This technique made it possible to significantly reduce the number of pipes that needed to be installed in the overhead section of the platform. Over 200 m of tunnel length were stabilized with separate post-grouting. The high grout consumptions always corresponded to the higher settlements at the surface (Fig.4).

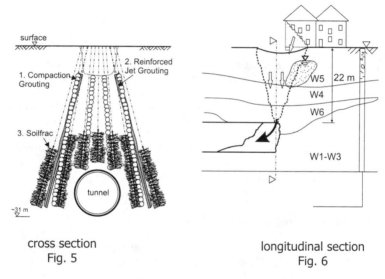

cross section
Fig. 5

longitudinal section
Fig. 6

FIG. 5. and FIG. 6   Consolidation scheme after a tunnel collapse combining compaction grouting, jet grouting and fracture grouting as well as chemical grouting in front of the cutting head

4.1.2 Remedial grouting from the surface

A further collapse happened in the same tunnel, when a water-filled lens of sandgrained weathered granite was encountered in a zone of more cohesive material. Before the counterpressure of the TBM could be regulated properly, large settlement reached the surface and caused the collapse of an ancient building. One person died in this incident. The tunneling operation was stopped for a few months and large remedial works took place to stabilize the affected area and to allow the restart of the TBM.

In the area of the incident, very loose sand was found as well as silt and some silty clay. It was determined that the best way to continue was to use a combination of grouting methods to reduce the risk of further settlements. Additionally it was recommended that it was important to maintain a safe distance away from the installed tunnel lining to avoid additional unfavorable pressures. After initial consolidation using compaction grouting a grid of jet grouting columns of small diameter was installed down to 25 m depth (triple system, 0,6 to 1,0 m diameter, relief holes to avoid heavings). The final consolidation was reached by fracture grouting with small quantities, similar to what was previously used in the tunnel.

Additional grouting measures including injection of resin, were used to stabilize the soil in front of the cutting head. This also allowed for some repair and modification of the cutting tools, and finally permitted the restart of the TBM. The following part of the tunnel in more compact granite was performed without problems. The last stretch, which again has a very low overburden above important historical buildings will be protected by means of preconsolidation.

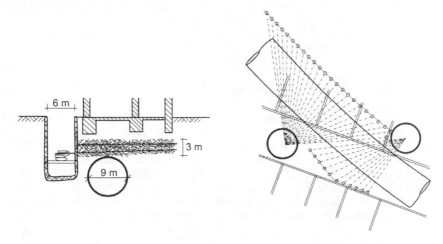

Fig.7 cross section   Fig.8 plan view

FIG. 7. and FIG. 8. Preconsolidation scheme using drilling shafts and two layers of 2" steel pipes to reinforce the ground between the tunnel and the foundations in a case where a settlement reduction is sufficient to avoid larger cracking

4.2 Preconsolidation as Settlement Reduction

Example shown in Fig.7 and Fig.8.

When active compensation grouting seems to be inadequate, it can still be useful to use passive measures to obtain a significant settlement reduction. The very low overburden of only 3 m between the tunnel and the ancient stone foundations requires additional safety measures, like reinforcement of the ground with steel tubes. Using these 3" steel tubes as sleeve pipes for repeated hydraulic fracturing until first lifting reactions are obtained, results in a considerable consolidation of the treated layer. In this case grouting is only foreseen prior to the passage of the TBM. Reductions of total settlements in a range of 10 to 45 % are likely to be obtained.

4.3 Compensation grouting

In situations where even a very controlled excavation method will cause unacceptable settlements or angular distortions, compensation can be foreseen as a continued operation before, during and after the tunnel excavation. In particular, phased excavations of large underground spaces can require a reduction of settlements over a long period.

4.3.1 Aspects of Compensation Grouting(Fig.10)

Effective settlement compensation can only be obtained if the stress conditions inside and around the treated zone are modified in a suitable way. Increased horizontal stresses compared to vertical stresses allow heaving operations, comprising high loaded single foundations. Basically, only geometrical relations limit the possibilities of settlement compensation. Different loading conditions of neighboring foundations, different foundation levels, or a combination of footings and selfbearing slabs can be handled if the thickness of the treated soil and the remaining natural soil allows an adequate distribution of the involved forces. It is important to state clearly that hydraulic fracturing includes significant deformations of and around the treated zone, and that these movements are to be controlled in the best way possible. Usually experts are more concerned about the pressures applied during grouting than about the implemented deformations. In that respect it is important to separate between pumping pressures and effective pressures in the ground, which always remain in the range of the overburden pressure. However, it may be necessary to use pumps with the capacity to achieve 100 bars, to open the primary space around the sleeve where the injection is done using a double packer. Specifications should always require equipment to provide the capacity of controlled hydrofracturing in all types of soil such as: high-shear mixers for grout up to a water solid ratio of 0,4; a sufficient number of pumps with computer control and automatic data acquisition, pumping rates from 1 to 20 l/min and a pressure range up to 100 bars; sleeve pipes/tubes a manchettes ("TAM's") with a sufficient durability to maintain usability also after a great number of grouting operations (double packers sometimes are inflated up to a pressure of 200 bars to remain in their position).

4.3.2 Elements of Compensation Grouting

The practical application of compensation grouting requires a number of working steps. Some of them involve the use of sensitive equipment that is usually not designed for operating in dusty and dirty environments.

4.3.2.1 Working Platform

A very limited space is sufficient to install a drill rig for subhorizontal or inclined drillings and to provide the access to the TAM's during the grouting work. Spaces of 15 to 30 m² are commonly utilized for drilling while the batching/mixing/pumping plant can be installed at a distance of more than 100 m away. The main objective of fracture grouting arrangements is to reach the zone to be treated with a minimum of drilling meters, and to install a sufficient number of sleeves in homogeneous grid distances. Drillings can be performed from the existing ground level or from excavated trenches. It can also be performed from shafts, existing excavation pits or abandoned tunnels.

4.3.2.2 Drilling

Boreholes for the installation of sleeve pipes are usually made with cased drilling methods of at least 100mm diameter, always providing a sufficient coverage of the 2" tubes with a medium-hard anular grout (Fig. 9).

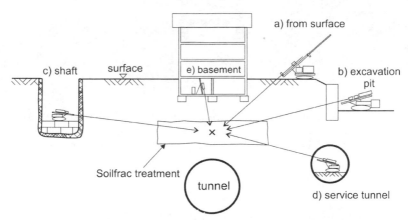

FIG. 9 Options for drilling arrangements a) from the surface; b) from a working trench; c) from a circular or rectangular shaft; d) from existing service tunnels; and e) from basements

Drilling lengths of 70 m have been reached in certain cases. However, this involves increased operational problems which can result in the loss of a number of sleeves. In some situations it is almost impossible to install additional tubes because of logistic reasons.

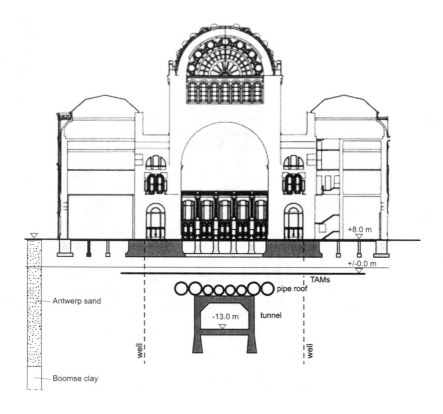

FIG. 10 Complex compensation grouting application: a high speed railway tunnel is excavated below a roof of 8 microtunnels and between diaphragm walls; a historical railway station was to protect in a two year operation

4.3.2.3 Grouting

A wide range of grouting can be used for compensation grouting. Water – solid ratios are varied between 0,4 and 1,0 according to the soil conditions and the working phase. Some granular soils have to be pretreated with permeation grouting to become suitable for fracture grouting. Quantities per single grouting operation are ranging from 20 to 120 l. Accelerators can be added to the mix consisting of water, binder, filler, bentonite to minimize the amount of bleeding water in cohesive soils.

4.3.2.4 Monitoring

An observational approach includes not only the real time-acquisition of all relevant grouting parameters but also the continuous observation of movements of structures. All important parts of buildings to be controlled are instrumented with automatically registered gauges such as: electronic water level systems, hydrostatic pressure levelling, automated geodetic survey, permanent vertical or horizontal inclinometers, tiltmeters etc. Traditional levelling is used as a reference system when technical problems are faced.

All data shall be represented in a three-dimensional graph (based on the CAD-drawings of the buildings to be protected) to allow a quick interpretation by the responsible site manager. The software should also contain the option to define limits for angular distortions between selected points to visualize the actual alert status.

4.3.2.5 Site organization

For complex compensation targets it is recommended that different levels of safety related to the development of differential settlements over time be defined. The frequency of readings in the monitoring system can be directly referred to the actual safety level. Increasing settlement or lifting rates are generally the most significant parameters to determine alert situations. The benefit from such a sophisticated system depends mainly on the reaction to the provided information. The site managers for the tunnel operation as well as for the grouting activities must be available 24 hours a day on short notice, to make decisions with the client's supervisor. Decision may include halting the excavation or to evacuating people from a building at risk. Examples have shown that only an immediate and careful interpretation of the large quantity of data provided by frequent monitoring assures guarantees to avoid possible damages.

4.3.2.6 Contractual concept

The specification of compensation grouting works should make allowance for unexpected developments that can impact the original schedule. For example, delays in tunnel excavation for various reasons have an impact on standby periods for equipment. Also, there may be a need to analyze readings of the monitoring system in a more economical way.

The following items allow to specify the elements of compensation works:

a) site installation
b) drilling

c) installation of Soilfrac-sleeve pipes
d) primary grouting
e) repeated fracture grouting
f) parameter acquisition
g) monitoring
h) documentation of the works and elaboration of grouting programs

The nature of the above mentioned elements differs regarding the phases of a complex application like pretreatment, preconditioning, preheaving, standby of people and/or equipment, monitoring during standby, "alert status" of staff an equipment, compensation, postmonitoring, etc.

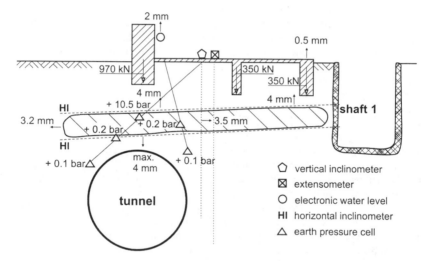

Fig. 11 Typical section of a large scale trial with extensive monitoring to get information about the ability to concentrate lifting actions under high-loaded foundations and to limit additional pressures to the tunnel lining

4.3.2.7 Field trials (Fig.11)

Large scale field trials should be part of the overall scope of work. The main objectives of trials are:

a) To show the ability to improve the soil and to lift buildings in an controlled way;
b) To assess the possible influence on a tunnel lining; and
c) To evaluate the efficiency factor of the grout during the compensation phase

It must be emphasized that the dimensions of a field trial shall be chosen according to the dimensions of the main work. Trials with less than 10 by 10 m extensions will only result in very limited conclusions, and they should not be taken into consideration when important data for the execution parameters need to be checked.

5. CONCLUSIONS

Grouting methods are used to reduce settlement before, while and after tunnel excavations. When required by a realistic risk assessment active compensation is the most effective way to benefit from the available technical resources.

REFERENCES

Chambosse G., and Otterbein R. (2001) „Compensation grouting under high loaded foundations", *CIRIA Conference*, London.

Drooff E., Tavares P.D., and Forbes J. (1995) "Soil fracture grouting to remediate settlement due to soft ground tunneling", *Proc. of the Rapid Excavation and Tunnelling Conference*, San Francisco.

Harris D. (2001) "Protective measures", *Chapter 11 in Building response to tunneling – Volume 1*, Projects and methods, CIRIA, Thomas Telford.

Melis M., Arnaiz M., Oteo C., and Álvarez J. (2001) "Inyecciones de compensación de la linea 1 del Metro de Madrid", Sociedad Espanola de Mecánica del Suelo.

Mair R.J., and Taylor R.N. (1997) "Bored tunnelling in the urban environment" *Proc. of the fourteenth international conference on soil mechanics and foundation engineering*, Vol. 4, Hamburg.

GROUTING TO MINIMIZE SETTLEMENTS PRIOR TO TUNNEL EXCAVATION – A CASE STUDY

Douglas, M. Heenan[1], P.Eng., Member, ASCE and Michael Xu[2], Ph.D., P.Eng.

ABSTRACT: Phase I of the Glen Echo Creek Culvert Reconstruction Project, involved lowering the invert of an existing concrete arch tunnel along with replacing an existing corrugated metal pipe (CMP) with a new tunnel between 28th and 29th street in Oakland, California. All construction work was performed within the tunnel by hand mining techniques. Due to the close proximity to residential buildings, roadways and utilities, various grouting techniques were applied to minimize ground settlement, stabilize subgrade soils and reduce water seepage to facilitate hand excavation. This paper describes the design and execution of the various grouting programs implemented, as well as the grouting performance.

INTRODUCTION

Glen Echo Creek flows through the City of Oakland, California. Channel hydraulic investigations indicated that the 25-year and 100-year storm events would cause short term, but widespread flooding along certain sections of the creek. The County of Alameda Public Works Agency, therefore, decided to conduct several improvements to the creek system. This project was a portion of Phase I the Glen Echo Creek Culvert Reconstruction Project, and involved lowering the invert of a concrete arch culvert tunnel and replacing an existing

---

[1] Vice President, Advanced Construction Techniques Ltd., 10495 Keele Street, Maple, Ontario L6A 1R7

[2] Geotechnical Engineer, Advanced Construction Techniques Ltd., 10495 Keele Street, Maple, Ontario   L6A 1R7

2.1 meter diameter CMP with a 3 meter diameter tunnel between 28th and 29th street, which is about two blocks east of Broadway in Oakland.

The existing concrete arch tunnel extended from station 23+99 to 26+76, was 85 m long with inside dimensions of 2.4 m high by 2.1 m wide was to be lowered by 0.9 m. The work in this section included removal of the existing concrete invert and subgrade materials and construction of new reinforced cast-in-place sidewalls and invert. Hand mining methods were used to perform the excavation work and grouting techniques were utilized for temporary support.

The existing 2.1 m diameter CMP extended from station 22+92 to 23+96, was 31.7 m long and was also to be lowered by 0.9 m. A small backhoe was used to remove the existing CMP. The open excavation was initially supported by stabilizing the ground using grouting techniques, followed by the installation of circular steel plates. The final support system consisted of a reinforced shotcrete lining.

SITE CONDITIONS AND GEOLOGY

There are several roadways, parking areas and numerous structures that are located adjacent to the culvert alignment as shown in Figure 1. Several one and two-story residential buildings are positioned along the west side of the alignment. Houses in close proximity to the culvert include a two-story wooden home at 4.6 m west of station 22+90 and a one-story home 3.0 m west of the culvert, between stations 24+80 and 25+70.

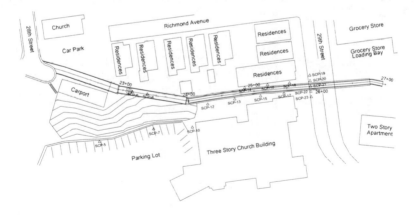

Figure 1. Site layout of line B improvement, Oakland.

A typical section at station 25+50 is shown in Figure 2. A three-story church was located 3 m to 6 m east of the culvert between stations 24+90 and 25+60, and the cross section of the structure relative to the tunnel alignment is shown in Figure

2, as well. An asphalt paved parking area associated with the three-story church overlaid the concrete arch section between stations 24+90 and 25+70. The alignment was intersected by 29th street that runs perpendicular to the project centerline at station 26+00.

The ground surface adjacent to the culvert alignment generally slopes east to west at a relatively moderate grade. Between stations 22+80 and 24+10 the culvert alignment is situated near the toe of a steep embankment fill slope to the east. This embankment supports a parking lot of the three-story church at about elevation 17.8 m. An asphalt drive/parking area associated with a small church to the east contains a carport canopy and is situated approximately 3 m east of station 22+80. A 1.4 m (4.5 ft) diameter tree is positioned near the slope toe and directly over the culvert at station 23+10.

Geotechnical investigation adjacent to the tunnel alignment indicated that the overburden materials above the CMP section consisted of silty sand and clayey sand fill, and then varied from clayey sand, silty sand, sandy clay to clayey gravel. At station 25+90, boring logs indicated that the top of overburden consisted of stiff silty clay fill to a depth of 4 m below ground surface.

Medium stiff to stiff silty clay with sand was present below the fill to a depth of approximately 7.9 m. The borings cored through the walls of the arch culvert encountered sandy clay and silty clay fill at the upstream portion of the culvert from station 26+55 to 25+34, and encountered silty sand from station 24+79 to 24+14. The borings cored through the floor of the arch culvert encountered sand and gravel. The grain size distributions of the subgrade soils are shown in Figure 3, indicating that the soils are not fully permeable at the top of CMP, and partially permeable at the invert of CMP and Arch tunnels.

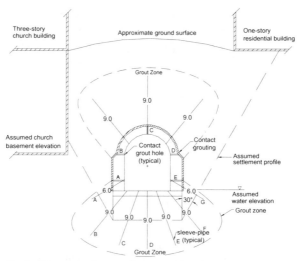

Figure 2. Typical layouts of grout pipes at station 25+50.

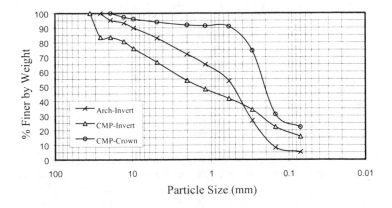

Figure 3. Soil particle size distributions.

DESIGN AND EXECUTION OF GROUTING PROGRAM

The grouting program was a design-build type of construction with performance based criteria. The main objectives of the grouting programs were:

1. To prevent settlement of buildings, utilities, and roadways in close proximity to the arch tunnel by means of ground stabilization.
2. To stabilize subgrade soils to facilitate open excavation.
3. To mitigate water seepage during hand mining operations.

To accommodate the above objectives, the following grouting programs were designed and performed at various locations in conjunction with different construction sequences.

Contact Grouting of Arch Culvert Concrete Tunnel

Given that the existing tunnel was more than 50 years old, it was anticipated that voids may exist directly above and beside the concrete arch walls and the concrete arch structure/soil interface. Contact grouting would provide a positive contact around the structure, resulting in reduced surface settlement while providing grout containment for the subsequent permeation or fracture grouting operations. Contact grouting was performed prior to any other grouting techniques.

The typical patterns of contact grout hole alignment are shown in Figure 2. Five 63.5 mm diameter holes, designated as A, B, C, D and E, were drilled through the concrete lining at a 1.2 m spacing along the alignment of the tunnel. Sequencing of the grouting operation was A, E, B, D, then C. Mechanical packers were installed and a neat cement grout was injected through the packer until a connection was established with adjacent holes or until a refusal pressure of 69 kPa was achieved.

During the contact grouting operation, extensive grout seepage occurred through the concrete structure at several locations. When excessive seepage was encountered grout injection was momentarily stopped, and fast setting hydraulic cement was used to seal off the excessive seepage and the grouting operation was then resumed. The total grout volume injected was 61 m^3, which exceeded the grout volume anticipated.

Pre-conditioning (Compensation) Grouting

Compensation grouting was to be performed during excavation in order to mitigate ground settlement caused by the hand excavation. This grouting was to be initiated by ground movement as indicated by surface monitoring. Three holes were drilled through the concrete tunnel lining at 1.2 m spacing. Custom fabricated 38.1 mm diameter steel sleeve pipes were then jacked into the roof at the arch tunnel, according to the alignment shown in Figure 2. Compensation grouting was not specified for the CMP, but was performed using a single sleeve pipe at the center of the roof (refer to Figure 2) in order to control the ground settlement due to shallow overburden coverage and close proximity to residential buildings.

Pre-conditioning grouting was carried out prior to permeation grouting and tunnel excavation at the arch tunnel, and was carried out simultaneously with the permeation grouting at the CMP. This operation was intended to condition the insitu soil prior to excavation to prevent large sudden settlements due to the existence of voids and loose soil. Ground heave was monitored during the pre-conditioning grouting. A limited amount of cement grout was injected at the concrete arch tunnel due to concerns of heaving the nearby structures. A considerable amount (19 m^3) of neat cement grout was injected at the CMP.

Soil Stabilization – Permeation and Fracture Grouting

As indicated by the soil particle distribution, the underlain soils were partially permeable at the invert of arch and CMP. The soil stabilization grouting was actually a combination of permeation and fracture grouting. Permeation and fracture grouting program was designed and performed, in order to stabilize the subgrade soils and facilitate lowering the culvert invert of the arch concrete tunnel, and the enlargement of the CMP. Sodium silicate solution grout was chosen for the intended purpose.

Typical grout hole layouts for the permeation/fracture grouting are shown in Figure 2. A series of holes designated as A, B, C, D, E, F and G were drilled through the floor slab and sidewalls of the arch culvert and CMP. Then various lengths of 38.1 mm diameter steel sleeve pipes were jacked into the pre-drilled holes to develop adequate grout zone coverage. After each sleeve pipe installation, the annular space between the steel pipe and floor slab was sealed for grout containment. The spacing for each array was 2.1 m along the tunnel alignment. Mechanical packers were used to seal individual sleeves with the grouting injection sequence proceeding from the side walls to the middle of the tunnel invert as A, G, B, F, C, E then D.

A sodium silicate grout plant (refer to Figure 4) was custom designed and commissioned for mixing and pumping the sodium silicate grout. A grout delivery system was installed throughout the tunnel with a header system used at point of placement. The grout plant was located outside the tunnel at the east end portal lay down area. The grout injection system consisted of a piston type two component pump with a 1 to 1 ratio. The hardener was pre-mixed with water as component A, and the sodium silicate solution as component B. Components A and B were pumped separately to the point of injection and combined at the header in a static in-line mixer prior to entering the soil.

The sodium silicate used was an N-grade silicate. It has a silica to soda ash molecular ratio (SiO_2/Na_2O) of 3.22. The silicate is composed of 28.7% SiO_2 and 8.9% Na_2O and has a density of 41°Be at 20°C, a specific gravity of 1.39, a viscosity of 177 cp at 20°C, a pH of 11.3, and solids of 37.6%. The hardener used was a Dibasic Ester (DBE) solution. The DBE is composed of 21% dimethyl adipate, 59% dimethyl glutarate, and 21% dimethyl succinate.

During the initial phase of the project, it was observed that the sodium silicate grout was not setting within the soil mass in a predictable manner. A quick set grout was required due to flowing ground water conditions. The sequencing of excavation immediately after grouting as well as ground water conditions required a modification to the grouting method in order to accelerate the setting time of the sodium silicate solution.   This was facilitated by the injection of a cement based grout.

The silicate based cement grout or cement based silicate grout have been used widely for applications where accelerated setting time of the injected grout is required (Shroff and Shah, 1992; Liao et al. 1992).   The modified permeation/fracture grouting method differs from the traditional Joosten process and overcomes many disadvantages of this method.   The Joosten process involves injecting concentrated sodium silicate into one hole and a strong calcium chloride solution under high pressure into an adjacent hole (Karol 1990). When a sodium silicate solution and a concentrated solution of appropriate salt are mixed, the reaction forms an instantaneous gel.   The Joosten process results in a strong gel, if properly used.   However, the high viscosity of the solution and the need for many closely spaced grout holes limits the use of this method.   Also, the nature of the process prohibits a complete reaction of the two liquids (Karol 1990).

The modified grouting operation involved multiple injections with a limited volume of grout material injected during each cycle, to facilitate a complete reaction of the two liquids. Mechanical packers were set near the top of the sleeve pipes and a predetermined volume of sodium silicate grout was injected into each hole according to the injection sequence. The sodium silicate grout was then immediately followed by injecting a predetermined volume of cement grout to accelerate the set of the sodium silicate grout. After all holes along a segment of tunnel had been injected (typically 6 to 10 m), a second cycle of injection was performed with a

Figure 4. Sodium silicate grout plant.

lower packer setting and higher injection pressures. This process was repeated until a predetermined total volume of grout for each section was reached or until the maximum injection pressure was reached with no flow.

The arch invert tunnel was divided into 69 sections, each covering 1.2 m of the tunnel. As a result of a stringent construction schedule, the excavation had to be preformed at five locations simultaneously. In order to provide stabilized soil for the open excavation, the grouting work was divided into 10 segments. Initially, 5 segments with 5-6 sections each were grouted at odd locations, with excavation proceeding with the 5 grouted segments. The remaining 5 segments, typically 8 sections, were grouted later along with the excavation. Figure 5 shows the grout volume injected at the arch tunnel. The average grout take is approximately 1.7 m^3 and 0.7 m^3 of sodium silicate grout and cement grout, respectively. It was observed during all excavation that the soils remained stable and no soil losses occurred.

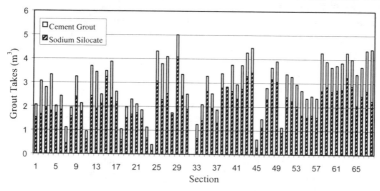

Figure 5. Permeation grout takes at Arch tunnel.

At the CMP, permeation/fracture and compensation grouting were carried out simultaneously as shown in Figure 6. Figure 7 shows the grout takes at the CMP, with an average of approx. 2 m^3 and 1 m^3 of sodium silicate grout and cement grout, respectively. After completion of all grouting work at the CMP, the excavation was started. It was observed that there was no soil loss occurring and the grouted soils were very stable, as shown in Figure 8. As a result, the open excavation could be carried out with longer segments during each cycle without the need of temporary support, typically 5.5 m, as compared with anticipated 1.8 m per cycle.

Figure 6. Permeation/fracture grouting and compensation grouting at CMP.

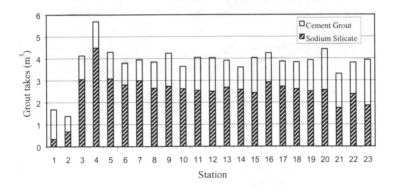

Figure 7. Permeation/fracture grout takes at CMP.

GROUTING PERFORMANCE

Prior to and during grouting and excavation, elevation changes of ground surface, buildings and sub-grade soil were closely monitored using various monitoring techniques. The layouts of surface control points are shown in Figure 1. The ground movements at Arch and CMP are shown in Figures 9 and 10.

Figure 8. Exposed soil surface after excavation at CMP.

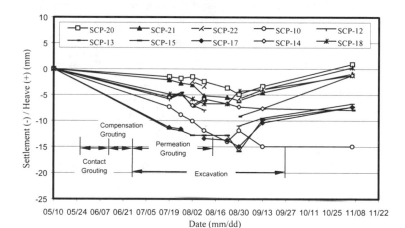

Figure 9. Movements during construction at Arch tunnel.

As indicated in Figure 10, the maximum ground settlement was 15.8 mm during construction at SCP-15 and 8 mm after construction at SCP-13 on the East side of the tunnel, in close proximity to the three-story church. The maximum ground settlement was 10 mm during construction and 3 mm after construction at SCP-16 on the West side of the tunnel. The maximum settlement on the roadway was 6 mm. No damage to the adjacent buildings were observed or recorded.

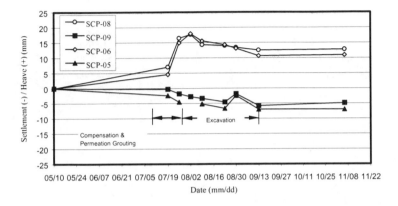

Figure 10. Movements during construction at CMP.

A maximum ground heave of 18 mm was intentionally created at the crown of CMP, which then settled to 12 mm after construction. A maximum settlement of 15 mm was observed at SCP-10 on the parking lot, and 7 mm at SCP-05, which was farther away from the CMP. No damage to adjacent buildings or roadway was observed or recorded.

Conclusions

The various grouting programs developed and instituted for the project were successful in minimizing ground settlements and facilitating open hand excavation. The grouting techniques utilized were practical and suited the site specific conditions, allowing the project to be completed on schedule. As is the case with most grouting projects, modifications were made to original grouting plans to best suit the existing site conditions.

References

Henn, R.W. (1996). "Practical guide to grouting of underground structures." ASCE New York, 191p.

Karol, R.H. (1990). *Chemical grouting,* Marcel Dekker Inc., New York, 465pp.

Liao, H.J., Krizek, R.J. & Borden, R.H. (1992). "Microfine cement/sodium silicate grout." *ASCE Geotechnical Special Publication No. 30,* New York, pp676-687.

Shroff, A.V. and Shah, D.L. (1992). "Chemical based cement grout system for rock grouting." *ASCE Geotechnical Special Publication No. 30,* New York, pp651-662.

DESIGN OF GROUTING PROCEDURES TO PREVENT GROUND SUBSIDENCE OVER SHALLOW TUNNELS

Ross T. McGillivray, P.E., Member, Geo-Institute[1]

ABSTRACT: Tunnels 3 m (10 feet) in diameter used for road and railroad crossings of water pipelines in Hillsborough County, Florida suffered ground surface subsidence problems due to their shallow depth and sand soil cover. The soil depth above the top of the tunnels was typically no more than 3 m (10 feet). A controlled low strength cementitious grout was designed to fill the void between the tunnel and the liner plates, and to maintain pressure around the tunnel boring machine (TBM) equal to the overburden pressure. Pre-grouting could not be done due to the low permeability of the fine sand soils and the lack of surface access. Therefore, the grout had to be designed for injection from ports inside the TBM without causing damage to the machine. This paper describes the design of the grout and the testing program to document the engineering properties of the grout, as well as the field grouting procedures developed for the project. The results of field measurements and laboratory test results are also presented.

INTRODUCTION

A series of large diameter pipelines were installed to distribute raw and treated drinking water as part of a water supply project for the west central area of Florida. The pipelines were 2.4 m (8 ft.) diameter steel. In order to limit disruption of traffic, all major road crossings were done using tunneling techniques. One contractor was using an Earth Pressure Balance Tunnel Boring Machine (TBM) to install tunnels.

---

[1] Chief Engineer, Tampa Branch, Ardaman & Associates, Inc., 3925 Coconut Palm Drive, Tampa, FL 33619, ross@ardaman.com

The TBM is a steel cylinder with an outside diameter of 3 m (10 ft.) and a closed face that can be forced against the soil as the tunnel is excavated to balance the soil pressure. Figure 1 shows the front of the TBM with the closed cutting head.

Figure 1: Closed Front End of TBM

The cuttings are removed by an auger to a Muck Car at the rear of the machine. The total length of the machine was 6.4 m (21 ft.). Liner plates are assembled inside of a shield at the tail of the machine. The liner plates have a nominal diameter when assembled of 2.7 m (9 ft.), leaving a circumferential void of about 150 mm (6 in.). Liner plates are assembled inside of the tail shield of the TBM behind a "Thrust Ring". The tunnel is advanced pushing the thrust ring against the liner plates, forcing the TBM into the soil or rock in front of the machine. Figure 2 shows the tail section of the TBM. The auger used to transfer soil from the cutting chamber to the Muck Car and the thrust ring can be seen in the photograph. The liner plates that are assembled behind the TBM, within the tail shield are used as a reaction for the thrust ring.

The operating procedure employed on other tunnels called for the injection of 5% bentonite clay/water slurry at the cutting head and around the top of the machine.

Figure 2: Rear View of TBM

The cutting head had a 38 mm (1.5 inch) over-cut. A cement/bentonite clay grout was pumped from ports behind the TBM to fill the gap between the liner plates and the tunnel excavation. The TBM was stopped after a "Step" of 0.9 m (3 ft.), a ring of liner plates was bolted together and to the ring of plates previously placed. The thrust ring pushed the TBM forward by pushing back against the liner plates to initiate the next cutting step. Grout was pumped through pipes discharging behind the TBM to fill the gap between the excavation and the assembled liner plates. A seal in the tail shield prevented leakage of grout into the TBM.

Tunnels installed with this method at other locations suffered significant surface subsidence. In an attempt to prevent the subsidence, the next tunnel was advanced using higher pressures on the grout. The result was a break-out of the grout at the ground surface, with no significant change in the surface subsidence.

The tunnel was to cross under a major rail line carrying high speed trains and a State Road, Broadway Avenue. The tunnel contractor was informed that the project was stopped until a procedure was submitted to assure that there would be no unacceptable settlement of the tracks or the road. The author was asked to develop new grouting procedures that would limit the subsidence to less than 13 mm (½ inch).

PROBLEM EVALUATION

The surface soils are typically sand, and some organic materials were found in the soil below the rails and above the tunnel. The soil profile at the tunnel site is shown in Figure 3.

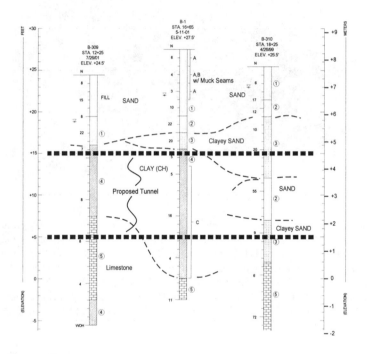

Figure 3: Soil Profile at the Tunnel

It was the opinion of the author that the subsidence was occurring due to the immediate collapse of the soil into the over-cut gap and the gap between the liner plates and the TBM diameter because the highly fluid grout could not provide reliable support of the soil. The shallow depth of the tunnel contributed to the problem. The soil depth was not great enough to allow stresses to arch around the tunnel. Any subsidence of the soil immediately above the tunnel was reflected at the ground surface.

If the sandy soils above the tunnel could be solidified by injection of grout, they would not collapse when the tunnel advanced. Therefore, consideration was given to pre-grouting using chemical grouts to solidify the soils. However, the permeability of the soil was low due small amounts of clay and organic material mixed with the very fine sand. Also, the soil was thinly layered with more clayey soil that would result in a tendency for hydro-fracture due to the grout pressure. Although grouting contractors may believe that the chemical grouting was feasible, it has been our experience that results were poor in other projects with similar soil conditions. Also, there was no access above the tunnel due to the right of way limitations by the railroad and the wetland protection problems associated with drainage ditches on each side of the tracks. Fore-pole grouting, grouting from inside of the TBM, was not possible because of the closed-face cutting system.

The solution that finally developed was to grout from inside of the TBM using the bentonite slurry ports and by modifying the grout injection ports that were located above the tail shield. However, the owner of the TBM was extremely concerned that the grout could lock the machine into the ground if the tunneling was stopped for any period time due to equipment break-downs or other problems. Therefore, the grout had to be very low strength.

DESIGN OF GROUT PROPERTIES

There are two typical cementitious grouts used in Florida; low slump sand/cement mixtures used for compaction grouting and void filling to solve sinkhole problems, and fluid cement/bentonite mixtures used for slurry walls and filling of small diameter voids around pipes or below structures. The sand/cement type grout would not be suitable for the tunneling application due to long pumping distances, and small diameter grout lines required inside the TBM. The TBM had grout lines with a diameter of only 25 mm (1 inch).

Both types of grout have high strengths due to the high cement contents required to create mixtures that can be pumped. Typical 2 day strengths of a sand/cement grout might be as high as several hundred psi, and 28 day strengths would exceed 1,500 psi. The cement/bentonite grout is more fluid, and can be pumped through small diameter lines. However, this grout also has relatively high strength. A typical mix design for Cement/grout for sealing wells consists of the following (SWFWMD, 2001):

Cement:	262	N	892	lbs.
Bentonite:	22	N	74	lbs
Water:	0.83	m3	167	gallons
Yield:	1.0	m3	1	cubic yard

This mix is very fluid, with a Flow Cone flow rate of 12 seconds (ASTM C 939). However, the mix has a relatively high shrinkage, as much as 10 percent, due to segregation of the solids from the water. Typically this is not a problem when the grout is injected in soil due to the loss of excess water by seepage into the soil. However, the resulting grout strengths are higher in situ than in the test cylinders due to the decrease in water cement ratio.

Strength tests were run on cylinders cast in plastic molds. The molds were 76 mm (3 inches) in diameter and 152 mm (6 inches) tall. The samples were cured in a moist room under controlled temperature and humidity. The attached Figure 4 is a summary of the strength test results for the cement/bentonite grout.

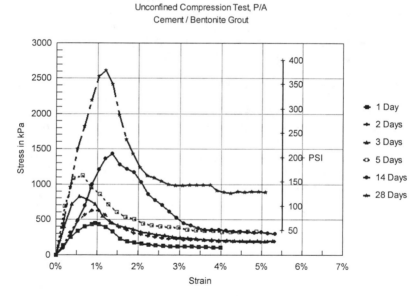

Figure 4

Figure 4 shows that the strength of the cement/bentonite grout mix listed above was too high to meet the needs of the owner of the TBM. He was concerned that this grout would be strong enough to lock the machine in the ground if the tunneling was stopped for any reason. Therefore, a lower strength grout had to be designed.

Also, if this grout is dropped through water, it immediately segregates, losing its consistency. Since the tunnel was below the water table, the void left by the excavation around the liner plates would be filled with water. Injection of the grout into the water filled voids could allow it to segregate. This may have contributed to the subsidence problems observed during the previous tunneling operations.

It was believed that substituting a non-cementitious, fine grained material for some of the cement in the grout would reduce its strength, but would not significantly change its characteristics with respect to mixing and pumping. The cement substitute had to be readily available, and of reasonable cost. The first trials were run substituting fly ash for cement in the mix. The result was a harsh mix, that separated very easily when dropped in water. In addition, the strength was still too high; 862 kN (125 psi) after only 2 days cure.

Metakaolin, a calcined kaolin clay, was being tested by associates of the author to document its effects as an add-mixture in concrete. The metakaolin was found to improve concrete workability at lower water cement ratios, and to improve strength. Therefore, the author decided to substitute about half the cement content in the grout with metakaolin. The mix appeared to be as fluid as the cement/bentonite grout, but it did not flow through an ASTM Flow Cone. However, its strength was still too high. The initial testing showed a 482 kPa (70 psi) in one day, and more than 1.03 Mpa (150 psi) in 2 days. Also, the metakaolin is an extremely expensive material, about 10-times the cost of cement, because it is a highly processed material designed to be a low volume add-mixture.

While researching the metakaolin, it was found that the same supplier had a dry kaolin powder that is used as a filler in plastics manufacturing. Other suppliers had similar dry kaolin that is used in paper manufacturing. The dry kaolin power could be obtained in 23 kg (50 lb) bags, and the cost was about the same cost as cement. Testing with kaolin in substitution for cement led to the following design mix:

Cement:	129	N	437	lbs.
Bentonite:	21	N	70	lbs
Kaolin:	134	N	455	lbs.
Water:	0.83	m^3	167	gallons
Yield:	1.0	m^3	1	cubic yard

The initial tests indicated that the mix had about the same consistency as the metakaolin mix. The kaolin/cement grout did not flow through a Flow Cone, and it did not separate when dropped through a cylinder of water. The strength of the kaolin/cement grout was tested in unconfined compression using 76 mm (3 inches)by 152 mm (6 inches) cylinders. Figure 5 shows the strength of the grout after curing.

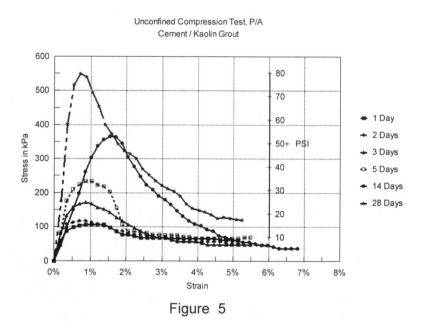

Figure 5

The peak strength reached only 100 kPa (15 psi) in one day and only 118 kPa (17 psi) in 2 days. This strength range was acceptable to the owner of the TBM as it was in the range of soil strengths that were typical in clay tunneling projects. In order to explore further the shear strength of the kaolin/cement grout, a Direct Shear test was run on a sample. This test was felt to be a model for the maximum level of adhesion between the grout and the skin of the TBM. The results of this test are summarized in Figure 6. After 18 days cure, the Direct Shear strength of the kaolin grout was only about 120 kPa (18 psi). In addition, the strength broke down at a strain of about 1% to a residual strength of less than 50 kPa (7 psi). Also, it was noted when testing the kaolin/cement grout that when grout was set only 1 to 3 days, it appeared to liquify if it was sheared. Demonstration of these properties provided assurance that the TBM would not be locked in the ground if tunneling had to be halted for any period of time.

Kaolin Based Low Strength Grout - 18 Days Cure
Direct Shear Test - Normal Stress = 7.0 PSI

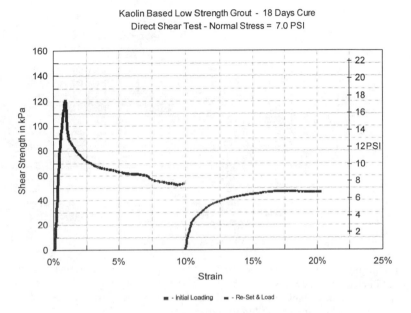

Figure 6

The high kaolin content appeared to act as a retarder in that it delayed setting time. Figure 5 shows little strength gain between 1 day and 2 days and about 60% strength gain between 1 day and 3 days.   This property was tested using the Modified Vicat Needle Test (ASTM C-807-99). The Vicat Needle test uses the penetration of a thin needle into the grout as it cures to measure the time to initial and final set. Figure 7 shows the Vicat Needle test results.   This test shows the delayed setting time for the kaolin cement grout.   More than 6 hours were required for initial set, and final set occurred after 9 hours. The typical "Step" of the TBM required 2 hours from completion of one cutting sequence, installing the liner plates and starting the next cutting sequence. As noted previously, the TBM had a seal in the tail shield so that the liner plates were assembled below the shield, and ahead of the seal. Therefore, grout was injected as the machine was pushed forward for the cut. The delayed set assured that the grout did not setup between the shield and the liner plates. Even then, the adhesion of the grout to the steel, probably half of the direct shear strength, was low enough to allow the machine to step after an extended delay between cutting sequences.

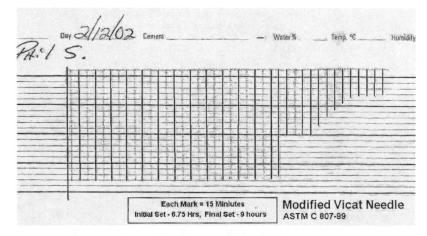

Figure 7: Vicat Needle Test Results

The results of the testing described above were used to determine that the kaolin/cement grout was suitable for use for the tunneling project. A report was then written describing the measured strength properties of the grout and the recommended grouting procedures for grout placement.

DESIGN OF FIELD MIXING PROCEDURES

Once the laboratory design of the grout was accepted, analyses were conducted to match the field mixing equipment with the grout mix design. The attached Figure 8 is a photograph of the grout pump that was used for the project.

The grout pump was a progressive cavity type pump driven by an electric motor. A second motor drove a paddle bar in the bottom of the hopper. The cement bentonite grout normally used for the tunneling operations was mixed in the pump/hopper using the paddle bar and grout recirculation. A reservoir of 5% water/bentonite slurry was used as the mixing fluid. As noted previously, the bentonite slurry was normally pumped to the cutting head and around the top of the

TBM through ports near the front of the shield to assist in removing the cuttings and to support the machine. In order to evaluate the potential problems with field mixing, the hopper was calibrated for volume, and trial field mixes were made up to test the pumping and to measure the properties of the field mixed grout.

A typical "step" by the TBM would leave a void of about 0.29 m3 (10 ft3) that had to be filled with grout. Since both the cement and the kaolin were available in bags, the mix was proportioned for even numbers of bags of each material: 2 bags of cement (43 kg. each bag) and 4 bags of kaolin (23 kg. each bag) with 0.057 m3 (15 gallons) of 5% bentonite slurry. This resulted in a mix volume for each batch of 0.34 m3 (12 ft3). The field test showed that the materials could be easily mixed in the field, and that the grout pumped as easily as the cement/bentonite grout that was originally used. The strength tests on the field mix produced about the same strengths as were produced in the controlled laboratory batches.

GROUTING PROCEDURES

In order to confirm that no excessive (more than 13 mm (½ inch)) settlements were occurring, a survey monitoring program was set up. A line of hubs was established over the centerline of the tunnel and at 4.6 meters (15 feet) on each side of the centerline. Each rail of the three sets of railroad tracks as well as each edge and the centerline of the Broadway Avenue pavement were also monitored. The surveying was scheduled for the start of each shift (8 hours).

The TBM originally had two aft grout pipes (25 mm (1 inch)) diameter tubes above the top of the shield and terminating at the edge of the shield. It was concluded that the soil could collapse before the grout was injected, creating an uncertain distribution of grout around the liner plates. A recommendation was made to the TBM owner to turn the grout pipes down to inject the grout inside of the shield where the cavity would be protected from ground collapse. As noted previously, the liner plates are erected inside of the tail shield and inside of a seal ring. When the cutting step was initiated, grout was injected at grout ports near the head of the TBM, and in the area below the shield and behind the seal. The pressure on the grout was maintained at a little above the overburden pressure as monitored using pressure gages mounted at the grout pipe connection inside of the TBM.

Each liner plate has a PVC pipe plug 50 mm (2 inches) in diameter for inspection and possible post-installation grouting. As part of the monitoring process, some grout plugs were opened. The attached Figure 9 shows the consistency of the grout after about a day at a typical plug.

Figure 9: Grout Inspection Port - 1 Day Cure

CONCLUSIONS

The tunneling operation was concluded with 0 to 10 mm (3/8 inch) heave at one location, and no settlement detected anywhere along the alignment of the tunnel.

In spite of several delays up to 2 days, due to equipment break-downs and, in one case, blockage of the discharge with a chert boulder, there were no problems with restarting the TBM.

Testing of the grout strength using samples taken from field mix batches showed that the field grout mix met the strength requirements established during the design phase.

ACKNOWLEDGMENTS

The author wishes to express appreciation to Mr. Tony DeAguial of Pipe Jacking Unlimited, the TBM owner and tunnel contractor as well as Philip Schlossnagle, EI for his assistance in conducting the tests and field inspections.

REFERENCES:

ASTM C 807-99 -   Standard Method for Time of Setting of Hydraulic Cement Mortar by Modified Vicat Needle.

ASTM D 4832 -   Standard Test Method for Preparation and Testing of Controlled Low Strength Material (CLSM) Test Cylinders

ASTM D 2166 -   Test Method for Unconfined Compressive Strength of Soil

ASTM D 3080 -   Test Method for Direct Shear Test of Soils Under Consolidated Drained Conditions

ASTM D 5971 -   Practice for Sampling Freshly Mixed Controlled Low Strength Material

ASTM C 939 -   Standard Test Method for Flow of Grout for Preplaced-Aggregate Concrete (Flow Cone Method)

SWFWMD -   Rules of the Southwest Florida Water Management District, Chapter 40D-3, "Regulation of Wells", January 2001

Use of Grouting to Reduce Deformations of an Existing Tunnel Underpassed by Another Tunnel

S.A.Mazek[1], K.T.Law[2], and D.T.Lau[2]

Abstract

The Greater Cairo metro and El-Azhar road tunnels, the major project of underground structures in Cairo city, Egypt, have been constructed. During the construction of the Greater Metro Line 2 (Shubra El-kheima- Mubarak) and El-Azhar road tunnels, which are installed by the tunneling boring machine (TBM), Geotechnical challenges are expected to occur related to soil stability around the tunnels. One of these problems arises when parts of the metro and the road tunnels cross under an existing sewage tunnel during the construction. To control the potential problem, the National Authority for Tunnels (NAT) has applied grouting to the soil around the sewage, the metro and the road tunnels. In the present study, one of these problems is highlighted and a model is proposed to provide a prediction of the soil structure interaction using a 3-D model of the multi-crossing tunnel incorporating the effect of grouting. The cement-bentonite grouting is considered in this study. The objective of this study is to evaluate the usefulness of permeation grouting in reducing deformation around the sewage tunnel when the metro and the road tunnels pass underneath it at different crossing zones.

To assess and understand effect of changing the soil stiffness by grouting on the behavior of the two tunnels, a parametric study has been performed. The study is conducted using a finite element method that models the three-dimensional behavior under both the effects of the tunnel crossings and effects of the grouting. A nonlinear stress-strain constitutive model is adopted to represent the soil surrounding the tunnel and a linear constitutive model is employed to represent the sewage tunnel liner. The effects are expressed in terms of the settlement and radial deformation of the sewage tunnel as the metro or the road tunnels pass underneath it for the cases with and without grouting. The 3-D model of predicting the behavior of the tunnel system under multi-crossing tunnels and grouting effect is described and the predictions are compared with the field measurements. The comparison reveals a good agreement between the computed and observed values.

---

[1] Ph.D. candidate, Civil and Environmental Engineering Department, Carleton University, 1125 Colonel By Dr., Ottawa, Ontario, Canada; smazek@ccs.carleton.ca
[2] Professor, Civil and Environmental Engineering Department, Carleton University, 1125 Colonel By Dr., Ottawa, Ontario K1S 5B6, Canada; Phone 1-613-520-7474; tlaw@ccs.carleton.ca.

Introduction

Metro Line 2 includes 6 km of surface and 13 km of underground construction. El-Azhar road tunnel is 2.7 km long. Many problems related to the soil stability were expected during the construction of the Greater Cairo metro and El-Azhar road tunnels in Egypt by NAT. One of these problems was located at north Massara station and at El Attaba district. Both the metro and the road tunnels pass four metres beneath the existing sewage tunnel that forms part the Greater Cairo wastewater project as shown in Figure 1. The lining of the sewage tunnel is built of reinforced concrete strengthened by steel lining on the outside surface of the concrete section. The thickness of the reinforced section and the steel lining is 0.25 m and 7 mm, respectively.

In order to investigate the settlement of the sewage tunnel due to the construction of the metro and the road tunnels, NAT in Cairo conducted a laboratory study. The results showed that the estimated settlement of the sewage tunnel would exceed the allowable limit of 10 mm. To control the potential problem, NAT grouted the soil around the sewage tunnel before the TBM crossed under it. Consequently, the measured settlement in the field when the metro and the road tunnels passed under the sewage tunnel was found to be significantly less than the estimated value without grouting and well within the allowable limit of 10 mm set by the Egyptian standards (Abdel Salam, 1998; Documented file issued by NAT, 1999).

Modeling such a problem must include the details of tunneling and the associated stresses around the tunnels. To assess and predict the behavior of the two tunnels due to grouting, a parametric study has been performed. The study is conducted using a finite element method that models and predicts the three-dimensional behavior under both the effects of the tunnel crossings and the grouting. A nonlinear stress-strain constitutive model is adopted for the soil surrounding the tunnel. Incremental nonlinear technique utilizing the hyperbolic model of Duncan and Chang (1970, 1974, and 1980) has been used. In addition, linear elastic behavior is assumed for the reinforced concrete lining.

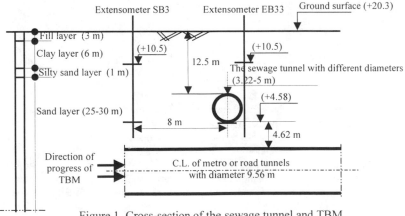

Figure 1. Cross-section of the sewage tunnel and TBM

The 3-D effects on the performance of the grouted sewage tunnel are examined. The effects and predictions are expressed in terms of the settlement and radial displacement of the sewage tunnel as well as the vertical displacement at different locations at the crossing zone when the metro and the road tunnels pass underneath it. The predicted results have been compared with the measured values in the field. A good agreement has been found.

Finite element model

The finite element computer program (COSMOS/M, 1996) has been used in this study. This finite element model takes into account the effects of the vertical overburden pressure, the lateral earth pressure, the nonlinear properties of the soils and the linear properties of the sewage tunnel. Figure 1 shows the configuration of the sewage tunnel and the TBM crossing of Line 2 of Cairo Metro and El-Azhar road tunnel. The soil, the tunnel lining and the interface medium are simulated using appropriate finite elements. Solid elements and thick shell elements are used for modeling the soil media and the sewage tunnel, respectively. One thick shell element is used to represent the composite behavior of the sewage tunnel reinforced concrete pipe section and the steel lining by following a section transformation procedure (Popov, 1998). The existing soil profile as shown in Figure 1 is simulated in the 3-D model, where the watertable is located at 3 m from the ground surface.

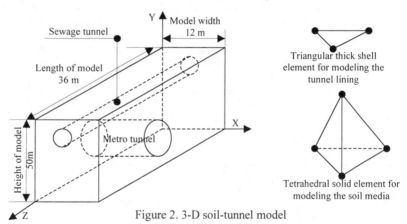

Figure 2. 3-D soil-tunnel model

Linear thick shell element (Figure 2) is used to model both the membrane (in-plane) and bending (out-of-plane) behavior of the sewage tunnel structure. A 3-node triangular thick shell element is used for modeling the sewage tunnel liner with each node having 6 degrees of freedom (three translation and three rotations). The solid element was chosen since it possesses in-plane and out-of-plane stiffness. The solid element allows for both in-plane and out-of-plane loads. The solid element is tetrahedral in shape and has 4 nodes with each node having 3 degrees of freedom (three translations). The tetrahedral solid element and the triangular shell element interface have been used between the soil media and the tunnel to ensure the compatibility conditions at the interface surface between them as well as the associated stress and strains along the interface surface. This type of finite element is used to link adjacent nodes characterized by stiffness components.

The 3-D finite element mesh used in the analysis represents a soil block height of model (y) and width of model (x) by length of model (z) in plan as shown in Figure 2. The vertical boundaries of the 3-D finite element model are restrained by roller supports to allow vertical displacement but prevent a movement normal to the boundaries. The horizontal plane at the bottom of the mesh represents a rigid bedrock layer and the movement at this plane is restrained in three directions. The movement at the upper horizontal plane is free to simulate a free ground surface. In the finite element analysis, the loading pertinent to the construction is considered.

The sewage tunnel was constructed in 1980. The construction of the metro and the road tunnels beneath the sewage tunnel later in 1996 and 1999, respectively, caused the soil beneath the sewage tunnel to unload. The nonlinear properties of soils, the depths of the sewage tunnel, and the confining pressure are varied to study their effects on the settlement and radial displacement of the sewage tunnel. The cases studied are subjected to overburden and lateral earth pressures. Studies have been made with the sewage tunnel, as it exists in the field at depth 12.5 m from the ground surface as well as at constant relative depth of 4.62 m, and TBM diameter of 9.56 m. In addition, the minimum length and width of the 3-D model have been determined to be 3.75 times the diameter of the tunnel in order to eliminate the size effect in the prediction of the performance of the sewage tunnel (Mazek et al., 2001b).

Different nonlinear properties of soil have been chosen to realistically simulate the behavior of the different soils along the metro tunnel Line 2 (Ezzeldin, 1999; Mazek et al., 2001b; Documented file issued by NAT, 1993 and 1999). Moreover, the soil-tunnel excavation and the construction procedure have been idealized using incremental nonlinear technique utilizing Duncan and Chang model (1970, 1974, 1980)

Properties of tunnel and soil

Since the sewage tunnel existed before the metro and the road tunnels were constructed, displacements would be induced in the sewage tunnel at the crossing zones with the construction of the metro and the road tunnels. The settlement and radial displacement of the sewage tunnel have been calculated for the non-grouted and grouted soil in this study.

The settlement is computed at the spring line and crown of the sewage tunnel as well as at different locations at the crossing zone. The sectional linear characteristics of the sewage, the metro, and the road tunnels are shown in Table 1.

Table 1. Characteristics of the sewage, the metro, and the road tunnels

Tunnel	D (m)	ν	E_b (t/m^2)	(t) cm	f_c (t/m^2)
Sewage	3.22- 5	0.18	2.1×10^6	25	4000
Metro or road	9.56	0.18	2.1×10^6	40	4000

In Table 1, D = the diameter of tunnel, v = Poisson's ratio of tunnel liner, E_b = the elastic modulus of the tunnel lining, t = thickness of tunnel lining, and f_c = the compressive strength of concrete.

The project area under analysis lies within the young alluvial plain, which covers the major area of the lowland portion of the Nile valley in Cairo vicinity (Campo and Richards, 1998; Mazek et al., 2001b; Documented file issued by NAT, 1993, 1999). Site investigations along the project alignment have indicated that the soil profile consists of a relatively thin surficial fill layer ranging from two to four metres in thickness. A natural deposit of stiff, overconsolidated silty clay underlies the fill. This deposit includes occasional sand and silt partings of thickness from four to ten metres. Beneath the clay layer, there is a thick alluvial sand that extends down to bedrock, which is well below the metro tunnel. The watertable varies between 2-4 m from the ground surface.

The upper few metres of this alluvial sand are parts of a transition layer of highly interbedded clay silt and fine sand. Below the transition layer, the alluvial sand layer is more uniform with coarse to fine sand, which occasionally contains layers of silt to clayey silt that varies in thickness from a few centimetres to several decimetres. Lenses of gravel and cobbles, up to several metres thick, may also be present at depths of 25 to 30 metres. Soil parameters for the grouted and non-grouted cases are shown in Table 2.

Table 2. Geotechnical properties

Soil parameter	Non-grouted sand	Grouted sand
γ_b (t/m^3)	2.0	2.2
k_o	0.40	0.293
υ_s	0.30	0.30
ϕ (Degree)	37	43

In Table 2, γ_b = bulk density, k_o = coefficient of lateral earth pressure = $1 - \sin \phi$, υ_s = Poisson's ratio, and ϕ = the angle of internal friction for the soil.

Duncan and his colleagues (1970, 1974, and 1980) developed a simplified and practical model for representing nonlinear stress-strain behavior. The behavior is expressed with a hyperbolic stress-strain relationship that is convenient for use in incremental finite element stress analyses. This model uses a tangent modulus (E_t) that varies with stress level as shown in Eq. 1.

$$E_t = [1 - \frac{R_f(1 - \sin\phi)(\sigma_1 - \sigma_3)}{(2c)\cos\phi + 2\sigma_3 \sin\phi}]^2 E_i \qquad (1)$$

where σ_1 is the major principal stress and σ_3 is the minor principal stress.

The friction angles ϕ adopted for the sand layer have been obtained using laboratory test results from reconstituted samples. The friction angles for non-grouted and grouted soil are 37° and 43°, respectively. The coefficient of earth pressure at rest K_0 of these layers is given by $K_0 = 1 - \sin\phi$. The variation of the initial tangent modulus (E_i) with confining pressure is related to the effective pressure based on Janbu's empirical equation (Canadian Foundation Engineering Manual, 1992), which is given by Eq. 2

$$E_i = mP_a (\frac{\sigma_3}{P_a})^n \tag{2}$$

where the modulus number (m) and the exponent number (n) are both pure numbers, P_a is the atmospheric pressure expressed in appropriate units, and σ_3 is the effective confining pressure as the initial minor principal stresses.

The nonlinear parameters for the non-grouted and grouted soil are shown in Table 3 (El-Nahhas et al., 1994; Ezzeldine, 1999). The hyperbolic model parameters are also discussed and analyzed using the documented report issued by NAT (1993 and 1999).

Table 3. Hyperbolic model parameters

Material	m	n	$\Delta\phi$	R_f	G	d	F
Fill	300	0.74	-	0.69	-	-	-
Clay	350	0.60	-	0.70	-	-	-
Transition soil	400	0.60	0	0.78	0.32	5	0.2
Sand	500-700	0.5-0.6	0	0.85–0.9	0.30	5	0.15
Grouted sand	1600-2500	0.21-0.4	0	0.9-0.95	0.30	5	0.15

In Table 3, c and ϕ are the Mohr-Coulomb shear strength parameters, E_i is the initial tangent modulus, ϕ_1 is the friction angle at confining stress of 1 atm (1 atm= 1 kg/cm^2), $\Delta\phi$ is the reduction in the friction angle for 10-fold increase in confining pressure, R_f is the failure ratio or the ratio between the compressive strength $(\sigma_1 - \sigma_3)_f$ and the value $(\sigma_1 - \sigma_3)_{ult}$ of the asymptotic stress for the hyperbolic stress-strain curve, G is the initial Poisson's ratio (v_i), F is the reduction of v_i for a 10-fold increase in σ_3, and d is the rate of increase of v_i with increase in residual strain.

The stress-strain relationship is defined by Eq. 3 (Duncan and Chang, 1970):

$$\varepsilon = \frac{\sigma_1 - \sigma_3}{E_i[1 - \frac{R_f(\sigma_1 - \sigma_3)}{(\sigma_1 - \sigma_3)_f}]} \tag{3}$$

where σ_1 is the major principal stress and σ_3 is the minor principal stress.

The curved nature of the failure envelope is also incorporated by considering ϕ for change with stress level as follows in Eq. 4:

$$\phi = \phi_1 - \Delta\phi \log\frac{\sigma_3}{P_a}$$ (4)

where σ_3 is the effective confining pressure.

The finite element analyses of the crossing of the tunnel system have been carried out to simulate the construction of both the sewage tunnel and the metro and the road tunnels. The construction of the sewage tunnel in 1980 was under the initial in-situ stress condition. The excavation of the metro and the road tunnels cause the soil around the sewage tunnel system to respond in an unloading manner, and unload moduli is appropriate during this stage. Under the unload-reload condition, Duncan et al. (1980) found that unload and reload moduli (E_{ur}) are similar and are 1.2-3 times the initial modulus (E_i). Byrne et al. (1987), based on tests on granular soils, found the E_{ur}/E_i is in the range of 2-4.

Wong and Duncan (1974) used the Young's modulus and Poisson's ratio or the bulk modulus to describe the effect of volume change. The volume change component refers to the variation of volumetric strain (ε_v). Therefore, Wong and Duncan used hyperbolic function to model the stress dependent characteristics of the volumetric strain.

The tangent Poisson's ratio implies that the volume-change is a hyperbolic function and stress dependent. The parameter G is the initial Poisson's ratio (v_i), F is the reduction of v_i for a 10-fold increase in σ_3 and d is the rate of increase of v_i with increase in residual strain as shown in Table 3 (Ahmed, 1994). The tangent Poisson's ratio (v_t) could be computed as shown in Eq. (5).

$$v_t = \left[\frac{G - F \log(\frac{\sigma_3}{P_a})}{\left[1 - d(\frac{\sigma_1 - \sigma_3}{E_i}) \right]^2} \right]$$ (5)

where σ_1 is the major principal stress and σ_3 is the minor principal stress.

Effective stress is used in the finite element analyses, as the tunnels are located in sand. The lateral earth pressure at the sewage tunnel has been calculated using the coefficient of lateral earth pressure at rest discussed earlier.

Stresses in soil

The stresses in the subsoil have undergone two steps of change. The first stage corresponds to the construction of the sewage tunnel and the second stage to the construction of the metro tunnel.

The loading steps have been simulated using the 3-D nonlinear finite element analysis for the non-grouted and the grouted cases. First, the initial principal stresses are computed with the absence of the sewage tunnel. Second, the excavation of the sewage tunnel is modeled by means of the finite element method. The excavation has been simulated by the removal of those elements and the associated stresses inside the boundary of the sewage tunnel surface to be exposed by the excavation. Third, the movements and stress changes induced in the soil media are calculated. Fourth, the calculated changes in stresses are then added to the initial principal stresses computed from the first step to determine the combined stresses resulting from the sewage tunnel construction. Fifth, the final stresses due to the construction of the metro tunnel are computed using the combined stresses obtained in the fourth step as the initial stresses. However, the construction sequence of the 12-metre section of the metro tunnel is modeled in four stages in 3 metres interval. The only difference between the grouted and the non-grouted cases lies in the properties of the soil used in the fifth steps. Appropriate soil properties have been assigned to the two different cases.

A simple stress level criterion is used for differentiating between primary loading and unloading-reloading behavior. The unloading-reloading condition is considered to have occurred when the current stress level (SL) is smaller than the maximum stress level achieved in the past ($SL_{\text{max past}}$). Subsequent investigations have indicated that this simple stress level criterion for assignment of unloading-reloading moduli (E_{ur}) moduli should be included in the finite element analysis. The stress level (SL) is defined as follows:

$$SL = \frac{(\sigma_1 - \sigma_3)}{(\sigma_1 - \sigma_3)_f} \qquad (6)$$

where σ_1 is the major principal stress and σ_3 is the minor principal stress.

The model assumes that for all stress level greater than or equal to the maximum previous stress level ($SL_{\text{max past}}$), the primary loading modulus (E_i) should be applied, and for all stress levels lower than $SL_{\text{max past}}$, the unloading-reloading moduli (E_{ur}) should be used. Because of unloading nature of the construction, the current stress level (SL) is smaller than the maximum previous stress level ($SL_{\text{max past}}$). Consequently, the finite element analysis uses the unloading and reloading modulus for modeling the soil behavior. The computed stresses resulting from analysis of the sewage tunnel construction is used for calculating the nodal force to simulate the construction of the metro tunnel.

Performance of the 3-D soil structure problem

The study evaluates the effectiveness of grouting to reduce deformation around the sewage tunnel. The effect of depth for the sewage tunnel is also studied by varying its value. With the relative depth of the metro tunnel from the sewage tunnel being constant 4.62 m, this change in depth changes the effective confining stresses and hence the initial modulus. For the grouted case, grouting has been applied to soil within the region around the sewage tunnel and the metro tunnel. The geotechnical properties of the soil for this part of the study have been shown in Table 3.

The numerical analysis is carried out using the unload reload modulus (E_{urn}) for the non-grouted soil ranging from 3,100 t/m^2 to 16,000 t/m^2 at different sewage tunnel depths: 12.5, 15, 17.5 and 20 metres based on Janbu's equation [Equation 2] using different nonlinear soil parameters, as the tunnels are located in non-grouted sand as shown in Table 3.

In addition, the numerical analysis has also been conducted using the unload reload modulus (E_{urg}) for the grouted soil ranging from 20,000 t/m^2 to 100,000 t/m^2 at different sewage tunnel depths based on Janbu's equation using different nonlinear soil parameters, as the tunnels are located in grouted sand as shown in Table 3.

The calculated maximum negative (shortened) radial displacement occurred at the crown of the sewage tunnel in each case and their values have been calculated for different overburden depths of the sewage tunnel as found in Table 5 (e.g. at depth 12.5 m from the original ground surface).

The calculated maximum settlements of the spring line of the sewage tunnel located at depths of 12.5, 15, 17.5, and 20 metres have also been estimated as found in Table 5 (e.g. at depth 12.5 m). The calculated maximum settlements of the crown of the sewage tunnel located at depths of 12.5, 15, 17.5, and 20 metres have also been computed as found in Table 5 (e.g. at depth 12.5 m).

The predicted deformations have been compared to the measured values. Figure 3 shows the computed and recorded settlements at the spring line level at Point EB33, which is right against the side of the sewage tunnel. The comparison of the predicted and measured values shows that there is a small difference of 1.6 mm. This is within the accuracy of the measurements.

Figure 4 compares the computed and recorded settlements at level +10.5 about 2.7 m above the crown of the sewage tunnel at Point SB3 (Figure 1). The comparison shows that there is a small difference of 0.9 mm, which is again within the accuracy of the measurements.

In order to assess the benefits of grouting, the settlement at the spring line of the sewage tunnel is examined. This is done by comparing the predicted and measured settlements at that point for the grouted and the non-grouted cases. The measured settlement for the non-grouted was deduced from laboratory model tests conducted by NAT. The measured settlement for the grouted case has been reported above at Point EB33. The comparison is shown in Table 4.

Table 4. Comparison of measured and predicted settlements at settlement point EB33

Condition	Recorded settlement (mm)	Computed settlement based on the nonlinear analysis (mm)
Non-grouted	37*	39.5
Grouted	6**	6.8

* Determined from laboratory modeling
** Actual measurements in the field.

The comparison shows; (1) there is good agreement between the computed and measured values; and (2) the settlement is substantially reduced with grouting. In fact, the grouting has helped achieved the objective of limiting the settlement within the specifications of the Egyptian standards.

Based on the good agreement between the computed and measured readings, one can proceed to use the computational model to explore other beneficial aspects of the grouting process. The study has been carried out for the case with the overburden depth 12.5 m from the ground surface and at the sewage tunnel diameter of 3.22 m. In the following, the effects of the relative values of the unloading and reloading moduli of the non-grouted and grouted soils are examined. The results are summarized in Table 6. Table 6 shows the percentage reduction in the negative (shortened) radial displacements occurred at the crown of the sewage tunnel, and the settlement of the spring line and the crown of the sewage tunnel with the grouted soil at different ratio of unloading/reloading moduli of the non-grouted and grouted soil (E_{urn} and E_{urg}, respectively). Based on the estimated readings as shown in Tables 5 and 6, grouting the soil reduces the radial displacement and the settlement of the spring line and crown of the sewage tunnel by up to 87, 85, and 86 percent, respectively. Therefore, grouting the soil increases the soil stiffness and then reduces the deformation of the soil around the tunnel system comparing to non-grouted soil.

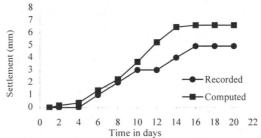

Figure 3. Recorded and computed settlemennt profile at settlement point EB33 at level (+ 2.3)

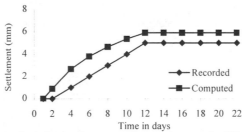

Figure 4. Recorded and calculated settlemennt profile at settlement point SB3 at level (+10.5)

Conclusions

A 3-D nonlinear finite element analysis has been used to analyze the crossing of two tunnels. The analysis takes into account the 3-D changes in stress and strain, the non-

linear behavior of the soil, and construction progress, etc. The following conclusions can be drawn regarding the performance of the tunnel under the effects of various factors.
• The proposed 3-D numerical model is applicable to analyze and predict the performance of the tunnel system under the crossing of two tunnels.
• The settlement of the spring line of the sewage tunnel increases with an increase of the advancing length of the metro tunnel under-passing the sewage tunnel. About 80% of the finial settlement occurs with the front of the metro tunnel reaching vertically below the center of the sewage tunnel.
• The radial displacement and settlement of the tunnel depend on the unload-reload modulus of soil, the overburden depth of the tunnel, and the properties of the tunnel cross-section, as well as the soil nonlinear parameters.
• Grouting the soil reduces the radial displacement and the settlement by up to 87% and 85%, respectively.
• Comparisons of the computed and measured results suggest that the 3-D model has a good potential in analyzing the 3-D soil-structure interaction between the two tunnel crossing each other.

Table 5. Predicted performance of the sewage tunnel at depth 12.5 m

E_{urn}	E_{urg}	Radial displacement (mm)		Settlement of the spring line (mm)		Settlement of the crown (mm)	
$\times 10^3 t/m^2$	$\times 10^3 t/m^2$	Non-grouted	Grouted	Non-grouted	Grouted	Non-grouted	Grouted
3.1	20	13.7	2.1	56.7	10.6	70.4	12.7
5.2	33	8.9	1.4	39.5	6.8	48.4	8.2
8	50	6.5	0.8	28.3	5.3	34.8	5.2
11	69	5	0.63	21.8	3.13	26.8	3.8
12	75	4.6	0.6	20.2	2.9	24.9	3.5
13	82	4.3	0.54	18.9	2.7	23.18	3.2
14.5	90	3.95	0.5	17.2	2.4	21.15	2.9
16	100	3.55	0.45	15.8	2.16	19.35	2.6

Table 6. Percentage reduction due to grouting of the soil

E_{urn}	E_{urg}	Reduction ratio, R*, (%) of the sewage tunnel diameter 3.22 m at depth 12.5 m		
$\times 10^3 t/m^2$	$\times 10^3 t/m^2$	Radial displacement	Settlement of spring line	Settlement of crown
3.1	20	84.6	81.3	82
5.2	33	84.3	82.7	83
8	50	87.7	81.3	85
11	69	87.4	85.6	85.8
12	75	87.4	85.7	85.9
13	82	87.4	85.7	86.2
14.5	90	87.4	86	86.3
16	100	87.3	86.3	86.5

[*R= (Group II – Group I)/ Group II], where, Group I is the result of the grouting case, Group II is the result of the non-grouted case.

Acknowledgment

The authors acknowledge the National Authority for Tunnels (NAT) and the Egyptian Tunneling Society for the te.chnical support and thank Mr. M.E. Abdel Salam, Former head of NAT and head of the Egyptian Tunneling Society and Dr. Ashraf Abu-Krusha, Consultant office at NAT, for their assistance in providing information for this study. Part of the funding for this research is provided by the Natural Science and Engineering Research Council of Canada.

References

Ahmed, A.A. (1994). Analysis of deck road tunnels. Proceedings of the international congress on tunneling and ground condition, Abdel Salam (ed.), Cairo, Egypt. Published by Balkema, Rotterdam.

Abdel-Salam, M.E. (1998). Urban constraints on underground works the Cairo metro-case histories. Egyptian society presentation, Cairo.

Byrne, P.M., Cheung, H., and Yan, L. (1987). Soil parameters for deformation analysis of sand masses. Canadian Geotechnical Journal. 24, 366-376.

Canadian Foundation Engineering Manual, Third edition. (1992). Canadian Geotechnical Society. BiTech Publishers Ltd. Canada.

COSMOS/M program. (1996). Structural Research and Analysis Corporation. Los Angeles, California. USA.

Compo, D.W., and Richards, D.P. (1998). Geotechnical challenges faced on line 2 of the Greater Cairo Metro System. ASCE, Big dig around the world. USA.

Duncan, J.M., and Chang, C.Y. (1970). Nonlinear Analysis of Stress and Strain in Soils. Journal of the Soil Mechanics and Foundation Div. ASCE, Vol. 96, No. SM5. September.

Duncan, J.M., Byrne, P.M., Wong, K.S., and Mabry, P. (1980). Strength, stress-strain and bulk modulus parameters for finite element analysis of stresses and movements in soil masses. University of California, Berkeley, CA. Report no. UCB/GT/80-01.

El-Nahhass, F.M., Ahmed, A.A., El-Gammal, M.A., and Abdel Rahman, A.H. (1994). Modeling Braced Excavation for Subway station. Proceedings of the international congress on tunneling and ground condition, Abdel Salam (ed.), Cairo, Egypt. Published by Balkema.

Ezzeldine, O.Y. (1999). Estimation of the Surface Displacement Field Due to Construction of Cairo Metro Line El-Khalafawy- St.Thereses. Tunneling and underground space technology, Vol. 14, No. 3, pp. 267-279. Published by Elsevier science Ltd.

Mazek, S.A., Law, K.T., and Lau, D.T. (2001). Effect of grouting on soil reinforcing and tunnel deformation. International Underground Infrastructure Research Conference, Waterloo University, Kitchener, Canada.

Mazek, S.A., Law, K.T., and Lau, D.T. (2001). 3-D Analysis on the Performance of a Grouted Tunnel. Canadian Geotechnical Conference. Calgary. Canada. Vol. 1, pp. 111-119.

Popov, E.P. (1998). Engineering mechanics of solid. Upper Saddle River, New Jersey, USA.

Wong, K. S., and Duncan, J. M. (1974). Hyperbolic stress-strain parameters for nonlinear finite element analyses of stresses and movements in soil masses. Report no. TE 74-3. University of California, Berkeley, California. USA.

SOIL STABILIZATION GROUTING UNDER A RAILWAY - FOR MICRO-TUNNELING FOR A SEWER CROSSING

M. Chuaqui[1] and RP. Traylor[2]

ABSTRACT

Soil stabilization grouting was performed to facilitate tunneling under an existing active railway line. The design consisted of creating a circular annulus of cemented soils around the trajectory of the future sewer tunnel by performing permeation grouting through sleeve-port pipes. Since the original soils investigation data indicated that the soils were well-graded, undisturbed native soils, grout volumes were estimated based upon 20% accessible void space. A single pass with sodium silicate grout was anticipated.

The sleeve-port pipes were installed by directional drilling methods. Careful monitoring of the drilling and casing grout injection processes revealed evidence of the presence of extensive fill along the tunnel trajectory. As a result of these observations, water pressure testing was added to the scope of work in order to better define the characteristics of the material to be grouted. The water pressure testing data revealed the material to be grouted had a very high permeability (greater than of 0.1 cm/sec). In order to manage the cost associated with the grouting work, a single pass of cement based suspension grouting was performed prior to the sodium silicate grouting. During the cement based grouting, high takes at low injection pressures were noted and as a direct result of these observations two passes with the sodium silicate grouts were performed.

During the tunneling process, direct verification of both the existence of extensive gravel layers and of successful full permeation and stabilization of the high and low permeability soils was provided.

The grouting methodology and the volume of grout used differed considerably from the original design. The final volume of grout used corresponded to an accessible void ratio of approximately 40%. The original design was consistent with the soils investigation data, however, if the program had not been modified for the actual ground conditions, it is unlikely that satisfactory ground treatment would have been achieved.

The project demonstrates that careful monitoring and analysis of drilling and grouting data and modification of the design based upon these data are necessary components of a successful geotechnical construction process.

[1]Principal, MC Grouting, Inc., 29 Haliburton Ave., Toronto, ON M9B 4Y5 Canada; Tel: (416) 695-2593; Fax: (416) 695 2399; email: marcelo.chuaqui@sympatico.ca.
[2]President, TRAYLOR,LLC, 17204 Hunter Green Rd, Upperco, MD 21155 U.S.A.; Tel/Fax (410) 239-9441; email: geobtraylor@msn.com;

1.0 Introduction

The project consisted of stabilizing soils in a circular pattern along the 240-cm (8') diameter tunnel trajectory. The tunnel intersected an active railway 150-cm (5') below grade and crossed perpendicular to railroad. There were three sets of tracks to be crossed and the length of the zone to be stabilized was 2400-cm (80'). The logs for the three soil borings available for the site, indicated that the soils to be stabilized were well-graded, native soils. The high standard penetration test blow counts further indicated that the soils were undisturbed. It is important to note these borings were located some distance on either side of the tunnel crossing.

2.0 Original Design

The soils described in the site investigation report were not anticipated to be permeable with a regular cement based grout, industry state-of-practice dictates that cement based particulate grouts are effective in soils with a minimum permeability of 1×10^{-1} to 1×10^{-2} cm/s depending on the silt content, and therefore either a solution or microfine grout was required (DePaoli, B. et al, 1992). Solution grouts, such as sodium silicates, are effective through soils with permeabilities as low as 1×10^{-4} cm/sec (DePaoli, B. et al, 1992). Sodium silicate grouts are considered temporary, due to degradation of the compounds over time, however, since only temporary ground support was required, sodium silicate was selected as the most economical grout type for this application.

The grouting program consisted of installing 3.8-cm (1-½") sleeve-pipes along the trajectory of the tunnel using directional drilling methods. Ten pipes would be installed to form a circular annulus around the tunnel trajectory.

The goal of the program was to create overlapping grouted soil cylinders around each sleeve port pipe to provide temporary excavation support during tunneling. The grout quantity was estimated by assuming a 20% assessable void space in these soils. The estimate quantity was approximately 15 gallons per sleeve or a total quantity of 9,850 gallons.

3.0 Execution: Anticipated and Modified Means and Methods

In this section for each process of the grouting operation the anticipated means and methods will be described. The unexpected conditions encountered during the execution of each process will be discussed and the modified means and methods will then be explained.

3.1 Directional Drilling, Sleeve-Pipe Design and Installation

Pilot holes were drilled with directional drilling methods by the tunneling contractor. Once the pilot hole was completed and the bit broke through the ground surface on the opposite side of the tracks, the sleeve-pipes were pulled back through the pilot hole as the bit was withdrawn.   Figure 1 shows the directional drilling rig.

Figure 1: Sleeve-Pipe Installation.

Each pipe was constructed as follows:
- a lead section with no sleeves (approximately 90' long);
- a pay zone section with sleeves every 16" (approximately 80' long);
- a tail section with (approximately 40' long).

Casing grout was a high viscosity, thixotropic, low strength cement based suspension grout, used to fill the annular space between the original pilot hole and sleeve-port pipe. If this annular space was not filled, the sodium silicate grout would travel along the pipe during injection, rather than into the soils requiring treatment.

The planned method of casing grout installation consisted of injecting the grout with the tunneling contractors mud pump, as the bit was being withdrawn from the pilot hole. After the installation of the first pipe, it was determined that type of pump on the rig was not well suited for grout injection; therefore an alternate method for casing grout installation was required. The alternate method involved inflating a pair geotextile bags at one end of the annular space between the pipe and the ground. These bags would provide a bulkhead so that the casing grout would travel along the pipe, filling the annular space. After the bags were filled, grout would be injected through sleeves adjacent to the geotextile bag until surfacing at the far end of pipe ensuring that the entire annular space was filled. The tail section of each sleeve-pipe was modified to have one sleeved section (150-cm long) wrapped with two geotextile bags that could be inflated to provide a bulkhead within the annular space between the existing hole (12.7-cm) and the pipe (3.8-cm) for installation of the casing grout.

The pilot hole drilling determined that the soils were not well graded, but were actually fill as evidenced by following:
• Holes encountering rocks or boulders that caused deviation in hole trajectories;
• One hole deviated sharply and encountered the side of a hand-dug stone well, which was within the zone of grouting influence;
• During the drilling of some holes, loss of circulation of the bentonite slurry used for flushing the drill hole was experienced.

As result of the hole deviations only eight of ten planned sleeve-pipes were installed. Based on the trajectories of the eight sleeve pipes with respect to new tunnel alignment, it was decided that reasonable ground treatment could be achieved with the reduced number of sleeved pipes by adjusting the grout quantities to ensure overlap between the theoretical grout/soil mass cylinders around each sleeve-pipe.

As a result of different ground conditions encountered, it was noted/decided that:
• The three investigation holes were not within the zone to be grouted and potentially were not representative of the existing ground conditions;
• The existence of the well was of concern and was likely indicative that there was fill within the zone to be grouted;
• The use of sodium silicate grout within fill was not an efficient form of ground treatment;
• Water testing in 20% to 33% of the sleeves should be performed to better define the characteristics of soils to be treated;
• A multiple-pass approach (i.e. each sleeve is injected more than once) with sodium silicate grouts or a combination of cement-based grouts, followed by sodium silicate grouts would be used to treat the soils.
• The type of multiple pass approach used would be determined by the insitu permeability of the soils, as evidenced by the water testing data.

3.2   Bag Inflation

A neat 0.8 water:cement ratio grout was selected for inflation of the bags. Prior to inflating production bags within the tunnel alignment, a test bag was inflated on surface to determine (a.) the actual volume required to fill the bag after water was pressure filtrated through the geotextile, (b.) the pressures required to open the sleeves and (c.) the pressures required to fail the bag.. The volume of each bag was approximately 12 liters, the pressure required to open the sleeves was 3 psi, and the bags failed at just over 50 psi.

A single fold double packer was used to straddle the sleeve corresponding to each bag. The bags were inflated to approximately 20 psi effective pressure. The bags were re-pressurized twice at 10-minute intervals, after initial filling to compensate for any time-dependant water bleed through the geotextile. Based on the pressure and flow data, 13 of the 14 bags were inflated successfully, while one bag was apparently damaged during pipe installation.

3.3   Installing Casing Grout

The purpose of casing grout is to provide a friable zone surrounding the sleeve pipe, which can be easily fractured to permit grouting of discrete zones of soil. The grout must be of high enough viscosity to prevent permeation of the soil and of low enough strength to allow fracturing with the same equipment to be used in the grouting program. The casing grout formulation is noted in Table 1.

Material	Quantity
Water	130 liters
Cement (type I/II)	43 kg (1 bag)
Bentonite	5% (bwow)

Table 1: Casing Grout Formulation

A single fold double packer was used to straddle the sleeves located outside the bags. Grout was continuously injected until grout flow was observed at the far end of the pipe, indicating that the grout was by-passing the geotextile bags. Since the effective grouting pressure was below the generally accepted hydro-fracture pressure for soils (1 psi/foot of depth), grout by-passing the bulkhead was either the result of the bulkhead not having been installed adequately or the ground being of such high permeability that the high viscosity casing grout was traveling around the bulkhead and back to surface. A surface test, along with careful observation of the pressure during inflation indicated that 13 of 14 bags were inflated successfully. The use of two bags on each pipe ensured the successful installation of at least one bulkhead along each pipe. Since the bags were inflated carefully and held pressure when injection was terminated, it was likely that the formation was very open.

In response, the casing grout was then injected by moving the packer along the pipe to different locations along the pipe. During casing grout injection, connection to surface was observed along the side and inside the stone well, as well as between the railroad tracks. Grout injection pressures were maintained below the expected hydrofracture pressure of the soils, indicating that there were several very high permeability zones within the soils to be treated. When connection to surface was observed, injection in that pipe was terminated. The packer was removed from the pipe and injection on a different pipe was initiated. After a 24-hour period, once the casing grout had gelled sufficiently, injection in the pipe that had connection to surface was reinitiated. This was a time consuming but necessary process. If the casing grout was not installed so that the entire annular space was filled along the length of the pipe, the water testing data would not be representative of the soil formation and excessive quantities of sodium silicate grout would be required. Approximately twice as much casing grout was installed than theoretically required to fill the annular space. This indicated that a considerable amount of consolidation and void filling was achieved with a high viscosity cement slurry grout, at low grouting pressures.

3.4   Water Testing

The water pressure testing was added to the scope of work in order to better characterize the soils that were to be treated and to determine if a pass of regular cement slurry grout would be performed prior to sodium silicate grouting. If the in situ permeability was below 0.1 cm/sec, sodium silicate grouting would be performed and if the in situ measured permeability was above 0.1 cm/sec, cement based suspension grouting would be performed prior to sodium silicate grouting.

The in situ permeability was calculated by applying Caron's equation:

$$K \, (cm/s) \; = \; \frac{Q}{4 \, \pi \, r \, p} \quad \frac{1 \, (meter)}{L \, (zone)}$$

where:

Q = grout flow in cm^3/sec
r = half the borehole diameter in cm
p = effective grouting pressure in cm water column
$L(zone)$ = the length of the test zone in meters.

In order to determine the effective pressure, the headloss pressure for the injection rate must be subtracted from the gage pressure. The headloss pressure was determined by performing a surface test. Water was injected through a sleeve pipe at various flow rates and the corresponding pressures were recorded. These data were then graphed to create a headloss curve. The headloss curve was then used during water testing to estimate the headloss at a specific injection rate.

The water testing procedure involved injecting water until stable flow and pressure readings were established. The tests were conducted for three minutes. The water testing was performed at effective pressures below the theoretical hydrofracture pressure (1 psi per foot of depth). Flow rates between 1 and 5 gpm were anticipated. Higher flow rates were used if no effective pressure could be measured.

Water testing was performed in approximately 33% of the sleeves. The average in situ permeability was greater than 0.1 cm/sec. The water testing data indicated the presence of high permeability fill within the zone to be grouted. Based on the water testing data, a single pass of cement based grout, utilizing 60 liters per sleeve (240 liters per meter) or 34,000 liters was anticipated.

3.5   Cement Based Grouting

The grout formulations selected for the regular cement based grouting had a 1.0 to 1.3 water:cement ratios with 2.0% to 3.5% bentonite by weight of water and with super plasticizer at 1% to 2% by weight of cement. The base mix had a 32 second marsh time and less than 2% bleed. The water:cement ratio, bentonite and super-plasticizer dosages were varied to increase the viscosity of the grout as needed.   The grout formulation was typically thickened since none of sleeves reached refusal.   The majority of the grout injected had a 60 second marsh time and no bleed. The bentonite was not pre-hydrated since only temporary support for the tunnel was required. When non-hydrated bentonite is added to grout, durability issues can arise due to cracking of the grout (Bruce et al, 1997).

The properties of the grouts were tested regularly to ensure batching was executed properly. The following tests were performed every 5th batch:
* Specific Gravity - The specific gravity of the grout was measured in accordance to the method described in API Recommended Practice 13B-1 with a Baroid Mud Balance (API Standard Specification)
* Apparent Viscosity - The marsh time of the grout was measured in accordance with the method described in API Recommended Practice 13B-1 (API Standard Specification).

During grouting operations the grouting pressure, flow-rate, take and mix being injected were recorded for each sleeve.   Two sleeve-pipes were injected simultaneously. Each line was color-coded. Double- fold, double packers were used for the cement based grouting (two sleeves were isolated for each pipe being grouted). These packer rods were marked with tape every 25-cm so that sleeves could be easily located. The equipment layout used for the cement slurry grouting is depicted in Figure 2

Figure 2 Cement Grouting Equipment Layout.

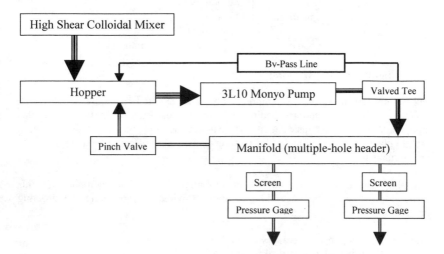

Each sleeve was injected with 60 liters of cement slurry grout for the entire length of the pipe. None of sleeves reached true refusal (zero grout take at maximum allowable grouting pressure) during initial injection.

The cement suspension grouting data indicated that the soils had zones of high permeability. The viscosity of the grout mix was increased to limit the spread of the grout as much as possible. Grout connection to the surface was noted during the injection of several pipes.

3.6   Sodium Silicate Grouting

Since the cement grouting data indicated that the formation was even more open than anticipated from the water testing, the sodium silicate grout was injected over two passes, rather than one single pass, in all but the two lowest elevation pipes.

3.6.1   Sodium Silicate Grout Formulation

A bench-scale testing program was executed to verify that the sodium silicate grout was compatible with the non-potable mix-water available at the site, and to determine mixing proportions to achieve a 20 to 30 minute set at various temperatures. The sodium silicate grout formulation is noted in Table 2.

Table 2: Sodium Silicate Grout Formulation

Component A	Component B
80 type N sodium silicate	80 water
20 water	20 Glyoxal

This mix design had a gel time of approximately 30 minutes.

3.6.2 Sodium Silicate Grouting Equipment

The sodium silicate grouting equipment layout is depicted in Figure 3. With the exception of the concentrated sodium silicate and hardener tanks the rest of this equipment was housed inside a heated sea-container that was configured to be a mobile plant. All mixing tanks and transfer pumps were electrical and connected to an internal panel.

The plant contained separate mixing and holding tanks for each component. This made it possible to mix A and B component without interrupting the grouting operation.

The A and B components were mixed by:
- o Zeroing the mechanical flowmeter;
- o Starting the paddle mixer;
- o Adding the appropriate amount of water (measured on the flowmeter);
- o Adding the appropriate amount of either sodium silicate for A component or hardener for B component (measured on the flowmeter).

The A and B components were then stored in the paddle mixer tank until the agitator holding tanks were sufficiently emptied to hold the mixing tank volume. The A and B components were then transferred to the agitated holding tanks.

The individual plant components consist of:

- Two Graco Bulldog Pumps - These pumps were configured with two pistons mounted on a single ram to ensure that equal volumes of A and B were being injected. These pumps provided constant pressure output and easily adjustable flow rates. These pumps were computer controlled by solenoid switches, if a cut-off volume was reached.

- Two Mixing Tanks – One for A-Component and one for B-Component. These tanks were manufactured from clear plastic 280 Gallon totes. They were calibrated along the front for verification of flowmeter readings and were enclosed to eliminate contamination from dust. A rotating paddle, driven by a variable speed electrical motor, provided a mixing action. The rotation speed and paddle design were designed to avoid air entrainment in the A and B components.

- Several Transfer Pumps - These pumps were electric powered centrifugal pumps.

- Two Mechanical Flowmeters For Mixing A and B Components. They could be easily zeroed after each batch is mixed.

- Two Agitator Holding Tanks - These tanks were manufactured from clear plastic 280 gallon totes. They were calibrated along the front for verification of equal pumping of A and B components by the Graco pumps. Low turbulence circulation pumps were used to keep the A and B components well mixed, without air entrainment.

- Inline Static Mixer - An inline static mixer was used for mixing of the A and B components.   Check valves were at the inlet of the static mixer to prevent backflow of A component in the B component line and vice versa.

Figure 3: Sodium Silicate Plant

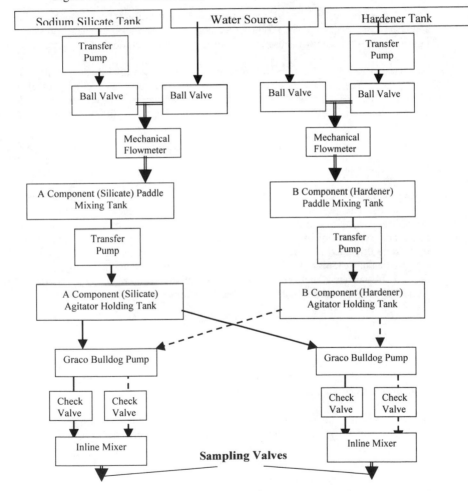

Each line had a mechanical flowmeter and pressure transducer with 4 to 20 mA outputs connected to datalogger and PC computer. The computer software monitored, stored and displayed in real-time the pressure, flow, and take. Additionally the software automatically stopped the pumps when the target volume had been reached through the use of electronically controlled solenoid valves.

Since different hoses and fittings, as opposed to those used for the cement grouting, were being utilized, the head-losses curve was determined for the new equipment configuration. This data was used to determine the effective pressure during subsequent grouting operations.

Two sleeve-pipes were injected simultaneously during the grouting. Double fold, double packers were used for the grouting operations. The grouting data (pressure, flow and trend-line data) were used to assess the grouting operation.

The six uppermost sleeve-pipes received two passes of sodium silicate grout. On average, the first pass consisted of 15 gallons per sleeve and the second pass of 5 gallons per sleeve. The total volume of sodium silicate was 12,000 gallons. A gradual reduction in permeability could be seen through the different phases and passes of the grouting.

3.7   Tunneling

During tunneling operations, the sleeve-pipe locations, casing grout installation, and treated soils were exposed and visually observed. The sleeve-pipe followed very closely the target trajectories. The casing grout could be seen encasing each sleeve pipe. The existence of high permeability gravel layers was verified. These gravel layers were cemented together with grout. All the soils that were encountered had been sufficiently stabilized with either sodium silicate or cement based grouts to allow for a safe and controlled tunneling operation. The gravel layers contained mainly cement-based grouts and the finer soils contained mainly sodium silicate.

4.0   Conclusions

The project demonstrates the need for good quality soil exploration within the area to be treated. The more remote exploration were misleading and resulted in under estimation of the grout quantities during initial design.

Careful planning, monitoring and analysis of drilling and grouting data and subsequent modification of the design based upon these data are necessary components of a successful geotechnical construction process involving grouting for excavation support.

The grouting methodology and the volume of grout used differed considerably from the original design. The final volume of grout used corresponded to an accessible void ratio of approximately 40%. Although the original grouting program design was appropriate for the conditions depicted in the soils investigation data, it is unlikely that satisfactory ground treatment would have been achieved. Extensive modification of the grouting program in response to the actual ground conditions, was singularly the defining turning point for the successful execution of the work.

Figure 4: Soils and sleeve-pipes encountered during tunnel.

References:

American Petroleum Institute (API) Standard Specification (1990). RP13B-1 Recommended Practice Stand. Proc. for Field Testing Water Based Drilling Fluids.

Bruce, D.A., G.S. Littlejohn, and A. Naudts. (1997). "Grouting Materials for Ground Treatment: A Practitioner's Guide," *Grouting – Compaction, Remediation, and Testing*, Proc. of Sessions Sponsored by the Grouting Committee of the Geo-Institute of the American Society of Civil Engineers, Logan UT, Ed. by C. Vipulanandan, Geotechnical Special Publication No. 66, July 16-18, pp. 306-334.

DePaoli, B., Bosco, B., Granata, R. and Bruce, D.A. (1992). "Fundamental Observations on Cement Based Grouts (2): Microfine Cements and the Cemill Process." Proc.   ASCE Conference, "Grouting, Soil Improvement and Geosynthetics", New Orleans, LA,. Feb. 25-28, 2 Volumes, pp. 486-499.

Principles of Ground Water Control through Pregrouting in Rock Tunnels

Orjan A Sjostrom[1]

Abstract

The principles of pregrouting as presented here are based on nearly 40 years of experience of pre- and post-grouting in rock tunnels in different parts of the world. Keywords: pregrouting, water control, ground treatment, stable grouts, ideal grout

Sealing of rock through grouting

Grouting is the process of injection under pressure of suspensions, emulsions and solutions into rock, soil and cavities in order to seal, strengthen and stabilize the ground, and to fill voids.

Sealing of rock in tunnels to control water inflow to the tunnel is done by pregrouting and by postgrouting.

By grouting it is normally not possible to get a completely watertight tunnel, only to reduce the water inflow to varying degrees of tightness depending on the particular circumstances. Rock quality may be improved by one – three Q-classes.

The degree of tightness *needed* depends on the sensitivity of the overlying ground and structures. In non-sensitive ground it depends on the level of water inflow that can be accepted without disrupting the advance and on the pumping costs. Also the accessibility for maintenance and maintenance cost during use of the tunnel has to be considered. Considerations must also be given to the effect of a continuous water inflow into a tunnel on the ground water balance and the environment.

The degree of tightness that *will be achieved* depends mainly on what type of grout material that is used, grout holes pattern, water pressure head, the characteristics of the rock, and of course on the knowledge and skill of the personnel undertaking the work.

By pregrouting it is possible to achieve a substantial reduction of water inflow. Postgrouting is time-consuming and costly and the effect on the water inflow varies with ground conditions, previous treatment, grouts used, pressure that can be applied, water pressure head, amount and length of drillholes, etc.

Often the volume of fissures in hard rock is less than 1%, occasionally higher. Grout takes in fault zones and decomposed rock varies within wide ranges.

Principal, MSc, MHKIE, ASCE, *OSCO,* Orjan Sjostrom Co Ltd, T23A-1B, South Horizons, Ap Lei Chau, Hong Kong, phone/fax +852 2234 5156; e-mail sjostrom@groutexpert.com

The tunneler's basic rules

There are two basic rules that have to be followed for successful tunneling: 1) Do not open the tunnel for rock movements, and 2) Do not open for water inflows.

If rock movements can be contained before and just after opening of the tunnel, before loosening pressures build up, and if water is controlled before opening the tunnel, the battle is won.

To contain and control water before opening of the tunnel by pregrouting is relatively easy. To contain water inflows after opening of the tunnel is very difficult and time-consuming, and it is a risk of opening up for heavy water inflows that cannot be controlled without time-consuming and costly measures. To reduce even moderate inflows by postgrouting is a tedious and frustrating job.

As the pregrouting often is on the critical path, it might be tempting to minimize the number of holes. But one or two more holes do not take long time, if it is planned and integrated in the grouting cycle. On the other hand, one or two holes too few may cause postgrouting for months.

"Do not save the small money and lose the big money". "Less haste, more speed"

Estimate of water inflow

Dr Littlejohn (1975) has presented a nomogram on the relation of water inflow and width of fissures. Using that nomogram we can get an indication of the water inflow from one fissure at 20m and 100m pressure head.

Opening of fissure	Water pressure head	
	20m	100m
0.1 mm	~1 l/min	~5 l/min
0.2 mm	5-10 l/min	25-50 l/min
0.4 mm	50-100 l/min	3-500 l/min

These figures should be seen as indications only. However, this illustrates that the inflow from fissures with apertures that are a fraction of a millimeter only, give quite substantial inflows of water.

Sealing effects achieved in practice

Current practice is to use Micro Fine Cement, MFC (< 30 microns), or finer cements for groundwater exclusion grouting of rock. With MFC the seepage into the tunnels can normally be kept under 5 l/min/100 m of tunnel at a water pressure head of usually some 10 – 30 meters. With a higher pressure head, inflows would be proportionally higher. If higher degree of sealing is required, or the rock is of low quality, finer cements than MFC as ultrafine cement, UFC, or chemicals have to be used.

Basic properties of grout

The ability of cement-based grouts to penetrate a fissure or the pores of a soil depends mainly on the size of the cement grains, the stability of the grout, viscosity and time before gelling, "open-time". The type of grout to be used and type of cement to be used in grouts depends on the required sealing effect, the prevailing rock conditions and existing water pressure head.

Chemical grouts, which do not contain particles, have basically the ability of penetrating fissures that water can go through. The main limiting factors for the use of chemical grouts are its viscosity and the strength of the ground, which limits the pressure that can be used. Chemical grouts are usually expensive, 10-100 times that of cement grouts.

Pregrouting in tunnel is done mostly with cement-based grouts, only to a small percentage with chemical grouts

The ideal grout. The ideal grout is a grout that has excellent flow properties and penetration into fine fissures initially, but that, as the traveling speed slows down a distance from the hole, starts to thicken and resist further displacement. This would allow quick grouting of a hole or a stage, without grout traveling to areas too far away from the intended grouting zone of 5 to 10m outside the tunnel. It should give a high early gelling and hardening time, allowing packers to be removed and tunneling to resume without delay. It should have high early strength and give a complete, competent and durable filling of the fissures. The ideal grout should give the required sealing effect at lowest total cost.

Cement grain size. For a cement grout of good flowability, it is commonly accepted that a fissure that may only be penetrated by a cement grout with cement grains about 3-5 times smaller than the aperture of the fissure.

Thus, OPC-grouts, with max grains 100-140μ, may penetrate fissures of aperture greater than 0.4 mm, and

Microfine cement (MFC) grouts, such as Rheocem 650 and Injection 30, with max grain size 30μ, may penetrate and treat fissures of greater aperture than 0.1 mm.

Ultrafine cement (UFC) grouts, with grain sizes less than 15 μ, may treat fissures less than 0.05 mm, and will have sealing and stabilizing effect also on granular soils.

Figure 1 illustrates the penetration of different grouts.

Please note that the best penetration grout Acrylamide is poisonous and should not be used, see below under Chemical grouts. On the other hand, grouts based on ultrafine cement have about the same penetration as the best chemical grouts, and cement grouts should then be preferred.

From the estimate of inflow from different fissure apertures above and the size of fissures that can be treated with a certain cement, it is clear that OPC cement has an effect only on the fissures that give high inflows. Many times inflows have to be controlled to less than 1 l/min per m of tunnel. In Scandinavia the requirement is often 5 l/min per 100 m of tunnel or less. For such water control, MFC grouts or finer are required, depending on rock conditions and the water pressure head.

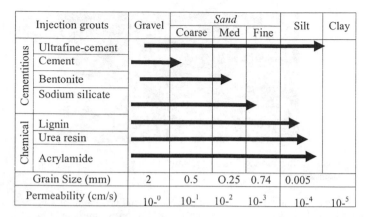

Injection grouts		Gravel	Sand			Silt	Clay
			Coarse	Med	Fine		
Cementitious	Ultrafine-cement						
	Cement						
	Bentonite						
	Sodium silicate						
Chemical	Lignin						
	Urea resin						
	Acrylamide						
Grain Size (mm)		2	0.5	O.25	0.74	0.005	
Permeability (cm/s)		10^{-0}	10^{-1}	10^{-2}	10^{-3}	10^{-4}	10^{-5}

Figure 1. Penetration of grouts. After Clarke

Normally, both relatively tight joint sets (< 0.1mm aperture) and more open joints are present at the same time in a drill hole that is to be treated.

The technique applied nowadays utilizes low water: cement ratio mixes with plasticizer that disperse the grains, giving a low viscosity for a certain time, enabling treatment of fine fissures, but at the same time not traveling too far in open joints, as the gelling and thixotropic effects sets in on the grout, as the rate of flow reduces further out from the hole.

Stability of grout. Stability of grout is important, as lack of this will cause the particles to come out of the suspension and cause incomplete grouting and/or clogging of pipes. Bleeding of unstable grouts will cause incomplete filling of fissures, see Figure 2, and may leave open paths for water and erosion of grout and soft ground.

A grout is usually considered stable if its bleed is less than 5%; sometimes the limit is 2%. The stability of a grout can be improved using additives such as bentonite, stabilizers, thixotropic agents, etc, or by reducing the water cement ratio. For MFC, as e g Rheocem 650 or Injection 30, 1:1 wcr mix normally gives a stable grout. A low water cement ratio mix usually provides a reliable filling of competent grout in the fissures. Such grouts are normally preferred in current practice.

Bentonite should not be used together with finer cements, as its grain size is bigger than that of fine cements. Tests have further indicated that a grout with bentonite is less stable under pressure.

Viscosity. To allow penetration into fine fissures the viscosity of the grout should be less than 35 seconds as measured by a Marsh cone.

Additives 1-3% in the form of plasticizers are needed to get an appropriate viscosity for a 1:1 mix as well as for ability to penetrate fissures without segregating.

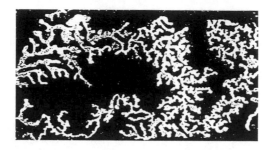

Figure 2. Incomplete filling of fissure due to bleed (seen from above). After P Hansson

Open time. To allow the grout to be placed, to avoid premature closing of fissures and to avoid blockage of grouting lines, an ideal open time would be about one hour. This would also enable tunneling to resume shortly after grouting is completed.

Chemical grouts. Examples of chemical grouts are Polyurethane (PU), Latex emulsions and silicate grouts.

PU grout can be very useful in stopping water inflows, as most PU-grouts used in tunneling reacts with water. It is however only a temporary measure, and should be followed by a more permanent grout.

Latex emulsion has been used successfully in reducing water inflows by postgrouting.

Silicate grout has very good penetration ability, but forms a rather soft gel in normal applications

The best chemical grouts for sealing of difficult ground are grouts based on acrylamide. Such grouts are however poisonous, and should not be used. Two such grouting material, marketed under the name of "Rocagil" and "Siprogel", was used on the Hallandsas tunnel in Sweden and the Romerike tunnel in Norway a few years ago. The supplier did not inform sufficiently on the environmental risks. Cows drinking water being discharged into a small creek got paralyzed and had to be slaughtered, and the grouting personnel got health problems as well. This led to the tunnels being stopped for long time, and a costly and time-consuming process was implemented, to eliminate the risks from already injected material and the chemicals leaching out from the tunnel.

Some types of chemical grouts should however be available on a tunnel site, as well as quick set cement grouts, to deal with unusual difficult ground and heavy water inflows.

Most grouting is however carried out with cementitious grouts, as they are both cheaper, easier to handle, virtually non-toxic and their durability is more dependable.

ROCK MASS	RMR	Q	GROUTING REQUIRED	TYPE OF GROUT MATERIAL
1.1 Massive, no joints	I	1000	no grouting	
1.2 Very few joints, < 0.1 joints/m3	I	100	spot/targeted grouting	MFC, if joints >0.5 mm, OPC
1.3 Few joints, < 1/m3, ≤ 2 joint sets	II	10	limited to continuous	MFC
1.4 Jointed rock, < 10/m3, >2 joint sets	III	1	continuous	MFC
1.5 Very jointed rock, ≥ 10 joints/m3	III-IV	< 0.1	continuous, closer spacing, in stages	MFC/UFC
2. FAULT ZONES				
2.1 Zones with clay	V	< 0.1	displace, wash out/replace, compact	OPC? MFC
2.2 Silty zones	V	< 0.1	penetrate, very close spacing, in stages	UFC, Chemicals
2.2 Sandy zones	V	< 0.1	penetrate, close spacing, in stages	MFC, UFC
2.3 Gravel zones/sugar cube rock	V-V	< 0.1	penetrate, quick set, stages	OPC, MFC
2.5 Mixed material	IV-V	< 0.1	penetrate, displace, compact, in stages, close spacing	OPC/MFC⇒UFC/ Chemicals

Note 1. Rock class, grouting required and type of grout materials are indicative only

Note 2. In rock classes 1.5, 2 and 3 repeated grouting and more than one grout cover could be needed

Note 3. Type of grout material to use is depending on degree of sealing effect needed, any filling in fissures, water pressure head, etc.

OPC = Ordinary Portland Cement. Max grain size 120-140 microns

MFC = Fine Cement, grain size 0-30 microns

UFC = Micro Fine Cement, grain size 0-15 microns

Chemical grout, contains no particles

Figure 3. Guide to Ground Treatment in rock Tunnels

Characteristics of rock masses

The degree of difficulty of grouting is dependent on the number of joints that have to be treated: single or few joints are easier to treat than sugar cube rocks. Of course the apertures of fissures and filling in them are important for the result.

In hard rock the sealing is accomplished by filling the fissures by grout, penetration-grouting. In fault zones and decomposed rock, pores in the decomposed rock have to be penetration-grouted, while bands and zones of clay needs to be either washed out/replaced or compacted and/or displaced, silty materials needs to be both penetration-grouted and compacted, and sandy/gravely materials need to be penetration-grouted. A high degree of tightness is normally required in fault zones and decomposed rock, to avoid erosion of sandy and silty layers.

Grouting of such materials is difficult and unpredictable, as the grout may go in any direction, giving an incomplete treatment. Treatment may have to be repeated, to get any degree of safety against inrush of water.

The enclosed Schedule "Guide to Ground Treatment with Pregrouting in Rock Tunnels", Figure 3, is an attempt to summaries and classify treatments needed for different types of rock. In rocks with a few joints the grout pattern may be more open, whilst in heavily fissured rock a regular drill pattern giving a good grout cover is needed.

Probing and prediction of rock quality

Pre-construction surveys give a general idea of rock conditions. However, the need of grouting is decided mainly if fissures are 0.1 or 1mm or closed, and such details are not possible to get from the pre-construction investigations. Pumping tests may give some indications, but to be able to decide on needs of pre-treatment of the ground, probe holes ahead of the tunnel face has to be carried out. Water inflows from holes and/or water pressure tests give an indication of conditions ahead and the need for pregrouting.

The information received by probing can be enhanced by using Measurement While Drilling, MWD, where torque, rotation, feed, penetration, flush water pressure, etc are continuously logged. The data received need to be fed into a computer program, calibrated and interpreted. The result can be presented as e.g. a tomographic map.

With the use of MWD presented as a tomographic map, the results are more readily available to be used by all tunnel personnel.

Pregrouting improves the rock quality.

Pregrouting will also improve the rock quality of the ground. "In dry conditions, pregrouting will improve the rock quality within the order of one quality class. In wet conditions, pregrouting will improve the rock quality by two or three quality classes. The overall result of efficient pregrouting will be reduced permeability, deformation and reinforcement requirement when tunneling, increased deformation modulus and seismic velocity. Each can be linked in a simple way to Q-value increases." From Barton et al 2001/02.

A very good example of this is the SSDS tunnels in Hong Kong. Tunnel F was driven on 150 m dept under the sea, with some 30-40m rock cover. At one point he open TBM had to pass a fault zone that showed to be 270m long with severely decomposed rock with inclusions of bands of silt and sand. The "rock" quality was improved 3-4 Q-classes by extensive pregrouting, by slowly "wedging" the grout into the ground, and the zone could be excavated successfully.

Pregrouting will thus will play an important role in avoiding unpleasant surprises. The excavation works will be more predictable, giving better chances in completing the tunnel according to schedule.

Continuous grout cover

As exemplified by Figure 3, in rock masses with few joints limited pregrouting may be successful, whilst in heavily fissured rock pregrouting giving a continuous cover has to be carried out.

As a general rule, when grout cover all around the tunnel periphery is needed, grouting in holes not more than 2.5-3.0m apart at the far end, in drillholes no longer than 20-25m, using stable MFC-grouts with plasticizers, will give a high probability of an adequate treatment.

Often the grouting of the drill holes is done in one stage, but in difficult rock, heavily fissured and with big variation of width of fissures, treatment has to be carried out in 3-5m long stages, or even less.

Grouting of fault zones and decomposed rock can be very demanding. Often the treatment has to be repeated a number of times to give an adequate water exclusion, making zones of soil less prone to erosion.

Drill hole lengths and deviations

Drilling for pregrouting in tunnels is usually carried out with percussion drills. Experience of measurements of deviation shows that drill holes deviate to a certain extent. Up to a drill length of about 20m, holes are normally within 1m of target, if guide rods are used. From about that length of drilling, the deviation normally increases. This explains the rule of thumb not to drill more than 20-25m long holes for successful pregrouting.

In broken rock the deviation has a tendency to increase. Sometimes, if drillholes collapse, drilling and grouting has to be carried out in successive descending stages, to enable drilling onwards.

Grouting pressures

For good penetration into fissures, it is beneficial to use a high pressure. For pregrouting in hard rocks pressures applied vary from 1 MPa to 6 MPa (10-60 bars), depending on thickness of overlaying rock and the strength of the rock, whilst the pressure for postgrouting from within the opened tunnel has normally to be limited to 0.5 – 1 MPa (5–10 bars).

Batch grouting/refusal

Grouting techniques employed earlier used thin grouts, starting with 5:1 and 3:1, water: cement by weight, successively reducing water content to mixes below 1:1, and continuing grouting to refusal.

With the advent of stable grouts, with plasticizers to give high flowability, batch grouting has come into use. A pre-decided amount of grout is injected into the hole, where after the grouting is stopped. It has been found that this normally gives the required sealing effect, and only in certain conditions the grouting has to be repeated.

This technique is very beneficial for the cycle time, using a stable grout in one mix, with a limited quantity. Also it is considered beneficial to use as high speed and pressure as possible from the start, which further reduces the time spent on any one hole. The technique can be used under fairly homogenous conditions, in rocks with not too big variation in fissure widths.

Test of grouting results

At the end of a grouting cycle, check holes are often drilled to control the result of the grouting. When the inflows from checkholes are less than the pre-set limit, the grouting is considered satisfactory and excavation may resume.

The result of the grouting over any excavated length of tunnel may have to be verified by measurements of the inflow of water. This is normally done by isolating a section of the tunnel and setting up a cofferdam or a stank board with an outlet, from which the water flow is collected and measured.

Test of grouts

As the grout is pumped into the rock, it is normally not accessible afterwards to sample or to check on the quality.

It is therefore more important to take samples and make tests of grouts than any other construction material that can be sampled after being placed in the structure.

There are some simple tests that should be performed on each grouting site: Cup-tests for gelling and hardening, Marsh Cone tests, Mud Balance tests and bleed tests. All are simple, site-adapted tests that should be carried out routinely on each grouting site.

   -Cup test consists of taking grout in a simple cup, and checking it for gelling time, the time when the grout thickens and pumpability is reduced. It is also used to indicate onset of hardening, and therefore when packers may be removed. Cup tests can also be used as an indication of bleed, but preferably graduated measuring glasses should be used for deciding the bleed.

   -Marsh Cone tests are done to check the viscosity of the grout, which is given in seconds. A good cement grout has normally a Marsh viscosity of 30-35 seconds.

   -Mud Balance tests are carried out to check on the volume weight of the grout, and are also an indirect check on the mix proportions.

All tests should be logged in a test book, recording date, time and section of grouting.

Cost aspects

As the relation of costs between OPC: MFC: UFC is about 1: 5: 10 it might be tempting to use OPC for the grouting. But OPC can seal off only the relatively high flows, and would leave the finer fissures untreated. Refer to 3 and 5.1.1 above. Those remaining fissures will usually give an intolerable inflow. As soon as there is a requirement for ground water control it is usually worthwhile to use finer cements. The cost of cement is usually marginal in a tunneling job, and by using cheaper cement it is a big risk that the result is not satisfactory. Just a few days postgrouting would quickly eat up any savings made on cheaper cement.

Any person that has an experience of postgrouting knows that it usually takes an unbelievable length of time, and many times it is not possible to achieve satisfactory results.

Many tunneling engineers have decided to try to speed up the pregrouting by using fewer holes than required, thinking that this can be dealt with later, only to find them spending a lot of time and money on post-grouting.

As the cost of a tunneling operation often is in the region of US\$ 1,000 to 1,500 per hour, a cost/benefit analysis would normally show that it is worth while to employ a more expensive material, if the grouting cycle is faster, and gives less risk for inadequate sealing. Stable grouts based on MFC or UFC have cohesion and are thixotropic to a much higher extent than OPC grouts, and will give closure of fissures with less amount of grout, giving less time for grouting. The higher cost for the cement material in a MFC grout is very often more than compensated by a shorter tunneling cycle.

Pregrouting as a contract provision

The author has come across quite a few numbers of tunneling projects, where high quantities of ground water came into the tunnel, causing big delays and/or settlements. There were no provisions of pregrouting in the contract, or it was a lump sum contract, in which the contractor had to estimate the need for ground water control and include it in his bid.

Even with elaborate investigations and hydro-geological evaluations, it is not possible to predict in detail what the tunneler will met, only to line up a plausible scenario. The reason for this is that investigations are only spot-checks. In most cases only a few parts per million or less of the ground to be tunneled through is sampled.

It is therefore prudent to include provisions for pregrouting and criteria when this should be implemented. The cost of such provisions is normally very marginal – if included in the bid documents - but the consequences of high water inflows (and low water inflows under sensitive ground) may be far reaching.

Closing remarks

Grouting was previously considered to be more an art than an engineering job, but with the advent of stable grouts and one-shot grouting, the job can and should be engineered. However, more than most engineering processes grouting requires a lot of understanding, skill and attention to details. Grouting is controlled by about 20 factors, and only about 5 of those can be controlled or investigated. Every grouting job is more or less unique, but by daily following the job, the grouting engineer develops a feel for the ground and what the grout does when it flows through the fissure and zones hidden in the ground ahead.

References

Barton et al (2001/02) "Strengthening the case for grouting" Tunnels & Tunnelling

Clarke WJ (1985) "Performance characteristics of microfine cement." ACSE, Preprint 84-023

Hansson P: Basic course in grout material technology. Vattenfall Utveckling AB

Littlejohn G S (1975) "Acceptable water flows for rock anchor grouting." Ground Engineering, 8(2) pp 46-48

Additional Material

Hakami E (1995) "Aperture Distribution of Rock Fractures." Division of Engineering Geology, Royal Institute of Technology, Stockholm, Sweden

Martinsen S (2000) "Success story at Bomlafjord." Tunnels & Tunnelling

Sjostrom O A (1988) Grouting, Sealing, Strengthening and Stabilising of Rock and Soil. Grouting Handbook

Sjostrom O A (1998) "Ground Water Control Through Pre-grouting for the Northern Link of the Stockholm Ring Road". Proceedings of the World Tunnel Congress '98 on Tunnels and Metropolises, Sao Paolo, Brazil

Sjostrom O A (1998) "Preliminary study of repair of earth fill dams using grouting". Swedish Water Power Board.

Sjostrom O A (1995) "Sealing of Rocks - Prerequisites and principles of design". Swedish Road Administration, NL2, May 1995

Sjostrom O A (1995) Development of Swedish Pre-grouting Technologies. Water Sealing and Prevention of Frost in Road Tunnels. Seminar, NTH, Trondheim, Norway

Injection of a Ventilation Tower of an Underwater Road Tunnel Using Cement and Chemical Grouts

Danielle Palardy[1], Gérard Ballivy[2] Jean-Philippe Vrignaud[3], and Caroline Ballivy[4]

Abstract

This paper presents the monitoring and the grouting works of an immersed road tunnel. As this underground structure is almost 30 years old, groundwater has gradually damaged and penetrated the concrete, bringing water problems in the winter as the water forms into ice that can fall on cars or damage electrical equipment. To understand the problem related to water infiltration and find solutions, instruments were installed to monitor the movement of a horizontal construction joint and exploratory drilling was conducted in the concrete ventilation tower structure. These investigations led to choosing two solutions: a first phase of cement grouting and a second phase of chemical grouting were carried out between 1998 and 1999. From the different available observation locations, both grouting methods gave satisfactory results for the stopping of water leakages in the ventilation towers.

Introduction

An urban road tunnel presented infiltration problems in one of its ventilation towers, promoting the formation of ice in winter. These water inflows were associated with the freeze-thaw period which caused the opening and closing of joints at that level. To confirm this phenomenon, crack measurement apparatus were installed to detect and follow all movements of the joints affected by the leak. In order to establish a relation between these temperature cycles, thermocouples were installed as controls in the construction joint and on the outside. As well, the starting-up of a pump at 0.0 elevation of this same tower was registered in order to evaluate the quantity of water circulating at this level.

Together with this instrumentation, an examination of the area and the carrying out of some drilling enabled to identify the causes for the water inflow and solutions that should be implemented to ensure the imperviousness of the structure at the level of damaged joints. Among the solutions considered, the injection of cement

---

[1] Materials Engineer, Labo S.M. inc., Varennes, PQ, J3X 1P7, Canada
[2] Professor, Civil Eng. Dept., Université de Sherbrooke, Sherbrooke, PQ, J1K 2R1, Canada
[3] M.A.Sc., Civil Eng. Dept., Université de Sherbrooke, Sherbrooke, PQ, J1K 2R1, Canada
[4] Materials Engineer, Construction Div., Hydro-Québec, Montréal, PQ, Canada

and chemical grouts was selected. Also, injection should allow the consolidation of the degraded concrete under the 0.0 level.

This paper presents the results obtained during various phases of the investigation of the ventilation tower, as well as those related to the injection of cement and chemical grouts.

Description of the site

The underground structure under examination is the Louis-Hyppolite-Lafontaine bridge-tunnel, a section of the Trans-Canada highway that crosses the St. Lawrence River between Boucherville and the Island of Montreal (Québec, Canada). It was built between 1962 and 1967. As illustrated in Figure 1, the tunnel was built using several concrete sections. The north part, made up of sections 1 to 8, was built in a plant and sunk to the bottom of the river while the south part was built in a dry dock.

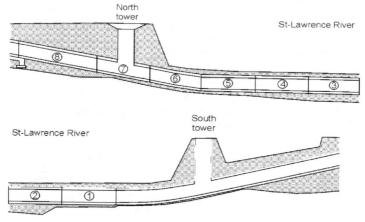

Figure 1. General view of the tunnel

The north ventilation tower, where this study was undertaken, is situated in section 7 of the bridge-tunnel on the banks of the Island of Montréal. It is made up of a double steel casement filled with concrete. The inside of this cylinder is formed of many concrete sections cast under water in dry conditions. Figure 2 gives a general view, as well as the dimensions of this tower.

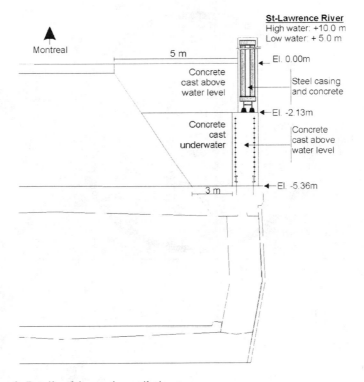

Figure 2. Details of the north ventilation tower

Instrumentation

Crack measurement apparatus and thermocouples

Four crack measurement apparatus (Nos. 1 to 3 and No. 5) were installed at the level of the main joint of the north ventilation tower, facing north (Palardy et al, 1997a). This is the joint between the concrete cast under water and that cast in dry conditions during construction of the tower. For each crack measurement apparatus, a Type T thermocouple was installed at a depth of 5 cm in the joint. Also, near crack measurement apparatus 3, a fourth thermocouple was installed to measure ambient temperature. Figure 3 represents the position of crack measurement apparatus and thermocouples installed at the level of the tower, as well as that of the data acquisition system (CR-10).

Between February and December 1997, a good correlation was established between variations in temperature and movements of the joint. Indeed, a decrease in temperature produces an opening of the joint due to a contraction of concrete while an increase in temperature produces the expansion of the concrete. Figure 4 presents these data.

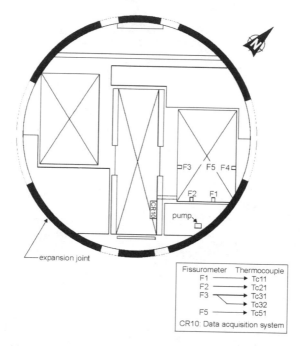

Figure 3. Position of the instrumentation and data acquisition system

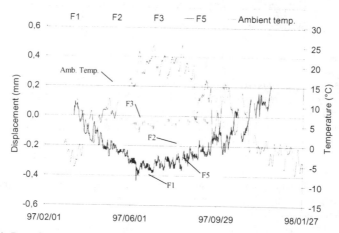

Figure 4. Data from thermocouples and crack meters. February to December 1997

First of all, temperatures increase during the month of May and then maintain an average value of approximately 20°C for the months of June to August 1997. For this same period, crack measurement apparatus indicated a gradual closing of the crack of 0.15 mm for the month of May; for the summer period, the variations are very low. Afterwards, the temperature decreased gradually until December, to reach an average temperature of – 2°C. According to the decrease in temperature, the crack opening gradually increased from September to December 1997, to reach 0.4 mm.

0.0 Level Pump

A pump at the 0.0 level of the ventilation tower was connected to the data acquisition system. This allowed the pump to register when switched on. Given the known 30 litre capacity of the pump sump, it was possible to calculate the volume of water circulating at this level.

According to data obtained since the beginning of the month of June 1997, the pump switched on approximately every two and a half hours. As indicated in Figure 5, this implies the pumping of nearly 300 litres a day. Following this, the carrying out of drilling operations at 0.0 elevation caused an important increase in the quantity of water pumped. The water rate shows an increase up to the end of August 1997, reaching more than 1,200 litres/day. Finally, the rate decreased and stabilized around 700 litres/day followed by a second increase at the beginning of September 1997. The rate stabilized at around 900 litres/day prior to the injection of cement grout.

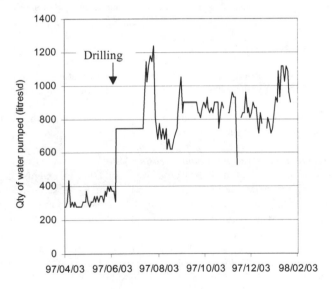

Figure 5. Quantity of water pumped per day – 0.0 elevation.

Investigation drilling

Two test holes (F1 and F2) were drilled at 0.0 elevation, as described in Table 1. These drillings made it possible to observe the deterioration of the concrete cast under water, at between –2.1 and –5.5 m in elevation. The concrete is delaminated and only the aggregate has been retrieved over a length of 3 metres. Also, artesian conditions have been detected. However, as soon as the drilling exceeds a depth of 3 metres, the water level in the holes dropped at approximately –1.2 m in elevation. In the zone between 0.0 and –2.1 m, an average compressive strength of 40 MPa was measured on the concrete cores obtained. Water inflows originate from construction joints circling the tower at –2.1 and 5.2 m in elevation.

Table 1. Description of test holes – 0.0 Elevation

Parameters	Test hole F1	Test hole F2
Type of drill :	Electrical	Electrical
Diametre of the drilling hole :	76 mm (3 po)	76 mm (3 po)
Drilling angle :	75 degrees	75 degrees
Depth aimed for :	±2.1 m (±7 pi)	±6.1 m (±20 pi)
Depth drilled :	2.2 m (7'4")	±3.0 m (± 10 pi)
Recovery :	±2.2 m (± 7'4")	±2.5 m (± 8'2")
Remarks :	At 2.2 m, the water arrives from the bottom of the drilling hole. It rises to the surface (artesian pressure above the 0.0 level).	The concrete starting from ± 2.4 m is damaged. When the hole is filled with water, it drains immediately up to approximately 1.2 m.

Cement grout injection

Injection work was carried out according to the specifications described in the quotation drawn up by the Laboratoire de mécanique des roches et de géologie appliquée de l'Université de Sherbrooke (Palardy et al, 1997b). This document indicates the injection equipment and method used, as well as the performance criteria of mortar mixes for such structures.

Primary, secondary and reconnaissance drilling holes
The holes used for injection of cement grout were drilled at an exterior angle to the structure in order to reach the joints at –2.1 and 5.2 m. Two series of holes were drilled: 25 primary long (L) holes, followed by 18 secondary short (S) holes (Figure 6). Four reconnaissance (R) holes were drilled after the injection work to verify the penetrability of the grouts.

A wireline coring system using an NQ (3-inch) calibre corer for the long drilling holes and an hydraulic drill with a HQ (4-inch) calibre corer for the short drilling holes was selected as drilling method. The holes were drilled continuously with a steel tube, then an MPSP (multiple packer sleeve pipe) was lowered into the hole. Each drill hole was sealed, on average at 3 meters from the surface, with the help of jute bags filled with thick cement grout.

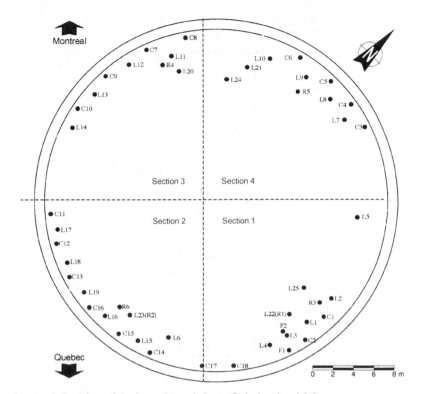

Figure 6. Position of the long (L) and short (S) holes, level 0.0.

Delamination of the concrete was evident in the holes drilled in the south part of the ventilation shaft (sections 1 and 2). Voids and fin sand were observed in these sections. The drilling of hole L3 confirmed the occurrence of an artesian condition previously identified in test holes F1 and F2 (Palardy et al., 1998).

Mixes used

Injection was carried out with Type 10 cement-based silica fume grout mixes (St. Lawrence Cement Type 10 HSF 8.5%). However, part of the drilling holes were injected with microfine cement-based mixes (Spinor A12 from Origny, France) due to the fineness of the sand encountered.

The mixes tested and used, as well as their rheological characteristics are found in Tables 2 and 3. Percentages of superplasticizer (Rheobuild 1000 from Master Builders Technologies) and colloidal agent (Welan gum from Multiurethanes) are expressed in terms of dry extract relative to cement mass.

Table 2. Type 10 Portland cement-based grout with silica fume

Mix B	Initial mix : W/C : 1.0 - SP : 0.5% - AC : 0.03% 1- Marsh cone : 33 sec - Gravity : 1.40 - Mini-cone : 168 mm Temperature : 12° C (temperature adjusted for injection) 2- Marsh cone: 31 sec - Gravity : 1,35 - Mini-cone : 203 mm Temperature : 21° C
Mix E	Initial mix : W/C : 0.8 - SP : 0.5% - AC : 0.03% 1- Marsh cone: 32 sec - Gravity : 1.43 - Mini-cone : 198 mm Temperature : 21° C– Static bleeding : < 5% after 2 hours 2- Marsh cone: 40 sec - Gravity : 1.44 - Mini-cone : 192 mm Temperature : 20° C
Mix F	Initial mix : W/C : 0.6 - SP : 0.5% - AC : 0.03% 1- Marsh cone : 44 sec - Gravity : 1.54 - Mini-cone : 178 mm Temperature : 20° C – Static bleeding : < 5% after 2 hours
Mix G	Initial mix : W/C : 0.6 - SP : 0.8% - AC : 0.03% 1- Marsh cone: 42 sec - Gravity : 1.55 - Mini-cone : 182 mm Temperature : 20° C – Static bleeding : < 5% after 2 hours
Mix H	Initial mix : W/C : 0.6 - SP : 1.0% - 0.03% 1- Marsh cone: 38 sec - Gravity : 1.50 - Mini-cone : 215 mm Temperature : 20° C – Static bleeding : < 5% after 2 hours 2- Marsh cone : 35 sec - Gravity : 1,53 - Mini-cone : 186 mm Temperature : 15° C – Static bleeding : < 5% after 2 hours
Mix I	Initial mix : W/C : 0.6 - SP : 1.0% - AC : 0.04% 1- Marsh cone: 36 sec - Gravity : 1.53 - Mini-cone : 189 mm Temperature : 20° C – Static bleeding : < 5% after 2 hours 2- Marsh cone: 44 sec - Gravity : 1.63 - Mini-cone : 169 mm Temperature : 20° C – Static bleeding : < 5% after 2 hours

Table 3. Spinor A12 microfine cement-based grout

K Mix	Initial mix : W/C : 0.8 - SP : 1.2% 1- Marsh cone : 30 sec - Gravity : 1.45 – Mini-cone : 222 mm Temperature : 22° C – Static bleeding : < 5% after 2 hours 2- Marsh cone : 31 sec - Gravity : 1.45 – Mini-cone : 218 mm Temperature : 20° C – Static bleeding : < 5% after 2 hours 3- Marsh cone : 32 sec - Gravity : 1,45 - Mini-cone : 209 mm Temperature : 19° C – Static bleeding : < 5% after 2 hours 4- Marsh cone : 31 sec - Gravity : 1.43 - Mini-cone : 210 mm Temperature : 19° C – Static bleeding : < 5% after 2 hours

Sequence of water and injection tests

Water tests were conducted on all holes before injection. As well as evaluating the site's absorption level, they helped in determining injection pressure, cement and

W/C ratio of the grout. According to practice, injection is impossible for Lugeon values inferior or equal to 3 LU (Houlsby, 1990).

Maximum pressure used for the water tests was 350 kPa. Primary and secondary drilling holes were tested according to the modified Lugeon test, that is, with three levels of pressure of two minutes each. Water tests in the reconnaissance drilling holes, on the other hand, were conducted according to the complete Lugeon test, which consists of five levels of pressure for a total duration of 30 minutes. These modified and complete Lugeon tests were carried out in one run over the effective length of drilling.

The highest Lugeon values (20 to 50 LU) come mainly from sections 1 and 2. For the sections situated north, most of the drilling holes indicate an absorption of 0 LU, with the exception of drilling holes C9/L11/L7/L9 that have values of 42, 37, 11, and 35 LU, respectively.

Following these results, it was decided to begin injection in the most deteriorated zone in order to block water inflows. Therefore, injection began at the lowest point (areas 1 and 2 to –6.1 m) at the level of drilling holes L4/L3/F2/L1/F1 and L15/L16/C16/L19 so as to push the grout toward the lowest point (areas 3 and 4 to –4.6 m). Also, as for the water test, injection was conducted in one run for all drilling holes. It should be noted that the rate of injection of the holes was generally very low with a maximum rate of 15 litres/minute and an average rate of 4 litre/minute. Maximum injection pressure of the cement mortar was fixed at 1 MPa.

Cement mortar injection should have sealed water leaks from the horizontal joints at –2.1 and –5.2 m of elevation, as well as provide renewed soundness to concrete below the –2.1 m level. The extent of penetration of the grout into the concrete could not be observed and therefore remained uncertain. However, grout was observed at various openings in the structure. Openings such as the ventilators and the electrical block located in the highest part of sections 3 and 4 were used to monitor the progressive sealing of the ventilation shaft during the injection process.

Table 4 indicates the total quantities of cement grout used during the process. Due to the fact that some zones were less damaged than previously thought, the quantities of mortar injected remained below the estimated values.

Table 4. Volume of cement grout injected in all drilling holes

Section	Primary holes		Secondary holes		Reconnaissance holes		Total quantity	
1	10:	2 211 litres	10:	2 936 litres	10:	0 litre	10:	5 147 litres
	S:	1 359 litres	S:	3 litres	S:	5 litres	S:	1 367 litres
2	10:	4 807 litres	10:	3 litres	10:	0 litres	10:	4 810 litres
	S:	285 litres	S:	14 litres	S:	2 623 litres	S:	2 922 litres
3	10:	0 litre	10:	889 litres	10:	0 litre	10:	889 litres
	S:	25 litres	S:	0 litre	S:	1 017litres	S:	1 042 litres
4	10:	258 litres	10:	0 litre	10:	0 litre	10:	258 litres
	S:	2 749 litres	S:	0 litre	S:	70 litres	S:	2 810 litres
* 10: type 10 / S: Spinor A12			Total cement grout injected :				10:	11 104 litres
							S:	8 141 litres

Chemical grout injection

Inspection of the site showed water inflows at the level of the horizontal joint in the
ventilation tower including traffic lanes, continued after completion of the cement
grout injection. A further round of sealing was performed using chemical mortar
injection. This section of the paper describes the products used as well as the
sequence of work (Vrignaud et al., 1999).

Product used

In accordance with the requirements of the tender call document, one product was
determined to be appropriate. This was a semi-rigid polyurethane resin known as
Universal H100, manufactured by the Multiurethanes company. The characteristics
of the product are presented in Table 5.

Table 5. Characteristics of UNIVERSAL H100 polyurethane

	Universal H100
Description	Aquareactive polyurethane resin injection Monocomponent that reacts with water Hydrophobic semi-rigid foam and chemical attack-proof Foam does not shrink and does not swell after final setting time
Composition	Resin : prepolymer with isocyanates (MDI) Accelerator : alkylamine
Characteristics	Resin : Gravity (20°C) 1.15 / Viscosity (25°C) 100 cPo Accel. : Gravity (20°C) 0.995 / Viscosity (25°C) 25 cPo
Application	Waterproofing of underground tunnels, subways, excavations and cracks in rocky walls Where there are important water inflows at high pressure

Work carried out

Injection was carried out according to the specification described in the quotation
drawn up by the Laboratoire de mécanique des roches et de géologie appliquée de
l'Université de Sherbrooke (Palardy et al., 1997b).

Holes for injection were drilled from the base of the horizontal joint spaced
25 cm apart. These intersected the joint at 40 cm depth. The joint was cleaned under
pressure using phosphoric acid and water to clean all particles from the joint that
could block distribution of the resin and prevent good adherence of the grout to the
concrete. Observing the quantity of water absorbed by the joint and the time it took
the water to return to the surface allowed evaluation of the future absorption of grout
as well as adjustment of reaction time. Heated containers helped maintain the
temperature of the mortar above 20 °C so as to obtain a quick reaction time and low
viscosity. The grout was injected successively in each sleeve installed along the
joint. Injection was carried out in two runs at approximately 15 minutes apart.
Generally, the first run was done with a pressure of 4 to 6 MPa (600 to 800 psi) and
the second at a pressure of 7 MPa (1000 psi). Finally, the sealing of the joint was not
necessary because loss of resin at the surface was not very important.

The percentage of accelerator was set at 5%, then 3%, to allow the grout to
move as far as possible. Injection pressure went progressively from 500 to 1000 psi.

It was maintained until the water was pushed back as far as possible on the south and east sides of the joint. When injection ended, it was observed that the grout had traveled over a length of 3 m on each side of the injection point, the water being pushed back in the left lane.

In all, 325 litres were injected for ventilators VE 152 and 154, in the south direction and 186 litres for VE 154, in the north direction, for a total of 511 litres of mortar for 66 linear meters of repairs. Absorption was relatively high (7.7 l/meter), which shows that a few voids could still be found at the level of this joint. It should be noted that 75% of the quantity of grout, 372 litres, were injected in both south-east and north-west corners of VE 152 and 153. The rest of the joint showed a lower absorption, as expected, that is 2.0 l/m in the south direction (89 litres over 44 m) and 2.3 l/m in the north direction (50 litres over 22 m).

Conclusion

Data collected for the period from February to December 1997 agree with the expected displacements and temperatures. A good correlation can be made between changes in temperature and the contraction and expansion of concrete that are evidenced by an opening and closing, respectively, of the crack at the level of a casting joint of the north ventilation tower in the north direction. Maximum movements registered occur during important temperature changes, that is in fall and in spring. Therefore, the arrival of milder temperatures produces a closing of the crack of approximately 0.4 mm while fall cooling causes an opening of approximately 0.4 mm. Water inflow, prior to sealing program, was calculated at 300 litres/day.

The concrete making up the north tower between levels –2.1 and –5.2 m was damaged throughout, to a greater or lesser extent. It still displayed binding properties overall, but was completely leached in some areas with many holes, drilled in sections 1 and 2, rich in fine sand.

The use of Type 10 cement-based silica fume grout with low W/C ratios (W/C = 0.8) was conclusive. However, due to the quantity of sand encountered, microfine cement-based grouts had to be considered. The total volume of grout injected was 11,104 litres of Type 10 HSF cement-based grout and 8,150 litres of microfine cement-base grout. Based on the available observation points, this injection stopped water inflows at the level of the ventilators through the sealing of the joints on the periphery of the north ventilation tower.

Overall, injection using chemical grout of the semi-rigid hydrophobic, prepolymer-type were deemed satisfactory. A great absorption level was observed at the level of the joint of the ventilation tower, in the amount of 511 litres were injected in 66 linear meters of the horizontal joint in the ventilation tower. This is a relatively high volume of injection. Water was pushed back outside the traffic lane, mainly toward the service tunnels situated in the central part of the tunnel where it can be evacuated by the trenches.

A first inspection of the tower in mid-May 1998 helped in evaluating the work carried out in winter 1998. It seems that the major water leaks were stopped, particularly at 0.0 level. Injection seems to have completely stopped the water leaking from ventilators VE-151, VE 153 in the direction of Montréal (north) and in ventilators VE-152 and VE-154 in the direction of Québec City (south). The cement grout that escaped through the cracks and joints during the injection is visible and the internal surfaces of the ventilators are dry. The electrical block, situated at the highest point of the tower, is now completely dry whereas it displayed important leaks when work began. Also the trench along the wall and the pump reservoir in section 1 (north-east) of the tower are completely dry. These observations have been confirmed by a new inspection in March 2002.

Acknowledgements

The authors gratefully acknowledge the collaboration of Ministère des Transports du Québec (Department of Transportation, Province of Quebec) participating in field work. They also appreciate the financial support of Natural Sciences and Engineering Research Council of Canada (NSERC) and the Quebec Provincial FCAR fund (Fonds pour la formation de chercheurs et l'aide à la recherche).

References

Houlsby, A.C. (1990). *Construction and design of cement grouting, A guide to grouting in rock foundations.* Wiley series of practical construction guides, John Wiley and Sons, inc.

Palardy, D. and Ballivy, G. (1997a). *Instrumentation de la tour de ventilation nord, direction nord, rapport d'installation.* Laboratoire de mécanique des roches et de géologie   appliquée   (LMRGA),   Univ.   de   Sherbrooke,   Report GR-97-03-01.

Palardy, D. and Ballivy, G. (1997b). *Devis technique - Travaux d'injection au coulis de ciment et au coulis chimique au pont-tunnel L.-H.-Lafontaine.* LMRGA, Univ. de Sherbrooke, Report GR-97-11-02.

Palardy, D. and Ballivy, G. (1998). *Instrumentation de la tour de ventilation nord, direction nord, Rapport final du suivi - Année 1997.* LMRGA, Univ. de Sherbrooke, Report GR-98-02-01.

Palardy, D., Ballivy, C., Vrignaud, J.-P. and Ballivy, G. (1998). *Suivi des travaux d'injection au coulis de ciment et au coulis chimique, pont-tunnel L.-H.-Lafontaine,* LMRGA, Univ. de Sherbrooke, Report GR-98-03-02.

Vrignaud, J.-P., Palardy, D. and Ballivy, G. (1999). *Suivi des travaux d'injection au printemps 1999 aux tunnels Georges-Vanier, Ville-Marie et L.-H.-Lafontaine et recommandations pour les interventions futures,* LMRGA, Univ. de Sherbrooke, Report GR-99-07-01.

THE TORONTO TRANSIT COMMISSION'S SUBWAY TUNNEL AND STATION LEAK REMEDIATION GROUTING PROGRAM

L. Narduzzo, P.Eng.*

ABSTRACT:

The Toronto Transit Commission (TTC), one of the largest public transportation systems in North America has been plagued by leaking tunnels since the time they were constructed. The water infiltration problems was causing both delays and concerns for passenger safety as well as causing accelerated aging of the rail and rail fastening systems, deterioration and malfunction of electrical systems and their components, and decay of the structure itself. A professionally engineered state-of-the-art solution grouting program has brought the problem under control. The grouting program at the TTC, which was started up in May 1997, is one of the largest continuous on-going leak remediation grouting projects using solution grouts in North America. The leakage remediation program originally focused on solving the tunnel leakage problems but has since expanded to include water infiltration problems in the stations. The success gained to date can be attributed to a combination of several key components: the assembly of an in-house team of grouting expertise – from design engineer to field technician and the selection and meticulous use of the most suitable sealing materials available in industry for this specific and extremely difficult application.

The unique challenge of performing all the leak remediation grouting work within the nightly two hour maintenance window without impacting on customer service was successfully accomplished using a strategically implemented, engineered grouting procedure. Time limitations and difficult ground conditions proved to be the two most difficult obstacles facing the grouting engineer. The grouting design had to take into account the multiple phase, multiple stage grouting, operations that were anticipated and required to successfully shut off the leakage problems.

----

*Senior Leak Remediation Engineer, Toronto Transit Commission, 1900 Yonge Street, Toronto, Ontario, Canada, M4S 1Z2, Phone 416 393-4063, luigi.narduzzo@ttc.ca

The sealing material selected to be the most suitable for this demanding and challenging application was AVANTI International's AV118 Duriflex; a low-viscosity aqueous solution of acrylic resins which, when properly catalyzed produces an impermeable and cohesive gel grout. Approaching five years from the time the grouting program was initiated, the program is on track in meeting its goals and objectives as originally set out in 1997.

This paper aims to provide some helpful insight on how the specific sealing system engineered and designed to solve the TTC's tunnel and station leakage problems has worked and continues to be the leak remediation solution of choice for this public transit authority.

Introduction

The City of Toronto has one of the largest public transportation systems in North America. More than 1,000,000 commuters rely on this public transportation service every day. Toronto's Transit system is comprised of approximately 52 kilometers of underground tunnels and numerous below-grade structures in different geological zones and hydrological conditions. There are two main arteries, one running north to south and the other running east to west.

The first subway line in Toronto was opened in 1954. It extended along Yonge Street from Union Station in the south to Eglinton Station in the north. Since then their have been several extensions linking the city of Toronto from McCowan Road to Kipling Avenue (East to West respectively) and Union Station to Downsview and Finch Stations (South to North respectively). The most recent extension is slated for completion in the fall of 2002. It extends eastward along Sheppard Avenue adding four new stations to the subway system.

Water infiltration has been a problem for the TTC since the time of construction. The water accelerates the life cycle of many operational components and systems affecting the safe operation of the system. Rail plates and anchors need replacement, electrical signal components require significant up-keep and replacement, drains become plugged etc.... Since their construction the tunnels had not received any major restoration of any kind. The requirement for a structure maintenance program was in great need due to the persistent and progressive water infiltration that was causing extensive concrete and steel deterioration. Up to the start of the remediation grouting program only emergency repairs had been performed on an as required basis. Given that water infiltration into the tunnels was the leading contributor to a large range of problems with electrical and mechanical systems and components including the structure itself, tunnel leak remediation was deemed a primary objective and focus of the Commission.

The implementation of a focused maintenance program, by skilled and knowledgeable professionals in their respective areas of expertise was launched early in 1997 with the formation of a tunnel leak remediation crew. An "in-house" design-build approach was selected due to the large, variable and complex scope of work needed to restore the integrity of the structures since localized remediation to control water leakage (the emergency repairs) had yielded minimal success and could not keep pace with the increasing severity of the problem. The tunnel leak remediation program was built around a solid foundation of expertise in the specialty field of grouting, restoration and rehabilitation.

This paper gives a 5 year update on some of the key aspects of the tunnel and station leak remediation program using solution grouts. It shares some of the experiences gained from this fast pace, successful state-of-the-art tunnel leak remediation grouting program.

Historical Program Background

In 1996 a detailed consultants study that analyzed the current leakage conditions within the existing subway tunnels clearly indicated that it is imperative that the Commission immediately allocate sufficient funding to implement a comprehensive maintenance and repair program. The program was to be geared to minimize the detrimental effects of leakage and ensure safe operation of the subway and rapid transit system. So was born the tunnel leak remediation program. On average, approximately 30 percent of the underground tunnel structures require some form of remediation.

The start-up of the present in-house specialty leak remediation crew was done from "ground zero". Prior to the spring of 1997 the TTC had no professional grouting equipment, no material and very limited in-house expertise in the field of grouting restoration and rehabilitation. The TTC went externally to hire the required expertise to head up and deliver the desired end results it was looking for.

The tunnel leak remediation crew is currently managed under the structure maintenance section of the Track & Structure Department. The condition surveys, investigations, design of leak repairs, product selection, grouting methodology, quality control, reporting, scheduling and analysis of results are the responsibility of the senior leak remediation grouting engineer. The Tunnel Leak Remediation Crew is made up of thirteen employees: a foreperson, two lead hands, two structure mechanics, two special vehicle operators and six structure repair persons. In 1999, the program was expanded to include station leak repairs.

Tunnel Type and Construction

There are three different types of tunnel construction in the Toronto Transit Commission:

- concrete box structure

- precast concrete circular lined tunnel
- cast iron circular lined tunnel

In locations of the subway system that allowed it, the cut and cover tunneling system was used whereby a concrete box tunnel construction was utilized. Approximately 80% of the subway tunnel was constructed by the cut and cover method. At some locations there is as little as 0.5 metres of fill covering the subway box tunnel.

In several locations throughout the city, the subway extended to greater depths below ground requiring the bored tunnel construction system using either a cast iron segmental liner or a precast concrete segmental liner. The bored subway tunnels are as deep as 25 metres.

Concrete Box Tunnel

The standard concrete box structure portion of tunnel comprises a rectangular concrete box of vertical walls, roof and floor slabs, with a central wall dividing the tracks which run in opposite directions. The structure is monolithic with keyed construction joints near the floor/wall joints and wall/roof joints.

The box structure sections are typically 12.2 metres long and each unit is separated by an expansion joint running transversely to the track direction. Lateral movement between adjacent units is restrained by means of a shear key. The wall/roof intersection is hunched. The clearance inside each running tunnel formed by the interior box section for passage of the trains is 4 metres high from the top of the rail to the underside of the roof and 4.2 metres between inside faces of walls.

Cast Iron Lined Tunnel

Cast iron liners were used to line the inside of the bored tunnel. The tunnel was excavated behind a shield and the liner was erected immediately behind the shield and the 40 millimeter annulus between the ground and liner was grouted with neat cement.

The cast iron liners were the type of liners used in the first bored tunnels in the Transit system and were used in conjunction with the precast concrete liners on later sections of the bored tunnels where the more economic precast concrete liners could not be used.

The tunnel formed by the cast iron liner is 4.9 metres in diameter. The liner is made up of 8 segments around the circumference plus a key section at the crown. Each ring is 610 millimetre (2 feet) wide. The thickness of the liner is 22 millimetres (7/8 inch) with flanges 35 millimetre (1-3/8 inch) wide. The overall thickness of the liner is 150 millimetres. This thickness coincides with the thickness of precast concrete liner. Each segment is bolted to the adjacent segment with 3 – 25 millimetre diameter bolts per 610 millimeter (2-foot) ring, and 4 – 25 millimeter diameter bolts per segment. The outside faces of all flanges were machined so that the segments were in contact over the full contact width of the flanges.

A caulking groove was machined on the inside of each flange forming a groove 8 millimetres wide and 25 millimetres deep when the segments are erected. This groove was caulked with an asbestos-cement ribbon, which was moistened and rammed into the caulking groove. In some places, lead caulking was used instead of asbestos cement ribbon. Plastic grommets were placed behind the head and nut of each bolt to seal the bolt holes. Grouting the annulus between the outside of the cast iron liner and the ground was achieved by pumping grout through grout holes in each segment and closing these holes with plugs.

Precast Concrete Lined Tunnel

The precast concrete liners were used to line the inside of the bored tunnels in a similar manner to that of the cast iron liner. These liners were usually used in conditions where the groundwater table was below the tunnel axis. The use of precast concrete liner was preferred over the cast iron liner as it was cheaper overall. The cast iron liner was only used in areas of the tunnel where groundwater table conditions prevented the use of the precast concrete liner.

The tunnel formed by the precast concrete liner is all 4.9 metres in diameter except at stations and the new Sheppard Subway line that is currently under construction. This new tunnel is 5.2 metres in diameter. The liner is made up of 8 segments plus a key section at the crown. Each ring is the same size and thickness as the cast iron liner, namely 610 millimetres (2 feet) and 150 millimetres (6 inches) respectively. Each segment is bolted to the adjacent section using two 23 millimetre diameter bolts in the circumferential direction. The radial joints between the segments was cast slightly convex, with a radius of 1.5 metres, to allow rotation within the joint. In some cases 8 millimetre plywood packing was inserted between the precast concrete liner rings to align the tunnel and to avoid stress concentrations during shield jacking. The precast concrete liner was made using 45 MPa concrete in a factory environment. Each unit was moist cured for 7 days. The liner was erected in the tail of the shield in a similar manner as the cast iron liners and grouted in a similar fashion.

Identification and Analysis of Seepage Problems

The mechanism for water infiltration into the tunnels and stations is predominantly via the expansion and construction joint systems and in the case of the bored tunnels the infiltration is primarily along the radial and longitudinal joints but has also been observed coming in via the bolt holes and also through cracks. The ground water table is typically above the tunnel roof and is the source for all the water infiltration problems. The water sits in behind the tunnel walls and/or liners and migrates along the backside of the tunnel lining/walls until it finds an entry point via an expansion joint, construction joint, crack, segmental tunnel joint or other point of ingress. The zone immediately in behind the concrete box tunnel walls or segmental tunnel liners that was disturbed during the tunnel construction is a zone of high hydraulic conductivity. For the most part, our hydraulic conductivity tests reveal an "infinite" permeability (infinite meaning our equipment cannot supply water fast enough to register an effective pressure after taking into account the headlosses in the system). The hydraulic conductivity tests are continually performed using the acrylamide solution grout as the testing fluid.

The geotechnical data is fairly consistent throughout all areas that the subway passes. The soils vary from fairly silty clay till to dense till. In general, the soils are un-injectable to at best marginally injectable by solution grouts. However, it is the preferential hydraulic pathway immediately adjacent to the back of the structure that is the zone that is the focus of our grouting efforts.

THE GROUTING PROGRAM

Time Limitations

All tunnel and station leak remediation work is performed at night when the subway system is not in service. On average the actual maintenance window from which work can be performed is approximately two hours per night. A six hour working window is available on Sunday mornings. Due to the short working window, it was necessary to develop and implement a strategic and innovative design solution to execute the leakage repairs in a swift, effective and efficient manner. Given the vast amount of other maintenance activities that are also performed at night the availability of work zones in the same area night after night is not always possible.

Self Sufficient Work Vehicle

The majority of the grouting operations are performed from a twenty four metre long work car that was custom built specifically for this program. The work car was designed to be self sufficient for power, storage space and enough water for the entire shift. The work car has two working platforms strategically separated to allow work to be performed on two expansion joints simultaneously. The work car is electrically powered and has been equiped with a diesel powered 250 CFM compressor equipped with scrubber to control emissions. The cost of the work car alone was one million dollars. The work car is loaded with grout material and other consumables at the start of every shift.

A second work car was made available for the tunnel and station leak remediation program on a permanent basis starting in mid 1999. The addition of the second work car increased the production and efficiency of the overall operation. The second work car became the "drilling" car, preparing areas for grouting allowing grouting to be carried on seven nights a week by the grouting work car.

Injection Methodology

The grouting repair techniques that has been successfully used to stop the water infiltration into the tunnels and stations has been the "back-wall" or exterior soil grouting approach using acrylamide based solution grouts. AVANTI's AV100 and AV118 have both been used. The manner in which these leak repairs are carried out is slightly different in the circular bored tunnel sections compared to the concrete box tunnel sections.

In the concrete box tunnel sections the grouting is targeted in the soils immediately behind the expansion joint. In essence a mini grout curtain or grout bulb is created immediately behind the expansion joint preventing water from entering the joint via the back of the structure. Grouting ("back-wall grouting") is carried out via mechanical injection packers installed at a 1 to 1.5 metre spacing along the expansion joint. The holes are drilled by means of rotary air percussion techniques using stoper and jack-leg drills. Drill holes are 35 millimetres in diameter. A ball valve in conjunction with a hydraulic connection system is then attached to the mechanical injection packers so that a grout hose can be hooked-up when grouting is to take place. Multiple hole grouting with up to six holes at a time is performed.

Figure 1: View of grouting work car during a "back-wall" grouting operation using AV118 Duriflex injection gel to seal a leaking expansion joint.

The grouting operations are slightly different for the circular bored tunnels. Given the nature of these circular tunnels (i.e. 610 millimetre wide segment rings bolted together) a larger scope external soil grouting operation is performed. Naturally, there are many more potential infiltration points in the circular tunnels compared to the concrete box tunnels. Again, through extensive hydraulic conductivity testing it was determined that a preferential hydraulic pathway exists immediately behind the liner. One explanation for this condition is the result of ungrouted or unsuccessfully grouted tunnel annulus during the original tunneling.

For the precast concrete segmental liner, 23 millimetre diameter grout holes, are drilled at a 1 to 2 metre staggered spacing. Mechanical injection packers with a ball valve and hydraulic connection system are then installed in the drill holes. The drill holes are drilled using rotary electric percussive drilling techniques. Multiple hole acrylamide grout injection is then performed working from the bottom upwards to the crown.

The depth of injection drill holes ranges from 600 millimetres to over 4200 millimetres for the concrete box tunnel structure. For the round tunnel areas the depth of holes range from 50 millimetres to greater than 900 millimetres.

In the case of the cast iron segmental tunnel liner, the original existing grout plug holes are used as grout access points. The new grout hole spacing is dependent on the locations of the old existing grout plugs. Apart from this fact, grouting procedures are the same as for the precast concrete segmental lined tunnel. External soil grouting is performed starting from the bottom upwards to the crown. Good hydraulic interconnection has always been observed during all grouting operations ideally creating overlapping grout cylinders.

The set times used in the acrylamide (AV100 and AV118) grouting is for the most part between 30 seconds to 3 minutes. However, set times of less than 30 seconds are sometimes required due to excessive leakage encountered during the grouting operations.

In all cases, the effective grout injection pressures are kept (on average) to a maximum of 15 psi above the hydrostatic pressure at that location. Due to time constraints, grout volumes have been pre-determined and fixed volumes of grout are placed at each location. In the box tunnel the typical volume of grout that is placed to successfully create a grout curtain is 800 litres of acrylamide based solution grout. By experience, the installation of this quantity of AV118 Duriflex grout optimizes the success in sealing problem leakage locations. On an average night a total of four box tunnel locations are completed. For the bored tunnels, an average of 250 litres of acrylamide based solution grout are injected per linear metre of tunnel, completing and average of 4.5 lineal metres of bored tunnel per grouting shift.

Grout Selection

The requirement for a technically suitable, durable, proven grout material that is applicable in the type of repairs to be executed lead to the use of an acrylamide based solution grout. Much of the success that has been achieved can be attributed to the use of the acrylamide based solution grout. At the start of the program, AVANTI INTERNATIONAL's AV100 was used. In April of 2000 the switch was made to AV118 Duriflex. The key reason made for switching to the AV118 Duriflex was regarding the high air quality test result readings that were obtained. The air quality sampling results were higher than the Ontario allowable Time Weighted Average Exposure Value level of 0.03 mg/cubic meter for acrylamide. For this reason the alternative AV118 Duriflex was selected due to the lower acrylamide content. The AV118 Duriflex has a free acrylamide content of less than 5% of concentrate solution as delivered on site from the plant. Since the switch was made air quality sampling results were quite satisfactory to the extent that respiratory protection could be eliminated if there would be no concern for mists or sprays during the grouting operation.

As it turns out, the switch proved to be much more advantageous to the overall program as it allowed for the elimination of a concentrate batching plant. AV118 Duriflex already comes in a 40% concentrate solution. This proved to be a favourable move since it eliminated the concentrate batching step in the grouting procedures. With higher volumes of acrylamide being used on a nightly basis, the elimination of the batching process stream-lined the overall grouting procedure. Naturally, because the AV118 Duriflex already comes in a 40% concentrate solution less handling is required. This simplification in the preparation stage made for a safer more efficient grouting operation overall.

Grout Formulations

The acrylamide is injected at a 10% to 20% final injection concentration depending on the in-situ conditions at the time of grouting. In areas where there is a high water presence (evident by flowing water from the grout packers) a final injection concentration of 20% AV118 Duriflix is selected. Locations that are dry at time of injection (due to time of season) are injected with a final injection concentration of 10% to 13% AV118 Duriflex.

All batches are prepared at the start of the shift prior to heading out to the work zone. Set times are adjusted at this time by the addition of a specific amount of initiator and activator to the respective A and B components. Quality control of each batch is checked for set time. Batch sizes are typically 2,000 litres at final injection concentration.

Safety Program

One important feature of the Toronto Transit Commission's tunnel leak remediation program is its stringent safety program. The T.T.C. has it's own Occupational Hygiene Department that independently monitors the safety aspects of the work. The Occupational Hygiene Department makes it's own assessment in personal protective equipment requirements for the job and then recommends the specific personal protective equipment for the job at hand.

A unique system of personal protective equipment was specially developed for the specific needs of the tunnel leak program. The Occupational Hygiene Department was responsible in selecting the personal protective equipment to meet the safety needs of the crew. One of the greatest safety concerns was having adequate protection while performing grout injection work overhead. The ultimate system of personal protective equipment selected consisted of the following:

• Powered air-purifying respirator with organic vapour cartridges
• Snap cap protective hood and visors
• Rain gear with tapered sleeves
• Rubber gloves with sealing attachment to rain coat
• Rubber boots

To avoid even the slightest concerns for acrylamide contamination the safety protocol has been established to replace all the above listed safety gear on a weekly basis with the exception of the rubber boots.

Experiences and Results

Since the start of the program many milestones have been achieved and many hurdles have been overcome. Over 1,700,000 litres of acrylamide (AVANTI INTERNATIONAL's AV100 and AV118) solution grout has been successfully injected throughout the tunnel structures to stop the water infiltration problems. In consideration of the difficult working conditions and the short nightly work window available, the success has been exceptional. The whole grouting operation has become extremely efficient over time, enabling the completion of approximately 4 locations per shift. Presently, the program continues to use two work cars: a drilling car and a grouting car and as a result, a steady level of production could be maintained.

TUNNEL LEAK GROUTING STATISTICS

Linear metres of expansion joints, construction joints and cracks sealed in Concrete Box Tunnel Structures	14,078 metres
Square metres (area) of Bored Tunnel Sealed (Pre-cast concrete lined lined/Cast Iron Lined)	30,000 square metres
Station structures: Linear metres of expansion joints, construction joints and cracks sealed	1,273 metres
Total volume of AV100/AV118 injected (litres): 1997 to 2001 inclusive	1,724,550 litres

Conclusions

The main reason for the success behind TTC's tunnel leak remediation program is the professional manner in which the program was developed. The right decision was made by assembling a team of knowledgeable professionals in their area of expertise along with an experienced crew in general grouting procedures.

The reduction in inflow that has been achieved with the sealing methods used has yielded positive results; specifically, a dry tunnel. Due to the time limitations during the nightly injection work, approximately ten percent of the areas addressed on the first attempt require some form of secondary grouting at achieve a completely dry end product.

The results attained to date with the acrylamide (AV100 and AV118) grouting have been very successful and the program is expected to continue in the years to come as an on-going maintenance program.

Large-scale Field Investigation of Grouting in Hard Jointed Rock

Thomas Dalmalm[1]

Thomas Janson[2]

ABSTRACT

During the construction of the rock tunnels at the South Link, Stockholm, Sweden, grout investigations were carried out. The aim was to develop alternative methods/concepts for sealing of tunnels compared to the methods/concepts described in the contract. The investigation started with a laboratory study, which outlined the most appropriate grouts for field trials. The field trials consisted of seven grout fans. The front of each fan was divided for equal pumping test values, and grouted with different concepts. The sealing result and the spreading of the grout were studied in 10 holes in the tunnel front and in 10 holes perpendicular to the tunnel. The holes were test pumped and studied with a borehole camera. From the investigation it was concluded that regardless the maximum grain size of cement (9.5, 16 or 30 μm) fine joints (< 100 μm) were still unsealed after grouting. In open joints or rock masses (>1 Lugeon) there was some indications that micro cement was more effective than the more coarse cement type. Higher stop pressure did not give a better sealing result and some parameters achieved at laboratory especially for micro cements were not possible to reproduce with ordinary field mixers.

INTRODUCTION

Preface

The project was a co-operation between Swedish National Road Administration, Royal Institute of Technology and the contractor NCC AB. The project was carried out during the construction of rock tunnels for the south part of the Stockholm Circle. The investigation was carried out between July –99 and February –00.

A working team from the Royal Institute planned, performed and analysed the trials, (Dalmalm et al. 2000). A reference group of representatives from the co-operation companies have, during the project, participated in the work.

---

[1] NCC AB / Royal Institute of Technology, S-100 44 Stockholm, Sweden; dalmalm@kth.se

[2] Ph. D. Golder Associates AB, Box 20127, S-104 60 Stockholm, Sweden; Thomas_janson@golder.se

Background and aim

The tunnel excavation at the South Link, Stockholm, Sweden was, due to the location in the Central Stockholm, connected to high demands regarding sealing and performance. The Swedish National Road Administration wished to develop methods for tunnel sealing, based on latest experiences and science results, to reach a high sealing efficiency and a low impact on the environment.

The aim was to develop alternative methods/concepts for tunnel sealing, in relation to the methods described in the contract, so that the need for post-grouting and chemical grouting could be reduced.

- The methods/concepts should be cement based and production efficient.
- The methods/concepts should be compared to what was described in the contract.

Geology

The geology varied strongly within the area of the grout trials. A general judgement, prior to the trials, was that the rock mass from a grouting point of view could be described as:

- A low hydraulic conductivity, approximately $1 \cdot 10^{-7}$ m/s (app.1 Lugeon) or lower.
- Single joints, between 0.1-0.5 joints per meter, with varying conductivity.
- Narrow apertures, < 0.5 mm, with varying joint filling.

The prior judgement was in accordance with the rock mass present at site of the trials. The measured average Lugeon values were very low, equivalent to a hydraulic conductivity between approximately $3 \cdot 10^{-7}$ and $2 \cdot 10^{-8}$ m/s.

Demands and assumptions

In order to fulfil the allowed inflow, which at the area of the trials was between 1-3 litres per 100 meter of tunnel, a sealing efficiency between 90 to 95 % was needed. The sealing efficiency was then defined according to equation 1.

$$Sealing\ efficiency\ (\%) = 1 - \frac{Inflow\ after\ grouting}{Inflow\ before\ grouting} \qquad (Equation\ 1)$$

The grout needed to both penetrate and fill the joints, and in addition, the grout needed to be long term resistant. To reach the allowed inflow it was necessary to obtain a hydraulic conductivity of the grouted zone of $0.5\text{-}0.4 \cdot 10^{-8}$ m/s, equal to seal joints down to an aperture of 0.1 mm partially. It was predicted that it would be difficult to reach a hydraulic conductivity below $0.5 \cdot 10^{-7}$ m/s. When sealing apertures less than 0.2-0.3 mm, filtration of the grout was expected to affect the result; the grout trials confirmed the predictions. At this point, a priority order between the different methods/concepts to seal the fracture system was done. The trials started with finding a proper grout mix, to be followed up by trials with high grout pressures.

A number of grout properties demands were set up. related to environment. function and production.

The specification of the grout was as follows:

*   Shear strength after 2 hours should for production purpose be at least 3 k Pa.
*   Separation and shrinkage should be less than 2 % after 2 hours.
*   At least 300 ml of grout should be able to pass through a 125 µm filter one hour after mixing, measured with a filter pump (Hansson. 1995).

GROUT TRIALS

Investigation of grouts

The trials started with a request to nine different producers of grout material. The geological situation was described and they were asked to recommend a cement based concept, based on the given information. Five producers answered with a number of concepts.

The initial laboratory investigation of grout were based on six different products; Rheocem 800, Cementa Injektering 30, Dyckerhoff Rock U, P-U, P-X and Thermax. The different products were tested for different WC-ratios and different additives, to achieve an optimal specification. In the laboratory, properties such as rheology, stability, filtration and shear strength development were studied. The different properties were then evaluated by means of a Marsh cone, viscometer, mud balance scales, filter pump, sand column, measure column and fall cone equipment. During the field trials, exactly the same set up of grout investigation was performed, except for the sand column, in order to make them comparable. After the laboratory investigation, three concepts were chosen for field investigation, (see Table 1).

Table 1.   The three grout mixes which where selected for field investigation.

Type	WC ratio	Additive %	Accelerator
Rheocem 800	1.2	Rheobuild 1.4 %	
Rock U	1.0	Compound B, 4 %	
Cementa IC 30	0.6	HPM 0.7 %	2 %

The strategy, during the trials was, firstly to mix a recipe in the lab in order to obtain reliable grout parameters, secondly to mix at the grout rig and use the same equipment for grout property measurement as in the lab. During the field trials. some recipes showed a large divergence from the laboratory study. An extra control of the dispersion ability of the field grout mixing equipment was. therefore. performed.

Field trials

The grout trials consisted of seven fans (see Table 2), each 20 meters of length; the fan layout is shown in Figure 1 and 2. The grout fans were carried out in three stages. 2+ 2 + (2 +1).

Table 2. Performed concept for the different trial fans and corresponding Q- and RQD values.

Fan	Concept A	Concept B	Chainage	Q-value	RQD-value
1*	RockU	Rheocem 800	313	62	92
	wc 1.0	wc 1.2	1/791		
2*	Rheocem 800	RockU	313	30	90
	wc 1.2	wc 1.0	1/804		
3	RockU	IC 30	313	1.5	22
	wc 1.0	wc 1.0	1/864		
4	IC 30	RockU	313	2	25
	wc 1.0	wc 1.0	1/878		
5A	IC 30	IC 30	313	7	70
	wc 0.6, high pressure	wc 1.0 – 0.5	1/921		
5B	IC 30	IC 30	313	5	60
	wc 0.6, high pressure	wc 1.0 – 0.5	1/936		
6	IC 30	IC 30	314	3.3	45
	wc 1.0 – 0.5	wc 0.6	1/306		

* In the two first fans, colored grouts were used. The measurements with the filter pump showed decreased penetration for the colored grouts. Therefore, it was decided not to use colored grouts in subsequent fans.

The fan was drilled in agreement with the ordinary drill plan for the area, which made the trials comparable to surrounding ordinary grouting. All holes were test pumped and individual Lugeon vales were noted. The fan was divided in two halves with equal Lugeon sums based on water loss measurement in grout holes, (Figure 1).

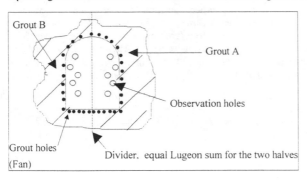

Figure 1. Principal layout. Divided fan and corresponding observation holes.

Within each fan, 10 observation holes, with a length of 15 meters, were drilled parallel to the tunnel, five in each half (see Figure 1 and 2) to measure the sealing effect. The holes were test pumped and studied with a borehole camera. Before grouting the observation holes were sealed with a sleeve and water filled.

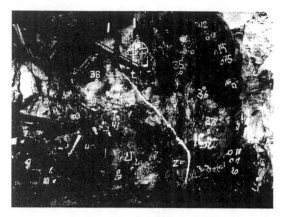

Figure 2.   The grout fan consisted of 32 holes located in the periphery, numbered 1-32. In addition 10 observation holes, numbered 33-42 and the divider line between equal Lugeon sum for grout holes is shown in the figure.

The fan was grouted hole by hole. Connections with other holes in the fan or with observation holes were noticed. After grouting and hardening, the observation holes were again test pumped, so that the sealing effect could be measured. The tunnel was then excavated forward; so that holes perpendicular to the tunnel could be installed to measure the perpendicular grout distribution, see Figure 3.

The spreading holes were also test pumped and studied with borehole camera. A number of times, during the tests, the rock mass was so tight, that no grouting were necessary, or that the pump test values before grouting were so low that a significant difference after grouting would not be possible to detect. Therefore, some tests were interrupted at an early stage. Investigation of the trial areas was carried out, both as ordinary geological survey and by means of digital photography.

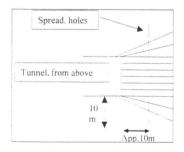

Figure 3.   Principal layout of spreading holes perpendicular to the tunnel.

RESULTS

Laboratory testing

The grout, which best fulfilled the demands were cement Rock U. closely followed by cement Rheocem 800 and cement IC 30. From laboratory three mixes were then presented as ready for field investigation, as shown in Table 1.

It may be noticed that both Rock U and Rheocem 800 are micro cements (grain size under 16 μm), while Cementa IC 30 was conventional injection cement, with coarser grains (30 μm). It may also be noticed that Rock U was gypsum-free and that the retarding gypsum content was added in the superplasticizer. In that way grout hardening could be directed, as desired. Furthermore, it was noticed that hardening of Rock U were strongly correlated to the amount of superplasticizer and the present temperature, which might be both an advantage and a disadvantage.

During grouting, grout consumption normally varies strongly. With low consumption, risk for grout hardening in the agitator was high. As a solution, adding a superplasticizer based on gypsum into the agitator was investigated. By doing so, the short hardening time needed in some situations could be combined with a grout that does not cure in the agitator.

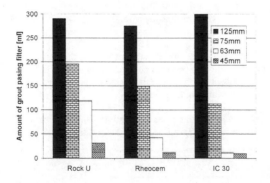

Figure 4.   Results from laboratory investigation with filter pump.

The penetration ability of the different grout mixes was compared by means of three different methods: filter pump, sand column and NES equipment (Sandberg, 1997). Generally it was noticed that a finer grain size distribution give a higher penetration (see Figure 4), but a number of exceptions was also noticed. The three different methods also showed some difference in the results. For the filter pump, penetration and grain size correspond well. Even for the sand column, penetration and grain size correspond, but change of filter (sand) was a troublesome operation, making the method a bit slow. A certain threshold value was also observed where some particles passed and others were trapped. For the NES equipment, grain size and penetration correspond initially. The threshold value was here a smallest column width, which no particles pass, without respect to grain size. This result was interesting in combination with the field trials. According to the field trials, the sealing efficiency was not always improved by using micro cement.

For rock mass purpose, a combination of the NES equipment and the filter pump should be appropriate for measuring the penetration ability. Passing.

Mixing trials with field mixer

During the trials at South Link, two micro cements were used. For both of them, but in particular, for Rock U, the pre-laboratory properties of the grout, were not achieved during the field trials. Therefore, so called "field mixing trials" were performed, without using the grout for grouting. To produce a correct mix takes between 7 and 8 minutes - much longer than what is common for ordinary production of grout mix. It is in addition very important that the different components are added in the correct order.

The purpose of field mixing trials was to examine the difference in properties achieved using a field mixer and a laboratory mixer, respectively, and, in addition, to strictly mix according to the producer's recommendations. One parameter, which clearly did not correspond between laboratory and the field mixing, was the filter pump values. In Table 5, a comparison is shown between field- and laboratory mixing for the filter pump.

Even though the mixing was carried out strictly according to the producer's recommendations, it was not possible to reproduce the parameters from the laboratory. It might therefore, be suggested that the laboratory mixer disperse grout better than the field mixer, which was what could be expected.

Table 5.   Filter pump values for the cement RockU. The left values are from the field mixer and the right values are from the laboratory mixer. (field/lab) [ml].

Time / Filter:	125 µm	75 µm	63 µm	45 µm
0 min	240 / 300	100 / 170	60 / 170	- / 30
10 min	300 / 300	120 / 140	50 / 130	- / 25
30 min	300 / 300	90 / 140	50 / 120	- / 25
60 min	280 / 300	75 / 140	30 / 90	- / 20
120 min	240 / 300	75 / 130	40 / 15	- / 20

Grout holes

An overview of the results from the field trials is presented in Figure 5.

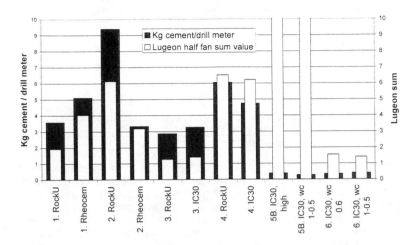

Figure 5.   Lugeon sum and grouted cement for the 12 fan halves.

In general, for all fans, the grout amount in kg was correlated to the Lugeon value, although some exceptions and explanations are here noticed. In Fan 4, two of the grout holes had contact with the soil layer, above the rock mass, therefore the maximum pressure could not be used as stop criteria, instead the maximum volume was used as stop criteria. A tendency during all trials was that the first grout used for the first half, consumed more grout than the second grout for the second half. In fan 5A, there was face leakage resulting in a far too high Lugeon value. Due to missing data Fan 5A is not shown in the figure. In Fan 5B, high face leakage occurred, resulting in abnormal Lugeon values, not corresponding to grout volume. The Lugeon values from Fan 6 were according to the criteria for trial grouting (at least 10 grout holes should have a Lugeon value above 0.4), too low. That means that no significant difference in observation holes, before and after grouting, would be possible to measure, as was confirmed by the trials.

Observation holes

The result from grouting has been measured in observation holes and is shown in Table 3. Holes with a value below 0.15 Lugeon was considered water tight, although it was not possible to measure differences of that scale. A hole was interpreted as influenced, if the Lugeon value was lower after grouting.

Table 3. Observation holes influenced by grouting. A observation hole was registered as influenced if the Lugeon value was decreased by grouting.

Fan	Rheocem 800	Cementa IC 30	Cementa IC30 high pressure	Rock U
		Influenced observation holes (total number of holes)		
1	3 (3)			0 (1)
2	1 (3)			0 (0)
3		0 (0)		1 (2)
4		2 (2)		1 (2)
5A		2 (2)	1 (2)	
5B		5 (5)	3 (4)	
6		- (0)	-(1)*	
Influenced %	67	100	67	40

*ordinary pressure and WC ratio 0.6

According to table 3, it may be noticed that Cement IC 30, have influenced all corresponding holes in all fans. Compared to the result in Table 4, it may be noticed that this influence was equal to a 49 % reduction of the Lugeon value. According to Table 4, Cement Rheocem 800 achieved a 66 % reduction of the Lugeon value, equal to that 67 % of the holes were influenced. The conclusion from above was, therefore, that Cementa IC 30 more often reaches the water bearing structure, but not seal so good and Rheocem 800 less often reaches the water bearing structure, but when doing so, the sealing effect was very good.

Table 4. Lugeon sum before and after grouting, as measured in the observation holes. T_{eff} = sealing efficiency.

Fan	Rheocem 800 Before (After)	T_{eff} %	Cementa IC 30 Before (After)	T_{eff} %	Cementa IC 30 high pressure Before (After)	T_{eff} %	RockU Before (After)	T_{eff} %
1	7.35 (0)	100					1.07 (0.59)	45
2	1.46 (1.01)	31					0.5 (0.12)	76
3			0.44 (0.23)	48			2.73 (1.96)	28
4			2.78 (0.23)	92			0.78 (0.7)	10
5A			1.33 (1.16)	13	0.96 (0.73)	24		
5B			10.1 (6.64)	44	2.70 (1.79)	34		
6			0.42 (-)		0.63 (-)*			
Mean		66		49		29		40

- = Missing value.   *ordinary pressure and WC ratio 0.6

From table 4 it could be understood that both micro cements (Rheocem 800 and Rock U) had a very good sealing effect in fan 1 and 2, located in a fairly jointed rock mass (Q=46, RQD=91). Cement IC 30 had a better sealing effect than Cement Rock U at fan 3 and 4, located in a dense jointed rock mass (Q=1.75, RQD=23). At fan 5A and 5B a higher grout pressure was used together with Cement IC 30, but the sealing effect was not here improved by increased pressure in the fairly jointed rock mass (Q=6, RQD=65).

Spreading holes

In order to be able to verify the conductivity of the grouted sections and to estimate perpendicular grout spreading, spreading holes were drilled in the middle of each trial fan, (see Figure 3).

In each fan there were 10 spreading holes with a length of 10 meters, which were test pumped and studied with a borehole camera. Results from pump tests are shown in Table 6. The table has been divided into grout 1 and grout 2, corresponding to the first and second grout mix of each fan. In the table, the average value for all spreading holes of each fan is also shown.

Table 6.   Measured Lugeon values in the spreading holes for the different fans.

Fan	Lugeon (mean)			Mean
	Grout 1		Grout 2	1 and 2
1	0.33	(Rheocem)	0.13 (RockU)	0.23
2	0.07	(Rock U)	0 (Rheocem)	0.04
3	0.04	(IC 30)	0.04 (RockU)	0.04
4	0.26	(RockU)	0.22 (IC 30)	0.24
5a	3.24	(IC30 high)	3.10 (IC30 contr.)	3.17
5b	0.01	(IC 30 contr.)	0.60 (IC30 high)	0.31
6	0	(IC30 contr.)	0.71 (IC30 vct 0.6)	0.35

As shown in the table, the sealing result measured from the spreading holes was very good for all except for one of the fans, indicating that a good sealing of the rock mass has been achieved as far out as 10 meters from the tunnel face in those fans. In Fan 5A, high Lugeon values were obtained which can be explained by high face leakage. In Fan 6 the rock mass was very tight even before the grouting, and very small amounts of grout were consumed. This indicates that the sealing achieved was local in the near field of the grout hole, and that sealing 5 to 10 meters from the drill hole not was achieved in this rock type.

Surveying of grouted areas

The trial grouted areas have been ocularly inspected by geologists and logged with a borehole camera. Damp, drip and presence of grout at tunnel face have been observed. If the different concepts at the different fans should be compared, all surrounding conditions have to be stable and equal. The digital photography was performed a couple of months after the last grouted fan At that time, all trial fans had achieved stable conditions after grouting and were therefore more comparable.

The ocular inspection of damp and drip after grouting on shotcrete areas was graded 1 to 5, where 1 means that the area was very wet, and 5 very dry. Fans 1 and 2 were by the ocular inspection graded 5. The trial grouted area was here much dryer than the surrounding areas, grouted with the ordinary concept for the tunnel. Fans 1 and 2 were grouted with micro cement (Rock U, Rheocem 800) and the surrounding areas with ordinary grout cement (IC 30). For this rock type (Q=46, RQD=91), micro cement was consequently more efficient.

Fans 3 and 4 were by the ocular inspection graded as 4. The areas are relatively dry and had less wet spots than the surrounding areas. Fans 3 and 4 where grouted with the grout cement IC 30 and the micro cement RockU. Fans number 5A, 5B and 6 was graded as 3. The areas were OK, but with the same intensity of wet spots on the shotcrete as surrounding areas.

The grouts used at Fans 1 and 2 contained red and yellow iron oxide, in order to follow the grout path ahead of the tunnel face. The coloured grout was observed in a nearby tunnel, confirming at least 20 meters' spread ahead, from the bottom of the grout hole. In a parallel tunnel, with 5 meters' of rock between the tunnels, the coloured grout was observed from the bottom to the top of that tunnel, confirming at least 15 meters' of spread perpendicular to the grout holes.

Conclusions

Our conclusions from the laboratory and field investigation have been summarised as:

- Joints with a hydraulic aperture of 0.1 mm or less were not possible to seal, neither with conventional injection cement, IC 30, (d_{95}= 30 μm) nor with micro cements (9.5 μm).

- The trials with a higher stop pressure, 45 bars, did not during these limited tests give a better sealing result than the trials with 25 bars.
- Based on survey of damp and drip on the tunnel face, micro cements (Fans 1 and 2) performed better than other concepts.

- Based on the water loss measurements, the coarser cement IC 30, sealed as well as the finer micro cements Rock U and Rheocem.

- Lugeon values below 0.15 are below the accuracy of the equipment.

- For the micro cement, Rock U, it was not possible to reproduce laboratory grout parameters, not even outside production, when mixing strictly to the recipe.

- A Lugeon value below approximately 0.3 gave, in general, a grouted quantity sufficient only to fill the hole (water pressure during pump tests were 20 bars).

- The available grout mixing equipment restrains the choice of suitable grout.

Recommendations for further grout works at South Link

- The investigation has shown that, with ordinary grouting equipment, and for tight rock mass situations, here equal to a joint intensity of about 2.3 joints per metre and apertures between 0.05 to 0.15 mm, micro cement do not perform better than conventional injection cement.

- Micro cement may be used for more jointed rock, with wider joint apertures (>0.5 Lugeon), and will then fill the joint better than conventional injection cement, and thereby give a better sealing. At such rock mass conditions, micro cement could be used, even with traditional mixing equipment, with poor dispersion ability. But naturally, the sealing effect will be much better if the equipment is exchanged to a high dispersion equipment or method.

- In some of the grout holes, which had been cleaned, large amount of debris was found during inspection with the borehole camera. An improved cleaning was therefore recommended before grouting.

- Train the personnel in grout mixing.

- Use sleeves, which can handle larger grout hole diameter variation.

REFERENCES

Dalmalm, T., Eriksson, M., Janson, T., Brantberger, M., Slunga, A., Delin, P., Stille, H. (2000). "Injekteringsförsök vid Södra Länkens bergtunnlar." Rapport 3075, Department of Soil and Rock mechanics, Royal Institute of Technology, Stockholm, Sweden.

Hansson P. (1995). "Filtration stability of cement grouts for injection of concrete structures." IABSE symposium, San Francisco.

Hellgren, T., Slunga, A. (2000). "Injekteringsförsök i Södra Länken." Examensarbete 99/14, Department of Soil and Rock mechanics, Royal Institute of Technology, Stockholm, Sweden.

Sandberg, P. (1997). "NES-metod för mätning av injekteringsbruks inträngningsförmåga." Svensk Bergs - & Brukstidning 5-6/97. Sweden.

Long-Distance Grouting, Materials and Methods
Grouting Conference 2003

Christopher R. Ryan [1]
Steven R. Day[2]
Donald W. McLeod[3]

Abstract

An abandoned 1600-meter (mile-long) rock tunnel had to be completely filled with grout. The total tunnel volume was approximately 4500 cubic meters (6000 cubic yards). The tunnel was water-filled with access only at each end through narrow, 25-meter deep (80 ft), vertical shafts. Access for pumping was feasible only from one end of the tunnel, thereby requiring unusually long distances for pumping.

Through an extensive laboratory testing and modeling program, different grouts were tested for suitability for this project. The ideal grout would have low viscosity, good stability and, after setting, low bleed, moderate strength and low permeability. Materials tested included cement-bentonite, cement-flyash and combinations including blast furnace slag cement. Data is presented on the various grout materials leading up to the choice of a cement-bentonite-slag cement blend as the optimal mix for the project.

The unusual conditions at this project required the use of divers and remote-operated vehicles to inspect the tunnel and to place the initial cable that would allow grout pipes to be drawn into the tunnel. Each component of the grout system was engineered to provide adequate capacity to fill the tunnel in three to four days, working around the clock. A backup system using a sleeve pipe to provide secondary grout was devised and installed.

The work in the field progressed more or less as planned, with a few unknowns cropping up to make for some difficult moments. As it turned out, the secondary grout line was necessary to complete the work. Grout samples were taken during the project for confirmation testing and borings were drilled into the tunnel after the work to verify that the tunnel was full. Data from this phase of the project are also presented.

This project presented an unusual opportunity to plan and test components pre-construction. While there is no way to verify, the distances that the grout was pumped may represent some kind of record.

[1] President, Geo-Solutions Inc., 201 Penn Center Blvd, Suite 401, Pittsburgh, PA 15235; phone 412-825-5164; cryan@geo-solutions.com
[2] Vice President, Geo-Solutions Inc., 26 West Dry Creek Circle, Suite 600, Littleton, CO 80120; phone 720-283-0505; sday@geo-solutions.com
[3] Project Manager, Miller Springs Remediation Mgmt. Inc, 2480 Fortune Drive, Suite 300, Lexington, KY 40509; phone 859-543-2174; don_mcleod@oxy.com

Project Outline

A former water intake tunnel extending under the Niagara River was contaminated with organic wastes from a nearby landfill and was to be filled and closed at the request of regulatory authorities. The two-meter (six-foot) tunnel is nearly 1600 m (one mile) long and accessible from just two 25-meter (80 ft) deep vertical shafts, one of which is in the river. Closure of the tunnel presented a unique remediation challenge because of the limited access, considerable volume of the tunnel, and because the tunnel was full of potentially contaminated water. A plan was developed and implemented that closed the tunnel by filling it with cementitious grout while simultaneously removing and treating the displaced water. The grout used to fill the tunnel had to meet demanding requirements for both regulatory acceptance and workability.

The project work plan had to take into account a number of unique complicating factors, including:

- The tunnel was level, making it difficult to displace water upwards with a heavier grout.
- Access for material placement was really only practical from the land end of the tunnel
- Access by divers into the flooded tunnel was limited to about 200 meters (600 ft) from each end.
- The tunnel could not be dewatered due to the nearly unlimited volume of water from both the tunnel and infiltration from the river that would need to be treated.
- Once work would begin to fill the tunnel, no further personnel access would be permitted, requiring a remote operation.
- Redundant systems would be required to account for multiple variations and breakdowns that might occur. It would be difficult to ever restart the work in the event of a disruption.
- The work plan had to account for the fact that the first stage of grouting might not be totally effective in sealing the tunnel up to the roof, so a secondary grouting system would need to be devised.
- Because of the dimensions of the project and the problems of grout setting, the system would be designed to operate continuously once work started until completion.

The key to success on the project was the selection of a grout with parameters that would fit the situation as well as the design of a placement system that could reliably place a large amount of grout over a period of a few days. Grout for an application like this had never been designed and it was necessary to go back to the laboratory to search out the ideal combination. Since the grout had to be mixed and pumped from shore and, based on the placement work plan, the initial grout would have to pass through almost 1600 m (5000 ft) of pipe to the point of placement and would have to flow back through the tunnel, displacing water, for a distance of at least 300 m (1000 ft) over a period of 30 hours or more before it would set too much to pump. The volumes were considerable. It would take approximately 4500 cubic meters (6000 cubic yards) of grout to fill the tunnel, so more than 1000 cubic meters (1300 cubic yards) would have to be placed before it would start to set. Based on these requirements, as well as regulatory requirements for the completed grout fill, parameters for the grout design were set as follows:

- Unconfined strength at 28 days in the range of 100 to 200 kPa (15-30 psi).
- Heavier and more viscous than water so that water would be displaced out of the tunnel as the grout was placed.
- The grout should be immiscible in water, so that it would form a face displacing the water, rather than a semi-mixed zone of water and grout.
- The grout should have an extended set time, 24 hours or more, to allow significant volumes of grout to be placed from a single point.
- The mixed grout had to have low viscosity, preferably less than 60 seconds Marsh Funnel to allow it to be placed through small diameter pipes over long distances without significant head losses.
- The permeability of the hardened grout had to be no higher than 1 x 10^{-6} cm/sec.

Laboratory Testing Program

Based on a review of the literature and previous experience, three basic types of grouts were selected for consideration in the laboratory testing program. The grout mixtures tested were divided into three groups labeled as Portland Cement-Bentonite with admixtures (CB); cement-bentonite with Blast Furnace Slag Cement and admixtures (BFSB); and Portland Cement-Fly ash with and without foam and other admixtures (CF). A variety of additives designed to improve grout workability were tested, including: super plasticizer, anti-wash, pre-formed foam, and lignosulfonate.

A total of 19 grout mixtures were formulated and tested. Seven grouts were CB, eight were BFSB, and four CF mixes. The proportions (all expressed as a percent by weight of water) and ingredients of six representative mixtures are provided in Table 1.

Table 1: Example Grout Proportions and Ingredients

Ingredients (% Wt of Water)	Grout Type & Mix Number					
	CF-4A	CF-4G	CB-5A	CB-5B	BFSB-(	BFSB-6D
Portland Cement	15	52	19	19	5.5	5.5
BFS	0	0	0	0	16.5	16.5
Fly Ash	35	115	0	0	0	0
Bentonite	5.5	0	5.5	4.5	4.0	4.5
Foam	0	2.5	0	0	0	0
Anti-Wash	0.14	0.26	0	0.13	0.13	0
Super plasticizer	0.06	0.11	0	0	0	0
Lignosulfonate	0	0	0.10	0.13	0	0.06

The grout mixtures were first subjected to a series of tests including: viscosity, density, set time, bleed, shrinkage, unconfined compressive strength, and permeability. The results of the tests on the six representative grout mixtures are provided in Table 2.

Table 2: Grout Properties

| Property | Grout Type & Mix Number | | | | | |
	CF-4A	CF-4G	CB-5A	CB-5B	BFSB-6A	BFSB-6D
Viscosity (MF sec.)	33	>90	49	60	55	43
Density (gm/cc)	1.27	1.22	1.14	1.14	1.15	1.15
Set Time (days)	5	1	3	3	5	6
Bleed (ml/1000 ml)	77	0	<5	0	<5	0
Shrinkage (%)	19.5	7	4.4	1.3	1.2	1.3
UCS – 7 day (kPa)	59	959	69	83	276	48
UCS – 28 day (kPa)	290	1884	159	179	1049	662
Permeability (cm/sec)	NR	4 E-7	5 E-7	5 E-7	8 E-8	6 E-8

With respect to viscosity, all of the grout mixtures were workable or could be made workable using additives. The set times of the CB and BFSB mixtures were acceptable, but some of the CF grouts set too quickly for the placement conditions (e.g. CF-4G) and BFSB grouts that did not include some Portland Cement did not set at all. The most significant finding was the variability in the bleed and shrinkage of some of the grouts. While the CB and BFSB had minimal shrinkage, the CF grouts performed poorly. No additive provided significant improvement in the bleed, so the CF mixtures were deleted from the program. While the strength and permeability of the CB and BFSB grouts were both acceptable, the BFSB grout had better properties. (See Figure 1 below)

Figure 1. CF samples on left show significant bleed. CF Samples on right with foam show significant shrinkage.

Three kinds of tests were performed to check the compatibility of the grout mixes still under consideration with site leachate, specifically DNAPL (dense non-aqueous phase liquid) and APL (aqueous phase liquid). In the first test, the fluid grout is poured into pans full of leachate and of water (for comparison). The grout is tested with a modified set test apparatus (ASTM C-403) as it hardens and a comparison is made between the times for the grout to set in leachate compared to times to set in water. The results showed no effect due to the leachate.

The hardened grout was subjected to an immersion test (ASTM C-267) designed to predict the long-term performance of cement products exposed to chemicals. The test is performed by soaking cured grout specimens in sealed jars filled with leachate and tap water (for comparison) for up to 45 days. No effect due to the leachate was observed

A limited number of mixtures that had been permeated with water were retained for continued permeation with DNAPL and APL. With the DNAPL, the material apparently creates a coating that stops all flow within a few days. These tests were started after the permeability tests with water were completed. The results of the tests are shown in Table 3.

Table 3: Permeability of Grout Mixtures to Leachate

	Grout Type & Number			
Mix Number	CB-5A	CB-5B	BFSB-6A	BFSB-6D
Water Permeability (cm/sec)	5.8 E-7	1.9 E-6	8.1 E-8	6.2 E-8
APL Permeability (cm/sec)	3.9 E-7	1.3 E-6	1.69 E-8	3.8 E-8
Pore Volumes APL	1.6	1.3	0.18	0.12
Time of APL permeation (days)	13	14	19	6
DNAPL Permeability (cm/sec)	Stop	3.0 E-8	1.75 E-8	Stop
Pore Volumes DNAPL		0.025	0.015	
Time of DNAPL perm (days)		26	2	

The final grout mix was then selected based on the testing to date. It actually was a slight variation on mix BFSB-6D shown in this paper. It had minimal bleed and shrinkage, so it would maintain good contact with the top of the tunnel and all of the other properties met the requirements of the project. The final mix design was 4% bentonite and 22% cement by weight of water. The cement was a pre-blended combination of 75% Blast Furnace Slag Cement and 25% Portland Cement.

Model Testing Program

The final step in the testing program was model testing. The model tests were devised to investigate the potential behavior of the grout as it was placed underwater, in a long tunnel.

The first test was a simple tremie test with the grout placed through a tube into a container full of water. The grout should not mix with the water and the bleed of the grout should still be acceptable as it set underwater. The selected grout passed this test with no problem. Even when it was placed in a manner so that it dropped through the water, the grout bulb remained intact until it rejoined the grout at the bottom of the container with essentially no mixing with the water. Subsequent bleed during the setting process was no more than it had been in the earlier testing.

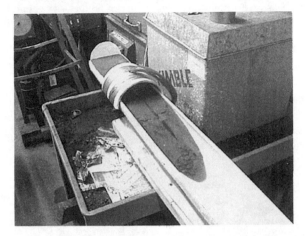

Figure 2.   Model test of grout placement in a tunnel full of water

The second bench-scale test was devised to model the horizontal displacement of the water in the tunnel as the grout is placed. The setup was a half-pipe full of water with a grout tube inserted at one end. A long slope of grout pushing the water forward was expected. As the photo in Figure 2 shows, there was actually a surprisingly steep face of grout (1:5 vertical : horizontal) that formed. Again, there was essentially no mixing of the leading edge of the grout with the water.

Field Implementation

An unusual feature of this project was the available time to plan and think through each step in the operation. This planning was critical to the success of the project because a failure at a critical stage in the preparations or operation could leave the tunnel blocked with no way to restart the work.

The first step was to prepare the tunnels by inserting the grout pipes. The difficulty here was, as stated earlier, that the tunnel had to remain flooded and was only accessible to divers for a short distance. The shore shaft was a 2-meter diameter riser pipe 25 meters deep. At the bottom there was an immediate transition into the horizontal tunnel that generally was about 1.5 meters wide and 2 meters high, lined with concrete. The only other access was on a small concrete platform out in the river, 1600 meters (one mile) away, where there was a 2.5-meter diameter riser shaft.

Fabricated rollers were installed at the base of the access shafts to allow pipes and cables to be pulled through without damage. This was accomplished using divers. Next a small remote-operated vehicle went through the tunnel carrying a small diameter cable. This cable was used to pull through a larger cable that was about 3200 meters (10,000 ft) long. This cable was wound onto cable drive pipe pullers at both end of the tunnel so that it could travel back and forth through the tunnel, dragging in the many components of the grouting system. Considerable attention was paid to the details of cable connections to avoid the possibility of snagging on a previously placed component.

Figure 3. Pipe puller and shaft at the shore end of the project

The first part of the grout placement system to be installed was the secondary grout line. Since the most likely place for voids to form would be above the grout as it settled, this line had to be at the top of the tunnel. It consisted of a 75mm (3 inch) diameter HDPE pipe with rubber sleeve valves placed over holes drilled in the pipe on approximate 15-meter (50 foot) centers. The pipe was dragged through then evacuated by blowing the water out of the line with compressed air to a sump pump at the opposite end, forcing the line to float to the top of the tunnel.

The cable was then rewound on the pipe puller at the shore and attached to the first and longest primary grout pipe that was then dragged through using the river-end pipe puller. This pipe reached all the way to the base of the river shaft and succeeding pipes were each about 300 m (1000 ft) shorter than the last. A total of five primary grout pipes would be placed, ranging in diameter from 150mm (6-inch) to 100mm (4-inch). To speed placement of the pipes, they were prefabricated into sections of approximately 150 meters (500 feet). Each pipe had a conical tip designed to keep out sediment as the pipe was dragged into the tunnel yet that would open when grout pressure was applied. An emergency sleeve valve was mounted a short distance back from the tip designed to provide an outlet in the event the tip was blocked.

Once the pipes were all placed, it was time to block the river shaft so that grout would be forced back towards the shore once pumping started. This was accomplished by placing a steel "stool" in the shaft, sealing around the edges and pouring tremie concrete on top of the structure to seal the shaft. After this was done, the only grout pipe left accessible from both ends was the secondary grout pipe.

Figure 4. Placing the steel "stool" in the river shaft in preparation for the tremie seal

The tunnel was now ready for the grouting operation to begin. To simplify the mixing process, all the bentonite slurry needed was premixed and stored in large ponds on the site. Cement and blast furnace slag cement were delivered to the site pre-blended, so only one dry component would have to be mixed, allowing for easier quality control and a faster placement rate.

The one-component mixing system was crucial to the project design since it would allow for a continuous as opposed to a batch mixing system. The mix plant consisted of a 4.5 cubic meter (6 cubic yard) capacity colloidal enclosed mixer fed by a variable flow liquid system and a variable speed dry material feed system. The level in the mixer was controlled by a sonar device; feed rates of the materials were adjusted based on real time continuous density measurements provided by a highly accurate coriolis density meter. Slurry was pumped in a rate sufficient to keep the level in the mixer constant as the material was pumped out and the rate of feed for the cement blend was controlled based on density.

Figure 5. Grout mixing plant

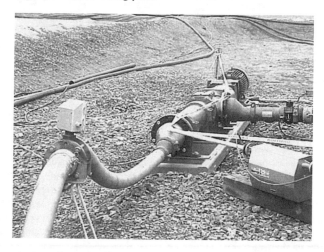

Figures 6. Pumping unit with density meter to the right and flow meter to the left.

This highly productive mix plant was designed to produce grout at a rate of more than 750 cubic meters (1000 cubic yards) in a 24-hour period. After some test runs where grout was mixed and pumped into a pit on site, the work was finally ready to begin.

There were a few moments that pushed the system to its limits. The grout could barely be pumped the long distance out to the end of the first pipe. Once it began to consistently flow, water started to rise in the shore shaft. Work continued on a 24 hour a day basis until completion. Problems arose during the first run when a violent nighttime thunderstorm forced a cessation of operations and the pipe was lost before the grout reached the next pipe end. Work continued from the second pipe but it was known that there might be a gap at this location. Grouting through the succeeding pipes went according to plan. At one location, there was a breakout of grout into the riverbed through an old shaft that was supposed to have been sealed. The grout came up the shore shaft on schedule.

Once the primary grout had been allowed to set, the secondary grout program began. The secondary grout pipe was pressurized, allowing the seals to pop open in any locations where there might be a weakness and theoretically no grout. After a period of time, the secondary line was flushed out by pumping water through it and the secondary grout allowed to set before starting another phase of secondary grouting. By the end of this process, grout was coming out of the ground at various locations, so it appeared that the tunnel was tightly sealed.

Field Verification

Throughout the grouting operation, samples of the grout mixture were collected for strength and permeability testing. The results, Shown in Figures 7 and 8, show that the quality of the grout consistently met the project objectives of greater than 100 kPa unconfined strength and less than 1×10^{-6} cm/sec permeability after 28 days of cure.

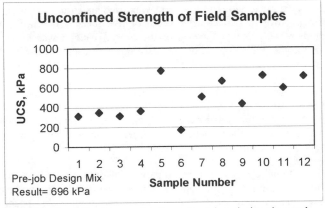

Figure 7. Unconfined strength of samples taken during the work

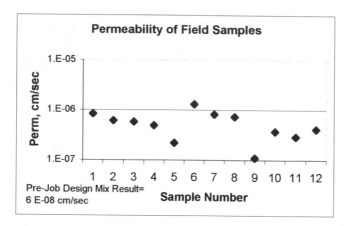

Figure 8. Permeability of samples taken during the work

An additional quality control requirement of this project was confirmatory drilling after placement and curing of the grout to visually confirm the effectiveness of the fill and to check for the presence of any voids above the grout. Because of problems that arose during the grouting, there were two locations in which grout was introduced into the tunnel from a tremie pipe that was in front of the grout face. This created the potential for a gap caused by hydraulically confined water. Three borings were advanced into the tunnel – one from land where the tunnel passed under an island and two from a barge in the river.

Locating the tunnel required great care, as none of the original construction plans were available. The drilling coordinates were calculated from known locations of the tunnel shafts. At each location, a well casing was carefully set vertically to the top of bedrock and a "Full-Hole" outer tube core barrel system was used to ensure a vertical hole through the bedrock down to the 25-meter depth of the tunnel. To check for the presence of a void in the grout at the top of the tunnel, caution was used at the appropriate depth and the drilling operation was videotaped to capture the drop of the drill stem if it were to occur.

The boring on land intercepted the tunnel on the second attempt and confirmed the presence of one of the suspected gaps (caused by the stoppage of grouting during the lightning storm). When the tunnel was penetrated, air erupted from the borehole and the drill bit dropped 2 meters to the tunnel bottom. A relief hole was bored into the gap some distance away so that additional grouting could fill the gap.

For the work in the 10-kph (6 mph) current of the river, the drilling rig was placed on a barge equipped with a 0.6-meter (2-ft) diameter hole for the drill bits. The hole was positioned directly over the calculated drilling coordinates using global positioning.

When the barge was close, one spud of the barge was sunk into the riverbed while a tug was used to rotate the barge to achieve the final location. Both of the borings in the river confirmed a complete fill with competent grout. The core in the upper part of Figure 9 (encased in a Lexan® sleeve) shows the integrity of the grout; in color, the grout is the green-black color that is typical of BFSB-based grouts. The cores in the lower part of the figure are samples of the concrete tunnel wall and the bedrock.

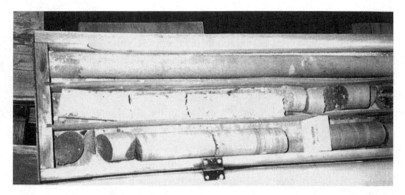

Figure 9. Confirmatory drilling core showing, from the top, grout from the tunnel, the concrete tunnel floor and bedrock.

Conclusions

The key to this project was finding a grout mixture that would meet the requirements of viscosity and set time to allow placement over the distances and time periods required as well as meeting the physical strength and permeability requirements set forth by regulatory agencies. The additional requirements of immiscibility with water and low decantation soon focused our search on combinations of blast furnace slag cements and Portland cement.

Blast Furnace slag cements need a percentage of Portland cement to perform at all in this application. Without it, they do not set. With a proper mix ratio, the grout will have a low viscosity, low bleed, low shrinkage and will form a grout of low strength (100-600 kPa) and low permeability (less than 1×10^{-6} cm/sec).

This project clearly tested the limits of the current knowledge of grout mix design as well as the technology of grout mixing and pumping. The project parameters and the design requirements made the job one of the most challenging imaginable. The combination of a far-sighted owner and a competent contractor to do the design, testing, and construction supervision made the project a success.

The Effect of TAS Method by a Supplementary Method to Tunnel

Byung-Sik Chun[1], Yoo-Hyeon Yeoh[2]

Abstract

Generally, it is known that urethane injection is an excellent method in long-term durability and environmental characteristics for ground improvement. However, urethane grouting has short rise-time that is, reaction time of solution A and solution B is short, and the injection distance is very short. Therefore, urethane injection cannot be used for the site where locates deeply from the initial injection point in a ground. Other injection materials such as cement cannot apply an alternative material when rapid hardening is required. From this study, disadvantages of urethane injection could be improved by TAS method. From the results of field tests, it was known that TAS increases injection distance over 10m, which is longer than that of original urethane grouting. In addition, the rise-time of TAS is shorter than that of cement grouting, so rapid improvement can be gained from TAS. Short rise-time and long injection distance of TAS can give excellent applicability for tunnel construction.

Introduction

Recently, a method considering improvement effect, workability, economy, and environmental influence is needed in many of constructions to solve geo-environmental problems. In Korea, general urethane methods have remarkable improvement effect and little environmental influence, but chemical grout is so expensive and the injection distance is limited because of short rise-time(Chun, 2001). Therefore, TAS(Tunneling method on Advanced reinforcing system)is the method that can expand the improvement range by the excellent urethane and

----

[1]Professor, Department of Civil Engineering Hanyang Univ. 17 Haengdang-dong Seongdong-ku, Seoul, R.O.Korea; phone 822-2290-0326; hengdang@unitel.co.kr

[2]President. Gaya Engineering CO., LTD. Seoul. R.O.Korea. phone 822-575-1735; gaya21@chollian.net

by injection mechanism. In this study, laboratory tests and field tests were performed to determine the improvement characteristics and the expanded range of injection.

The properties of Urethane chemical grout

The physical properties of Urethane chemical grout. The rise-time of urethane chemical grouts is so short and the viscosity is changed because of high heat generation in mixing process. Liquid type chemical grout whose viscosity controls its permeability emits high heat after reaction and the viscosity is decreased continuously until the time of gelation(Chun, 1997). The efficiency of injection is remarkably superior to the case of which is considered the viscosity of urethane chemical grout (Table 1) is about 50cps.

The mechanical properties of Urethane chemical grout.

Homo-Gel strength. Table 2 shows the strength tests results of completely hardened Homo-Gel made out of Urethane solution A and B, mixing ratio is 1(10 g) : 3(30 g), in cubic mold(5 × 5 × 5 cm).

Table 1. The specific gravity & viscosity of urethane chemical

Chemical grouts		Standard value		Measured value	
		Grout material	Caulking material	Grout material	Caulking material
Viscosity (cps)	Main material (Solution A)	50±10	400±50	44	340
	Hardening material (Solution B)	50±10	200±20	42	230
	Mixing ratio	1 : 2.7 ~ 3.0	1 : 1	1 : 3	1 : 1
Specific gravity	Main material (Solution A)	1.1 ± 0.05	1.1±0.05	1.05	1.05
	Hardening material (Solution B)	1.25 ± 0.05	1.25±0.05	1.23	1.23

Table 2. Homo-Gel strength of urethane

Chemical grouts		Standard value		Measured value	
		Grout material	Caulking material	Grout material	Caulking material
Strength (kg/cm^2)	Bending	Over 40	Over 40	92	72
	Unconfined compression	Over 50	Over 50	73	88
	Shearing adhesion	Over 18	Over 18	52	-

Sand-Gel strength. Table 3, 4 show Sand-Gel strength. From Table 3, 4, it is known that strength improvement and cut-off effect is very effective.

Table 3. 5-36block, line No. 5, Seoul subway (Chun, 1991)

R.M	γ_t (t/m^3)	Gs	q_u (kg/cm^2)	k (cm/sec)	V* (km/sec)	C (kg/cm^2)	ϕ (°)
Before injection	-	-	-	7×10^{-3} (Estimated)	-	0.0	35.79
After injection	1.93	2.55	41.92	6.75×10^{-5}	1.043	Not measured	Not measured

* : Elastic wave velocity

Table 4. 4-8block, line No. 1, Busan subway (Jung, 1993)

R.M	γ_t (t/m^3)	Gs	q_u (kg/cm^2)	k (cm/sec)	V (km/sec)	C (kg/cm^2)	ϕ (°)
Before injection	1.77	2.60	1.44	1.841×10^{-3}	-	0.092	34.91
After injection	2.18	-	54.29	2.599×10^{-7}	-	-	-

The Stabilization and Durability of hardening materials. To evaluate the leakage rate of grouted soil, ϕ 5cm × L 10cm homogel specimens were made and soaked in the water tank during 28days. The tests results showed that weathering of homogel specimens were rarely happened. Therefore, the durability of urethane is excellent.

Generally, when waterglass is injected, 10~50% of waterglass is leaked, but when urethane is injected, the leakage quantity is less than 0.1% of urethane quantity.

The secondary penetration test by swelling pressure. To check properties of the secondary penetration by swelling pressure, cylindrical mold that is 70 mm (2.8 in.) in height and 60 mm (2.4 in.) in diameter was made(Figure 1) At the center of the mold, steel bar that is same in height and 15 mm (0.6 in.) in diameter was placed and standard sands were filled in the mold. Finally, steel bar was extracted at the center of the mold and urethane was injected in the center hole.

As the result, urethane grouts could substantially ensure a reinforcement area larger than the general injection used only pressure. Jiacai(1982) used the properties of swelling of urethane to stabilize the foundation of dam built on clay layers of sand stone ground.

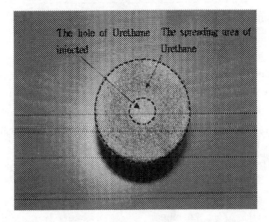

Figure 1. The secondary swelling of urethane grout

Evaluation of the environmental characteristics of urethane

Evaluation of the environmental characteristics by leaching test. ICP test and refuse leachate test(KS M 0111) for urethane which was used for grouting were performed. The tests results are shown in the Table 5. From the test results, heavy metals such as Pb, Cd, As, Hg, CN, Phenol were not found. Therefore, urethane has no environmental problems.

Table 5. Leaching test of Urethane chemical and soil standard of natural environment

Test item		Unit	Test result	Method of test	Norm
Leaching Test	Pb	mg/ℓ	N.D	I.C.P	Less than 0.01 mg/ℓ
	Cd	mg/ℓ	N.D	I.C.P	Less than 0.01 mg/ℓ
	Cr	mg/ℓ	N.D	I.C.P	Less than 0.05 mg/ℓ
	As	mg/ℓ	N.D	I.C.P	Less than 0.01 mg/ℓ
	Hg	mg/ℓ	N.D	I.C.P	Less than 0.0005 mg/ℓ
	Phenol	mg/ℓ	N.D	I.C.P	Less than 0.005 mg/ℓ
	CN	mg/ℓ	N.D	Uv/VIS analysis	N.D

* result of test: test results of Korea Testing and Research Institute for chemical industry(2002. 2.25).
*Norm: Soil standard of natural environment(Mika, 2000).
*N.D: not detected.

Evaluation of the environmental characteristics by fish toxicity test. Fish toxicity tests were performed for urethane which was used for grouting. From the tests results, LC50 that is, a level of concentration of toxicity that make 50 of 100 fishes die in 96 hours, of solution A was 65,300mg/ℓ (6.5%), and that of solution B was 5,000mg/ℓ. Therefore, urethane has less pollution characteristics than other chemical grouts (ex: Detergent:22mg/ℓ, Cement grout :several hundreds mg/ℓ).

Table 6. Lethality rate after 24, 48, 96 hours of each grouts

Injection liquid		24 hours		48 hours		96 hours	
Solution A	Solution B	pH	lethality rate	pH	lethality rate	pH	lethality rate
Sodium silicate No. 3	Ordinary Portland Cement	10.9	85%	10.9	100%	10.9	100%
Sodium silicate No. 3	Micro Cement	10.68	5%	10.58	15%	10.38	65%
Silica-Sol	Ordinary Portland Cement	8.37	0%	8.37	0%	8.35	0%
Silica-Sol	Micro Cement	8.1	0%	8.3	0%	8.3	0%
Urethane Solution A	Urethane Solution B	-	0%	-	0%	-	0%

Consideration about exiting urethane

Applicability, economical efficiency, and workability were considered about 3m short pipe or 6m double pipe.

Application. (A) 3m injection bolt (short pipe): The length of injection bolt is short, so the bolt is applied to a short sectional tunnel such as duct for electrical utilities and tele-communication utilities etc. Because the hardness of injection bolt is low, it couldn't be applied to soil tunnel. (B) 6m injection bolt (a double pipe): If width D is wide in large sectional tunnel that cross the soft ground, construction angle θ of injection bolt must great and it is difficult to consider the bolt as a beam element.

Economical efficiency. (A) 3m injection bolt (short pipe): When reinforcing depth is thick, a injection bolt must be executed every cutting face. So, workability and economical efficiency are reduced. (B) 6m injection bolt (a double pipe): When reinforcing depth is thick, because the construction space of urethane injection becomes narrow, workability and economical efficiency are reduced. When the length of injection bolt increases, the loss of urethane increases.

Workability. (A) 3m injection bolt (short pipe): Because main area of the injection bolt is small, it is difficult to expect enough frictional resistance at high pressure urethane injection. So, injection bolt does occurrence projection to outside, and cause problems of hazardous factors on construction and function of bolt. (B) 6m injection bolt (a double pipe): When pressure is interacted with injection bolt, injection bolt is projected to injection hole outside happens and functional problem of bolt.

Consideration about applicability

Injection Mechanism. Urethane injection has a short rise-time. Therefore, when urethane injection is performed during longer time than rise-time at underground, urethane penetration at spherical shape at first, then urethane is gelling to outside part, and chemical grout which impenetrate part by driven chemical grout gradually become hardening(cabbaging effect) in cauliflower form to infiltrate into weakness section(Christian Kutzner, 1996). According to tests results, continuous injection in weathered soil layer was possible during 45 times of rise-time, but the inner part of pipe had been stopped longer than rise time. Therefore, it is important first of all that arrival time of urethane must be keep to outlet by within rise-time inner part of pipe.

In addition, to form the uniform and homogeneous injection bulb, it is important that the quantity of discharges at both sides of injection bolt must be keep almost equally during injection.

Injection tests for conventional injection bolt. Conventional injection bolt consists of double steel tube that has length, outside and inside diameters of 6m, 50.8mm and 27mm, respectively, as shown in Figure 2. From the results of field tests, urethane not jetted at the same time through the whole block in maximum jetted discharge. But the urethane was jetted only 3.5m of front block as shown in Figure 3.

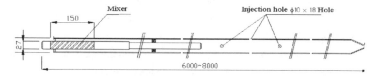

Figure 2. Conventional injection bolt

Figure 3. Urethane injection by conventional bolt

Because conventional injection bolt is grouted simultaneously in outside (in borehole) and inside of bolt, form pipe interior to a certain outlet arrival time depends on the injection velocity of urethane in outside of injection bolt. Therefore, when object ground has many large voids or was heavily weathered, arrival time of urethane to the objective spot exceed the rise-time because of much leakage of urethane. Therefore, injection bolt is clogged and the state of injection at the edge of borehole is much worse than in front. Three conventional injection bolts of 10m were examined in silty sand ground by field injection tests. From the test result, it was confirmed that urethane is injected only to 5.0~5.7m from borehole.

TAS injection bolt.   TAS injection bolt consists of double steel tube that has length, outside and inside diameters of 6~12m, 50.8mm and 34mm, respectively, as shown in Figure 4.

The characteristics of TAS injection bolt are as follows.

1.   Progressive enlarged entrance to minimize the head loss in entrance.

2.   Installation of the latticed caulking which is increased friction to control the protrusion of injection bolts to outside of borehole in high-pressure injection.

3.   Installation of spacer to separate injection bolts from the side of borehole and improve the performance of injection in construction.

4.   Determination of the size and spacing of borehole to insure the optimum conditions of injection.

5.   Installation of the hollow double-steel tube considering the effective injection of urethane and improvement of performance.

6.   Minimization of the use of urethane and installation of the inner hollow-tube with conical shape at both ends of installed spacer to make the injection of urethane smooth.

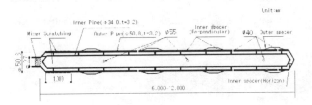

Figure 4. Diagram of TAS injection bolt

Constitution of TAS injection bolt. Outer tube comprises; 1 Entrance : As shown in Figure 5(a), the entrance has a progressive enlarged shape to minimize an energy loss and eddy that occur when the entrance has rapid enlarged shape in injection. 2 Caulking : The maximum pressure over 40kg/㎠ acts in injection bolts when the urethane is injected in ground. At this time, to prevent the protrusion of injection bolts to outside of borehole and to increase friction, latticed figures are installed as shown in Figure 5(b). 3 Nozzle : As shown in Figure 5(c), the variety of the size and space of nozzle is possible the equal jet of chemical grout. 4 Edge of injection bolts : It is produced to easy the insert of bolts in borehole as shown in Figure 5(d).

Inner pipe comprises; 1 Inner spacer (horizon, perpendicular) : The spacer is installed to separate injection bolts from the side of borehole and to improve the performance of injection in construction as shown in Figure 5(e).

The insert of the inner pipe in outer pipe : The two parts (L=5m×2) of injection bolts are bolted as shown in Figure 5(f).

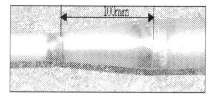 

a) Progressive enlarged entrance   (b) Caulking

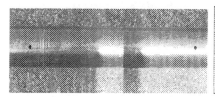 

(c) Nozzle   (d) Edge

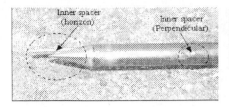 

(e) Horizontal and perpendicular spacer   (f) Insert of inner pipe in outer pipe

Figure 5. The constitution of TAS injection bolt

Results of tests for performance of TAS injection bolt. The sample injection bolt has the holes of 4mm and 5.5mm alternated each 5cm of injection bolt. The injection bolt is installed an angle of about 15 degree and its holes are taped. The performance tests of injection bolt were performed according to the size of hole, the injection quantity, and the injection location. After the beginning of injection, arrival time of urethane and the discharge during rise-time were determined according to each holes. From the tests results, the optimum outlet form was determined at the position and size of holes as shown in Table 7.

Table 7. The results of the performance tests of injection bolt

No.	Position of injection hole (m)	Arrival time of chemical grout (sec)	Jetted discharge (ℓ)	Discharge at the front of pipe (ℓ)	Discharge at the end of pipe (ℓ)
①	2.50	5.16	5.43		
②	3.80	9.43	2.17		
③	5.80	12.07	1.90	7.60	7.87
④	7.55	18.61	2.71		
⑤	8.75	26.31	2.17		
⑥	9.35	30.17	1.09		

Improvement by TAS method.

Pilot tests. Pilot tests were performed for weathered rock. The pilot tests were performed at slope of weathered zone located in Seojong-ri, Yangpyeong-gun, Gyeonggido, Korea. Three holes that leave a space of 60cm are drilled as shown in Figure 6. During the injection, the urethane drained out through joint. This is the

---

outflow of surplus urethane that is infiltrated a joint and voids in ground. Therefore, it is determined that the effect of urethane injection is excellent. The core of hardened soil of urethane in Figure 7 could be gathered from the center of grouting in Figure 6.

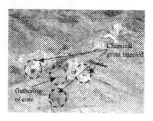

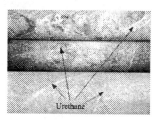

Figure 6. The shape of slope after applying TAS method

Figure 7. The core sample at a point blow 8m

From the results of pilot tests for the above TAS method, the urethane is infiltrated to a micro void around the injection bolt and makes a round solidity homogeneous and high strength in a regular range. Therefore, it could be confirm that the application of TAS method as the supplementary method of tunnel for a soft ground like the weathered zone is very good.

Conclusions

The results about evaluation for application of TAS method in field are summarized as follows.

1. Generally, the urethane is a material that has physically and mechanically excellent properties and a long-term durability. However, urethane has no pollution characteristics. Therefore, the injection effect is very excellent for soil improvement. However, general application is not easy because of high cost. That is, after consider economical efficiency, construction application may be proper in field.
2. The injection distance of TAS is longer than 10m. When TAS is compared with the existing urethane injection, TAS is the method that can extend the injection distance over 6m of the existing urethane injection. In addition, TAS can be used as supplementary method of tunnel.

3. Because the ground applied TAS method is improved immediately after injection, it is evaluated that the application of TAS is proper to the construction of a soil tunnel and large cross-section tunnel.
4. TAS method has the effect of increase of soil strength by high permeability, durability. In addition, TAS has no pollution. Therefore, TAS method will be expected to apply in field of cut-off of water and ground reinforcement.

References

Christian Kutzner. (1996). *Grouting of Rock and Soil*, A.A. Balkema, 203-204.

Chun, B. S. (2001). "Chemical Grouting and Geo-environmental Issue", *Journal of the Korean Geo-environmental Society*, 12-23.

Chun, B. S. (1997). "The Engineering Properties and Reinforcement of Poly-Urethane for Grouting", *Journal of the Korean Society of Civil Engineers*, 476-478.

Jung, S. K. (1993). "The Report of Soil Test Results with Execution of PU-IF Method for the 4~8 Section of Works in Busan Subway Line Number One", 1-2.

Lee W. Abramson, Chair., (1997). *Ground Improvement, Ground Reinforcement Ground Treatment Developments 1987-1997*, ASCE, 353-360.

Mika. (2000). *Purification system for geoenvironmental pollution*, Gihodo Shuppan. , 44.

Yonekuraetal. (2000). *Grouting of permanency grout*, Sankaido Publishing. 103-114.

Subject Index

Page number refers to the first page of paper

Admixtures, 1180
Airports, 464
Anchors, 772, 791, 1010
Aqueducts, 893

Bearing capacity, 354, 707
Bedrock, 377
Binders, materials, 25, 145
Bitumen, 1293
Bored piles, 707, 716
Boreholes, 752
Boston, 236, 670
Bridges, highway, 575
Buildings, high-rise, 377
Buildings, residential, 377
Bulkheads, 330

Cables, 1010
California, 413, 893, 1392, 1546
Canada, 869, 1338, 1617
Case reports, 218, 294, 401, 452,
    813, 824, 953, 1010, 1020, 1130,
    1546
Cement grouting, 740, 1089, 1115,
    1153, 1180, 1200, 1208, 1221,
    1243, 1360, 1605
Cements, 501, 562, 681
Chemical grouting, 837, 1115, 1454,
    1605
Chicago, 389, 464
China, 837
Classification, 489
Clays, 236, 575, 634, 772, 845, 1192,
    1314
Coal mine wastes, 1103
Cofferdams, 1383

Columns, 365, 389, 527, 540, 634,
    681, 695, 1044
Compaction grouting, 869, 991,
    1020, 1032, 1044, 1056
Compressive strength, 1208
Computer analysis, 1440
Cooling towers, 365
Core walls, 1141
Curtain grouting, 881, 917, 929,
    1405, 1417
Cutoffs, 967

Dam design, 917
Dam foundations, 857, 881
Dam safety, 869
Dams, 905, 929, 1293, 1417
Deformation, 1570
Design, 1071, 1347
Dikes, 1429
Drainage, 791
Drilling, 303, 752, 1617

Earthquakes, 1305
Egypt, 1570
Embankment stability, 803
Embankments, 893, 991
Europe, 1534
Excavation, 257, 269, 515, 813, 837,
    1546
Expansion joints, 1617

Facility expansion, 1032
Field investigations, 1628
Finland, 552, 780
Florida, 1557
Fly ash, 1169, 1192

Foundation construction, 248, 540, 752
Foundation settlement, 354, 389
Foundations, 428, 893, 1490
Freeze-thaw, 1338

Germany, 248, 740, 1372
Gravel, 1383
Ground water, 1594
Grout, 1141, 1235

Highway construction, 318, 575
History, 1, 50, 100, 218, 294, 401, 452, 610, 857, 1130, 1266
Hong Kong, 342

Illinois, 389, 464
In situ tests, 695
Inflow, 1200, 1293, 1628
Injection, 50, 845, 1221, 1372
Intake structures, 1383

Japan, 25, 1056
Jet grouting, 198, 218, 236, 257, 269, 281, 294, 303, 318, 330, 365, 377, 389, 401, 413, 428, 440, 452, 464, 515, 527, 1490
Joints, 1208, 1254, 1466

Karst, 941, 953, 967, 979

Lateral loads, 428
Leaching, 658
Limestone, 967
Liquefaction, 501, 1305, 1478, 1507

Manufacturing facilities, 452
Mapping, 1082
Masonry, 1243
Massachusetts, 236, 670
Measurement, 1326
Methodology, 979, 1082, 1347, 1652

Mexico, 354
Microtunneling, 303
Mining, 1115, 1180, 1200, 1293
Mixing, 25, 145, 474, 489, 515, 634, 646, 658, 670, 681, 695, 1020, 1314, 1628
Models, 1466

Netherlands, 728
Nevada, 791
Nuclear power plants, 1392
Numerical models, 1089, 1454

Offshore structures, 330
Ohio, 905
Oregon, 1010

Penetration, 1326
Permeability, 622
Peru, 917, 929
Philippine Islands, 1383
Piers, 740
Pile foundations, 728
Piles, 780, 791
Pipes, 1208
Polyurethane, 1254, 1266, 1281, 1338
Porous media, 1454
Portugal, 401
Powerplants, 428, 440
Pull-out resistance, 772
Pumps, 1640

Quality control, 236, 281, 474, 489, 695, 991, 1153

Railroad tracks, 803
Railroad trains, 562
Reconstruction, 248
Rehabilitation, 330, 791, 869, 905, 1243, 1392, 1605
Reliability, 772

Remedial action, 1103
Resins, 1254, 1338
Retrofitting, 413
Rheology, 1221
Rock masses, 164, 979, 1089
Rock mechanics, 1089
Rocks, 50, 100, 772, 1082, 1383,
    1405, 1466, 1628

Safety factors, 330
Sampling, 489
Sand, 1235, 1305, 1372
Sea walls, 330
Sealing, 1254, 1338, 1594, 1628
Seepage, 881, 967, 1180, 1200, 1293,
    1417
Seepage control, 1347, 1429
Settlement, 501, 813, 837, 1534,
    1546
Sewers, 1582
Shallow foundations, 941
Shrinkage, 1281
Silty soils, 1507
Simulation, 1454
Sinkholes, 869
Slopes, 318
Slovakia, 1417
Soft soils, 257, 269, 440, 552, 586,
    622, 1314
Soil cement, 527, 540, 586, 598, 646,
    658, 670
Soil conditions, 634, 752
Soil grouting, 303, 1180, 1192, 1200,
    1293, 1347
Soil improvement, 377, 440, 452,
    991, 1020, 1347, 1478, 1490,
    1518, 1652
Soil mixing, 527, 575, 586, 598, 610
Soil pollution, 610
Soil stabilization, 269, 318, 464, 501,
    527, 552, 562, 575, 586, 598, 610,
    991, 1192, 1314, 1360, 1478, 1582

Soils, 50, 100
Solidification, 1360
South Africa, 1115
South Carolina, 1032
Spain, 1518
Stability analysis, 330
Stabilization, 1010
Stadiums, 389
Strain, 1044
Structure reinforcement, 824
Sweden, 145, 552, 622, 681, 780

Taiwan, 1305
Temperature, 1281
Tests, 1056
Texas, 330, 440
Time factors, 1466
Towers, 1010
Tunnel construction, 236, 342, 1518,
    1652
Tunneling, 257, 464, 1180, 1200,
    1534, 1582
Tunnels, 413, 791, 1020, 1338, 1546,
    1557, 1570, 1594, 1605, 1617,
    1640
Turkey, 281, 365, 377, 452

Underground construction, 527, 837
Underground structures, 1605
Underwater structures, 1338
United Kingdom, 610
United States, 501, 857

Verification, 474, 1044
Vibration, 562
Virginia, 575
Voids, 1, 1103

Water content, 1208
Water infiltration, 1605
Water levels, 1254
Water pipelines, 1557

Water pressure, 1417

West Virginia, 824

Water treatment plants, 1032

Author Index

Page number refers to the first page of paper

Abedi, Hasan, P.E., 1429
Akbulut, S., 1192
Ali, Rajiv, 1010
Al-Tabbaa, A., 610
Altugu, T., 452
Amaya, F., 917, 929
Ånberg, Helen, 622
Anderson, Andy, 1044
Anderson, Randy, P.E., 413
Andromalos, Kenneth B., P.E., 515
Aronsson, Stefan, 780
Au, S. K. A., 845
Axelsson, Morgan, 681

Bachand, Michael L., 1032
Bahner, Eric W., P.E., 515
Bahrekazemi, Mehdi, 562
Baker, Hayward, 1534
Ballivy, Caroline, 1605
Ballivy, G., 1243, 1338
Ballivy, Gérard, 1605
Barata, Carlos, 401
Bares, F. A., 1490
Bell, Kenneth R., 354
Berry, Richard M., 1130
Blanco, Miguel Angel, 1518
Bodare, Anders, 562
Boehm, D., 440
Boehm, Dennis W., 330
Boes, N., 610
Boghart, Ray (Alireza), 1020
Bolisetti, Tirupati, 1454
Bolton, M. D., 845
Boys, I. E., 452
Breitsprecher, Georg, 740
Brengola, Andrew F., P.E., 428
Brill, Gary T., P.E., 218

Bruce, D. A., 1153
Bruce, Donald A., 474, 752, 967
Bruce, Mary Ellen C., 474
Brunner, Wolfgang G., 248
Burke, George, 1534
Burke, George K., P.E., 218
Byle, Michael J., P.E., 1071

Cadden, Allen, P.E., 905
Cadden, Allen W., P.E., 941, 979
Camper, Kyle E., P.E., 389
Carter, T. G., 917
Cebola, Duílio, 401
Chao, B. S., 1305
Chitambira, B., 610
Chlapik, Dusan, 1417
Chuaqui, M., 1153, 1582
Chuaqui, Marcelo, 1347
Chun, Byung-Sik, 1314, 1652
Clemente, José L. M., 354
Cockburn, James, 1440

Dahlström, L.-O., 1089
Dalmalm, Thomas, 1466, 1628
Dash, Umakant, P.E., 413
Davie, J. R., 440
Davie, John, 365
Day, Steven R., 1640
Doven, Ata G., 1169
Dreese, Trent L., 1440
Dreese, Trent L., P.E., 824, 881
Druss, David L., 527
Durgunoglu, H. T., 377, 452
Düzceer, Rasin, 281

Eldridge, T. L., 917
El-Kelesh, Adel M., 1056

Emrem, C., 452
Eriksson, Magnus, 1326
Esrig, Melvin I., 501

Falcão, João, 401
Falk, Eduard, 1534
Fernagu, E., 1338
Ferreira, Sandra, 401
Filz, George M., 489
Fischer, Joseph A., P.E., 953
Fischer, Joseph J., 953
Fondazioni, Keller, 1534
Forte, Edward P., 501
Francis, Mathew, P.E., 303
Fransson, Åsa, 1082
Franz, Raymond J., P.E., 389
Fu, Xudong, 707

Gallagher, Patricia M., 1478
Ganse, Margaret A., 575
Garcia, J. P., 929
Garner, Steve, 869
Gemmi, Bruno, 1490
Geraci, Jeffrey, 1010
Glynn, E., 1405
Gökalp, Alp, 281
Gómez, Jesús, P.E., 905
Gómez, Jesús E., P.E., 941
Griffin, H. Clay, 1130
Gularte, Francis B., 354, 791
Gustafson, Gunnar, 1082

Hamby, James A., P.E., 967
Han, Jie, 634
Harris, Joe, 991
Hayashi, Hirochika, 598
Heenan, Douglas M., 824, 1440,
    1546
Hegazy, Yasser A., 772
Heinz, W. F., 1115
Hennings, Steve, 1281
Hill, Jeffrey R., 1020

Ho, Chu Eu, 269, 716
Hok, Robert, 1417
Holm, Göran, 145
Holmquist, Darrel V., P.E., 1103
Hooey, Stephen, 1180, 1293
Huang, C. C., 1305
Hulla, Jozef, 1254, 1417
Hundley, Paul S., 1020

Jagello, Jon J., 1044
Janicek, Drahomir, 1254
Janson, Thomas, 1628
Jefferies, M. G., 917, 929
Jefferies, Michael, 869
Jefferis, Stephan A., 1141
Jelisic, Nenad, 552
Jia, W., 1507
Johnsen, Lawrence F., 1044

Kapadia, R. J., 1383
Kasali, Gyimah, 540
Kim, Jin-Chun, 1314
Kitazume, Masaki, 586
Kocak, D., 377
Koch, Alyssa J., 1478
Koga, Hirofumi, 634
Kulac, H. F., 377, 452
Kummerer, Clemens, 813
Kwong, James, P.E., 303

Lambrechts, James R., 575, 670
Landry, Eric, 1180, 1200
Laporte, R., 1243
Larsson, Stefan, 681
Lau, D. T., 1570
Law, K. T., 1570
Layhee, Carrie A., 575
Lee, Thomas S., P.E., 413
Lees, Daniel, 1347
Lehtonen, Jouko, 780
Leppänen, Mikko, 552
Lewis, Dwayne A., 464

Liao, H. J., 1305
Licuanan, D. O., 342
Littlejohn, Stuart, 50, 100
Liu, Jinyuan, 837
Lohman, David P., 1429
Lombardi, Giovanni, 164
Lopez, Roberto A., 354

Mac Kenna, Peter E., 501
Magill, David, 1281
Martin, Lucas, 1518
Maswoswe, Justice J. G., 236
Matsui, Tamotsu, 1056
Mattey, Y., 1208, 1281
Mazek, S. A., 1570
McGillivray, Ross T., P.E., 1557
McLeod, Donald W., 1640
Meyers, John, P.E., 318
Millar, Gerry, 791
Mittag, Jens, 1372
Miura, Norihiko, 634
Mollamahmutoglu, Murat, 1235
Monroe, A. Sarah, 1518
Morelli, Gianfranco, 1490
Myers, Tim, 318

Nagel, Scott, 670
Nakamura, Takeshi, 586
Narduzzo, L., 1617
Naudts, Alex, 1180, 1200, 1266,
    1293
Naudts, Ward, 1180
Nelson, E., 342
Nishida, Takahiro, 658
Nishikawa, Jun'ichi, 598

Oakland, Michael W., P.E., 1032
Ohishi, Kanta, 586, 598, 658
Oruc, K., 452
Otsuki, Nobuaki, 658
Otterbein, Reiner, 813
Ottoson, Richard S., 953

Pacheco, Joana, 401
Palardy, D., 1243
Palardy, Danielle, 1605
Pekrioglu, Ayse, 1169
Pelnik, Thomas W., III, 489
Perera, R., 610
Perkins, Steven W., 991
Perret, S., 1243, 1338
Petrasic, Kerry, P.E., 318
Pinto, Alexandre, 401
Piyal, Mehmet, 365
Porbaha, Ali, 695
Posey, Thomas A., P.E., 330
Puppala, Anand J., 695

Ramachandra, K., 1383
Reitsma, Stanley, 1454
Rennie, David C., P.E., 893
Richely, A., 342
Ringen, Alan R., P.E., 218
Ritchie, D. G., 929
Roberts, Bradford W., P.E., 428
Rourke, Monica M., 1392
Ryan, Christopher R., 1640

Saglamer, A., 1192
Samavedam, Gopal, 803
San Juan, Raúl, 1518
Santagata, E., 1221
Santagata, M. C., 1221
Sanver, Armagan, 365
Savidis, Stavros A., 1372
Scherer, Steven D., 1020
Schweiger, Helmut F., 813
Senapathy, H., 440
Shen, Shui-Long, 634
Shibazaki, Mitsuhiro, 198, 294
Shiells, David P., 489
Shim, S. K., 1383
Shirlaw, J. Nick, 257
Shuttle, D. A., 1405
Sickling, J., 452

Simard, Gary, P.E., 1429
Simon, Kent, P.E., 1103
Sjostrom, Orjan A., 1594
Slastan, Peter, 1254
Sluz, Andrew, 803
Smith, Graham C. G., 905
Soga, K., 845
Sola, Pedro R., 1518
Stille, Håkan, 1326, 1466
Storry, R. B., 342
Sussmann, Theodore R., 803
Swedenborg, S., 1089

Taki, Osamu, 540, 646
Tan, Chin Gee, 269
Taube, Martin G., P.E., 464
Tekinturhan, Bahattin, 365
Terashi, Masaaki, 25, 586, 598, 658
Thevanayagam, S., 1507
Thomas, Damon B., P.E., 1103
Tóth, Paul Stefan, 740
Traylor, R. P., 1582
Traylor, Robert, 905
Tumay, Mehmet T., 1169

van der Stoel, Almer E. C., 728
Vataja, Janne W., 824
Vipulanandan, C., 1208, 1281
Vrignaud, J. P., 1338
Vrignaud, Jean-Philippe, 1605

Walz, Arthur H., Jr., P.E., 967
Wargo, Richard H., P.E., 979
Warner, James, 1, 869, 1360
Weaver, K. D., 857
Wehling, Timothy M., 893
Wern, A., 1383
Wilson, David B., 1440
Wilson, David B., P.E., 881, 967

Xu, Michael, 1546

Yeoh, Yoo-Hyeon, 1652
Yildiz, R., 452
Yilmaz, S., 377
Yokoo, Mitsuru, 294
Yoshida, Hiroshi, 294

Zhou, Zhengbing, 707